# Multisensor Data Fusion

## From Algorithms and Architectural Design to Applications

# Devices, Circuits, and Systems

**Series Editor**
*Krzysztof Iniewski*
CMOS Emerging Technologies Research Inc.,
Vancouver, British Columbia, Canada

**PUBLISHED TITLES:**

## PUBLISHED TITLES:

## PUBLISHED TITLES:

**Testing for Small-Delay Defects in Nanoscale CMOS Integrated Circuits**
*Sandeep K. Goel and Krishnendu Chakrabarty*

**VLSI: Circuits for Emerging Applications**
*Tomasz Wojcicki*

**Wireless Technologies: Circuits, Systems, and Devices**
*Krzysztof Iniewski*

**Wireless Transceiver Circuits: System Perspectives and Design Aspects**
*Woogeun Rhee*

## FORTHCOMING TITLES:

**Advances in Imaging and Sensing**
*Shuo Tang, Dileepan Joseph, and Krzysztof Iniewski*

**Analog Electronics for Radiation Detection**
*Renato Turchetta*

**Cell and Material Interface: Advances in Tissue Engineering, Biosensor, Implant, and Imaging Technologies**
*Nihal Engin Vrana*

**Circuits and Systems for Security and Privacy**
*Farhana Sheikh and Leonel Sousa*

**CMOS Time-Mode Circuits and Systems: Fundamentals and Applications**
*Fei Yuan*

**Ionizing Radiation Effects in Electronics: From Memories to Imagers**
*Marta Bagatin and Simone Gerardin*

**Magnetic Sensors: Technologies and Applications**
*Kirill Poletkin*

**MRI: Physics, Image Reconstruction, and Analysis**
*Angshul Majumdar and Rabab Ward*

**Multisensor Attitude Estimation: Fundamental Concepts and Applications**
*Hassen Fourati and Djamel Eddine Chouaib Belkhiat*

**Nanoelectronics: Devices, Circuits, and Systems**
*Nikos Konofaos*

**Nanomaterials: A Guide to Fabrication and Applications**
*Sivashankar Krishnamoorthy and Gordon Harling*

**Physical Design for 3D Integrated Circuits**
*Aida Todri-Sanial and Chuan Seng Tan*

**Power Management Integrated Circuits and Technologies**
*Mona M. Hella and Patrick Mercier*

**Radio Frequency Integrated Circuit Design**
*Sebastian Magierowski*

# FORTHCOMING TITLES:

**Silicon on Insulator System Design**
*Bastien Giraud*

**Semiconductor Devices in Harsh Conditions**
*Kirsten Weide-Zaage and Malgorzata Chrzanowska-Jeske*

**Smart eHealth and eCare Technologies Handbook**
*Sari Merilampi, Lars T. Berger, and Andrew Sirkka*

**Structural Health Monitoring of Composite Structures Using Fiber Optic Methods**
*Ginu Rajan and Gangadhara Prusty*

**Terahertz Sensing and Imaging: Technology and Devices**
*Daryoosh Saeedkia and Wojciech Knap*

**Tunable RF Components and Circuits: Applications in Mobile Handsets**
*Jeffrey L. Hilbert*

**Wireless Medical Systems and Algorithms: Design and Applications**
*Pietro Salvo and Miguel Hernandez-Silveira*

# Multisensor
# Data Fusion

## From Algorithms and Architectural Design to Applications

EDITED BY

## Hassen Fourati

GIPSA-LAB
DEPARTMENT OF CONTROL SYSTEMS
UNIVERSITY GRENOBLE ALPES
GRENOBLE, FRANCE

## Krzysztof Iniewski MANAGING EDITOR

CMOS EMERGING TECHNOLOGIES RESEARCH INC.
VANCOUVER, BRITISH COLUMBIA, CANADA

**CRC Press**
Taylor & Francis Group
Boca Raton  London  New York

CRC Press is an imprint of the
Taylor & Francis Group, an **informa** business

MATLAB® is a trademark of The MathWorks, Inc. and is used with permission. The MathWorks does not warrant the accuracy of the text or exercises in this book. This book's use or discussion of MATLAB® software or related products does not constitute endorsement or sponsorship by The MathWorks of a particular pedagogical approach or particular use of the MATLAB® software.

CRC Press
Taylor & Francis Group
6000 Broken Sound Parkway NW, Suite 300
Boca Raton, FL 33487-2742

First issued in paperback 2020

© 2016 by Taylor & Francis Group, LLC
CRC Press is an imprint of Taylor & Francis Group, an Informa business

No claim to original U.S. Government works

ISBN-13: 978-1-4822-6374-9 (hbk)
ISBN-13: 978-0-367-65628-7 (pbk)

**Visit the Taylor & Francis Web site at**
**http://www.taylorandfrancis.com**

**and the CRC Press Web site at**
**http://www.crcpress.com**

*Dedicated to my wife, Emna, whose Zen-like patience continues to amaze;*
*to my parents, without whom I would not be where I am today.*

**Hassen Fourati**

# Contents

## SECTION I   Novel Advances in Multisensor Data Fusion Algorithm Design

## SECTION II  Multisensor Data Fusion Showcases Advancements

# Preface

The technology of multisensor data fusion seeks to combine information coming from multiple and different sources and sensors, resulting in an enhanced overall system performance with respect to separate sensors and sources. Multisensor data fusion has gained in importance over the last decades and found applications in an impressive variety of areas within diverse disciplines: navigation, sensor networks, intelligent transportation systems, security, medical diagnosis, biometrics, environmental monitoring, remote sensing, measurements, robotics, and so forth. Different concepts, techniques, and architectures have been developed to optimize the overall system output in applications for which sensor fusion might be useful and enables development of concrete solutions.

The idea for this book arose as a response to the immense interest and strong activities in the field of multisensor data fusion during the last few years, both in theoretical and practical aspects. This book is targeted toward researchers, academics, engineers, and graduate students working in the field of sensor fusion, estimation and observation, filtering, and signal processing.

This book captures the latest data fusion concepts and techniques drawn from a broad array of disciplines. With contributions from the world's leading fusion researchers and academicians, this book has 34 chapters, divided roughly into two sections, and covers the fundamental theory and recent theoretical advances, as well as showcasing applications of multisensor data fusion. Each chapter is complete in itself and can be read in isolation or in conjunction with other chapters of the book. Chapters 1 through 23 in Section I are devoted to the state of the art and novel advances in multisensor data fusion algorithm design. New materials and achievements in optimal fusion and multisensor filters are provided. In Section II, Chapters 24 through 34 mostly showcase multisensor data fusion advancements in fields such as medical applications, navigation, traffic analysis, and so on.

We are grateful to all the contributors for sharing their valuable knowledge and we expect this book to offer a good balance between academic and industrial research throughout the different chapters. We sincerely hope that this book will be a source of inspiration for new concepts and applications and stimulate further the development of data fusion architecture. We would also like to acknowledge CRC Press and its staff for technical and editorial assistance that improved the quality of this book and resulted in its publication. Finally, we hope readers will enjoy this book and that it will prove to be a useful addition to the increasingly important and expanding field of data fusion.

**Hassen Fourati**
*Univ. Grenoble Alpes, Gipsa-Lab, F-38000 Grenoble, France*
*CNRS, Gipsa-Lab, F-38000 Grenoble, France*
*Inria, Grenoble, France*

MATLAB® is a registered trademark of The MathWorks, Inc. For product information, please contact:

The MathWorks, Inc.
3 Apple Hill Drive
Natick, MA 01760-2098 USA
Tel: 508 647 7000
Fax: 508-647-7001
E-mail: info@mathworks.com
Web: www.mathworks.com

# Editors

**Hassen Fourati, PhD**, is currently an associate professor in the Electrical Engineering and Computer Science Department at the University Grenoble Alpes, Grenoble, France and a member of the Networked Controlled Systems Team (NeCS), affiliated with the Automatic Control Department of the Laboratoire Grenoble Images Parole Signal Automatique (GIPSA-LAB) and the Institut National de Recherche en Informatique et en Automatique (INRIA). He received the B.Eng. in Electrical Engineering from the National Engineering School of Sfax, Tunisia, a MS in Automated Systems and Control from the University of Claude Bernard, Lyon, France, and a PhD in Automatic Control from the University of Strasbourg, France, in 2006, 2007, and 2010, respectively. His research interests include nonlinear filtering, estimation, and multisensor fusion with applications in navigation, inertial and magnetic sensors, robotics, and traffic management. He has published several research journal articles, papers in international conferences, and book chapters. He can be reached at hassen.fourati@gipsa-lab.fr.

**Krzysztof (Kris) Iniewski** is managing R&D at Redlen Technologies Inc., a start-up company in Vancouver, Canada. Redlen's revolutionary production process for advanced semiconductor materials enables a new generation of more accurate, all-digital, radiation-based imaging solutions. Kris is also a president of CMOS Emerging Technologies Research Inc. (www.cmosetr.com), an organization of high-tech events covering communications, microsystems, optoelectronics, and sensors. In his carrier, Dr. Iniewski held numerous faculty and management positions at University of Toronto, University of Alberta, SFU, and PMC-Sierra Inc. He has published over 100 research papers in international journals and conferences. He holds 18 international patents granted in the United States, Canada, France, Germany, and Japan. He is a frequently invited speaker and has consulted for multiple organizations internationally. He has written and edited several books for CRC Press, Cambridge University Press, IEEE Press, Wiley, McGraw-Hill, Artech House, and Springer. His personal goal is to contribute to healthy living and sustainability through innovative engineering solutions. In his leisurely time, Kris can be found hiking, sailing, skiing, or biking in beautiful British Columbia. He can be reached at kris.iniewski@gmail.com.

# Contributors

**Andrea Abrardo**
University of Siena
Siena, Italy
abrardo@dii.unisi.it

**Nisar R. Ahmed**
University of Colorado Boulder
Boulder, Colorado, USA
Nisar.Ahmed@Colorado.EDU

**T. Ajitha**
Indian Institute of Technology Madras
Chennai, Tamilnadu, India
tajitha98@gmail.com

**Murat Akcakaya**
University of Pittsburgh
Pittsburgh, Pennsylvania, USA
akcakaya@pitt.edu

**Ienkaran Arasaratnam**
Apple Inc.
Cupertino, California, USA
haran@ieee.org

**Erik Blasch**
Air Force Research Laboratory
Rome, New York, USA
erik.blasch@us.af.mil

**Jean-Marc Le Caillec**
Telecom Bretagne
Brest Cedex 3, France
jm.lecaillec@telecom-bretagne.eu

**Mark E. Campbell**
Cornell University
Ithaca, New York, USA
mc288@cornell.edu

**Kumar Pakki Bharani Chandra**
University of Exeter
Exeter, UK
b.c.k.pakki@exeter.ac.uk

**Sébastien Changey**
GNC Department
ISL–French-German Research Institute of
    Saint-Louis
Saint-Louis Cedex, France
sebastien.changey@isl.eu

**Chee-Yee Chong**
Independent Consultant
Los Altos, California, USA
cychong@ieee.org

**Thyagaraju Damarla**
US Army Research Laboratory
Adelphi, Maryland, USA
thyagaraju.damarla.civ@mail.mil

**Belur V. Dasarathy**
Independent Consultant
Huntsville, Alabama, USA
fusion-consultant@ieee.org

**François Delmotte**
Université Lille Nord de France
Béthune, France
francois.delmotte@univ-artois.fr

**Jacques Demongeot**
Laboratoire AGIM
Université Grenoble-Alpes
La Tronche, Grenoble, France

and

Institut Universitaire de France
Paris, France

and

Mines Douai
IA
Douai, France
jacques.demongeot@yahoo.fr

**Bruno Diot**
Laboratoire AGIM
Université Grenoble-Alpes
Grenoble, France

and

IDS
Montceau-les-Mines, France
bruno.diot@ids-assistance.com

**Gregorio E. Drayer**
Georgia Institute of Technology
Atlanta, Georgia, USA
drayer@gatech.edu

**Zhansheng Duan**
Center for Information Engineering Science
    Research
Xi'an Jiaotong University
Xi'an, China
zduan@uno.edu

**Deniz Erdogmus**
Northeastern University
Boston, Massachusetts, USA
erdogmus@ece.neu.edu

**António J. Falcão**
Uninova-CA3
Monte da Caparica, Portugal
ajf@uninova.pt

**Nour-Eddin El Faouzi**
Transport and Traffic Engineering Laboratory
Bron, France

and

ENTPE
LICIT
Vaulx-en-Velin, France

and

University of Lyon
Lyon, France
nour-eddin.elfaouzi@ifsttar.fr

**Gianluigi Ferrari**
University of Parma
Parma, Italy
gianluigi.ferrari@unipr.it

**Anthony Fleury**
Mines Douai
IA
Douai, France
anthony.fleury@mines-douai.fr

**José M. Fonseca**
Uninova-CA3
Monte da Caparica, Portugal
jmf@uninova.pt

**Céline Franco**
Laboratoire AGIM
Université Grenoble-Alpes
La Tronche, Grenoble, France

and

Institut Universitaire de France
Paris, France

and

Mines Douai
IA
Douai, France
celine.franco@imag.fr

**Shrikant Fulari**
Indian Institute of Technology Madras
Chennai, India
shrikant.f@gmail.com

**Robert X. Gao**
Department of Mechanical and Aerospace
    Engineering
Case Western Reserve University
Cleveland, Ohio, USA
robert.gao@case.edu

**Wei Gao**
College of Automation
Harbin Engineering University
Harbin, China
gaow@hrbeu.edu.cn

**Jean-François Grandin**
THALES Systèmes Aéroportés
Elancourt, France
jean-francois.grandin@fr.thalesgroup.com

**Samir Hachour**
Université Lille Nord de France
Béthune, France
samirhachour@yahoo.fr

**Uwe D. Hanebeck**
Karlsruhe Institute of Technology (KIT)
Karlsruhe, Germany
uwe.hanebeck@ieee.org

**Hao He**
Department of EECS
Syracuse University
Syracuse, New York, USA
hhe02@syr.edu

**Matt Higger**
Northeastern University
Boston, Massachusetts, USA
higger@ece.neu.edu

**Ayanna M. Howard**
Georgia Institute of Technology
Atlanta, Georgia, USA
ayanna.howard@ece.gatech.edu

**Fangming Huang**
Nanjing Research Institute of Electronics
    Engineering
Nanjing, China
HFM3000@sina.com

**Zhiliang Huang**
Nanjing Research Institute of Electronics
    Engineering
Nanjing, China
zhiliangh28@163.com

**Radoslav Ivanov**
University of Pennsylvania
Philadelphia, Pennsylvania, USA
rivanov@seas.upenn.edu

**Satish G. Iyengar**
General Electric Global Research Corporation
Niskayuna, New York, USA
iyengar@ge.com

**Alex Pappachen James**
Department of Electrical and Electronics
    Engineering
Nazarbayev University
Astana, Kazakhstan
apj@ieee.org

**Zhou Jian**
College of Computer
Nanjing University of Posts and
    Telecommunications
Nanjing, China
zhoujian@njupt.edu.cn

**Simon J. Julier**
Department of Computer Science
University College of London
London, UK
s.julier@cs.ucl.ac.uk

**Fakhri Karray**
Department of Electrical  and Computer
    Engineering
University of Waterloo
Waterloo, Ontario, Canada
karray@uwaterloo.ca

**Bahador Khaleghi**
IMS Inc.
Waterloo, Ontario, Canada
bkhalegh@uwaterloo.ca

**Alaa Khamis**
Vestec Inc.
and
Suez University
Waterloo, Ontario, Canada
akhamis@pami.uwaterloo.ca

**Ali Khenchaf**
ENSTA Bretagne
Brest Cedex 9, France
ali.khenchaf@ensta-bretagne.fr

**Lawrence A. Klein**
Klein & Associates
Santa Ana, California, USA
larry@laklein.com

**Wolfgang Koch**
Fraunhofer/University of Bonn
Wachtberg, Germany
wolfgang.koch@fkie.fraunhofer.de

**Laurent Lecornu**
Telecom Bretagne
Brest Cedex 3, France
Laurent.lecornu@telecom-bretagne.eu

**Insup Lee**
University of Pennsylvania
Philadelphia, Pennsylvania, USA
lee@cis.upenn.edu

**X. Rong Li**
Department of Electrical Engineering
University of New Orleans
New Orleans, Louisiana, USA
xli@uno.edu

**Shaopeng Liu**
Distributed Intelligent Systems Lab
GE Global Research
Niskayuna, New York, USA
victorlsp@gmail.com

**Zheng Liu**
Toyota Technological Institute
Nagoya, Japan
zheng.liu@ieee.org

**James Llinas**
Center for Multisource Information Fusion
University at Buffalo
Buffalo, New York, USA
llinas@buffalo.edu

**Jing Ma**
Department of Automation
Heilongjiaang University
Harbin, China
jingma427@163.com

**Ronald Mahler**
Intelligent Robotics Laboratory
Lockheed Martin Advanced Technology
    Laboratories
Eagan, Minnesota, USA
MahlerRonald@comcast.net

**Marco Martalò**
University of Parma
Parma, Italy

and

E-Campus University
Novedrate (CO), Italy
marco.martalo@unipr.it

**David Mercier**
Université Lille Nord de France
Béthune, France
david.mercier@univ-artois.fr

**Luís Miranda**
Uninova-CA3
Monte da Caparica, Portugal
lmm@ca3-uninova.org

**André D. Mora**
Uninova-CA3
Monte da Caparica, Portugal
atm@uninova.pt

**Shozo Mori**
Systems & Technology Research
Sunnyvale, California, USA
shozo.mori@stresearch.com

**Manohar Murthi**
University of Miami
Coral Gables, Florida, USA
mmurthi@miami.edu

**Benjamin Noack**
Karlsruhe Institute of Technology (KIT)
Karlsruhe, Germany
benjamin.noack@ieee.org

**Ruixin Niu**
Department of Electrical and Computer
    Engineering
Virginia Commonwealth University
Richmond, Virginia, USA
rniu@vcu.edu

**Umut Orhan**
Honeywell Aerospace
Redmond, Washington, USA
uorhan@cu.edu.tr

**Miroslav Pajic**
University of Pennsylvania
Philadelphia, Pennsylvania, USA
pajic@seas.upenn.edu

**Emmanuel Pecheur**
GNC Department
ISL–French-German Research Institute of
    Saint-Louis
Saint-Louis, France
emmanuel.pecheur@isl.eu

**Fangfang Peng**
Department of Automation
Heilongjiaang University
Harbin, China
pengfangfang2013@163.com

**Kamal Premaratne**
University of Miami
Coral Gables, Florida, USA
kamal@miami.edu

**Marc Reinhardt**
Karlsruhe Institute of Technology (KIT)
Karlsruhe, Germany
marc.reinhardt@ieee.org

**Rita A. Ribeiro**
Uninova-CA3
Monte da Caparica, Portugal
rar@uninova.pt

**Max Mauro Dias Santos**
Department of Electronics
Federal University of Technology–Paraná
    (UTFPR)
Ponta Grossa, Brazil
maxsantos@utfpr.edu.br

**Jonathan R. Schoenberg**
Arzentech, Inc.
Fishers, Indiana, USA
jon@arzentech.com

**Joris Sijs**
TNO Technical Sciences
The Hague, The Netherlands
joris.sijs@tno.nl

**Enbin Song**
College of Mathematics
Sichuan University
Chengdu, China
e.b.song@163.com

**Arun Subramanian**
Department of EECS
Syracuse University
Syracuse, New York, USA
arsubram@syr.edu

**Shankar C. Subramanian**
Indian Institute of Technology Madras
Chennai, India
shankarram@iitm.ac.in

**Qian Sun**
College of Automation
Harbin Engineering University
Harbin, China
sunsl@hlju.edu.cn

**Shuli Sun**
Department of Automation
Heilongjiaang University
Harbin, China
sunsl@hlju.edu.cn

**Ion-George Todoran**
Telecom Bretagne
Brest Cedex 3, France
iongeorge.todoran@telecom-bretagne.eu

**Lelitha Vanajakshi**
Indian Institute of Technology Madras
Chennai, India
lelitha@iitm.ac.in

**Pramod K. Varshney**
Department of EECS
Syracuse University
Syracuse, New York, USA
varshney@syr.edu

**Nicolas Vuillerme**
Laboratoire AGIM
Université Grenoble-Alpes
Grenoble, France

and

Institut Universitaire de France
Paris, France
nicolas.vuillerme@agim.eu

**Kun Wang**
Nanjing Research Institute of Electronics
    Engineering
Nanjing, China
kun.wang1981@gmail.com

**Thanuka Wickramarathne**
University of Notre Dame
Notre Dame, Indiana, USA
twickram@nd.edu

**Jian Xu**
Nanjing Research Institute of Electronics
    Engineering
Nanjing, China
xujian2001-1@163.com

**Liguo Zhang**
School of Electronic Information and Control
    Engineering
Beijing University of Technology
Beijing, China
zhangliguo@bjut.edu.cn

**Ya Zhang**
College of Automation
Harbin Engineering University
Harbin, China
yzhang@hrbeu.edu.cn

# Section I

---

## Novel Advances in Multisensor Data Fusion Algorithm Design

# 1 Challenges in Information Fusion Technology Capabilities for Modern Intelligence and Security Problems

*James Llinas*

## CONTENTS

## 1.1 HETEROGENEITY OF SUPPORTING INFORMATION

Experiences in dealing with intelligence and security problems in Iraq and Afghanistan and other places in the world have required the (ongoing) formulation of new paradigms of intelligence analysis and dynamic decision making. Broadly, these problems fall into the categories of counterterrorism and counterinsurgency (COIN) as well as stability operations. Depending on the phases of COIN or other operations, the nature of decision making ranges from conventional military-like to sociopolitical. Because of this wide spectrum of action, the nature of information support required for analysis has an equally wide range. As automated information fusion (IF) processes provide some of the support to such decision making, requirements for IF process design must address these varying requirements, resulting in considerable challenges in IF process design.

### 1.1.1 OBSERVATIONAL DATA

Further, these experiences have also shown that some of the key observational and intelligence data in such operations come not only from traditional sensor systems, but also from dismounted soldiers or other human observers reporting on their patrol activities. These data are naturally communicated in language in the form of various military and intelligence reports and messages. Such "soft" data finds its way into IF processes as both structured and unstructured digitized text, and this input modality creates new challenges to IF process designs, contrasted with more traditional IF applications involving the use of highly calibrated, numerically precise observational data from sensors. Combined with the data from the usual repertoire of "hard" or sensor data from various radio frequency (RF) sensors, video and other imaging systems, as well as signals intelligence (SIGINT) and satellite imagery, the observational data stream is a composite of data of highly different quality, sampling rates, content, and structure.

### 1.1.2 Open Source and Social Media Data

Soft or hard data can also find their way into modern IF processes in the form of monitored open source and social media feeds such as newswire feeds, Twitter, and blog sources judged to be possibly helpful. Getting such data into an IF system will require automated Web crawlers and related capabilities, as well as subsequent natural language processing capabilities.

### 1.1.3 Contextual Data

Modern problems also afford (and demand) the use of additional data and information beyond just observational data. A major category of such data and information is Contextual Information (CI). CI is that information that can be said to "surround" a situation of interest in the world (many definitions and characterizations exist but we do not address such issues here). It is information that aids in understanding the (estimated) situation and also in reacting to the situation, if a reaction is required. CI can be relatively or fully static or can be dynamic, possibly changing along the same timeline as the situation (e.g., weather). It is also likely that it may not be possible to know the full characterization and specification of CI at system/algorithm design time, except in very closed worlds. Thus, we envision an "a priori" framework of exploitation of CI that attempts to account for the effects on situational estimation of that CI that is known at design time. Even if such effects are known at design time, there is a question of the ease or difficulty involved in integrating CI effects into a fusion system design or into any algorithm designs. This issue is influenced in part by the nature of the CI and the manner of its native representation, for example, as numeric or symbolic, and the nature of the corresponding algorithm; for example, cases can arise that involve integrating symbolic CI into a numeric algorithm. Strategies for a priori exploitation of CI may thus require the invention of new hybrid methods that incorporate whatever information an algorithm normally employs in estimation (usually observational data) with an adjunct CI exploitation process. Note too that CI may, like observational data, have errors and inconsistencies itself, and accommodation of such errors is a consideration for hybrid algorithm design. Similarly, we envision the need for an "a posteriori" CI exploitation process, owing to at least two factors: (1) that all relevant CI may not be able to be known at system/ algorithm design time and may have to be searched for and discovered at runtime, as a function of the current situation estimate, and (2) that such CI may not be of a type that was integrated into the system/algorithm designs at design time and so may not be able to be easily integrated into the situation estimation process. In this case we then envision that at least part of the job of a posteriori CI exploitation would involve checking the consistency of a current situational hypothesis with the newly discovered (and situationally relevant) CI.

There are yet other system engineering issues. The first is the question of accessibility; CI must be accessible to use it, but accessibility may not be a straightforward matter in all cases. One question is whether the most current CI is available; another may be that some CI is controlled or secure and may have limited availability. The other question is one of representational form. CI data can be expected to be of a type that has been created by "native" users; for example, weather data, important in many fusion applications as CI, are generated by meteorologists, for meteorologists (not for fusion system designers). Thus, even if these data are available, there is likely to be a need for a "middleware" layer that incorporates some logic and algorithms both to sample these data and shape them into a form suitable for use in fusion processes. In even simpler cases, this middleware may be required to reformat the data from some native form to a usable one. In spite of some a priori mapping of how CI influences or constrains the way in which situational inferences or estimates can be developed, which may serve certain environments, the defense and security type applications, with their various dynamic and uncertain types of CI, demand a more adaptive approach. Given a nominated situational hypothesis Hf from a fusion process or "engine," the first question is: What CI type information is relevant to this hypothesis? Relevant CI is only that information that influences our interpretation or understanding of Hf. Presuming a "relevancy filter" can be crafted, a search function would explore the available CI and make this CI available to an "a posteriori" reasoning engine. That reasoning engine would then use (1) a CI-guided subset of Domain Knowledge and (2) the retrieved CI to reason over Hf to first determine consistency of Hf with the relevant CI. If it is inconsistent, then some type of adjudication logic will need to be applied to reconcile this inconsistency between (1) the fusion process that produced Hf and (2) the a posteriori reasoning process that judges it as inconsistent. If, however, Hf is judged as consistent with the additional CI, an expanded interpretation of Hf could be developed, providing a deeper situational understanding. This overall process, which can be considered a "Process Refinement" operation, would be a so-called "Level 4" process in the context of the Joint Directors of Laboratories (JDL) Data Fusion Process Model [1], that is, as an adaptive operation for fusion process enhancement. The overall ideas discussed here are elaborated in Ref. [2].

### 1.1.4  ONTOLOGICAL DATA

IF processes and algorithms historically have been developed in a framework that has assumed the a priori availability of a reliable body of procedural and dynamic knowledge about the problem domain, that is, knowledge that supports a more direct approach to temporal reasoning about the unfolding patterns of interest in the problem domain. In COIN and other complex problems, such a priori and reliable knowledge is most often not available—the Tactics, Techniques, and Procedures of modern-day adversaries are highly adaptive and extremely hard to model with confidence. The US DARPA COMPOEX Program [3] attempted to develop such models but achieved only partial success, experiencing gaps in the overall modeling space of such desired behavioral models. We label these types of problems as "weak knowledge" problems, implying that only fragmentary a priori behavioral model type knowledge is available to aid in IF-based reasoning, inferencing, and estimation.

Ontological information, however, that does not attempt to form such comprehensive behavioral and temporal models overtly but does include temporal primitives along with structural/syntactic relations among entities can be specified a priori with reasonably good confidence, and thus provides a declarative knowledge base to support IF reasoning and estimation. Note that such knowledge is also represented in language and is available as digital text, in the same way as data from messages, documents, Twitter, and so forth. The use of ontological information in IF systems can be varied; ontological information can augment observed data and can aid in asserting possible relationships, directing search and also sensor management (to acquire expected information based on ontological relations), and yet in other ways. Importantly, specified ontologies can also serve as providing consistent and grounded semantic terminology for any given system. In our current research, we employ ontologies primarily for augmenting observational data with asserted ontological data whose relevance is algorithmically determined using "spreading activation" and then integrated to enrich the evidential basis for reasoning [4]. The broader implications of ontologies for intelligence analysis are described in Ref. [5], which come from the University of Buffalo's National Center for Ontological Research (see http://ncorwiki.buffalo.edu/index.php/Main_Page).

### 1.1.5  LEARNED INFORMATION

Finally, there is the class of information that could be learned (online) from all of the aforementioned sources if the IF process is designed with a Data Mining/Inductive or Abductive Learning

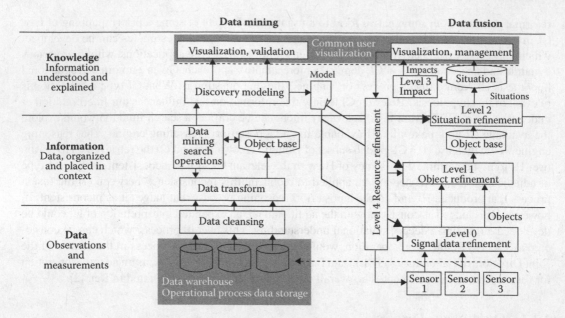

**FIGURE 1.1** Notional fusion process architecture combining data mining and data fusion. (From *ISCAS '98—Proceedings of the 1998 IEEE International Symposium on Circuits and Systems*, Ann Arbor, 1998.)

functional component. Very little research and prototyping of such dual-process type IF systems has been done although the conceptualization of such IF schemes and architectures was put forward some time ago (e.g., Ref. [6]), as shown in Figure 1.1. The runtime integration of learned information raises a number of both algorithmic issues as well as architectural issues. For example, if meaningful patterns of behavior can be learned and can be measured/judged as persistent or enduring, such patterns could be incorporated in a dynamically modifiable knowledge base to be reused (as a Level 4 Process refinement function). Such learning processes will also not be perfect and have some uncertainty that also needs to be factored into the traditional Common Referencing and Data Association functions of the target fusion process.

## 1.2 COMMON REFERENCING AND DATA ASSOCIATION

Common Referencing (CR) is that traditional IF system function that is sometimes called "Alignment" and is the function that normalizes these input sources for any given fusion application or design. CR addresses such issues as coordinate system normalization, temporal alignment, and uncertainty alignment across the input streams, among others. With the highly disparate input streams described earlier, the design of required CR techniques is a nontrivial challenge. There are at least two major CR issues that these heterogeneous data represent temporal alignment and uncertainty alignment. Consider a textual input message whose free text, in just a few lines, could have past–present–future tense expressions, for example, "3 days ago I saw…", "past precedents lead me to believe that tomorrow I should see…" and so forth. Other sources can also have varied temporal structures regarding their input. Such data lead to the issue of what the IF community has called "OOSM: out-of-sequence measurements" for hard/sensor data but the issue carries over to all sources as well and requires complex temporal alignment techniques for CR; it also raises the issue of retrospective fusion processing operations to correct for delayed inputs (if warranted; this is a design choice). Temporal alignment methods we have used for soft data are described in Ref. [7].

The uncertainty alignment requirement evolves as a result of the high likelihood that any uncertainty in the widely various sources will be represented in disparate forms. Consider the basic differences between the uncertainty in sensor (hard) data and textual (soft) data; sensor data uncertainty is sensibly always expressed in probabilistic form whereas, owing to the problem of imprecise adjectives and adverbs in language, linguistic uncertainty is often expressed in possibilistic (fuzzy) terms. It can be expected that uncontrolled open source or social media data may use yet other uncertainty formalisms to express or tag inputs. Transformation and normalization of disparate forms of uncertainty is a specialized topic in the uncertainty/statistical literature (e.g., Ref. [8]), and is among the high-priority issues in the IF community [9]. It should be noted that such transformations largely can be developed only by invoking some statistical type qualities that are preserved across the transform, such as some form of total uncertainty; that is, the transform does not create an "equivalent" value of a probability in say a possibilistic space; seminal papers on the probability–possibility transformation issue are in Refs. [10–12]. In our research, we have addressed the probabilistic–possibilistic transformation issue in an approach that satisfies the consistency and preference preservation principles [13], while resulting in the most specific distribution for a specified portion of a probabilistic representation, resulting in the use of a truncated triangular transformation in our case [14].

Regarding the Data Association (DA) function, which some consider the heart of a fusion process, these varied data raise the level of DA complexity in significant ways. The soft data category, which inherently reports about Entities and (judged) Relationships, and is inherently in semantic format (language/words), raises the important issue of how to measure semantic similarity of such elements as reported in these various input streams. Such scores are needed in the Hypothesis Evaluation step of the DA process (see Ref. [15] on these DA subfunctions). But there are further DA complications that arise due to the soft data: linguistic phrases have verbs that reflect inter-Entity (noun) relationships; also of note is that the Natural Language Processing (NLP) community has employed graphical methods for the representation of linguistic structures. As a result, the DA process now involves interassociation of both Entities (nouns) and Relations (verbs), that is, of graphical structures. This requirement extends to the hard data as well because that data need to be cast in a semantic framework to enable the overall DA process for the combined hard and soft data. Developing DA methods for graphical structures represents an entirely new challenge for the DA function. In such approaches for these applications, a scoring approach also needs to be developed to assess Relational similarity as well as Entity similarity, and a composite association scheme for these graphical substructures needs to be evolved. Historical approaches to DA have often employed solution methods drawn from assignment problems in operations research. When association is required between many nongraphical data sources, this can be handled by the multidimensional assignment problem [16,17]. The main difference between the multidimensional assignment problem and graph association is how topological information from the graphs is used. Our research center has attacked this problem and has developed research prototype algorithms, as described in Ref. [18], where the graph association problem is formulated as a binary linear program and a heuristic for solving the multiple graph association is developed using a Lagrangian relaxation approach to address issues with a between-graph transitivity requirement.

## 1.3 SEMANTICS

The introduction of linguistic information, as well as the transformation of sensor + algorithm estimation process outputs into a semantic frame, also adds to the complexity of IF process design and development. Semantic complexity is also added by the very nature of modern intelligence and security problems wherein the situations of interest relate to both military operations and also sociopolitical behaviors and entities. Clear meanings of such notions of interest as "patterns of life," "rhythm of the city," and "radicalization" as patterns or situations of interest—to be estimated by IF systems—have proven difficult to specify in clear semantic terms. Although the use of ontologies helps in this regard, standardization issues remain when considering networked and distributed systems, which are typical in the modern era. For example, in distributed intelligence or military systems there is typically no single point of architectural authority that can mandate a single ontological framework for the network. For large-scale real systems there is also the problem of large legacy systems that were never designed with ontological formalisms in mind; this creates a "retrofit" problem of adjusting the semantic framework of that system to some new ontological standard, a costly and complex operation.

It must also be noted that the way in which all textual/linguistic information gets into an IF system is through processing in some type of NLP or text extraction system. Such systems serve as a front-end filter for the admission of fundamental entity and relationship data, the raw soft data of the system, and so any imperfections in such extractions bound the capture of evidential information for the subsequent reasoning and estimation processes. Whereas errors in hard sensor data are typically known with reasonable accuracy because of sensor calibrations, the errors in text extraction and NLP systems are either weakly known or unknown, sometimes as a result of proprietary constraints.

Other strategies to deal with the complexities of semantics involve the use of controlled languages, to bound the grammatical structures and also the extent of the vocabulary that has to be dealt with. A good example for military/intelligence applications is the Battle Management Language (BML) [19] that has been under development since about 2003 for both Command and Control simulation studies but also for IF applications (e.g., Refs. [20,21]).

There is a corresponding need to understand better the nature of semantic (and syntactic) complexity in language, and also to develop measures and metrics that aid in developing better NLP processes and controlled languages. There is a reasonably rich literature on these topics (e.g., Ref. [22]) that should be exploited in regard to the integrated design of IF systems that today have to deal with a wide range of semantic difficulties.

## 1.4 GRAPHICAL REPRESENTATIONS AND METHODS

There are a number of reasons that, for COIN and asymmetric warfare-type problems, graphs are becoming a dominant representational form for the information in and the processes involved in IF systems. In the information domain, many of the components discussed in Section 1.1 are textual/linguistic and to capture this information in digital form, graphs are the representational form of choice. The problem domain is also described in the ontologies that are also typically couched in graphical forms. Note that ontologies describe inter-Entity relations of various types. Note too that the inferences and estimates of interest in these problems are of the higher level type in the sense of the JDL Model of Information Fusion, that is, estimates of situations and threat states. These higher level states—the conditions of interest for intelligence and security applications—are also best described as graphs, as situations can in the most abstract sense be considered as a graph of entities and relations.

As a result, it is not unexpected to see that the core functions of IF, such as DA as previously described, are employing graphical methods in these fusion function operations. The US Army's primary intelligence support system, the Distributed Common Ground Station-Army (DCGS-A), employs a global graph approach to capture all of the evidentiary information that supports IF and other intelligence analysis operations; see Ref. [23] and Figure 1.2, which shows the top-level structure of this graphical concept.

Developing a comprehensive understanding of these problems thus involves a logical synthesis of the many situational substructures or subgraphs in these problem domains. The subgraphs are somewhat thematic and can be thought of as revolving about the Political, Military, Economic,

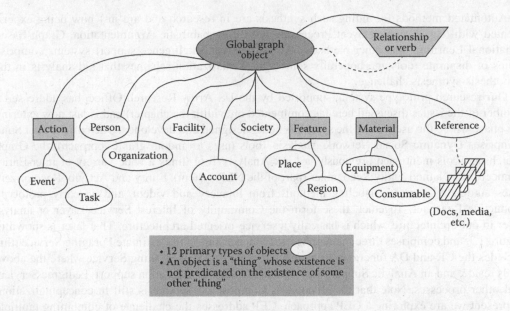

- 12 primary types of objects
- An object is a "thing" whose existence is not predicated on the existence of some other "thing"

**FIGURE 1.2** US Army's "global graph" concept for DCGS-A. (From Walsh, D. Relooking the JDL Model for fusion on a global graph. In *National Symposium on Sensor and Data Fusion*, Las Vegas, July 2010.)

Social, Infrastructure, and Information (PMESII) notion of the heterogeneity of the classes of information of interest in such problems. Thus, it is not surprising to see social network analysis tools—which are by the way graph-theoretic and graph-centric—employed in support of intelligence analysis, here with the focus on the social and infrastructure patterns and subgraphs of the problem space.

In our own work for such problems, we considered that it would be broadly helpful in analysis to enable a subgraph-querying capability as a generalized analysis tool. In such an approach, the analyst forms a query in text that can be transformed to a graph (we call these template graphs in that they are subgraph structures of interest) that is then searched for in the associated-evidence graph that is formed by the DA process. This search operation is in effect an stochastic inexact graph-matching problem, as the nodes and arcs of the evidential (or perhaps the template graph) have uncertainty values associated with them, and also because what is sought is the best match to the query, not an exact match, as there may be no exact match in such unpredictable problem situations. Other complexities arise in trying to realize such capability, such as executing such operations incrementally for streaming data, and also doing them in a computationally efficient way because the graphs can get quite large. As a consequence of several PhD efforts, we have realized today a rather mature graph-matching capability for intelligence analysis that is implemented in a cloud-based process; see Refs. [24–26], among other of our works.

## 1.5 OVERALL SYSTEM ARCHITECTURES AND ANALYSIS FRAMEWORKS

It can perhaps be appreciated from the preceding discussion that the major challenge for intelligence analysis in these modern problems is the synthesis of a total situational picture. In the face of highly heterogeneous data of varying uncertainty and of a problem domain that has many substructures and relations and entities of interest, and has a varying temporal operational tempo, these problems—even with state of the art automated support/analysis systems such as IF systems—create a cognitive challenge even for the best analysts.

Automated methods for aiding such synthesis are in research and are just now being experimented with (e.g., Complex Event Processing [CEP], Probabilistic Argumentation, Graph-Based Relational Learning, and other methods). At the moment, intelligence support systems comprise suites of disparate tools and hopefully some agile visualization schemes that aid analysts in the hypothesis-synthesis challenge.

Our research prototype system, supported by the US Army Research Office, has addressed a number of the issues discussed here (as commented on within the chapter) and is just now entering the phase where the user end of the system is being designed and developed. The current Tool Suite comprises Dynamic Social Network Analysis Tools (uses a random graph approach), the Graph Matching Tools mentioned previously, a Link Analysis Tool (finds a wide variety of inter-Entity connections), Named Entity Recognizer (part of the NLP system), Entity and Activity Recognizers (uses automated semantic labeling methods from imagery and video), and an early prototype Abductive Reasoner. Together, these form the Community of Interest Service Layer or analyst layer in our architecture, which is basically a service-oriented architecture. That layer is shown in Figure 1.3, and comprises three main services: Evidence and Entity-estimate Foraging Service (this includes the CR and DA functions previously described), a Sensemaking Service where the above tools reside, and an Analytic Support Service that includes Visualization support, Pedigree Service, and other processes. Note that the Hypothesis Composition Service is still in conceptualization; at present we are exploring a CEP approach. CEP addresses the challenge of combining multiple heterogeneous data streams into a hierarchical structure that can represent higher order events and semantic meaning through the application of rules and filters at multiple levels of information (e.g., Ref. [27]).

The Core Enterprise Services that do all of the front-end data processing and conditioning are shown in Figure 1.4; this figure does not show much of the detail but the flavor of these operations can be appreciated. Multiple hard data streams (in our case these are LIDAR, EO/IR, and Visible imagery, video sources, and acoustic devices) are processed individually to the point where semantic information is developed from various estimation algorithms. Multiple soft message streams as arising from multiple soldier reports enter the NLP-based soft processing stream and the primary

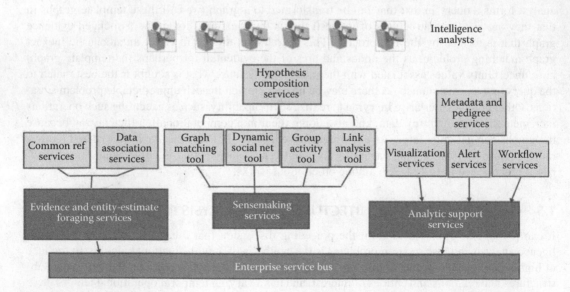

**FIGURE 1.3**  Analyst layer in our service-oriented architecture for counterinsurgency analysis support.

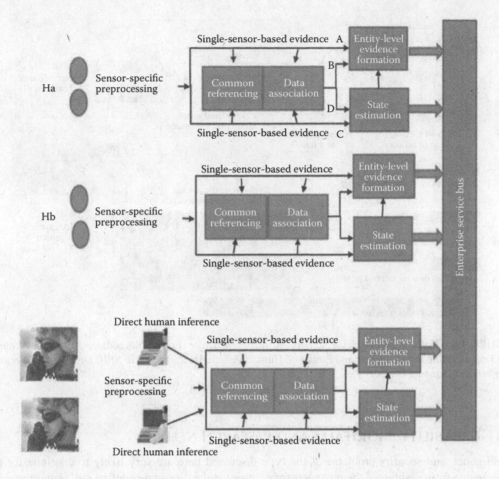

**FIGURE 1.4** Core enterprise services in our service-oriented architecture for counterinsurgency analysis support.

entities and relations are extracted. All of this semantic information flows to the Enterprise Bus, where it is accessible to the Community of Interest (Analysts) Service Layer for inferencing and estimation operations.

Space permits us to describe only one of our tools and we choose to show our Activity Recognizer Tool system; this is developed by our colleagues at Tennessee State University and is described in Ref. [28]. This tool focuses on human–vehicle interactions and activities, a type of activity that is critically important for the problem of improvised explosive devices (IEDs). It is assumed that video and acoustic sensing of the human–vehicle settings is feasible. An approach that is based on human–vehicle activity ontology is used, wherein discrete activities (e.g., door opening as detected by acoustics, human entering vehicle as detected by video) are detected by each sensing modality. Spatiotemporal and semantic association of these discrete activities, along with the activity-class ontology, allows fusion-based inferencing of aggregated activity classes of interest. Further, as previously described, these inferences are framed into "pseudo-messages" that are sent to the Enterprise Bus for access by the hard + soft DA service, to allow combination with soft message data on the same activities. These operations are depicted in Figure 1.5.

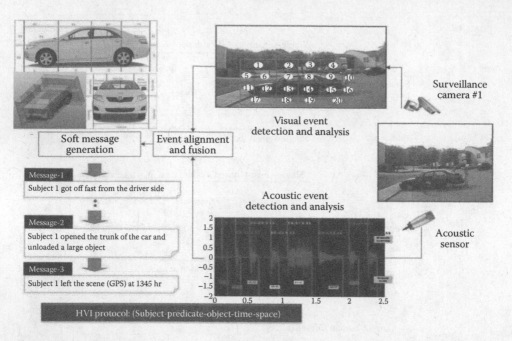

**FIGURE 1.5** Fused human–vehicle activity estimation scheme. (From Shirkhodaie, A. et al., "Acoustic and Imagery Semantic Labeling and Fusion of Human-Vehicle Interactions," in *SPIE Defense and Security Conference*, Orlando, FL, 2011.)

## 1.6 CAPABILITY SHORTFALLS AND RESEARCH NEEDS

Intelligence and security problems of the type discussed here are very likely to continue for the foreseeable future, although the prospects for conventional nation-state conflict still remain as well, and will drive yet other technology requirements, and it should be noted that there are some overlaps in such requirements. Although the IF community is reacting to the COIN and asymmetric/irregular warfare and stability operations needs of the type described here, there are very few well-tested capabilities that have been transitioned to operational systems. The IF community also has to make judgments about research investments that are peculiar to IF process needs and those that are supportive of IF processes; a good example is in NLP and text extraction—this is a core capability supportive of IF but that is actively being matured by the NLP community. For IF processing in particular, there are needs to improve CR methods for temporal and uncertainty alignment, and to improve capabilities for retrospective and reliable temporal estimation when meaningful amounts of input are continually shifting in time. Improvements in DA techniques are similarly required, both for new ideas in measures for semantic similarity and scoring in support of DA, but equally in graphical or other methods that in addition exhibit computational efficiency, as DA is typically the computational bottleneck in IF systems. Regarding state estimation, we see the major challenge being in the creation of automated methods to support synthesis of the disparate hypotheses emanating from data and theme-specific inferencing/estimation tools, to aid intelligence analysts in forming more comprehensive assessments of the "story" of interest suggested by the evidence. New studies and implementations of information foraging theory [29] are needed as one means to achieve more informative and efficient examinations of both associated evidence and inferences. Deeper examinations are also needed of the various Sensemaking paradigms [30,31] and ways that IF technologies can support them.

## 1.7  CONCLUSION

Evolving international sociopolitical events and dynamics, coupled with rapid growth of a wide variety of informational technologies, has given rise to a marked increase in the complexity of intelligence analysis and efforts to design and develop technological capabilities to aid such analyses. As IF technologies have been a major contributor for intelligence analysis, these complexities have carried over to create significant challenges in IF system design. This chapter has reviewed what are considered to be many of these challenges and, by referencing a major academic research program at our research center on such challenges, provided some examples of candidate methods to deal with these complexities. Much of the research on modern IF system design is still in the basic research domain, and far from being proven, much remains to be done and explored.

## ACKNOWLEDGMENT

This research activity has been supported in part by a Multidisciplinary University Research Initiative (MURI) grant (No. W911NF-09-1-0392) for Unified Research on Network-based Hard/Soft Information Fusion, issued by the US Army Research Office (ARO) under the program management of Dr. John Lavery.

## REFERENCES

1. D. L. Hall and J. Llinas, Introduction to multisensor data fusion. *Proceedings of IEEE*, 85(1):6–23, 1997.
2. J. Gómez-Romero et al., Strategies and techniques for use and exploitation of contextual information in high-level fusion architectures. *13th International Conference on Information Fusion (Fusion 2010)*, Edinburgh, UK, July 2010.
3. A. Kott and P. S. Corpac, COMPOEX technology to assist leaders in planning and executing campaigns in complex operational environments. *12th International Command and Control Research and Technology Symposium*, Newport, RI, 2007.
4. M. Kandefer and S. C. Shapiro, Evaluating spreading activation for soft information fusion. *14th International Conference on Information Fusion*, Chicago, July 5–8, 2011.
5. B. Smith et al., Ontology for the intelligence analyst. *Crosstalk: The Journal of Defense Software Engineering*, 25(6):18–25, 2012.
6. E. Waltz, Information understanding: Integrating data fusion and data mining processes. *ISCAS '98—Proceedings of the 1998 IEEE International Symposium on Circuits and Systems*, Ann Arbor, MI, 1998.
7. D. McMaster, Temporal alignment in soft information processing. *14th International Conference on Information Fusion*, Chicago, July 2011.
8. G. J. Klir, A principle of uncertainty and information invariance. *International Journal of General Systems*, 17:249–275, 1990.
9. E. Blasch et al., High level information fusion developments, issues, and grand challenges. *Fusion10 Panel Discussion, 13th International Conference on Information Fusion (Fusion 10)*, 2010.
10. M. Oussalah, On the probability/possibility transformations: A comparative analysis. *International Journal of General Systems*, 29(5):671–718, 2000.
11. J. F. Geer and G. J. Klir, A mathematical analysis of information-preserving transformations between probabilistic and possibilistic formulations of uncertainty. *International Journal of General Systems*, 20:143–176, 1992.
12. G. Klir and B. Parviz, Probability-possibility transformations: A comparison. *International Journal of General Systems*, 21(1):291–310, 1992.
13. D. Dubois and H. Prade, Unfair coins and necessity measures: A possibilistic interpretation of histograms. *Fuzzy Sets and Systems*, 10:15–20, 1983.
14. G. Gross, R. Nagi and K. Sambhoos, A fuzzy graph matching approach in intelligence analysis and maintenance of continuous situational awareness. *Journal of Information Fusion*, 18:43–61, 2014.
15. D. L. Hall and J. Llinas, *Handbook of Multisensor Data Fusion*. CRC Press, Boca Raton, FL, 2001.

16. A. Poore, S. Lu and B. Suchomel, Data association using multiple-frame assignments. In *Handbook of Multisensor Data Fusion*, 2nd ed., M. Liggins, D. Hall and J. Llinas (eds.). CRC Press, Boca Raton, FL, 2009, pp. 299–318.

17. A. Poore and N. Rijavec, A Lagrangian relaxation algorithm for multidimensional assignment problems arising from multitarget tracking. *SIAM Journal on Optimization*, 3:544–563, 1993.

18. G. Tauer, R. Nagi and M. Sudit, The graph association problem: Mathematical models and a lagrangian heuristic. *Naval Research Logistics (NRL)*, 60(3):251–268, 2013.

19. J. M. Pullen et al., Joint Battle Management Language (JBML)—US Contribution to the C-BML PDG and NATO MSG-048 TA. *IEEE European Simulation Interoperability Workshop*, Genoa, Italy, June 2007.

20. U. Schade, J. Biermann, M. Frey and K. Kruger, From Battle Management Language (BML) to automatic information fusion. *Proceedings of Information Fusion & Geographic Information Systems (GIS)*, 2007, pp. 84–95.

21. H. Lee and B. P. Zeigler, SES-based ontological process for high level information fusion. *The Journal of Defense Modeling and Simulation: Applications, Methodology, Technology*, 7(4), Art. No. 129, 2010.

22. S. Pollard and A. W. Biermann, A measure of semantic complexity for natural language systems. *Proceedings of NLP Complexity Workshop: Syntactic and Semantic Complexity in Natural Language Processing Systems*, Stroudsburg, PA, 2000.

23. D. Walsh, Relooking the JDL model for fusion on a global graph. *National Symposium on Sensor and Data Fusion*, Las Vegas, NV, July 2010.

24. A. Stotz, R. Nagi and M. Sudit, Incremental graph matching for situation awareness. *12th International Conference on Information Fusion*, Seattle, WA, July 6–9, 2009.

25. K. Sambhoos, R. Nagi, M. Sudit and A. Stotz, Enhancements to high level data fusion using graph matching and state space search. *Information Fusion*, 11(4):351–364, 2010.

26. G. Gross, Continuous preservation of situational awareness through incremental/stochastic graphical methods. *14th International Conference on Information Fusion*, Chicago, July 5–8, 2011.

27. A. Buchmann and B. Koldehofe, Complex event processing. *IT-Information Technology*, 51:241–242, 2009.

28. A. Shirkhodaie, A. Rababaah and V. Elangovan, Acoustic and imagery semantic labeling and fusion of human-vehicle interactions. *SPIE Defense and Security Conference*, Orlando, FL, 2011.

29. P. Pirolli, *Information Foraging Theory: Adaptive Interaction with Information*. Oxford University Press, New York, 2007.

30. P. Pirolli and S. K. Card, The sensemaking process and leverage points for analyst technology. *Proceedings of the 2005 International Conference on Intelligence Analysis*, McLean, VA, 2005.

31. G. Klein, B. Moon and R. F. Hoffman, Making sense of sensemaking I: Alternative perspectives. *IEEE Intelligent Systems*, 21(4):70–73, 2006.

# 2 Multisensor Data Fusion
## A Data-Centric Review of the State of the Art and Overview of Emerging Trends

*Bahador Khaleghi, Alaa Khamis, and Fakhri Karray*

## CONTENTS

## 2.1 INTRODUCTION

All living organisms have the ability to gain information about their environment, as well as to interpret this information to take appropriate decisions. Building a complete picture of the environment could be achieved using a single sensing element or by the fusion of the data gathered from multiple sensing elements. The operation of the human brain is probably the best analogy to a multisensor data fusion system, where the brain acts as the fusion node and makes sense of input provided by our five sense organs, as illustrated in Figure 2.1.

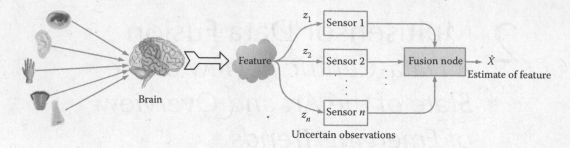

**FIGURE 2.1** Analogy between the brain operation and the fusion node in a data fusion system.

The goal of this chapter is to provide readers with a comprehensive review of contemporary data fusion methodologies, as well as an overview of the most recent developments and emerging trends in this field. The existing data fusion methodologies are examined according to a data-centric taxonomy, that is, the specific data-related issue they aim to tackle. Inspired by recent advances in mobile and ubiquitous sensing, cloud storage and computing, and prevalence of social networks, the new and emerging directions in data fusion research, such as social data fusion, cloud-enabled and big data fusion, and fusion of streaming data, are briefly discussed. The rest of this chapter is organized as follows. A brief discussion of multisensor data fusion definitions and most common applications are presented in Section 2.2. Relying on a data-centric taxonomy, a review of the existing data fusion literature is provided in Section 2.3. Section 2.4 is dedicated to a short discussion on research on data fusion evaluation methodologies and frameworks. The emerging research trends within the data fusion community are enumerated and briefly examined in Section 2.5. Lastly, in Section 2.6 the concluding remarks are presented.

## 2.2  MULTISENSOR DATA FUSION

### 2.2.1  What Is Multisensor Data Fusion?

From a data integration perspective, Luo defines multisensor fusion as any stage in the integration process in which there is an actual combination (or fusion) of different sources of sensory information into one representational format [1]. The boundary between sensor fusion and sensor integration is quite fuzzy and the terms are used interchangeably sometimes. Joshi and Sanderson describe multisensor fusion as part of the multisensor integration process [2]. This process refers to the synergistic use of multiple sensors to improve operation of the system as a whole and it also includes sensor planning and sensor architecture. Multisensor planning deals with the acquisition of sensor data while multisensor architecture is responsible for the organization of data processing and data flow in the system. Joshi and Sanderson define multisensor fusion as a process that deals with the combination of data from multiple sensors into one coherent and consistent internal representation or action [2].

Many other definitions for data fusion exist in the literature. Joint Directors of Laboratories (JDL) [3] defines data fusion as a "multi-level, multifaceted process handling the automatic detection, association, correlation, estimation, and combination of data and information from several sources." Klein [4] generalizes this definition, stating that data can be provided either by a single source or by multiple sources. Both definitions are general and can be applied in different fields including remote sensing. In Ref. [5], Bostrom et al. present a review and discussion of many data fusion definitions. Based on the identified strengths and weaknesses of previous work, a principled definition of information fusion is proposed as "Information fusion is the study of efficient methods for automatically or semi-automatically transforming information from different sources and different points in time into a representation that provides effective support for human or automated decision making." Data fusion is a multidisciplinary research area borrowing ideas from many diverse fields such as signal processing, information theory, statistical estimation and inference, and artificial intelligence. This is indeed reflected in the variety of techniques presented in Section 2.3.

Various conceptualizations of the fusion process exist in the literature. The most common and popular conceptualization of fusion systems is the JDL model [3]. The JDL classification is based on the input data and produced outputs, and originated in the military domain. The original JDL model considers the fusion process in four increasing levels of abstraction: object, situation, impact, and process refinement. Despite its popularity, the JDL model has many shortcomings, such as being too restrictive and especially tuned to military applications. The JDL formalization is focused on data (input/output) rather than processing. An alternative is Dasarathy's framework [6], which views the fusion system, from a software engineering perspective, as a data flow characterized by input/output as well as functionalities (processes). Another general conceptualization of fusion is the work of Goodman et al. [7], which is based on the notion of *random sets*. The distinctive aspects of this framework are its ability to combine decision uncertainties with decisions themselves, as well as presenting a fully generic scheme of uncertainty representation. One of the most abstract fusion frameworks was proposed by Kokar et al. [8]. This formalization is based on category theory and is claimed to be sufficiently general to capture all kinds of fusion, including data fusion, feature fusion, decision fusion, and fusion of relational information. It can be considered as the first step toward development of a formal theory of fusion. The major novelty of this work is the ability to express all aspects of multisource information processing, that is, both data and processing. Further, it allows for consistent combination of the processing elements (algorithms) with measurable and provable performance. Such formalization of fusion paves the way for the application of formal methods to standardized and automatic development of fusion systems.

## 2.2.2 APPLICATIONS OF MULTISENSOR DATA FUSION

Multisensor data fusion aims to overcome the limitations of individual sensors and produce accurate, robust, and reliable estimates of the world state based on multisensory information [9]. Multisensor data fusion has attracted many researchers from academia and industry because of its foreseen benefits in many applications. These benefits include, but are not limited to, enhanced confidence and reliability of measurements, extended spatial and/or temporal coverage, and reduced data imperfection aspects. Mitchell listed four main advantages of multisensor data fusion [10]: a greater granularity in the representation of information; greater certainty in data and results; elimination of noise and errors, producing a greater accuracy; and allowing a more complete view on the environment. These foreseen benefits of multisensor data fusion result in its wide applicability in a variety of military and civilian applications. As part of a comprehensive survey on multisensor integration and fusion in intelligent systems, Luo and Kay described a number of military and industrial applications in this area [11].

Data fusion is an established military technology and available for numerous military applications. These military applications include, but are not limited to, surveillance [12], anomaly detection [13] and behavior monitoring [14], target tracking [15,16], target engageability improvement [17], fire control [18], and landmine detection [19]. For example, modern military Command & Control (C2) systems are making increasing use of data fusion and resource management technology and tools [20]. By reducing uncertainty in the existing pieces of information and providing a means to infer about the missing pieces, data fusion supports the decision makers in compiling and analyzing the tactical/operational picture, and ultimately improving their situation awareness [21].

Examples of nonmilitary applications of data fusion include air traffic control [22], healthcare [23,24], speaker detection and tracking [25], mobile robot navigation [26], mobile robot localization [27], intelligent transportation systems [28], remote sensing [29,30], environment monitoring [31,32], and situational awareness [33]. For example, a Bayesian approach with pre- and postfiltering to handle data uncertainty and inconsistency in mobile robot local positioning is described in Ref. [27]. Mobile robot positioning provides an answer for the question: Where is the robot? The robot positioning solutions can be roughly categorized into relative position measurements (dead reckoning) and absolute position measurements. In the former, the robot position is estimated by applying to a previously determined position the course and distance traveled since. In the latter, the absolute position of the robot is computed by measuring the direction of incidence of three or more actively transmitting beacons, using artificial or natural landmarks, or using model matching to estimate the absolute location of the robot. There will always be an error in the readings provided by these techniques, and therefore the notion of multisensor data fusion is commonly used to tackle various imperfection aspects of data and yield a more accurate estimate for the robot position [27].

## 2.3 A DATA-CENTRIC TAXONOMY FOR MULTISENSOR DATA FUSION ALGORITHMS

Regardless of how different components (modules) of the data fusion system are organized, which is specified by the given fusion architecture, the underlying fusion algorithms must ultimately process (fuse) the input data. Real-world data fusion applications have to deal with several data-related challenges. As a result, we explore data fusion algorithms according to a data-centric taxonomy [34]. Figure 2.2 illustrates an overview of data-related challenges that are typically tackled by data fusion algorithms. The input data to the fusion system may be imperfect, correlated, inconsistent, and/or in disparate forms/modalities. Each of these four main categories of challenging problems can be further subcategorized into more specific problems, as shown in Figure 2.2 and discussed in the following.

Various classifications of imperfect data have been proposed in the literature [35–37]. Our classification of imperfect data is inspired by the pioneering work of Smets' [36] as well as an elaboration by Dubois and Prade [38]. Three aspects of data imperfection are considered in our classification: uncertainty, imprecision, and granularity.

Data are uncertain when the associated degree of confidence about what is stated by the data is less than 1. On the other hand, imprecise data are those data that refer to several, rather than only one, object(s). Finally, data granularity refers to the ability to distinguish among objects, which are

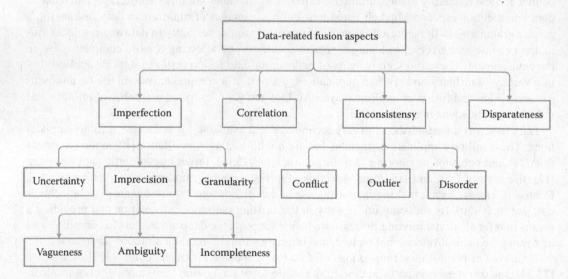

**FIGURE 2.2** A taxonomy of data fusion methodologies: Different data fusion algorithms can be roughly categorized based on one of the four challenging problems of input data that are mainly tackled: data imperfection, data correlation, data inconsistency, and disparateness of data form.

described by data, being dependent on the provided set of attributes. Mathematically speaking, assume the given data $d$ (for each described object of interest) to be structured as the following:

| object $O$ | attribute $A$ | statement $S$ |
| --- | --- | --- |

representing that the data $d$ is stating $S$ regarding the relationship of some attribute(s) $A$ to some object $O$ in the world. Further assume $C(S)$ to represent the degree of confidence we assign to the given statement $S$. Then, data are regarded to be uncertain if $C(S) < 1$ while being precise, that is, a singleton. Similarly, data are deemed as imprecise if the implied attribute $A$ or degree of confidence $C$ are more than 1, for example, an interval or set. Please note, the statement part of the data is almost always precise.

The imprecise $A$ or $C$ may be well defined or ill defined and/or miss some information. Thus, imprecision can manifest itself as ambiguity, vagueness, or incompleteness of data. The ambiguous data refers to those data where the $A$ or $C$ is exact and well defined yet imprecise. For instance, in the sentence "Target position is between 2 and 5" the assigned attribute is the well-defined imprecise interval [2 5]. The vague data is characterized by having ill-defined attributes, that is, the attribute is more than 1 and not a well-defined set or interval. For instance, in the sentence "The tower is large" the assigned attribute "large" is not well defined as it can be interpreted subjectively, that is, have different meaning from one observer to the other. The imprecise data that has some information missing is called incomplete data. For instance, in the sentence "It is possible to see the chair," only the upper limit on the degree of confidence $C$ is given, that is, $C < \tau$ for some $\tau$ [39].

Consider an information system [40] in which a number of (rather than one) objects $O = \{o_1,...,o_k\}$ are described using a set of attributes $A = \{V_1, V_2, ..., V_n\}$ with respective domains $D_1, D_2, ..., D_n$. Let $F = D_1 \times D_2 \times ... \times D_n$ to represent the set of all possible descriptions given the attributes in $A$, also called the frame. It is possible for several objects to share the same description in terms of these attributes. Let $[o]_F$ to be the set of objects that are equivalently described (thus indistinguishable) within the frame $F$, also called the equivalence class. Now, let $T \subseteq O$ represent the target set of objects. In general, it is not possible to exactly describe $T$ using $F$, because $T$ may include and exclude objects that are indistinguishable within the frame $F$. However, one can approximate $T$ by the lower and upper limit sets that can be described exactly within $F$ in terms of the induced equivalence classes. Indeed, the rough set theory provides a systematic approach to this end. In summary, data granularity refers to the fact that the choice of data frame $F$ (granule) has a significant impact on the resultant data imprecision. In other words, different attribute subset selections $B \subseteq A$ will lead to different frames, and thus different sets of indiscernible (imprecise) objects.

Correlated (dependent) data are also a challenge for data fusion systems and must be treated appropriately. We consider inconsistency in input data to stem from (highly) conflicting, spurious, or out of sequence data. Finally, fusion data may be provided in different forms, that is, in one or several modalities, as well as generated by physical sensors (hard data) or human operators (soft data).

We believe such categorization of fusion algorithms is beneficial as it enables explicit exploration of popular fusion techniques according to the specific data-related fusion challenge(s) they target. Further, our taxonomy is intended to facilitate ease of development by supplying fusion algorithm designers with an outlook of the appropriate and established techniques to tackle the data-related challenges their given application may involve. Finally, such exposition would be more intuitive and therefore helpful to nonexperts in data fusion by providing them with an easy-to-grasp view of the field.

## 2.3.1 FUSION OF IMPERFECT DATA

The inherent imperfection of data is the most fundamental challenging problem of data fusion systems, and thus the bulk of research work has been focused on tackling this issue. A number of

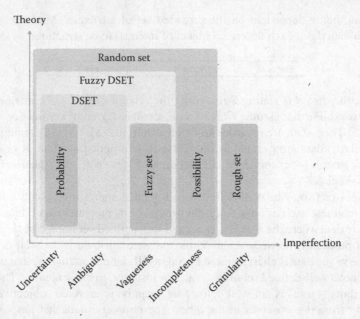

**FIGURE 2.3** Overview of theoretical frameworks of imperfect data treatment (note: the fuzzy rough set theory is omitted from the diagram to avoid confusion).

mathematical theories are available to represent data imperfection [41], such as probability theory [42], fuzzy set theory [43,44], possibility theory [45], rough set theory [46], and Dempster–Shafer evidence theory (DSET) [47]. Most of these approaches are capable of representing specific aspect(s) of imperfect data. For example, a probabilistic distribution expresses data uncertainty, fuzzy set theory can represent vagueness of data, and evidential belief theory can represent uncertain as well as ambiguous data. Historically, the probability theory was used for a long time to deal with almost all kinds of imperfect information, because it was the only existing theory. Alternative techniques such as fuzzy set theory and evidential reasoning have been proposed to deal with perceived limitations in probabilistic methods, such as complexity, inconsistency, precision of models, and uncertainty about uncertainty [42]. There are also hybridizations of these approaches that aim for a more comprehensive treatment of data imperfection. Examples of such hybrid frameworks are fuzzy rough set theory (FRST) [48] and fuzzy Dempster–Shafer theory (fuzzy DSET) [49]. Lastly, there is the fairly new field of fusion using *random sets*, which could be used to develop a unified framework for treatment of data imperfections [50]. Figure 2.3 provides an overview of the aforementioned mathematical theories of dealing with data imperfections. On the *x*-axis, various aspects of data imperfection, introduced in Figure 2.3, are depicted. The box around each of the mathematical theories designates the range of imperfection aspects targeted mainly by that theory. Interested readers are referred to Refs. [39] and [34] for a comprehensive review of the classical theories of representing data imperfections, describing each of them along with their interrelations.

## 2.3.2 Fusion of Correlated Data

Many data fusion algorithms, including the popular Kalman filter (KF) approach, require either independence or prior knowledge of the cross covariance of data to produce consistent results. Unfortunately, in many applications fusion data are correlated with potentially unknown cross covariance. This can occur as a result of common noise acting on the observed phenomena [51] in centralized fusion settings, or the rumor propagation issue, also known as the data incest or double counting problem [52], in which measurements are inadvertently used several times in distributed fusion settings [53]. If not addressed properly, data correlation can lead to biased estimation, for example, artificially high confidence value, or even divergence of fusion algorithm [54]. Most of the proposed solutions to correlated data fusion attempt to solve it by either eliminating the cause of correlation [55,56] or tackling the impact of correlation in fusion process [57–59].

### 2.3.3 FUSION OF INCONSISTENT DATA

#### 2.3.3.1 Spurious Data

Data provided by sensors to the fusion system may be spurious as a result of unexpected situations such as permanent failures, short duration spike faults, or slowly developing failure [60]. If fused with correct data, such spurious data can lead to dangerously inaccurate estimates. For instance, KF would easily break down if exposed to outliers. The majority of work on treating spurious data has been focused on identification/prediction and subsequent elimination of outliers from the fusion process. Indeed, the literature work on sensor validation is partially aiming at the same target [61–63]. The problem with most of these techniques is the requirement for prior information, often in the form of specific failure model(s). As a result, they would perform poorly in a general case in which prior information is not available or unmodeled failures occur [64]. In Refs. [60,65] a general framework for detection of spurious data has been proposed that relies on stochastic adaptive modeling of sensors and is thus not specific to any prior sensor failure model. Extensive experimental simulations have shown the promising performance of this technique in dealing with spurious data [64].

#### 2.3.3.2 Out-of-Sequence Data

The input data to the fusion system are usually organized as discrete pieces each labeled with a timestamp designating its time of origin. Several factors such as variable propagation times for different data sources as well as having heterogeneous sensors operating at multiple rates can lead to data arriving out of sequence at the fusion system. Such out-of-sequence measurements (OOSM) can appear as inconsistent data to the fusion algorithm. The main issue is how to use these, usually old, data to update the current estimate while taking care of the correlated process noise between the current time and the time of the delayed measurement [66].

Most of the early work on OOSM assumed only single-lag data. For example, an approximate suboptimal solution to OOSM called Algorithm B [67], as well as its famous optimal counterpart Algorithm A [68], both assume single-lag data. Some researchers have proposed algorithms to enable handling of OOSM with arbitrary lags [69–71]. Among these methods the work in Ref. [71] is particularly interesting as it provides a unifying framework for treating OOSM with Algorithm A as a special case. Nonetheless, it was shown in Ref. [72] that this approach, along with many other multilag OOSM methods, is usually very expensive in terms of computational complexity and storage. The same authors proposed an extension to the Algorithm A and Algorithm B called Algorithm A/1 and Algorithm B/1, respectively. They further showed that these new algorithms have requirements similar to their single-lag counterparts and are therefore recommended for practical applications; Algorithm B/1 especially is preferred because it is almost optimal and very efficient. Research work also investigates the OOSM problem in the case of having both single-lag and multiple-lag data, termed the mixed-lag OOSM problem. The proposed algorithm is claimed to handle all three types of OOSM data and is shown to be suboptimal in the linear MMSE sense under one approximation [73].

#### 2.3.3.3 Conflicting Data

Fusion of conflicting data, when, for instance, several experts have very different ideas about the same phenomenon, has long been identified as a challenging task in the data fusion community. In particular, this issue has been heavily studied for fusion within the Dempster–Shafer evidence theory framework. As shown in a famous counterexample by Zadeh [74], naive application of Dempster's rule of combination to fusion of highly conflicting data results in unintuitive results. Since then, Dempster's rule of combination has been subject to much criticism for rather counterintuitive behavior [75]. Most of the solutions proposed alternatives to Dempster's rule of combinations [76–79]. On the other hand, some authors have defended this rule, arguing that the counterintuitive results are due to improper application of this rule [50,80,81]. For example, in Ref. [50] Mahler shows that the supposed unintuitive result of Dempster's combination rule can be resolved using a simple corrective strategy, i.e. to assign arbitrary small but nonzero belief masses to hypotheses deemed extremely unlikely.

Fusion of conflicting data within the Bayesian probabilistic framework has also been explored by some authors. For example, the Covariance Union (CU) algorithm is developed to complement the Contextual Information (CI) method, and enable data fusion where input data is not just correlated but may also be conflicting [82]. Furthermore, a new Bayesian framework for fusion of uncertain, imprecise, as well as conflicting data was proposed in Ref. [83].

### 2.3.4 FUSION OF DISPARATE DATA

The input data to a fusion system may be generated by a wide variety of sensors, humans, or even archived sensory data. Fusion of such disparate data to build a coherent and accurate global view or the observed phenomena is a very difficult task. Nonetheless, in some fusion applications such as human–computer interaction (HCI), such diversity of sensors is necessary to enable natural interaction with humans. Our focus of discussion is on fusion of human generated data (soft data) as well as fusion of soft and hard data, as research in this direction has attracted attention in recent years. This is motivated by the inherent limitations of electronic (hard) sensors and recent availability of communication infrastructure that allow humans to act as soft sensors [84]. Further, although a tremendous amount of research has been done on data fusion using conventional sensors, very limited work has studied fusion of data produced by human and nonhuman sensors. An example of preliminary research in this area includes the work on generating a dataset for hard/soft data fusion intended to serve as a foundation and a verification/validation resource for future research [85,86]. Also in Ref. [84], Hall et al. provide a brief review on ongoing work on dynamic fusion of soft/hard data, identifying its motivation and advantages, challenges, and requirements. A Dempster–Shafer theoretic framework for soft/hard data fusion is presented that relies on a novel conditional approach to updating as well as a new model to convert propositional logic statements from text into forms usable by Dempster–Shafer theory [87]. Another recent example of data fusion systems capable of leveraging soft data is presented in Ref. [88], where Seifzadeh et al. describe a solution to the problem of agile target tracking using fuzzy inference applied to soft data reports that characterize the target agility level.

## 2.4 EVALUATION OF DATA FUSION SYSTEMS

Performance evaluation aims at studying the behavior of a data fusion system operated by various algorithms and comparing their pros and cons based on a set of measures or metrics. The outcome is typically a mapping of different algorithms into different real values or partial orders for ranking [89]. Generally speaking, the obtained performance of a data fusion system is deemed to be dependent on two components: the quality of input data and the efficiency of fusion algorithm. As a result, the literature work on (low-level) fusion evaluation can be categorized into the following groups:

- *Evaluating the quality of input data to the fusion system.* The target here is to develop approaches that enable quality assessment of the data, which are fed to the fusion system, and calculation of the degree of confidence in data in terms of attributes such as reliability and credibility [90]. The most notable work in this group is perhaps the standardization agreements (STANAG) 2022 of the North Atlantic Treaty Organization (NATO). STANAG adopts an alphanumeric system of rating, which combines a measurement of the reliability of the source of information with a measurement of the credibility of that information, both evaluated using the existing knowledge. STANAG recommendations are expressed using natural language statements, which makes them quite imprecise and ambiguous. Some researchers attempted to analyze these recommendations and provide a formal mathematical system of information evaluation in compliance with the NATO recommendations [90,91]. The proposed formalism relies on the observation that three notions underline an information evaluation system: the number of independent sources supporting a piece of information, their reliability, and that the information may conflict with some available/prior information. Accordingly, a model of evaluation is defined and its fusion method, which accounts for the three aforementioned notions, is formulated. The same authors have extended their work to enable dealing with the notion of degree of conflict, in contrast to merely conflicting or non-conflicting information [92]. Nonetheless, the current formalism is still not complete as there are some foreseen notions of the STANAG recommendations, such as total ignorance about the reliability of the information source, that are not being considered. Another important aspect related to input information quality, which is largely ignored, is the rate at which it is provided to the fusion system. The information rate is a function of many factors, including the revisit rate of the sensors, the rate at which data sets are communicated, and also the quality of the communication link [93]. The effect of information rate is particularly important in decentralized fusion settings where imperfect communication is common.

- *Assessing the performance of the fusion system.* The performance of fusion systems itself is computed and compared using a specific set of measures referred to as measures of performance (MOPs). The literature work on MOP is rather extensive and includes a wide variety of measures. The choice of the specific MOP(s) of interest depends on the characteristics of the fusion system. For instance, there is more to evaluate in a multiple-sensor system than there is in a single-sensor system. Further, in the case of multitarget problems, the data/track association part of the system also needs to be evaluated along with the estimation part. The commonly used MOPs may be broadly categorized into the metrics computed for each target and metrics computed over an ensemble of targets. Some of the MOPs belonging to the former category are track accuracy, track covariance consistency, track jitter, track estimate bias, track purity, and track continuity. Examples of measures in the latter category are average number of missed targets, average number of extra targets, average track initiation time, completeness history, and cross-platform commonality history [94,95]. There are also other less popular measures related to the discrimination and/or classification capability of the fusion system that can be useful to collect in some applications. Aside from the conventional approaches for performance measurement, there is some notable work on development of MOPs for multitarget fusion systems within the finite set theory framework [96,97]. The key observation is that a multitarget system is fundamentally different from a single-target system. In the former case, the system state is indeed a finite set of vectors rather than a single vector. This is due to the appearance/disappearance of targets, which leads to the number of states varying with time. In addition, it is more natural to mathematically represent the collection of states as a finite set, as the order in which the states are listed has no physical significance [98]. This approach is especially useful in fusion applications in which the number of targets is not known and has to be inferred along with their positions. Finally, it is worth pointing out some of the fusion evaluation tools and testbeds that have recently become available. The Fusion Performance Analysis (FPA) tool from Boeing is software that enables computation of technical performance measures (TPMs) for virtually any fusion system. It is developed in Java (thus is platform-independent) and implements numerous TPMs in three main categories: state estimation, track quality, and discrimination [99]. Another interesting development is the multisensor–multitarget tracking testbed [100], which has been lately introduced and is the first step toward the realization of a state-of-the-art testbed for evaluation of large-scale distributed fusion systems.

To the best of our knowledge, there is no standard and well-established evaluation framework to assess the performance of data fusion algorithms. Most of the work is being done in simulation and based on sometimes idealized assumption(s), which make it difficult to predict how the algorithm would perform in real-life applications. A review of literature on data fusion performance evaluation is presented in Ref. [101], where the challenging aspects of data fusion performance evaluation, in practice, are discussed. Having analyzed more than 50 of the related literature work, it has been shown that only very little (i.e., about 6%) of the surveyed research work treats the fusion evaluation problem from a practical perspective. Indeed, it is demonstrated that most of the existing work is focused on performing evaluation in simulation or unrealistic test environments, which is substantially different from practical cases. Regarding the preceding discussions, there appears to be a serious need for further research on development and standardizing measures of performance applicable to the practical evaluation of data fusion systems.

## 2.5 NEW DIRECTIONS IN MULTISENSOR DATA FUSION

### 2.5.1 SOCIAL DATA FUSION

The advent of social network services such as Facebook and Twitter has enabled users to share their social data—pictures, videos, news, ideas—on the Web. The social data are highly rich in content and thus provide an unprecedented opportunity for both researchers in social sciences and practitioners in industry to study human behavior, identify customer preferences, and much more. Within the context of data fusion, the conventional sensory data and the recently availability social data can be deemed as mutually compensatory in numerous data fusion and processing applications [102]. For instance, social network services can be leveraged in a participatory sensing manner, that is, human as a social sensor [103], to collect data in areas where physical sensors are not available. Similarly, data provided by conventional sensors can be used to construct context regarding the available social data, thus enabling them to be analyzed more effectively. On the other hand, social data are typically provided as streams of massive unstructured data. Accordingly, exploiting social data in fusion applications involves tackling challenges in stream data processing, scalable and distributed data storage and processing, and ability to model and interpret unstructured data. The aforementioned advantages along with the inevitable theoretical and practical challenges has made social data fusion a highly attractive and promising area of research within the fusion community, as reflected in the plethora of recent publications [103–106].

### 2.5.2 OPPORTUNISTIC DATA FUSION

Regarding the limitations of traditional data fusion systems, which are designed mostly to use dedicated sensor and information resources, and the availability of new ubiquitous computing and communication technologies, the opportunistic data fusion paradigm considers the possibility of treating sensors as shared resources and performing fusion in an opportunistic manner [107]. New challenging problems associated with such fusion systems are identified and novel approaches to tackle them are explored. Some of the distinctions of the opportunistic information fusion model (OIFM) compared to the conventional approach are the need for on-the-fly discovery of sensors, ad hoc computational load, and dynamic (not predefined) fusion rules. The key enabling component required to realize an OIFM is a new approach toward middleware development called opportunistic middleware model (OMM). This is because the existing middleware platforms do not scale to the device diversity, size, and runtime dynamics required by OIFM applications [107]. Unfortunately, current specifications for the OMM do not address many issues related to its implementation and thus future research is still needed to make OIFM viable. Nonetheless, some preliminary research work is reported in the literature. For instance, in Ref. [108] an opportunistic fusion of data across time, space, and feature level is performed in a visual sensor network to achieve human gesture analysis. In Ref. [109], the authors study the problem of optimal camera placement in a visual sensor network designed to serve multiple applications (each to be operated in an opportunistic manner). The problem is formulated as a multiobjective optimization problem and solved efficiently using a multiobjective genetic algorithm.

### 2.5.3 ADAPTIVE FUSION AND LEARNING

Early work on adaptive data fusion dates back to the early 1990s [110]. Nonetheless, this problem has rarely been explored in the fusion literature until recently. Some of the existing work is focused on incorporation of adaptivity into the KF algorithm. In Ref. [111] an adaptive fusion system capable of intelligent allocation of limited resources is described that enables efficient tracking of moving targets in three dimensions. An adaptive variant of KF called FL-AKF that relies on fuzzy inference based on covariance matching to adaptively estimate the covariance matrix of measurement noise is proposed in Ref. [112]. In a similar approach, in Ref. [113] Tafti and Sadati present a novel adaptive Kalman filter (NAKF) that achieves adaptation using a mathematical function termed degree of matching (DoM), which is based on covariance matching. An adaptive UKF algorithm with multiple fading factors-based gain correction is proposed and applied to the picosatellite attitude estimation problem [114]. Another trend of work investigates explicit integration of machine learning algorithms into the fusion process to accomplish adaptation. For example, machine learning methods are deployed in Ref. [115] to achieve online adaptation to users' multimodal temporal thresholds within a human–computer interaction application framework. Some other works study application of reinforcement learning to adaptive fusion systems to perform dynamic data reliability estimation [116,117]. Another research work also proposed using kernel-based learning methods to achieve adaptive decision fusion rules [118].

### 2.5.4 DATA RELIABILITY AND TRUST

The majority of data fusion literature work is based on an optimistic assumption about the reliability of underlying models producing the beliefs associated with imperfect data. For instance, sensory data are commonly considered as equally reliable and play a symmetrical role in the fusion process [119]. Nonetheless, different models usually have different reliabilities and are valid only for a specific range. A recent trend in data fusion has addressed this issue mostly by attempting to account for reliability of beliefs. This has been accomplished through introduction of the notion of a second level of uncertainty, that is, uncertainty about uncertainty, represented as reliability coefficients. The main challenges are first to estimate these coefficients and then to incorporate them into the fusion process. A number of approaches to estimate reliability coefficients have been proposed that rely on domain knowledge and contextual information [120], learning through training [121], possibility theory [122], and expert judgments [123]. Further, the problem of reliability incorporation has been studied within several fusion frameworks such as Dempster–Shafer theory [124], fuzzy and possibility theory [125], transferable belief model [126], and probability theory [127]. Another research work also investigates the impact of belief reliability on high-level data fusion [128]. The issue of reliability in data fusion is still not well established, and several open questions such as interrelationship between reliabilities, reliability of heterogeneous data, and a comprehensive architecture to manage data fusion algorithm and reliability of data sources remain part of future research [119,124]. Taking research to the next level, more recent work attempts to address the overarching issues of information quality and higher level quality [129].

### 2.5.5 DATA FUSION IN THE CLOUD AND BIG DATA FUSION

Recent advances in cloud computing technologies have led to the availability of interesting new capabilities for data fusion systems. Some of the most notable benefits of cloud-based computing include scalable and flexible data storage and processing, while maintaining a high level of reliability and security. Enabled by the power of the cloud, modern data fusion algorithms can now be performed over vast amounts of entities across multiple databases [130]. Efficient implementation of such big data fusion systems, however, requires attending to data management, system design, and real-time execution. For instance, the Google fusion table can be deemed as a preliminary cloud-enabled data management and fusion service that allows for uploading, sharing, filtering, and visualization. In particular, it supports the fusion of data from multiple sources through joining across tables sourced by different users [131]. The notion of cloud robotics, initially proposed by James Kuffner at Google [132], is another example of the potential of cloud-enabled data fusion and processing to revolutionize modern robotics. A cloud-enabled team of robots is capable of offloading expensive computational and/or storage robotic tasks, hence allowing users to focus on the application at hand rather than worrying about the underlying IT infrastructure. A recent exemplary work in cloud robotics is described in Ref. [133], where a holistic robotic system is able to leverage cloud services to enhance the performance of video tracking. The experimental results demonstrate the feasibility of offloading computation to the cloud, which is especially beneficial when there are a large number of robot networks demanding image processing tasks.

### 2.5.6 Fusion of Data Streams

The so-called tidal wave of big data is typically characterized by its three-V properties: volume, velocity, and variety. In particular, advances in mobile technologies have led to the proliferation of numerous online data-intensive applications where data streams are being collected continuously in large volume and high speed. When it comes to time-sensitive big data fusion applications that involve processing such extremely large data feeds produced at high speeds by multiple sources, the conventional static database technologies are not sufficient. Examples of such modern real-time applications are traffic monitoring and management, stock price prediction, and flight schedule checking. An alternative solution is to fuse data streams on-the-fly as soon as they are available. Research work on to the stream data fusion problem is gradually gaining popularity. Schueller and Behrend argue [134] in favor of the Reactive Programming (RE) paradigm and the Language Integrated Query (LINQ) language as a promising solution to the problem of storing and fusing real-time streams of data. Dynamic trust assessment over data-in-motion is another important challenge, which has been addressed in Ref. [135]. Their proposal is perhaps the first attempt to develop a dynamic trust assessment framework applicable to data streams with subjective logic as the underlying computational toolset. In a related study, Zhao et al. present a probabilistic model to transform the problem of truth discovery over data streams into a probabilistic inference problem [136]. The proposed approach is claimed to possess advantages such as requiring only a single pass over data, limited memory usage, and short response time, which are backed by preliminary experimental results.

### 2.5.7 Low-Level versus High-Level Data Fusion

The discussion on high-level data fusion may appear to be outside the scope of this chapter. However, as argued in Ref. [137], as soft human-generated data, in the form of complex natural language statements, play an ever-increasing role in modern fusion systems, the clear distinction between low-level and high-level fusion processes might no longer be applicable. The key observation is that the interpretation and analysis of soft data necessitate development of complicated models not restricted to the immediate time frame, similar to those used by the high-level data fusion processes. In particular, knowledge of the context within which a piece of soft data is uttered is crucial in our ability to exploit soft data [138]. A similar line of thought is presented in Ref. [139], where Dragos examines the main challenges involved in dealing with various forms of uncertainties potentially expressed by soft data and the need for high-level ontological analysis to assess them properly.

## 2.5.8 EVOLUTION OF THE JDL MODEL

In an attempt to unify the terminology associated with data and information fusion, the JDL Data Fusion Group developed the JDL fusion model [94]. With subsequent revisions, it is the most widely used system for understanding data fusion processes. The goal of this model is to facilitate understanding and communication among researchers, designers, developers, evaluators, and users of data and information fusion techniques to permit cost-effective system design, development, and operation. The JDL model differentiates data fusion functions into a set of fusion levels and provides a useful distinction among data fusion processes that relate to the refinement of objects, situations, threats, and processes at several levels.

Ever since its conception, the original JDL model has been revised several times to alleviate some of its shortcomings [140,141]. The most recent extensions aim at enabling the JDL model to accommodate emerging trends in the data fusion community. In Ref. [142], Snidaro et al. discuss the important role of contextual information in all JDL levels of modern fusion processes. The authors also present several design and engineering challenges that should be tackled to incorporate contextual knowledge effectively into a data fusion system. Similarly, in Ref. [138], Blasch et al. investigate ways to extend the original JDL model to support exploitation functions and information management for situation awareness and massive data analytics for contextual awareness. There are even studies that present an argument about the adequacy of the JDL model when it comes to describing data fusion systems capable of processing complex human-generated data [137].

## 2.6 CONCLUSION

Data fusion is a multidisciplinary research area with a wide range of potential applications in numerous domains including defense, robotics, automation, intelligent system design, and machine learning. This has driven, and will continue to do so, an ever-increasing interest in the research community in developing more sophisticated data fusion methodologies and architectures. This chapter provided a short yet comprehensive view of the current state of affairs in the data fusion community. Moreover, it examined the effect of recent advances in sensing, storage, and computation on data fusion research trends and consequently presented several areas of work that are promising avenues of future research.

## REFERENCES

1. R. C. Luo, and M. G. Kay. Data fusion and sensor integration: State-of-the-art 1990s. In *Data Fusion in Robotics and Machine Intelligence*, M. A. Abidi and R. C. Gonzalez (eds.), pp. 7–135, Academic Press, San Diego, CA, 1992.
2. R. Joshi, and A. C. Sanderson. *Multisensor Fusion: A Minimal Representation Framework*, vol. 11. World Scientific, Singapore, 1999.
3. A. N. Steinberg, C. L. Bowman, and F. E. White. Revisions to the JDL data fusion model. In *AeroSense'99*, pp. 430–441. International Society for Optics and Photonics, Bellingham, WA, 1999.
4. L. A. Klein. *Sensor and Data Fusion Concepts and Applications*. Society of Photo-Optical Instrumentation Engineers (SPIE), Bellingham, WA, 1993.
5. H. Boström, S. F. Andler, M. Brohede, R. Johansson, A. Karlsson, J. Van Laere, L. Niklasson, M. Nilsson, A. Persson, and T. Ziemke. On the definition of information fusion as a field of research, Informatics Research Centre, University of Skövde, Tech. Rep. HS-IKI-TR-07-006, 2007.
6. B. V. Dasarathy. *Decision Fusion*, vol. 1994. IEEE Computer Society Press, Los Alamitos, CA, 1994.
7. I. R. Goodman. *Mathematics of Data Fusion*, vol. 37. Springer, New York, 1997.
8. M. M. Kokar, J. A. Tomasik, and J. Weyman. Formalizing classes of information fusion systems. *Information Fusion*, 5(3):189–202, 2004.
9. H. Hu, and J. Q. Gan. Sensors and data fusion algorithms in mobile robotics. *Report CSM-422*, Department of Computer Science, University of Essex, Tech. Rep. CSM-422, 2005.
10. H. B. Mitchell. *Multi-Sensor Data Fusion*. Springer, New York, 2007.

11. R. C. Luo, and M. G. Kay. Multisensor integration and fusion in intelligent systems. *Systems, Man and Cybernetics, IEEE Transactions on,* 19(5):901–931, 1989.

12. Y. Fischer, and A. Bauer. Object-oriented sensor data fusion for wide maritime surveillance. In *Waterside Security Conference (WSS), 2010 International,* pp. 1–6. IEEE, 2010.

13. Y. Wu, A. Patterson, R. D. C. Santos, and N. L. Vijaykumar. Topology preserving mapping for maritime anomaly detection. In *Computational Science and Its Applications–ICCSA 2014,* pp. 313–326. Springer, New York, 2014.

14. H. Oh, S. Kim, H.-S. Shin, A. Tsourdos, and B. A. White. Behaviour recognition of ground vehicle using airborne monitoring of unmanned aerial vehicles. *International Journal of Systems Science,* 45(12):2499–2514, 2014.

15. R. Du, J. Liu, Y. Wang, Z. Li, and C. Gao. Fast data fusion algorithm for tracking maneuvering target by vehicle formation. In *Radar Conference (RADAR), 2013 IEEE,* pp. 1–6. IEEE, 2013.

16. W. Koch. *Tracking and Sensor Data Fusion: Methodological Framework and Selected Applications.* Springer, New York, 2013.

17. A. R. Benaskeur, F. Rheaume, and S. Paradis. Target engageability improvement through adaptive tracking. *Journal of Advances in Information Fusion,* 2(2):99–112, 2007.

18. F. Rheaume, and A. R. Benaskeur. Fire control-based adaptation in data fusion applications. In *Information Fusion, 2007 10th International Conference on,* pp. 1–8. IEEE, 2007.

19. H. Frigui, L. Zhang, P. Gader, J. N. Wilson, K. C. Ho, and A. Mendez-Vazquez. An evaluation of several fusion algorithms for anti-tank landmine detection and discrimination. *Information Fusion,* 13(2):161–174, 2012.

20. M. Liggins II, D. Hall, and J. Llinas. *Handbook of Multisensor Data Fusion: Theory and Practice.* CRC Press, Boca Raton, FL, 2008.

21. M. R. Endsley. Toward a theory of situation awareness in dynamic systems. *Human Factors: The Journal of the Human Factors and Ergonomics Society,* 37(1):32–64, 1995.

22. P. Park, H. Khadilkar, H. Balakrishnan, and C. Tomlin. High confidence networked control for next generation air transportation systems. *IEEE Transactions on Automatic Control,* 59(12):3357–3372, 2014.

23. Z. Jin, X. Wang, Q. Gui, B. Liu, and S. Song. Improving diagnostic accuracy using multiparameter patient monitoring based on data fusion in the cloud. In *Future Information Technology,* pp. 473–476. Springer, New York, 2014.

24. G. Fortino, S. Galzarano, R. Gravina, and W. Li. A framework for collaborative computing and multisensor data fusion in body sensor networks. *Information Fusion,* 22:50–70, 2015.

25. A. Rae, A. Khamis, O. Basir, and M. Kamel. Particle filtering for bearing-only audio-visual speaker detection and tracking. In *Signals, Circuits and Systems (SCS), 2009 3rd International Conference on,* pp. 1–6. IEEE, 2009.

26. S. Parasuraman, and B. Shirinzadeh. Fuzzy logic based sensors data fusion for mobile robot navigation. *Key Engineering Materials,* 467:794–799, 2011.

27. W. A. Abdulhafiz, and A. Khamis. Bayesian approach with pre-and post-filtering to handle data uncertainty and inconsistency in mobile robot local positioning. *Journal of Intelligent Systems,* 23(2):133–154, 2014.

28. N.-E. El Faouzi, H. Leung, and A. Kurian. Data fusion in intelligent transportation systems: Progress and challenges—A survey. *Information Fusion,* 12(1):4–10, 2011.

29. T. Ranchin, and L. Wald. *Data Fusion in Remote Sensing of Urban and Suburban Areas.* Springer, New York, 2010.

30. A. Brook, E. Ben-Dor, and R. Richter. Modelling and monitoring urban built environment via multisource integrated and fused remote sensing data. *International Journal of Image and Data Fusion,* 4(1):2–32, 2013.

31. D. Chen, Z. Liu, L. Wang, M. Dou, J. Chen, and H. Li. Natural disaster monitoring with wireless sensor networks: A case study of data-intensive applications upon low-cost scalable systems. *Mobile Networks and Applications,* 18(5):651–663, 2013.

32. Y. Dingcheng, W. Zhenghai, X. Lin, and Z. Tiankui. Online Bayesian data fusion in environment monitoring sensor networks. *International Journal of Distributed Sensor Networks,* 2014:Article ID 945894, 2014.

33. M. P. Jenkins, G. A. Gross, A. M. Bisantz, and R. Nagi. Towards context aware data fusion: Modeling and integration of situationally qualified human observations to manage uncertainty in a hard + soft fusion process. *Information Fusion,* 21:130–144, 2015. doi: http://dx.doi.org/10.1155/2014/945894.

34. B. Khaleghi, A. Khamis, F. O. Karray, and S. N. Razavi. Multisensor data fusion: A review of the state-of-the-art. *Information Fusion,* 14(1):28–44, 2013.

35. P. Krause, and D. Clark. *Representing Uncertain Knowledge: An Artificial Intelligence Approach.* Kluwer Academic Publishers, Dordrecht, 1993.

36. P. Smets. Imperfect information: Imprecision and uncertainty. In *Uncertainty Management in Information Systems*, A. Motro and P. Smets (eds.), pp. 225–254. Springer, New York, 1997.

37. G. J. Klir, and M. J. Wierman. *Uncertainty-Based Information: Elements of Generalized Information Theory*, vol. 15. Springer, New York, 1999.

38. D. Dubois, and H. Prade. Formal representations of uncertainty. In *Decision-Making Process: Concepts and Methods*, D. Bouyssou, D. Dubois, H. Prade, and M. Pirlot (eds.), pp. 85–156. John Wiley & Sons, New York, 2009.

39. M. Florea, A.-L. Jousselme, and E. Bosse. Fusion of imperfect information in the unified framework of random sets theory: Application to target identification. Technical Report, *Defence R&D Canada*. Valcartier, Tech. Rep. ADA475342, 2007.

40. J. Komorowski, Z. Pawlak, L. Polkowski, and A. Skowron. Rough sets: A tutorial. In *Rough Fuzzy Hybridization: A New Trend in Decision-Making*, S. K. Pal and A. Skowron (eds.), pp. 3–98. Springer, New York, 1999.

41. F. K. J. Sheridan. A survey of techniques for inference under uncertainty. *Artificial Intelligence Review*, 5(1–2):89–119, 1991.

42. H. Durrant-Whyte, and T. C. Henderson. Multisensor data fusion. In *Springer Handbook of Robotics*, B. Siciliano and O. Khatib (eds.), pp. 585–610. Springer, New York, 2008.

43. L. A. Zadeh. Fuzzy sets. *Information and Control*, 8(3):338–353, 1965.

44. F. O. Karray, and C. W. De Silva. *Soft Computing and Intelligent Systems Design: Theory, Tools, and Applications*. Pearson Education, Upper Saddle River, NJ, 2004.

45. C. V. Negoita, L. A. Zadeh, and H. J. Zimmermann. Fuzzy sets as a basis for a theory of possibility. *Fuzzy Sets and Systems*, 1:3–28, 1978.

46. Z. Pawlak. *Rough Sets: Theoretical Aspects of Reasoning about Data*, vol. 9. Springer, New York, 1991.

47. G. Shafer. *A Mathematical Theory of Evidence*, vol. 1. Princeton University Press, Princeton, NJ, 1976.

48. D. Dubois, and H. Prade. Rough fuzzy sets and fuzzy rough sets. *International Journal of General System*, 17(2–3):191–209, 1990.

49. J. Yen. Generalizing the Dempster-Schafer theory to fuzzy sets. *Systems, Man and Cybernetics, IEEE Transactions on*, 20(3):559–570, 1990.

50. R. P. S. Mahler. *Statistical Multisource-Multitarget Information Fusion*, vol. 685. Artech House, Boston, 2007.

51. S. J. Julier, and J. K. Uhlmann. A non-divergent estimation algorithm in the presence of unknown correlations. In *Proceedings of the American Control Conference*. Albuquerque, NM, 1997.

52. A. Makarenko, A. Brooks, T. Kaupp, H. Durrant-Whyte, and F. Dellaert. Decentralised data fusion: A graphical model approach. In *Information Fusion, 2009. FUSION'09. 12th International Conference on*, pp. 545–554. IEEE, 2009.

53. D. Smith, and S. Singh. Approaches to multisensor data fusion in target tracking: A survey. *Knowledge and Data Engineering, IEEE Transactions on*, 18(12):1696–1710, 2006.

54. P. S. Maybeck. *Stochastic Models, Estimation, and Control*, vol. 3. Academic Press, New York, 1982.

55. S. P. McLaughlin, R. J. Evans, and V. Krishnamurthy. A graph theoretic approach to data incest management in network centric warfare. In *Information Fusion, 2005 8th International Conference on*, vol. 2, pp. 1162–1169. IEEE, 2005.

56. T. Brehard, and V. Krishnamurthy. Optimal data incest removal in Bayesian decentralized estimation over a sensor network. In *Acoustics, Speech and Signal Processing, 2007. ICASSP 2007. IEEE International Conference on*, vol. 3, pp. III-173–III-176. IEEE, 2007.

57. W. Niehsen. Information fusion based on fast covariance intersection filtering. In *Information Fusion, 2002. Proceedings of the 5th International Conference on*, vol. 2, pp. 901–904. IEEE, 2002.

58. D. Franken, and A. Hupper. Improved fast covariance intersection for distributed data fusion. In *Information Fusion, 2005 8th International Conference on*, vol. 1, pp. 154–160. IEEE, 2005.

59. A. R. Benaskeur. Consistent fusion of correlated data sources. In *IECON 02. Industrial Electronics Society, IEEE 2002 28th Annual Conference of the*, vol. 4, pp. 2652–2656. IEEE, 2002.

60. M. Kumar, D. P. Garg, and R. A. Zachery. A generalized approach for inconsistency detection in data fusion from multiple sensors. In *American Control Conference, 2006*, pp. 2078–2083. IEEE, 2006.

61. S. J. Wellington, J. K. Atkinson, and R. P. Sion. Sensor validation and fusion using the nadaraya-watson statistical estimator. In *Information Fusion, 2002. Proceedings of the 5th International Conference on*, vol. 1, pp. 321–326. IEEE, 2002.

62. P. H. Ibarguengoytia, L. E. Sucar, and S. Vadera. Real time intelligent sensor validation. *Power Systems, IEEE Transactions on*, 16(4):770–775, 2001.

63. J. Frolik, M. Abdelrahman, and P. Kandasamy. A confidence-based approach to the self-validation, fusion and reconstruction of quasi-redundant sensor data. *Instrumentation and Measurement, IEEE Transactions on*, 50(6):1761–1769, 2001.

64. M. Kumar, D. P. Garg, and R. A. Zachery. A method for judicious fusion of inconsistent multiple sensor data. *Sensors Journal, IEEE*, 7(5):723–733, 2007.

65. M. Kumar, D. Garg, and R. Zachery. Stochastic adaptive sensor modeling and data fusion. In *Smart Structures and Materials*, M. Tomizuka, C. B. Yun, and V. Giurgiutiu (eds.), pp. 100–110. International Society for Optics and Photonics, San Diego, CA, 2006.

66. U. Orguner, and F. Gustafsson. Storage efficient particle filters for the out of sequence measurement problem. In *Information Fusion, 2008 11th International Conference on*, pp. 1–8. IEEE, 2008.

67. S. Blackman, and Artech House. *Design and Analysis of Modern Tracking Systems*. Artech House, Boston, 1999.

68. Y. Bar-Shalom. Update with out-of-sequence measurements in tracking: Exact solution. *Aerospace and Electronic Systems, IEEE Transactions on*, 38(3):769–777, 2002.

69. M. Mallick, S. Coraluppi, and C. Carthel. Advances in asynchronous and decentralized estimation. In *Aerospace Conference, 2001, IEEE Proceedings*, vol. 4, pp. 1873–1888. IEEE, 2001.

70. K. Zhang, and X. R. Li. Optimal update with out-of-sequence measurements for distributed filtering. In *Information Fusion, 2002. Proceedings of the 5th International Conference on*, vol. 2, pp. 1519–1526. IEEE, 2002.

71. K. Zhang, X. R. Li, and Y. Zhu. Optimal update with out-of-sequence measurements. *Signal Processing, IEEE Transactions on*, 53(6):1992–2004, 2005.

72. Y. Bar-Shalom, H. Chen, and M. Mallick. One-step solution for the multistep out-of-sequence-measurement problem in tracking. *Aerospace and Electronic Systems, IEEE Transactions on*, 40(1):27–37, 2004.

73. Y. Anxi, L. Diannong, H. Weidong, and D. Zhen. A unified out-of-sequence measurements filter. In *Radar Conference, 2005 IEEE International*, pp. 453–458. IEEE, 2005.

74. L. A. Zadeh. Review of a mathematical theory of evidence. *AI Magazine*, 5(3):81, 1984.

75. M. C. Florea, A.-L. Jousselme, E. Bosse, and D. Grenier. Robust combination rules for evidence theory. *Information Fusion*, 10(2):183–197, 2009.

76. R. R. Yager. On the Dempster-Shafer framework and new combination rules. *Information Sciences*, 41(2):93–137, 1987.

77. P. Smets. The combination of evidence in the transferable belief model. *Pattern Analysis and Machine Intelligence, IEEE Transactions on*, 12(5):447–458, 1990.

78. J. Dezert. Foundations for a new theory of plausible and paradoxical reasoning. *Information and Security*, 9:13–57, 2002.

79. E. Lefevre, O. Colot, and P. Vannoorenberghe. Belief function combination and conflict management. *Information Fusion*, 3(2):149–162, 2002.

80. F. Voorbraak. On the justification of Dempster's rule of combination. *Artificial Intelligence*, 48(2):171–197, 1991.

81. R. Haenni. Are alternatives to Dempster's rule of combination real alternatives?: Comments on "About the belief function combination and the conflict management problem"—Lefevre et al. *Information Fusion*, 3(3):237–239, 2002.

82. J. K. Uhlmann. Covariance consistency methods for fault-tolerant distributed data fusion. *Information Fusion*, 4(3):201–215, 2003.

83. S. Maskell. A Bayesian approach to fusing uncertain, imprecise and conflicting information. *Information Fusion*, 9(2):259–277, 2008.

84. D. L. Hall, M. McNeese, J. Llinas, and T. Mullen. A framework for dynamic hard/soft fusion. In *Information Fusion, 2008 11th International Conference on*, pp. 1–8. IEEE, 2008.

85. M. A. Pravia, R. K. Prasanth, P. O. Arambel, C. Sidner, and C.-Y. Chong. Generation of a fundamental data set for hard/soft information fusion. In *Information Fusion, 2008 11th International Conference on*, pp. 1–8. IEEE, 2008.

86. M. A. Pravia, O. Babko-Malaya, M. K. Schneider, J. V. White, C.-Y. Chong, and A. S. Willsky. Lessons learned in the creation of a data set for hard/soft information fusion. In *Information Fusion, 2009. FUSION'09. 12th International Conference on*, pp. 2114–2121. IEEE, 2009.

87. K. Premaratne, M. N. Murthi, J. Zhang, M. Scheutz, and P. H. Bauer. A Dempster-Shafer theoretic conditional approach to evidence updating for fusion of hard and soft data. In *Information Fusion, 2009. FUSION'09. 12th International Conference on*, pp. 2122–2129. IEEE, 2009.

88. S. Seifzadeh, B. Khaleghi, and F. Karray. Soft-data-constrained multi-model particle filter for agile target tracking. In *Information Fusion (FUSION), 2013 16th International Conference on*, pp. 564–571. IEEE, 2013.

89. H. Chen, G. Chen, E. P. Blasch, P. Douville, and K. Pham. Information theoretic measures for performance evaluation and comparison. In *Information Fusion, 2009. FUSION'09. 12th International Conference on*, pp. 874–881. IEEE, 2009.

90. L. Cholvy. Information evaluation in fusion: A case study. In *Proceedings of the Conference IPMU 2004*. Perugia, Italy, 2004.

91. V. Nimier. Information evaluation: A formalisation of operational recommendations. *NATO Science Series Sub Series III: Computer and Systems Sciences*, 198:81, 2005.

92. L. Cholvy. Modelling information evaluation in fusion. In *Information Fusion, 2007 10th International Conference on*, pp. 1–6. IEEE, 2007.

93. A. E. Gelfand, C. Smith, M. Colony, and C. Bowman. Performance evaluation of decentralized estimation systems with uncertain communication. In *Information Fusion, 2009. FUSION'09. 12th International Conference on*, pp. 786–793. IEEE, 2009.

94. O. E. Drummond. Methodologies for performance evaluation of multitarget multisensor tracking. In *SPIE's International Symposium on Optical Science, Engineering, and Instrumentation*, pp. 355–369. International Society for Optics and Photonics, 1999.

95. R. L. Rothrock, and O. E. Drummond. Performance metrics for multiple-sensor multiple-target tracking. In *AeroSense 2000*, pp. 521–531. International Society for Optics and Photonics, 2000.

96. T. Zajic, J. L. Hoffman, and R. P. S. Mahler. Scientific performance metrics for data fusion: New results. In *AeroSense 2000*, pp. 172–182. International Society for Optics and Photonics, 2000.

97. D. Schuhmacher, B.-T. Vo, and B.-N. Vo. A consistent metric for performance evaluation of multi-object filters. *Signal Processing, IEEE Transactions on*, 56(8):3447–3457, 2008.

98. B.-T. Vo. *Random Finite Sets in Multi-Object Filtering*. PhD dissertation, University of Western Australia, 2008.

99. P. Jackson, and J. D. Musiak. Boeing fusion performance analysis (FPA) tool. In *Information Fusion, 2009. FUSION'09. 12th International Conference on*, pp. 1444–1450. IEEE, 2009.

100. D. Akselrod, R. Tharmarasa, T. Kirubarajan, Z. Ding, and T. Ponsford. Multisensor-multitarget tracking testbed. In *Computational Intelligence for Security and Defense Applications, 2009. CISDA 2009. IEEE Symposium on*, pp. 1–6. IEEE, 2009.

101. J. van Laere. Challenges for if performance evaluation in practice. In *Information Fusion, 2009. FUSION'09. 12th International Conference on*, pp. 866–873. IEEE, 2009.

102. S. R. Yerva, H. Jeung, and K. Aberer. Cloud based social and sensor data fusion. In *Information Fusion (FUSION), 2012 15th International Conference on*, pp. 2494–2501. IEEE, 2012.

103. A. Rosi, M. Mamei, and F. Zambonelli. Integrating social sensors and pervasive services: Approaches and perspectives. *International Journal of Pervasive Computing and Communications*, 9(4):294–310, 2013.

104. A. Beach, M. Gartrell, X. Xing, R. Han, Q. Lv, S. Mishra, and K. Seada. Fusing mobile, sensor, and social data to fully enable context-aware computing. In *Proceedings of the 11th Workshop on Mobile Computing Systems & Applications*, pp. 60–65. ACM, 2010.

105. T. Lovett, E. O'Neill, J. Irwin, and D. Pollington. The calendar as a sensor: Analysis and improvement using data fusion with social networks and location. In *Proceedings of the 12th ACM International Conference on Ubiquitous Computing*, pp. 3–12. ACM, 2010.

106. C. C. Aggarwal, and T. Abdelzaher. Integrating sensors and social networks. In *Social Network Data Analytics*, C. C. Aggarwal (ed.), pp. 379–412. Springer, New York, 2011.

107. S. Challa, T. Gulrez, Z. Chaczko, and T. N. Paranesha. Opportunistic information fusion: A new paradigm for next generation networked sensing systems. In *Information Fusion, 2005 8th International Conference on*, vol. 1, pp. 720–727. IEEE, 2005.

108. C. Wu, and H. Aghajan. Model-based human posture estimation for gesture analysis in an opportunistic fusion smart camera network. In *Advanced Video and Signal Based Surveillance, 2007. AVSS 2007. IEEE Conference on*, pp. 453–458. IEEE, 2007.

109. R. Al-Hmouz, and S. Challa. Optimal placement for opportunistic cameras using genetic algorithm. In *Intelligent Sensors, Sensor Networks and Information Processing Conference, 2005. Proceedings of the 2005 International Conference on*, pp. 337–341. IEEE, 2005.

110. L. Hong. Adaptive data fusion. In *Systems, Man, and Cybernetics, 1991. Decision Aiding for Complex Systems, Conference Proceedings, 1991 IEEE International Conference on*, pp. 767–772. IEEE, 1991.

111. G. Loy, L. Fletcher, N. Apostoloff, and A. Zelinsky. An adaptive fusion architecture for target tracking. In *Automatic Face and Gesture Recognition, 2002. Proceedings, 5th IEEE International Conference on*, pp. 261–266. IEEE, 2002.

112. P. J. Escamilla-Ambrosio, and N. Mort. Hybrid kalman filter-fuzzy logic adaptive multisensor data fusion architectures. In *Decision and Control, 2003. Proceedings, 42nd IEEE Conference on*, vol. 5, pp. 5215–5220. IEEE, 2003.

113. A. D. Tafti, and N. Sadati. Novel adaptive Kalman filtering and fuzzy track fusion approach for real time applications. In *Industrial Electronics and Applications, 2008. ICIEA 2008. 3rd IEEE Conference on*, pp. 120–125. IEEE, 2008.

114. H. E. Soken, and C. Hajiyev. Adaptive unscented Kalman filter with multiple fading factors for pico satellite attitude estimation. In *Recent Advances in Space Technologies, 2009. RAST'09. 4th International Conference on*, pp. 541–546. IEEE, 2009.

115. X. Huang, and S. Oviatt. Toward adaptive information fusion in multimodal systems. In *Machine Learning for Multimodal Interaction*, A. Popescu-Belis, S. Renals, and H. Bourlard (eds.), pp. 15–27. Springer, New York, 2006.

116. N. Ansari, E. S. H. Hou, B.-O. Zhu, and J.-G. Chen. Adaptive fusion by reinforcement learning for distributed detection systems. *Aerospace and Electronic Systems, IEEE Transactions on*, 32(2):524–531, 1996.

117. M. A. Hossain, P. K. Atrey, and A. El Saddik. Learning multisensor confidence using a reward-and-punishment mechanism. *Instrumentation and Measurement, IEEE Transactions on*, 58(5):1525–1534, 2009.

118. G. Fabeck, and R. Mathar. Kernel-based learning of decision fusion in wireless sensor networks. In *Information Fusion, 2008 11th International Conference on*, pp. 1–7. IEEE, 2008.

119. G. L. Rogova, and V. Nimier. Reliability in information fusion: Literature survey. In *Proceedings of the 7th International Conference on Information Fusion*, pp. 1158–1165. IEEE, 2004.

120. V. Nimier. Introducing contextual information in multisensor tracking algorithms. In *Advances in Intelligent Computing IPMU'94*, pp. 595–604. Springer, New York, 1995.

121. B. Yu, and K. Sycara. Learning the quality of sensor data in distributed decision fusion. In *Information Fusion, 2006 9th International Conference on*, pp. 1–8. IEEE, 2006.

122. F. Delmotte, L. Dubois, and P. Borne. Context-dependent trust in data fusion within the possibility theory. In *Systems, Man, and Cybernetics, 1996, IEEE International Conference on*, vol. 1, pp. 538–543. IEEE, 1996.

123. S. A. Sandri, D. Dubois, and H. W. Kalfsbeek. Elicitation, assessment, and pooling of expert judgments using possibility theory. *Fuzzy Systems, IEEE Transactions on*, 3(3):313–335, 1995.

124. R. Haenni, and S. Hartmann. Modeling partially reliable information sources: A general approach based on Dempster–Shafer theory. *Information Fusion*, 7(4):361–379, 2006.

125. D. Dubois, and H. Prade. Possibility theory in information fusion. In *Information Fusion, 2000. FUSION 2000. Proceedings of the 3rd International Conference on*, vol. 1, pp. PS6–P19. IEEE, 2000.

126. Z. Elouedi, K. Mellouli, and P. Smets. Assessing sensor reliability for multisensor data fusion within the transferable belief model. *Systems, Man, and Cybernetics, Part B: Cybernetics, IEEE Transactions on*, 34(1):782–787, 2004.

127. E. J. Wright, and K. B. Laskey. Credibility models for multi-source fusion. In *Information Fusion, 2006 9th International Conference on*, pp. 1–7. IEEE, 2006.

128. A. Karlsson. Dependable and generic high-level information fusion: Methods and algorithms for uncertainty management, University of Skovde, Tech. Rep. HS-IKI-TR-07-003, 2007.

129. G. L. Rogova, and E. Bosse. Information quality in information fusion. In *Information Fusion (FUSION), 2010 13th Conference on*, pp. 1–8. IEEE, 2010.

130. E. Blasch, Y. Chen, G. Chen, D. Shen, and R. Kohler. Information fusion in a cloud-enabled environment. In *High Performance Cloud Auditing and Applications*, K. J. Han, B. K. Choi, and S. Song (eds.), pp. 91–115. Springer, New York, 2014.

131. H. Gonzalez, A. Halevy, C. S. Jensen, A. Langen, J. Madhavan, R. Shapley, and W. Shen. Google fusion tables: Data management, integration and collaboration in the cloud. In *Proceedings of the 1st ACM Symposium on Cloud Computing*, pp. 175–180. ACM, 2010.

132. J. J. Kuffner. Cloud-enabled robots. In *IEEE-RAS International Conference on Humanoid Robotics*. Nashville, TN, 2010.

133. B. Liu, Y. Chen, E. Blasch, K. Pham, D. Shen, and G. Chen. A holistic cloud-enabled robotics system for real-time video tracking application. In *Future Information Technology*, J. J. Park, I. Stojmenovic, M. Choi, and F. Xhafa (eds.), pp. 455–468. Springer, New York, 2014.

134. G. Schueller, and A. Behrend. Stream fusion using reactive programming, linq and magic updates. In *FUSION*, pp. 1265–1272. IEEE, 2013.

135. S. Arunkumar, M. Srivatsa, D. Braines, and M. Sensoy. Assessing trust over uncertain rules and streaming data. In *Information Fusion (FUSION), 2013 16th International Conference on*, pp. 922–929. IEEE, 2013.

136. Z. Zhao, J. Cheng, and W. Ng. Truth discovery in data streams: A single-pass probabilistic approach. In *International Conference on Information and Knowledge Management*, pp. 1589–1598. 2014.

137. K. Rein, and J. Biermann. Your high-level information is my low-level data-a new look at terminology for multi-level fusion. In *Information Fusion (FUSION), 2013 16th International Conference on*, pp. 412–417. IEEE, 2013.

138. E. Blasch, A. Steinberg, S. Das, J. Llinas, C. Chong, O. Kessler, E. Waltz, and F. White. Revisiting the JDL model for information exploitation. In *Information Fusion (FUSION), 2013 16th International Conference on*, pp. 129–136. IEEE, 2013.

139. V. Dragos. An ontological analysis of uncertainty in soft data. In *Information Fusion (FUSION), 2013 16th International Conference on*, pp. 1566–1573. IEEE, 2013.

140. J. Llinas, C. Bowman, G. Rogova, A. Steinberg, E. Waltz, and F. E. White. Revisiting the JDL data fusion model II. In *International Conference on Information Fusion*, pp. 1218–1230, 2004.

141. J. J. Salerno. Where's level 2/3 fusion—A look back over the past 10 years. In *Information Fusion, 2007 10th International Conference on*, pp. 1–4. IEEE, 2007.

142. L. Snidaro, I. Visentini, J. Llinas, and G. L. Foresti. Context in fusion: Some considerations in a JDL perspective. In *Information Fusion (FUSION), 2013 16th International Conference on*, pp. 115–120. IEEE, 2013.

# 3 Information Fusion
## *Theory at Work*

*Jean-François Grandin*

## CONTENTS

## 3.1 INTRODUCTION

### 3.1.1 RELEVANCE OF DATA FUSION FOR MILITARY SYSTEMS

The volume and complexity of data handled by modern defense systems have taken several quantum leaps as a result of advances made in the electronics sector. Further, every decision has to be taken very quickly and its consequences can be increasingly costly. There is also a lack of experts able to analyze these data. The equation is simple: the volume of data is multiplied by 10,000; the complexity is multiplied by 10 or more; and the number of clever experts is reduced by a factor of 2 or 3. So it is of prime importance to develop automatic systems to help operators and reduce their stress. Fortunately, the considerable growth in the power of data processing resources in all areas (data acquisition, communication, storage, computation) now enables reasonably priced hardware to use real-time interpretation and analysis techniques. Data fusion is applied to many defense systems. In the domain major references are Refs. [1–4].

### 3.1.2   Major Functions of Data Fusion

The functions of surveillance systems are to detect, localize, track, classify, and identify different objects that cross the area of interest [1]. To assess the requirements of such a system, one has to combine the information given by several sensors or information sources that are, or are not, co-localized and often deliver heterogeneous information. These functions always optimize some criteria; for example, in probability the criterion that is often used is the maximum likelihood criterion. It means that the measurement errors knowing the decision (detection, association, estimation, identification) have a maximum likelihood.

#### 3.1.2.1   Association

The purpose of association is to cluster similar measurements coming from different sensors. To be able to do so, it is very important that the sensors deliver, at least in part, redundant information, because the association is based on correlation of common characteristics. Association is essentially a comparison of common data characteristics. This comparison is not trivial because it can use several characteristics that are not corrupted in the same way. Further, the association operator sometimes has to take into account a veto partial decision that cannot be represented by a metric. It is important to know that the association performance is dominated by the sensor that is the less effective.

#### 3.1.2.2   Estimation

The purpose of estimation is to combine measurements coming from different sensors. The combination of information will deliver

- A more and more accurate description based on complementary information. Two pieces of information are complementary if the estimation of characteristics of an object requires both. The integration of several complementary measures increases the specificity or accuracy of the description.
- More and more reliable information and a measure of the reliability based on supplementary information. Two pieces of information are supplementary or redundant if the knowledge of one eliminates the necessity to know the other. This redundancy is useful because the confrontation of redundant information will deliver some knowledge on data reliability: integration on several redundant measures increases the certainty of the measure.

The estimation performance is dominated by the sensor that is the most effective.

#### 3.1.2.3   Identification

The purpose of identity estimation is to recognize an object. The identity estimation is of crucial importance in military systems, especially for electronic intelligence (ELINT) or electronic support measure (ESM) systems [5–8]. Identity estimation relies not only on measurements but also on a database where the knowledge of object characteristics are stored. Data fusion methods for identification have been presented in Ref. [2] and in Chapter 7 of Ref. [3], where Section 7.2 presents one of the first comparisons between Bayesian and Dempster–Shafer methods. These types of comparisons have been detailed in Refs. [9,10]. In this chapter, we focus on the identification function and on comparisons of Bayesian and Dempster–Shafer methods for identity estimation in data fusion.

### 3.1.3 IMPORTANCE OF INFORMATION QUALITIES

A fusion process improves the perception of the environment thanks to the multiple interpretations provided by different sensors. However, any information processing is based on information content. If there is no information embedded in the processing the result will be very poor: "garbage in, garbage out." Hence it is of primary importance to describe all the information the system may process before discussing the processing itself. It is always verified that adding more information to "simple" processing is better than making processing more complex. Information that is of prime importance is the quality of the measurement sources. In data fusion this was highlighted in Ref. [11]. The fusion process basically consists in the propagation of inaccuracies and uncertainties over the sensors and processing that make up the whole system [12]. Without these considerations, no real information fusion is performed. In signal processing, the process of filtering while considering known statistical characteristics of signal and noise process is commonplace. Probabilistic filtering consists in computing posterior probability thanks to the knowledge of sensor characteristics. Here, this filtering process is fundamental because it is applied to decision events, whose semantic level is higher than those of signal samples in signal processing. In multiple sensor systems, it is necessary to take advantage of the multiplicity of the measured parameters to mitigate the temporary deficiency of a part of the measurements. It is thus very important to filter the probabilities of the events by use of quality factors. These quality factors might be estimated instantaneously or through a statistical process.

### 3.1.4 TYPES OF INFORMATION QUALITIES

There are several major information imperfections:

- The *inaccuracy*, which represents a quantitative lack of knowledge. The variance reflects the precision on a set of measures.
- The *incompleteness*, which represents a lack of observability. Often sensors have insufficient time or field of view to achieve a complete description of objects. But often, observation conditions make it possible to describe the incompleteness.
- The *reliability* expresses the proportion of the measures realized within the framework of a normal functioning of the sensor [12]. The sensors measure the environment and interpret the measurements with a priori models. The sensors misread the situation if the a priori models are too different from the real models. Any information can be characterized by a degree of validity with regard to an a priori model or by immediate criteria. If the sensors do not provide reliability information, it might be possible to evaluate it through cross-entropy calculation, which basically measures the concordance of the sources to each other, and to themselves. As shown by Grandin and Marques [13,14], the use of reliability coefficients significantly improves the results thanks to a probabilistic filtering process for different methods taking into account these coefficients, although these reliability coefficients are very roughly specified.
- The *conflict* between partial elements of information is also a source of uncertainty. This situation can result from noisy measurements, false information, and poor reliability but can also be due to the specificity of some combination of information. In the first case, the presence of a conflict must decrease the estimated belief. In the second case, the high degree of conflict is admissible because it corresponds to an actual object.

Information imperfections can further be classified in two kinds of uncertainty [15]:

- *Objective uncertainty* or *stochastic uncertainty* corresponds to the variability that emerges randomly. It is the case of accuracy and reliability.
- *Subjective uncertainty* corresponds to knowledge deficiency; this is the case of incompleteness.

## 3.2  THEORIES FOR DATA FUSION

Uncertainty propagation techniques are needed in systems in which several sensors contribute to the analysis of a situation. It is necessary to take into account in the fusion process the uncertainty of each sensor. It has been demonstrated [13,14] that the profits of a model informed by the reliability are consistent and moreover robust to partially unknown reliabilities. Several techniques have been suggested to do this: Probabilities of Bayes [16], Possibility–Necessity of Dubois and Prade [17,18]. Upper and Lower Probabilities of Dempster [19], and Belief-Plausibility of Dempster–Shafer (also called Evidence Theory) [20,21]. We also mention the Transferable Belief Model (TBM), which is another model to represent qualified uncertainties based on belief functions. The TBM is presented in Refs. [22–25]. It is therefore essential for engineers to appreciate the advantages and drawbacks of each one, through theoretical but also through some concrete elements. The combination operators in the theories rely on axiomatic properties that make them appropriate or not for different situations. This chapter presents some important differences that exist between Bayesian and Dempster–Shafer theories, which allow the uncertainty management in data fusion. This was previously highlighted by Grandin and Moulin [26]. In the case of Bayesian mass allocation, Probability and Evidence theories are in full concordance. It is when non-Bayesian allocation is provided that differences between the two theories can be highlighted [26]. This chapter illustrates and explains these aspects. It is generally agreed that the two theories do not address the same problems. This does not mean that these theories are contradictory, but rather that they can be used as complementary to solve complex problems of the type that will be presented here. The performance of a fusion process also depends on the general level of conflict between sources. According to the degree of conflict, it will be necessary to privilege the use of certain operators, adaptive in nature, that take into account the conflict between sources. In a previous paper the author [12] demonstrated how to build such operators using the different theories. We can see that the conflict is often the consequence of the qualities of the measures. Little reliable or too vague sources will supply more scattered and thus conflicting measures.

### 3.2.1  Credibility Mass Assignment

It is important to start with the definition of a credibility mass assignment. The sets from $\wp(\Omega)$ are stable for Union and Complement operators:

$$\varnothing, \Omega \in \wp(\Omega) \quad E, F \in \wp(\Omega) \Rightarrow E \cup F \in \wp(\Omega) \quad E \in \wp(\Omega) \Rightarrow \bar{E} \in \wp(\Omega)$$

A credibility mass assignment is a function $\mu$ from the elements of $\wp(\Omega)$ to the interval $[0, +\infty]$ or $[0, 1]$ when normalized.

$$\mu : \wp(\Omega) \rightarrow [0, +\infty], E \rightarrow \mu(E)$$

The function $\mu$ has three fundamentals properties:

- Monotony: $\forall E, F \in \wp(\Omega), \quad E \subseteq F \Rightarrow \mu(E) \leq \mu(F)$
- Limit conditions: $\mu(\varnothing) = 0, \quad \mu(\Omega) = 1$

The interpretation is simple: If one is confident of the fact that an event belongs to a subset, one is necessarily more confident that this event belongs to any subset that includes this previous subset. A first consequence of these axioms is

$$\forall E, F \in \wp(\Omega), \quad \mu(E \cap F) \leq \min(\mu(E), \mu(F))$$

and

$$\mu(E \cup F) \geq \max(\mu(E), \mu(F))$$

Further to these three axioms, one can add the additivity axiom:

$$\forall E, F \in \wp(\Omega), \quad \text{if} \quad E \cap F = \emptyset \quad \mu(E \cup F) = \mu(E) + \mu(F)$$

When the additivity axiom is verified, the credibility mass assignment is a probability assignment. When the mass assignment is not additive, we call it a capacity.

The dual of any capacity is defined by

$$\bar{\mu}(A) \equiv 1 - \mu(\bar{A})$$

We have other properties:

- The overadditivity:

$$\forall E, F \in \wp(\Omega), \quad \text{if} \quad E \cap F = \emptyset \quad \mu(E \cup F) \geq \mu(E) + \mu(F)$$

A consequence of overadditivity is

$$\forall A \in \wp(\Omega), \quad \mu(A) + \mu(\bar{A}) \leq 1$$

- The underadditivity:

$$\forall E, F \in \wp(\Omega), \quad \text{if} \quad E \cap F = \emptyset \quad \mu(E \cup F) \leq \mu(E) + \mu(F)$$

A consequence of underadditivity is

$$\forall A \in \wp(\Omega), \quad \mu(A) + \mu(\bar{A}) \geq 1$$

If a capacity $\mu$ is overadditive, its dual capacity $\bar{\mu}$ is underadditive.

For a nonadditive capacity we define the kernel of the capacity by the set of probability distributions (additive):

$$K(\mu) = \left\{ P \, / \, \forall A \in \wp(\Omega), \mu(A) \leq P(A) \leq \bar{\mu}(A) \right\}$$

It clearly states that a capacity is able to represent an interval for a probability.

### 3.2.2 PROBABILITY

In probability theory [16], the mass assignation consists of the assignation of a probabilities density $p$ on the elements of $\Omega$ such as

$$p : \Omega \to [0,1], \quad \sum \left\{ p(\omega) | \omega \in \Omega \right\} = 1$$

The belief degrees on the subsets of $\Omega$ are quantified by a probability distribution $P$ as

$$P : \wp(\Omega) \to [0,1], \quad \forall \omega \in \Omega, P(\{\omega\}) = p(\omega)$$

$$\forall A, B \subseteq \Omega \,/\, A \cap B = \varnothing, P(A \cup B) = P(A) + P(B)$$

Additivity implies

$$P(A) = 1 - P(\bar{A})$$

and

$$P(A) = \sum \big\{ p(\omega) \big|\, \omega \in A \big\}$$

Those properties describe the static aspect. The dynamic aspect is introduced by the Bayesian conditional rule. Let $A$ be an arbitrary event such as $P(A) > 0$. The conditional probability of $B$ given that the event $A$ has been realized, is

$$P(B/A) = \frac{P(A \cap B)}{P(A)}$$

When a knowledge or event $B$ occurs, the knowledge on the event $A_i$, $P(A_i)$ is updated: $P(A_i/B)$ is calculated and will become $P(A_i)$ for the next iteration. At this end, it is necessary to know the prior probabilities: $P(A_i)$ prior probability of the event $A_i$ and $P(B/A_i)$ probability of the event $B$ given $A_i$. Let $P_k(A_i)$ be the probability of $A_i$ at the moment $k$, $P_k(B)$ the probability of $B$ at the moment $k$, and $P_k(B/A_i)$ the conditional probability of $B$ given $A_i$ at the moment $k$, the probability of $A_i$ at the moment $k + 1$ is given by

$$P_{k+1}(A_i) = P_k(A_i/B) = \frac{P_k(A_i) \cdot P_k(B/A_i)}{P_k(B)}$$

In this case, $P(B/A_i)$ defines the probability distribution of the measure in the class $A_i$; that is often characterized by a known model. The validity of the knowledge on this term has a direct impact on the validity of the final result.

### 3.2.3 POSSIBILITY–NECESSITY

In Possibility theory [17], uncertainty of an event $A$ is described by two degrees slightly linked: the degrees of Possibility $\Pi(A)$ and Necessity $N(A)$ of the event $A$. A measure of Possibility, denoted $\Pi$, is defined on $\wp(\Omega)$ by

$$\Pi : \wp(\Omega) \to [0,1], \quad \forall \omega \in \Omega, \Pi(\{\omega\}) = \pi(\omega)$$

$$\forall A, B \subseteq \Omega \quad \Pi(A \cup B) = \max(\Pi(A), \Pi(B))$$

Moreover

$$\Pi(\varnothing) = 0, \quad \Pi(\Omega) = 1$$

As

$$\Pi(A \cup B) = \text{Max}(\Pi(A), \Pi(B)) \leq \Pi(A) + \Pi(B)$$

$\Pi$ is underadditive.

A dual measure of Necessity is defined by

$$N(A) \equiv 1 - \Pi(\overline{A})$$

$$N(\varnothing) = 0, \quad N(\Omega) = 1 \quad \forall A, B \subseteq \Omega \quad N(A \cap B) = \min(N(A), N(B))$$

$$N(A \cap B) = 1 - \Pi(\overline{A \cap B}) = 1 - \Pi(\overline{A} \cup \overline{B}) = 1 - \max(\Pi(\overline{A}), \Pi(\overline{B}))$$

$$= 1 - \max(1 - N(E), 1 - N(F)) = \min(N(A), N(B))$$

$N$ is overadditive. We have

$$\text{Max}(\Pi(A), \Pi(\overline{A})) = \text{Max}(\Pi(A), 1 - N(A)) = 1 \quad \Rightarrow [N(A) \neq 0 \Rightarrow \Pi(A) = 1]$$

and also

$$\Rightarrow [\Pi(A) \neq 1 \Rightarrow N(A) = 0]$$

It is important to note that when necessity is not null, the possibility value is 1 (a necessary event is totally possible) and when the possibility value is not equal to 1, the necessity value is null (a nontotally possible event has no necessity). The dynamic assignation is realized by a conditioning model usually defined by

$$\Pi(A \cap B) = \Pi(A/B)*\Pi(B)$$

where the combination operator * is generally the minimum or the multiplication.

### 3.2.4   BELIEF–PLAUSIBILITY

Belief theory, also called Evidence theory [20,21], describes a *generalization of the theory of Probabilities and Possibilities*. A total belief mass equal to 1 is distributed between the elements of a finite subset $V$ of $\wp(\Omega)$ by the function $m$ called mass assignation. The sets of $\wp(\Omega)$ for which $m(A)$ is strictly positive are called focal elements, and the focal set $F$ is defined by $F = \{V \in \wp(\Omega) / m(V) > 0\}$. When the mass assignation is made on singletons of $\wp(\Omega)$ the mass assignation is probabilistic and there is no difference with probability theory. This belief mass assignation allows one to deduce a Belief and Plausibility degree evaluation of any element defined by

$$\text{Bel}: \wp(\Omega) \to [0,1], \quad \text{Bel}(A) = \sum \left\{ m(V) \middle| V \in F, V \subseteq A \right\}$$

$$\text{Pl}: \wp(\Omega) \to [0,1], \quad \text{Pl}(A) = \sum \left\{ m(V) \middle| V \in F, V \cap A \neq \varnothing \right\}$$

The belief of an event $A$ is the sum of the masses associated with the focal elements which imply $A$. This clarifies that belief is a generalization of necessity in possibility theory. Belief is overadditive. The plausibility of $A$ is the sum of the masses associated with the focal elements that "do not contradict" $A$. This clarifies that plausibility is a generalization of possibility. Plausibility is underadditive.

The belief and the plausibility are dual:

$$\text{Pl}(A) = 1 - \text{Bel}(\bar{A}).$$

We have then two pieces of information: the degree of belief on $A$, Bel($A$) and the degree of belief on $\bar{A}$, $\text{Bel}(\bar{A})$. Both pieces of information can be defined independently, at the opposite of the probabilistic frame. Hence we can increase the belief on $A$ without decreasing the belief on $\bar{A}$. In the static frame, we have

$$N(A) \leq \text{Bel}(A) \leq P(A) \leq \text{Pl}(A) \leq \Pi(A)$$

$[\text{Bel}(A), \text{Pl}(A)] \left( \left[ \text{Bel}(A), 1 - \text{Bel}(\bar{A}) \right] \right)$ is called the uncertainty interval. See Figure 3.1.

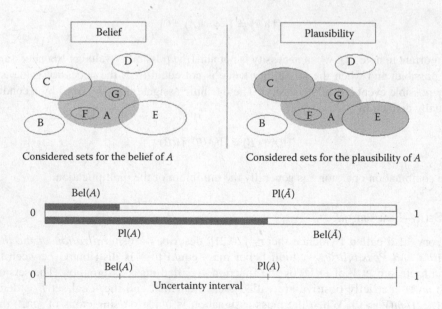

FIGURE 3.1   Uncertainty interval [Belief, Plausibility].

Apart from the conditioning, Dempster defines a rule named the combination rule of Dempster; the aim of this rule is to combine two belief assignations based on two distinct sources of information.

$$\text{Bel}_1 \oplus \text{Bel}_2(A_i \cap B_j) = \frac{\text{Bel}_1(A_i) \cdot \text{Bel}_2(B_j)}{1 - K} \quad \text{with} \quad K = \sum_r \sum_s \left\{ \text{Bel}_1(A_r) \cdot \text{Bel}_2(B_s) \middle| A_r \cap B_s = \varnothing \right\}$$

Conditioning formulas can be used, which depends on the definition of conditional beliefs:

- Dempster–Shafer: $\mu_Y^{DS}(X) = 1$ if $X \cap Y \neq \varnothing$, $\quad \mu_Y^{DS}(X) = 0$ if $X \cap Y = \varnothing$

  which gives the same combination as the Dempster rule.

- Modified Dempster–Shafer definition:

$$\mu_Y^{M-DS}(X) = \text{card}(X \cap Y)/(\text{card}(X) \cdot \text{card}(Y)), \quad \text{if } Y \neq \varnothing, \mu_Y^{M-DS}(X) = 0 \text{ if } Y = \varnothing$$

- Estimated Bayes definition: $\mu_Y^B(X) = \text{card}(X \cap Y)/\text{card}(X)$
- Pearl definition: $\mu_Y^P(X) = \text{card}(X \cap Y)/\text{card}(X \cap \bar{Y})$, $\quad$ if $X \cap \bar{Y} \neq \varnothing, \mu_Y^P(X) = 1$ if $X \cap \bar{Y} = \varnothing$

## 3.3   COMPARISON OF BEHAVIORS

In Ref. [26], Grandin and Moulin point out, in illustrative examples, divergent results between the theory of Possibilities, theory of Beliefs, and theory of Probabilities. Several authors have highlighted similar results [4,9,10]. In a static framework the following relation holds: $\text{Bel}(A) \leq P(A) \leq \text{Pl}(A)$. Because of this relation it is attractive to think that we will use Credibilities or Possibilities to frame Probabilities by a kind of confidence interval, in the dynamic framework. Unfortunately it is not correct (except if we consider a particular definition of the conditional belief; see modified Dempster–Shafer approach in Ref. [27]). Because Bel and Pl are not additive, in the general case (non-Bayesian masses) Credibilities combination leads to reinforcing the mass given to some combination of events [12,26].

### 3.3.1   SYNERGY DUE TO NONADDITIVITY

Several authors [3,9,10] underlined the superior speed of convergence of the probabilistic in front of credibilistic approach. But we can find examples in which the convergence of the probabilistic model is lower. Consider the following example: The universe is the set of events $\Omega = \{T_1, T_2, T_3, T_4\}$. The situation is that event $T_4$ is the one that really occurs, so it is the one that will be observed. Sensor propositions are the following:

- Time even $(t = 2P)$: $m(T_1) = 0.2$, $m(T_2) = 0.2$, $m(T_3 \cup T_4) = 0.6$
- Time $2P + 1$: $m(T_1) = 0.2$, $m(T_3) = 0.2$, $m(T_2 \cup T_4) = 0.6$

Probability, Belief, and Plausibility for the event $T_4$ are plotted during time (Figure 3.2).

The product of the masses of the disjunction supplied by sensors 1 and 2 is reported to $T_4$: $m_{12}(T_4) = m_{12}((T_2 \cup T_4) \cap (T_4 \cup T_3)) = m_1(T_4 \cup T_2) \times m_2(T_4 \cup T_3)$.

To not take into account the proportion of the intersection between both disjunctions is questionable. An important mass can be allocated to $T_4$ if the disjunction is strong while the intersection can be very weak. Following both plans will make the same contribution in $T_4$ (see Figure 3.3).

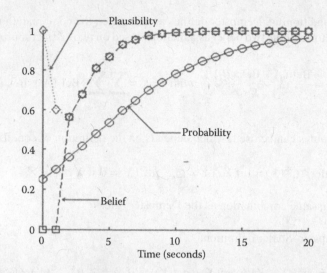

**FIGURE 3.2** Comparison of convergence speed for Bayes and Dempster–Shafer theories.

Strong intersection                                          Weak intersection

**FIGURE 3.3** Cause of the synergy effect in Dempster–Shafer combination.

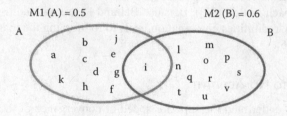

**FIGURE 3.4** Synergy effect in Dempster–Shafer combination.

A consequence in this example is that the Dempster–Shafer combination has a positive synergy for $T_4$. The hypothesis ($T_4$) is advantaged to the detriment of the other hypotheses. It is a direct consequence of nonadditivity. Consider now the following example. We consider two sets $A$ and $B$ of 10 events. With probability the mass is shared on each single event (see Figure 3.4).

With probabilities: $P_{12}(A \cap B) = P_1(\{i\}) \cdot P_2(\{i\}) = 0.5/10*0.6/10 = 0.003$
With credibilities: $m_{12}(A \cap B) = m_1(A) \cdot m_2(B) = 0.5*0.6 = 0.3$

Applying the theory depends on the problem itself. With probability each elementary event has the same chance to occur. It is a bet theory. The combination of events proposed by credibilities is a logical demonstration. The effective intersection here is event $i$. The credibility assigned to each event is a mass of probability. And the probability of the logical conclusion is the product of elementary degrees of probability.

### 3.3.2   UNCOMMITTED BELIEF AND RELIABILITY

In Belief theory, a belief can be assigned to the entire frame of discernment. It is written $m(\Omega)$. In the case of *singleton masses*, it has the effect of producing results such as

$$\text{Pl}(A) = \sum_{B \cap A \neq \varnothing} m(B) = m(A) + m(\Omega) \qquad \text{Bel}(\overline{A}) = 1 - \text{Pl}(A) = 1 - m(A) - m(\Omega)$$

$$\text{Bel}(A) = \sum_{B \subset A} m(B) = m(A) \qquad \text{Bel}(A) + \text{Bel}(\overline{A}) = 1 - m(\Omega) \leq 1$$

With $m(\Omega) = 0$, Bel resumes to classical probabilities.

In Refs. [13,14], Grandin and Marques suggest introducing the notion of reliability in probability. The Bayesian combination rule weighted by the reliability of the sensors is derived:

$$P(H_i)_t = \frac{P(H_i)_{t-1} \prod\limits_{j=1}^{n} (1 - f_j + f_j P(m_j / H_i)_t)}{\sum\limits_{i=1}^{P} P(H_i)_{t-1} \prod\limits_{j=1}^{n} (1 - f_j + f_j P(m_j / H_i)_t)}$$

where $P(H_i)_t$ = probability of $H_i$ at time $t$ and $f_j$ = reliability of measurement $m_j$.

The relations between $m(\Omega)$ and $F$ (reliabilities in probability) are described in Ref. [12]. When reliabilities on the sensors are used (which is recommended for an efficient fusion), it is taken into account in the Dempster–Shafer theory by a reinforcement of the mass $m(\Omega)$, that is to say a weakening of all the other masses by the factor $1 - m(\Omega)$, and in Bayesian theory by a weakening of all the individual probabilities by the factor $F$. If probabilistic and credibilistic fusion are "informed by uncertainty on sensor measures" they produce similar convergences on Bayesian masses [12]. Reliability is like a filter that weakens extracted information and the predictivity is affected (convergence less fast). In exchange, we gain robustness to absurd values. So, it is logical to observe in examples presented in Refs. [3,9,10] a faster convergence of a Bayesian method that does not take into account the reliability, ahead of Dempster–Shafer methods, which take it into account. If this reliability is taken into account, at the same level that the uncommitted, for example, $F = 1 - m(\Omega)$, we observe an equivalent predictivity. Whichever theory is used, it is essential to model the sensor well, particularly the confidence or the reliability of the measures [11,12]. Any information can be characterized by a degree of validity with regard to an a priori model or by immediate criteria. The reliability expresses the proportion of the measures realized within the framework of a normal functioning of the sensor. In the following example the information from four sensors is combined. Each sensor has outliers (40%, 70%, 70%, 80%). The reliabilities are estimated. We compare the result from Bayes combination without and with reliability weighting. The true hypothesis is $H_1$. Figure 3.5 shows the result of one run. If reliability is not used, the estimated credibilities are chaotic and sometimes with large errors. If reliability is used, credibilities are smoothed and converge toward the true values.

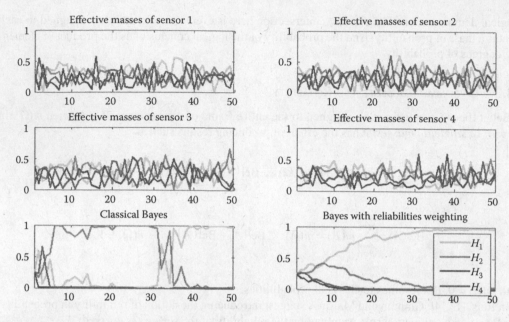

**FIGURE 3.5**   Benefits of reliabilities weighting.

### 3.3.3   Conclusion on Behaviors

In case of probabilistic masses, there are no differences between Bayes and Dempster–Shafer results. In this case, credibilities resume to probabilities. Reliabilities can be defined in both theories and induce the same behavior. *A large difference in behavior is the effect caused by the relaxation of the additivity axiom when considering non-Bayesian masses.* In this case, it is generally not possible to converge on compatible decisions between probabilities and credibilities.

## 3.4   EXPERIMENTS

As the two theories do not have the same behavior, we have to select which is the best one for an application. To do this, we have to know whether we are in an application where noise integration is dominant or where logical demonstration is dominant. In the first case, we combine objective uncertainty; in the second case, we combine subjective uncertainty. To highlight this conclusion, we relate in this section two experiments, the first one with objective uncertainty and the second with subjective uncertainty.

### 3.4.1   Experiment with Objective Uncertainty

This type of experiment was first presented in Ref. [26]. Let us consider the fusion of two sources, $S_1$ and $S_2$. We are interested in three hypotheses ($H_1$, $H_2$, and $H_3$). Assume $m(H_1)$, $m(H_2)$, and $m(H_3)$ obey Normal probability density functions:

$$m_1(H_1) \sim N(0.25, 0.01) \quad m_1(H_2) \sim N(0.35, 0.01) \quad m_1(H_3) \sim N(0.4, 0.01)$$
$$m_2(H_1) \sim N(0.2, 0.01) \quad m_2(H_2) \sim N(0.3, 0.01) \quad m_2(H_3) \sim N(0.5, 0.01)$$

Nonprobabilistic masses on two sensors are simulated:

$$m_1(H_1 \cup H_2) = \frac{m_1(H_1) + m_1(H_2)}{m_1(H_1) + m_1(H_2) + m_1(H_3)} \qquad m_2(H_1 \cup H_3) = \frac{m_2(H_1) + m_2(H_3)}{m_2(H_1) + m_2(H_2) + m_2(H_3)}$$

$$m_1(H_3) = \frac{m_1(H_3)}{m_1(H_1) + m_1(H_2) + m_1(H_3)} \qquad m_2(H_2) = \frac{m_2(H_2)}{m_2(H_1) + m_2(H_2) + m_2(H_3)}$$

In the simulation, hypothesis $H_3$ is the class that originates the measurements. Figure 3.6 shows information to fuse.

Each sensor delivers masses on union of elementary hypotheses (Dempster–Shafer masses). In Figure 3.6, effective masses are represented on upper curves and non-Bayesian masses on lower curves. The information provided by sensors are non-Bayesian masses. To adapt the information to probabilistic combination, it is necessary to share the masses on each union between all elementary hypotheses that compose it. This can be criticized because this sharing is abusive (in fact it does not correspond to the actual effective masses; in the simulation we have the information on how to share the mass, but we do not use it).

The first hypothesis $(H_1)$ is reinforced to the detriment of the other hypotheses with the Dempster–Shafer combination rule $(m_{12}(H_1) = m_1(H_1 + H_2) \times m_2(H_1 + H_3))$. The following curves represent the evolution of the masses after pignistic transformation [28] for the Dempster–Shafer operator and the a posteriori probabilities for the Bayes combination. Despite the fact that Bayes has to share the masses abusively $(P_1(H_1) = m(H_1 + H_2)/2)$ Bayes converges toward the right hypotheses $(H_3)$ while

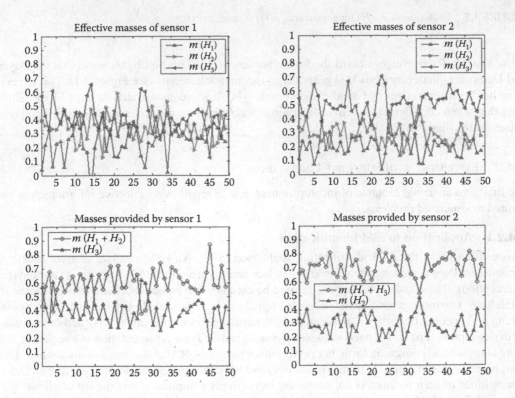

**FIGURE 3.6** Effective masses and non-Bayesian masses for the sensors.

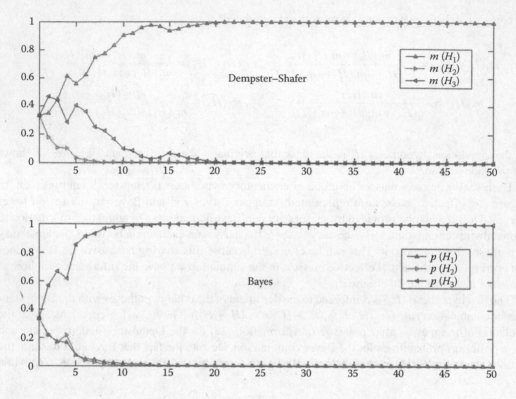

**FIGURE 3.7**   Evolution of $m_{12}(H)$ (upper curves), $p(H)$ (lower curves).

Dempster–Shafer converges toward the first hypotheses ($H_1$). This highlights again that the Bayes and Dempster–Shafer methods lead generally to incompatible results (see Figure 3.7). This experiment has been generalized by Grandin and Marques [14]. It was demonstrated on large Monte Carlo runs that when considering objective uncertainty with non-Bayesian masses the probability gives better results than credibilities.

### 3.4.2   EXPERIMENT WITH SUBJECTIVE UNCERTAINTY

We first give an actual example of incompleteness, just to recall that subjective information is not limited to simulated problems.

#### 3.4.2.1   Application to ESM Identification

This example is in the ESM identification application [5–8]. An ESM receiver is used to detect, measure, and segregate streams of incoming pulses coming from RADAR. This gives rise to radar interceptions. These radar interceptions have to be compared to emissions characteristics stored in a database. The more generic model of a pulse signal can be represented by a sequence description, taking into account frequency $F$, pulse width PW, pulse repetition intervals (PRIs), and other waveform descriptors. The aim of pulse sequence characterization is the representation of a sequence in a more compact and convenient form. For performance purposes, RADAR emissions are summarized first, based on parameter histograms. To simplify, we limit the application description to $F$ and PRI. The agilities of such parameters are becoming increasingly complicated and the list of all possible values of $F$ and PRI for a waveform can be quite long. For such RADAR emission, the complete characterization of the waveform requires a long interception time, which is not compatible with

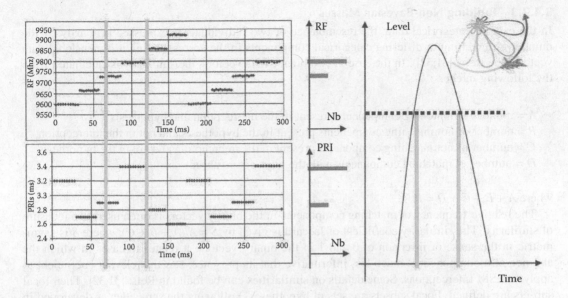

**FIGURE 3.8**  Incompleteness in ESM interception.

other ESM requirements and cost. Because of sensitivity limitation the ESM intercepts only a small part of the waveform, on the rotating beam of the RADAR. It means a small part of the RF and PRI values. In Figure 3.8 the two histograms (one for RF, one for PRI) shows the cell values measured on the interception. It is common to observe only a small number of all possible cell values for a RADAR emission. Then processing errors (due to low intercept duration, pulse loss, and noise) lead to mixture of several coincident emissions giving rise to some mix of emission patterns. RADAR signatures (RADAR emissions in the database) describe the possibility for the radar to emit particular patterns. The database itself is corrupted by the lack of knowledge due to the partial knowledge of RADAR emissions. RADAR signatures are also frequently overlapped on several parameters. The perturbations of the interception, at a given date, of one of these patterns are not a question of noise. The list of intercepted features is due to the fact that at the date of interception the RADAR selected one of its numerous possibilities. The perturbations are essentially due to incompleteness of the measurement (due to ESM interception), and incompleteness of the signature's knowledge. Signatures describe possible interceptions and not the density of probability of potential measurements. For these reasons, Dempster–Shafer theory is preferred to Bayes theory. This remark on subjective uncertainty in the knowledge base was previously reported in Ref. [29]. But to provide measurable proof, the author designed an experiment comparing Bayes and Dempster–Shafer combinations with the support of two students. This has been detailed in Ref. [30]. In this chapter, we just summarize this experiment.

### 3.4.2.2  Description of the Experiment

RADAR signatures (from the database) are first simulated. The number of signatures is the number of hypotheses. These hypotheses have a large number of similar values. ESM interceptions are then simulated with a percentage of correct values (chosen at random from the true hypothesis), with a percentage of interceptions with false values and in this case with a percentage of false values coming from other signatures and with a percentage of false values (noise values) that do not belong to any signature. Four parameter histograms for each signature are simulated. The parameters are $F$ and PRI for two of them.

### 3.4.2.3  Building Non-Bayesian Masses

In the case of categorical data, the resemblance of two individuals is expressed primarily by the number of matching or differing categorical components in the corresponding $p$-dimensional data vectors (see Refs. [31,32]). In the case of two binary data vectors, the similarity is calculated from the following integers:

$A$ = number of matching components present in the interception and hypothesis
$B$ = number of nonmatching components present in the hypothesis and not in the interception
$C$ = number of nonmatching components present in the interception and not in the hypothesis
$D$ = number of matching components with the two vectors absent

where $A + B + C + D = P$.

The relative frequency of matching components in the binary vectors is often used as a measure of similarity. The similarity coefficient of Jaccard is given by $S = A/(A + B + C)$, where $S$ is asymmetric in the sense of inversion of 0 and 1 in the binary vectors. It relates to cases in which the absence of a component is much less informative that its presence. $S$ is the relevant coefficient to apply to ESM interceptions. Some details on similarities can be found in Refs. [31,32]. Then focal subsets are defined. Focal subsets are sets of hypotheses. Following the same idea as developed in Ref. [33], for each interception and each parameter, we build the focal subset with all the hypotheses that share the same set of values with the interception; that means the same $A$ and $B$ coefficients. Then we get the mass of the focal subset for each parameter of the interception based on the Jaccard coefficient using normalization (the masses of all the subsets for each parameter have to sum to 1). To convert these masses to Bayesian masses, the mass of the focal element is divided by the number of hypotheses in the focal subset [28]. Then Bayes and Dempster–Shafer combination rules are applied concurrently.

Conceptually we distinguish two levels as proposed by the TBM [22]:

- At the creedal level beliefs are entertained.
- At the pignistic level beliefs are transformed to decisions.

### 3.4.2.4  Experimental Results

A Monte Carlo simulation has been realized with 2000 elementary lists of 100 events each. Each iteration corresponds to a full identification process over 100 events. To generate these lists, a number of parameters are chosen at random (with a uniform law) in an interval. These intervals are as follows:

| Parameter Name | Value (Interval) |
| --- | --- |
| Number of hypotheses | [8, 12] |
| Percentage of similar values | [50%, 90%] |
| Percentage of correct values | [2%, 15%] |
| Percentage of false values | [10%, 25%] |
| Percentage of interceptions with false values | [35%, 55%] |

These values were chosen to create difficult identification problems. Even if thousands of signatures are in an actual RADAR database for ESM identification, only a local mix of interleaved signatures can give rise to identification problems, so to consider 10 interleaved signatures in the simulation is very representative. In fact, on easy cases the two methods give in a few steps a correct answer.

One typical result is the following.

| Number of Interceptions | Dempster–Shafer Result | Bayes Result |
| --- | --- | --- |
| 25 | 1341/2000 | 542/2000 |
| 50 | 1492/2000 | 837/2000 |
| 75 | 1528/2000 | 1000/2000 |
| 100 | 1617/2000 | 1122/2000 |

The two methods are convergent: According to time they give better and better results. As expected, the Dempster–Shafer combination delivers the best performance. Above all, we see that the Dempster–Shafer method is able to give a better performance (percentage of success multiplied by 2.5) in a shorter time. The latency of the RADAR identity estimation is a critical requirement.

## 3.5 WEIGHTING DECISIONS WITH CONFLICT MEASUREMENT

We now add a last but important aspect to the discussion: How to manage the conflict between measurements? A taxonomy of combination rules has been presented in Ref. [34]. It opposes conjunctive operators to disjunctive operators. Conjunctive operators have to be used when elements of information are concordant (with no conflict); here, focal subsets have a large intersection and the intersection provides a more specific subset with a small risk. Disjunctive operators have to be used when elements of information are discordant (with a high degree of conflict); in this case focal subsets have a small intersection and the union provides a less specific subset (loss of information) but the risk is minimized. In addition, it is possible to build adaptive operators where the degree of conflict is used as a weight between conjunctive and disjunctive operators. See Ref. [12] for details. According to the different theories these operators are described below.

Conjunctive operators are

- Product with probabilities: $P_F(A \cap B) = P_{S1}(A) \cdot P_{S2}(B)$
- Minimum with possibilities: $\Pi_F(A \cap B) = \min[\Pi_{S1}(A), \Pi_{S2}(B)]$
- DS combination: $m_F(C) = \displaystyle\sum_{A \cap B \subseteq C} m_{S1}(A)m_{S2}(B)$

Disjunctive operators are

- Sum with probabilities: $P_F(A \cup B) = P_{S1}(A) + P_{S2}(B)$
- Maximum with possibilities: $\Pi F(A \cup B) = \max[\Pi_{S1}(A), \Pi_{S2}(B)]$
- With DS: $m_F(C) = \displaystyle\sum_{A \cup B \subseteq C} m_{S1}(A)m_{S2}(B)$ if $\displaystyle\sum_{A \cap B \neq \varnothing} m_{S1}(A)m_{S2}(B) = 0$

The degrees of conflict are

- With probabilities: $K_{Pr} = 1 - \displaystyle\sum_{A,B} P_{S1}(A) \cdot P_{S2}(B)$
- With possibilities: $K_{po} = 1 - \sup_{A,B}[\min[\Pi_{S1}(A), \Pi_{S2}(B)]]$
- With DS: $K_{DS} = 1 - \displaystyle\sum_{A \cap B \neq \varnothing} m_{S1}(A)m_{S2}(B)$

Adaptive operators are

- With probabilities: $P_F(A) = (1 - K_{Pr})P_F(A \cap B) + K_{Pr}P_F(A \cup B)$
- With possibilities: $\Pi_F(s) = \max(\min(1 - K_{po}, \Pi_F(A \cap B)), \min(K_{po}, \Pi_F(A \cup B)))$

- With DS: If $K = 1\, m_F(C) = \displaystyle\sum_{A \cup B \subseteq C} m_{S1}(A) m_{S2}(B)$

$$\text{if } K \neq 1\, m_F(C) = \frac{\displaystyle\sum_{A \cap B \subseteq C} m_{S1}(A) m_{S2}(B)}{1 - K}$$

These operators are convenient in the case of objective uncertainties. In the case of subjective uncertainties, a high degree of conflict can simply be the result of the high specificity of the combination of two sources of information; for example, there is only one RADAR signature with frequency $F_1$ and PRI $P_1$, but there are many signatures with $F_1$ and many with $P_1$. The fact that one signature with this specific combination exists validates the result. How to manage the conflict has to be selected according to the physics of the application.

## 3.6 CONCLUSION

To combine uncertainties in information fusion, Bayes and Dempster–Shafer theories were used. They differ by their behaviors on non-Bayesian masses. Bayes theory gives better results on objective uncertainty and Dempster–Shafer theory gives better results on subjective uncertainties. On the RADAR identity estimation in ESM, the uncertainty is mainly subjective and Dempster–Shafer theory demonstrates a superior performance.

## REFERENCES

1. D.L. Hall, J. Llinas, *Handbook of Multisensor Data Fusion*. CRC Press, Boca Raton, FL, 2001.
2. C.J. Harris, *Application of Artificial Intelligence to Command and Control Systems*. Peter Peregrinus, London, 1988.
3. E. Waltz, J. Llinas, *Multisensor Data Fusion*. Artech House, Boston, MA, 1990.
4. D.L. Hall, *Mathematical Technique in Multisensor Data Fusion*. Artech House, Boston, MA, 1992.
5. R.G. Willey, *Electronic Intelligence: The Analysis of Radar Signals*. Artech House, Boston, MA, 1982 (new augmented edition, 2006).
6. D. Adamy, *A First Course in Electronic Warfare*, pp. 75–77. Artech House, Boston, MA, 2001.
7. A. de Martino, *Introduction to Modern EW Systems*, pp. 206–208. Artech House, Boston, MA, 2012.
8. J.F. Grandin, D. Neveu, Principes d'identification de signaux radars utilisant imprécisions, incertitudes, informations contextuelles et fiabilités. *GRETSI 99*.
9. E.L. Waltz, D.M. Buede, Data fusion and decision support for command and control. *IEEE Transactions on Man, Systems and Cybernetics*, Vol. 6, No. 6, pp. 865–879, 1986.
10. D.M. Buede, P. Girardi, A target identification comparison of Bayesian and Dempster-Shafer multisensor fusion. *IEEE Transactions on Systems, Man and Cybernetics*, Vol. 27, No. 5, pp. 569–577, 1997.
11. D.L. Hall, A.K. Garga, *Pitfalls in Data Fusion (and How to Avoid Them)*. Proceedings of the 2nd International Conference on Information Fusion—FUSION 99, Vol. 1, pp. 429–436, Sunnyvale, CA. July 8, 1999.
12. J.F. Grandin, Fusion de données: Théorie et méthodes. *Techniques de l'Ingénieur*, Mars 2006.
13. J.F. Grandin, M. Marques, Robust data fusion. *Third International Conference on Information Fusion, Fusion 2000*, Paris, France, July 10–13, 2000.
14. J.F. Grandin, M. Marques, Robust fusion with reliabilities weights. *SPIE AeroSense*, 2002.
15. J.C. Helton, Uncertainty and sensitivity analysis in the presence of stochastic and subjective uncertainty. *Journal of Statistical Computation and Simulation*, Vol. 57, Issue 1–4, pp. 3–76, 1997.
16. T. Bayes, Essay Toward Solving a Problem in the Doctrine of Changes. *Philosophical Transactions of the Royal Society*, London, 1763.
17. D. Dubois, H. Prade, *Possibility Theory*. Plenum Press, New York, 1985.
18. D. Dubois, H. Prade, Fusion d'informations imprécises. *Revue Traitement du Signal*, Vol. 11, No. 6, pp. 447–458, 1994.

19. A.P. Dempster, Upper and lower probabilities induced by a multi-valued mapping. *Annals of Mathematical Statistics*, Vol. 38, No. 2, pp. 325–339, 1967.
20. A.P. Dempster, A generalization of Bayesian inference. *Journal of the Royal Statistical Society*, Vol. 30, No. 2, pp. 205–247, 1968.
21. G. Shafer, *A Mathematical Theory of Evidence*. Princeton University Press, Princeton, NJ, 1976.
22. P. Smets, The combination of evidence in the Transferable Belief Model. *IEEE Transactions on Pattern Analysis Machine Intelligence, PAMI*, Vol. 12, pp. 447–458, 1990.
23. P. Smets, Belief functions: The disjunctive rule of combination and the generalized Bayesian theorem. *International Journal of Approximate Reasoning*, Vol. 9, pp. 1–35, 1993.
24. P. Smets, R. Kennes, The transferable belief model. *Artificial Intelligence*, Vol. 66, pp. 191–234, 1994.
25. P. Smets, *Advances in the Dempster-Shafer Theory of Evidence*. Wiley, New York, 1994.
26. J.F. Grandin, C. Moulin, What practical differences between probabilities, possibilities and credibilities. *SPIE AeroSense*, 2002.
27. D. Fixsen, R.P.S. Mahler, The modified Dempster-Shafer approach to classification. *IEEE Transactions on Systems, Man and Cybernetics*, Vol. 27, No. 1, pp. 96–104, 1997.
28. P. Smets, Decision making in the transferable Belief model: The necessity of the pignistic transformation. *International Journal of Approximate Reasoning*, Vol. 38, pp. 133–147, 2005.
29. F. Campos, A. Neves, E.V. Neto, *Dealing with Subjective Uncertainty in Knowledge Based Systems*. IEEE, 2007.
30. D. Nayigizente, *Comparaison entre l'approche bayésienne, la théorie de Dempster-Shafer et la logique floue pour l'identification d'émissions radar en guerre électronique*. Mémoire Master Recherche Ecole Centrale de Paris, 2014.
31. M. Rifqi, *Mesures de similarité, raisonnement et modélisation de l'utilisateur*. Habilitation Thesis, 2010.
32. F. Esposito, D. Malerba, V. Tamma, H.H. Bock. F.A. Lisi, H. Bacelar Nicolau. Chapter 8: Similarity and dissimilarity. In *Analysis of Symbolic Data: Exploratory Methods for Extracting Statistical Information from Complex Data*, H.H. Bock and E. Diday, (eds.), pp. 139–197. Springer, New York, 2000.
33. B. Yu, K. Sycara, J. Giampapa, S. Owens, *Uncertian Information Fusion for Force Aggregation and Classification in Airborne Sensor Networks*. American Association for Artificial Intelligence, Palo Alto, CA, 2004.
34. I. Bloch, Information combination operators for data fusion: A comparative review with classification. *IEEE Transactions on Systems, Man and Cybernetics*, Vol. 26, No. 1, pp. 52–67, 1996.

# 4 JDL Model (III) Updates for an Information Management Enterprise

*Erik Blasch*

## CONTENTS

## 4.1 INTRODUCTION

This chapter is intended to serve as a reexamination of the use of the Joint Directors of Laboratories (JDL) model [1–5] to meet the needs for fusion processing that is responsive to new technologies of big data, cloud computing, and machine analytics. The inception of the original JDL Data Fusion model aimed at providing a process flow for sensor and data fusion [1]. Subsequent revisions were made in 1998 [2] to incorporate a new understanding of the issues involved in exploiting information for diverse purposes such as impacts versus threats and process refinement versus sensor management. There also were needs to clarify terms, broaden the model from its initial focus on tactical military applications, and apply it to different users [3]. With the establishment of the International Society of Information Fusion (ISIF), the notion of "information" became an important concept in information fusion designs and analysis. In 2004, the JDL model was revised (aka JDL II) as per the issues discussed from the JDL Data Fusion Group [4,5]. At the same time, the working group proposed the Data Fusion Information Group (DFIG) model, which included suggested updates from the constituent member discussions such as a focus on user functions [6]. The focused application of this chapter deals with information exploitation of distributed multimedia content that requires revisions of the JDL model for machine analytics, information management, and user contextual reasoning over enterprise applications. Hence, this chapter is a further assessment of the JDL model (i.e., JDL III).

Since the 1990 publication of *Multisensor Data Fusion* [7], the process flow for "fusion" has been about the same. However, with the popularity of the Web and the formation of ISIF (http://www.isif.org), new developments have resulted from information exploitation and "management" [8]. These developments affect the existing models. An accepted modeling distinction has been made between high-level information fusion (HLIF) and low-level information fusion (LLIF) that was first described in Ref. [7]. The low-level functional processes support target classification, identification, and tracking, while high-level functional processes support situation, impact, and fusion process assessment. LLIF concerns numerical data (e.g., target locations, kinematics, and attribute types). HLIF concerns abstract symbolic information (e.g., threat, intent, and goals) [9]. Over the last decade, there have been numerous panel discussions on HLIF that demonstrate the ever changing issues in systems-level information fusion [10]. Some recent systems issues include new methods for evidential reasoning [11], situation awareness logic [12], enterprise architectures [13], interface management [14], and distributed architectures [15]. A complete list of interesting modifications to the use of the JDL functional model can be found in hundreds of papers in the ISIF conferences.

The focus of the chapter is aimed at stakeholders who are building information fusion systems that support information exploitation (InfoEx) [16]. InfoEx from many sensors and users distributed across a variety of locations requires a revisit to the JDL model for system-based applications. For a closed-loop sensor system, the tenets of the original JDL process flow are applicable with real-world parameters (e.g., a Kalman filter). For InfoEx systems, which are envisaged to have multiple users, data sources, intelligence data collections, and missions; the JDL model can be used for guidance, but has to be adapted to meet these needs. A large-scale system is difficult to model comprehensively, and thus a model provides a notional representation of the process flows across parameters and conditions as in Ref. [4].

## 4.2 INFORMATION FUSION PROCESS MODELS

The JDL II model [4] focused on the internodal processing as extensions and discussions of the JDL I model [1]. A representation of the discussion is shown in Figures 4.1 and 4.2. From the JDL I model, the goal was to determine the processes of an information fusion system. The updates focused on the common attributes of information fusion that focused on state refinement and assessment.

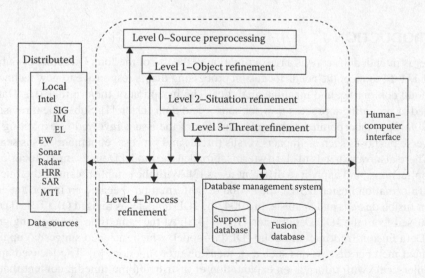

**FIGURE 4.1**   JDL I model (1992).

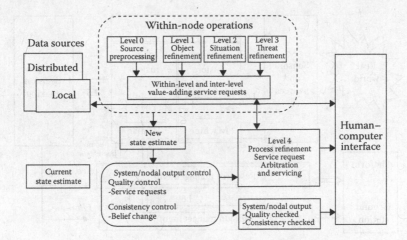

**FIGURE 4.2** JDL II model (2004).

Common to the Level 0–3 processing was state assessment. However, within each node; there is a complementary process refinement function of updating the state and inherently presenting to the human–computer interface.

The key goal of JDL II was to clarify definition of terms and establish a clear and generalized basis for the JDL model. It defined data fusion as *the process of combining multiple pieces of data to estimate the state of some aspect of the world*. Partitioning into data fusion "levels" was stated in terms of the types of such aspects: L0: features, L1: individual entities, L2: structures (e.g., relationships and situations), L3: scenarios and outcomes, and L4: aspects of the system itself. Sensor and other information management functions were partitioned into a series of analogous levels. The model is a functional model and therefore agnostic as to whether implementation of any of these functions is by people or automated processes [2]. Extending the JDL model to encompass resource management with data fusion supports functions of problem-solving under uncertainty. System designers and process managers may want a tight integration of diverse fusion, data mining, adaptive modeling, and sensing and diverse resource management processes.

Information exploitation (InfoEx) considers the utility of information, probability of acquiring such information via candidate action plans, and costs of plans, all of which are time dependent (utility and cost are also user variables). InfoEx involves considerations of the characterization and representation of utility and cost, multiagent and multiuser systems, and adaptive management and adaptive context exploitation.

Steinberg states [17,18]:

"Context provides expectations, constrains processing, and can infer or refine desired information ('problem variables') on the basis of other available information ('context variables')."

A key issue of utilizing context relates to the original distinction of low-level information fusion (LLIF) and high-level information fusion (HLIF). In the Data Fusion Information Group (DFIG) model (see Figure 4.3), LLIF (L0–1) composes explicit object assessment. HLIF (L2–6) composes much of the discussions and changes in the last decade [6].

In the DFIG model, the goal is to utilize the Data Fusion and Resource Management (DF/RM) functions proposed in the revisions, while highlighting the user's involvement. Humans are active observers with the ability to reason over contextual information [19]. RM is divided into fusion process control (L4), user refinement (L5), and platform placement/resource collection to meet mission objectives (L6). L2 Situation Assessment (SA) includes structures and relationships which are inferred from L1 Object Assessment of individual entities from L0 Data Alignment (i.e., registered

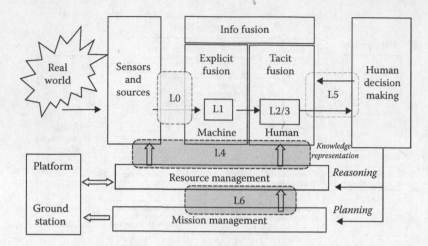

**FIGURE 4.3** DFIG Information Fusion model (2006). (From Blasch, E. et al., *Journal of Advances in Information Fusion*, 1(2):122–139, 2006.)

features). Because unobserved SA events are difficult for a computer to assess, user knowledge and reasoning are necessary. L3 (Impact Assessment) includes scenarios, outcomes, sensemaking of threats, courses of action, optimal decisions obtained using game theory [20], estimation of intent, and so forth to help refine the SA estimation and information needs for different actions.

Table 4.1 provides a comparison of the various models that include elements of a user-fusion system. The Observe, Orient, Decide, Act (OODA) model serves as a starting point and updates include multiple OODA loops such as the Technology, Emotion, Culture, and Knowledge (TECK-OODA) [21]; Cognitive (C-OODA) [22]; and (Us versus Them) OODA loops [23]. For example, *orient* is an assessment function that includes the DFIG processes of object, situation, and impact assessment. These are mapped to a situation awareness model of perception, comprehension, and projection in the C-OODA. The key aspect of the chapter for information management is *Decide-Act* which is the DFIG elements of sensor, user, and mission (SUM) refinement. SUM is further determined by the evaluation and ion, recall, action selection and implementation in the C-OODA. To perform information fusion refinements, there is a need for scalable, efficient, and effective information management.

Current definitions of the DFIG levels include:

*Level 0—Data Acquisition*: Estimation and prediction of signal/object observable states on the basis of pixel/signal level data association (e.g., information systems collections)

**TABLE 4.1**
**Comparisons of Information Fusion Models**

| OODA | DFIG Model | | Activity | C-OODA |
|---|---|---|---|---|
| Act | Level 6 | Mission Refinement | Command Execution | Implementation |
| | | | Resource Tasking | |
| Decide | Level 5 | User Refinement | Decision Making | Recall |
| | Level 4 | Sensor Refinement | Process Management | Evaluate |
| Orient | Level 3 | Impact Assessment | Pattern Learning | Projection |
| | Level 2 | Situation Assessment | Context Awareness | Comprehension |
| | Level 1 | Object Assessment | Feature Analysis | Feature Matching |
| Observe | Level 0 | Signal Alignment | Signal Processing | Perception |
| | | Data Acquisition | Sensing | Data Gathering |

*Level 1—Object Assessment*: Estimation and prediction of entity states on the basis of data association, continuous state estimation, and discrete state estimation (e.g., data processing)

*Level 2—Situation Assessment*: Estimation and prediction of relations among entities, to include force structure and force relations, communications, and so forth (e.g., information processing)

*Level 3—Impact Assessment*: Estimation and prediction of effects on situations of planned or estimated actions by the participants; to include interactions between action plans of multiple players (e.g., assessing threat actions to planned actions and mission requirements, performance evaluation)

*Level 4—Process Refinement* (an element of Resource Management): Adaptive data acquisition and processing to support sensing objectives (e.g., sensor management and information systems dissemination, command/control)

*Level 5—User Refinement* (an element of Knowledge Management): Adaptive determination of who queries information and who has access to information (e.g., information operations) and adaptive data retrieved and displayed to support cognitive decision making and actions (e.g., human computer interface)

*Level 6—Mission Management* (an element of Platform Management): Adaptive determination of spatial-temporal control of assets (e.g., airspace operations) and route planning and goal determination to support team decision making and actions (e.g., theater operations) over social, economic, and political constraints.

Key attributes of the DFIG model include

*Motivations*: The DFIG model suggestions were revisits to the original JDL model during the 2004 discussions, essentially constituting the JDL III version. JDL III focuses on users and information needs. For example, the user requires different tasks utilizing various skills (perception), rules (tasks), and cognition (knowledge) [24]. Also, many fusion systems are designed to meet a specific mission need. Thus, L5/L6 are distinct as to represent the different interactions, tailor the information fusion design to different systems, and address the advancements in the use of information fusion by the general community. However, there are further insights to the DFIG/JDLIII model.

    *Insight 1*: The L1 (explicit machine fusion), L2/L3 (implicit machine fusion) and interface to human (L5) are not unique (although they are tailored at the design level for any system development). Processes at any of these levels can include explicit, implicit, human, or machine approaches [6]. A better distinction than explicit/implicit is between (1) *filtering/refinement*, which converts an input estimate of a random variable to a more accurate or less uncertain estimate of that variable and (2) *inference/abstraction*, which estimates (frequently latent) variables from those available as inputs.

    *Insight 2*: From Figure 4.3, the choice to combine L2/L3 based on common functions needs to be separated based on variables extracted for exploitation. Distinguishing (L1, L2, L3) as data fusion functions (blocks) and (L0, L4, L5, L6) as resource management functions (interfaces) helps to characterize that these blocks and interfaces represent fusion functions and require control exploitation processes. For example, cognitive reasoning (L5) to constrain control could also be evidential reasoning (L2) for fusion such as belief changes. Planning, typically done by humans through automation, is now afforded through autonomy with enterprise resources. Finally, sensor platforms and ground stations could encompass other enterprise resources (e.g., World Wide Web) with varying degrees of controllability over not just the implied physical "real world," but perhaps also the "external universe of discourse" to cover virtual realities.

Major suggestions since the 2004 revisions are the focus on information management, to include advanced visualizations, data mining, and mission focus with teams, priorities, and coordination.

All of these functions help to *establish context* for the HLIF unsolved issues of fusion and resource management. For example, RM can be aided by information management enterprise computing (aspects of data acquisition, access, recall, and storage services). Key developments include context assessment and context management [25].

The rest of the chapter is organized as follows. Section 4.3 discusses the information management enterprise with system level (or processing) context information. Section 4.4 describes machine analytics, and Section 4.5 discusses user refinement in contextual analysis. Section 4.6 presents a notional example of data processing using physics-derived (e.g., video) and human-derived (e.g., text) fusion for both a JDL-type instantiation and suggestions for the JDL-based information exploitation. Discussions and conclusions are presented in Section 4.7.

## 4.3 INFORMATION MANAGEMENT ENTERPRISE

In this section, we detail the enterprise, information management, and layered service constructs to support management and exploitation of information of context.

### 4.3.1 INFORMATION FUSION MODELING IN THE ENTERPRISE

The current trends in information fusion (IF) are *data mining*, *enterprise architectures*, and *communications* [26]. Different mission applications require coordination over (1) data (e.g., models, storage/accesses control, and process and transport flow), (2) architectures (e.g., service-oriented architecture); and (3) the enterprise (e.g., service bus, computing environment, and the cloud). Figure 4.4 highlights the needs of the user, elements of data mining [27], and data flow in the enterprise.

*Information exploitation* involves two processes: (1) integrating and managing the data and (2) analyzing the data through IF and data mining. The JDL model deals only with part of data analysis as IF. Services and clouds can deal with the implementation of these two steps, as represented in Figure 4.4.

Context is both input and output of data analysis. For example, context (e.g., terrain and background traffic) is used to support fusion (e.g., finding usual behavior) and at the same time is the output of

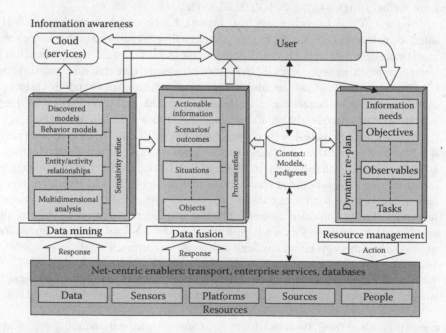

**FIGURE 4.4**  Information Fusion Enterprise model.

fusion (e.g., terrain from fusing video). Recently, Solano and Jernigan [13] presented an enterprise architecture to manage intelligence products for mission objectives highlighting data formats (e.g., schemas, unstructured, and metadata); data processes (e.g., access, ingest, cleansing, profiling, and ontology workflows); and database management services (DBMS). DB resources include contextual information such as terrain models. Cloud technology can serve as a basis for access to DB resource information but requires access (e.g., service-oriented architecture [SOA]) to enterprise services.

### 4.3.2 Information Management Model

The goal of information management (IM) is to maximize the ability (effectiveness) of a user to act on information that is available to, produced, or consumed within the enterprise. There are several means by which this can be accomplished.

- Reducing barriers to effective information use by providing notification, mediation, access control, and persistence services
- Providing an information space wherein information is managed directly, rather than del-egating IM responsibilities to applications that produce and consume information
- Focusing on consumer needs rather than producer preferences to ensure that information can be effectively presented and used
- Providing tools to assess information quality and suitability
- Exploiting producer-provided characterization of information to support automated man-agement and dissemination of information [9]
- Tools for goal-driven search, discovery, and exploitation of information to meet dynamic information needs

Optimal users' ability/effectiveness achieved by any of these means can make applications less compli-cated and enable the enterprise to be more agile to adapt to changing requirements and environments.

There are several best practices that help achieve the goals of information management. Organizations will greatly improve the interoperability and agility of their future net-centric infor-mation fusion (and command and control) systems by

1. Adopting dedicated information management infrastructures (e.g., cloud computing);
2. "Packaging" information for dissemination and management;
3. Creating simple, ubiquitous services that are independent of operating system and pro-gramming language;
4. Using a common syntax and semantics for common information attributes such as loca-tion, time, and subject; and
5. Adopting interfaces among producers, consumers, and brokers that are simple, effective, and well documented.

If appropriately employed, these best practices can reduce the complexity of information fusion systems, allow for effective control of the information space, and facilitate more effective sharing of information over an enterprise environment.

Figure 4.5 presents an IM model that illustrates the extended relation of the actors coordinating through the enterprise with the various layers and inner circles providing the protocol for informa-tion service access and dissemination [9].

People or autonomous agents interact with the managed information enterprise environment by producing and consuming information or by managing it. Various actors and their activities/services within an IM enterprise surround the IM model that transforms data into information. Within the IM model, there are various services that are needed to process the managed informa-tion objects (MIOs).

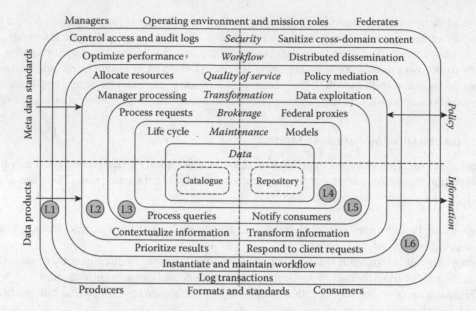

**FIGURE 4.5**  Information Management (IM) model.

**TABLE 4.2**

**Service Layers**

| | |
|---|---|
| Security | Control access, Log transactions, Audit logs, Negotiate security policy with federated information spaces, Transform identity, and Sanitize content |
| Workflow | Manage workflow model configurations, Instantiate and maintain workflows, Assess and optimize workflow performance |
| QoS | Respond to client context, Allocate resources to clients, QoS policy mediation, Prioritize results, and Replicate information |
| Transformation | Contextualize information, Transform MIOs, Support state and context-sensitive processing, Support user-defined processing functions, Support manager defined processing functions |
| Broker | Process queries, Support browsing, Maintain subscriptions, Notify consumers, Process requests for information and advertisements, Support federated information space proxies |
| Maintenance | Post MIOs, Verify Adherence to standards, Manage MIO lifecycle, Manage information space performance, Retrieve specific MIOs from repositories, Support configuration management of information models |

A set of service layers is defined that uses artifacts to perform specific services. An artifact is a piece of information that is acted upon by a service or that influences the behavior of the service (e.g., a policy). The service layers defined by the model are Security, Workflow, Quality of Service (QoS), Transformation, Brokerage, and Maintenance, as shown in Table 4.2. These services are intelligent agents that utilize the information space within the architecture, such as cloud computing and machine analytics.

One recent technology development is cloud computing, which supports an enterprise analysis [28], use of cloud computing in information fusion [29], and video tracking [30].

### 4.3.3  LAYERED VIEW OF THE CLOUD

Using elements of the JDL/DFIG, the enterprise, and the IM model for sensing, networking, and reporting can be realized. Figure 4.6 presents the layered information where the end-user (operator

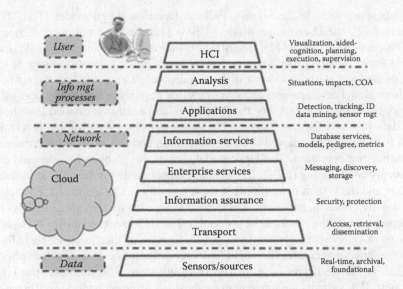

**FIGURE 4.6** Layered information services.

or machine) desires quality information as fused products from data, which requires various methods and services from sensor collections to information delivery. "Sensors/Sources" can be viewed as a general term as it relates to physical sensors, humans, and database services (e.g., data mining) that seek data from the environment.

Current trends in information fusion share common developments with cloud computing such as agent-based network service architectures [31], ontologies [32], and metrics [33] to combine physics-derived sensing and human-derived reporting using fusion products.

### 4.3.4 SYSTEMS-LEVEL MANAGEMENT OF CONTEXT

The JDL model is a functional model that seeks to identify and organize mathematical fusion functions; with management functions evident in all levels of the model. Three issues that have been addressed as inherent at each level are

- *Uncertainty management through contextual awareness*, particularly management of second-order uncertainly. This involves issues of source characterization (particularly with human sources), representation of diverse flavors of uncertainty, as well as computational issues (e.g., in probability hypothesis density-based methods) [34–36].
- *Adaptive context exploitation*, considering the utility of information, probability of acquiring such information via candidate action plans and costs of plans, all of which are time-variable (utility and cost are also user-variable). This involves considerations analogous to the preceding: consistent characterization and representation of utility and cost. Additional factors include multiagent and multiuser systems. Aspects of this problem are the integration of fusion; mining and general problem-solving methods; adaptive process, model, and goal management; and adaptive context exploitation [37,38].
- *Methods of user involvement through contextualization*, to include the interface and control issues for specific tasks that are operationally divided among different users such as information collection, real-time control, forensic analysis, and visualizations. These functions are local functions distributed over a large team for specific tasks versus a single user that is responsible for the entire information fusion system [39–41].

The JDL variations include the Signals, Feature, Decision (SFD) model [7], (DF/RM) dual-node architecture [2], the User-Fusion Model (UFM) [42], Transformation of Requirements for Information Process (TRIP) model [43], and State-Transition Data Fusion (STDF) [9]. Each of these JDL variations seek common processes across many levels and a more robust way of implementation by reuse of data and methods to serve different purposes. The method of *context* has also been a subject of these model developments and variations that have influenced developers using the JDL as a framework for implementation. For the DF/RM model, context is divided into those elements that support data fusion (L0–3) and those for management (L4). An example is terrain information in which tracking can be adapted to road information and sensors can be managed as related to zones of authorized operation. The TRIP model divides the problem into demand conditioning (L4–6) and supply conditioning (L0–3). The demand/supply duality is meant to serve users requesting information and the system determining if these requests can be answered. For demand conditioning, decision, exploitation, and observability context are used to decompose the user request into machine functions (i.e., fusion) that can be processed. The UFM seeks to utilize the user's cognitive requests (L5–6) in establishing context in their coordination of the fusion levels (L0–4). Finally, the STDF utilizes the same control (L4–6) to look over different functions such as physics-derived (e.g., radar) (L0–1), human-derived (e.g., text) (L0–1), situations (L3), and scenarios (L4). The key attributes of these JDL variants are inherent in establishing different methods of fusion and control with common issues described at each level.

The future of the use of JDL will require developments in uncertainty analysis, exploitation, and user involvement based on the context, such as mission management (L6). Although these related models (SDF, UFM, TRIP, DF/RM, STDF) all have elements of context, they also try to solve real problems for real users. Hence, a current trend in information processing is machine analytics that seeks methods of data compression and presentation for user interaction.

## 4.4   MACHINE ANALYTICS

In the JDL model revisits in 2004, little attention was paid to the enormous amount of types of data (e.g., e-mail and sensor), distributed locations, and various connections to different applications (e.g., finance to surveillance) that have resulted from the expansion of the World Wide Web. Related concepts recently emerging are machine, descriptive, prescriptive, predictive, visual, and other analytics yet to be coined. There are three issues of importance *hardware* (e.g., Apache Hadoop data intensive distributed architecture), *software* (e.g., machine analytics), and *user/domain* applications (e.g., visual analytics, text analytics). As the JDL is a functional model, we focus on the last two. Data mining (DM) and data fusion (DF) are analysis functions supported by machines.

### 4.4.1   Modeling Analytics

The commercial business industry has been managing large volumes of transactional data using powerful databases systems and, most recently, using cloud infrastructures. Traditional business analytics has focused mostly on descriptive analyses of structured historical data using statistical techniques. The current trend in DM is toward predictive analytics of unstructured data such as documents, video, image sets, multimedia data, network data, matrices, tensors, and graphs and tensors [44]. The field of data fusion [45] has been making use of cutting edge artificial intelligence (AI) and machine learning (ML) techniques to perform situation and threat assessments. Given that analytics and data fusion are two sides of the same coin, an effective revision of the JDL model warrants a close cooperation between the two communities in terms of technology enrichment for managing and intelligent processing of large volumes of data. The model extensions should be able to exploit such enriched technologies seamlessly in areas such as massive data analytics, hybrid reasoning, text analytics, and distributed processing.

**TABLE 4.3**

**Approaches to Modeling Analytics**

| Paradigm | Approach | Technologies |
|---|---|---|
| Statistical | Nondeterministic relationships between variables are captured in the form of mathematical equations and probability distributions. | Test hypothesis, regression analyses, probability theory, sampling, inferencing |
| Artificial intelligence | Domain experts provide knowledge of system behavior, and knowledge engineers develop computational models using an underlying ontology. | Logic-based expert systems, fuzzy logic, Bayesian networks |
| Temporal | Linear/nonlinear equations specify behavior of stochastic processes or of dynamic systems as state transitions and observations. | Autoregression, survival analysis, Kalman filters, hidden Markov models, dynamic Bayesian networks |
| Machine learning | System input/output behavior is observed, and techniques extract system behavior models. | Clustering, neural network, and various linear, nonlinear, and symbolic approaches to learning |

Traditional statistical approaches are invaluable in data-rich environments, but there are areas where AI and ML approaches provide better analyses, especially where there is an abundance of subjective knowledge. Benefits of such augmentation include mixing of numerical and categorical variables in algorithms, "what-if" or explanation-based reasoning, explainable results of inferences easily understood by human analysts, and efficiency enhancement incorporating knowledge from domain experts as heuristics to deal with the "curse of dimensionality." Though early AI reasoning was primarily symbolic in nature (i.e., manipulation of linguistics symbols with well-defined semantics), it has moved toward a hybrid of symbolic and numerical, and therefore one is expected to find probabilistic and statistical foundations in many AI approaches. Conversely, business analytics traditionally has powerful customer or other related segmentation techniques via various powerful clustering algorithms such $k$-means, hierarchical clustering, and $k$-nearest neighbor. Table 4.3 depicts some of the well-known techniques categorized along the statistics, AI, and ML paradigms along with a special category for temporal reasoning given the dynamic nature of analytics and fusion problems. The following are some examples of augmentation and enrichment of business analytics:

- Enrich principal component and factor analyses with subspace methods (e.g., latent semantic analyses);
- Meld regression analyses with probabilistic graphical modeling;
- Extend autoregression and survival analysis techniques with Kalman filter and dynamic Bayesian networks, embed decision trees within influence diagrams; and
- Augment "nearest-neighbor" and $k$-means clustering techniques with support vector machines and neural networks.

*Machine analytics* (MA) covers the broad spectrum of applications and provides a direct link to the traditional JDL model. Processes that require machine data analysis include physics-derived sensor (e.g., video), human-derived (e.g., text), and machine (e.g., Web files) data. MA is also based on the processes of human–machine and machine–machine interactions. Inside MA are the emerging concepts of (1) descriptive, prescriptive, predictive analytics and (2) scientific, information, and visual analytics. These concepts mirror discussions in the current JDL model revisits of information exploitation.

The business-oriented definitions of analytics complement elements of data fusion.

**TABLE 4.4**

**Machine Analytics Mapped to Information Fusion Levels**

| Fusion | Machine | Concept |
|---|---|---|
| Level 0 | Scientific | Access to data and pedigree of information and issues of structured/ unstructured data |
| Level 1 | Information (visual, text) | Development of graphical methods for data analysis |
| Level 2 | Descriptive | Uses data mining to estimate the current state (i.e., machine learning) over different reasoning of trends for modeling |
| Level 3 | Predictive | Future options from current estimates |
| Level 4 | Prescriptive | Sequencing of selected actions |
| Level 5 | Visual | Sensemaking and reasoning |
| Level 6 | Activity-based | Policy instantiation of desired outcomes as to a focused mission |

- *Descriptive analytics* looks at an organization's historical and current performances that can be used to diagnose the situation.
- *Predictive analytics* forecasts future trends, behavior, and events for decision support such as to suggest courses of action profile and trending.
- *Prescriptive analytics* determines alternative courses of actions or decisions options (using predictive information), given the historical, current, and projected situations and a set of objectives, requirements, and constraints [46].

*Visual Analytics* (VA) [47] seeks scientific, information, and cognitive representations. Visualization supports analytical reasoning, planning, and decision making through effective data representations and transformations over physical- and human-derived data (sometimes referred to hard and soft data fusion). Finally, user interaction with machines is important for the collection, exploitation, and dissemination of data.

- Scientific visualization deals with data that have a natural geometric structure (e.g., magnetic resonance imaging data, wind flows).
- Information visualization handles abstract data structures such as trees or graphs.
- Visual analytics is especially concerned with sensemaking and reasoning.

If we look at "analytics" (Table 4.4), it mirrors the JDL (and other proposed models – DFRM and DFIG) in having both the data fusion *reasoning* (e.g., Bayes) and systems-level *management* (e.g., control) functions. Thus, the MA is like reasoning, whereas VA is about management.

### 4.4.2 Data Fusion Analytics

Descriptive and Predictive Analytics together establish current and projected situations of an organization, but do not recommend actions. An obvious next step is *Prescriptive Analytics*, which is a process to determine alternative courses of actions or decision options, given the situation along with a set of objectives, requirements, and constraints. Automation of decision making of routine tasks is ubiquitous, but subjective processes within organizations are still used for complex decision making. This current use of subjectivity should not prohibit the fusion and analytics community from pursuing a computational approach to the generation of decision options by accounting for various nonquantifiable subjective factors together with numerical data. The analytics-generated options can then be presented, along with appropriate explanations and backing, to the decision makers of the organization.

Systems routinely collect and store large volumes of data on a continuous basis from a variety of disparate and heterogeneous sources. Though such distribution is coherent with recent thrusts toward net-centric warfare, analysts often face a daunting task when searching for specific data or for series of correlated data residing in distributed sources. One solution is to build a large centralized data storage area in advance, such as a cloud infrastructure. However, the proprietary nature of some of the sources requires that they operate autonomously and hence a distributed fusion approach [48] is vital.

There are also other types of analytics. (1) *Web analytics* is Internet usage data for purposes of understanding and optimizing Web usage and business and market research. (2) *Image analytics* uses real-world videos and images to extract information with machine performance comparable to that of humans. (3) *Cross-lingual text analytics* has contents in multiple languages that enable a system to discover and maximize the value of information within large quantities of text (open-source or internal). The growing trend in analytics serves at DF/DM for both autonomy (machine) and automation (machine to user).

## 4.5  USER INVOLVEMENT

The *Observe–Orient–Decide–Act (OODA) model* serves as another common widely used model and is referenced in the information fusion community. As differentiated from the IM model, which deals with the enterprise and distribution of the data, the OODA is focused on a local control loop. Both of these models are useful for information fusion and can be enhanced with InfoEx capabilities such as visual analytics. For example, in Figure 4.7, a real-time operator is reactive, whereas an analyst has more time for data mining and discovery.

When using multiple OODA loops [49,50], there are two types of cases in which an operator is making real-time decisions and an analyst is making forensics non-real-time assessments. The motivation is based on the paradigms of (A) getting inside the enemy's control loop (L3) and (B) team decision making (L5). Obviously, the selection of the competitor, such as in game theory [20], requires determining a focused threat (L6). What is extended from the JDL and variations [8] is highlighting real-time versus the non-real time data mining and machine analytics for adaptive information exploitation.

## 4.6  ANALYSIS OF INFORMATION EXPLOITATION

With the JDL and advances in computation, there is a need to focus on the HLIF management functions such as managing the amount of information available. Current developments in enterprise

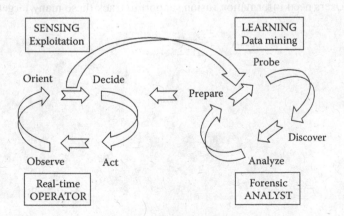

**FIGURE 4.7**  Multiple OODA loop human–machine interactions.

architectures support information management, cloud computing, user refinement through queries, and machine analytics. For *contextual reasoning*, much of the ancillary data are available on a distributed enterprise system to provide situational analysis focused on mission needs. One example is that a user would like to refine the estimates of many target tracks and uses semantic queries (ontology) to access the database through machine analytics/data mining.

### 4.6.1 CONTEXT TRACKING WITH PHYSICS AND HUMAN-DERIVED DATA

We designate two types of data: video and user text (e.g., documents). The physics-derived video can be processed by a machine; however, the human-derived text can be processed by a machine, but still needs user involvement to understand the meaning of the connected words.

Information fusion developments include large data (e.g., imagery), flexible autonomy (e.g., from moving airborne platforms over communication systems), and human coordination for situation awareness [51]. Figure 4.8 demonstrates an imagery data collection example using electro-optical (EO) cameras and wide-area motion imagery (WAMI). Using information fusion for situation awareness based on imagery includes (1) tracking targets in images (fusion over *time*) [52], (2) identifying targets using different sensors (fusion over *frequency*) [53], and (3) linking target measurements over wide areas (fusion over *space*) [54]. Enterprise information includes stored data of the physical (terrain), resource (sensors), and social context (objects) that are easily accessible from cloud services.

Given WAMI data [55,56] shown in Figure 4.8, we seek the benefits that are enabled from an enterprise network. For example, when the user designates an area of interest, the machine can then detect and track targets (L1 fusion) [57,58] and access contextual information through data mining (L2/L3 fusion) to enhance understanding (e.g., target type, identification [ID], and activity). These results can be used for sensor management [59]. Finally, the results are used to query the sensors to get more information (L4 fusion), store the results, present to the user (L5 fusion), and disseminate back to the cloud for mission awareness (L6 fusion). Recent efforts include information management for video- and text-based multimodal fusion [60,61].

### 4.6.2 INFORMATION MANAGEMENT CONTEXTUAL TRACKING

#### 4.6.2.1 Information Fusion Example

In the first notional scenario, raw (images, text), filtered (tracks, keywords), and fused (tracks with classification and ID labels) are sent to user stations. For the five cases (1–5), shown in Table 4.5, we model data scaled to the relative volume values. We assume that there are five distributed operators maintaining surveillance. The complexity from Case 1 to Case 5 is akin to the difference between tracking a few targets (8 = 40/5) and trying to maintain tracks over the entire context of a city (200 tracks = 1000/5). Users need information fusion support to track these many targets.

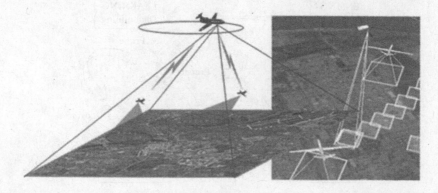

**FIGURE 4.8**   Wide area motion imagery (WAMI) data.

**TABLE 4.5**

**Scenario Data**

| Case | Video | Text | Tracks |
|------|-------|------|--------|
| 1 | 3 | 10 | 40 |
| 2 | 10 | 500 | 40 |
| 3 | 10 | 50 | 200 |
| 4 | 20 | 500 | 1000 |
| 5 | 50 | 1000 | 1000 |

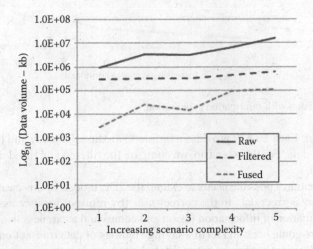

**FIGURE 4.9** Traditional JDL data fusion.

Assuming raw data messaging, it is intuitive that the filtered information can reduce the volume by ruling out unnecessary data (see Figure 4.9). The results indicate that from the traditional JDL methods, there is little to no information management from the user, mission, or machine analytics and fusion is the correlation of text with tracks.

### 4.6.2.2 Information Management Example

In the second notional scenario, we are interested in seeing the benefits of user involvement, machine analytics, and information management to support distributed users. Not all users require the same collection information; however, they can query the information they need from the cloud to be presented on their displays. In this case, we use a publish/subscribe (pub/sub) architecture to afford the interactions of distributed users.

When using pub/sub information management, it is important to note that over a different set of users desiring situational context, with two data feeds (video to tracks and text to keywords) and 40 to 200 tracks, there is a break point in operations. For this case, if one tries to normalize over the time duration of the scenario, the fused information accounts to three pieces of information for each image. If there are two operators, they can follow more objects by specializing in the data source and area of designated interest to highlight tracks. The interesting part is the difference between 5 and 10 operators in that there is an order of magnitude more information being passed around, as many users (assumed) are requesting the same information.

The second notional scenario speaks to team management. A situation of too many operators looking at similar data could result in an overlap in functions. By using contextual information and user-defined operating pictures (UDOP) displays, information can deliver machine analytic results

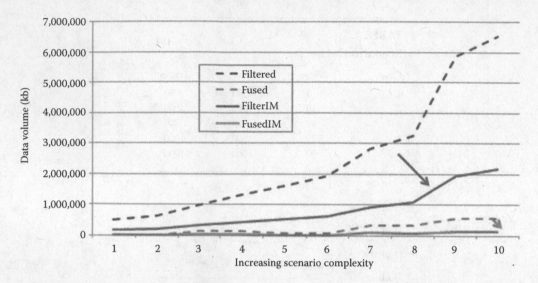

**FIGURE 4.10**  Data fusion with information exploitation.

to the appropriate user. Each UDOP provides a tailored situational display and the user can call the raw, filtered, or fused data. For this scenario, we compare the filtered and fused data as the user does not want the raw data.

Figure 4.10 demonstrates possible needs to extend the JDL model. In this case, using information management sends the correct data to the correct user (by request) for user assessment. However, when using the combination of information fusion and contextual awareness in a pub/sub enterprise, only the fused data are going back and forth – saving volumes of data transactions. Here, we assume such aspects as visualization (versus the entire filtered image and text), resident overhead imagery to overlay tracks and text reports at the distributed site, and command and control interfaces that allow each user to know the updates from the other users.

From Figure 4.10, we see that information management provides a reduction in the volume of information being passed around in the information fusion system. Likewise, there is time/space uncertainty reduction as target track and ID information association is refined by multiple users against terrain, feature, and attribute clarifications. Going from the raw data to the filtered information determines the relevant data. Information Management (IM) processes the filtered information and sends only the needed updates. Again, if there is a correlation of data in time as part of the fusion process, the association of the key information from information management (based on quality, workflow, and maintenance as services), then a further reduction on information being passed is possible, which is labeled as FusedIM in Figure 4.10. To determine the fusion gain from data and uncertainty reduction, we determined the analysis over the notional cases. Table 4.6 presents the information gain at each level where for more complex scenarios, information management provides a 75% gain in uncertainty reduction to the salient information needed for users to determine the target tracks.

### TABLE 4.6
### Fusion Gain Analysis

| Scenario | 1 | 2 | 3 | 4 | 5 | 6 | 7 | 8 | 9 | 10 |
|---|---|---|---|---|---|---|---|---|---|---|
| Information gain from IM Filtering | 0.68 | 0.69 | 0.70 | 0.71 | 0.71 | 0.72 | 0.73 | 0.74 | 0.74 | 0.75 |
| Information gain from IM Fused | 0.55 | 0.58 | 0.62 | 0.64 | 0.67 | 0.69 | 0.71 | 0.72 | 0.74 | 0.75 |

## 4.7 CONCLUSIONS

Since its inception, the JDL model has served as a process model to explain information fusion. This chapter revisits the JDL model based on technology developments in the last decade as discussed in the community. The chapter overviewed the major developments in the JDL model, with information exploitation being the third major development in the information fusion modeling process. We highlight three information exploitation technology developments affecting the JDL model that include (1) an enterprise architecture that supports information management to store and access data through cloud computing, (2) machine analytics for data mining, and (3) human–machine interfaces to support users as active observers of contextual reasoning (L5). Through notional examples, we demonstrated an information gain through data and uncertainty reduction using information management to present only the useful updates with target tracking, situation assessment, resource management, and user refinement. Information management utilizes dynamic data-driven applications systems processes of measurement assessment, statistical modeling, and software strategies to provide real-time support to users.

## ACKNOWLEDGMENT

Erik Blasch was supported under an AFOSR grant in Dynamic Data-Driven Applications Systems. The author appreciates the discussions with Alan Steinberg, Subrata Das, and Chris Bowman. The views and conclusions contained herein are those of the author and should not be interpreted as necessarily representing the official policies, either expressed or implied, of Air Force Research Laboratory, or the U.S. government.

## REFERENCES

1. O. Kessler, K. Askin, N. Beck et al. "Functional description of the data fusion process. Technical report for the Office of Naval Technology Data Fusion Development Strategy," Naval Air Development Center, November 1991.
2. A. N. Steinberg, C. L. Bowman, and F. E. White, "Revisions to the JDL model," *Joint NATO/IRIS Conference*, 1998.
3. E. Blasch, and S. Plano, "JDL Level 5 fusion model 'user refinement' issues and applications in group tracking," *Proceedings of SPIE*, Vol. 4729, 2002.
4. J. Llinas, C. Bowman, G. Rogova, A. Steinberg, E. Waltz, and F. White, "Revisiting the JDL data fusion model II," *Proceedings of the 7th International Conference on Information Fusion*, 2004.
5. E. Blasch, and S. Plano, "DFIG Level 5 (user refinement) issues supporting situational assessment reasoning," *Proceedings of the 8th International Conference on Information Fusion*, 2005.
6. E. Blasch, I. Kadar, J. Salerno et al. "Issues and challenges in situation assessment (Level 2) fusion," *Journal of Advances in Information Fusion*, Vol. 1, No. 2, pp. 122–139, 2006.
7. E. Waltz, and J. Llinas, *Multisensor Data Fusion*, Artech House, Boston, MA, 1990.
8. E. Blasch, I. Kadar, K. Hintz, J. Biermann, C. Chong, and S. Das, "Resource management coordination with Level 2/3 fusion issues and challenges," *IEEE Aerospace and Electronic Systems Magazine*, Vol. 23, No. 3, pp. 32–46, 2008.
9. E. Blasch, E. Bossé, and D. A. Lambert, *High-Level Information Fusion Management and Systems Design*, Artech House, Norwood, MA, 2012.
10. E. Blasch, D. A. Lambert, P. Valin et al. "High level information fusion (HLIF) survey of models, issues, and grand challenges," *IEEE Aerospace and Electronic Systems Magazine*, Vol. 27, No. 9, pp. 4–20, 2012.
11. J. Dezert, D. Han, Z.-G. Liu, and J.-M. Tacnet, "Hierarchical DSmP transformation for decision-making under uncertainty," *Proceedings of the 15th International Conference on Information Fusion*, 2012.
12. A. Jøsang, and R. Hankin, "Interpretation and fusion of hyper opinions in subjective logic," *Proceedings of the 15th International Conference on Information Fusion*, 2012.

13. M. A. Solano, and G. Jernigan, "Enterprise data architecture principles of high-level Mult-INT fusion: A pragmatic guide for implementing a heterogeneous data exploitation," *Proceedings of the 15th International Conference on Information Fusion*, 2012.

14. L. Scholz, D. Lambert, D. Gossink, and G. Smith, "A blueprint for command and control: Automation and interface," *Proceedings of the 15th International Conference on Information Fusion*, 2012.

15. D. L. Hall, C.-Y. Chong, J. Llinas, and M. Liggins, *Disrtibuted Data Fusion for Network-Centric Opertaions*, CRC Press/Taylor & Francis, Boca Raton, FL, 2012.

16. E. Blasch, A. Steinberg, S. Das et al. "Revisiting the JDL model for information exploitation," *Proceedings of the 16th International Conference on Information Fusion*, 2013.

17. A. N. Steinberg, and G. L. Rogova, "Situation and context in data fusion and natural language understanding," *Proceedings of the 11th International Conference on Information Fusion*, 2008.

18. A. N. Steinberg, "Context-sensitive data fusion using structural equation modeling," *Proceedings of the 12th International Conference on Information Fusion*, 2009.

19. J. García, L. Snidaro, and I. Visentini, "Exploiting context as binding element for multi-level fusion," *Proceedings of the 15th International Conference on Information Fusion*, 2012.

20. G. Chen, D. Shen, C. Kwan, J. Cruz, M. Kruger, and E. Blasch, "Game theoretic approach to threat prediction and situation awareness," *Journal of Advances in Information Fusion*, Vol. 2, No. 1, pp. 1–14, 2007.

21. E. Blasch, P. Valin, E. Bosse, M. Nilsson, J. Van Laere, and E. Shahbazian, "Implication of culture: User roles in information fusion for enhanced situational understanding," *Proceedings of the 12th International Conference on Information Fusion*, 2009.

22. E. Blasch, R. Breton, P. Valin, and E. Bosse, "User information fusion decision making analysis with the C-OODA model," *Proceedings of the 14th International Conference on Information Fusion*, 2011.

23. E. Blasch, J. J. Salerno, and G. Tadda, "Measuring the worthiness of situation assessment," *IEEE National Aerospace Electronics Conference*, 2011.

24. J. Rasmussen, "Skills, rules, knowledge; signals, signs, and symbols, and other distinctions in human performance models," *IEEE Transactions on Systems, Man and Cybernetics*, Vol. 13, pp. 257–266, 1983.

25. A. N. Steinberg, C. L. Bowman, G. Haith, E. Blasch, C. Morefield, and M. Morefield, "Adaptive context assessment and context management," *Proceedings of the 17th International Conference on Information Fusion*, 2014.

26. O. Kessler, and F. White, "Data fusion perspectives and its role in information processing, Chapter 2," in *Handbook of Multisensor Data Fusion*, 2nd Ed., M. E. Liggins, D. Hall, and J. Llinas (eds.), CRC Press, Boca Raton, FL, 2008.

27. E. L. Waltz, "Information understanding: Integrating data fusion and data mining processes," *International Symposium on Circuits and Systems*, Vol. 6, pp. 553–556, 1998.

28. E. Blasch, O. Kessler, F. E. White, J. Morrison, and J. F. Tangney, "Information fusion management and enterprise processing," *IEEE National Aerospace and Electronics Conference*, 2012.

29. E. Blasch, Y. Chen, G. Chen, D. Shen, and R. Kohler, "Information fusion in a cloud-enabled environment," in *High Performance Semantic Cloud Auditing*, B.-Y. Choi, K. Han, and S. Song (eds.), Springer, New York, 2013.

30. B. Liu, E. Blasch, Y. Chen et al. "Information fusion in a cloud computing era: A systems-level perspective," *IEEE Aerospace and Electronic Systems Magazine*, Vol. 29, No. 10, pp. 16–24, 2014.

31. S. Das, J. Llinas, G. Pavlin, D. Snyder, A. Steinberg, and K. Sycara, "Agent based information fusion: Panel discussion," *Proceedings of the 10th International Conference on Information Fusion*, 2007.

32. E. Blasch, "Ontological issues in higher levels of information fusion: User refinement of the fusion process," *Proceedings of the 6th International Conference on Information Fusion*, 2003.

33. E. Blasch, M. Pribilski, B. Daughtery, B. Roscoe, and J. Gunsett, "Fusion metrics for dynamic situation analysis," *Proceedings of SPIE*, Vol. 5429, 2004.

34. E. Blasch, P. C. G. Costa, K. B. Laskey et al. "Issues of uncertainty analysis in high-level information fusion—Fusion2012 Panel Discussion," *Proceedings of the 15th International Conference on Information Fusion*, 2012.

35. E. P. Blasch, P. Valin, A.-L. Jousselme, D. Lambert, and E. Bosse, "Top ten trends in high-level information fusion," *Proceedings of the 15th International Conference on Information Fusion*, 2012.

36. P. C. G. Costa, K. B. Laskey, E. Blasch, and A.-L. Jousselme, "Towards unbiased evaluation of uncertainty reasoning: The URREF ontology," *Proceedings of the 15th International Conference on Information Fusion*, 2012.

37. A. N. Steinberg, and C. Bowman, "Adaptive context exploitation," *Proceedings of SPIE*, Vol. 8758, 2013.
38. A. N. Steinberg, and C. Bowman, "Adaptive context discovery and exploitation," *Proceedings of the 16th International Conference on Information Fusion*, 2013.
39. E. Blasch, and S. Plano, "Cognitive fusion analysis based on context," *Proceedings of SPIE*, Vol. 5434, 2004.
40. E. Blasch, "Level 5 (user refinement) issues supporting information fusion management," *Proceedings of the 9th International Conference on Information Fusion*, 2006.
41. E. Blasch, "User refinement in information fusion, Chapter 19," in *Handbook of Multisensor Data Fusion*, 2nd Ed., M. E. Liggins, D. Hall, and J. Llinas (eds.), pp. 503–536, CRC Press/Taylor & Francis, Boca Raton, FL, 2009.
42. E. Blasch, and P. Hanselman, "Information fusion for information superiority," *IEEE National Aerospace and Electronics Conference*, 2000.
43. W. Fabian, Jr., and E. P. Blasch, "Information architecture for actionable information production (a process for bridging fusion and ISR management)," *National Symposium on Sensor and Data Fusion*, 2002.
44. T. Li, C. Ding, and F. Wang, "Guest editorial: Special issue on data mining with matrices, graphs and tensors," *Data Mining and Knowledge Discovery*, Vol. 22, pp. 337–339, 2011.
45. Available at http://en.wikipedia.org/wiki/Visual_analytics.
46. S. Das, *Computational Business Analytics*, CRC Press/Chapman & Hall, Boca Raton, FL, 2013.
47. S. Das, *High-Level Data Fusion*, Artech House, Norwood, MA, 2008.
48. S. Das, "A framework for distributed high-level fusion," in *Net-Centric Distributed Fusion*, D. L. Hall, J. Llinas, M. Liggins, and C. Chong (eds.), CRC Press/Taylor & Francis, Norwood, MA, 2012.
49. E. Blasch, and S. Plano, "Proactive decision fusion for site security," *Proceedings of the 8th International Conference on Information Fusion*, 2005.
50. A. N. Steinberg, "Foundations of situation and threat assessment," in *Handbook of Multisensor Data Fusion*, 2nd Ed., M. E. Liggins, D. Hall, and J. Llinas (eds.), pp. 437–502, CRC Press/Taylor & Francis, Boca Raton, FL, 2009.
51. E. Blasch, R. Breton, and P. Valin, "Information fusion measures of effectiveness (MOE) for decision support," *Proceedings of SPIE*, Vol. 8050, 2011.
52. H. Ling, L. Bai, E. Blasch, and X. Mei, "Robust infrared vehicle tracking across target change using $L_1$ regularization," *Proceedings of the 13th International Conference on Information Fusion*, 2010.
53. Y. Wu, E. Blasch, G. Chen, L. Bai, and H. Ling, "Multiple source data fusion via sparse representation for robust visual tracking," *Proceedings of the 14th International Conference on Information Fusion*, 2011.
54. K. Palaniappan, F. Bunyak, P. Kumar et al. "Efficient feature extraction and likelihood fusion for vehicle tracking in low frame rate airborne video," *Proceedings of the 13th International Conference on Information Fusion*, 2010.
55. H. Ling, Y. Wu, E. Blasch, G. Chen, H. Lang, and L. Bai, "Evaluation of visual tracking in extremely low frame rate wide area motion imagery," *Proceedings of the 14th International Conference on Information Fusion*, 2011.
56. R. Pelapur, S. Candemir, F. Bunyak, M. Poostchi, G. Seetharaman, and K. Palaniappan, "Persistent target tracking using likelihood fusion in wide-area and full motion video sequences," *Proceedings of the 15th International Conference on Information Fusion*, 2012.
57. X. Mei, H. Ling, Y. Wu, E. Blasch, and L. Bai, "Efficient minimum error bounded particle resampling L1 tracker with occlusion detection," *IEEE Transactions on Image Processing (T-IP)*, Vol. 22, No. 7, pp. 2661–2675, 2013.
58. E. Blasch, Z. Wang, H. Ling et al. "Video-based activity analysis using the L1 tracker on VIRAT data," *IEEE Applied Imagery Pattern Recognition Workshop*, 2013.
59. C. Yang, L. M. Kaplan, E. Blasch, and M. Bakich, "Optimal placement of heterogeneous sensors for targets with Gaussian priors," *IEEE Transactions on Aerospace and Electronics Systems*, Vol. 49, No. 3, pp. 1637–1653, 2013.
60. E. Blasch, J. Nagy, A. Aved et al. "Context aided video-to-text information fusion," *Proceedings of the 17th International Conference on Information Fusion*, 2014.
61. R. I. Hammoud, C. S. Sahin, E. P. Blasch, and B. J. Rhodes, "Multi-source multi-modal activity recognition in aerial video surveillance," *International Computer Vision and Pattern Recognition (ICVPR)*, 2014.

# 5  Elements of Random Set Information Fusion

*Ronald Mahler*

## CONTENTS

## 5.1  INTRODUCTION

This chapter summarizes some of the central ideas of *finite-set statistics* (FISST), a systematic, unified approach to multisensor–multitarget detection, tracking, identification, and information fusion. FISST has been the subject of considerable worldwide research interest during the last decade, as evidenced by a thousand research publications from well over a dozen nations. This interest can be attributed to the fact that finite-set statistics (1) employs explicit, systematic, and unified statistical models of multisensor–multitarget systems; (2) Bayes optimally unifies multitarget detection and state-estimation; (3) results in innovative multitarget tracking algorithms (probability hypothesis density [PHD] filters, cardinalized PHD [CPHD] filters, multi-Bernoulli filters, etc.) that do not require measurement-to-track association, while still achieving tracking performance comparable to, or better than, that of conventional approaches; (4) results in promising multitarget tracking filters for unknown, dynamically changing backgrounds; and (5) has been a fertile source of fundamentally new approaches.

In particular, finite-set statistics has led to significant advances such as (1) "classical" CPHD filters that can successfully perform in clutter with clutter rates of 70 measurements per frame and higher (regimes in which conventional algorithms tend to suffer combinatorial breakdown) [1]; (2) CPHD filters for superpositional sensors that significantly outperform conventional Markov Chain Monte Carlo (MCMC) methods, while also being 30 to 87 times faster (depending on the specific application) [1]; (3) multi-Bernoulli track-before-detect (TBD) filters for tracking in images, which have been shown to significantly outperform the previously best algorithm, the histogram-probabilistic multi-hypothesis tracker (PMHT) [1]; (4) PHD filter-based algorithms that have been shown to significantly outperform conventional simultaneous localization and mapping (SLAM) algorithms such as MH-FastSLAM in heavy clutter conditions [2]; and (5) the first (and apparently only) provably Bayes-optimal approach for fusing both "hard" data and "soft" data (such as attributes, features, natural-language statements, and inference rules) [1].

The purpose of the chapter is to answer the following questions: (1) How can multiple, noncooperative targets be Bayes optimally detected and tracked using one or more imperfect sensors? (2) How can multisensor–multitarget systems be statistically modeled so that Bayes optimality is possible? (3) How can faithful Bayesian models of multisensor–multitarget systems be constructed? (4) How can the optimal multisource–multitarget Bayes recursive filter be approximated in a principled statistical manner—meaning that the underlying models and their relationships are preserved as faithfully as possible?

The chapter is organized into the following major sections: The Philosophy of Finite-Set Statistics; Single-Sensor, Single-Target Systems; Multisensor–Multitarget Systems; Principled Approximate Multitarget Filters; The Finite-Set Statistics Approximation Methodology; Conclusions; and References. More complete information can be found in tutorial articles [3–6] and books [1,2,7,8]. This chapter is essentially a condensed version of Ref. [4].

## 5.2 THE PHILOSOPHY OF FINITE-SET STATISTICS

Multisensor, multitarget systems differ from single-sensor, single-target systems in a major respect: They are composed of *randomly varying numbers* of (1) targets of various kinds; (2) disparate sensors and other information sources; (3) measurements of various kinds; and (4) various sensor-carrying platforms. A rigorous mathematical foundation for stochastic multiobject problems—*point process theory*—is a half-century old, but has been formulated with the requirements of mathematicians rather than practitioners in mind. One fundamental motivation for finite-set statistics is this: *Tracking and information fusion R&D practitioners should not have to be virtuoso experts in point process theory to produce meaningful engineering innovations.* As was emphasized in Refs. [3,4], engineering statistics is a tool and not an end in itself. It must have two qualities: (1) *trustworthiness*—that is, it is constructed upon a systematic, reliable mathematical foundation, to which we can appeal when the going gets rough; and (2) *fire and forget*—that is, this foundation can be safely neglected in most situations, leaving a serviceable mathematical machinery in its place.

These two goals are inherently in conflict. Foundations are shackles if they are so mathematically complex that they cannot be taken for granted in most situations. But they are simplistic if they are so simple that they repeatedly result in practical blunders. This gap between trustworthiness versus engineering pragmatism is what finite-set statistics attempts to bridge. Four objectives are paramount: (1) directly generalize familiar single-sensor, single-target Bayesian "Statistics 101" concepts to the multisource–multitarget realm; (2) avoid all avoidable abstractions; (3) as much as possible, replace theorem-proving with "turn-the-crank," purely algebraic procedures; and (4) despite this fact, retain all mathematical power necessary.

Toward this end, let us begin by comparing the "random finite set" (RFS) paradigm of finite-set statistics with the conventional multitarget tracking approach: *report-to-track association* (MTA). The most familiar tracking algorithms presume a radar-type sensor model. Suppose that, for a given range-bin $r$, azimuth $\alpha$, and elevation $\theta$, a radar amplitude-signature is thresholded. If the amplitude exceeds the threshold, there are two possible reasons. First, it was produced by a target—a "target detection" is at $\mathbf{z} = (r, \alpha, \theta)$. Second, it was caused by noise—it is a "false detection." A third possibility: a target was present but was not detected—a "missed detection."

For target-generated detections, the "small target" model is presumed. Targets are (1) distant enough (relative to radar resolution) that any single target generates at most a single detection but (2) also near enough that any detection is caused by at most a single target. A divide-and-conquer strategy can then be applied to the multitarget detection and tracking problem (see Refs. [8] and [3, pp. 321–335]). Suppose that, at time $t_k$, there are $n$ "target tracks" $\left(\ell_1^{k|k}, \mathbf{x}_1^{k|k}, P_1^{k|k}\right),\dots,\left(\ell_n^{k|k}, \mathbf{x}_n^{k|k}, P_n^{k|k}\right)$. That is, these are $n$ hypothesized targets where, for the $i$th track, $\mathbf{x}_i^{k|k}$ is its state (position, velocity), $P_i^{k|k}$ is its error covariance matrix, and $\ell_i^{k|k}$ is the "track label" that uniquely distinguishes it. The Gaussian distribution $f_i(\mathbf{x}) = N_{P_i^{k|k}}\left(\mathbf{x} - \mathbf{x}_i^{k|k}\right)$ is the "track density" of the $i$th track.

Now suppose that the radar collects $m$ detections $Z = \{\mathbf{z}_1,\dots,\mathbf{z}_m\}$ at the next time $t_{k+1}$. Because of false alarms, usually $m > n$. Employ the EKF (extended Kalman filter) time-update equations to construct the *predicted tracks* $\left(\ell_1^{k+1|k}, \mathbf{x}_1^{k+1|k}, P_1^{k+1|k}\right),\dots,\left(\ell_n^{k+1|k}, \mathbf{x}_n^{k+1|k}, P_n^{k+1|k}\right)$. Given this, consider the following hypothesis $H$: for each $i$, if the track $\left(\ell_i^{k+1|k}, \mathbf{x}_i^{k+1|k}, P_i^{k+1|k}\right)$ generates a detection then it generated $\mathbf{z}_{\tau(i)}$. The "excess measurements" $Z - \{\mathbf{z}_{\tau(1)};\dots,\mathbf{z}_{\tau(n)}\}$ are then interpreted as due to either false alarms or to previously undetected targets. The hypothesis $H$ is an MTA. Taking all such hypotheses into account, we get a list $H_1^{k+1|k+1},\dots,H_\nu^{k+1|k+1}$ of MTAs. For each $H_i^{k+1|k+1}$, we can employ the EKF measurement-update equations and $\mathbf{z}_{\tau(i)}$ to construct a list of revised tracks, $\left(\ell_1^{k+1|k+1}, \mathbf{x}_1^{k+1|k+1}, P_1^{k+1|k+1}\right),\dots,\left(\ell_n^{k+1|k+1}, \mathbf{x}_n^{k+1|k+1}, P_n^{k+1|k+1}\right)$. The dominant algorithms using the MTA approach are *multi-hypothesis trackers* (MHTs).

In contrast to MTA, finite-set statistics employs a top-down paradigm grounded in the random finite set (RFS) version of point process theory. The hypothesis-list $H_1^{k+1|k+1},\dots,H_\nu^{k+1|k+1}$ is replaced by a *multitarget probability distribution* $f_{k|k}(X|Z^{(k)})$, defined on the finite-set variable $X = \{\mathbf{x}_1,\dots,\mathbf{x}_n\}$ with $n \geq 0$, where $Z^k : Z_1,\dots,Z_k$ is the time history of measurement sets at time $t_k$. The radar-type measurement model is used to construct, and is replaced by, a *multitarget likelihood function* $L_Z(X) = f_{k+1}(Z|X)$. The value $L_Z(X)$ is the likelihood that a measurement set $Z$ will be generated if targets with state set $X$ are present. Given this, MTA propagation is replaced by the *multitarget recursive Bayes filter* [1,3,4,7]. This filter is computationally intractable in most situations. It must be approximated, resulting in the PHD, CPHD, multi-Bernoulli, and other filters [1,4,7].

Finite sets are order independent. It is thereby often mistakenly asserted that RFS algorithms are inherently incapable of constructing time sequences of labeled tracks. But as has been explained in Refs. [7, pp. 505–508] and [4, p. 10], target states in general have the form $\mathbf{x} = (\ell, \mathbf{x}')$, where $\ell$ is a identifying label unique to each track. Thus, the multitarget Bayes filter, along with any RFS approximation of it can (at least in principle) maintain temporally connected tracks.

## 5.3 SINGLE-SENSOR, SINGLE-TARGET SYSTEMS

The purpose of this section is to motivate the concepts that follow by illustrating them for the simplest special case: a single target being observed by a single sensor, with no false alarms (false targets) and no missed detections (undetected targets).

### 5.3.1 SINGLE-SENSOR/TARGET SYSTEMS: THE RECURSIVE BAYES FILTER

This is the theoretical foundation for optimal single-sensor, single-target tracking. It requires two a priori items: the *Markov transition density* and the *sensor likelihood function*. The Markov density $f_{k+1|k}(\mathbf{x}|\mathbf{x}')$ is constructed from a *single-target motion model*. It is the probability that the target will have state $\mathbf{x}$ at time $t_{k+1}$ if it had state $\mathbf{x}'$ at time $t_k$. The sensor likelihood function $f_{k+1}(\mathbf{z}|\mathbf{x})$ is constructed from a *single-target measurement model*. It is the probability that the target will generate measurement $\mathbf{z}$ at time $t_{k+1}$ if its state is $\mathbf{x}$.

The recursive Bayes filter propagates a single-target probability distribution $f_{k|k}(\mathbf{x}|Z^k)$ through time:

$$\ldots \to f_{k|k}(\mathbf{x}|Z^k) \to f_{k+1|k}(\mathbf{x}|Z^k) \to f_{k+1|k+1}(\mathbf{x}|Z^{k+1}) \to \ldots$$

First, $f_{k+1|k}(\mathbf{x}|\mathbf{x}')$ is used to propagate $f_{k|k}(\mathbf{x}|Z^k)$ from time $t_k$ to time $t_{k+1}$ via the *prediction integral*:

$$f_{k+1|k}\left(\mathbf{x}|Z^k\right) = \int f_{k+1|k}\left(\mathbf{x}|\mathbf{x}'\right) \cdot f_{k|k}\left(\mathbf{x}|Z^k\right) d\mathbf{x}' \tag{5.1}$$

Second, suppose that the sensor collects a new measurement $\mathbf{z}_{k+1}$ at time $t_{k+1}$. Then $f_{k+1}(\mathbf{z}|\mathbf{x})$ and Bayes' rule are used to modify $f_{k+1|k}(\mathbf{x}|Z^k)$ into

$$f_{k+1|k+1}\left(\mathbf{x}|Z^{k+1}\right) = \frac{f_{k+1}\left(\mathbf{z}_{k+1}|\mathbf{x}\right) \cdot f_{k|k}\left(\mathbf{x}|Z^k\right)}{f_{k+1|k}\left(\mathbf{z}_{k+1}|Z^k\right)} \tag{5.2}$$

where the Bayes normalization factor is

$$f_{k+1|k}\left(\mathbf{z}_{k+1}|Z^k\right) = \int f_{k+1}\left(\mathbf{z}_{k+1}|\mathbf{x}\right) \cdot f_{k|k}\left(\mathbf{x}|Z^k\right) d\mathbf{x}. \tag{5.3}$$

Third, to extract information of interest (position, velocity, identity, etc.) requires a *Bayes-optimal state estimator*, such as the maximum a posteriori (MAP) estimator:

$$\hat{\mathbf{x}}_{k+1|k+1} = \arg\sup_{\mathbf{x}} f_{k+1|k+1}\left(\mathbf{x}|Z^{k+1}\right) \tag{5.4}$$

## 5.3.2  SINGLE-SENSOR/SINGLE-TARGET SYSTEMS: MOMENT APPROXIMATIONS

The single-target Bayes filter is usually implemented using *moment approximations*. That is, let $f_i(\mathbf{x}) = N_{P_0}(\mathbf{x} - \mathbf{x}_0)$ denote a Gaussian distribution with mean $\mathbf{x}_0$ and covariance matrix $P_0$. Assume that SNR (signal-to-noise ratio) is large enough that the track distributions can be approximated by their first-order and second-order moments:

$$f_{k|k}(\mathbf{x}) = N_{P_{k|k}}(\mathbf{x} - \mathbf{x}_{k|k}), \quad f_{k+1|k}(\mathbf{x}) = N_{P_{k+1|k}}(\mathbf{x} - \mathbf{x}_{k+1|k}) \tag{5.5}$$

Then the Bayes filter can be replaced by a filter—the EKF—which propagates first- and second-order moments rather than the full track distributions:

$$\ldots \to (\mathbf{x}_{k|k}, P_{k|k}) \to (\mathbf{x}_{k+1|k}, P_{k+1|k}) \to (\mathbf{x}_{k+1|k+1}, P_{k+1|k+1}) \to \ldots$$

Given a fixed covariance matrix $P$, a less accurate but faster filter results if the $f_{k|k}(\mathbf{x})$ can be approximated by their first-order moments:

$$f_{k|k}(\mathbf{x}) \cong N_P(\mathbf{x} - \mathbf{x}_{k|k}), \quad f_{k+1|k}(\mathbf{x}) \cong N_P(\mathbf{x} - \mathbf{x}_{k+1|k}) \tag{5.6}$$

The Bayes filter can then be replaced by one that propagates the first-order moment:

$$\ldots \to \mathbf{x}_{k|k} \to \mathbf{x}_{k+1|k} \to \mathbf{x}_{k+1|k+1} \to \ldots$$

The $\alpha$–$\beta$ filter and the $\alpha$–$\beta$–$\gamma$ filters are the most common examples of such a filter.

### 5.3.3  Single-Sensor/Target Systems: Target Modeling

For a single-target system, target modeling begins with a *state transition function* $\phi_k(\mathbf{x}')$. That is, the target is hypothesized to have state $\phi_k(\mathbf{x}')$ at time $t_{k+1}$ if it had state $\mathbf{x}'$ at time $t_k$. Uncertainty in this hypothesis is modeled by appending a zero-mean random noise vector $\mathbf{W}_k$ (the "plant noise") with noise distribution $f_{\mathbf{W}_k}(\mathbf{x})$:

$$\mathbf{X}_{k+1|k} = \phi_k(\mathbf{x}') + \mathbf{W}_k \tag{5.7}$$

The information in Equation 5.7 is equivalent to that contained in the *probability mass function* (p.m.f.) of $\mathbf{X}_{k+1|k}$ (otherwise known as a *probability measure*):

$$p_{k+1|k}(S|\mathbf{x}') = \Pr(\mathbf{X}_{k+1|k} \in S|\mathbf{X}_{k|k} = \mathbf{x}') \tag{5.8}$$

where $S$ is a measurable subset of the space $\mathcal{X}$ of all possible target states. This is equivalent to the Markov density

$$f_{k+1|k}(\mathbf{x}|\mathbf{x}') = f_{\mathbf{W}_k}(\mathbf{x} - \varphi_k(\mathbf{x}')) \tag{5.9}$$

which, in turn, is a consequence of the Radon–Nikodým equation

$$p_{k+1|k}(S|\mathbf{x}') = \int_S f_{k+1|k}(\mathbf{x}|\mathbf{x}') \, d\mathbf{x} \tag{5.10}$$

The Markov density in Equation 5.9 is "true" because it satisfies the relationship in Equation 5.10—and thus contains exactly the same information as in the original motion model, Equation 5.7. No information has been lost and no spurious information has been inadvertently introduced (which could produce an unrecognized statistical bias).

### 5.3.4  Single-Sensor/Target Systems: Sensor Modeling

In a single-sensor, single-target system, the modeling of the sensor begins with a *measurement function* $\eta_{k+1}(\mathbf{x})$. This indicates that the sensor will collect the measurement $\eta_{k+1}(\mathbf{x})$ at time $t_{k+1}$ if a target with state $\mathbf{x}$ is present. Because any real sensor is noisy, this model must be extended to the form

$$\mathbf{Z}_{k+1} = \eta_{k+1}(\mathbf{x}) + \mathbf{V}_{k+1} \tag{5.11}$$

where $\mathbf{W}_{k+1}$ is a zero-mean noise vector with noise distribution $f_{\mathbf{V}_{k+1}}(\mathbf{z})$. The information contained in the noisy model is equivalent to that contained in the p.m.f.

$$p_{k+1}(T|\mathbf{x}) = \Pr(\mathbf{Z}_{k+1} \in T|\mathbf{X}_{k+1|k} = \mathbf{x}) \tag{5.12}$$

where $T$ is a measurable subset of the space $\mathcal{Z}$ of all possible sensor measurements. This is equivalent to the likelihood function

$$f_{k+1}(\mathbf{z}|\mathbf{x}) = f_{\mathbf{V}_{k+1}}(\mathbf{z} - \eta_{k+1}(\mathbf{x})) \tag{5.13}$$

which, in turn, is a consequence of the equation

$$p_{k+1}(T|\mathbf{x}) = \int_T f_{k+1}(\mathbf{z}|\mathbf{x})\,d\mathbf{z} \tag{5.14}$$

The likelihood function in Equation 5.13 is "true" because of Equation 5.14. No information in Equation 5.11 has been lost and no spurious information has been introduced.

### 5.3.5 Single-Sensor/Target Systems: "True" Markov Densities and Likelihoods

Equations 5.9 and 5.13 are derived using *constructive Radon–Nikodým derivatives* [9]. Let $E_\mathbf{x}$ be a small region centered at $\mathbf{x}$ of size $|E_\mathbf{x}|$. Then from Equation 5.10

$$p_{k+1|k}(E_\mathbf{x}|\mathbf{x}') = \int_{E_\mathbf{x}} f_{k+1|k}(\mathbf{y}|\mathbf{x}')\,d\mathbf{y} \cong |E_\mathbf{x}| \cdot f_{k+1|k}(\mathbf{x}|\mathbf{x}') \tag{5.15}$$

from which follows the *Lebesgue differentiation theorem* and a formula for $f_{k+1|k}(\mathbf{x}|\mathbf{x}')$:

$$f_{k+1|k}(\mathbf{x}|\mathbf{x}') = \lim_{|E_\mathbf{x}|\searrow 0} \frac{p_{k+1|k}(E_\mathbf{x}|\mathbf{x}')}{|E_\mathbf{x}|} = \lim_{|E_\mathbf{x}|\searrow 0} \frac{\Pr(\mathbf{W}_k \in E_\mathbf{x} - \varphi_k(\mathbf{x}'))}{|E_\mathbf{x}|} = \lim_{|E_\mathbf{x}|\searrow 0} \frac{\int_{E_\mathbf{x}-\varphi_k(\mathbf{x}')} f_{\mathbf{W}_k}(\mathbf{y})\,d\mathbf{y}}{|E_\mathbf{x}|} \tag{5.16}$$

## 5.4 MULTISENSOR–MULTITARGET SYSTEMS

The purpose of this section is to show how the previous concepts, formulated for single-sensor and single-target systems, can be extended to multisensor–multitarget systems.

### 5.4.1 Multisensor–Multitarget Systems: Random Finite Sets

The state of a multitarget system is most accurately represented as a finite set of the form $X = \{\mathbf{x}_1,...,\mathbf{x}_n\}$. Both the number $n \geq 0$ of targets and their states $\mathbf{x}_1,...,\mathbf{x}_n$ are random. This includes the possibility $n = 0$ (no targets are present), in which case we write $X = \varnothing$ (the null set). The finite-set representation is "natural." Each target already has its own unique identity, as indicated by a discrete identity state variable $c$. But there is nothing in the physics of the situation that endows the targets with an inherent or natural *ordering*. Thus a random multitarget state is a random finite set (RFS), $\Xi$.

Similar comments apply to the measurements collected from the targets by a sensor, which also (usually) have no inherent physical ordering. They have the form $Z = \{\mathbf{z}_1,...,\mathbf{z}_m\}$, where the number $m \geq 0$ of measurements, and the measurements $\mathbf{z}_1,...,\mathbf{z}_m$ themselves, are random. Thus a random multitarget measurement is an RFS, $\Sigma$.

### 5.4.2 Multisensor–Multitarget Systems: Probability Distributions of RFSs

Finite-set statistics is based on the *stochastic geometry* formulation of random set theory [1,7]. Just as the statistics of a random vector $\mathbf{W}$ are characterized by its p.m.f. $p_\mathbf{W}(S) = \Pr(\mathbf{W} \in S)$, so the statistics of an RFS $\Xi$ are characterized by its *belief-mass function* (b.m.f.) (also known as a *belief measure*):

$$\beta_\Xi(S) = \Pr(\Xi \subseteq S) \tag{5.17}$$

Just as $\mathbf{W}$ has a probability distribution $f_{\mathbf{W}}(\mathbf{x})$ such that $p_{\mathbf{W}|}(S) = \int_S f_{\mathbf{W}}(\mathbf{x})d\mathbf{x}$, so $\Xi$ has a probability distribution $f_{\Xi}(X)$ such that

$$\beta_{\Xi|}(S) = \int_S f_{\Xi}(X)\delta X \tag{5.18}$$

Here, the integral is a *set integral* that accounts for the randomness in both the number of objects and the randomness of the objects themselves:

$$\int_S f_{\Xi}(X)\delta X = f(\varnothing) + \sum_{n=1}^{\infty} \frac{1}{n!} \int_{S \times ... \times S} f\left(\{\mathbf{x}_1,...,\mathbf{x}_n\}\right)d\mathbf{x}_1 \cdots d\mathbf{x}_n. \tag{5.19}$$

The probability distribution of the number $n$ of objects in $\Xi$ (the "cardinality distribution" of $\Xi$) is

$$p_{\Xi}(n) = \Pr\left(|\Xi| = n\right) = \int_{|X|=n} f_{\Xi}(X)\delta X = \frac{1}{n!} \int f\left(\{\mathbf{x}_1,...,\mathbf{x}_n\}\right)d\mathbf{x}_1 \cdots d\mathbf{x}_n \tag{5.20}$$

### 5.4.3  Multisensor–Multitarget Systems: The Multisensor-Multitarget Bayes Filter

This is the theoretical foundation for optimal multisensor–multitarget detection, tracking, and identification. As in the single-target case, it requires two a priori items. The *multitarget Markov density* $f_{k+1|k}(X|X')$ is constructed from a *multitarget motion model*. It is the probability that the multitarget system will have multitarget state set $X$ at time $t_{k+1}$ if it had multitarget state-set $X'$ at time $t_k$. The *multisensor–multitarget likelihood function* $f_{k+1}(Z|X)$ is constructed from a *multisensor–multitarget measurement model*. It is the probability that a set $X$ of targets will generate a set $Z$ of measurements at time $t_{k+1}$. Given this, the multisensor–multitarget recursive Bayes filter propagates a multitarget probability distribution $f_{k|k}(X|Z^k)$:

$$... \to f_{k|k}(X|Z^{(k)}) \to f_{k+1|k}(X|Z^{(k)}) \to f_{k+1|k+1}(X|Z^{(k+1)}) \to ...$$

where $Z^{(k)} : Z_1,..., : Z_k$ denotes the time sequence of measurement sets collected through time $t_k$. First, $f_{k+1|k}(X|X')$ is used to propagate $f_{k|k}(X|Z^{(k)})$ from time $t_k$ to time $t_{k+1}$ via the *multitarget prediction integral*:

$$f_{k+1|k}\left(X|Z^{(k)}\right) = \int f_{k+1|k}\left(X|X'\right) \cdot f_{k|k}\left(X'|Z^{(k)}\right)\delta X' \tag{5.21}$$

Second, suppose that the sensors collect a new measurement set $Z_{k+1}$ at time $t_{k+1}$. Then $f_{k+1}(Z|X)$ and Bayes' rule result in

$$f_{k+1|k+1}\left(X|Z^{(k+1)}\right) = \frac{f_{k+1}\left(Z_{(k+1)}|X\right) \cdot f_{k|k}\left(X|Z^{(k)}\right)}{f_{k+1|k}\left(Z_{(k+1)}|Z^{(k)}\right)} \tag{5.22}$$

$$f_{k+1|k}\left(Z_{k+1}|Z^{(k)}\right) = \int f_{k+1}\left(Z_{k+1}|X\right) \cdot f_{k|k}\left(X|Z^{(k)}\right)\delta X \tag{5.23}$$

Third, we must extract information of interest: the number of targets and their positions, velocities, identities, and so forth. Multitarget versions of the maximum a posteriori (MAP) and expected a posteriori (EAP) estimators do not exist [3,7]. One must use alternative estimators [7, pp. 494–508] such as the joint multitarget (JoM) estimator:

$$\hat{X}_{k+1|k+1} = \arg\sup_X \frac{c^{|X|} \cdot f_{k+1|k+1}\left(X|Z^{(k+1)}\right)}{|X|!} \tag{5.24}$$

where $c$ is a constant with the same units of measurement as $\mathbf{x}$.

### 5.4.4 Multisensor–Multitarget Systems: Multitarget Motion Modeling

In multitarget systems, besides individual target motions one must also model *target disappearance* (target "death") and *target appearance* (target "birth"). The process begins with a *multitarget motion model* of the form

$$\Xi_{k+1|k} = T_k\left(\mathbf{x}_1'\right) \cup \ldots \cup T_k\left(\mathbf{x}_{n'}'\right) \cup B_k \tag{5.25}$$

Here, (1) $X' = \left\{\mathbf{x}_1', \ldots, \mathbf{x}_{n'}'\right\}$ is the multitarget state set at time $t_k$; (2) $T_k(\mathbf{x}')$ is the RFS of targets at time $t_{k+1}$ that originated with a target with state $\mathbf{x}'$ at time $t_k$; (3) $B_k$ is the RFS of newly appearing targets at time $t_{k+1}$; and (4) $\Xi_{k+1|k}$ is the RFS of all targets at time $t_{k+1}$. The RFS $T_k(\mathbf{x}')$ is usually assumed to be *Bernoulli*—that is, a target either disappears or does not disappear:

$$T_k(\mathbf{x}') = \begin{cases} \varnothing & \text{if} \quad \mathbf{x}_i' \quad \text{vanishes} \quad \left(\text{probability} \quad 1 - p_{S,k+1|k}(\mathbf{x}')\right) \\ \{\varphi_k(\mathbf{x}') + \mathbf{W}_k\} & \text{if} \quad \mathbf{x}_i' \quad \text{persists} \quad \left(\text{probability} \quad p_{S,k+1|k}(\mathbf{x}')\right) \end{cases} \tag{5.26}$$

Also, the $B_k$ is usually assumed to be *Poisson*—that is, its probability distribution is

$$f_{B_k}(X') = e^{-N_{k+1|k}^B} \prod_{\mathbf{x}' \in X'} \left(N_{k+1|k}^B \cdot s_{k+1|k}(\mathbf{x}')\right) \tag{5.27}$$

where $N_{k+1|k}^B$ is the expected number of newly appearing targets and $s_{k+1|k}(\mathbf{x}')$ is the spatial distribution of the target appearances.

Assuming independence of all the target RFSs, the b.m.f. of $\Xi_{k+1|k}$ is

$$\beta_{k+1|k}(S|X') = \beta_{\Xi_{k+1|k}}(S|X') = \beta_{T_k(\mathbf{x}_1')}(S) \cdots \beta_{T_k(\mathbf{x}_{n'}')}(S) \cdot \beta_{B_k}(S) \tag{5.28}$$

where

$$\beta_{T_k(\mathbf{x}')}(S) = 1 - p_{S,k+1|k}(\mathbf{x}') + p_{S,k+1|k}(\mathbf{x}') \int_S f_{k+1|k}\left(\mathbf{x}|\mathbf{x}'\right) d\mathbf{x} \tag{5.29}$$

$$\beta_{B_k}(S) = \exp\left(-N_{k+1|k}^B + N_{k+1|k}^B \int_S s_{k+1|k}(\mathbf{x}) d\mathbf{x}\right) \tag{5.30}$$

From Equation 5.18 we know that $f_{k+1|k}(X|X')$ and $\beta_{k+1|k}(S|X')$ are related by

$$\beta_{k+1|k}\left(S|X'\right)=\int_S f_{k+1|k}\left(X|X'\right)\delta X \tag{5.31}$$

A central aspect of finite-set statistics is a procedure for inverting this equation to derive an explicit formula for the "true" $f_{k+1|k}(X|X')$ from the formula for $\beta_{k+1|k}(S|X')$, Equation 5.28.

### 5.4.5 MULTISENSOR–MULTITARGET SYSTEMS: MULTITARGET MEASUREMENT MODELING

For clarity, only the single-sensor case is considered here. In the multitarget case, one must also address not only sensor noise but also the possibility that targets may not be detected and that measurements may be due to clutter rather than targets. The process begins with a *multitarget measurement model* of the form

$$\sum_{k+1} = \Upsilon_{k+1}(\mathbf{x}_1)\cup...\cup\Upsilon_{k+1}(\mathbf{x}_n)\cup C_{k+1} \tag{5.32}$$

Here, (1) $X = \{\mathbf{x}_1,...,\mathbf{x}_n\}$ is the multitarget state set at time $t_{k+1}$; (2) $\Upsilon_{k+1}(\mathbf{x})$ is the RFS of measurements at time $t_{k+1}$ generated by a target with state $\mathbf{x}$; (3) $C_{k+1}$ is the RFS of clutter measurements at time $t_{k+1}$; and (4) $\Sigma_{k+1}$ is the RFS of all measurements at time $t_{k+1}$. The RFS $\Upsilon_{k+1}(\mathbf{x})$ is usually assumed to be Bernoulli—that is, a target either generates a measurement or it does not:

$$\Upsilon_{k+1}(\mathbf{x})=\begin{cases}\varnothing & \text{if}\quad \mathbf{x}\quad\text{undetected}\quad\left(\text{probability}\quad 1-p_{D,k+1}(\mathbf{x})\right)\\\{\eta_{k+1}(\mathbf{x})+\mathbf{V}_{k+1}\} & \text{if}\quad \mathbf{x}\quad\text{detected}\quad\left(\text{probability}\quad p_{D,k+1}(\mathbf{x})\right)\end{cases} \tag{5.33}$$

The RFS $B_k$ is usually assumed to be Poisson—that is,

$$f_{C_{k+1}}(Z)=e^{-\lambda_{k+1}}\prod_{\mathbf{z}\in Z}\left(\lambda_{k+1}\cdot c_{k+1}(\mathbf{z})\right) \tag{5.34}$$

where $\lambda_{k+1}$ is the expected number of clutter measurements ("clutter rate") and $c_{k+1}(\mathbf{z})$ is the spatial distribution of the clutter.

Given independence assumptions, the b.m.f. $\beta_{k+1}(T|X)$ of $\Sigma_{k+1}$ is constructed as in the previous subsection. Also, $f_{k+1}(Z|X)$ and $\beta_{k+1}(T|X)$ are related by

$$\beta_{k+1}(T|X)=\int_T f_{k+1}(Z|X)\delta Z \tag{5.35}$$

This equation can be inverted to derive the "true" $f_{k+1}(Z|X)$ from $\beta_{k+1}(T|X)$.

### 5.4.6 MULTISENSOR–MULTITARGET SYSTEMS: MULTITARGET CALCULUS FOR MODELING

Equations 5.31 and 5.35 do not tell us how to construct formulas for $f_{k+1|k}(X|X')$ and $f_{k+1}(Z|X)$ from formulas for $\beta_{k+1|k}(S|X')$ and $\beta_{k+1}(T|X)$, respectively. This is the purpose of *multitarget calculus*. Let $E_{\mathbf{x}}$ and $E'_{\mathbf{x}}$ be two very small regions centered at $\mathbf{x}$. For any $S$, let $S_{E'_{\mathbf{x}}} = S - E'_{\mathbf{x}}$ where "$-$" is the set-theoretic difference. For any $\beta(S)$, define its *generalized constructive Radon–Nikodým derivative* as

$$\frac{\delta\beta}{\delta\mathbf{x}}(S) = \lim_{|E'_\mathbf{x}|\searrow 0} \lim_{|E_\mathbf{x}|\searrow 0} \frac{\beta\left(S_{E'_\mathbf{x}} \cup E_\mathbf{x}\right) - \beta(S_{E'_\mathbf{x}})}{|E_\mathbf{x}|} \tag{5.36}$$

More generally, let $X = \{\mathbf{x}_1, \ldots, \mathbf{x}_n\}$ with $|X| = n \geq 0$ and define the *set derivative* as

$$\frac{\delta\beta}{\delta X}(S) = \frac{\delta^n\beta}{\delta\mathbf{x}_1 \cdots \delta\mathbf{x}_n}(S), \quad \frac{\delta\beta}{\delta\varnothing}(S) = \beta(S). \tag{5.37}$$

The set derivative and set integral are inverse operations:

$$\int_S \frac{\delta\beta}{\delta X}(\varnothing)\,\delta X = \beta(S), \quad \frac{\delta}{\delta X}\int_S f(Y)\,\delta Y = f(X). \tag{5.38}$$

The Lebesgue differentiation theorem, Equation 5.16, involves a first-order set derivative. More generally, the $f_{k+1|k}(X|X')$ and $f_{k+1}(Z|X)$ can be constructed using set derivatives:

$$f_{k+1|k}\left(X|X'\right) = \frac{\delta\beta_{k+1|k}}{\delta X}(\varnothing|X'), \quad f_{k+1}(Z|X) = \frac{\delta\beta_{k+1}}{\delta Z}(\varnothing|X). \tag{5.39}$$

## 5.5 PRINCIPLED APPROXIMATE MULTITARGET FILTERS

The multisensor–multitarget recursive Bayes filter is computationally intractable in general. How can it be approximated in a manner that preserves, as faithfully as possible, the underlying multi-target motion and measurement models as well as their statistical interrelationships? This question is answered by assuming that $f_{k|k}(X|Z^{(k)})$ and/or $f_{k+1|k}(X|Z^{(k)})$ have a simplified form. Five such simplifications are currently being investigated: *Poisson*, *independent identically distributed cluster* (i.i.d.c.), *multi-Bernoulli*, and *generalized labeled multi-Bernoulli* (GLMB).

### 5.5.1 POISSON APPROXIMATION: PHD FILTERS

As noted earlier, $\alpha$–$\beta$ filters are based on *first-moment approximation* of the single-sensor, single-target Bayes filter as in Equation 5.6. By analogy, assume that SNR is so large that the multitarget Bayes distribution can be approximated by its first-order statistical moment. What, however, is a multitarget first-order moment? The naïve definition,

$$X_{k|k} = \int X \cdot f_{k|k}(X|Z^{(k)})\,\delta X \tag{5.40}$$

is mathematically undefined because addition and subtraction of finite sets cannot be usefully defined. Instead we must replace $X$ in Equation 5.40 with a substitute:

$$D_{k|k}(\mathbf{x}|Z^{(k)}) = \int \left( \sum_{\mathbf{x}' \in X} \delta_{\mathbf{x}'}(\mathbf{x}) \right) \cdot f_{k|k}(X|Z^{(k)})\delta X = \int f_{k|k}(\{\mathbf{x}\} \cup W)\delta W \qquad (5.41)$$

where $\delta_{\mathbf{x}'}(\mathbf{x})$ is the Dirac delta function concentrated at $\mathbf{x}'$; and where the density function $D_{k|k}(\mathbf{x}|Z^{(k)})$ is called a *probability hypothesis density* (PHD), *first-moment density*, or *intensity function*. It is uniquely characterized by the following property:

$$\int_S D_{k|k}(\mathbf{x}|Z^{(k)})d\mathbf{x} = E\left[|S \cap \Xi_{k|k}|\right] \qquad (5.42)$$

where $E[|S \cap \Xi_{k|k}|]$ is the expected number of targets in $S$. Its value $N_{k|k}$, when $S$ is the entire state space $X$, is the expected total number of targets. Thus $D_{k|k}(\mathbf{x}|Z^{(k)})$ is the *target density* at $\mathbf{x}$. A target with state $\mathbf{x}$ is more likely to be present when $D_{k|k}(\mathbf{x}|Z^{(k)})$ is large. Thus the number and states of the targets can be estimated as follows. Let $\nu$ be the nearest integer to $N_{k|k}$ and let $\mathbf{x}_1,...,\mathbf{x}_\nu$ be the states corresponding to the $\nu$ largest values of $D_{k|k}(\mathbf{x}|Z^{(k)})$. Then $X_{k|k} = \{\mathbf{x}_1,...,\mathbf{x}_\nu\}$ is the multitarget state estimate.

The PHD filter propagates the PHD rather than the multitarget distribution:

$$... \to D_{k|k}(\mathbf{x}|Z^{(k)}) \to D_{k+1|k}(\mathbf{x}|Z^{(k)}) \to D_{k+1|k+1}(\mathbf{x}|Z^{(k+1)}) \to ...$$

It must be assumed that the target-appearance and clutter processes are Poisson and, as a simplifying assumption, the predicted multitarget distribution is approximately Poisson:

$$f_{k+1|k}(X|Z^{(k)}) \cong e^{-N_{k|k}} \prod_{\mathbf{x} \in X} D_{k+1|k}(\mathbf{x}|Z^{(k)}) \qquad (5.43)$$

The specific formulas for the PHD filter will not be reproduced here. Their computational order is very attractive: $O(mn)$, where $m$ is the current number of measurements and $n$ is the current number of tracks; see Refs. [2,5,6].

### 5.5.2 INDEPENDENT IDENTICALLY DISTRIBUTED CLUSTER APPROXIMATION: CPHD FILTERS

The CPHD filter generalizes the PHD filter. It has better performance but greater computational complexity $O(m^3n)$. It propagates both the PHD and the distribution $p_{k|k}(n|Z^{(k)})$ of the target number $n$ (cardinality distribution):

$$... \to \quad D_{k|k}(\mathbf{x}|Z^{(k)}) \quad \to \quad D_{k+1|k}(\mathbf{x}|Z^{(k)}) \quad \to \quad D_{k+1|k+1}(\mathbf{x}|Z^{(k+1)}) \quad \to ...$$

$$\updownarrow$$

$$... \to \quad p_{k|k}(n|Z^{(k)}) \quad \to \quad p_{k+1|k}(n|Z^{(k)}) \quad \to \quad p_{k+1|k+1}(n|Z^{(k+1)}) \quad \to ...$$

It must be assumed that the target-appearance and clutter processes are independent, identically distributed cluster (i.i.d.c.) processes; and as a simplification, that both $f_{k|k}(X|Z^{(k)})$ and $f_{k+1|k}(X|Z^{(k)})$ are i.i.d.c.—for example,

$$f_{k|k}(X|Z^{(k)}) \cong |X|! \cdot p_{k|k}(|X|) \prod_{\mathbf{x} \in X} D_{k|k}(\mathbf{x}|Z^{(k)}). \qquad (5.44)$$

The specific formulas for the CPHD filter will not be reproduced here.

The PHD and CPHD filters can be extended to the multisensor case ([1], Chapter 10) and to unknown, dynamically changing clutter ([1], Chapter 18).

### 5.5.3 Multi-Bernoulli Approximation: Multi-Bernoulli Filters and GLMB Filters

A multi-Bernoulli filter propagates a *track table* in place of the multitarget distribution:

$$... \rightarrow \left\{\left(q_i^{k|k}, s_i^{k|k}(\mathbf{x})\right)\right\}_{i=1}^{v_{k|k}} \rightarrow \left(q_i^{k+1|k}, s_i^{k+1|k}(\mathbf{x})\right)\right\}_{i=1}^{v_{k+1|k}} \rightarrow \left(q_i^{k+1|k+1}, s_i^{k+1|k+1}(\mathbf{x})\right)\right\}_{i=1}^{v_{k+1|k+1}} \rightarrow ...$$

Here, $s_i^{k|k}(\mathbf{x})$ is the track distribution of the $i$th track and $q_i^{k|k}$ is the track probability—that is, the probability that the $i$th track is an actual target. It must be assumed that the multitarget distributions are "multi-Bernoulli." The formulas for the multi-Bernoulli filter will not be reproduced here.

The GLMB filter is a generalization of the multi-Bernoulli filter based on the following presumption: every track has a unique identifying label. It propagates a GLMB distribution in place of the multitarget distribution. The GLMB filter provides an *algebraically closed-form solution* of the general multitarget Bayes filter. It is the first implementable, provably Bayes-optimal multitarget detection and tracking filter and has the first provably Bayes-optimal track management scheme. For more details, see Refs. [1], Chapter 15, and [10].

### 5.6 THE FINITE-SET STATISTICS APPROXIMATION METHODOLOGY

The approximation methodology used for PHD, CPHD, and multi-Bernoulli filters consists of the following steps: (1) construct RFS motion and measurement models for the targets and sensor; (2) use multitarget calculus to convert these models into multitarget Markov densities and likelihood functions; (3) from these, construct the optimal approach for the application: a multitarget Bayes filter; (4) convert the multitarget Bayes filter into probability generating functional (p.g.fl.) form; and (5) use simplifying approximations (Poisson, i.i.d.c., multi-Bernoulli) and *multitarget functional calculus* to derive PHD, CPHD, and/or multi-Bernoulli filters for the particular application.

Multitarget functional calculus generalizes the concepts of belief-mass function, and set derivatives of belief-mass functions (described earlier) to the concepts of *probability generating functional* (p.g.fl.), and the *functional derivatives of p.g.fl.s.* For more details, see Refs. [1,4,7].

### 5.7 CONCLUSIONS

Finite-set statistics provides a unified, systematic, and theoretically rigorous approach to multisource–multitarget information fusion. This chapter has provided a brief introduction to some of its basic ideas.

### REFERENCES

1. R. Mahler, *Advances in Statistical Multisource-Multitarget Information Fusion*, Artech House, Norwood, MA, 2014.
2. J. Mullane, B.-N. Vo, M. Adams, and B.-T. Vo, *Random Finite Sets in Robotic Map Building and SLAM*, Springer, New York, 2011.
3. R. Mahler, "'Statistics 101' for multisensor, multitarget data fusion," *IEEE Aerospace and Electronics Systems, Part 2: Tutorials*, 19(1): 53–64, 2004.
4. R. Mahler, "'Statistics 102' for multisensor-multitarget detection and tracking," *IEEE Journal of Selected Topics in Signal Processing*, 7(3): 376–389, 2013.
5. B.-N. Vo, B.-T. Vo, and D. Clark, "Bayesian multiple target filtering using random finite sets" in M. Mallick, V. Krishnamurthy, and B.-N. Vo (eds.), *Integrated Tracking, Classification, and Sensor Management*, Wiley, Hoboken, NJ, 2013.
6. B. Ristic, B.-T. Vo, B.-N. Vo, and A. Farina, "A tutorial on Bernoulli filters: Theory, implementation, and applications," *IEEE Transactions on Signal Processing*, 61(13): 3406–3430, 2012.

7. R. Mahler, *Statistical Multisource-Multitarget Information Fusion*, Artech House, Norwood, MA, 2007.

8. B. Ristic, *Particle Filters for Random Set Models*, Springer, New York, 2013.

9. G. Shilov, and B. Gurevich, *Integral, Measure and Derivative: A Unified Approach*, Prentice-Hall, Englewood Cliffs, NJ, 1966.

10. S. Reuter, B.-T. Vo, and K. Dietmayer, "Multi-object tracking using labeled multi-Bernoulli random finite sets," *Proceedings of the 17th International Conference on Information Fusion*, Salamanca, Spain, July 7–10, 2014.

# 6 Optimal Fusion for Dynamic Systems with Process Noise

*Chee-Yee Chong and Shozo Mori*

## CONTENTS

## 6.1  INTRODUCTION

Multiple sensors can provide better state estimation performance over a single sensor because each additional sensor contributes more information. An important example is target tracking, where multiple sensors provide better geographical or spectral coverage than a single sensor, and improve detection and false alarm rejection performance. Improved location or classification estimation accuracy also results from geometric or phenomenological diversity provided by multiple sensors.

Centralized fusion of the measurements from all sensors at a single fusion node is theoretically optimal because the information contained in the measurements is not degraded by any intermediate processing. However, centralized fusion is not always feasible when communication bandwidth is limited. Thus, fusion systems with geographically distributed sensors frequently use a distributed estimation or information fusion architecture where the individual sensors process their measurements to generate local state estimates and estimation error covariances, which are then sent to a fusion node to be combined into global state estimates and estimation error covariances.

In most distributed estimation systems, the local state estimates to be fused are optimal estimates given the local sensor measurements and computed using only local sensor model information [1]. When the local estimation errors are not conditionally independent given the state, fusion algorithms have to address this cross-sensor dependence or correlation. The most common approach [2–4], frequently called tracklet fusion or the equivalent measurement approach, uses the current and predicted local estimates to find the new information received since the last fusion time. However, it does not generate the optimal global estimate when process noise is present and the fusion rate is lower than the sensor observation rate. Optimality can be regained if the state is augmented to be the entire state trajectory for all observation times since the most recent fusion [5–8]. Then tracklet fusion with augmented state produces the same result as centralized fusion (measurement fusion). Although augmented state fusion was discovered a while ago [5], the algorithm was not completely developed until very recently [8].

Other fusion algorithms address dependence that can be characterized by cross-sensor covariances of the local estimation errors. Examples include maximum likelihood estimate [9,10], best linear unbiased estimate (BLUE) [11,12], and minimum variance (MV) or maximum a posteriori estimate [13,14]. Because cross-covariances are used in fusion, the local sensor models have to be communicated to the fusion site along with the local estimates. Furthermore, the fused estimate may not be globally optimal because it is only the best estimate given the local estimates characterized by local and cross-sensor estimation error covariances.

The recently developed distributed Kalman filter (DKF) [15–18] has taken a different approach to distributed estimation. Local sensor measurements are processed using global sensor models to produce pseudo state estimates that are not locally optimal. The fusion site then combines the local pseudo estimates to generate global estimates that are optimal even when the process noise is present. Various approaches [19,20] have been developed to address the degradation in DKF performance when the actual measurement error covariances do not match the assumed model. In particular, covariance debiasing [20] computes a correction matrix that allows a consistent covariance matrix to be reconstructed.

The accumulated state density (ASD) used for exact memoryless track fusion [21,22] is similar to the augmented state, and the fusion algorithm also produces the same results as optimal centralized fusion. However, there are significant differences between the two approaches. Each local estimator does not require model knowledge of other sensors to generate the local ASD, but local prediction uses a relaxed evolution model that involves the number of sensors in the system. Thus the local estimate is again a pseudo estimate.

The rest of this chapter is structured as follows. Section 6.2 discusses a general class of distributed estimation problems, including the state and measurement models, the centralized solution for both linear and nonlinear problems, and the distributed estimation architecture. Section 6.3 presents optimal fusion for nonlinear systems with process noise. Section 6.4 presents optimal fusion for linear systems with process noise using augmented state. Section 6.5 discusses alternative fusion approaches for linear systems with process noise, including maximum likelihood (or Bar-Shalom–Campo) fusion with cross-covariance, MV estimate, DKF, and ASD fusion. Section 6.6 compares the performance of augmented state tracklet fusion with standard tracklet fusion, Bar-Shalom–Campo, and minimum variance fusion, and Section 6.7 contains the conclusions. An appendix presents the derivations for the equations used in augmented state tracklet fusion.

## 6.2  DISTRIBUTED ESTIMATION PROBLEM

This section presents the state and measurement models for the estimation problem, algorithms for centralized measurement fusion, and architecture for distributed estimation or fusion.

### 6.2.1  STATE AND MEASUREMENT MODELS

The state $x_k \in R^n$ to be estimated at time $t_k$, for $k = 0,1,2,...$, is modeled by

$$x_{k+1} = f_k(x_k, w_k) \tag{6.1}$$

For linear systems, the dynamic model becomes

$$x_{k+1} = F_k x_k + G_k w_k \tag{6.2}$$

where $F_k$ and $G_k$ are matrices representing the system dynamics, and $w_k$ is the process noise modeled by a zero-mean white random process with covariance $Q_k$.

The state $x_k$ is observed by $S$ sensors. For $s = 1,...,S$, and $k = 1,2,...$, the measurement $z_k^s$ of the $s$th sensor at time $t_k$ is modeled by

$$z_k^s = h_k^s\left(x_k, v_k^s\right) \tag{6.3}$$

For linear systems, the observation model becomes

$$z_k^s = H_k^s x_k + v_k^s \tag{6.4}$$

where $H_k^s$ is the measurement matrix, and $v_k^s$ is the measurement error process modeled by a zero-mean random vector with covariance $R_k^s$. The measurement noises are assumed to be independent with each other and the process noise. The initial state $x_0$ is independent of the noises, with mean $\bar{x}_0$ and covariance $\bar{P}_0$. For simplicity, we assume synchronous observations by all sensors but the results can be generalized to non-synchronous measurements with appropriate modifications.

### 6.2.2 CENTRALIZED FUSION

Define the collective measurements by all sensors at time $t_k$ to be $z_k$, that is, $z_k = \left( z_k^s \right)_{s=1}^{S}$, the cumulative measurements of sensor $s$ to be $Z_k^s = \left( z_j^s \right)_{j=1}^{k}$, and the cumulative measurements of all sensors at $t_k$ to be $Z_k$, or $Z_k = \left( z_j \right)_{j=1}^{k} = \left( Z_k^s \right)_{s=1}^{S}$. Then centralized fusion for the nonlinear models of equations 6.1 and 6.3 is provided by a nonlinear filter with the following two steps:

Prediction

$$p\left( x_{k+1} \middle| Z_k \right) = \int p_k \left( x_{k+1} \middle| x_k \right) p\left( x_k \middle| Z_k \right) dx_k \tag{6.5}$$

Update

$$p\left( x_{k+1} \middle| Z_{k+1} \right) = C^{-1} \prod_{s=1}^{S} p\left( z_{k+1}^s \middle| x_{k+1} \right) p\left( x_{k+1} \middle| Z_k \right) \tag{6.6}$$

where $p(x_{k+1}|x_k)$ and $p\left( z_{k+1}^s \middle| x_{k+1} \right)$ are the transition probability densities defined by Equations 6.1 and 6.3, respectively, and $C$ is a normalizing constant.

For linear systems, centralized fusion is provided by Kalman filtering. Let $x_{k|l}$ and $P_{k|l}$ denote the optimal estimate of $x_k$ and its error covariance given $Z_l$. The information matrix form of the centralized Kalman filter (CKF) involves the following two steps:

Prediction

$$x_{k|k-1} = F_{k-1} x_{k-1|k-1} \tag{6.7}$$

$$P_{k|k-1} = F_{k-1} P_{k-1|k-1} F_{k-1}^T + G_{k-1} Q_{k-1} G_{k-1}^T \tag{6.8}$$

Update

$$P_{k|k}^{-1} x_{k|k} = P_{k|k-1}^{-1} x_{k|k-1} + \sum_{s=1}^{S} i_k^s \tag{6.9}$$

$$P_{k|k}^{-1} = P_{k|k-1}^{-1} + \sum_{s=1}^{S} I_k^s \tag{6.10}$$

where $i_k^s \triangleq \left( H_k^s \right)^T \left( R_k^s \right)^{-1} z_k^s$ and $I_k^s \triangleq \left( H_k^s \right)^T \left( R_k^s \right)^{-1} H_k^s$. The initial conditions are $x_{0|0} = \bar{x}_0$ and $P_{0|0} = \bar{P}_0$.

The Kalman filter estimates are optimal according to several criteria. If the random variables are Gaussian, they are the conditional mean and minimum variance estimates. When they are not Gaussian but the estimates are constrained to be linear, then they are the linear minimum variance or best linear unbiased estimates.

### 6.2.3  DISTRIBUTED ESTIMATION ARCHITECTURE

The centralized estimate is by definition optimal because it uses all sensor measurements directly without possible degradation due to intermediate processing. However, the communication networks in many real applications may not have enough bandwidth to communicate all the measurements to a central fusion node. Instead, local processors compress the sensor measurements to generate local state estimates, which are communicated typically at a rate lower than the observation rate to the fusion node. The fusion node then combines the local estimates to generate the global state estimate. This global estimate can be sent back to the local sensor processors if necessary. Figure 6.1 shows a distributed estimation system with no feedback from the fusion node, which is the architecture considered in this chapter.

The problem of fusing state estimates is sometimes called track-to-track fusion, when the local estimates are generated by local trackers processing sensor measurements in a tracking problem. Then the fusion problem is combining the estimates of local tracks that are assumed to have originated from the same target. The term "estimate fusion" is more general because the state to be estimated may not be the state of a moving target.

A key issue in distributed estimation is choice of the local state estimates to be communicated for fusion by the fusion node. Ideally, the local estimates should be the optimal state estimates given the sensor measurements because these estimates can also be used to support local operations such as sensor management. However, these optimal local estimates may not contain sufficient information for reconstructing the optimal global estimates. This is certainly the case when process noise is present and the communication rate is lower than the observation rate.

Distributed estimation without feedback from the fusion node to sensor nodes consists of the following steps:

1. *Local processing*: Each sensor processor uses the local sensor measurements to compute local estimates and estimation error covariances using prior knowledge on sensor models. For fusion algorithms such as DKF, the prior knowledge also includes models for other sensors. For ASD, the local processor also knows the number of sensors. The system dynamic model is assumed known by all sensor and fusion nodes.
2. *Communication*: Each local processor communicates the local estimates and estimation error covariances to the fusion node at given fusion times. For fusion algorithms using cross-sensor local state estimation error covariances, the local sensor models also have to be communicated, or assumed to be known to all sensors.

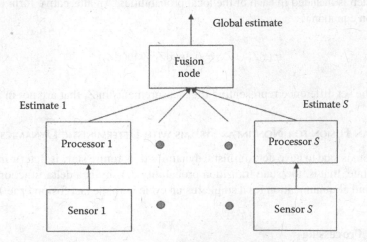

**FIGURE 6.1**  Distributed estimation or fusion architecture.

3. *Fusion processing*: The fusion node combines all local estimates and estimation error covariances to generate the global estimate. The fusion node is assumed to have no prior knowledge of local sensor models except for those communicated by the local nodes. However, it may also need to compute the cross-covariances for some algorithms such as Bar-Shalom–Campo or minimal variance estimates.

Fusion processing may utilize results from the last fusion time in addition to the local estimates. This is sometimes known as repeated fusion, in contrast to memoryless fusion, when the fusion processor uses only the most recent local estimates. Memoryless fusion can be viewed as repeated fusion at the first fusion time with no prior communication.

## 6.3  OPTIMAL FUSION FOR NONLINEAR SYSTEMS

Optimal fusion uses the Bayesian distributed fusion equation, which optimally combines local conditional probabilities of a state when the individual sensor measurements are conditionally independent given the state. Thus the main problem for optimal fusion is finding the state that satisfies the conditional independence assumption.

### 6.3.1  BAYESIAN DISTRIBUTED FUSION EQUATION

Let $x$ be the state to be estimated and $p(x)$ be its prior probability density or distribution function. Suppose two measurement sets $Z_1$ and $Z_2$ are conditionally independent given $x$.

Let the posterior probabilities for the individual data sets be $p(x|Z_1)$ and $p(x|Z_2)$. Then the posterior conditional probability $p(x|Z_1 \cup Z_2)$ given all the measurements $Z_1 \cup Z_2$ can be computed from the following Bayesian distributed fusion equation [6]:

$$p\left(x\middle|Z_1 \cup Z_2\right) = C^{-1}\frac{p\left(x\middle|Z_1\right)p\left(x\middle|Z_2\right)}{p\left(x\middle|Z_1 \cap Z_2\right)} \tag{6.11}$$

where $C$ is a normalizing constant.

This Bayesian distributed fusion equation states that the fused posterior probability $p(x|Z_1 \cup Z_2)$ is the product of the local probabilities $p(x|Z_1)$ and $p(x|Z_2)$, divided by the common probability $p(x|Z_1 \cap Z_2)$, which is included in each of the local probabilities. An alternative form of the Bayesian distributed fusion equation is

$$p(x|Z_1 \cup Z_2) = C^{-1}p(x|Z_1 \backslash Z_2)p(x|Z_2) \tag{6.12}$$

where $Z_1 \backslash Z_2$ is the set difference representing the measurements in $Z_1$ that are not in $Z_2$.

### 6.3.2  BAYESIAN FUSION FOR NONLINEAR SYSTEMS WITH DETERMINISTIC DYNAMICS

A dynamic system is said to have deterministic dynamics if its future state is determined completely by the current state, that is, the state transition probability density is a delta function. We consider the processing and communication for a single fusion cycle with the last fusion time $t_{\bar{K}}$ and current fusion time $t_K$.

#### 6.3.2.1  Local Processing

Local processing at sensor $s$ consists of the following prediction and update steps for $k = \bar{K}+1,...,K$.

Prediction

$$p\left(x_{k+1}\middle|Z_k^s\right) = \int p_k\left(x_{k+1}\middle|x_k\right)p\left(x_k\middle|Z_k^s\right)dx_k \qquad (6.13)$$

Update

$$p\left(x_{k+1}\middle|Z_{k+1}^s\right) = C^{-1}\left(Z_{k+1}^s\right)^{-1} p\left(z_{k+1}^s\middle|x_{k+1}\right)p\left(x_{k+1}\middle|Z_k^s\right) \qquad (6.14)$$

with initial condition $p^s(x_{\bar{K}})$. Note that local processing uses only the local sensor model.

### 6.3.2.2 Communication

At the fusion time $t_K$, each local node $s$ sends its local posterior probability density $p\left(x_K\middle|Z_k^s\right)$ to the fusion node.

### 6.3.2.3 Fusion Processing

The fusion node first computes $p\left(x_K\middle|Z_{\bar{K}}\right)$ by extrapolating $p\left(x_{\bar{K}}\middle|Z_{\bar{K}}\right)$ from the previous fusion time $t_{\bar{K}}$ to the current fusion time $t_K$. Then it combines the sensor conditional probabilities $p\left(x_K\middle|Z_K^s\right)$, $s = 1,...,S$ with $p\left(x_K\middle|Z_{\bar{K}}\right)$. Define $Z_K^s \backslash Z_{\bar{K}}^s = Z_K^s \backslash Z_{\bar{K}} \triangleq \left(z_{\bar{K}+1}^s,...,z_K^s\right)$. For deterministic dynamics, $Z_{\bar{K}}$ and $Z_K^s \backslash Z_{\bar{K}}$, $s = 1,...,S$, are conditionally independent given $x_K$. Thus, recursive application of Equation 6.12 results in

$$p\left(x_K\middle|Z_k\right) = C^{-1}\prod_{s=1}^{S} p\left(x_K\middle|Z_K^s \backslash Z_{\bar{K}}\right)p\left(x_K\middle|Z_{\bar{K}}\right)$$

$$= C^{-1}\prod_{s=1}^{S} p\left(x_K\middle|Z_K^s \backslash Z_{\bar{K}}^s\right)p\left(x_K\middle|Z_{\bar{K}}\right) \qquad (6.15)$$

$$= C^{-1}\prod_{s=1}^{S} \frac{p\left(x_K\middle|Z_K^s\right)}{p\left(x_K\middle|Z_{\bar{K}}^s\right)}p\left(x_K\middle|Z_{\bar{K}}\right)$$

where $C^{-1}$ is a normalizing constant. Equation 6.15 states that the fused conditional probability is the product of the conditional probabilities predicted from the last fusion time and the conditional probabilities of the state given the local data received since the last communication/ fusion. This fusion equation is optimal independent of the number of sensor revisits between fusion times.

### 6.3.3 BAYESIAN FUSION FOR NONDETERMINISTIC DYNAMICS

When the state has nondeterministic dynamics, $Z_{\bar{K}}$ and $Z_K^s \backslash Z_{\bar{K}}^s$, $s = 1,...,S$, are no longer conditionally independent given $x_K$ at a single time $t_K$ unless fusion takes place at every observation time, that is, $t_K$ and $t_{\bar{K}}$ are consecutive observation times $(K = \bar{K} + 1)$. Thus the fusion equation 6.15 may not be valid for nondeterministic cases. However, the measurements $\left(Z_K^s \backslash Z_{\bar{K}}^s\right)_{s=1}^{S}$ for the $S$ sensors are conditionally independent given the augmented state $X_K = \left(x_j\right)_{j=\bar{K}+1}^{K}$. This conditional independence motivates the augmented state fusion algorithm [5–8].

#### 6.3.3.1   Local Processing

Local processing now computes the conditional probabilities of the augmented state $X_K \triangleq \left(x_j\right)_{j=\bar{K}+1}^{K}$. For $k = \bar{K}+1,...,K,$

Prediction

$$p\left(X_{k+1}\middle|Z_k^s\right) = p\left(x_{k+1}\middle|x_k\right)p\left(X_k\middle|Z_k^s\right) \tag{6.16}$$

Update

$$p\left(X_{k+1}\middle|Z_{k+1}^s\right) = C^{-1}p\left(z_{k+1}^s\middle|x_{k+1}\right)p\left(X_{k+1}\middle|Z_k^s\right) \tag{6.17}$$

with initial condition $p^s(x_{\bar{K}})$. Note that local processing uses only the local sensor model.

#### 6.3.3.2   Communication

At the fusion time $t_K$, each local node $s$ sends its local posterior probability density $p\left(X_K\middle|Z_K^s\right)$ of the augmented state to the fusion node.

#### 6.3.3.3   Fusion Processing

Because $Z_K^s \backslash Z_{\bar{K}}^s$, $s = 1,...,S$, are conditionally independent given the augmented state $X_K$, the fusion Equation is 6.15 with the state replaced by augmented state, that is,

$$p\left(X_K\middle|Z_K\right) = C^{-1}\prod_{s=1}^{S} p\left(X_K\middle|Z_K^s\backslash Z_{\bar{K}}^s\right)p\left(X_K\middle|Z_{\bar{K}}\right)$$

$$= C^{-1}\prod_{s=1}^{S} \frac{p\left(X_K\middle|Z_K^s\right)}{p\left(X_K\middle|Z_{\bar{K}}^s\right)}p\left(X_K\middle|Z_{\bar{K}}\right) \tag{6.18}$$

Optimal fusion for nonlinear systems with nondeterministic dynamics is conceptually simple. However, implementation is difficult because the probability density involves high dimensional random vectors.

## 6.4   OPTIMAL FUSION FOR LINEAR SYSTEMS WITH PROCESS NOISE

The most popular optimal fusion algorithm for linear systems is sometimes called tracklet fusion because the new information of each sensor is contained in the tracklet of measurements received since the last fusion time. The tracklet is also called equivalent measurement or report because it is frequently treated as a (aggregated) measurement or report. As in fusion for nonlinear systems, the main problem is finding the state variables that satisfy the conditional independence assumption.

### 6.4.1   BAYESIAN DISTRIBUTED FUSION FOR GAUSSIAN RANDOM VECTORS

Suppose the state $x$ is a Gaussian random vector, and the measurements are linear functions with additive zero mean Gaussian errors. The local state estimates are conditional means $\hat{x}_i$ given the local measurements. Let the error covariance of the local estimate $\hat{x}_i$ be $P_{i=1,2}$. Then the fused estimate can be defined as the conditional mean $\hat{x}_{1\cup 2}$ given the global set of measurements, that is, the union of the two local sets of measurements. Let the error covariance of the fused estimate $\hat{x}_{1\cup2}$ be $P_{1\cup2}$. Then the fusion Equation 6.11 becomes

$$P_{1\cup2}^{-1}\hat{x}_{1\cup2} = P_1^{-1}\hat{x}_1 + P_2^{-1}\hat{x}_2 - P_{1\cap2}^{-1}\hat{x}_{1\cap2} \tag{6.19}$$

$$P_{1\cup2}^{-1} = P_1^{-1} + P_2^{-1} - P_{1\cap2}^{-1} \tag{6.20}$$

where $\hat{x}_{1\cap2}$ is the conditional mean of the state $x$ given the common information with the estimation error covariance $P_{1\cap2}$.

Equations 6.19 and 6.20 are the information matrix form of the fusion equations because the inverse of the covariance matrix can be called the information matrix. Equation 6.20 states that the information matrix of the fused estimate is the sum of the information matrices of the local estimates minus the information matrix of the common estimate. Equation 6.19 states that the information of the fused estimate is the sum of the local information minus the common information.

## 6.4.2 STANDARD TRACKLET OR EQUIVALENT REPORT FUSION

Standard tracklet fusion represents the new information by a state estimate and error covariance matrix at a single time.

### 6.4.2.1 Local Processing

Let $x_{k|l}^s$ and $P_{k|l}^s$ be the optimal (local) estimate of $x_k$ and its error covariance given $Z_l^s$. Assuming no feedback from the fusion node to each sensor node, local processing at sensor $s$ consists of the following prediction and update steps for $k = 0,1,2,....$

Prediction

$$x_{k|k-1}^s = F_{k-1}x_{k-1|k-1}^s \tag{6.21}$$

$$P_{k|k-1}^s = F_{k-1}P_{k-1|k-1}^s F_{k-1}^T + G_{k-1}Q_{k-1}G_{k-1}^T \tag{6.22}$$

Update

$$\left(P_{k|k}^s\right)^{-1} x_{k|k}^s = \left(P_{k|k-1}^s\right)^{-1} x_{k|k-1}^s + i_k^s \tag{6.23}$$

$$\left(P_{k|k}^s\right)^{-1} = \left(P_{k|k-1}^s\right)^{-1} + I_k^s \tag{6.24}$$

with initial conditions $x_{0|0}^s$ and $P_{0|0}^s$. Note that local processing uses only the local sensor model.

### 6.4.2.2 Communication

At the fusion time $t_K$, each local node $s$ sends its state estimate $x_{K|K}^s$ and error covariance $P_{K|K}^s$ to the fusion node.

### 6.4.2.3  Fusion Processing

The fusion node computes $x^s_{K|\bar{K}}$ and $P^s_{K|\bar{K}}$ using $x^s_{K|\bar{K}}$ and $P^s_{\bar{K}|\bar{K}}$ received at the last communication time $t_K$ for each sensor, and $x_{K|\bar{K}}$ and $P_{K|\bar{K}}$ from its own $x_{\bar{K}|\bar{K}}$ and $P_{\bar{K}|\bar{K}}$. Then the global estimate $x_{K|K}$ and its error covariance $P_{K|K}$ are computed using the fusion equations

$$P^{-1}_{K|K} x_{K|K} = P^{-1}_{K|\bar{K}} x_{K|\bar{K}} + \sum_{s=1}^{S} \left( \left( P^s_{K|K} \right)^{-1} x^s_{K|K} - \left( P^s_{K|\bar{K}} \right)^{-1} x^s_{K|\bar{K}} \right) \tag{6.25}$$

$$P^{-1}_{K|K} = P^{-1}_{K|\bar{K}} + \sum_{s=1}^{S} \left( \left( P^s_{K|K} \right)^{-1} - \left( P^s_{K|\bar{K}} \right)^{-1} \right) \tag{6.26}$$

Basically the new global estimate $x_{K|K}$ combines the global estimate $x_{K|\bar{K}}$ predicted from the last fusion time $t_{\bar{K}}$ with the new information received by the individual sensors as represented by $\left( P^s_{K|K} \right)^{-1} x^s_{K|K} - \left( P^s_{K|\bar{K}} \right)^{-1} x^s_{K|\bar{K}}$. If fusion takes place after every observation time, this new information is $i^s_k = \left( H^s_k \right)^T \left( R^s_k \right)^{-1} z^s_k$, thus the name of equivalent measurement or information.

Tracklet fusion produces the optimal global estimate when there is no process noise, where "optimal" is in the sense of the minimum variance or conditional mean. When there is process noise, the global estimate is optimal only when the fusion rate is the same as the observation rate. Despite this restriction, equivalent measurement or tracklet fusion is the most common approach for track fusion because of its easy implementation, use of optimal local estimates that can be computed from local sensor models, and good performance as the communication or fusion rate increases.

### 6.4.3  Tracklet Fusion with Augmented State

As in nonlinear systems, when the communication (and fusion) rate is lower than the observation rate, the new measurements $\left( Z^s_K \backslash Z^s_{\bar{K}} \right)^S_{s=1}$ collected by the sensors since the last fusion time are no longer conditionally independent given the state at a single time because each $Z^s_k$ includes the common process noise in addition to the independent measurement noise. Thus equivalent measurement or tracklet fusion no longer produces the optimal global estimate. However, the measurements $\left( Z^s_K \backslash Z^s_{\bar{K}} \right)^S_{s=1}$ for the $S$ sensors are conditionally independent given the augmented state $X_K = \left( x_j \right)^K_{j=\bar{K}+1}$. Thus optimal fusion is achieved by using the augmented state instead of the state at a single time.

### 6.4.3.1  Local Processing

The local processor performs the prediction and update functions to estimate the local state and its error covariance. In addition, it also computes $X^s_{k|k} = \left[ \left( x^s_{k|k} \right)^T, ..., \left( x^s_{\bar{K}+1|k} \right)^T \right]^T$, the estimate of the augmented state vector $X_k$ given the measurements $Z^s_k$, and $\mathbf{P}^s_{k|k}$, its error covariance by the following equations.

Prediction (see Appendix for derivations)

$$X^s_{k|k-1} = \begin{bmatrix} F_{k-1} x^s_{k-1|k-1} \\ X^s_{k-1|k-1} \end{bmatrix} \tag{6.27}$$

$$\mathbf{P}_{k|k-1}^s = \begin{bmatrix} P_{k|k-1}^s & F_{k-1}\mathbf{P}_{k-1|k-1}^s \\ \mathbf{P}_{k-1|k-1}^s F_{k-1}^T & \mathbf{P}_{k-1|k-1}^s \end{bmatrix} \tag{6.28}$$

where $\mathbf{F}_{k-1} = \begin{bmatrix} F_{k-1} & 0_{n \times n(k-2-\overline{K})} \end{bmatrix}$ and the initial conditions are $X_{\overline{K}|\overline{K}}^s = x_{\overline{K}|\overline{K}}^s$ and $\mathbf{P}_{\overline{K}|\overline{K}}^s = P_{\overline{K}|\overline{K}}^s$.

Update

$$\left(\mathbf{P}_{k|k}^s\right)^{-1} X_{k|k}^s = \left(\mathbf{P}_{k|k-1}^s\right)^{-1} X_{k|k-1}^s + J_k i_k^s \tag{6.29}$$

$$\left(\mathbf{P}_{k|k}^s\right)^{-1} = \left(\mathbf{P}_{k|k-1}^s\right)^{-1} + J_k I_k^s J_k^T \tag{6.30}$$

where $J_k = \begin{bmatrix} I_n, 0_{n \times n(k-\overline{K}-1)} \end{bmatrix}^T$ is a $n(k-\overline{K}) \times n$ matrix that selects the $x_k$ in $X_k$ to generate the measurement $z_k^s$.

Note that local processing only uses the local sensor model. Because the local prediction in distributed ASD (DASD) [21,22] uses a relaxed evolution model that depends on the number of sensors, the local estimates and covariance in DASD are only pseudo estimates and covariances and are not optimal.

### 6.4.3.2 Communication

At the fusion time $t_K$, each local node $s$ sends its augmented state estimate $X_{K|K}^s$ ($n(K - \overline{K})$ vector) and error covariance $\mathbf{P}_{K|K}^s$ ($n(K - \overline{K}) \times n(K - \overline{K})$ matrix) to the fusion node.

### 6.4.3.3 Fusion Processing

The fusion node computes recursively the predictions $X_{K|\overline{K}}^s$ and $\mathbf{P}_{K|\overline{K}}^s$ for sensor $s$ using the following equations (see Appendix for derivation)

$$X_{k|\overline{K}}^s = \begin{bmatrix} F_{k-1} x_{k-1|\overline{K}}^s \\ X_{k-1|\overline{K}}^s \end{bmatrix} \tag{6.31}$$

$$\mathbf{P}_{k|\overline{K}}^s = \begin{bmatrix} F_{k-1} P_{k-1|\overline{K}}^s F_{k-1}^T + G_{k-1} Q_{k-1} G_{k-1}^T & F_{k-1} \mathbf{P}_{k-1|\overline{K}}^s \\ \mathbf{P}_{k-1|\overline{K}}^s F_{k-1}^T & \mathbf{P}_{k-1|\overline{K}}^s \end{bmatrix} \tag{6.32}$$

with initial conditions $X_{\overline{K}|\overline{K}}^s$ and $\mathbf{P}_{\overline{K}|\overline{K}}^s$ received at the last communication time. It also computes $X_{K|\overline{K}}$ and $\mathbf{P}_{K|\overline{K}}$ from the last fused estimate $X_{\overline{K}|\overline{K}}$ and error covariance $\mathbf{P}_{\overline{K}|\overline{K}}$ using similar equations.

The global estimate $X_{K|K} = \begin{bmatrix} x_{K|K}^T, ..., x_{\overline{K}+1|K}^T \end{bmatrix}^T$ of the augmented state and its error covariance $\mathbf{P}_{k|k}$ are given by

$$\mathbf{P}_{K|K}^{-1} X_{K|K} = \mathbf{P}_{K|\overline{K}}^{-1} X_{K|\overline{K}} + \sum_{s=1}^{S} \left( \left(\mathbf{P}_{K|K}^s\right)^{-1} X_{K|K}^s - \left(\mathbf{P}_{K|\overline{K}}^s\right)^{-1} X_{K|\overline{K}}^s \right) \tag{6.33}$$

$$\mathbf{P}_{K|K}^{-1} = \mathbf{P}_{K|\overline{K}}^{-1} + \sum_{s=1}^{S} \left( \left(\mathbf{P}_{K|K}^s\right)^{-1} - \left(\mathbf{P}_{K|\overline{K}}^s\right)^{-1} \right) \tag{6.34}$$

The global estimate $x_{K|K}$ and error covariance $P_{K|K}$ can be extracted from the augmented estimate and error covariance.

The augmented state fusion algorithm computes the optimal global estimate is where the number of states equals the number of observations for all sensors between fusion times (note the difficulty for nonsynchronous observations). Reducing the length or dimension of the augmented state produces a suboptimal global estimate but requires less communication bandwidth. The numerical results will show that augmented state with very short length such as 2 or 3 has performance similar to full augmented state under normal conditions.

## 6.5  ALTERNATIVE FUSION APPROACHES FOR LINEAR SYSTEMS WITH PROCESS NOISE

This section presents alternative fusion approaches for linear systems with process noise. The first type of fusion algorithms address dependence that can be characterized by cross-sensor covariances of the local estimation errors. Section 6.5.1 presents the maximum likelihood estimate [9,10]. Section 6.5.2 presents the minimum variance (MV) estimate [13,14], which is equivalent to the best linear unbiased estimate (BLUE) [11,12]. Because cross-covariances are used in fusion, the local sensor models have to be communicated to the fusion node along with the local estimates. Further, the fused estimate may not be globally optimal because it is only the best estimate given the local estimates characterized by local and cross-sensor estimation error covariances.

The second type of fusion algorithms computes the optimal global estimates from centralized fusion. The recently developed DKF and ASD fusion algorithms both compute the optimal global estimate for linear systems with process noise. However, the local estimates require global model knowledge for their computation and the local estimates are not optimal local estimates given the local sensor measurements.

### 6.5.1  MAXIMUM LIKELIHOOD CROSS-COVARIANCE FUSION

This approach first appeared in Ref. [9], and we may call it Bar-Shalom–Campo fusion rule after the names of the authors of [9].

#### 6.5.1.1  Local Processing

Each local processor $s$ computes an estimate $x_{K|K}^s$ with error covariance $P_{K|K}^s$ from the local measurements using Kalman filters.

#### 6.5.1.2  Communication

The local processors send the estimates $x_{K|K}^s$ and error covariances $P_{K|K}^s$ to the fusion node. In addition, they also send the local sensor models for computing the cross-covariances between the local estimates by the fusion node.

#### 6.5.1.3  Fusion Processing

The fusion node computes the fused estimate by a convex combination of the local estimates, that is,

$$x_{K|K} = \sum_{s=1}^{S} W_{K|K}^s x_{K|K}^s \tag{6.35}$$

where the weights $W_{K|K}^s$, $s = 1,...,S$ satisfy $\sum_{s=1}^{S} W_{K|K}^s = I$ and are selected to minimize the classical likelihood function [10].

For two sensors, with $\hat{x}_i = x_{K|K}^i$ and $P_i = P_{K|K}^i$, $i = 1, 2$, and $P_{ij}$ as the cross-covariance between the local estimates $\hat{x}_i$ and $\hat{x}_j$, the fusion rule becomes

$$\hat{x} = W_1\hat{x}_1 + W_2\hat{x}_2 \tag{6.36}$$

where

$$W_i = (P_j - P_{ji})(P_1 + P_2 - P_{12} - P_{21})^{-1} \tag{6.37}$$

for $i = 1, 2$ with $j = 3-i$. The cross-covariance is not communicated from the local processors and has to be computed from the fusion node from the sensor models.

### 6.5.2 Minimum Variance (MV) Fusion

MV fusion uses the prior estimate $x_{K|\bar{K}}$ and error covariance $P_{K|\bar{K}}$ at the fusion node predicted from the last fusion time, in addition to the local estimates and error covariances.

#### 6.5.2.1 Local Processing

Each local processor $s$ computes an estimate $x_{K|K}^s$ with error covariance $P_{K|K}^s$ from the local measurements using a Kalman filter.

#### 6.5.2.2 Communication

Each local processor sends the local estimate $x_{K|K}^s$ and error covariances $P_{K|K}^s$ to the fusion node. It may also send other estimates $x_{k|k}^s$ and error covariances $P_{k|k}^s$, $\bar{K} < k \le K$ in a full memory form. As in all fusion approaches that use cross-covariances, the local sensor models are also communicated.

#### 6.5.2.3 Fusion Processing

The main difference between minimum variance fusion and maximum likelihood fusion is the use of the predicted estimate $x_{K|\bar{K}}$ and error covariance $P_{K|\bar{K}}$ in addition to the local estimates $x_{K|K}^s$ and error covariances $P_{K|K}^s$. Then the minimum variance fusion algorithm is

$$x_{K|K} = W_{K|\bar{K}}x_{K|\bar{K}} + \sum_{s=1}^{S} W_{K|K}^s x_{K|K}^s \tag{6.38}$$

where the weighting matrices $W_{K|\bar{K}}$ and $W_{K|K}^s$ are selected to minimize an error covariance. Since they depend on the local covariances and cross-covariances, these have to computed from the sensor models.

The MV fusion is a special case of the BLUE estimate when the estimate depends only on the last global estimate $x_{K|\bar{K}}$ and the local estimates $\left(x_{K|K}^s\right)_{s=1}^{S}$. The Bar-Shalom–Campo fusion rule is a BLUE estimate when the global estimate is restricted to be a function of only the local estimates $\left(x_{K|K}^s\right)_{s=1}^{S}$.

### 6.5.3 Distributed Kalman Filter

The DKF [15–18] is developed recently to handle fusion at arbitrary times for linear systems with process noise. The following presents the basic DKF that assumes perfect knowledge of the global sensor models at each local processor.

#### 6.5.3.1 Local Processing

Each local processor has knowledge of the global sensor model, that is, models of other sensors in addition to its own. The local pseudo-estimates $\tilde{x}_{k|k}^s$ and $\tilde{x}_{k|k-1}^s$, and pseudo error covariances $\tilde{P}_{k|k}^s$ and $\tilde{P}_{k|k-1}^s$, are computed by the following prediction and update equations. The estimates are only pseudo because they are not local estimates of the state due to the use of other sensor model parameters.

Prediction

$$\tilde{x}_{k|k-1}^s = F_{k-1}\tilde{x}_{k-1|k-1}^s \tag{6.39}$$

$$\tilde{P}_{k|k-1}^s = SP_{k|k-1} = S\left(F_{k-1}P_{k-1|k-1}F_{k|k-1}^T + Q_{k-1}\right) \tag{6.40}$$

with initial conditions $\tilde{x}_{0|0}^s$ and $\tilde{P}_{0|0}^s$.

Update

$$\left(\tilde{P}_{k|k}^s\right)^{-1}\tilde{x}_{k|k}^s = \left(\tilde{P}_{k|k-1}^s\right)^{-1}\tilde{x}_{k|k-1}^s + i_k^s \tag{6.41}$$

$$\left(\tilde{P}_{k|k}^s\right)^{-1} = SP_{k|k} = S\left(P_{k|k-1}^{-1} + I_k\right)^{-1} \tag{6.42}$$

#### 6.5.3.2 Communication

At the fusion time $t_K$, each local node sends the pseudo estimate $\tilde{x}_{K|K}^s$ and pseudo error covariance $\tilde{P}_{K|K}^s$ to the fusion node.

#### 6.5.3.3 Fusion Processing

The global estimate $x_{K|K}$ and error covariance $P_{K|K}$ are obtained by the following fusion equations:

$$P_{K|K}^{-1}x_{K|K} = \sum_{s=1}^{S}\left(\tilde{P}_{K|K}^s\right)^{-1}\tilde{x}_{K|K}^s \tag{6.43}$$

$$P_{K|K}^{-1} = \sum_{s=1}^{S}\left(\tilde{P}_{K|K}^s\right)^{-1} \tag{6.44}$$

The global estimate is optimal given all sensor measurements if the global sensor models assumed by the local processors are correct. If the sensor models are time varying, for example, they are the results of linearization, then the models have to be communicated between sensors. For time-varying fusion architectures, each local processor also has to know the identities of other sensors generating estimates so that their models are included in local processing. Because the fused estimate quality degrades if the information model is not matched, a debiasing methodology has been introduced. This methodology is based on a *global information hypothesis* [19] and iteratively calculates a local deviation parameter.

These fusion equations do not use the global estimate from the last fusion time. Alternative equations use the prior global estimate and local estimates that are based only on measurements since the last fusion time [23].

### 6.5.4  TRACK FUSION WITH ACCUMULATED STATE DENSITY (ASD)

The accumulated state $x_{k:n}$ is defined as the joint vector of states within a given time window $[t_n, t_k]$, that is, $x_{k:n} = \left[ x_k^T, ..., x_n^T \right]^T$. Thus the ASD [24] is the posterior density of $x_{k:n}$ with mean $x_{k:n|k}$ and covariance $P_{k:n|k}$. Track fusion with ASD [21,22], is similar to track fusion with augmented state and produces the optimal global estimate from centralized fusion. However, the local estimates are not locally optimal because local prediction uses a relaxed evolution model that depends on the number of sensors.

#### 6.5.4.1  Local Processing
Local processing involves prediction and update steps similar to local augmented state processing but the local estimate is not locally optimal because it uses a different model for prediction.

Prediction
The prediction equations are similar to Equations 6.27 and 6.28 except that it uses a relaxed evolution model where the process noise has covariance $SQ_k$ instead of $Q_k$.

Update
The update equations are similar to Equations 6.29 and 6.30.

#### 6.5.4.2  Communication
The local processor sends the local ASD as represented by the estimate $x_{k:n|k}^s$ and error covariance $P_{k:n|k}^s$.

#### 6.5.4.3  Fusion Processing
The fused ASD estimate and error covariance are given by

$$P_{k:n|k}^{-1} x_{k:n|k} = \sum_{s=1}^{S} \left( P_{k:n|k}^s \right)^{-1} x_{k:n|k}^s \tag{6.45}$$

$$\left( P_{k:n|k} \right)^{-1} = \sum_{s=1}^{S} \left( P_{k:n|k}^s \right)^{-1} \tag{6.46}$$

Note that these equations are similar to the augmented state fusion Equations 6.33 and 6.34 except that they do not consider the prior information.

## 6.6  PERFORMANCE EVALUATION

This section compares the performance of augmented state tracklet fusion with cross-covariance fusion algorithms and sensitivity to augmented state with different lengths. The performance is compared by varying the intensity of the process noise as in Refs. [25,26].

### 6.6.1 TARGET AND MEASUREMENT MODELS

The target moves according to the two-dimensional almost constant course and speed model

$$dx(t) = \begin{bmatrix} 0 & \mathbb{I}_2 \\ 0 & 0 \end{bmatrix} x(t)dt + \begin{bmatrix} 0 \\ \sqrt{q}\mathbb{I}_2 \end{bmatrix} dw(t) \tag{6.47}$$

in continuous time, where $\mathbb{I}_2$ is the $2 \times 2$ identity matrix. Thus

$$F_k \triangleq \begin{bmatrix} \mathbb{I}_2 & \Delta t \mathbb{I}_2 \\ 0_{22} & \mathbb{I}_2 \end{bmatrix} \text{ and } G_k Q_k G_k^T \triangleq q \begin{bmatrix} (\Delta t^3/3)\mathbb{I}_2 & (\Delta t^2/2)\mathbb{I}_2 \\ (\Delta t^2/2)\mathbb{I}_2 & \Delta t \mathbb{I}_2 \end{bmatrix} \tag{6.48}$$

with $\Delta t \equiv t_k - t_{k-1}$, where $0_{22}$ is the $2 \times 2$ zero matrix. The initial condition is

$$\mathcal{E}\left(x(t_0)x(t_0)^T\right) = \begin{bmatrix} 10^2 \mathbb{I}_2 & 0_{22} \\ 0_{22} & 9\mathbb{I}_2 \end{bmatrix} \tag{6.49}$$

There are two complementary sensors with measurements

$$z_k^s = \begin{bmatrix} \mathbb{I}_2 & 0_{22} \end{bmatrix} x_k + v_k^s \tag{6.50}$$

for $s = 1, 2$, and $k = 1, 2, \dots$ with measurement noise covariances

$$R_k^1 = \mathcal{E}\left(v_k^1 \left(v_k^1\right)^T\right) = \begin{bmatrix} 1 & 0 \\ 0 & 4 \end{bmatrix} \tag{6.51}$$

$$R_k^2 = \mathcal{E}\left(v_k^2 \left(v_k^2\right)^T\right) = \begin{bmatrix} 4 & 0 \\ 0 & 1 \end{bmatrix} \tag{6.52}$$

### 6.6.2 PERFORMANCE RESULTS

We compare the performance of augmented state tracklet fusion with Bar-Shalom–Campo (maximum likelihood) fusion and minimum variance estimate fusion. Comparison with DKF and DASD fusion is not meaningful because both produce the optimal global estimate if the assumed global model knowledge is correct. Reference [8] compares the performance of DKF with augmented state tracklet fusion when there a mismatch in global sensor model knowledge.

Each sensor makes 10 observations of the target state and computes the local estimates. At the end of the 10 scans, the estimates of the state (at a single time or multiple times) and their error covariances are sent to the fusion node for fusion. Figures 6.2 and 6.3 compare the root mean square (RMS) position and velocity errors relative to those of centralized fusion. For each fusion rule, the RMS errors are calculated analytically. All fusion algorithms have performance that is very close to that of centralized fusion, so that the differences in performance can be clearly visible only with its percentage increase of the RMS errors. Augmented state tracklet fusion exhibits almost the same performance as centralized fusion. The performance degradation of augmented state with

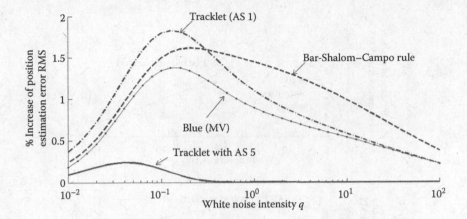

**FIGURE 6.2** Position estimation performance degradation.

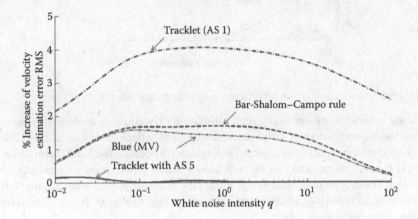

**FIGURE 6.3** Velocity estimation performance degradation.

length 5 is negligible. Standard tracklet fusion does not perform as well as minimum variance or Bar-Shalom–Campo but the difference is very small. The simple implementation and robustness to sensor knowledge explains the popularity of standard tracklet fusion.

It is interesting to note that the performance of all suboptimal fusion algorithms degrades initially and then improves as the white noise intensity increases. Performance improvement with large process noise is somewhat counterintuitive because the algorithms should have become even more "suboptimal." However, as the process noise increases, the optimal global estimate depends mostly on the most current local estimates, which are included even in the standard tracklet fusion algorithm. The same behavior is observed in Refs. [25,26] with the Ornstein-Uhlenbeck model.

Figure 6.4 compares the performance of augmented state fusion for different lengths in the augmented state. Increase in state length results in better performance but satisfactory performance can be achieved even with a small length.

## 6.7 CONCLUSIONS

Optimal fusion of local state estimates for dynamic systems with nonzero process noise is difficult because the sensor measurements are not conditionally independent given the state at a single time. Fusion algorithms based on cross-covariances do not attempt to compute the optimal global estimate and require communication of sensor models to compute the cross covariances. Tracklet fusion with

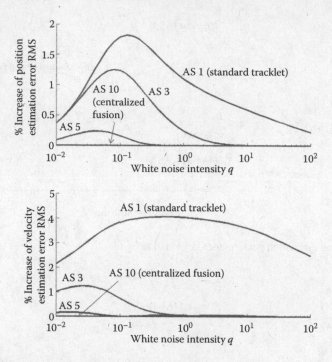

**FIGURE 6.4**  Performance comparison for different augmented state lengths.

augmented state computes the optimal global estimate for dynamic systems with nonzero process noise because the augmented state satisfies the conditional independence condition. It has advantages over optimal fusion algorithms such as DKF or DASD because the local estimates are optimal given the local measurements and can be computed without global model knowledge. Numerical results show that full augmented state is not necessary because partial state augmentation with a small number of state history can achieve performance close to that of full augmentation, that is, optimal system state estimation fusion.

## APPENDIX

A. Computation of $\mathbf{P}^s_{k|k-1}$

Define $e^s_{i|j} \triangleq x_i - x^s_{i|j}$, and $E^s_{i|j} \triangleq X_i - X^s_{i|j}$. Then

$$\mathbf{P}^s_{k|k-1} = \begin{bmatrix} P^s_{k|k-1} & \mathcal{E}\left\{ e^s_{k|k-1}\left(E^s_{k-1|k-1}\right)^T \right\} \\ \mathcal{E}\left\{ E^s_{k-1|k-1}\left(e^s_{k|k-1}\right)^T \right\} & \mathbf{P}^s_{k-1|k-1} \end{bmatrix} \tag{A6.1}$$

$$\mathcal{E}\left\{ \left(E^s_{k-1|k-1}\right)\left(e^s_{k|k-1}\right)^T \right\} = \mathcal{E}\left\{ \begin{bmatrix} e^s_{k-1|k-1} \\ E^s_{k-2|k-1} \end{bmatrix} \left(e^s_{k|k-1}\right)^T F^T_{k-1} \right\}$$

$$= \begin{bmatrix} P^s_{k-1|k-1} F^T_{k-1} \\ \mathcal{E}\left\{ E^s_{k-2|k-1}\left(e^s_{k-1|k-1}\right)^T \right\} F^T_{k-1} \end{bmatrix} = \mathbf{P}^s_{k-1|k-1}[F_{k-1} \; 0_{n \times n(k-2-\bar{K})}]^T \tag{A6.2}$$

where $\mathcal{E}$ is the mathematical expectation operator. Thus

$$\mathbf{P}^s_{k|k-1} = \begin{bmatrix} P^s_{k|k-1} & F_{k-1}\mathbf{P}^s_{k-1|k-1} \\ \mathbf{P}^s_{k-1|k-1}F^T_{k-1} & \mathbf{P}^s_{k-1|k-1} \end{bmatrix} \tag{A6.3}$$

B. Computation of $\mathbf{P}^s_{k|\bar{K}}$

$$\mathbf{P}^s_{k|\bar{K}} = \begin{bmatrix} \mathcal{E}\left\{e^s_{k|\bar{K}}\left(e^s_{k-1|\bar{K}}\right)^T\right\} & \mathcal{E}\left\{e^s_{k|\bar{K}}\left(E^s_{k-1|\bar{K}}\right)^T\right\} \\ \mathcal{E}\left\{E^s_{k-1|\bar{K}}\left(e^s_{k|\bar{K}}\right)^T\right\} & \mathcal{E}\left\{E^s_{k-1|\bar{K}}\left(E^s_{k-1|\bar{K}}\right)^T\right\} \end{bmatrix} \tag{A6.4}$$

Since

$$\mathcal{E}\left\{\left(E^s_{k-1|\bar{K}}\right)\left(e^s_{k|\bar{K}}\right)^T\right\} = \mathcal{E}\left\{\begin{bmatrix} e^s_{k-1|\bar{K}} \\ E^s_{k-2|\bar{K}} \end{bmatrix}\left(e^s_{k-1|\bar{K}}\right)^T F^T_{k-1}\right\}$$

$$= \begin{bmatrix} P^s_{k-1|\bar{K}}F^T_{k-1} \\ \mathcal{E}\left\{E^s_{k-2|\bar{K}}\left(e^s_{k-1|\bar{K}}\right)^T\right\}F^T_{k-1} \end{bmatrix} = \mathbf{P}^s_{k-1|\bar{K}}[F_{k-1}\ 0_{n\times n(k-2-\bar{K})}]^T \tag{A6.5}$$

we have

$$\mathbf{P}^s_{k|\bar{K}} = \begin{bmatrix} F_{k-1}P^s_{k-1|\bar{K}}F^T_{k-1} + G_{k-1}Q_{k-1}G^T_{k-1} & F_{k-1}\mathbf{P}^s_{k-1|\bar{K}} \\ \mathbf{P}^s_{k-1|\bar{K}}F^T_{k-1} & \mathbf{P}^s_{k-1|\bar{K}} \end{bmatrix} \tag{A6.6}$$

# REFERENCES

1. C. Y. Chong, S. Mori, K. C. Chang, and W. H. Barker, "Architectures and algorithms for track association and fusion," *IEEE Aerospace and Electronic Systems Magazine*, vol. 15, no. 1, pp. 5–13, 2000.
2. C. Y. Chong, "Hierarchical estimation," *Proceedings of the MIT/ONR Workshop on C3*, pp. 205–220, Monterey, CA, July 16–27, 1979.
3. O. Drummond, "Tracklets and a hybrid fusion with process noise," *Proceedings of SPIE*, vol. 3163, 1997.
4. X. Tian, and Y. Bar-Shalom, "Exact algorithms for four track-to-track fusion configurations: All you wanted to know but were afraid to ask," *Proceedings of the 12th International Conference on Information Fusion*, pp. 537–544, Seattle, WA, July 6–9, 2009.
5. C. Y. Chong, and S. Mori, "Graphical models for nonlinear distributed estimation," *Proceedings of the 7th International Conference on Information Fusion*, Stockholm, Sweden, June 28–July 1, 2004.
6. C. Y. Chong, K. C. Chang, and S. Mori, "Fundamentals of distributed estimation and tracking," *Proceedings of SPIE*, vol. 8392, 2012.
7. C. Y. Chong, K. C. Chang, and S. Mori, "Fundamentals of distributed estimation," in D. Hall, C. Y. Chong, J. Llinas, and M. Liggins II, editors, *Distributed Data Fusion for Network-Centric Operations*, pp. 95–124, CRC Press, Boca Raton, FL, 2012.
8. C. Y. Chong, S. Mori, F. Govaers, and W. Koch, "Comparison of tracklet fusion and distributed Kalman filter for track fusion," *Proceedings of the 17th International Conference on Information Fusion*, Salamanca, Spain, July 7–10, 2014.

9. Y. Bar-Shalom, and L. Campo, "The effects of the common process noise on the two-sensor fused-track covariance," *IEEE Transactions on Aerospace and Electronic Systems*, vol. 22, no. 6, pp. 803–805, 1986.

10. K. C. Chang, R. K. Saha, and Y. Bar-Shalom, "On optimal track-to-track fusion," *IEEE Transactions on Aerospace and Electronic Systems*, vol. 33, no. 4, pp. 1271–1276, 1997.

11. Y. Zhu, and X. R. Li, "Best linear unbiased estimation fusion," *Proceedings of the 2nd International Conference on Information Fusion*, Sunnyvale, CA, July 6–8, 1999.

12. X. R. Li, Y. Zhu, J. Wang, and C. Han, "Optimal linear estimation fusion—Part I: Unified fusion rules," *IEEE Transactions on Information Theory*, vol. 49, no. 9, pp. 2192–2208, 2003.

13. S. Mori, W. H. Barker, C. Y. Chong, and K. C. Chang, "Track association and track fusion with non-deterministic target dynamics," *IEEE Transactions on Aerospace and Electronic Systems*, vol. 38, no. 2, pp. 659–668, 2002.

14. K. C. Chang, T. Zhi, S. Mori, and C. Y. Chong, "Performance evaluation for MAP state estimate fusion," *IEEE Transactions on Aerospace and Electronic Systems*, vol. 40, no. 2, pp. 706–714, 2004.

15. W. Koch, "Exact update formulae for distributed Kalman filtering and retrodiction at arbitrary communication rates," *Proceedings of the 12th International Conference on Information Fusion*, pp. 2209–2216, Seattle, WA, July 6–9, 2009.

16. F. Govaers, and W. Koch, "Distributed Kalman filter fusion at arbitrary instants of time," *Proceedings of the 13th International Conference on Information Fusion*, Edinburgh, July 26–29, 2010.

17. F. Govaers, and W. Koch, "On the globalized likelihood function for exact track-to-track fusion at arbitrary instants of time," *Proceedings of the 14th International Conference on Information Fusion*, Chicago, IL, July 5–9, 2011.

18. F. Govaers, and W. Koch, "An exact solution to track-to-track-fusion at arbitrary communication rates," *IEEE Transactions on Aerospace and Electronic Systems*, vol. 48, no. 3, pp. 2718–2829, 2012.

19. M. Reinhardt, B. Noack, and U. D. Hanebeck, "Advances in hypothesizing distributed Kalman filtering," *Proceedings of the 16th International Conference on Information Fusion*, Istanbul, Turkey, July 9–12, 2013.

20. F. Govaers, A. Charlish, and W. Koch, "Covariance debiasing for the distributed Kalman filter," *Proceedings of the 16th International Conference on Information Fusion*, Istanbul, Turkey, July 9–12, 2013.

21. W. Koch, F. Govaers, and A. Charlish, "An exact solution to track-to-track fusion using accumulated state densities," *2013 Workshop on Sensor Data Fusion: Trends, Solutions, Applications (SDF)*, Bonn, Germany, Oct. 9–11, 2013.

22. W. Koch, and F. Govaers, "On decorrelated track-to-track fusion based on accumulated state densities," *Proceedings of the 17th International Conference on Information Fusion*, Salamanca, Spain, July 7–10, 2014.

23. C. Y. Chong, and S. Mori, "Optimal fusion for non-zero process noise," *Proceedings of the 16th International Conference on Information Fusion*, Istanbul, Turkey, July 9–12, 2013.

24. W. Koch, and F. Govaers, "On accumulated state densities with applications to out-of-sequence measurement processing," *IEEE Transactions on Aerospace and Electronic Systems*, vol. 47, no. 4, pp. 2766–2778, 2011.

25. S. Mori, K. C. Chang, and C. Y. Chong, "Comparison of track fusion rules and track association metrics," *Proceedings of the 15th International Conference on Information Fusion*, Singapore, July 9–12, 2012.

26. S. Mori, K. C. Chang, and C. Y. Chong, "Essence of distributed target tracking—Track fusion and track association," in D. Hall, C. Y. Chong, J. Llinas, and M. Liggins II, editors, *Distributed Data Fusion for Network-Centric Operations*, pp. 125–160, CRC Press, Boca Raton, FL, 2012.

# 7 A Fuzzy Multicriteria Approach for Data Fusion

*André D. Mora, António J. Falcão, Luís Miranda, Rita A. Ribeiro, and José M. Fonseca*

## CONTENTS

## 7.1 INTRODUCTION

This chapter discusses a recent data fusion algorithm (Ribeiro et al. 2014), Fuzzy Information Fusion (FIF), that provides a general method, based on decision matrices, for fusing heterogeneous sources, into a single composite, with the final goal of rating and ranking alternatives. In addition, FIF has the ability to handle imprecise information, either from lack of confidence in the input data or from its imprecision (allowing deviation errors). In general, FIF can be viewed as a decision support method for handling multiple criteria (i.e., input sources) classification problems under uncertain environments.

We start this chapter by presenting a comprehensive overview of existing data fusion types and models, to set up the context of the discussed FIF algorithm. Then, we discuss in detail the four steps of the FIF algorithm: data normalization, uncertainty filtering, assignment of relative importance to the data sources, and data aggregation/fusion. After introducing FIF, we illustrate its potential with two applications in the aerospace domain, one for selecting the safest planet landing sites for spacecraft and another for safe landing of drones. Finally, we present some concluding remarks.

## 7.2  DATA FUSION OVERVIEW

Data fusion is essentially a means of combining information from multiple sources (possibly hetero-geneous) into a single unified view of the various data. Khaleghi et al. (2013) and Castanedo (2013) provide an in-depth overview of past proposals and recent advances in multisource fusion.

To set up the context of this work in the next two subsections we summarize the main types of data fusion and models that have been proposed for performing data fusion.

### 7.2.1  MAIN TYPES

In general, data fusion includes three main types: multisensor, image, and information fusion.

#### 7.2.1.1  Multisensor Fusion

The main objective of multisensor fusion is to integrate data measurements extracted from different sensors and combine them into a single representation. Within this context, most approaches focus on multisensor data fusion using statistical methods (Kalman filters) and probabilistic techniques (Bayesian networks) (Waltz and Llinas 1990; Manyika and Durrant-Whyte 1994; Goodman et al. 1997; Epp et al. 2008; Lee et al. 2010). Statistical methods focus on minimizing errors between measured and predicted values, whereas probabilistic methods rely on using weighting factors based on how accurate the sensor data is. Hybrid strategies combine different multisensor fusion techniques taking the advantages of the individual approaches and mitigating their flaws. Because this type of data fusion is not applicable to this work, we will not discuss it any further.

#### 7.2.1.2  Image Fusion

The main objective of image fusion is to reduce uncertainty and redundancy while maximizing relevant information particular to an application by combining different image representations of the same scene (Goshtasby and Nikolov 2007). The general approach is to use multiple images of the same scene, provided by different sensors, and combine them to obtain a better understanding of the scene, not only in terms of position and geometry, but also in terms of semantic interpretation.

Image fusion algorithms are usually divided into pixel, feature, and symbolic levels (and starting at the signal level) (Esteban et al. 2005). Feature and symbolic algorithms have not received the same level of attention as pixel-level algorithms. Pixel-level algorithms are the most common and they work either in the spatial domain or in the transformation domain. Although pixel-level fusion is a local operation, transformation domain algorithms create the fused images globally. Feature-based algorithms (symbolic) typically segment the images into regions and fuse them using their intrinsic properties (Piella 2003; Hsu et al. 2009). A summary of characteristics for this type of data fusion is depicted in Table 7.1.

---

**TABLE 7.1**

**Characteristics of Data Fusion Levels**

| Characteristics | Representation Level of Information | Type of Sensory Information | Model of Sensory Information |
|---|---|---|---|
| Signal level | Low | Multidimensional signal | Random variable with noise |
| Pixel level | Low | Multiple images | Random process across the pixel |
| Feature level | Medium | Features extracted from signals/images | Noninvariant form of features |
| Symbol level | High | Decision logic from signals/images | Symbol with degree of uncertainty |

*Source:* Adapted from Esteban, J., A. Starr, R. Willetts, P. Hannah, and P. Bryanston-Cross. *Neural Computing and Applications*, 14(4):273–281, 2005.

---

There are many suitable algorithms and techniques proposed in the literature to deal with the three levels of image fusion processes: multiresolution analysis, hierarchical image decomposition, pyramid techniques, wavelet transform, artificial neural networks, biological inspired models, fuzzy rules, principal component analysis (PCA), and so forth (Goodman et al. 1997; Piella 2003; Serrano 2006; Goshtasby and Nikolov 2007; Dong et al. 2009; Hsu et al. 2009). Good overviews on image fusion and its applications can be seen in (Piella 2003; Smith and Heather 2005; Goshtasby and Nikolov 2007).

### 7.2.1.3 Information Fusion

As mentioned before, there is a tenuous line between image fusion and information fusion; for example, feature and symbolic levels of fusion are sometimes considered image fusion but they can also be considered as information-based fusion (Piella 2003; O'Brien and Irvine 2004). There are many interesting definitions for fusion in general, and, specifically, for information fusion as a multilevel process of integrating information from multisources to produce fused information (Lee et al. 2010). Moreover, Lee et al. (2010) discuss a list with six important factors for a successful information fusion: (1) robustness, (2) extended spatial-temporal coverage, (3) high confidence, (4) low ambiguity, (5) reliability and validity, and (6) low vulnerability.

The most traditional and well-known frameworks for information fusion are based on statistical methods (e.g., Kalman filters, optimal theory, distance methods) and probabilistic techniques (e.g., Bayesian networks, evidence theory) (Waltz and Llinas 1990; Manyika and Durrant-Whyte 1994; Goodman et al. 1997; Akhoundi and Valavi 2010; Lee et al. 2010). Statistical methods focus on minimizing errors between actual and predicted values, whereas probabilistic methods rely on using weighting factors based on how accurate the data is. Interesting applications of these techniques for fusion are (Serrano 2006; Epp and Smith 2007; Rogata et al. 2007; Epp et al. 2008; Johnson et al. 2008; Shuang and Liu 2009; Li et al. 2010).

Recently, computational intelligent approaches for information fusion have been proposed, particularly using Dempster–Shafer evidential theory, fuzzy set theory, and neural networks (Akhoundi and Valavi 2010; Lee et al. 2010). Some interesting applications using these novel approaches are in (Howard 2002; Serrano and Seraji 2007; Devouassoux et al. 2008; Serrano and Quivers 2008; Hsu et al. 2009; Bourdarias et al. 2010; Pais et al. 2010). In summary, information fusion models are crucial as a mean to present a coherent and uniform view to support decision makers and the respective decision-making process.

## 7.2.2 EXISTING MODELS

Bedworth and O'Brien (2000) define a process model to be a description of a set of processes, and several models have been developed to deal with data fusion issues (Veloso et al. 2009). Considering data fusion cases, usually there is a module to deal with the sensors and their output (input module), other to make some preprocessing or full processing of the data and a module to output the data processed on the environment. The so-called adjustments/calibrations of the sensors and system itself are normally done in a closed loop interface, common in these cases (Veloso et al. 2009).

### 7.2.2.1 JDL Model

Developed in 1985 by the United States Joint Directors of Laboratories (JDL) Data Fusion Group, the JDL data fusion model became the most common and popular conceptualization model of fusion systems (Steinberg et al. 1999). The group was intended to assist in coordinating activities in data fusion and improve communication and cooperation between development groups with the purpose of unifying research (Macii et al. 2008; Hall and Steinberg 2000; Khaleghi et al. 2013). The result of that effort was the creation of a number of activities (White 1991; Hall and Garga 1999; Steinberg et al. 1999; Hall and Steinberg 2000) such as (1) development of a process model for data fusion, (2) creation of a lexicon for data fusion, (3) development of engineering guidelines for building data fusion systems, and (4) organization and sponsorship of the Tri-Service Data Fusion Conference (e.g., from 1987 to

1992). The JDL Data Fusion Group has continued to support community efforts in data fusion, leading to the annual National Symposium on Sensor Data Fusion and the initiation of a Fusion Information Analysis Center. In the initial JDL data fusion lexicon (dated 1985), the group defined data fusion as "a process dealing with the association, correlation, and combination of data and information from single and multiple sources to achieve refined position and identity estimates, and complete and timely assessments of situations and threats, and their significance. The process is characterized by continuous refinements of its estimates and assessments, and the evaluation of the need for additional sources, or modification of the process itself, to achieve improved results" (Steinberg et al. 1999, p. 431).

According to this model, the sources of information used for data fusion can include both local and distributed sensors (those physically linked to other platforms), or environmental data, a priori data, and human guidance or inferences (Hall and Steinberg 2000; Macii et al. 2008; Khaleghi et al. 2013). Using these sources of information, the original JDL data fusion process consists of four increasing levels of abstraction, namely, "object," "situation," "threat," and "process" refinements, which were described in (Hall and Garga 1999; Steinberg et al. 1999; Hall and Steinberg 2000; Llinas and Hall 2008). To manage the entire data fusion process through control input commands of information requests, the system includes a user interface as well as a Data Management unit, a lightweight database to provide access and management of the dynamic data (Hall and Steinberg 2000; Macii et al. 2008). There are, however, some concerns with the ways in which these model levels have been used in practice. The JDL levels have frequently been interpreted "as a canonical guide for partitioning functionality within a system" (Steinberg et al. 1999); do level 1 fusion first, then levels 2, 3, and 4. The original JDL titles for these levels appear to be focused on tactical targeting applications (e.g., threat refinement), so that the extension of these concepts to other applications is not obvious (Steinberg et al. 1999). Therefore, despite its popularity, the JDL model has many shortcomings, such as being too restrictive and especially tuned to military applications, which have been the subject of several extension proposals (Hall and Llinas 1997; Steinberg et al. 1999; Llinas et al. 2004) attempting to alleviate them. In summary, the JDL formalization is focused on data (input/output) rather than processing (Khaleghi et al. 2013).

### 7.2.2.2 Waterfall Model

As mentioned in Veloso et al. (2009), the waterfall model is a hierarchical architecture where the information output by one module will be input to the next module. This model focuses on the processing functions on the lower levels. The stages relate to the source preprocessing and levels 1, 2, and 3 of the JDL model as follows: Sensing and signal processing correspond to source preprocessing, feature extraction and pattern processing match object refinement (level 1), situation assessment is similar to situation refinement (level 2), and decision making corresponds to threat refinement (level 3) (Veloso et al. 2009). Being this similar to the JDL model, the waterfall model suffers from the same drawbacks. Although being more exact in analyzing the fusion process than other models, the major limitation is the lack of description of the feedback data flow (Bedworth and O'Brien 2000). However, Esteban et al. (2005) and Zegras et al. (2008) proposed that the sensor system is continuously updated with feedback information arriving from the decision-making module. The main aspects of the feedback element are the recalibration, reconfiguration, and data gathering alerts to the multisensor system (Esteban et al. 2005).

The waterfall model does not clearly state that the sources should be parallel or serial (though processing is serial), assumes centralized control, and allows for several levels of representation (Zegras et al. 2008). This model has been widely used in the U.K. defense data fusion community but has not been significantly adopted elsewhere (Bedworth and O'Brien 2000).

### 7.2.2.3 Luo and Kay Model

Luo and Kay (1988) presented a generic data fusion structure based in a hierarchical model, yet different from the waterfall model. In this system, data from the sensors are incrementally added on different fusion centers in a hierarchical manner, thus increasing the level of representation from the raw data or signal level to more abstract symbolic representations at the symbol level (Esteban et al. 2005; Zegras et al. 2008; Veloso et al. 2009). This model has a parallel input and processing of data sources interface, which may enter the system at different stages and levels of representation. The model is based on decentralized architecture and does not assume a feedback control (Veloso et al. 2009).

### 7.2.2.4 Thomopoulos Model

Esteban et al. (2005) and Veloso et al. (2009) mention that Thomopoulos proposed a three-level model, formed by signal level fusion, where data correlation takes place; evidence level fusion, where data are combined at different levels of inference; and dynamics level fusion, where the fusion of data is done with the aid of an existing mathematical model. Depending on the application, these levels of fusion can be implemented in a sequential manner or interchangeably (Esteban et al. 2005). At each level exists data integration preserving a given order that may cause communication problems with data transmission, and factors such as spatial (Esteban et al. 2005; Veloso et al. 2009). There will also be a database component, responsible for data gathering that will be used in a learning process (Veloso et al. 2009).

### 7.2.2.5 Intelligence Cycle

The intelligence cycle is an approach that falls in a fusion model subgroup with cyclic character (Elmenreich 2007), with the Boyd control loop (see Section 7.2.2.6) an integral part. Intelligence processing involves both information processing and information fusion as observed in (Bedworth and O'Brien 2000). Although the information is often at a high level the processes for handling intelligence products are broadly applicable to data fusion in general (Bedworth and O'Brien 2000). The intelligence cycle model (Bedworth and O'Brien 2000; Elmenreich 2007) comprises four stages:

1. *Collection*: Appropriate raw intelligence data (intelligence report at a high level of abstraction) are gathered, for example, through sensors.
2. *Collation*: Associated intelligence reports are correlated and brought together.
3. *Evaluation*: The collated intelligence reports are fused and analyzed.
4. *Dissemination*: The fused intelligence is distributed to the users.

An additional stage is enumerated by Elmenreich (2007) regarding planning and directions where the intelligence requirements are determined. However, in the particular case of the United Kingdom Intelligence Community model, this stage is subsumed in the dissemination process, unlike in the U.S. model (Bedworth and O'Brien 2000).

### 7.2.2.6 Boyd Control Loop Model

In 1987, John Boyd proposed a cycle containing four stages that was first used for modeling the military command process (Bedworth and O'Brien 2000; Elmenreich 2007). The Boyd control loop represents the classic decision support mechanism in military information operations (Elmenreich 2007). Because decision support systems for situational awareness are tightly coupled with fusion systems, the Boyd model has also been used for sensor fusion (Bedworth and O'Brien 2000).

The Boyd model has some similarities between the intelligence cycle and the JDL model. Bedworth and O'Brien (2000) and Elmenreich (2007) compared the stages of those models. Although the Boyd model represents the stages of a closed control loop system and gives an overview on the overall task of a system, the model structure is significantly limited in identifying and separating different sensor fusion tasks (Elmenreich 2007).

### 7.2.2.7 Dasarathy Model

The Dasarathy fusion model is categorized in terms of the types of data/information that are processed and the types that result from the process (Steinberg and Bowman 2008). Its data flow is characterized

**TABLE 7.2**

**The Five Levels of Fusion in the Dasarathy Model**

| Input | Output | Analogues |
|---|---|---|
| Data | Data | Data-level fusion |
| Data | Features | Feature selection and feature extraction |
| Features | Features | Feature-level fusion |
| Features | Decisions | Pattern recognition and pattern processing |
| Decisions | Decisions | Decision-level fusion |

*Source:* Adapted from Bedworth, M., and J. O'Brien. *IEEE Aerospace and Electronic Systems Magazine*, 15:30–36, 2000.

by input/output (I/O) processes (Khaleghi et al. 2013), and three main levels of abstraction during the data fusion process are identified (Bedworth and O'Brien 2000) as:

1. *Decisions*: symbols or belief values
2. *Features*: or intermediate-level information
3. *Data*: or more specifically sensor data

Dasarathy states that fusion may occur both within these levels and as a means of transforming between them (Bedworth and O'Brien 2000). In the Dasarathy fusion model, there are five possible categories of fusion, illustrated in Table 7.2 as types of I/O considered (Steinberg and Bowman 2008).

In conclusion, several models and techniques have already been proposed in the literature for data fusion; however, most of them just apply to one type of data fusion: multisource, image fusion, or information fusion. The FIF algorithm, discussed in this chapter, is a general method that can be applied to any of those types. The main difference between FIF and the described models is that FIF uses a fuzzy multicriteria approach, which, we believe, can simplify and generalize any fusion process.

## 7.3 FIF ALGORITHM

### 7.3.1 CONTEXT AND SCOPE

There are a number of issues that make information fusion a challenging task. The development of FIF (Ribeiro et al. 2014) started with trying to answer the important question posed by Maître and Bloch (1997): "How to obtain a final value for each potential solution (e.g., pixel, hypothesis) from combining different heterogeneous measurements?" In those authors' point of view, it can be solved in three stages, as follows:

1. Transform the measures in such a way that it is possible to combine them.
2. Combine the data as transformed by the representation according to the allowed rules for the chosen framework (e.g., Bayes rule).
3. From the resulting combination make a decision in agreement with the problem.

Information fusion is related only to the first two steps: representing, transforming, cleaning, and aggregating information. The last one is already in the realm of ranking or selecting candidate alternatives (decision) and is outside the goal of obtaining a single composite of fused information.

In addition, information fusion should comply with important critical success factors (Lee et al. 2010), such as (1) robustness, (2) extended spatial-temporal coverage, (3) high confidence, (4) low ambiguity, (5) reliability and validity, and (6) low vulnerability. Other authors (Khaleghi et al. 2013) also pointed out different critical factors, such as data correlation, data alignment, static versus dynamic phenomena, and so forth. However, these later challenges are mostly related to image fusion and therefore not totally applicable in the FIF context. Summarizing, the FIF algorithm incorporates the first two stages and critical factors (1) to (6). FIF is a general information fusion algorithm suitable for tackling uncertain environments, particularly when there is a lack of confidence and accuracy in the input data, as well as when the criteria are heterogeneous and require intelligent normalization (Ribeiro et al. 2014).

### 7.3.2 FIF Architecture

In Figure 7.1 a complete architecture for the FIF algorithm is proposed. Each of the input data sources ($i_1, ..., i_n$) will run a customized data transformation process, preparing them ($w_1, ..., w_n$) for the multisource data fusion.

Observing Figure 7.1, it can be seen that the filtering uncertainty step is independent and should be performed before the assignment of relative importance of criteria to ensure that the data imprecision is taken into consideration when performing data fusion. Further, in FIF we strongly support that one of the best aggregation operators for fusing information is a mixture operator with weighting functions (Ribeiro et al. 2014), because it allows rewarding or penalizing poorly satisfied criteria (i.e., input variables). By extending this operator to include the uncertainty filtering the steps after normalization are jointly taken into consideration. Summarizing, the FIF algorithm includes four main steps:

1. Normalization, which includes a mathematical transformation (fuzzification) of input sources to ensure numerical and comparable data for fusion
2. Filtering uncertainty from data regarding inaccuracies and lack of confidence in input data
3. Assigning relative importance to each criteria membership value, which depends on the satisfaction/suitability of criteria for a specific alternative
4. Aggregation/fusion method (i.e., aggregation operator) for combining all fuzzified inputs (criteria) into a single composite (fused information)

We next describe all the steps in detail.

Step 1. Normalization

      Considering that inputs (criteria) can be originated by heterogeneous data sources, the inputs are normalized using fuzzy membership functions, that is, they undergo a

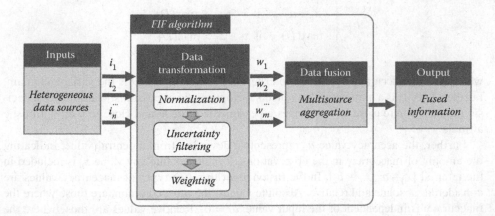

**FIGURE 7.1**  FIF algorithm architecture.

fuzzification process (Ross 2005). Besides guaranteeing normalized and comparable data, a fuzzification method allows representing data with semantic concepts, which facilitates problem understanding.

An important problem on the "variables fuzzification" is the choice of the best topology for the membership functions, because they must take into account the context objective. Hence, topologies such as triangular, trapezoidal, or Gaussian membership functions should be chosen depending on the nature of the criterion and the desirable representation and interpretation (semantic) for each criterion.

Step 2. Filtering uncertainty

The FIF algorithm enables dealing with the intrinsic uncertainty that can be found in the input information to be fused. The reasoning behind this step is that for any given alternative, each criterion must be adjusted by decreasing its membership value accordingly to the lack of confidence and inaccuracy (deviation interval) in the input data. This step is performed with a twofold filtering function: (1) combines metrics to deal with both types of uncertainty in the input values and (2) reflects the attitude of the decision maker (Pais et al. 2010).

Lack of confidence affects all input values regarding their membership values and inaccuracy creates an interval with left and right deviations from the initial value. This function enables to adjust (decrease) the membership functions to reflect the embedded input information uncertainty and also to incorporate a pessimist or optimist view of a developer. Formally, the accuracy and confidence parameters ($a_{ij}$ and $wc_j$ respectively) will modify the membership functions' values using the following expression for the filtered uncertainty (Pais et al. 2010):

$$fu_{ij} = wc_j^* \left( 1 - \lambda^* \max_{x \in [a;b]} \left\{ \left| \mu(x) - \mu(x_{ij}) \right| \right\} \right)^* \mu(x_{ij}) \qquad (7.1)$$

where $x_{ij}$ is the value of the $j$th criterion for site $i$; $\mu()$ is a membership degree in a fuzzy set; $wc_j$ is the confidence (percentage) associated to criterion $j$; and $[a, b]$ is the inaccuracy deviation interval, defined as follows:

$$a = \begin{cases} \min(D) & \text{if } x_{ij} - a_{ij} \leq \min(D) \\ x_{ij} - a_{ij} & \text{if } x_{ij} - a_{ij} > \min(D) \end{cases}$$

$$b = \begin{cases} x_{ij} + a_{ij} & \text{if } x_{ij} + a_{ij} \leq \max(D) \\ \max(D) & \text{if } x_{ij} + a_{ij} > \max(D) \end{cases}$$

where $a_{ij}$ is the accuracy associated to criterion $j$ for site $i$, and $D$ is the variable domain. Further, $\lambda$ [0, 1] is a parameter that reflects the optimistic or pessimistic attitude of a decision maker, where close to 1 indicates a pessimistic attitude and close to 0 an optimistic attitude.

Further, the accuracy value $a_{ij}$ represents a deviation from a central value, indicating the amount of inaccuracy in the observations. It indicates that any value $x_{ij}$ is included in the interval $[x_{ij} - a_{ij}, x_{ij} + a_{ij}]$. In the fusion algorithm two types of inaccuracy values are considered: absolute and relative. Absolute values, the most common, are those where the inaccuracy is independent of the input value: $a_{ij} = a_j$. Relative values are those where the accuracy is a function of the input value $x_{ij}$. These will take the form of $a_{ij} = a_j * x_{ij}$.

In this fashion all input values affected by any kind of uncertainty (inaccuracy or confidence) can be taken into consideration without loss of robustness. Obviously these parameters can be customized for any information fusion problem.

Step 3. Weighting functions

In FIF, linear weighting functions (Pereira and Ribeiro 2003; Ribeiro and Pereira 2003) are used to express the relative importance of criteria. The rationale of these weighting functions is that the satisfaction value of a criterion should influence its predefined relative importance.

FIF algorithm uses a modified linear function $L(x)$, to increase the computational efficiency and understandability, as follows (Ribeiro et al. 2014):

$$L(fu_{ij}) = \alpha \frac{1 + \beta fu_{ij}}{1 + \beta}, \text{ where } 0 \leq \alpha \leq 1 \text{ and } 0 \leq \beta \leq 1 \tag{7.2}$$

where $\alpha_j$, $\beta_j \in [0, 1]$ and $fu_{ij}$ is the *accuracy and confidence* membership value from Equation 7.1 for the $j$th criterion of alternative $i$.

We can see that the parameter $\alpha$ is used to express the relative importance of the different criteria. The parameter $\beta$ controls the ratio $L(1)/L(0) = 1 + \beta$ between the maximum and minimum values of the effective generating function. When we have $\beta = 0$ (the weighting function does not depend on criteria satisfaction) it falls in the classical weighted average aggregation operator.

The definition of the weighting functions morphologies is given by the parameters $\alpha$ and $\beta$. The $\alpha$ parameter provides the semantics for the weighting functions as for example (very important, important, low importance, etc.). The $\beta$ parameter provides the required slope for the weighting functions to enable more or less penalization or rewarding. The $\alpha$ and $\beta$ parameters will be set depending on the initial assigned semantic importance (e.g., important = 0.8) and on the sharpness of the function decrease, which depends on how much we want to penalize the poorly satisfied criteria.

The motivation for using weighting functions in the FIF algorithm is to enable rewarding or penalizing input criteria and also to remove (filter) the imprecision in the input data regarding lack of confidence and possible inaccuracies.

Step 4. Multisource aggregation

The last step on FIF algorithm is to fuse the transformed input information from diverse sources. The information fusion aggregation method proposed for fusing information is based on the mixture of operators with weighting functions (Pereira and Ribeiro 2003; Ribeiro and Pereira 2003), and its general formulation is

$$r_i = \oplus (W(fu_{i1}) \otimes ac_{i1}, \ldots, W(fu_{in}) \otimes fu_{in}) \tag{7.3}$$

where
- $\oplus$ is an aggregation method (e.g., sum, max, parametric operators)
- $\otimes$ is a conjunction operator (e.g., multiplication, min)
- $fu_{ij}$ is the filtered uncertainty *accuracy and confidence* membership value of the $j$th input for solution $i$ (Equation 7.1)
- $W(fu_{ij}) = \dfrac{L(fu_{ij})}{\sum\limits_{k=1}^{n} L(fu_{ik})}$ where $L(fu_{ij})$ is the weighting function above (Equation 7.2)

As can be observed, in FIF the weighting function of this mixture operator was extended to include dealing with imprecision. The result of this step concludes the information fusion steps of the FIF algorithm.

## 7.4  ILLUSTRATIVE APPLICATIONS

The preliminary research for devising this algorithm was done in the context of past research projects, whose main goals were to recommend a safe interplanetary spacecraft target-landing site (Intelligent Planetary SIte Selection [IPSIS]) and also its adaptation for autonomously landing unmanned aerial vehicles (Intelligent Landing of Unmanned (Aerial) Vehicles [ILUV]) (more detail about these two projects is available at http://www.ca3-uninova.org). The following subsections describe how the FIF algorithm was used in these two experimental applications.

### 7.4.1  IPSIS—Safe Landing with Hazard Avoidance

The main objective of IPSIS was to develop a model to choose the safest landing site for spacecraft with hazard avoidance and to determine when retargetings should occur. A preliminary version of the FIF algorithm was devised in IPSIS, to fuse information from shadows, textures, slopes, reachability, and fuel maps (Figure 7.2). The main publications on this project are Bourdarias et al. (2010), Pais et al. (2010), and Simoes et al. (2012). In these publications hazard maps were modeled as fuzzy functions to become the inputs for a dynamic fuzzy multicriteria model with feedback and an optimization process to answer real-time requirements. The preliminary works implied dynamic environments, several iterations, and feedback information, while here the scope is restricted to fusing information at a certain time $t$, without any concerns about past information. Here, we only discuss details about the fusion process and use two hazard maps (slope and texture), from the aforementioned project, to illustrate the FIF algorithm.

Considering the normalization of the "low-slope," the membership function relates to slopes lower than 20°. This criterion is normalized ("fuzzified") with an open triangular membership function defined as follows:

$$\mu(x) = \begin{cases} 1 & \text{if } x \leq c \\ \dfrac{20 - x}{20 - c} & \text{if } c < x \leq 20 \end{cases}$$

where
$x \in [0, 20]$
$c = \min(x) + \alpha \cdot (\max(x) - \min(x)) = 0.05 * 20 = 1$
$\alpha$ is the range of the function plateau

In this case we consider $\alpha = 0.05$, which results in a function plateau in the range [0, 5], although the parameter can be adjusted. The min and max values are extracted from the input matrix. Figure 7.3 depicts the input hazard map, its normalizing function, and the resulting output matrix.

To illustrate the uncertainty filtering, Figure 7.4 shows its application to a membership function of "low texture." As can be observed, we have in curve (1) the membership function representing

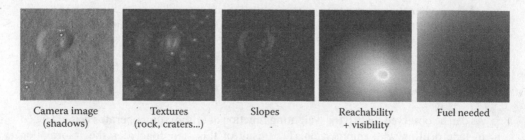

| Camera image (shadows) | Textures (rock, craters...) | Slopes | Reachability + visibility | Fuel needed |

**FIGURE 7.2**   IPSIS heterogeneous input data sources.

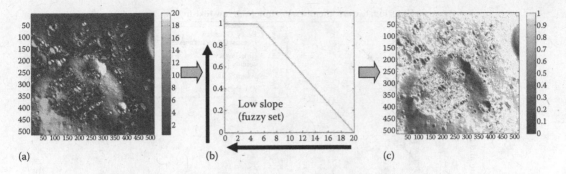

**FIGURE 7.3** Example of normalization (fuzzification) of criteria low-slope. (a) Original hazard map. (b) Membership function topology. (c) Normalized map.

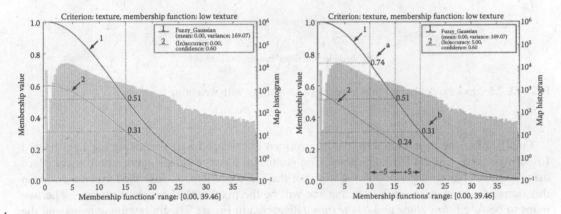

**FIGURE 7.4** Example of filtering "low texture" with lack of confidence and inaccuracy.

the fuzzy set "low texture" and in curve (2) the adjusted function after using the *Filter*. Note that the bar charts in both graphics represent the histogram of all values in the input matrix that were used to define the membership function topology.

In the left figure, we see the decrease caused in the membership function by having a confidence on the input data of only 60% but without any inaccuracy. In the right figure, we see the further decrease in the membership function by considering an inaccuracy, that is, deviation interval, of 5. In both examples we used a pessimistic attitude from the decision maker, that is, $\alpha = 1$. Illustrating, to a "low-texture" value of $x = 15$ corresponds a membership value of $\mu(15) = 0.51$. When we filter the information with a confidence of 60% the membership value decreases to $\mu(15) = 0.51$ (left graphic). If we further filter the information with an (in)accuracy deviation of 5, the possible slope range is [10, 20] and the corresponding membership value decreases to $\mu(15) = 0.24$ (right graphic).

Using the same "low variance texture" example (Figure 7.4), we now exemplify in Figure 7.5 the third step of the IPSIS weighting process to express the relative importance of this criterion. The curve (3) represents the defined weighting function with parameters $\alpha = 0.8$ (important criteria) and $\beta = 0.33$ (low decrease for the weight). Hence, for a texture of 15 units ($x$-axis) the initial membership value is 0.51. After filtering (step 2) its membership degree decreases to 0.30 and after weighting, that is, considering the relative importance of the criteria, the final value to be used for the fusion is 0.20. As can be observed, we highly penalized the input variable satisfaction value (0.51) because it displayed relatively low performance and our confidence in the data was relatively low.

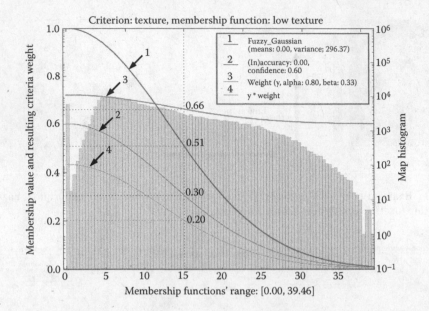

FIGURE 7.5   Example of hazard map importance assignment with weighting functions.

Finally, after the weighting step, the transformed inputs are fused generating the final hazard fused map (step 4 of the FIF algorithm). The multisource fusion (maps aggregation) takes place using the calculations described in the step 4 of the FIF algorithm, as described previously. From this output-fused map, the safest landing site will be the pixel with the highest ranking. The two maps to be fused (*low slope* and *low texture*), displayed in Figure 7.6, are raw input maps and the respective scale on the right shows dark for *low* values (objective for the fusion) and clear for *good* values. The fused hazard map (right) uses the same logical scale, that is, "good" alternatives are clearer and "bad" ones are in darker. The latter scale corresponds to the membership values within the interval [0, 1].

It should also be noticed that the fused hazard map (Figure 7.6, right) represents the input for the decision-making process, that is, it is the complete search space including all the potential alternatives for safe landing sites. In this fused map redder pixels correspond to best landing spots; bluer pixels represent worst landing alternatives.

With this last (fusing process) we conclude the illustration of FIF algorithm in a real case study of selecting the safest landing place with hazard avoidance.

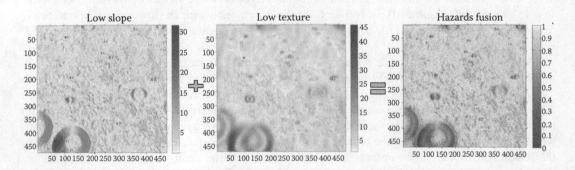

FIGURE 7.6   Example of fusion of low-slope and low-variance texture hazard maps.

## 7.4.2  ILUV—UAV LANDING SITE SELECTION

The ILUV was a technology transfer project to evaluate the applicability of the FIF algorithm for autonomous landing situations of a micro-UAV (Unmanned Aerial Vehicle). This project was developed within the context of the Portuguese Technology Transfer Initiative (PTTI; see http://www.ca3-uninova.org/project_iluv for more details).

The main objective of ILUV was to evaluate the transfer of a space technology—IPSIS technology described previously—to "earth-based" applications. The focus was on safety landing issues of a UAV, in particular for emergency landing situations. In situations caused by motor or battery failure, FIF can be used to select a safe landing site, minimizing the risks to people and property, as well as maximizing the chances of safely recovering the aircraft.

' Transferring the space-oriented site selection algorithm to landing a UAV on Earth required certain adaptions. Mainly, the landing profile is different: A spacecraft will have a near vertical descent on the landing phase, while for a fixed-wing UAV the landing profile is much more horizontal. Environmental factors also come into play—in particular wind direction has a major influence on the UAV landing, whereas it is not an existing criterion on a moon landing, for example. Water features such as rivers and lakes, which pose a major hazard for a UAV, were nonexistent on the IPSIS landing on the moon and Mars. Manmade features such as roads, buildings, electrical pylons, and so forth also had to be considered.

In this case the main requirements were presence of obstacles, adequate size for the landing profile, landing surface as smooth as possible, reduced slope levels, considering new criteria such as wind orientation, and safe distances from civilization—manmade structures, and so forth. Finally four criteria were selected and the respective hazard maps created: illumination, texture, shadow, and reachability. With these hazard maps the first step of FIF was applied and we defined the normalized criteria. Figure 7.7 shows a visualization of the one normalized criteria for a given iteration in the flight.

Step 1 of FIF, the normalization process of constructing the memberships for the four criteria, consists of the following. *Illumination* (or visibility) maps specify whether a certain site (image pixel) is visible or not. Because only visible sites are desired for site selection, this criterion served

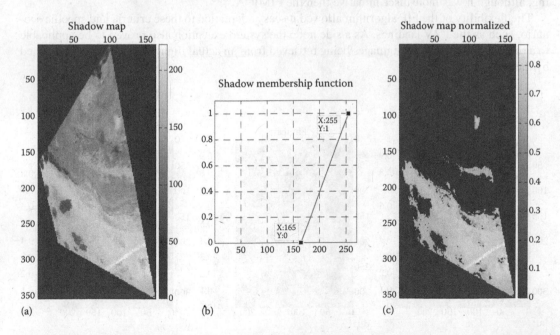

**FIGURE 7.7**  Example of normalization (fuzzification) of criteria shadow. (a) Original hazard map. (b) Membership function topology. (c) Normalized map.

as a "mask"; only pixels with a "1" value (visible) were considered by the algorithm for suitable landing sites and therefore a crisp membership function {0, 1} was defined. *Texture* maps had the information related to the details of the scenario's surface. Here the created membership function (normalization) represents "Low texture," where near zero regions had higher ratings because they were similar to flat regions (better for landing). The *Shadow* map was an actual frame of the video images from the UAV flights. This set of hazard maps had a domain in the interval [0, 255]. To normalize these maps we used a simple open triangle membership function to express the transition of black (0) to white (255), within the interval [0, 1] (see Figure 7.7 for an example of one iteration). For rating purposes, brighter regions were preferred. *Reachability* maps provided the areas in the map where the UAV was able to reach, considering the current flight status and taking into account its flight dynamics. The normalized maps have a range within [0, 1], with 1 the desired value, which is the site with the highest probability of being reached during flight.

In this case study, we skipped steps 2 and 3 of FIF because it was only a proof of concept study and the main goal was to assess the suitability of different aggregation operators for determining the safest place for landing UAVs, that is, step 4 of FIF. Three aggregation operators were used in the region aggregation process (step 4 of FIF) to find the most reliable. The two operators, Uninorm and Fimica, belong to the class of reinforcement operators (Ribeiro et al. 2010) and for comparative purposes we used the most pessimistic intersection operator, the Min operator. It should be noted that Uninorm and Fimica are full reinforcement operators, that is, they benefit "good" values in the criteria, but penalize any "bad" values, whereas the Min operator simply outputs the lowest criteria value. Validation consisted of multiple executions of the algorithm over various test scenarios. Figure 7.8 shows the outcome using the three aggregation operators. These images represent the "safety" values for each pixel if it were considered as a landing site, with values closer to 1.0 considered safer.

It can be observed in Figure 7.8 that the Uninorm operator (a) is much more discriminative in terms of the aggregation process, which is in fact a good property because it favors only very good landing sites. The FIMICA operator (b) was unable to discriminate positive selecting good safety landing sites and provided too many options. The min operator also provided too many options for landing, although it was more discriminative then the FIMICA.

The flexibility of the FIF algorithm allowed its easy adaptation to these criteria and specific scenario, such as the UAV landings. As a side note, the system execution also proved to be applicable to a real-world scenario, with images being retrieved from an actual flight test campaign performed by the UAV.

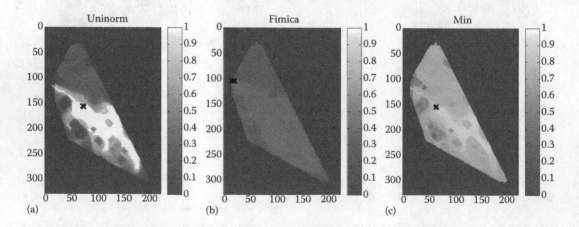

**FIGURE 7.8**  Safety maps for each aggregation operator; from left-to-right: (a) Uninorm, (b) Fimica, and (c) Min.

## 7.5    CONCLUDING REMARKS

In this chapter, a fuzzy information fusion algorithm (FIF) was presented that uses both computational intelligence and multicriteria decision-making techniques for combining heterogeneous data sources into one fused information output. FIF fully addresses the three main challenges of any information fusion process: Data must be numerically comparable; imprecision and uncertainty must be taken into consideration; and a suitable aggregation operator must be selected to combine the information. Moreover, FIF is general enough to be applied to any problem, where the inputs may be from many different sources, as long as they can be modeled as fuzzy sets representing a semantic concept. FIF's versatility also allows customization and tuning of the chosen parameters for expressing relative importance, as well as for the aggregation function (fusion).

The FIF algorithm is divided into two main phases: the data transformation and the data fusion process. The first phase includes steps 1, 2, and 3 (normalization, filtering uncertainty, and relative criteria weighting) and transforms the heterogeneous data sources into homogeneous inputs, which then can be combined together in step 4 (fusion process). The FIF normalizes (step 1) each input data source by using fuzzy membership functions, which can be individually customized to the data. The inputs' uncertainty (lack of confidence or imprecision) can influence a system's behavior. In FIF an uncertainty filtering is available (step 2) for penalizing imprecise and unconfident inputs. The third step of FIF data transformation phase is the assignment of relative importance of the input criteria using penalizing/rewarding weighting functions. After the three steps of the data transformation phase, the resulting transformed inputs are finally combined using aggregation operators with weighting functions to obtain the final output ratings (e.g., a fused rated image).

The algorithm's potential was illustrated with two experimental applications in the aerospace domain, one for selecting safest planet landing sites for spacecraft and another for safe landing of drones. It should also be pointed out that FIF has the potential to be used in diverse fields including selecting suppliers, remote sensing monitoring (i.e., by combing spatial-temporal satellite images), and medical diagnoses. Some preliminary work on those topics, using parts of FIF but in a dynamic decision environment, is already published (Campanella and Ribeiro 2011; Jassbi et al. 2014). Work on other topics is already underway.

## ACKNOWLEDGMENTS

This work was partially supported by ESA contract (ESTEC Contract No, 21744/08/NL/CBI) and IPN contract (PTTI—National Technology Transfer Initiative in Portugal Application for a Feasibility Study-contract with UNINOVA), referring to the illustrative applications.

## REFERENCES

Akhoundi, M. A. A., and E. Valavi. 2010. "Multi-Sensor Fuzzy Data Fusion Using Sensors with Different Characteristics." *The CSI Journal on Computer Science and Engineering*, submitted, arXiv preprint arXiv:1010.6096.

Bedworth, M., and J. O'Brien. 2000. "The Omnibus Model: A New Model of Data Fusion?" *IEEE Aerospace and Electronic Systems Magazine* 15: 30–6.

Bourdarias, C., P. Da-Cunha, R. Drai, L. F. Simões, and R. A. Ribeiro. 2010. "Optimized and Flexible Multi-Criteria Decision Making for Hazard Avoidance." In *Proceedings of the 33rd Annual AAS Rocky Mountain Guidance and Control Conference*. February 5–10, 2010, Breckenridge, CO: American Astronautical Society.

Campanella, G., and R. A. Ribeiro. 2011. "A Framework for Dynamic Multiple-Criteria Decision Making." *Decision Support Systems* 51 (1): 52–60.

Castanedo, F. 2013. "A Review of Data Fusion Techniques." *The Scientific World Journal* 2013: 704504.

Devouassoux, Y., S. Reynaud, and G. Jonniaux. 2008. "Hazard Avoidance Developments for Planetary Exploration." In *Proceedings of the 7th International ESA Conference on Guidance, Navigation & Control Systems*. June 2–5, 2008, Tralee, Ireland.

Dong, J., D. Zhuang, Y. Huang, and J. Fu. 2009. "Advances in Multi-Sensor Data Fusion: Algorithms and Applications." *Sensors* 9: 7771–84.

Elmenreich, W. 2007. "A Review on System Architectures for Sensor Fusion Applications." *Software Technologies for Embedded and Ubiquitous System, Lecture Notes in Computer Science* 4761: 547–59.

Epp, C. D., and T. B. Smith. 2007. "Autonomous Precision Landing and Hazard Detection and Avoidance Technology (ALHAT)." In *2007 IEEE Aerospace Conference*, March 3–10, 2007, Big Sky, Montana: IEEE, pp. 1–7.

Epp, C. D., E. A. Robertson, and T. Brady. 2008. "Autonomous Landing and Hazard Avoidance Technology (ALHAT)." In *2008 IEEE Aerospace Conference*, March 1–8, 2008, Big Sky, Montana: IEEE, pp. 1–7.

Esteban, J., A. Starr, R. Willetts, P. Hannah, and P. Bryanston-Cross. 2005. "A Review of Data Fusion Models and Architectures: Towards Engineering Guidelines." *Neural Computing and Applications* 14 (4): 273–81.

Goodman, I. R., R. P. Mahler, and H. T. Nguyen. 1997. *Mathematics of Data Fusion*. Dordrecht: Kluwer Academic Publishers.

Goshtasby, A. A., and S. Nikolov. 2007. "Image Fusion: Advances in the State of the Art." *Information Fusion* 8 (2): 114–8.

Hall, D. L., and A. K. Garga. 1999. "Pitfalls in Data Fusion (and How to Avoid Them)." In *Proceedings of the 2nd International Conference on Information Fusion—FUSION'99*, July 6–8, 1999, Sunnyvale, CA, vol. 1, pp. 429–36.

Hall, D. L., and J. Llinas. 1997. "An Introduction to Multisensor Data Fusion." *Proceedings of the IEEE* 85 (1): 6–23.

Hall, D. L., and A. N. Steinberg. 2000. "Dirty Secrets in Multisensor Data Fusion." In *National Symposium on Sensor Data Fusion, (NSSDF)*, June 2000, San Antonio, TX, p. 14.

Howard, A. 2002. "A Novel Information Fusion Methodology for Intelligent Terrain Analysis." In *2002 IEEE World Congress on Computational Intelligence. 2002 IEEE International Conference on Fuzzy Systems. FUZZ-IEEE'02. Proceedings (Cat. No. 02CH37291)*, May 12–17, 2002, Honolulu, HI: IEEE. vol. 2, pp. 1472–5.

Hsu, S. L., P. W. Gau, I. L. Wu, and J. H. Jeng. 2009. "Region-Based Image Fusion with Artificial Neural Network." *World Academy of Science, Engineering and Technology* 3 (5): 144–7.

Jassbi, J. J., R. A. Ribeiro, and L. R. Varela. 2014. "Dynamic MCDM with Future Knowledge for Supplier Selection." *Journal of Decision Systems* 23 (3): 232–48.

Johnson, A. E., A. Huertas, R. A. Werner, and J. F. Montgomery. 2008. "Analysis of On-Board Hazard Detection and Avoidance for Safe Lunar Landing." In *2008 IEEE Aerospace Conference*, March 1–8, 2008, Big Sky, Montana: IEEE, pp. 1–9.

Khaleghi, B., A. Khamis, F. O. Karray, and S. N. Razavi. 2013. "Multisensor Data Fusion: A Review of the State-of-the-Art." *Information Fusion* 14 (1): 28–44.

Lee, H., B. Lee, K. Park, and R. Elmasri. 2010. "Fusion Techniques for Reliable Information: A Survey." *International Journal of Digital Content Technology and Its Applications* 4 (2): 74–88.

Li, S., Y. Peng, Y. Lu, L. Zhang, and Y. Liu. 2010. "MCAV/IMU Integrated Navigation for the Powered Descent Phase of Mars EDL." *Advances in Space Research* 46 (5): 557–70.

Llinas, J., and D. L. Hall. 2008. "Multisensor Data Fusion." In *Handbook of Multisensor Data Fusion*, edited by M. E. Liggins, D. L. Hall, and J. Llinas. Boca Raton, FL: CRC Press, pp. 1–14.

Llinas, J., C. Bowman, G. Rogova, A. Steinberg, E. Waltz, W. Frank, and F. White. 2004. "Revisiting the JDL Data Fusion Model II." In *Fusion 2004: 7th International Conference on Information Fusion*, edited by P. Svensson, and J. Schubert, June 28 to July 1, 2004, Stockholm, Sweden, vol. 2, pp. 1218–30.

Luo, R. C., and M. G. Kay. 1988. "Multisensor Integration and Fusion: Issues and Approaches." *Proceedings of 1988 Orlando Technical Symposium*. April 4, 1988, Orlando, FL, pp. 42–9.

Macii, D., A. Boni, M. De Cecco, and D. Petri. 2008. "Tutorial 14: Multisensor Data Fusion." *Instrumentation Measurement Magazine, IEEE* 11 (3): 24–33.

Maître, H., and I. Bloch. 1997. "Image Fusion." *Vistas in Astronomy* 41 (3): 329–35.

Manyika, J., and H. F. Durrant-Whyte. 1994. *Data Fusion and Sensor Management: A Decentralized Information-Theoretic Approach*. Hemel Hempstead: Ellis Horwood.

O'Brien, M. A., and J. M. Irvine. 2004. "Information Fusion for Feature Extraction and the Development of Geospatial Information." In *Fusion 2004: Seventh International Conference on Information Fusion*, June 28 to July 1, 2004, Stockholm, Sweden, pp. 976–82.

Pais, T. C., R. A. Ribeiro, and L. F. Simões. 2010. "Uncertainty in Dynamically Changing Input Data." In *Computational Intelligence in Complex Decision Systems*, edited by D. Ruan, pp. 47–66. Paris: Atlantis Press.

Pereira, R. A. M., and R. A. Ribeiro. 2003. "Aggregation with Generalized Mixture Operators Using Weighting Functions." *Fuzzy Sets and Systems* 137 (1): 43–58.

Piella, G. 2003. "A General Framework for Multiresolution Image Fusion: From Pixels to Regions." *Information Fusion* 4: 259–80.

Ribeiro, R. A., and R. A. M. Pereira. 2003. "Generalized Mixture Operators Using Weighting Functions: A Comparative Study with WA and OWA." *European Journal of Operational Research* 145 (2): 329–42.

Ribeiro, R. A., T. C. Pais, and L. F. Simões. 2010. "Benefits of Full-Reinforcement Operators for Spacecraft Target Landing." In *Preferences and Decisions*, edited by S. Greco, R. A. M. Pereira, M. Squillante, R. R. Yager, and J. Kacprzyk, 257: 353–67. Studies in Fuzziness and Soft Computing. Berlin, Heidelberg: Springer.

Ribeiro, R. A., A. Falcão, A. Mora, and J. M. Fonseca. 2014. "FIF: A Fuzzy Information Fusion Algorithm Based on Multi-Criteria Decision Making." *Knowledge-Based Systems* 58: 23–32.

Rogata, P., E. Di Sotto, F. Câmara, A. Caramagno, J. M. Rebordão, B. Correia, P. Duarte, and S. Mancuso. 2007. "Design and Performance Assessment of Hazard Avoidance Techniques for Vision-Based Landing." *Acta Astronautica* 61 (1–6): 63–77.

Ross, T. J. 2005. *Fuzzy Logic with Engineering Applications*, 2nd Edition. New York: John Wiley & Sons.

Serrano, N. 2006. "A Bayesian Framework for Landing Site Selection during Autonomous Spacecraft Descent." In *2006 IEEE/RSJ International Conference on Intelligent Robots and Systems*, October 9–15, 2006, Beijing, China: IEEE, pp. 5112–17.

Serrano, N., and A. Quivers. 2008. "Evidential Terrain Safety Assessment for an Autonomous Planetary Lander." In *International Symposium on Artificial Intelligence, Robotics and Automation in Space*, September 4–6, 2012, Turin, Italy.

Serrano, N., and H. Seraji. 2007. "Landing Site Selection Using Fuzzy Rule-Based Reasoning." In *Proceedings 2007 IEEE International Conference on Robotics and Automation*, April 10–14, 2007, Rome, Italy: IEEE, pp. 4899–904.

Shuang, L., and Z. Liu. 2009. "Autonomous Navigation and Guidance Scheme for Precise and Safe Planetary Landing." *Aircraft Engineering and Aerospace Technology* 81 (6): 516–21.

Simoes, L. F., C. Bourdarias, and R. A. Ribeiro. 2012. "Real-Time Planetary Landing Site Selection—A Non-Exhaustive Approach." *Acta Futura* 5: 39–52.

Smith, M. I., and J. P. Heather. 2005. "A Review of Image Fusion Technology in 2005." In *SPIE 5782—Thermosense XXVII*, edited by R. S. Harmon, J. T. Broach, and J. H. Holloway, Jr., Orlando, FL, March 28, 2005, vol. 5782, pp. 29–45.

Steinberg, A. N., and C. L. Bowman. 2008. "Revisions to the JDL Data Fusion Model." In *Handbook of Multisensor Data Fusion*, edited by M. E. Liggins, D. L. Hall, and J. Llinas, Boca Raton, FL: CRC Press, pp. 45–67.

Steinberg, A. N., C. L. Bowman, and F. E. White. 1999. "Revisions to the JDL Data Fusion Model." *Proceedings of SPIE—Sensor Fusion: Architectures, Algorithms, and Applications III*, April 5, 1999, Orlando, FL, vol. 3719, pp. 430–41.

Veloso, M., C. Bentos, and F. Câmara Pereira. 2009. *Multi-Sensor Data Fusion on Intelligent Transport Systems*. MIT Portugal Transportation Systems Working Paper Series.

Waltz, E. L., and J. Llinas. 1990. *OLD Multisensor Data Fusion*. Norwood, MA: Artech House.

White, F. E. 1991. "Data Fusion Lexicon." In *The Data Fusion Subpanel of the Joint Directors of Laboratories, Technical Panel for C3*, 15: 15.

Zegras, C., F. Pereira, A. Amey, M. Veloso, L. Liu, C. Bento, and A. Biderman. 2008. "Data Fusion for Travel Demand Management: State of the Practice & Prospects." In *4th International Symposium on Travel Demand Management, (TDM'08)*, July 16–18, 2008, Vienna, Austria.

# 8 Distributed Detection and Data Fusion with Heterogeneous Sensors

Satish G. Iyengar, Hao He, Arun Subramanian,
Ruixin Niu, Pramod K. Varshney, and Thyagaraju Damarla

## CONTENTS

## 8.1  INTRODUCTION

Our lives today are constantly aided and enriched by various types of sensors, which are deployed ubiquitously. Multimodal or heterogeneous signal processing refers to the joint analyses and fusion of data from a variety of sensors (e.g., acoustic, seismic, magnetic, video, and infrared) to solve a common inference problem. Such a system offers several advantages and new possibilities for system improvement in many practical applications. For example, speech perception is known to be a bimodal process that involves both auditory and visual inputs [1]. Visual cues such as lip movements of the speaker have shown to improve speech intelligibility significantly, especially in environments where the auditory signal is compromised. In addition, much useful information can be extracted from the joint analysis of the different modalities. The use of multiple modalities may provide complementary information and thus increase the accuracy of the overall decision-making process, for example, fusion of "functional" images from positron emission tomography (PET) with "structural" data from magnetic resonance imaging (MRI).

In each of the preceding applications, sensors are used to make complex inferences about an underlying observed process or *phenomenon*. This is similar, in many ways, to how we, as humans, combine or *fuse* different streams of information originating from our sense organs. Whereas humans have a natural ability to handle and process such diverse streams of information, teaching machines to do the same remains a challenge. Over the past two decades, the field of information fusion has been extensively studied and researched. Although a rich body of literature exists, the increasing complexity of systems as well as the vast diversity of applications require constant revision to existing technologies and continued research in this area. Specifically, for the fusion of heterogeneous information, the data provided by different sensors are generally incommensurate and are characterized by considerably different features. Thus, it is unclear how one would weight each modality in the fusion process.

For many of these applications, the sensors deployed are such that some amount of local processing or sensor-level quantization or compression of data is performed, as the output of a local decision-making process. Eventually these decisions are combined at a central unit called the fusion center, which can provide a unified decision. This method of inference is also called distributed detection, and is described in more detail in Section 8.2. Coarse sensor-level quantization of data is a commonly employed technique used to overcome bandwidth and energy limitations in wireless sensor networks.

When sensors observe a common phenomenon, their measurements often exhibit spatial statistical dependence. This dependence may emerge in spite of sensors observing the phenomenon of interest as independent observers. For example, signals may be modeled as being embedded in additive *correlated* noise. The issue of statistical dependence is even more complex when different sensor modalities are used and linear correlation models are often insufficient to characterize statistical dependence. For the case in which acoustic and video modalities simultaneously observe a person talking, features extracted from voice data and image sequences of lip movements are statistically dependent; this dependence is induced by the phenomenon. Distributed detection using statistically dependent heterogeneous sensor data is discussed in Section 8.3. This chapter employs copula theory to address the issue of modeling heterogeneous dependence. Data modeling issues are discussed in more detail in Section 8.3.1. Distributed detection using statistically dependent quantized data is known to be computationally intractable. A computationally feasible approximation is possible using copula theory. We discuss the theory and application of distributed detection using dependent quantized data in Sections 8.3.2 and 8.3.3. Concluding remarks are provided in Section 8.4.

## 8.2 PRIMER ON DISTRIBUTED DETECTION THEORY

The problem of signal detection can be formulated as a binary hypothesis testing problem where the hypotheses $H_0$ and $H_1$ represent the absence and presence of a signal, respectively. Assume that $N$ sensors are deployed in the region of interest (ROI) to collect observations $Z_n$, for $n = 1, ..., N$. In traditional centralized detection, each sensor node transmits a sequence of $L$ observations to a fusion center for deciding the true state of nature. However, owing to limited transmission resources, including channel bandwidth, and energy, many applications such as wireless sensor networks require local compression/processing of the raw observations before transmission. In a distributed decision-making system, various forms of sensor-level compression, $u_n = \gamma_n(\mathbf{z}_n)$, can be employed. For example, the local sensor output can be a hard decision so that $\gamma_n(\mathbf{z}_n) \in \{0, 1\}$ or a soft decision, where $\gamma_n(\mathbf{z}_n)$ can take multiple values, as in a multilevel quantizer. Based on the compressed data $\mathbf{u} = [u_1, ..., u_n]$, the fusion center makes a global decision $u_0 = \gamma_0(\mathbf{u})$ that either favors $H_1$ ($u_0 = 1$) or $H_0$ ($u_0 = 0$).

The distributed network described in the preceding text is a parallel network (see Figure 8.1a). Another frequently used distributed detection topology, shown in Figure 8.1b, is serial/tandem topology [2–5]. In this system, the first sensor makes a decision based on its own observation and transmits to the second sensor. Combining the incoming decision and its own observation, the second sensor computes a decision and transmits it to the next sensor. Such a process continues until the last sensor in the network makes the final decision. A variety of other additional detection network topologies can be envisaged, such as a tree topology [6–8]. Among all the topologies

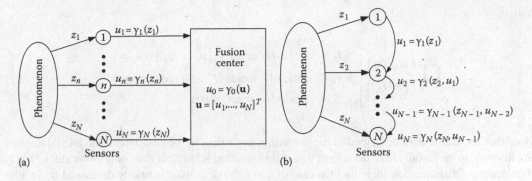

**FIGURE 8.1** Typical sensor network topologies. (a) Parallel configuration and (b) serial configuration.

considered in the literature, the parallel network topology has received the most attention, which is also what we focus on in this chapter.

From the signal processing perspective, two different problems need to be considered for the distributed detection system: (1) the design of local sensor signal processing rules, $[\gamma_1, \ldots, \gamma_N]$ and (2) the design of $\gamma_0$, the decision rule at the fusion center, often referred to as the fusion rule. In the most general setting, the design of the set of decision rules $\Gamma = [\gamma_1, \ldots, \gamma_N, \gamma_0]$, is an NP-complete problem [9–11]. However, it becomes tractable by assuming conditionally independent sensor observations, that is,

$$f\left(z_1, \ldots, z_N \middle| H_i\right) = \prod_{n=1}^{N} f_n\left(z_n \middle| H_i\right), \quad \forall i = 0,1$$

where $f_n(\cdot|H_i)$ represents the probability density function (PDF) of sensor $n$ under hypothesis $H_i$. Some well established results are reviewed next under the assumption of independence.

## 8.2.1 Conditionally Independent Observations

A common framework for solving decision problems is to maximize the probability of detection for a predetermined constraint on the probability of false alarm. This is known as the Neyman–Pearson framework of hypothesis testing.* It can be formulated more precisely as follows. Find optimal decision rules $\Gamma$ that maximize the probability of detection $P_D = P(u_0 = 1|H_1)$ given the false alarm constraint $P_F = P(u_0 = 1|H_0) \le \alpha$. For conditionally independent sensor observations the local sensor rules and fusion rule are likelihood ratio tests [12]. Recall that $u_n = \gamma_n(z_n)$ for $n = 1, 2, \ldots, N$, and, therefore,

$$\frac{f_n\left(z_n \middle| H_1\right)}{f_n\left(z_n \middle| H_0\right)} \begin{cases} > t_n, & \text{then } u_n = 1 \\ = t_n, & \text{then } u_n = 1 \text{ with probability } \epsilon_n \\ < t_n, & \text{then } u_n = 0 \end{cases} \tag{8.1}$$

---

* An alternative approach to developing decision rules, the Bayesian approach, considers that each hypothesis is a random entity and, hence, has a probability mass associated with it. The theory of distributed detection, for conditionally independent observations, using Bayesian decision rules has been derived in Ref. [4].

and

$$\prod_{n=1}^{N} \frac{P(u_n|H_1)}{P(u_n|H_0)} \begin{cases} > \eta, & \text{decide } H_1 \text{ or set } \gamma_0(u_1,\dots,u_N) = u_0 = 1 \\ = \eta, & \text{randomly decide } H_1 \text{ with probability } \epsilon \\ < \eta, & \text{decide } H_0 \text{ or set } u_0 = 0 \end{cases} \qquad (8.2)$$

Note that the framework described in the preceding text refers to the case where the local detectors are allowed to make only hard decisions, that is, in Equation 8.1, $u_n$ can take only two values, 0 or 1. The design of fusion rules, including the case of soft local decisions, is briefly discussed in Section 8.2.1.1.

Although the local and fusion center test statistics are based on likelihood ratios, determining the thresholds, $\{t_n\}_{n=1}^{N}$ and $\eta$, is a difficult problem as they are coupled with each other and affect the system performance in an interdependent manner. Therefore, the thresholds are usually obtained using an iterative person-by-person optimization (PBPO) approach, where each threshold of the sensor is optimized assuming fixed decision rules at all other sensors and the fusion center. The PBPO approach can only guarantee a locally optimal solution and not necessarily a solution that is globally optimal.

Simpler solutions can be achieved by assuming identical local decision rules. Interestingly, the use of identical decision rules was shown to be optimal in terms of the error exponent in the asymptotic regime (i.e., as the number of sensors $N \to \infty$) [13]. In Ref. [14], the exact asymptotics of the minimum error probabilities achieved by the optimal parallel fusion network and the system obtained by imposing the identical decision rule constraint was investigated. It was shown analytically that the restriction of identical decision rules leads to little or no loss of performance. Chamberland and Veeravalli [15] have shown that, in the asymptotic regime, binary sensors are optimal if there exists a binary quantization function whose Chernoff information exceeds half of the information contained in an unquantized observation. This requirement is fulfilled in many practical applications [16] such as the problem of detecting deterministic signals in a Gaussian noise and the problem of detecting fluctuating signals in a Gaussian noise using a square-law detector. In these scenarios, the gain offered by having more sensor nodes outperforms the benefits of getting detailed information from each sensor.

### 8.2.1.1 Design of Fusion Rules

In the case when local sensor probabilities of false alarms and detections are known, the optimal decision rule at the fusion center is given by the Chair–Varshney fusion rule [17],

$$\sum_{n=1}^{N} \left[ u_n \log \frac{P_{d,n}}{P_{f,n}} + (1-u_n)\log \frac{1-P_{d,n}}{1-P_{f,n}} \right] \underset{u_0=0}{\overset{u_0=1}{\gtrless}} \log \eta, \qquad (8.3)$$

where $P_{f,n}$ and $P_{d,n}$ are the local false alarm and detection probabilities for sensor $n$. Thus, the optimum fusion rule is essentially a weighted sum of the incoming local decisions, $u_n$, and compares this weighted sum with the threshold, $\eta$. The weights and $\eta$ are determined by the local probabilities of detection and false alarm. If the local decisions have the same statistics, that is, $P_{f,n} = P_{f,m}$, and $P_{d,n} = P_{d,m}$, for $n \neq m$, the Chair–Varshney fusion rule reduces to a majority ($K$-out-of-$N$) rule.

When $\gamma_n$ are multilevel quantizers, the optimum fusion rule should be derived over the $M$-dimensional observation space of local decisions, as $u_n \in \{0, 1, \dots, M-1\}$. This fusion rule has the form

$$\sum_{i=0}^{M-1} \sum_{S_i} \log \frac{\beta_n^i}{\alpha_n^i} \underset{u_0=0}{\overset{u_0=1}{\gtrless}} \log \eta, \qquad (8.4)$$

where $\alpha_n^i = P\left(u_n = i \middle| H_0\right)$, $\beta_n^i = P\left(u_n = i \middle| H_1\right)$ for quantization levels $i = 0, 1, \ldots, M - 1$ and $\mathcal{S}_i$ is the set local decisions $u_n$ that are equal to $i$. This is a generalization of the case where sensors make hard decisions, that is, $M = 2$.

## 8.2.2 CORRELATED OBSERVATIONS

Optimal schemes for distributed detection with linearly correlated observations have been investigated over the past two decades. It has been shown that the likelihood ratio-based quantizer, which was optimal under the assumption of conditional independence, is no longer optimal when correlation is taken into account. Examples of the consequent loss in performance are provided by Aalo and Viswanathan [18]. In fact, earlier work by Tsitsiklis and Athans [11] has shown that the distributed detection problem with dependent observations is NP-complete. Willett et al. [19] study the problem of distributed detection of a mean shift in correlated Gaussian noise and establish how the nature of correlation affects the optimum fusion rule. They conclude that even for a simple two-sensor and linear correlation formulation the distributed detection problem "exhibits apparently very complicated behavior." For this mean shift in correlated Gaussian noise problem, local quantizers designed using the likelihood ratio test (LRT) are, in general, not optimal. They show that determining the parameter regions where this optimality may hold is itself a challenging task: Although the optimality of the LRT can be determined for certain parameter regions, the problem is mostly intractable in other regions. Chen et al. [20] have recently proposed a new framework for this problem. They introduce a hidden variable that induces conditional independence among the sensor observations and thus unifies distributed detection with dependent or independent observations.

The problem is usually simplified by constraining the local sensors to be binary quantizers: Drakopolous and Lee [21] derive the fusion rule for distributed detection under dependence by assuming that the correlation coefficients between the sensor decisions are known and local sensor thresholds are given, whereas Kam et al. [22] employ another approach, namely the Bahadur–Lazarsfeld expansion of PDFs to derive the optimal fusion rule. It was, however, assumed that the joint distribution of sensor observations was completely known. Recently, we have considered a scenario in which the dependence structure, and hence the joint distribution between sensor observations, may be unknown [23]. Such problems are typical of networks with heterogeneous sensors, that is, sensors with disparate sensing modalities.* Our findings in this area of distributed detection with heterogeneous sensors are presented in the next section.

## 8.3 DISTRIBUTED DETECTION WITH HETEROGENEOUS SENSORS

Consider, again, the parallel distributed detection system of Figure 8.1a, however, now with sensors of disparate sensing modalities. Observations at each sensor $n$ are assumed to be independent and identically distributed (i.i.d.) over time with PDFs $f_n(z_n; \psi_n)$ and $g_n(z_n; \lambda_n)$ under $H_1$ and $H_0$ respectively, where $\psi_n$ and $\lambda_n$ are distributional parameters. The marginal PDFs are assumed to be well specified under both hypotheses (see Definition 8.1).

**Definition 8.1:** *Well-Specified Model (White 1994 [24])*

A parametric model $\{f(x; \Theta)\}$ is well specified for a random variable $X$ if there exists a unique $\theta' \in \Theta$ such that $f(x; \theta') \in \{f(x; \Theta)\}$ corresponds to the true density of $X$. Otherwise, $\{f(x; \Theta)\}$ is said to be *misspecified* for $X$. ∎

---

* For example, it is not immediately clear how one could model the complex relationship between observations of an audio and a video sensor monitoring a common region of interest.

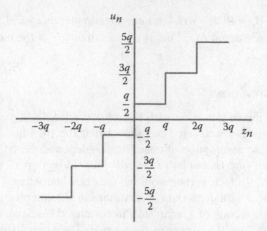

**FIGURE 8.2** Input-output transfer function of a uniform scalar quantizer. (From Iyengar S. G. et al., *IEEE Transactions on Signal Processing*, 60(9):4888–4897, 2012.)

However, no knowledge is assumed regarding the dependence structure between the heterogeneous data streams. Statistical dependence between heterogeneous sensor observations is approximated using copula functions (see Section 8.3.1.2). Copulas are more general descriptors of statistical dependence and possess all the ingredients necessary to model heterogeneous data [25].

Sensor observations are further passed through uniform multilevel quantizers (Figure 8.2) before their transmission to a remotely located fusion center. Quantizer output, during any time interval $1 \leq l \leq L$, can be given as

$$u_{nl} = \gamma_n(z_{nl}) = \begin{cases} -m_n q_n - \dfrac{q_n}{2}, & z_{nl} < -m_n q_n, \\[2ex] q_n \left\lfloor \dfrac{z_{nl}}{q_n} \right\rfloor + \dfrac{q_n}{2}, & -m_n q_n < z_{nl} \leq m_n q_n, \\[2ex] m_n q_n + \dfrac{q_n}{2}, & z_{nl} \geq m_n q_n, \end{cases}$$ (8.5)

where, $q_n$ and $2(m_n + 1)$ correspond to the quantizer step size and the number of quantization levels respectively, at sensor $n$. Further, $\lfloor x \rfloor$ stands for the floor operation that denotes an integer smaller than or equal to $x$. The quantized value at sensor $n$ can be represented with an integer $i_n = -m_n - 1$, $-m_n, \ldots, m_n$. In the analysis to follow, it is assumed that the dynamic range of the (analog) signal input to the quantizer is well within the lower and upper limits of the quantizer so that there are no quantizer saturation errors. The quantized observations thus received at the fusion center are used to estimate the unknown model parameters, and a generalized likelihood ratio test (GLRT)-based fusion rule is employed for global decision making.

### 8.3.1 MODELING HETEROGENEOUS DATA

The first and foremost challenge when designing heterogeneous fusion systems is to adequately model the joint distribution of sensor observations, **Z**. Heterogeneous data streams are not always commensurate. The physics governing each modality may be different and so may be their dimensionality, support, and sampling rates during data acquisition. These differences require formulation of multivariate statistical models that allow for disparate marginal distribution functions. A formal definition for heterogeneous observations follows.

**Definition 8.2:** (*Heterogeneous Random Vector*)

A random vector, $\mathbf{Z} = \left\{ Z_n \right\}_{n=1}^{N}$, governing the joint statistics of an $N$-variate dataset is termed as heterogeneous if the marginals, $Z_n$, follow nonidentical distributions.*  ∎

Further, in most applications, the signals share a common origin and thus may exhibit statistical dependence. Consider, for example, an acoustic sensor and a video camera monitoring a region for trespassers. Presence of a target may result in an increase in both the acoustic energy and the pixel intensities of the images acquired by the video camera. Both sensors provide information about the same event (and hence are statistically dependent) but in different *domains*. Thus, the following two requirements are expected of a *good* model for heterogeneous random variables:

1. The model should allow for disparate marginal distribution functions.
2. The model must also account for any statistical dependence that may be present between the disparate marginal distribution functions.

Section 8.3.1.2 reviews the theory of copulas and shows how copula densities possess properties that allow us to satisfy the preceding requirements. We first provide a brief discussion on some commonly adopted models. Although these approaches are attractive because of their analytical tractability, each of them leads to suboptimal solutions. Assessment of these shortcomings will motivate our copula based solution.

### 8.3.1.1  Commonly Adopted Models for Heterogeneous Data

Often statistical independence is assumed when dealing with heterogeneous datasets for mathematical tractability. That is, the joint PDF, $f(\mathbf{z}; \psi)$, is approximated as a product of the marginal densities,

$$\hat{f}\left(\mathbf{z}; \psi_p\right) = f_p\left(\mathbf{z}; \psi_p\right) = \prod_{n=1}^{N} f_n\left(z_n; \psi_n \in \Psi_n\right). \tag{8.6}$$

As is evident, the product model allows for disparate marginals but completely ignores the dependence across the disparate marginals.

Alternatively, one may model $\mathbf{Z}$ using a multivariate Gaussian PDF with covariance matrix $\Sigma$. Two limitations plague this approach: (1) Marginals are constrained to follow Gaussian distributions and (2) $\Sigma$ characterizes only the *linear* relationship, and is thus a weak measure of dependence [26]. Intermodal dependence and interactions are usually much more complex (than just being linear). Copula functions (see Section 8.3.1.2) are more general descriptors of dependence, and can accurately describe the functional relationship between multiple random variables.

### 8.3.1.2  Joint PDF Approximation Using Copula Theory

Copula functions *couple* multivariate joint distribution functions to their component marginal distribution functions [27]. We begin with the definition of a copula function.

---

* The definition is, of course, inclusive of the special case when the marginals are identically distributed. It also encompasses the case when the signals, although sharing a common modality (e.g., two acoustic sources or two video sources), may exhibit different statistics (due to different locations, signal strengths, etc.). Hence, we prefer the term *heterogeneous* in place of *multimodal* as used by some researchers.

**Definition 8.3**

A function $C : [0, 1]^N \to [0, 1]$ is an $N$-dimensional copula if $C$ is a joint cumulative distribution function (CDF) of an $N$-dimensional random vector on the unit cube $[0, 1]^N$ with uniform marginals [27–29]. ∎

The following theorem by Sklar is central to the statistical theory of copulas.

**Theorem 8.1: (*Sklar's Theorem*)**

Let $F$ be an $N$-dimensional CDF with continuous marginal CDFs $F_1, F_2, \ldots, F_N$. Then there exists a unique copula C such that for all $z_1, z_2, \ldots, z_n$ in $[-\infty, \infty]$,

$$F(z_1, z_2, \ldots, z_N) = C(F_1(z_1), F_2(z_2), \ldots, F_N(z_N)). \tag{8.7}$$

∎

Note that the copula function $C(u_1, u_2, \ldots, u_N)$ is a CDF with uniform marginals as $U_n = F_n(Z_n) \sim \mathcal{U}(0,1)$ (by probability integral transform).

Theorem 8.1 also admits the following converse, which is especially useful in practice when the true distribution $F$ (and hence the true copula $C$) is unknown. It allows one to construct a statistical model by considering separately the univariate behavior of the components of a random vector and the dependence structure defined by some copula function, say, $K$.

**Theorem 8.2**

If $F_1, F_2, \ldots, F_N$ are univariate marginal CDFs and if $K$ is an $N$-dimensional copula, then the function $\Xi : \mathbb{R}^N \to [0,1]$,

$$\Xi(z_1, \ldots, z_N) = K(F_1(z_1), \ldots, F_N(z_N)), \tag{8.8}$$

is a valid, $N$-variate CDF with marginals $F_1, F_2, \ldots, F_N$. ∎

The joint density can now be obtained by differentiating both sides of Equation 8.8 to obtain

$$\hat{f}(\mathbf{z}; \psi) = f_k(\mathbf{z}; \psi) = \underbrace{\left( \prod_{n=1}^{N} f_n(z_n; \psi_n) \right)}_{f_p(\mathbf{z}; \psi_p)} \underbrace{k\left( F_1(z_1; \psi_1), \ldots, F_N(z_N; \psi_N); \psi_d \right)}_{\text{dependence model}}, \tag{8.9}$$

where $k$ is called the copula density given by

$$k(\cdot) = \frac{\partial^N}{\partial u_1 \cdots \partial u_N} \dot{K}(u_1, u_2, \ldots, u_N), \tag{8.10}$$

with $u_n = F_n(z_n; \psi_n)$. Using Equation 8.9, we can construct a joint density function with specified marginal densities. This is well-suited for modeling heterogeneous random vectors where a different distribution might be needed to model each marginal $Z_n$.

A variety of copula functions, with different dependence properties, have been described in the literature [27]. Some examples are the Gaussian, Student's $t$, Clayton, Frank, Gumbel, and independence or product copula functions. Although there is a wide range of flexible copulas available for constructing models in two dimensions, multivariate copula modeling can be more challenging. Subramanian et al. [30,31] have addressed the issue of multisensor dependence modeling for detection applications through the use of vines [29].

A vine is a nested set of trees, where the edges of the $k$th tree are the nodes of the $(k + 1)$th tree, and each tree has a maximum number of edges. The trees are called dependence vines when they are used to encode dependence structures in multivariate distributions. There are several vine architectures possible; the most tractable methodologies use a graphical model that focuses on pairwise interactions of dependent variables using *regular vines*. Because pairwise interactions are used, bivariate copulas are used as building blocks.

Two types of regular vines have been analyzed in the literature in the context of expressing multivariate copulas: the canonical vines or C-vines and the drawable vines or D-vines. We use the D-vine architecture as an example because the D-vines easy to use for modeling; the C-vines are useful when it is known that a particular sensor plays a key role in governing intersensor dependencies. A vine, regular vine, and D-vine are formally defined below.

### Definition 8.4

$\mathcal{V}$ is a vine on $K$ elements if

1. $\mathcal{V} = (T_1, \ldots T_{K-1})$
- 2. $T_1$ is a connected tree with nodes $N_1 = \{1, \ldots, K\}$ and edges $E_1$; $T_k$ is a connected tree with nodes $N_k = E_{k-1}$ for $k = 2, 3, \ldots, K - 1$.
   $\mathcal{V}$ is a **regular vine** on $K$ elements if it satisfies the additional proximity condition,
3. For $k = 2, \ldots, K - 1$, if $a$ and $b$ are nodes of $T_k$ connected by an edge in $T_k$, where $a = \{a_1, a_2\}$ and $b = \{b_1, b_2\}$ are edges in $T_{k-1}$, then exactly one of $a_1, a_2$ equals one of $b_1, b_2$. ∎

A regular vine is called a **D-vine** if each node in $T_1$ has a degree of at most 2. A D-vine over four elements is shown in Figure 8.3. When a four-variate joint distribution is defined over this vine, we are essentially establishing a hierarchical, pairwise dependency relation, which can be expressed through copulas. Each tree in the vine represents a decomposition obtained by successively conditioning the variables. It can be shown that the joint PDF for a four-variable D-vine is [32]

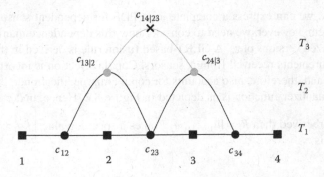

**FIGURE 8.3** D-vine over 4 elements. Labels indicate the copula density evaluated at each tree in the vine. (From Subramanian, A. et al., Fusion for the detection of dependent signals using multivariate copulas. In *Proceedings of the 14th International Conference on Information Fusion (FUSION)*, pp. 1–8, 2011.)

$$
\begin{aligned}
f(z_1, z_2, z_3, z_4) = {} & f_1(z_1) f_2(z_2) f_3(z_3) f_4(z_4) \cdot c_{12}\big(F_1(z_1), F_2(z_2)\big) \\
& \cdot c_{23}\big(F_2(z_2), F_3(z_3)\big) \cdot c_{34}\big(F_3(z_3), F_4(z_4)\big) \\
& \cdot c_{13|2}\Big(F_{1|2}\big(z_1|z_2\big), F_{3|2}\big(z_3|z_2\big)\Big) \cdot c_{24|3}\Big(F_{2|3}\big(z_2|z_3\big), F_{4|3}\big(z_4|z_3\big)\Big) \\
& \cdot c_{14|23}\Big(F_{1|23}\big(z_1|z_2,\,z_3\big), F_{4|23}\big(z_4|z_2,\,z_3\big)\Big)
\end{aligned}
\tag{8.11}
$$

The labels in Figure 8.3 indicate the copula density evaluated at each tree in the vine. The density of an $N$-dimensional distribution expressed in terms of a D-vine decomposition is given by Bedford and Cooke [33],

$$
\prod_{n=1}^{N} f_n(z_n) \prod_{j=1}^{N-1} \prod_{k=1}^{N-j} c_{j,j+k|\bar{j}}\Big(F_{j|\bar{j}}\big(z_j\big|\mathbf{z}_{\bar{j}}\big),\ F_{j+k|\bar{j}}\big(z_{i+j}\big|\mathbf{z}_{\bar{j}}\big)\Big)
\tag{8.12}
$$

where $\bar{j} = \{j+1,\ldots,j+k-1\}$ and $\mathbf{z}_{\bar{j}} = \big[z_{j+1},\ldots,z_{j+k-1}\big]^T$.

It is clear that different copula functions would associate the same set of marginals to different joint distributions. With such one-to-many possible mappings of the marginals, an important question that arises is: *How does one choose k from a finite set (say $\mathcal{A}_k$) of copula densities?* This is essentially a model selection problem.

Approaches for copula selection have typically focused on data fitting, that is, select the copula that best fits the given data. We refer readers to Refs. [34,35] for a review on goodness-of-fit tests for copulas. Recently, we have considered a more application-specific paradigm, and have developed copula selection methods that directly optimize performance metrics such as the area under the receiver operating characteristic curve (AUC) [25]. These methods, however, assume the availability of training data so that copula selection can be carried out offline. However, offline selection of copulas may not be possible for some applications. For such applications, we propose to embed the copula selection step in the (misspecified) GLRT-based fusion rule [25].

## 8.3.2 The (Misspecified) GLRT-Based Fusion Rule

Using copula theory, we can express a complete joint PDF for dependent sensor observations, $z_n$. For the fusion problem, however, we need to consider how this dependence manifests itself among all $u_n$, that is, quantized versions of $z_n$. A GLRT-based fusion rule is derived in this section for fusing quantized measurements received from $N$ sensors. Copula selection is incorporated within the GLRT formulation, and, therefore, also accounts for copula misspecification.

Recall that the quantizer function is as depicted in Figure 8.2. Hence, under hypothesis $H_1$, the probability that the received data $R_l = [u_{1l}, \ldots, u_{Nl}]$ takes a specific value $\left[i_1 q_1 + \dfrac{q_1}{2}, \ldots, i_N q_N + \dfrac{q_N}{2}\right]$ at the time instant $l$ is

$$
P_{i_1 \ldots i_N} = \int_{i_1 q_1}^{(i_1+1) q_1} \ldots \int_{i_N q_N}^{(i_N+1) q_N} f(z_1, \ldots, z_N)\, dz_N \ldots dz_1,
\tag{8.13}
$$

where $f(z_1,\ldots,z_N)$ is the true but unknown joint PDF of unquantized sensor observations under $H_1$. By approximating the dependence structure using a copula density function, $k_1(\cdot; \psi_d)$, contained in some set of valid copula densities $\mathcal{A}_k$, we have

$$\hat{P}_{i_1\ldots i_N}(\psi) = \int_{i_1 q_1}^{(i_1+1)q_1} \cdots \int_{i_N q_N}^{(i_N+1)q_N} \left( \prod_{n=1}^{N} f_n(z_n; \psi_n) \right)$$
$$\times k_1\left( F_1(z_1; \psi_1),\ldots,F_N(z_N; \psi_N); \psi_d \right) dz_N \ldots dz_1, \tag{8.14}$$

where $\psi = (\psi_1,\ldots,\psi_N,\psi_d)^T \in \Psi \subset \mathbb{R}^a$ is the $a$-dimensional unknown parameter vector that will be estimated from the received data and $F_n(\cdot; \psi_n)$ is the CDF of $Z_n$ under hypothesis $H_1$. $\hat{P}_{i_1\ldots i_N}(\psi)$ depends on $k_1(\cdot)$, but the relationship is not made explicit for notational convenience.

The likelihood function of the data $R_l$ under hypothesis $H_1$ can be written as

$$\hat{P}(R_l; \psi, H_1) = \prod_{i_1} \cdots \prod_{i_N} \left[ \hat{P}_{i_1\ldots i_N}(\psi) \right]^{\delta\left( u_{1l} - i_1 q_1 - \frac{q_1}{2},\ldots,u_{Nl} - i_N q_N - \frac{q_N}{2} \right)}, \tag{8.15}$$

where $\delta(\cdot)$ is the $N$-dimensional Kronecker-delta function defined as

$$\delta(x_1,\ldots,x_N) = \begin{cases} 1, & x_1 = \cdots = x_N = 0 \\ 0, & \text{otherwise.} \end{cases} \tag{8.16}$$

Similarly, the likelihood function of $R_l$ under $H_0$ can be derived as

$$\hat{P}(R_l; \lambda, H_0) = \prod_{i_1} \cdots \prod_{i_N} \left[ \hat{Q}_{i_1\ldots i_2}(\lambda) \right]^{\delta\left( u_{1l} - i_1 q_1 - \frac{q_1}{2},\ldots,u_{Nl} - i_N q_N - \frac{q_N}{2} \right)}, \tag{8.17}$$

where

$$\hat{Q}_{i_1\ldots i_2}(\lambda) = \int_{i_1 q_1}^{(i_1+1)q_1} \cdots \int_{i_N q_N}^{(i_N+1)q_N} \left( \prod_{n=1}^{N} g_n(z_n; \lambda_n) \right)$$
$$\times k_0\left( G_1(z_1; \lambda_1),\ldots,G_N(z_N; \lambda_N); \lambda_d \right) dz_N \ldots dz_1. \tag{8.18}$$

$\lambda = (\lambda_1,\ldots,\lambda_N,\lambda_d)^T \in \Lambda \subset \mathbb{R}^b$, is the $b$-dimensional unknown parameter vector, $k_0(\cdot; \lambda_d)$ is the copula density function used to approximate the joint distribution under $H_0$ and $G_n(\cdot; \lambda_n)$ is the CDF of $Z_n$ under hypothesis $H_0$.

With Equations 8.15 and 8.17, we have the following test at the fusion center [23],

$$\mathcal{T}_k(\mathbf{u}_1,\ldots,\mathbf{u}_N) \triangleq \frac{\displaystyle\max_{k_1(\cdot)\in\mathcal{A}_k,\Psi} \prod_l \hat{P}(R_l; \psi, H_1)}{\displaystyle\max_{k_0(\cdot)\in\mathcal{A}_k,\Lambda} \prod_l \hat{P}(R_l; \lambda, H_0)} \underset{H_0}{\overset{H_1}{\gtrless}} \eta, \tag{8.19}$$

where $\eta$ is the detector threshold.

The maximization in Equation 8.19 is over the copula densities, as well as the unknown marginal parameters and copula dependence parameters. Unlike the classical composite hypothesis testing formulation, which would have required the knowledge of the true PDFs under both hypotheses with possibly unknown parameters, it could very well happen that the set $\mathcal{A}_k$ may not be inclusive of the *true* copula models. Thus, the copula functions chosen after maximization may still be misspecified. We, therefore, call the test a misspecified GLRT (mGLRT). We refer readers to Ref. [23], where we discuss a method to determine the threshold, $\eta$, based on the asymptotic distribution of the detector statistic, $\mathcal{T}_k(\cdot)$.

### 8.3.3 A Computationally Efficient Fusion Rule

The fusion rule in Equation 8.19 involves evaluation of $N$-dimensional integrals and optimization over multiple dimensions to obtain maximum likelihood estimates of the unknown parameters. Application of mGLRT is, therefore, prohibitive as the number of sensors increases due to the increased computational complexity. An interesting alternative to the "exact" mGLRT (Equation 8.19) is the so-called LPF-noise based mGLRT which involves deliberate injection of noise to the quantized observations before fusion (see Figure 8.4) [11]. This approach is based on Widrow's quantization theory, which we review next.

#### 8.3.3.1 Widrow's Statistical Theory of Quantizaton: A Review

The statistical theory of quantization was developed by Widrow and co-workers [36–38]. They interpreted quantization of a random variable as sampling of its PDF, and showed that the PDF of the quantized signal is the convolution of the input signal PDF with a rectangular pulse function followed by conventional sampling. Thus, the PDF of the quantizer output, $u_{nl}$, at sensor $n$ and at any time instant, $l$, can be given as

$$p_{U_n}(z) = \left( p_{W_n}(z) \star p_{Z_n}(z) \right) \cdot c_{\delta_n}(z), \tag{8.20}$$

where $p_{Z_n}(z)$ is the PDF of the random variable at the input $Z_n$, $p_{W_n}(z)$ denotes the rectangular pulse function,

$$p_{W_n}(z) = \begin{cases} \dfrac{1}{q_n}, & -q_n/2 < z < q_n/2 \\ 0, & \text{elsewhere,} \end{cases} \tag{8.21}$$

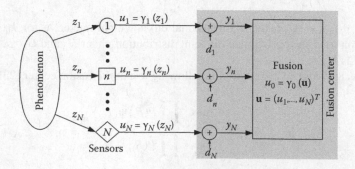

**FIGURE 8.4** A *controlled* noise $d_n$ is added at the output of each sensor $n$. The approach greatly simplifies the fusion rule by avoiding the need to compute multidimensional integrals.

whose width depends on the quantizer step-size ($q_n$), and $c_{\delta_n}(z)$ denotes the impulse train,

$$c_{\delta_n}(z) = \sum_{i_n \in \mathbb{Z}} q_n \delta'\left(z - i_n q_n - \frac{q_n}{2}\right). \tag{8.22}$$

The '$\star$' in Equation 8.20 denotes the convolution operation, and $\delta'(\cdot)$ in Equation 8.22 is the Dirac-delta function. This process of convolution followed by conventional sampling is popularly known as "area sampling" [38]. Also, note that $p_{W_n}(\cdot)$ is also the PDF of a uniform random variable, $W_n \sim \mathcal{U}\left(-\frac{q_n}{2}, \frac{q_n}{2}\right)$. Thus, quantization introduces two "types" of distortions or errors: (1) the additive uniform noise (AUN) error and (2) the aliasing error due to sampling.

The two errors introduced due to quantization can be better visualized in the characteristic function (CF) domain. Taking the Fourier transform of $p_{U_n}(z)$ (Equation 8.20), one obtains the CF of output variable $U_n$,

$$\phi_{U_n}(v) = \sum_{i_n=-\infty}^{\infty} \phi_{Z_n}\left(v + i_n \frac{2\pi}{q_n}\right) \operatorname{sinc}\left(\frac{q_n\left(v + i_n \frac{2\pi}{q}\right)}{2}\right) e^{-ji_n \frac{2\pi}{q_n}\frac{q_n}{2}}$$

$$= \sum_{i_n=-\infty}^{\infty} (-1)^{i_n} \phi_{Z_n}\left(v + i_n \frac{2\pi}{q_n}\right) \operatorname{sinc}\left(\frac{q_n\left(v + i_n \frac{2\pi}{q}\right)}{2}\right) \tag{8.23}$$

where $\phi_{Z_n}(v)$ is the CF of the input $Z_n$ and $\operatorname{sinc}(v) = \frac{\sin(v)}{v}$. Note that the central lobe ($i_n = 0$ in Equation 8.23),

$$\phi_{Z_n + W_n}(v) = \phi_{Z_n}(v) \cdot \operatorname{sinc}\left(\frac{q_n v}{2}\right), \tag{8.24}$$

corresponds to the CF one would obtain by adding an independent and uniformly distributed random variable $W_n$ to the input $Z_n$. It is clear from Figure 8.5 that, in addition to the error introduced due to the addition of uniform noise, quantization also causes an aliasing error due to overlapping (and phase shifted) lobes of $\phi_{Z_n + W_n}(v)$. However, if the input PDF is band-limited so that $\phi_{Z_n}(v) = 0$ for $|v| > \frac{\pi}{q_n}$, then the "frequency"-shifted versions of $\phi_{Z_n + W_n}(v)$ do not overlap and, in principle, the original PDF can be reconstructed from the knowledge of $p_{U_n}(\cdot)$. This is Widrow's first quantization theorem.

**Theorem 8.3: (*Widrow's Quantization Theorem I*)**

If the CF of the input variable $Z_n$ is bandlimited so that

$$\phi_{Z_n}(v) = 0, \quad |v| > \frac{\pi}{q_n}, \tag{8.25}$$

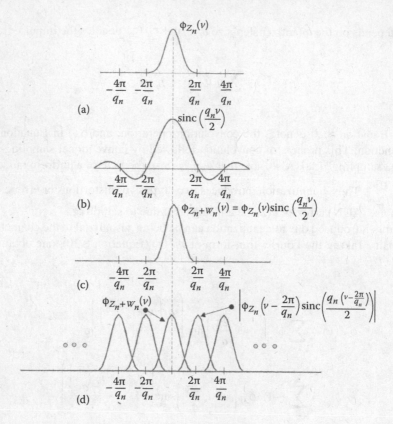

**FIGURE 8.5** Illustration of the quantization process in the CF domain: (a) CF of $Z_n$; (b) CF of $W_n$, the sinc function; (c) CF of $Z_n + W_n$; (d) repetition of CF of $Z_n + W_n$; the CF of the quantized variable is given by the summation of these repetitions after weighting each appropriately. (From Iyengar, S. G. et al., *IEEE Transactions on Signal Processing*, 60(9):4888–4897, 2012.)

then the different lobes in $\phi_{U_n}(v)$ do not overlap, and in principle, the orignal *PDF* $p_n(z_n)$ (before quantization) can be recovered from the PDF of $U_n$.   ∎

The LPF-noise based fusion rule (discussed next) assumes that observations, $Z_n$, at each sensor $n$ satisfies Widrow's first quantization theorem (Theorem 8.3).

### 8.3.3.2   LPF Noise-Based mGLRT

The LPF noise-based distributed detection system is shown in Figure 8.4. An externally generated noise, $d_n$, with PDF $p_{D_n}(d_n)$ is added to the quantized observations from each sensor $n$ before fusing them to make a global decision. Denote the new observations by $y_n = u_n + d_n$ whose CF is given by

$$\phi_{Y_n}(v) = \phi_{U_n}(v) \cdot \phi_{D_n}(v). \tag{8.26}$$

One can choose the noise source with a band-limited CF to filter out the repeated and phase-shifted CF lobes in $\phi_{U_n}(v)$. This is analogous to low pass filtering in signal processing. The noise, $D_n$, is therefore, termed the LPF noise. As shown in Figure 8.5d, an ideal noise source would be one with a rectangular CF in the pass-band, $-\dfrac{\pi}{q_n} \le v \le \dfrac{\pi}{q_n}$, (also see Figure 8.6). However, a rectangular

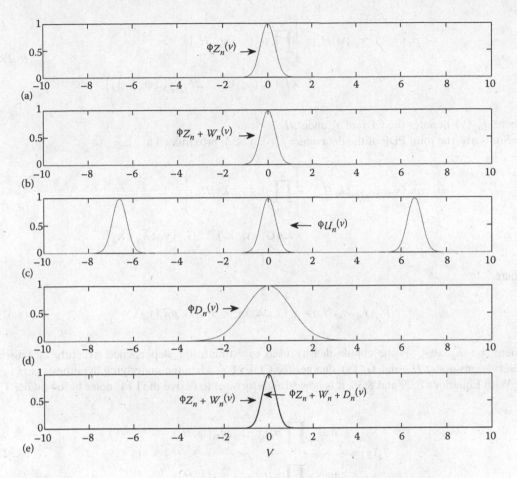

**FIGURE 8.6**  "Filtering" the quantized signal with LPF-noise. The quantization step size, $q_n$, is set to $0.3\sigma_n$: (a) CF of $Z_n$; (b) CF of $Z_n + W_n$; (c) CF of $U_n$; (d) CF of the external LPF-noise, $D_n$; (e) CF of $Y_n = Z_n + W_n + D_n$. (From Iyengar, S. G. et al., *IEEE Transactions on Signal Processing*, 60(9):4888–4897, 2012.)

function in the CF domain corresponds to a PDF whose shape corresponds to a sinc function, an invalid PDF. Note that this is similar to the nonrealizability of an ideal low-pass filter in signal processing. One, therefore, needs to design $D_n$ carefully so that it causes minimal distortion while transforming the discrete-valued random variable, $U_n$, into a continuous variable, $Y_n$. As long as the input variable $Z_n$ satisfies Widrow's first quantization theorem (Theorem 8.3) under both $H_1$ and $H_0$, we have

$$Y_n = Z_n + W_n + D_n. \tag{8.27}$$

Thus, under hypothesis $H_1$, the PDF of data, $y_{nl}$, at time instant $l$ is

$$p_{Y_n}(y_{nl}; \psi_n, H_1) = f_n(y_{nl}; \psi_n) \star p_{W_n}(y_{nl}) \star p_{D_n}(y_{nl}). \tag{8.28}$$

Using a copula density (say) $k_1(\cdot; \psi_d) \in \mathcal{A}_k$ to estimate the dependence structure between sensor observations, the joint PDF of the data $\mathbf{y}_l = (y_{1l}, y_{2l}, ..., y_{Nl})$ can now be approximated as

$$
\hat{p}_{\mathbf{Y}}(y_{1l},\ldots,y_{Nl};\psi,H_1) = \left\{ \prod_{n=1}^{N} p_{Y_n}(y_{nl};\psi_n,H_1) \right\}
$$
$$
\times k_1\big( F_{Y_1}(y_{1l};\psi_1),\ldots,F_{Y_N}(y_{Nl};\psi_N);\psi_d \big),
\tag{8.29}
$$

where $F_{Y_n}(y)$ denotes the CDF of $Y_n$ under $H_1$.

Similarly, the joint PDF of the data under $H_0$ can be approximated as

$$
\hat{p}_{\mathbf{Y}}(y_{1l},\ldots,y_{Nl};\lambda,H_0) = \left\{ \prod_{n=1}^{N} p_{Y_n}(y_{nl};\lambda_n,H_0) \right\}
$$
$$
\times k_0\big( G_{Y_1}(y_{1l};\lambda_1),\ldots,G_{Y_N}(y_{Nl};\lambda_N);\lambda_d \big),
\tag{8.30}
$$

where

$$
p_{Y_n}(y_{nl};\lambda_n,H_0) = g_n(y_{nl};\lambda_n) \star p_{W_n}(y_{nl}) \star p_{D_n}(y_{nl}),
\tag{8.31}
$$

where $k_0(\cdot;\lambda_d) \in \mathcal{A}_k$ is the copula density used to estimate the dependence structure of sensor observations under $H_0$, and $G_{Y_n}(y)$ denotes the CDF of $Y_n$ when the underlying hypothesis is $H_0$.

With Equations 8.29 and 8.30, it is now straightforward to derive the LPF-noise based mGLRT,

$$
T_k'(\mathbf{y}) = \frac{\displaystyle\max_{k_1(\cdot)\in\mathcal{A}_k,\Psi} \prod_{l=1}^{L} p_{\mathbf{Y}}(y_{1l},\ldots,y_{Nl};\psi,H_1)}{\displaystyle\max_{k_0(\cdot)\in\mathcal{A}_k,\Lambda} \prod_{l=1}^{L} p_{\mathbf{Y}}(y_{1l},\ldots,y_{Nl};\lambda,H_0)} \overset{H_1}{\underset{H_0}{\gtrless}} \eta.
\tag{8.32}
$$

The test derived above involves only continuous-valued variables and does not involve multi-dimensional integrals. This greatly simplifies the test. The reduced complexity is, however, at the expense of decreased signal-to-noise ratio due to the injection of noise to the quantized sensor observations. The addition of LPF noise facilitates *filtering* of the baseband CF, $\phi_{Z_n+W_n}(v)$, from the received quantized observations $\phi_{U_n}(v)$. This noise should be designed so that as little information as possible is lost while filtering the required signal.

### 8.3.4 AN ILLUSTRATIVE EXAMPLE

Consider a detection problem using a two-sensor network. It is known that the observations received at the local quantizers each follow a Gaussian distribution. That is,

$$
\begin{aligned}
H_0 &: \quad Z_1 \sim \mathcal{N}(0, 10), \quad Z_2 \sim \mathcal{N}(0, 10) \\
H_1 &: \quad Z_1 \sim \mathcal{N}(0.5, 10), \quad Z_2 \sim \mathcal{N}(0.5, 10)
\end{aligned}
\tag{8.33}
$$

where $\mathcal{N}(\mu,\sigma^2)$ is the usual univariate Gaussian density function with mean $\mu$ and variance $\sigma^2$. The marginal PDFs of both $Z_1$ and $Z_2$ are assumed to be known under both $H_1$ and $H_0$. The observations are statistically dependent under $H_1$; however, no knowledge about the dependence structure (and hence the joint distribution) is available at the fusion center. The observations, $\left\{z_{1l}, z_{2l}\right\}_{l=1}^{L}$, at the two local sensors are passed through uniform scalar quantizers before their transmission to the fusion center. Thus, the fusion center has access only to the quantized measurements, $\mathbf{u} = \left\{u_{1l}, u_{2l}\right\}_{l=1}^{L}$, to make a global decision in favor of one of the two hypotheses. The mGLRT-based fusion rule for this problem is the same as the one derived in Equation 8.19 with $N = 2$.

An alternative computationally efficient test was derived in Section 8.3.3.2, which involves injection of LPF noise before fusion. We evaluate its performance using the example presented here. Although Gaussian CFs are not perfectly band-limited, a property necessary for using the LPF noise-based fusion rule, they are very close to being band-limited for all practical purposes. Figure 8.6 shows quantization and the effect of LPF noise in the CF domain. The quantization step size, $q_n$ is set to 0.3 of the input standard deviation, that is, $q_n = 0.3\sigma_n$. The CF of the input variable $z_n$ is shown in Figure 8.6a. Addition of the quantization noise, $w_1$, is equivalent to multiplication of $\phi_{Z_n}(v)$ (shown in Figure 8.6a) with a sinc function, sinc $\left(\dfrac{q_n v}{2}\right)$. The resultant CF, $\phi_{Z_n+W_n}(v)$, is shown in Figure 8.6b. This CF is repeated and summed in Figure 8.6c, which represents the CF of the quantized signal, $u_n$ (see Equation 8.23). The CF of the LPF noise, $D_n$, a standard Gaussian distributed variable in this example, is shown in Figure 8.6d.* It is clear that multiplication of $\phi_{U_n}(v)$ with $\phi_{D_n}(v)$, which is equivalent to addition of $d_n$ to $z_n + w_n$ in the random variable domain, "filters" the signal so that only the main lobe ($v = 0$) of $\phi_{U_n}(v)$ is retained (Figure 8.6e). Because the LPF noise is different from the ideal one with rectangular CF, the signal, $z_n + w_n$, undergoes some distortion while being "filtered." However, this distortion is almost imperceptible as evident from Figure 8.6e.

The PDF of the transformed variable $Y_n = Z_n + W_n + D_n$ under $H_0$ and $H_1$ can be derived using Equations 8.28 and 8.31. The LPF noise-based fusion rule (Equation 8.32) is now applied for distinguishing the two hypotheses. We include the Gumbel and the Gaussian copula functions in the set $\mathcal{A}_k$ of candidate copula models for characterizing dependence between observations under $H_1$.

For the data generation under $H_1$, we use a Clayton copula with Kendall's $\tau$ set to 0.31. Detection performance of the LPF noise-based mGLRT is evaluated using this synthetic dataset. The data generating copula is deliberately excluded from $\mathcal{A}_k$ so that the detection performance can be evaluated even when the true underlying copula is unavailable. In other words, the test is defined over misspecified set of copula functions.

In Figure 8.7, we plot the ROC curves using 25,000 Monte Carlo trials. The decision window, $L$, is set to 50 samples. That is, we assume that the sensors observe the phenomenon over $L = 50$ time intervals before the fusion center makes a decision in favor of either hypothesis. It is evident from the Figure 8.7, that the performance of the LPF noise-based fusion rule is very close to the "exact" mGLRT, albeit with reduced computational complexity. This is true for both the quantizers, $q_n = 0.3\sigma_n$ and $q_n = 0.9\sigma_n$, considered here.

Another approach that is often adopted to address the issue of computational complexity is to deliberately neglect statistical dependence between sensor observations while designing the test. The test, so designed, would require computation of $N$ one-dimensional integration operations as opposed to $N$-dimensional integrations. However, such an approach severely degrades the detection performance as is evident from Figure 8.7.

We have also investigated the detection performance as a function of the quantization step size $q_n$ for a fixed false alarm probability ($P_F = 0.05$). A plot of $P_D$ versus $q_n/\sigma_n$ is shown in Figure 8.8. We observe that $P_D$ for the LPF-noise-based mGLRT decreases rather noticeably in the region $q_n/\sigma_n > 0.9$. We conjecture that this sharp decrease is due to the significant contribution of aliasing (in CF) errors, in addition to LPF noise.

---

* It is important to note that a standard Gaussian noise may not be the "best" LPF noise. It is used here to provide a simple illustrative example.

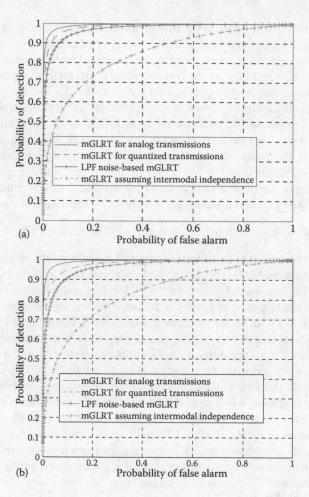

FIGURE 8.7 Monte Carlo-based receiver operating characteristic curves. Performance of the fusion rule based on LPF noise is very close to the mGLRT and signicantly outperforms the test that assumes intermodal independence. (a) Quantizer step size $q_n = 0.3\sigma_n$ and (b) Quantizer step size $q_n = 0.9\sigma_n$.

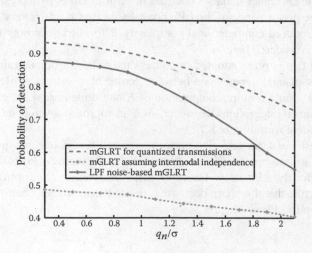

FIGURE 8.8 Probability of detection as a function of quantization step size (measured by its ratio to $\sigma_n$) for given $P_F = 0.05$.

## 8.4 CONCLUSION

Extension of conventional detection theory to the distributed setting with multiple sensors is non-trivial and introduces several new dimensions to the problem such as sensor network topology, intersensor relationships, optimization criteria, and quantization levels. For general network structures, the optimal solution to the distributed detection problem is NP-complete. Initial research in this area achieved tractability by limiting the local sensors to suboptimal quantizers. For the heterogeneous fusion problem, it is also important to model appropriately the statistical dependence and heterogeneity of sensor observations while designing the fusion rule. One such approach, discussed in this chapter, is based on copula theory. However, straightforward application of copula-based methodologies will lead to fusion rules that are not scalable with the number of sensors in a given network. A fusion algorithm, based on Widrow's additive quantization noise model, yields a scalable, computationally efficient, yet accurate, approximation to fusion results that otherwise would have been obtained using direct computation of the mGLRT. This approach uses the idea that the deliberate injection of a controlled amount of external noise before fusion can behave in a manner similar to frequency-selective filtering in the characteristic function domain. Designing a noise source that introduces minimal distortion while filtering is an important issue that remains to be addressed in a systematic manner.

## ACKNOWLEDGMENTS

This material is based on work supported by, or in part by, the U.S. Army Research Laboratory and the U.S. Army Research Office under contracts/grant numbers W911NF-13-2-0040 and W911NF-14-1-0339.

## REFERENCES

1. H. McGurk, and J. MacDonald. Hearing lips and seeing voices. *Nature*, 264:746–748, 1976.
2. P.F. Swaszek. On the performance of serial networks in distributed detection. *IEEE Transactions on Aerospace and Electronic Systems*, 29(1):254–260, 1993.
3. Z.B. Tang, K.R. Pattipati, and D.L. Kleinman. Optimization of detection networks. I. Tandem structures. *IEEE Transactions on Systems, Man and Cybernetics*, 21(5):1044–1059, 1991.
4. P.K. Varshney. *Distributed Detection and Data Fusion*. Springer-Verlag, New York, 1997.
5. W.P. Tay, J.N. Tsitsiklis, and M.Z. Win. On the subexponential decay of detection error probabilities in long tandems. *IEEE Transactions on Information Theory*, 54(10):4767–4771, 2008.
6. Z.-B. Tang, K.R. Pattipati, and D.L. Kleinman. Optimization of detection networks. II. Tree structures. *IEEE Transactions on Systems, Man and Cybernetics*, 23(1):211–221, 1993.
7. W.P. Tay, J.N. Tsitsiklis, and M.Z. Win. On the impact of node failures and unreliable communications in dense sensor networks. *IEEE Transactions on Signal Processing*, 56(6):2535–2546,.2008.
8. W.P. Tay, J.N. Tsitsiklis, and M.Z. Win. Bayesian detection in bounded height tree networks. *IEEE Transactions on Signal Processing*, 57(10):4042–4051, 2009.
9. N.S.V. Rao. Computational complexity issues in synthesis of simple distributed detection networks. *IEEE Transactions on Systems, Man and Cybernetics*, 21(5):1071–1081, 1991.
10. J.N. Tsitsiklis. Decentralized detection. In H.V. Poor, and J.B. Thomas, editors, *Advances in Statistical Signal Processing*. JAI Press, Greenwich, CT, 1993.
11. J.N. Tsitsiklis, and M. Athans. On the complexity of decentralized decision making and detection problems. *IEEE Transactions on Automatic Control*, 30(5):440–446, 1985.
12. A.R. Reibman. Performance and fault-tolerance of distributed detection networks. PhD thesis, Duke University, Durham, NC, 1987.
13. J.N. Tsitsiklis. Decentralized detection with a large number of sensors. *Mathematics of Control, Signals, and Systems*, 1:167–182, 1988.
14. P. Chen, and A. Papamarcou. New asymptotic results in parallel distributed detection. *IEEE Transactions on Information Theory*, 39(6):1847–1863, 1993.
15. J.F. Chamberland, and V.V. Veeravalli. Decentralized detection in sensor networks. *IEEE Transactions on Signal Processing*, 51:407–416, 2003.

16. J.F. Chamberland, and V.V. Veeravalli. Asymptotic results for decentralized detection in power constrained wireless sensor networks. *IEEE Journal of Selected Areas in Communications*, 22(6):1007–1015, 2004.

17. Z. Chair, and P.K. Varshney. Optimal data fusion in multiple sensor detection systems. *IEEE Transactions on Aerospace and Electronic Systems*, 22:98–101, 1986.

18. V. Aalo, and R. Viswanathan. On distributed detection with correlated sensors: Two examples. *IEEE Transactions on Aerospace and Electronic Systems*, 25(3):414–421, 1989.

19. P. Willett, P.F. Swaszek, and R.S. Blum. The good, bad and ugly: Distributed detection of a known signal in dependent gaussian noise. *IEEE Transactions on Signal Processing*, 48(12):3266–3279, 2000.

20. H. Chen, B. Chen, and P.K. Varshney. A new framework for distributed detection with conditionally dependent observations. *IEEE Transactions on Signal Processing*, 60(3):1409–1419, 2012.

21. E. Drakopoulos, and C.-C. Lee. Optimum multisensor fusion of correlated local decisions. *IEEE Transactions on Aerospace and Electronic Systems*, 27(4):593–606, 1991.

22. M. Kam, Q. Zhu, and W.S. Gray. Optimal data fusion of correlated local decisions in multiple sensor detection systems. *IEEE Transactions on Aerospace and Electronic Systems*, 28(3):916–920, 1992.

23. S.G. Iyengar, R. Niu, and P.K. Varshney. Fusing dependent decisions for hypothesis testing with heterogeneous sensors. *IEEE Transactions on Signal Processing*, 60(9):4888–4897, 2012.

24. H. White. *Estimation, Inference and Specification Analysis*. Cambridge University Press, Cambridge, 1994.

25. S.G. Iyengar, P.K. Varshney, and T. Damarla. A parametric copula-based framework for hypothesis testing using heterogeneous data. *IEEE Transactions on Signal Processing*, 59(5):2308–2319, 2011.

26. D. Mari, and S. Kotz. *Correlation and Dependence*. Imperial College Press, London, 2001.

27. R.B. Nelsen. *An Introduction to Copulas*. Springer-Verlag, New York, 1999.

28. H. Joe. *Multivariate Dependence and Related Concepts*. Chapman and Hall, New York, 1997.

29. D. Kurowicka, and R. Cooke. *Uncertainty Analysis with High Dimensional Dependence Modeling*. John Wiley & Sons, Hoboken, NJ, 2006.

30. A. Subramanian, A. Sundaresan, and P.K. Varshney. Fusion for the detection of dependent signals using multivariate copulas. In *Proceedings of the 14th International Conference on Information Fusion (FUSION)*, pp. 1–8, 2011.

31. A. Subramanian. Hypothesis testing using spatially dependent heavy-tailed multisensor data. PhD thesis, Syracuse University, Syracuse, NY, 2014.

32. K. Aas, C. Czado, A. Frigessi, and H. Bakken. Pair-copula constructions of multiple dependence. *Insurance: Mathematics and Economics*, 44(2):182–198, 2009.

33. T. Bedford, and R. Cooke. Probability density decomposition for conditionally dependent random variables modeled by vines. *Annals of Mathematics and Artificial Intelligence*, 32:245–268, 2001.

34. D. Berg. Copula goodness-of-fit testing: An overview and power comparison. *The European Journal of Finance*, 15:675–701, 2009.

35. C. Genest, M. Gendron, and M. Bourdeau-Brien. The advent of copulas in finance. *The European Journal of Finance*, 15(7–8):609–618, 2009.

36. B. Widrow. A study of rough amplitude quantization by means of Nyquist sampling theory. *IRE Transactions on Circuit Theory*, 3(4):266–276, 1956.

37. B. Widrow, and I. Kollar. *Quantization Noise: Roundoff Error in Digital Computation, Signal Processing, Control, and Communications*. Cambridge University Press, Cambridge, 2008.

38. B. Widrow, I Kollar, and M.-C. Liu. Statistical theory of quantization. *IEEE Transactions on Instrumentation and Measurement*, 45(2):353–361, 1996.

# 9 Fusion Systems Evaluation
## An Information Quality Perspective

*Ion-George Todoran, Laurent Lecornu,*
*Ali Khenchaf, and Jean-Marc Le Caillec*

## CONTENTS

## 9.1 INTRODUCTION

The performance evaluation of information processing algorithms has been studied for a long time. Recent technological advances continue to increase the efficiency of these algorithms in specific application contexts. Meanwhile, the evaluation of complex information systems, for example, information fusion systems (IFS), remains a very difficult task even today. One of the main causes of this difficulty is the need to integrate and process a continuously increasing number of heterogeneous information sources. From an information consumer point of view, that is, the end user of the IFS, the information and the quality of information are of great importance. Therefore, he or she needs information accompanied by a degree of confidence, corresponding to the capacity of the IFS to deliver the right information, at the right time, and in the right format. In this chapter, we propose the tracking of the information quality evolution through the system, as a possible solution to evaluate the information proposed to the end user.

## 9.2 DATA AND INFORMATION QUALITY

In this section, we define data and information quality using as support the research literature from three different domains of application. In Section 9.2.1, we present the state of the art of data and information quality evaluation. Next, in Section 9.2.2, we present data and information notions in the case of an IFS and how their semantic characteristics influence the quality definition and evaluation.

### 9.2.1 INFORMATION QUALITY: A REVIEW

The "fitness for use" is the most common definition of quality [1,2]. Therefore, to be able to define data and information quality, we need to start by defining these two notions. Only then, after clearly presenting their desired properties, can quality be defined and assessed.

Usually the notion of data has a strong connection with numbers, measurements, mathematics, and science [3]. Although data are collected with a clear purpose, the most important characteristic of data is its context independence. On the other hand, information is obtained from data by a process of semantic enrichment in a well-defined context. Thus, whenever data are interpreted, that is, getting a meaning in the current context, they are transformed into information [4]. Following [5], by context we understand "any information that can be used to characterize the situation of an entity. An entity is a person, place or object that is considered relevant to the interaction between a user and an application, including the user and applications themselves" (p. 5).

Therefore, the inner value of data is the possible information that could be extracted from it. This is represented in Figure 9.1. On the left are the data that are available to the system. These data contain **N** *useful* pieces of information. The term "useful information" characterizes each element that has a certain utility for the end user in the current context. Therefore, if the existing data contain all the information needed by the end user, these data can be characterized as being of good quality. Nonetheless, these data might contain "non-information," which is not pertinent in the current context (but that might become valuable in another context).

The IFS is viewed as a process of information extraction from multiple data sources. It has the role of discerning between useful information and non-information. The output of this process generates $M$ pieces of information. If the end user expects the $N$ pieces of information and $M = N$, then the IFS satisfies the user's needs. In the other two cases, $N > M$ or $N < M$, the IFS's performance is not maximum.

From this illustrative example, two very important conclusions can be drawn:

1. Data quality needs to be assessed based on the information that can be extracted.
2. Information generated by an IFS is dependent on the data being used and on the IFS's performance, that is, its adaptation to extracting precisely the desired information (not less, not more).

The most common practice in defining and evaluating data and information quality is the use of quality attributes or criteria. In Table 9.1, we present the quality criteria from three different research domains: management information systems (MIS), Web information systems (WIS), and IFS. MIS

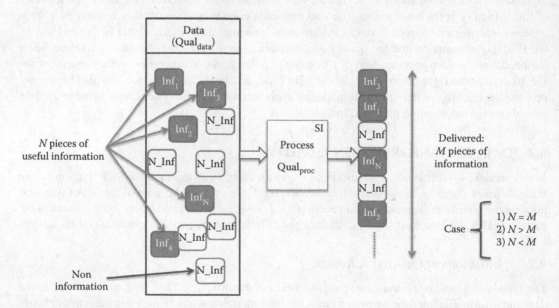

**FIGURE 9.1**  Data to information transformation.

**TABLE 9.1**

**Comparison of Quality Criteria Issued from Three Different Application Domains**

| | Management IS | Web IS | Information Fusion Systems |
|---|---|---|---|
| Quality Criteria | Accuracy | | ⊘ |
| | Believability | | Reliability Credibility Truthfulness |
| | Objectivity | | |
| | Reputation | | |
| | | | Level of expertise |
| | Value-added | | ⊘ |
| | Relevancy | | |
| | Timeliness | | |
| | | Latency | |
| | | Response | |
| | Completeness | | |
| | Amount of data | | ⊘ |
| | Interpretability | | |
| | Understandability | | |
| | Consistency | | Integrity |
| | Conciseness | | ⊘ |
| | Usability | Customer support documentation | ⊘ |
| | Accessibility | Availability | Availability Accessibility |
| | Security | | ⊘ |
| | ⊘ | Verifiability | ⊘ |
| | ⊘ | Price | ⊘ |
| | ⊘ | Quality of service | ⊘ |

has addressed the study of data and information quality for several years [6]. On the other hand, in the fusion community, defining a methodology for information quality evaluation is a more recent topic [7]. Most of the quality criteria from Table 9.1 are common to the three domains. Depending on the application's specifics, some quality criteria are unique to one domain, for example, quality of service in WIS. Others are developed in multiple quality criteria; for example, believability is confined to MIS and WIS and developed in three different criteria in IFS: reliability, credibility, and truthfulness. The explanation is that in the case of IFS, the believability and the source of information are very important and multiple aspects need to be assessed.

The information quality frameworks, like the ones presented in Table 9.1, have a series of deficiencies making them very difficult to implement in practice. First, they do not consider how the information was produced, that is, they have a vision of the IFS like that in Figure 9.1. Thus, they consider the IFS as a black box and study the quality evaluation at its input, that is, data quality, and at its output, that is, information quality. Second, they consider data and information as equivalent and therefore are using the same tools (criteria) to evaluate the quality of data and information. Third, they define these lists of quality criteria in natural language, without any formalism. As a consequence, they do not indicate what the applicable quality criteria are for any given data/information and how these criteria should be quantified.

We are persuaded that there exists a strong connection between the value and semantics of data/information and the process of selecting the applicable quality criteria. Therefore, we propose to study the semantic evolution in an IFS.

### 9.2.2  DATA AND INFORMATION IN A FUSION SYSTEM

Figure 9.2 represents a generic IFS architecture, adapted from Refs. [8] and [9]. This system employs three processing stages. The first, *data association*, is responsible for gathering and fusing data from multiple sources (e.g., databases, sensor reports, human experts). The second, *data processing*, has the role of extracting different elements of information. The third, *information fusion*, is responsible for fusing the information from the previous stage and for adapting the results to the end user(s)'s needs.

Figure 9.2 clearly illustrates that the passage from data to final information does not comprise only one step. In fact, there is a semantic enrichment of data in multiple stages to finally derive the information needed by the user(s). In this particular case, we identified raw data (as unprocessed data, for example, generated by sensors, stored in data bases), preprocessed data, primary information, and final information. These different types of data and information have specific characteristics, directly implying the necessity of using personalized quality definitions. Raw and preprocessed data represent the IFS's perception of the environment under observation. Usually, the quality of sensed data is evaluated considering the reliability of the source and the uncertainty measures associated to data values, for example, imprecision, conflict, fuzziness, and ambiguity [10]. On the other hand, primary and final pieces of information have an associated meaning, as they are responsible for the IFS's comprehension of the environment, for example, detection and identification of entities of interest and establishing the relationships between different entities. Moreover, the final information needs to be adapted to the end user to be easily projected and assimilated by its cognitive structures. If the quality of the primary information is generally evaluated by a reliability degree [11], to evaluate the final information it is necessary to consider multiple dimensions, for example,

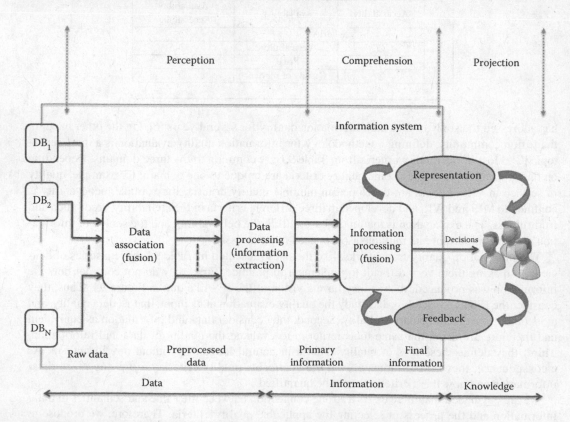

**FIGURE 9.2**  Data, information and knowledge in an information fusion system.

completeness, timeliness, relevancy. Also, if the end user is not satisfied with the current information quality, he or she might prefer to wait for new data to be available, to obtain final information of improved quality [12].

Evaluating the quality by means of forms delivered to users, the case of MIS and WIS gives only an average performance of the system. Instead, the user needs a quality evaluation for each piece of information proposed to him or her. In the next section we propose a new methodology for information quality evaluation based on a recent article [13].

## 9.3 METHODOLOGY FOR INFORMATION QUALITY EVALUATION

Unlike the existing methodologies, we propose to decompose the IFS into its elementary modules and to study the quality at the input and output of each processing module, that is, the local quality. This is presented in Section 9.3.1. Then, in Section 9.3.2, we present a possible solution for modeling the influence of each processing module on the quality. As a result, the output quality can be determined based on the input information and its quality. Finally, in Section 9.3.3, we present the evaluation of the entire IFS making use of the local quality.

### 9.3.1 SYSTEM DECOMPOSITION: LOCAL QUALITY

Most of today's IFS have a modular design architecture allowing evolution perspectives by operations such as module mix and match. In this paradigm, we propose to decompose the IFS into its elementary modules. As a result, we gain access to the different data and information semantic transformations. Therefore, we propose to study the input and output quality for each module. This is called local quality [13] (see Figure 9.5). $Q_{in}^i$ and $Q_{out}^i$ denote, respectively, the local quality at the input, and the output of the $i$th processing module. Of course, in the case of two consecutive processing modules, $i$ and $i + 1$, the output quality of the first is identical to the input quality of the second.

In Ref. [14] it was proposed to formalize the information quality definition by modeling it as a UML class diagram (Figure 9.3). An association between two classes of the type 1 to 1..* represents that the first class contains one or more instances of the second. This model puts each information element, *Info_element*, in relation with its quality, *QoI*. This association is directly dependent on the context of application *Context*, defined in our model by its spatial-temporal characteristics. The quality of information *QoI* class is defined by multiple quality criteria *Qual_Criteria*, as in the frameworks presented in Table 9.1. Depending on the processing module and the information element characteristics, that is, value and semantic, only some quality criteria are applicable. In Ref. [13], the quality criteria from Table 9.1 were classified based on their adequacy to capture data or information quality.

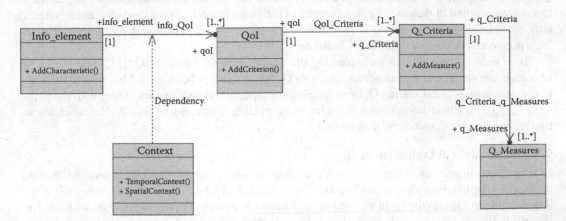

**FIGURE 9.3** UML diagram for information quality.

Unlike previous studies, we are persuaded that these criteria need to be quantified by quality measures *Qual_Measures*. The process of assigning quality measures to each quality criterion depends on the criterion definition and characteristics: quality criteria can be subjective or objective, context dependent or independent. Based on this, an adequate mathematical framework can be used to develop these measures. For example, the subjective quality criteria are more adapted to quantification by fuzzy-like measures [10], because they are usually evaluated by experts using natural language. Some other quality criteria are evaluated by statistics, and in this case probability theory is the most appropriate tool. For more details concerning the development of quality measurement in specific mathematical frameworks, we recommend readers consult Ref. [15].

To illustrate the notion of local quality, we consider the case of a module performing a general classification. It receives as its input a piece of data and outputs the data's corresponding class. In the general case of classification problems, the input quality is characterized by the quality criteria {*Accuracy and Data amount*}, and the output quality by the quality criterion {*Reliability*}:

$$
\begin{pmatrix}
Accuracy & : & \{\mu_{Acc}^1, \mu_{Acc}^2\} \\
Data\ amount & : & \{\mu_{DAm}^1\}
\end{pmatrix}
\overset{Q_f}{\Rightarrow}
\Big( Reliability : \{\mu_{Rel}^1, \mu_{Rel}^2\} \Big)
\tag{9.1}
$$

As explained in Section 9.2.2, the semantic changes determine a different quality definition. Therefore, in this particular case, the input and output qualities are evaluated using different quality criteria. Moreover, depending on the information to be described, these quality criteria could be evaluated by different quality measures $\mu_{Cr}^j$, denoting the *j*th quality measure employed for quality criterion *Cr*. Hence, taking the case of the *Reliability* criterion, it can be quantified by an *F*-measure [16] (as a result of a learning process using a set of training data, e.g., a neural network) and/or by a fuzzy measure representing an expert opinion.

In this subsection, we proposed to use the existing quality frameworks to evaluate the local quality. In the next subsection, we want to automatically determine the value(s) of the output quality based on the value(s) of the input information and its quality.

### 9.3.2 Quality Transfer Function

In Ref. [13], we proposed to model each processing module by a quality transfer function, $\mathbf{Q_f}$, making the connection between the output quality and the input information and its quality: $Q_{out} = \mathbf{Q_f}(I_{in}, Q_{in})$. Estimating the output reliability of a module based on its input is an old problem that has been extensively studied in the case of control systems [17]. Unfortunately, in the case of control systems only the reliability criterion case is considered. Therefore, we propose to extend these studies to incorporate all the quality criteria, not only one.

There are two possibilities for evaluating the quality transfer function, $\mathbf{Q_f}$ [13]. The first is to obtain an analytical mathematical expression for $\mathbf{Q_f}$. The second, when a closed form is too difficult to obtain, consists in estimating $\mathbf{Q_f}$ by measuring the input and output quality. Then, with the pairs $((Q_{in}, I_{in}), Q_{out})$ a linear interpolation provides an acceptable estimation of the $\mathbf{Q_f}$. Hereafter, these two techniques are detailed and exemplified.

#### 9.3.2.1 Analytical Evaluation of $\mathbf{Q_f}$

In some particular cases, when we have a complete knowledge of the processing module, that is, its functioning behavior can be mathematically modeled, and when its complexity is not too high, it is possible to express the $\mathbf{Q_f}$ in a closed form. Numerous analytical examples can be found in the literature. We present two of them; the first is from control systems and the second, a Bayesian classification system.

## Example 9.1

We consider the reliability modeling problem, as presented in Ref. [17]. In this case, the quality criterion *Reliability* completely describes the input quality, $R_{in}$ and the output quality, $R_{out}$. Furthermore, the control module is characterized by the following parameters:

- $P_w$, the probability that the control module is in operation
- $P_c$, the probability that the control module makes a correct decision given that the input information is correct
- $P_e$, the probability that the control module makes a correct decision given that the input information is erroneous, that is, detects and corrects the error
- If both reliabilities have a probabilistic meaning, then the analytic expression of the output reliability is given by Ref. [17]:

$$R_{out} = R_{in}(1 - P_w + P_w(P_c - P_e)) + P_wP_e \tag{9.2}$$

## Example 9.2

We now consider the case of a two-class Bayesian classifier indicating the presence or absence of a signal, taken from Ref. [13]. For simplicity in presenting the example, let us suppose that each input observation **x** has a Gaussian probability distribution given by the relation **x = S + N**, with **S** a binary variable taking the values 0 (signal is absent) or 1 (signal is present) with equal probability; **N** is the noise described by a standard Gaussian probability distribution: $N \sim \mathcal{N}(0,1)$. Using the previous notations, the classifier receives as input data a real number **x**, and it delivers, as output information, its class. The signal classification is performed by comparing it to a threshold. If the signal value is inferior to this threshold, then the observation belongs to class 0 (no signal detected); otherwise class 1 is designated (signal detected).

The classical measure for the input quality $Q_{in}$ is the signal-to-noise (SNR) level, corresponding to the quality criterion *Accuracy*. The output quality $Q_{out}$ is evaluated by the quality criterion *Correctness* and the quality measure for this criterion is the probability of correct classification, that is, the detection probability. In Bayesian mathematical theory the detection probability has the following analytic formula:

$$1 - P_{err} = 1 - \frac{1}{2}(P_{e0} + P_{e1}) = 1 - Q\left(\frac{1}{2}\right) \tag{9.3}$$

$$\text{with } Q(u) = \frac{1}{\sqrt{2\pi}} \int_u^\infty \exp\left(-\frac{x^2}{2}\right)dx \tag{9.4}$$

In conclusion, we have an analytical formula to evaluate the output information quality based on input quality and knowledge on the processing module, that is, two-class Bayesian classifier and the value of the threshold.

### 9.3.2.2  Nonanalytical Evaluation of $Q_f$

In some cases, analytical formulas are not available. This is the situation when the input and output quality have a large number of dimensions, with dependencies between them. Another situation is when we do not have complete knowledge of the module functioning behavior. In each case, we state that at least the knowledge of the input–output characteristics and the module general functioning behavior can be obtained. By means of this knowledge, the qualitative behavior of the processing module can be studied by varying the input quality and measuring the corresponding output quality. In this way, we obtain pairs of the type $((Q_{in}, I_{in}), Q_{out})$ for each processing module.

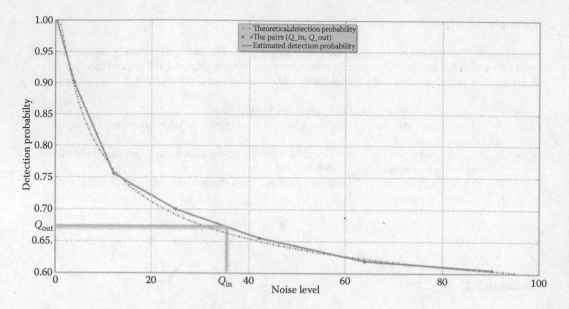

**FIGURE 9.4** Nonanalytical $\mathbf{Q}_f$ evaluation for a two-class Bayesian classifier (detection probability).

These pairs make it possible to determine relations between the input and output quality by using learning algorithms, for example, linear or nonlinear regression, neural networks.

We give an illustrative example of obtaining the function $\mathbf{Q}_f$ by measuring the pairs (($Q_{in}$, $I_{in}$), $Q_{out}$) in Figure 9.4. For simplicity of presentation we consider the one-dimensional case, analyzing the Bayesian classification example, presented in the analytical case. We have the following knowledge of the module: it works as a two-class Bayesian classifier; it receives at its input a one-dimensional signal susceptible to be affected by additive Gaussian noise; it delivers at its output the signal's estimated class. Without knowledge of the threshold value, the analytical determination of $\mathbf{Q}_f$ is not possible. In this case, the input quality is represented by the noise level and the output quality by the probability of correct classification (the detection probability). The output quality evaluation was carried out using a Monte Carlo simulation consisting in the generation of 10,000 data samples taken from a Bernoulli distribution with success probability $p = 0.5$. The noise follows a standard normal distribution $N \sim \mathcal{N}(0, \sigma^2)$, with $\sigma^2$ varying from 0.25 to 100. The points in Figure 9.4 represent the probability of correct detection. Small biases can be observed between the simulated $\mathbf{Q}_f$ and the theoretical one, determined by Equation 9.3. With $\mathbf{Q}_f$ determined, for each new piece of input data the output quality can be directly evaluated using a linear interpolation between the two points closest to the input noise level.

### 9.3.3 INFORMATION FUSION SYSTEM EVALUATION

In the previous two subsections, the quality evaluation was studied in the case of a single elementary module. We are now interested in evaluating the entire IFS, making use of the local quality evaluation and the quality transfer function.

Consider the case of an IFS composed by a number of elementary modules, such as the one presented in Figure 9.5. Let us assume that for each module $i$ of this system we were able to determine the applicable input and output quality criteria, and its corresponding quality transfer function, $Q_f^i$. For two successive modules, the output of the first module needs to be adapted to the input of the second one, w.r.t. the information exchange. Therefore, we believe that this condition should also be respected by the quality exchange. We state this quality condition by the following principle [15]:

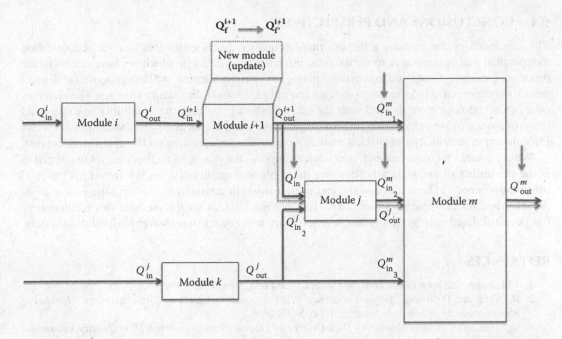

**FIGURE 9.5** Quality propagation and global quality evaluation (the doted lines designate the quality changes as a consequence of the module update).

**Principle 9.1** *In the case of two successive modules i and i + 1, the output quality of the first $Q_{out}^i$ engenders the input quality of the second $Q_{in}^{i+1}$ with respect to the quality criteria set, that is, $Q_{in}^{i+1} \subseteq Q_{out}^i$.*

The explanation of the stated principle is straightforward: the output information of the first module is identical to the input information of the second module. Therefore, the two pieces of information have the same characteristics and thus imply the use of the same quality criteria. As some quality criteria might not be available for use by the second module, we used the subset notation instead of the set equality.

Being able to locally evaluate the information quality, the quality transfer function allows us to automatically obtain the information quality in the further point of analysis in the system. Hence, the evaluation of the entire IFS is done by propagating the quality through the system, as stated by the following principle:

**Principle 9.2** *The local variations of the information quality propagate through the IFS to its output by means of $\mathbf{Q_f}$.*

This principle is illustrated in Figure 9.5. The upper branch contains two processing modules and we assume a situation when the second processing module is updated. As a consequence its quality transfer function changes:

$$\mathbf{Q_f^{i+1} \rightarrow Q_{f'}^{i+1}} \tag{9.5}$$

As a consequence, downstream of the $(i + 1)$th module, the information quality changes. We place an arrow to indicate the places where the quality of information changed. Moreover, we indicate with a dotted line how this quality change propagates. At the same time, on the second branch and upstream from the updated module, the quality remains unchanged. Thus, when a subpart of the system is updated or changed, we only need to analyze the new subpart to determine its quality transfer function.

## 9.4  CONCLUSIONS AND PERSPECTIVES

The methodology for evaluating the information quality described in this chapter is application independent and is applicable to all modular information systems in which we have access to the processing module. Compared to previous studies on system evaluation, we have proposed to decompose the system into its elementary modules and to locally study the quality changes. The elements of this methodology were defined with the idea of offering the user the possibility to understand the provenance of the information quality. At the same time, the local quality evaluation allows an information system analyst to check a module's performance depending on the application context.

We are aware that more research is needed to improve the proposed methodology. One direction is the possibility of automatically selecting the applicable quality criteria. Moreover, as the final information needs to be adapted to the end user, it is worth considering studying the format of its presentation, for example, as a dashboard. Last but not least, as for the moment this methodology has been validated only by simulation, it is necessary to apply it to a real-world information system.

## REFERENCES

1.  J.M. Juran, *Juran on Leadership for Quality*, Free Press, New York, 1989.
2.  R. Wang and D. Strong, "Beyond accuracy: What data quality means to data consumers," *Journal of Management of Information Systems*, 12(4), 5–33, 1996.
3.  L. Sebastian-Coleman, *Measuring Data Quality for Ongoing Improvement: A Data Quality Assessment Framework*, Morgan Kaufmann, Burlington, MA, 2013.
4.  A. Aamodt and M. Nygard, "Different roles and mutual dependencies of data, information, and knowledge—An AI perspective on their integration," *Data and Knowledge Engineering*, 16, 191–222, 1995.
5.  A.K. Dey, "Understanding and using context," *Personal Ubiquitous Computing*, 5(1), 4–7, 2001.
6.  R. Mason, "Measuring information output: A communication systems approach," *Information & Management*, 1(1), 219–234, 1978.
7.  G. Rogova and E. Bosse, "Information quality in information fusion," *13th International Conference on Information Fusion*, pp. 1–8, Edinburgh, UK. 2010.
8.  E. Lefebvre, M. Hadzagic and E. Bosse, "On quality of information in multi-source fusion environments," In *Advances and Challenges in Multisensor Data and Information Processing*, Ed. E. Lefebvre, IOS Press, Amsterdam, pp. 69–77, 2007.
9.  E. Waltz and J. Llinas, *Multisensor Data Fusion*, Artech House, Norwood, MA, 1990.
10. G.J. Klir and M.J. Wierman, *Uncertainty Based Information*, 2nd ed., Physica-Verlag, New York, 1999.
11. G. Rogova and V. Nimier, "Reliability in information fusion: Literature survey," *Proceedings of the 7th International Conference on Information Fusion*, pp. 1158–1165, Stockholm, Sweden, 2004.
12. G. Rogova, M. Hadzagic, M. St-Hilaire, M. Florea and P. Valin, "Context-based information quality for sequential decision making," *CogSIMA'13*, pp. 16–21, San Diego, 2013.
13. I.G. Todoran, L. Lecornu, A. Khenchaf and J.M. Le Caillec, "Information quality evaluation in fusion systems," *Proceedings of the 16th International Conference on Information Fusion*, pp. 906–916, Istanbul, Turkey, 2013.
14. I.G. Todoran, L. Lecornu, A. Khenchaf and J.M. Le Caillec, "Assessing information quality in information fusion systems," *NATO Symposium on Analysis Support to Decision Making in Cyber Defence & Security*, Tallinn, Estonia, 2014.
15. I.G. Todoran, L. Lecornu, A. Khenchaf and J.M. Le Caillec, "Toward the quality evaluation in complex information systems," *SPIE Signal Processing, Sensor/Information Fusion and Target Recognition XXIII*, Baltimore, MD, 2014.
16. D.M.W. Powers, "Evaluation: From precision, recall and F-measure to ROC, informedness, markedness and correlation,"*Journal of Machine Learning Technologies*, 2(1), 37–63, 2011.
17. R. Srivastava and B.H. Ward, "Reliability modeling of information systems with human elements: A new perspective," *IEEE Transactions: Total System Reliability Symposium*, pp. 30–39, 1983.

# 10 Sensor Failure Robust Fusion

*Matt Higger, Murat Akcakaya,
Umut Orhan, and Deniz Erdogmus*

## CONTENTS

## 10.1 INTRODUCTION

Many sensor fusion methods rely on the assumption that the sensors that generate the data operate according to a predetermined characteristic at all times [1]. When a sensor fails, and this assumption is challenged, detection systems demonstrate behaviors different from the expected. In such occurrences, one possible approach is to identify the failed sensors and update the fusion rule accordingly. A first intuition to identify a failed sensor may be to detect a change in the sensor's dependence on the target variable of interest. However, because sensors fail online, without access to ground truth of target variable, we cannot directly observe any shift in the sensor's relationship with the target variable. For this reason, many sensor failure detection schemes rely on quantifying the relationship between sensors in expected operating modes and declaring a failure when such a relationship changes. For example, if during training sessions we notice that sensors A and B fit a certain joint distribution and during operation they seem to behave in a highly unlikely way, according to the trained joint distribution, we would suspect sensor A or B may have failed. Once a particular failure has been detected, it is another challenge to isolate which sensor is failing, A or B?

One common method of characterizing normal sensor operation is training a neural network. Then, during operation, the system can use the same neural network to declare a sensor failed if the relationship changes [2–7]. Other approaches manually configure sensor relationships using a consistency measure [8], fuzzy set membership [9], or entropy-based measures [10]. It is also possible to construct sensor sets such that knowing which sets have failed one can deduce which particular sensors have failed [11].

Each of these methods identifies the existence of a failure during online operation of a system. Given the existence of a failure, one can seek to identify a failed sensor under the assumption that as few sensors as possible have failed. For example, if one sensor disagrees with the other nine sensors in a system, it is either the case that the one sensor has failed, or that the other nine have failed. Some voting procedure of consistency can be applied to isolate a failed sensor.

Once the failed sensor is detected, it is of interest to amend the fusion rule to account for the shift in sensor characteristics. One way of achieving this is by weighing the importance of sensor data, using fuzzy sets [12,13] or Dempster–Shafer theory [14] according to a sensor's agreement with all other sensor data. In the case of optical sensors with varying background light, models can be built for all lighting conditions and fused as appropriate [15,16]. If the failure follows a particular time dynamic then we can track the shift in characteristic using a Kalman [17] or particle filter [18,19].

Many of the aforementioned methods seek to identify failed sensors. Those methods that go a step beyond in reincorporating failed sensors back into fusion do so by leveraging some known dynamics of a failure. However, not all problems have a particular failure dynamic that is known to the designer. In this chapter, we describe a sensor fusion approach that builds a model of failure and derives a fusion rule that is sensor failure robust by minimizing Bayesian risk under the failure model [20]. We examine both the cases of dependent and independent sensor failures.

Of course, sensors fail in ways that are not always anticipated and it may not be possible to model the characteristic of the new "failed" state in an offline setting. To mitigate this, in Section 10.4.2 we demonstrate how one can model the failed sensor as drawing from some distribution of characteristics. In other words, we impose some prior knowledge of how one expects a failed sensor to act in lieu of knowledge of its characteristic behavior. We extend the minimum Bayesian risk fusion to include this unknown characteristic. In all the cases described, our method is shown to decrease risk. This decrease depends heavily on the probability of sensor failure; the greater the chance of failure, or the greater the shift in characteristic, the more worthwhile it becomes to model for these changes.

We demonstrate our methods on synthetic models of sensor failure as well as managing electroencephalogram (EEG) artifacts, modeled as sensor failures in a brain–computer interface [21].

## 10.2   SENSOR AND SENSOR FAILURE MODELS

We examine the problem of estimating the value of some discrete target variable $T$, where $t \in \{1, \ldots, N_T\}$. To do so we leverage sensor measurements $S_{1:N}$ where each $s_i \in \{1, \ldots, N_{S_i}\}$. We also include a variable $R_i$ that explicitly describes whether the sensor is functional, failed or in some other operating mode. For example $r = 1$ denotes the expected operating mode, $r = 0$ denotes a failed sensor characteristic, and other states of $r$ can describe other states of interest (low battery, visual occlusion, high temperature, etc.). (This notation is adopted from the "observation model" of Ref. [18].)

We assume the sensors are independent of each other given the target variable, $t$, and each sensor's operating mode, $r_{1:N}$. Dependencies between these variables can be seen in Figure 10.1.

In the remainder of the chapter, we use "operating mode" to refer to the state of $r_i$. We use the term "operating characteristic" to refer to how sensor data are generated based on ground truth $t$ as well as its operating condition $r$; in this way an operating characteristic is defined by the distribution

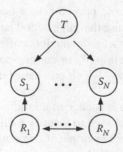

**FIGURE 10.1**   Sensor and sensor failure graphical model: $T$ is the target variable of interest, $S_i$ are sensors, and $R_i$ represent the operating mode (functional, failed, etc.) of each sensor.

$P_{S_i|T,R_i}$. We also use the common convention of using capital letters to denote random variables and lowercase letters to denote their values.

## 10.3   MINIMUM BAYESIAN RISK FUSION

We seek to minimize some expected risk value $E[r_{i|j}]$ that represents the risk associated with estimating $\hat{t} = i$ while in reality $t = j$. Given a prior distribution over our target variable, $P_T$, an associated risk with every decision-ground-truth pair, $r_{\hat{t}|t}$, and some classification rule, $F : s_{1:N} \mapsto \hat{t}$, we compute

$$
\begin{aligned}
E[\text{Risk}] &= \sum_{\hat{t},t} P_{\hat{T},T}(\hat{t},t) r_{\hat{t}|t} \\
&= \sum_{\hat{t},t} P_{\hat{T}|T}(\hat{t}|t) P_T(t) r_{\hat{t}|t} \\
&= \sum_{\hat{t},t,s_{1:N}} P_{\hat{T},S_{1:N}|T}(\hat{t},s_{1:N}|t) P_T(t) r_{\hat{t}|t} \\
&= \sum_{\hat{t},t,s_{1:N}} P_{\hat{T}|S_{1:N}}(\hat{t}|s_{1:N}) P_{S_{1:N}|T}(s_{1:N}|t) P_T(t) r_{\hat{t}|t} \\
&= \sum_{t,s_{1:N}} P_{S_{1:N}|T}(s_{1:N}|t) P_T(t) r_{F(s_{1:N})|t}
\end{aligned}
\tag{10.1}
$$

where the fourth equality is due to $\hat{t}$ being independent of $t$ given $S_{1:N}$ and the fifth equality is due to $P_{\hat{T}|S_{1:N}}(\hat{t}|s_{1:N}) = 1$ when $F(s_{1:N}) = \hat{t}_{s_{1:N}}$ and is 0 otherwise. Given a particular set of sensor measurements $s_{1:N}$, it follows that we can minimize the expected risk by choosing $\hat{t}$ as

$$
F_{\text{opt}}(s_{1:N}) = \arg\min_i \sum_t P_{S_{1:N}|T}(s_{1:N}|t) P_T(t) r_{i|t}
\tag{10.2}
$$

Of course, Equation 10.2 does not yet account for the various sensor failure states, $r_i$. We consider sensor failure robust fusion methods in Section 10.4.

## 10.4   MINIMUM BAYESIAN RISK FUSION WITH SENSOR FAILURE

Although it is convenient to assume that sensors fail independently, it may not always be the case. Consider if two sensors share a battery and are sensitive to fluctuations in their power source.

Further, when sensors fail we may not have knowledge of its new characteristic. Without access to some offline training data set, it is not fruitful to attempt to learn the new characteristic online (see Section 10.6). Without leveraging any problem specific insights, it is challenging to come up with a well-founded fusion of these new, unknown sensors.

We begin in Section 10.4.1 with the most idealistic case, that we have knowledge of the sensor's characteristics for each of its possible operating modes. In Section 10.4.2 we extend these methods to handle sensors with unknown characteristics. We examine the case of both dependent and independent failures in both sections below.

### 10.4.1   KNOWN SENSOR CHARACTERISTICS

In this section, we assume knowledge of all sensor characteristics under all operating modes. Namely, we are given $P_{S_{1:N}|T,R_{1:N}}$ for all possible states of $r_{1:N}$.

If we further assume that sensors fail independently ($r_i \perp\!\!\!\perp r_j$ for $i \neq j$) then it follows that sensor outputs are conditionally independent given the target variable $T$ (see Figure 10.1). Under these conditions, we compute

$$
\begin{aligned}
P_{S_{1:N}|T}\left(s_{1:N}|t\right) &= \prod_{j=1}^{N} P_{S_j|T}\left(s_j|t\right) \\
&= \prod_{j=1}^{N}\left[\sum_{r_j} P_{S_j|T,R_j}\left(s_j|t,r_j\right) P_{R_j}(r_j)\right]
\end{aligned}
\tag{10.3}
$$

where the first equality takes advantage of the fact that independent $r_i$ yield independent $s_i$ (see Figure 10.1). By plugging Equation 10.3 into 10.2 our independent failure minimum Bayesian risk fusion rule becomes

$$
F_{\text{opt}}(s_{1:N}) = \arg\min_i \sum_t P_T(t) r_{i|t} \prod_{j=1}^{N}\left[\sum_{r_j} P_{S_j|T,R_j}\left(s_j|t,r_j\right) P_{R_j}(r_j)\right]
\tag{10.4}
$$

In the case that our sensors do not fail independently our fusion computation becomes more computationally expensive:

$$
\begin{aligned}
P_{S_{1:N}|T}\left(s_{1:N}|t\right) &= \sum_{r_{1:N}} P_{S_{1:N},R_{1:N}|T}\left(s_{1:N},r_{1:N}|t\right) \\
&= \sum_{r_{1:N}} P_{S_{1:N}|T,R_{1:N}}\left(s_{1:N}|t,r_{1:N}\right) P_{R_{1:N}}(r_{1:N}) \\
&= \sum_{r_{1:N}}\left[\prod_{j=1}^{N} P_{S_j|T,R_j}\left(s_j|t,r_j\right)\right] P_{R_{1:N}}(r_{1:N})
\end{aligned}
\tag{10.5}
$$

where the last equality comes from the fact that $s_i$ and $s_j$ are conditionally independent given $t$ and $r_i$ for $i \neq j$; see Figure 10.1. By plugging Equation 10.5 into 10.2 our dependent failure minimum Bayesian risk fusion rule becomes

$$
F_{\text{opt}}(s_{1:N}) = \arg\min_i \sum_t P_T(t) r_{i|t} \sum_{r_{1:N}}\left[\prod_{j=1}^{N} P_{S_j|T,R_j}\left(s_j|t,r_j\right)\right] P_{R_{1:N}}(r_{1:N})
\tag{10.6}
$$

Before continuing to the case of unknown characteristics, it would be instructive to review our current framework. We start with a discrete naive Bayes sensor model and add a "operating mode" variable $r_i$ that describes the operating characteristic of the sensor. For example, possible sensor modes could include sensor failure, visually occluded, low battery, expected operation, and so forth. We develop minimum Bayesian risk fusion rules under the cases of dependent and independent sensor failures. The fusion rule makes the reasonable assumptions that we have knowledge of some prior distribution of target states, $P_T$, a risk associated with every possible type of missed decision, $r_{i|j}$, knowledge of the sensor characteristic under the expected state and the prior distribution of sensor states, $P_{R_{1:N}}$. It makes the potentially unreasonable assumption that we also have knowledge of a sensor's operating characteristic under all sensor states (including the "failed" sensor states). An example of this type is shown in Section 10.5.2.

## 10.4.2 UNKNOWN SENSOR CHARACTERISTICS

In this section, we extend the fusion rules of Section 10.4.1 to account for uncertainty in the sensor characteristics. To do so, we build a prior distribution over the expected sensor characteristics. This is equivalent to assuming that when a sensor fails, it draws some new characteristic from a known distribution.

Consider an unknown sensor characteristic for sensor $i$ by fixing $t = j$ and $r_i = k$. While we do not have explicit knowledge of $P_{S_i|T,R_i}(s_i|j,k)$ we define the random variable $\alpha_{i,j,k}$ such that

$$\alpha_{i,j,k}[l] = P_{S_i|T,R_i}(l|j,k) \tag{10.7}$$

where $\alpha_{i,j,k} \in \mathcal{R}^{N_{S_i}}$ and $\sum_l \alpha_{i,j,k}[l] = 1$. Of course, $\alpha_{i,j,k}$ is equivalent to the distribution of $s_i$ where $t$ and $r_i$ are fixed, but we define $\alpha_{i,j,k}$ as a random variable to emphasize that this sensor characteristic itself is not known with certainty, it is drawn from some distribution. Specifically,

$$\alpha_{i,j,k} \sim P_{\alpha_{i,j,k}} \tag{10.8}$$

As an example, the Dirichlet distribution is a natural choice for $P_{\alpha_{i,j,k}}$, as it is commonly employed as a distribution over discrete distributions [22]. From these definitions it follows:

$$
\begin{aligned}
P_{S_i|T,R_i}(s_i|j,k) &= \int P_{S_i|T,R_i,\alpha_{i,j,k}}(s_i|j,k,\alpha)P_{\alpha_{i,j,k}}(\alpha)d\alpha \\
&= \int \alpha[s_i]P_{\alpha_{i,j,k}}(\alpha)d\alpha \\
&= E\big[\alpha_{i,j,k}[s_i]\big]
\end{aligned}
\tag{10.9}
$$

In many cases, there will still be some operating modes for which we know the sensor's operating characteristic. For sensor $j$ we define the set of operating modes for which we know the characteristic as $K_j$. Then, assuming independent sensor failure we can plug Equation 10.9 into 10.4 only for those operating modes that we need to estimate:

$$
\begin{aligned}
F_{opt}(s_{1:N}) &= \arg\min_i \sum_t P_T(t)r_{ilt} \\
&\times \prod_{j=1}^{N}\left[\sum_{r_j \notin K_j} E\big[\alpha_{j,t,r_j}[s_i]\big]P_{R_j}(r_j) + \sum_{r_j \in K_j} P_{S_j|T,R_j}(s_j|t,r_j)P_{R_j}(r_j)\right]
\end{aligned}
\tag{10.10}
$$

and similarly, if sensors do not fail independently we plug Equation 10.9 into 10.6:

$$
\begin{aligned}
F_{opt}(s_{1:N}) &= \arg\min_i \sum_t P_T(t)r_{ilt} \\
&\sum_{r_{1:N}}\left[\prod_{j=1}^{N} P_{S_j|T,R_j}(s_j|t,r_j)^{I(r_j \in K_j)} E[\alpha_{j,t,r_j}[s_i]]^{I(r_j \notin K_j)}\right]P_{R_{1:N}}(r_{1:N})
\end{aligned}
\tag{10.11}
$$

## 10.5  NUMERICAL EXPERIMENTS

In this section, we show three numerical experiments that demonstrate our robust fusion methods. In the first example, we examine how robust fusion tempers the trust of a stronger classifier in a pool of weaker classifiers because of its failure rates. In this first example, we assume independent sensor failures with known characteristics. In the second example, we simulate the placement of a grid array of sensors to detect a target. This example demonstrates dependent sensor failures as well as unknown failure characteristics. As a final example, we model artifact conditions as sensor failures to make better decisions in a brain–computer interface using real data.

### 10.5.1  EXAMPLE: INDEPENDENT FAILURES AND KNOWN CHARACTERISTICS

In this example, we simulate the fusion of multiple sensors that have uneven discriminative power. To keep things simple, we use binary sensor states with a binary target (such an example is common in high level detection fusions). In our example, sensor 1 correctly identifies the target (present or not) 99% of the time while working ($r_1 = 1$), and only 60% of the time while failed ($r_0 = 0$). This sensor is fused with nine other sensors that are accurate only 85% of the time while functional and have an identical failed fusion characteristic when failing:

$$
\begin{aligned}
&P_{S_1|T,R_1}(0\,|\,0,1) = 0.99 \qquad && P_{S_i|T,R_i}(0\,|\,0,1) = 0.60 \\
&P_{S_1|T,R_1}(1\,|\,1,1) = 0.99 \qquad && P_{S_i|T,R_i}(0\,|\,0,1) = 0.60 \\
&P_{S_1|T,R_1}(0\,|\,0,0) = 0.85 \qquad && P_{S_i|T,R_i}(0\,|\,0,0) = 0.60 \\
&P_{S_1|T,R_1}(1\,|\,1,0) = 0.85 \qquad && P_{S_i|T,R_i}(0\,|\,0,0) = 0.60
\end{aligned}
\tag{10.12}
$$

for $i \in \{2,\dots,10\}$. We assume that we are detecting a highly unlikely event $P_T(0) = 0.9$ and that it is much more preferable to have a false positive than a missed target ($r_{1|0} = 1$ while $r_{0|1} = 100$). We compare robust fusion (Equation 10.4) with naive sensor fusion, which is equivalent to Equation 10.4 under the assumption that the sensors are always working in their optimal state.

Figure 10.2 shows the calculation of expected risk (Equation 10.1) as we vary the probability a sensor has failed, $P_{R_i}(0)$, from 0 to 0.25. As expected, increased "failure" probability increases the

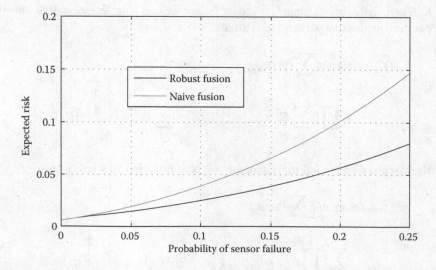

**FIGURE 10.2**  Expected risk vs. probability that a sensor has failed.

expected risk as the failure mode is less discriminative than the functional mode. Because failure robust fusion mitigates, as much as possible, the increase in expected risk by using the failure model to make an appropriate decision.

## 10.5.2　Example: Dependent Failures and Unknown Characteristics

In this example, we demonstrate an example of failure robust fusion in the case of dependent sensor failures with unknown characteristics. Consider the classification of a bird's species based on a video of the bird's flight. We are given classifiers that operate on a single image of the sequence. Given an appropriately segmented image of the bird, each classifier has an accuracy of 95%. Let us denote the event that the classifier is given a correctly segmented image as $r_i = 1$ such that:

$$P_{S_i|T,R_1}(0|0,1) = 0.95$$

$$P_{S_i|T,R_1}(1|1,1) = 0.95 \tag{10.13}$$

for each image $i$. Unfortunately, tracking the bird in the video is a challenge and we are not guaranteed an accurate segmentation. Further complicating matters is that we are unsure as to how our classifier performs when given an incorrectly segmented image. Given an incorrectly segmented image we model the classifier's characteristic $\alpha_{i,j,k}$ (see Equation 10.7) from a Dirichlet distribution with concentration parameter 1. (This is equivalent to choosing the accuracy uniformly from [0,1].)

Because tracking accuracy is dependent between consecutive images, we model segmentation accuracy as a Markov model:

$$P_{R_i|R_{i-1}}(0|0) = 0.9$$

$$P_{R_i|R_{i-1}}(1|1) \text{ varies}$$

If image $i$ in a sequence is incorrectly segmented then image $i + 1$ has a 90% of also being incorrectly segmented. This temporal significance introduces bursts of sensor failures in our sequences of outputs. In the numerical experiment below we vary $P_{R_i|R_{i-1}}(1|1)$, the probability that an image is incorrectly segmented given the previous image was correctly segmented. For simplicity, we assume that "image 0" was correctly tracked, but unavailable for classification, such that our initial image also has a $P_{R_i|R_{i-1}}(1|1)$ chance of being segmented correctly. We compare robust fusion (Equation 10.11) with naive sensor fusion, which is equivalent to (Equation 10.4) under the assumption that the sensors are always working in their optimal state.

As can be seen in Figure 10.3, leveraging the sensor failures, especially their temporal dependence, robust fusion outperforms naive fusion. This can be explained by considering a sequence of sensor outputs such as $S_{1:N} = \{1, 1, 2, 2, 2, 2, 1, 2, 1, 2\}$. In this case, naive fusion may choose species 2, as it is more common among all the classifiers which have identical characteristics. However, robust fusion considers that the first two decisions may be due to a tracking failure and the rest of the sequence may simply be drawn from the unknown characteristic. This seems plausible because the sequence switches so quickly between classifying species 1 and 2. Many of the sensor states which naive fusion and failure robust fusion disagree are of a similar form.

## 10.5.3　RSVP-Keyboard

We demonstrate our robust fusion methods on the RSVP-Keyboard [21,23–25], a machine designed to help paralyzed people type using EEG. In Rapid Serial Visual Presentation (RSVP), letters are quickly shown in a sequence of single-letter presentations. The user makes a selection from among

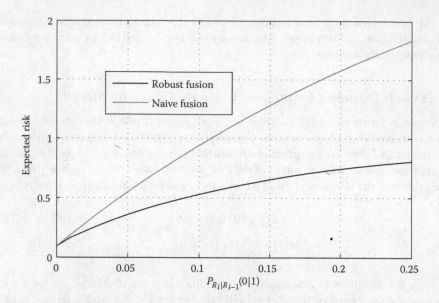

**FIGURE 10.3** Expected risk vs. probability of sensor failure given that the previous sensor was functional.

the letters by generating a particular brain response (P300) when their selection is shown. To spell, the system analyzes EEG data associated with each letter presentation and determines if the P300 signal was present or not; presence of the P300 implies that the user is selecting the associated letter.

Under traditional operating conditions, RSVP-Keyboard generates $s_1$ and makes a decision about whether the P300 is present ($t = 1$) or not ($t = 0$) for the EEG associated with each letter shown (see Refs. [23–25] for full details).

In this context, risk is not well motivated so we set $r_{ilj}$ to a constant risk for all types of errors yielding a traditional MAP classifier.

Unfortunately, a user's EEG is subject to muscle movements (jaw movements, eye blinks, smiling) that dramatically alter the EEG and offer no additional insights into user intent. Artifacts such as these are commonly detected and removed in other brain–computer interfaces. We take a different approach, modeling these artifacts as sensor failures as they shift the distribution of our features on the target class. Specifically, we assign each artifact type its own operating mode ($r_1 = 1 \Rightarrow$ no artifact, $r_1 = 2 \Rightarrow$ smiling, …) and learn the operating characteristic under each artifact type $\left( P_{S_1|T,R_1} \left( s_1 | t, r_1 \right) \right)$ from training data. We decide as

$$F_{\mathrm{opt}}(s_{1:N}) = \arg\max_i \sum_{r_j} P_{S_1|T,R_1}\left( s_1 | i, r_1 \right) P_{R_1}(r_1) P_T(i) \tag{10.14}$$

We ran experiments between a classifier that assumed a constant "no artifact" operating mode $\left( P_{r_1}(1) = 1 \right)$ compared with a classifier that explicitly accounted for artifacts. To put the results in context we include the area under the curve (AUC) values between the single sensor decision and the target variable for each user under all artifact conditions.

As can be noted in Table 10.1, typing accuracy changes dramatically between users. In the simulation, as with other trials we have performed, user 3 struggles to produce accurate classifications. In addition, in Figure 10.4 we note that robust classification consistently outperforms nonrobust methods. The performance advantage of our method is correlated to the magnitude of the difference in AUC between the control and artifact classes. In other words, the stronger the drop in AUC when an artifact is introduced (see Table 10.1), the greater the performance benefit of using robust fusion.

**TABLE 10.1**

**AUC Values between a Single Sensor Decision and the Target Variable**

|  | User 1 | User 2 | User 3 | User 4 |
|---|---|---|---|---|
| No artifact | 0.7644 | 0.8298 | 0.6488 | 0.8103 |
| Jaw movement | 0.6079 | 0.8026 | 0.6370 | 0.6527 |
| Smile | 0.7105 | 0.8423 | 0.6506 | 0.7023 |
| Eye blink | 0.6561 | 0.7641 | 0.4710 | 0.7373 |

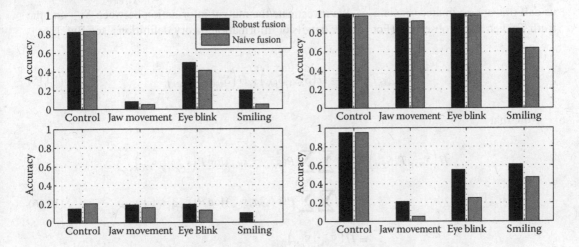

**FIGURE 10.4**    Robust vs. naive classification performance in RSVP.

## 10.6    LEARNING SENSOR CHARACTERISTICS ONLINE

Given the problem of sensor failure, a first intuition might be to identify failed sensors and learn their characteristics online. Ideally, once $P_{S_i|T,R_i}\left(s_i|t,r_i\right)$ was found, it could be inserted into the appropriate fusion rule to maintain minimum risk. Unfortunately, under the model given in Figure 10.5 such a method cannot yield an improvement in performance. We prove this is the case by examining the equivalent problem of adding a new sensor, $S_{N+1}$, and learning its characteristics online.

To demonstrate the intuition of our proof we first look at the graphical model in Figure 10.5. We seek to incorporate $S_{N+1}$ by learning the dotted line, $P_{S_{N+1}|T}$. Knowledge of this connection

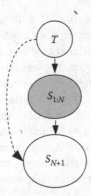

**FIGURE 10.5**    Graphical model of adding a sensor into fusion online.

(distribution) could certainly improve performance. However, we can incorporate node $S_{N+1}$ only by connecting it to observed quantities. We also note that $T$ isn't available online. Therefore, the "dotted line" connection is not feasible. Because of this, we can connect $S_{N+1}$ only by learning $P_{S_{N+1}|S_{1:N}}$. Incorporating the node in this way means when $S_{1:N}$ is observed $S_{N+1}$ is independent of $T$. As a result, $S_{N+1}$ cannot refine our fusion beyond the content available in $S_{1:N}$.

More rigorously, given a full joint distribution of $N$ sensor decisions and some target variable $T$, we would like to add sensor $S_{N+1}$ into our fusion rule through some online training process. However, because we never have access to ground truth of $T$ online, we restrict ourselves to using only samples from $S_{1:N}$ to accomplish this task. At best, if only because no other information is available, one could only ever incorporate $S_{N+1}$ into the decision by learning $P_{S_{N+1}|S_{1:N}}\left(s_{N+1}|S_{1:N}\right)$. Let us call this approximated sensor $\tilde{S}_{N+1}$. We investigate the mutual information between $\tilde{S}_{N+1}$ and the truth $T$ when $S_{1:N}$ is observed. (For further details of information theory definitions, see Ref. [26].)

$$I\left(\tilde{S}_{N+1},T\middle|S_{1:N}\right)=H\left(\tilde{S}_{N+1}\middle|S_{1:N}\right)-H\left(\tilde{S}_{N+1}\middle|T,S_{1:N}\right) \tag{10.15}$$

Note that

$$
\begin{aligned}
H\left(\tilde{S}_{N+1}\middle|T,S_{1:N}\right) &= \sum_{\tilde{s}_{N+1}}\sum_{t}\sum_{s_{N+1}}P(\tilde{s}_{N+1},t,s_{1:N})\log P\left(\tilde{s}_{N+1}\middle|t,s_{1:N}\right) \\
&= \sum_{\tilde{s}_{N+1}}\sum_{t}\sum_{s_{N+1}}P(\tilde{s}_{N+1},t,s_{1:N})\log P\left(\tilde{s}_{N+1}\middle|s_{1:N}\right) \\
&= H\left(\tilde{S}_{N+1}\middle|S_{1:N}\right)
\end{aligned} \tag{10.16}
$$

where the second equality is due to the essential fact that $\tilde{S}_{N+1}$ has no dependence on $T$ outside the influence of $S_{1:N}$. By plugging Equation 10.16 into 10.15 we see that:

$$I\left(\tilde{S}_{N+1},T\middle|S_{1:N}\right)=0 \tag{10.17}$$

The lack of mutual information between $\tilde{S}_{N+1}$ and $T$ shows that $\tilde{S}_{N+1}$ cannot add additional information to be used in the classification of $T$ once we are given $S_{1:N}$.

## 10.7  CONCLUSION

Accurate multimodal or multisensor detection of a target phenomenon requires knowledge of probabilistic sensor characteristics to determine an appropriate fusion rule which optimizes the expected Bayesian risk. However, a particular sensor characteristic can change online, introducing unaccounted additional risk to the fusion rule that was based on assumed sensor specifications. To mitigate such changes, we have shown a robust fusion rule that explicitly models alternate operating modes of a sensor (including different types of failures). We have constructed the fusion rule for dependent or independent sensor failures. Of course, in real scenarios one may not always have the operating characteristics of the sensor in its "failed" state. We present an argument for why learning such a characteristic online is not fruitful. As an alternative, we present a method that models a "failed" sensor as drawing a new characteristic from some known prior distribution. We develop fusion rules that minimize the expected Bayesian risk, accounting for these unknown sensor operating modes.

# REFERENCES

1. D. L. Hall, and J. Llinas, "Multisensor data fusion." In *Handbook of Multisensor Data Fusion*, The Electrical Engineering and Applied Signal Processing Series, D. L. Hall and J. Llinas, Eds. CRC Press, Boca Raton, FL, 2001.
2. Z. Shen, and Q. Wang, "Failure detection, isolation, and recovery of multifunctional self-validating sensor," *IEEE Transactions on Instrumentation and Measurement*, vol. 61, no. 12, pp. 3351–3362, 2012.
3. A. V. Bogdanov, "Neuroinspired architecture for robust classifier fusion of multisensor imagery," *IEEE Transactions on Geoscience and Remote Sensing*, vol. 46, no. 5, pp. 1467–1487, 2008.
4. C. Bachmann, M. Bettenhausen, R. Fusina, T. Donato, A. L. Russ, J. Burke, G. Lamela, W. Rhea, B. Truitt, and J. Porter, "A credit assignment approach to fusing classifiers of multiseason hyperspectral imagery," *IEEE Transactions on Geoscience and Remote Sensing*, vol. 41, no. 11, pp. 2488–2499, 2003.
5. F. Aminian, M. Aminian, and H. W. Collins, Jr., "Analog fault diagnosis of actual circuits using neural networks," *IEEE Transactions on Instrumentation and Measurement*, vol. 51, no. 3, pp. 544–550, 2002.
6. T.-H. Guo, and J. Nurre, "Sensor failure detection and recovery by neural networks," *IJCNN-91-Seattle International Joint Conference on Neural Networks*, vol. I, pp. 221–226, 1991.
7. T. Long, "Sensor fusion and failure detection using virtual sensors," *Control Conference, 1999*, June 1999.
8. M. Soika, "A sensor failure detection framework for autonomous mobile robots," *Proceedings of the 1997 IEEE/RSJ International Conference on Intelligent Robot and Systems. Innovative Robotics for Real-World Applications. IROS '97*, vol. 3, pp. 1735–1740, 1997.
9. Z. Zhang, and Q. Wang, "A new fuzzy neural network architecture for multisensor data fusion in non-destructive testing," *FUZZ-IEEE'99. 1999 IEEE International Fuzzy Systems. Conference Proceedings (Cat. No. 99CH36315)*, vol. 3, pp. 1661–1665, 1999.
10. A. Agogino, and K. Tumer, "Entropy based anomaly detection applied to space shuttle main engines," *2006 IEEE Aerospace Conference*, pp. 1–7, 2006.
11. S. Ahuja, S. Ramasubramanian, and M. M. Krunz, "Single-link failure detection in all-optical networks using monitoring cycles and paths," *IEEE/ACM Transactions on Networking*, vol. 17, no. 4, pp. 1080–1093, 2009.
12. S. Park, and C. Lee, "Fusion-based sensor fault detection," *Proceedings of the 1993 IEEE International Symposium on Intelligent Control*, pp. 156–161, 1993.
13. M. Oosterom, "Soft computing applications in aircraft sensor management and flight control law reconfiguration," *IEEE Transactions on Systems, Man, and Cybernetics*, vol. 32, no. 2, pp. 125–139, 2002.
14. R. Murphy, "Robust sensor fusion for teleoperations," *Proceedings IEEE International Conference on Robotics and Automation*, pp. 572–577, 1993.
15. S. Barotti, L. Lombardi, and P. Lombardi, "Multi-module switching and fusion for robust video surveillance," *Proceedings of 12th International Conference on Image Analysis and Processing*, pp. 260–265, 2003.
16. B.-K. Dan, Y.-S. Kim, J.-Y. Jung, and S.-J. Ko, "Robust people counting system based on sensor fusion," *IEEE Transactions on Consumer Electronics*, vol. 58, no. 3, pp. 1013–1021, 2012.
17. X. Gao, J. Chen, D. Tao, and X. Li, "Multi-sensor centralized fusion without measurement noise covariance by variational Bayesian approximation," *IEEE Transactions on Aerospace and Electronic Systems*, vol. 47, no. 1, pp. 718–722, 2011.
18. F. Caron, M. Davy, E. Duflos, P. Vanheeghe, and S. Member, "Particle filtering for multisensor data fusion with switching observation models: Application to land vehicle positioning," *IEEE Transactions on Signal Processing*, vol. 55, no. 6, pp. 2703–2719, 2007.
19. A. Chilian, H. Hirschmuller, and M. Gorner, "Multisensor data fusion for robust pose estimation of a six-legged walking robot," *2011 IEEE/RSJ International Conference on Intelligent Robots and Systems (IROS)*, pp. 2497–2504, 2011.
20. M. Higger, M. Akcakaya, and D. Erdogmus, "A robust fusion algorithm for sensor failure," *Signal Processing Letters, IEEE*, vol. 20, no. 8, pp. 755–758, 2013.
21. M. Higger, M. Akcakaya, U. Orhan, and D. Erdogmus, "Robust classification in RSVP keyboard," in *Foundations of Augmented Cognition*, Lecture Notes in Computer Science, vol. 8027, D. Schmorrow and C. Fidopiastis, Eds. Springer, Berlin Heidelberg, pp. 443–449, 2013.
22. C. Bishop, *Pattern Recognition and Machine Learning*. Springer, New York, 2006.
23. U. Orhan, "RSVP keyboard: An EEG based BCI typing system with context information fusion," *Electrical Engineering Dissertations*, Paper 85, 2014. Available at http://iris.lib.neu.edu/elec_eng_diss/85/.

24. U. Orhan, D. Erdogmus, B. Roark, B. Oken, and M. Fried-Oken, "Offline analysis of context contribution to ERP-based typing BCI performance," *Journal of Neural Engineering*, vol. 10, no. 6, 066003, 2013.

25. U. Orhan, K. E. Hild, D. Erdogmus, B. Roark, B. Oken, and M. Fried-Oken, "RSVP keyboard: An EEG based typing interface," *2012 IEEE International Conference on Acoustics, Speech and Signal Processing (ICASSP)*, pp. 645–648, 2012.

26. T. M. Cover, and J. A. Thomas, *Elements of Information Theory*. Wiley-Interscience, Hoboken, NJ, 2006.

# 11 Treatment of Dependent Information in Multisensor Kalman Filtering and Data Fusion

*Benjamin Noack, Joris Sijs, Marc Reinhardt, and Uwe D. Hanebeck*

## CONTENTS

## 11.1 INTRODUCTION

Distributed and decentralized processing and fusion of sensor data are becoming increasingly important. In view of the Internet of Things and the vision of ubiquitous sensing, designing and implementing multisensor state estimation algorithm have already become a key issue. A network of interconnected sensor devices is usually characterized by the idea to process and collect data locally and independently on the sensor nodes. However, this does not imply that the data are independent of each other, and the state estimation algorithms have to address possible interdependencies so as to avoid erroneous data fusion results.

Dependencies among local estimates generally can be traced back to common sensor information and common process noise. A wide variety of Kalman filtering schemes allow for the treatment of dependent data in centralized, distributed, and decentralized networks of sensor nodes, but making the right choice is itself dependent upon analyzing and weighing up the different advantages and disadvantages. This chapter discusses different strategies to identify and treat dependencies among Kalman filter estimates while pointing out advantages and challenges.

The study commences with an analysis of the sources of dependencies between state estimates and with the optimal estimation strategy, which consists of a centralized processing of sensor data. Although an efficient preprocessing of sensor data is possible, the central node must have access to each sensor measurement at each processing step. Therefore, centralized schemes may lead to unacceptably high rates and volumes of data transfers. Distributing storage and computation over the network often proves to be an alternative. With the concept of federated Kalman filtering, dependencies stemming from a common process noise model can be treated; also an optimal distributed Kalman filter can be achieved when strict prerequisites are ensured to be met. These requirements can successively be relaxed by employing hypotheses or by striving for a consensus among sensor nodes. In a fully decentralized network with nodes operating autonomously, an optimal fusion strategy is attainable if the underlying dependencies are accessible. However, bookkeeping of dependencies is often unacceptably expensive in terms of both performance and memory usage. In this regard, the focus lies on suboptimal estimation and fusion strategies that do not require precise information about the underlying dependency structure but employ conservative bounds on the dependencies. The concept of ellipsoidal intersection can be employed to counteract the problem of double-counting sensor data. Fusion of estimates by means of covariance intersection is most conservative but also most insusceptible to unmodeled dependencies and does not require any information about the underlying dependencies. Estimation quality and performance of these estimation concepts are considered, and the question of when and how to use them is discussed.

### 11.1.1 PRELIMINARIES

Underlined variables $\underline{x}$ denote vectors or vector-valued functions, and lowercase boldface letters $\underline{\mathbf{x}}$ are used for random quantities. Matrices are written in uppercase boldface letters $\mathbf{C}$. By $(\hat{\underline{x}}, \mathbf{C})$, we denote an estimate with mean $\hat{\underline{x}}$ and covariance matrix $\mathbf{C}$. The notation $\hat{\underline{x}}$ is used for the mean of a random variable, an estimate of an uncertain quantity, or an observation. The matrix $\mathbf{I}$ denotes the identity matrix of appropriate dimension. The inequality

$$\tilde{\mathbf{C}} \geq \mathbf{C} \quad \text{or} \quad \tilde{\mathbf{C}} - \mathbf{C} \geq 0$$

states that $\tilde{\mathbf{C}} - \mathbf{C}$ is a positive semidefinite matrix.

## 11.2  STATE ESTIMATION IN NETWORKED SYSTEMS

State estimation methods are utilized to provide insights into a system's behavior. An estimate for the system's state is dynamically computed based on prior information, a process model, and measurements stemming from sensor devices. The system is characterized by a discrete-time linear process model

$$\underline{\mathbf{x}}_{k+1} = \mathbf{A}_k \underline{\mathbf{x}}_k + \mathbf{B}_k \hat{\underline{u}}_k + \underline{\mathbf{w}}_k, \tag{11.1}$$

where the system matrix $\mathbf{A}_k \in \mathbb{R}^{n_x \times n_x}$ maps the state vector $\underline{x}_k \in \mathbb{R}^{n_x}$ from time step $k$ to the subsequent step $k + 1$. The temporal evolution can further be affected by a control input $\hat{\underline{u}}_k \in \mathbb{R}^{n_u}$ with control-input matrix $\mathbf{B}_k \in \mathbb{R}^{n_x \times n_u}$ and a white Gaussian process noise $\underline{w}_k \sim \mathcal{N}\left(\underline{0}, \mathbf{C}_k^w\right)$ with covariance matrix $\mathbf{C}_k^w \in \mathbb{R}^{n_x \times n_x}$.

To derive an estimate for the system's state, measurement data from sensor devices have to be acquired and processed. A multisensor system typically collects a vast number of measurements $\hat{\underline{z}}_k^i \in \mathbb{R}^{n_z}, i \in \{1, ..., N\}$. An observation $\hat{\underline{z}}_k^i$ provided by sensor device $i$ at time $k$ is related to the state through the linear measurement model

$$\underline{z}_k^i = \mathbf{H}_k^i \underline{x}_k + \underline{v}_k^i, \tag{11.2}$$

that is, $\hat{\underline{z}}_k^i$ is a realization of $\underline{z}_k^i$, where $\mathbf{H}_k^i \in \mathbb{R}^{n_z \times n_x}$ is the measurement matrix and $\underline{v}_k^i \sim \mathcal{N}\left(\underline{0}, \mathbf{C}_k^{z,i}\right)$ denotes a zero-mean white sensor noise with error covariance matrix $\mathbf{C}_k^{z,i} \in \mathbb{R}^{n_z \times n_z}$.

### 11.2.1 KALMAN FILTERING

With the model Equations 11.1 and 11.2 having a linear structure and being affected by Gaussian noise terms, state estimates for $\underline{x}_k$ can be computed recursively and in closed form in terms of the Kalman filter formulas. The Kalman filter algorithm [1], moreover, embodies an optimal solution to the state estimation problem providing estimates that minimize the mean-squared estimation error. The Kalman filter calculates the parameters $\left(\hat{\underline{x}}_k^e, \mathbf{C}_k^e\right)$, where $\hat{\underline{x}}_k^e \in \mathbb{R}^{n_x}$ is the estimate at time $k$ given all previous and current observations and $\mathbf{C}_k^e \in \mathbb{R}^{n_x \times n_x}$ is the corresponding error covariance matrix

$$\mathbf{C}_k^e = \mathrm{E}\left[\left(\hat{\underline{x}}_k^e - \underline{x}_k\right)\left(\hat{\underline{x}}_k^e - \underline{x}_k\right)^{\mathrm{T}}\right].$$

Its trace yields the mean-squared estimation error. The recursive processing of measurement data is carried out by combining the measurement information with prior information at each time $k$. The Kalman filter scheme is therefore composed of prediction and filtering steps. In the prediction step, the process model Equation 11.1 is employed to provide prior information for the subsequent time step $k + 1$. The prediction result $\left(\hat{\underline{x}}_{k+1}^p, \mathbf{C}_{k+1}^p\right)$ is computed according to

$$\hat{\underline{x}}_{k+1}^p = \mathbf{A}_k \hat{\underline{x}}_k^e + \mathbf{B}_k \hat{\underline{u}}_k \tag{11.3}$$

and

$$\mathbf{C}_{k+1}^p = \mathbf{A}_k \mathbf{C}_k^e \mathbf{A}_k^{\mathrm{T}} + \mathbf{C}_k^w. \tag{11.4}$$

The predicted estimate serves as prior information in the filtering step where it is combined with a current measurement vector $\hat{\underline{z}}_k^i$, which is related to the state by model Equation 11.2. The updated formulas are

$$\begin{aligned} \hat{\underline{x}}_k^e &= \hat{\underline{x}}_k^p + \mathbf{K}_k^i\left(\hat{\underline{z}}_k^i - \mathbf{H}_k^i \hat{\underline{x}}_k^p\right) \\ &= \left(\mathbf{I} - \mathbf{K}_k^i \mathbf{H}_k^i\right)\hat{\underline{x}}_k^p + \mathbf{K}_k^i \hat{\underline{z}}_k^i \end{aligned} \tag{11.5}$$

for the state estimate and

$$
\begin{aligned}
\mathbf{C}_k^{\mathrm{e}} &= \left(\mathbf{I} - \mathbf{K}_k^i \mathbf{H}_k^i\right) \mathbf{C}_k^{\mathrm{p}} \left(\mathbf{I} - \mathbf{K}_k^i \mathbf{H}_k^i\right)^{\mathrm{T}} + \mathbf{K}_k^i \mathbf{C}_k^{z,i} \left(\mathbf{K}_k^i\right)^{\mathrm{T}} \\
&= \mathbf{C}_k^{\mathrm{p}} - \mathbf{K}_k^i \mathbf{H}_k^i \mathbf{C}_k^{\mathrm{p}}.
\end{aligned}
\tag{11.6}
$$

for the corresponding error covariance matrix. In both formulas, the Kalman gain

$$
\mathbf{K}_k^i = \mathbf{C}_k^{\mathrm{p}} \left(\mathbf{H}_k^i\right)^{\mathrm{T}} \left(\mathbf{C}_k^{z,i} + \mathbf{H}_k^i \mathbf{C}_k^{\mathrm{p}} \left(\mathbf{H}_k^i\right)^{\mathrm{T}}\right)^{-1}
\tag{11.7}
$$

$$
= \left(\left(\mathbf{C}_k^{\mathrm{p}}\right)^{-1} + \left(\mathbf{H}_k^i\right)^{\mathrm{T}} \left(\mathbf{C}_k^{z,i}\right)^{-1} \mathbf{H}_k^i\right)^{-1} \left(\mathbf{H}_k^i\right)^{\mathrm{T}} \left(\mathbf{C}_k^{z,i}\right)^{-1}
\tag{11.8}
$$

is employed, which ensures an optimal combination in terms of a minimum mean-squared error. These formulas are apparently not restricted to a single measurement per time step; Equations 11.5 to 11.7 can simply be repeated for each available measurement. However, major difficulties arise when multiple sensors provide measurements over a network or multiple state estimation systems provide estimates to be fused.

### 11.2.2 MULTISENSOR SYSTEMS AND NETWORKS

In line with the rapid advances made in sensor and network technologies, there is a rising demand for distributed implementations of Kalman filter algorithms. The intended benefits include scalability, mobility, and robustness to failures. Sometimes, it is simply the need to use networked systems, for example, to monitor a large-scale phenomenon, that calls for distributed state estimation algorithms. From a visionary perspective, a network of sensor systems is capable of managing vast amounts of data, automatically responding to unpredictable changes of network conditions and environment, and reorganizing itself. In general, the design of state estimation algorithms cannot reflect these expectations without endangering reliability and compromising quality of estimates.

Apparently, distributed state estimation algorithms have to take into account the network architecture and are subject to network effects such as packet losses and delays. However, even a fully reliable network poses challenges to the task of estimating the system's state. Section 11.3 reveals dependencies between locally processed data as a major challenge for networked state estimation. Solutions to this challenge have to be oriented toward the underlying network architecture. More precisely, the state estimation algorithm depends on where and when measurement data is to be processed. On the basis of different, basic estimation architectures, Sections 11.4, 11.5, and 11.6 discuss several approaches to overcome these challenges.

### 11.3 SOURCES OF DEPENDENT INFORMATION

The standard formulation of the Kalman filter in Section 11.2.1 assumes conditional independence of measurements given the state, white process and measurement noise, and just-in-time processing. Many practical applications prevent these assumptions from being satisfied. In the last decades, much effort has been spent on making the Kalman filter robust to realistic violations of these assumptions. Extensions toward the treatment of dependencies due to colored noise terms, for instance, have been addressed in Ref. [2] by state augmentations. A processing of out-of-sequence measurements that may arise from packet delays in a network can be achieved by accumulated state densities [3] or delayed-state representations [4]. In general, the violated assumptions lead to correlations, that is, dependencies, that have to be properly treated. For example, an out-of-sequence measurement cannot be deemed to be conditionally independent of the current state anymore.

If dependencies across noise terms are known, the Kalman filter algorithm can be altered so as to still provide an optimal estimate. However, dependencies across different pieces of information in a networked system are often difficult to keep track of and to trace back. In particular, identifying dependencies becomes a serious problem if the network collects not only sensor data, but also the Kalman filter algorithm itself is distributed over the network [5]. Although distributed and decentralized Kalman filter implementations offer the advantage of handing over workload to the sensor nodes and of reducing the overall communication load, strong correlations among the outputs of the nodes can occur. The reason for this lies in the local acquisition and (pre-)processing of sensor data that is required to distribute the processing steps of the Kalman filter. In essence, the path data take to traverse the network and the redundant use of models are sources of underlying correlations.

### 11.3.1 DOUBLE-COUNTING OF COMMON SENSOR DATA

Double counting of data can be viewed as the most apparent reason for dependencies between data sets of different sensor nodes [6]. To establish efficient and communication-saving data transfers, the nodes do not simply transmit raw sensor data but usually preprocess and fuse received data before forwarding them. This leads to the problem that already processed information is concealed in the processing results and might erroneously be processed multiple times. Such a situation is depicted in Figure 11.1 where node A sends its data to node B and node C. Both node B and node C further process the received data and transmit their results to node D. Node D is not in the position to sort out that data from node A are included in both received data sets. Processing in node D possibly leads to double counting of data from node A. As shown in the following example, double counting leads to correlations.

### Example 11.1

Suppose that each node observes the state $\underline{x}$ with identity measurement matrix $\mathbf{H} = \mathbf{I}$, where each measurement serves as an estimate, i.e., $\hat{\underline{x}} = \hat{\underline{z}}$. Sensor node A provides a measurement $\hat{\underline{x}}_A$ with error covariance matrix $\mathbf{C}_A$. This information is transmitted to nodes B and C, which themselves have the observations $\hat{\underline{x}}_B$ and $\hat{\underline{x}}_C$, respectively. For the purpose of reducing the amount of data to be transferred, both nodes combine the received measurement with their own

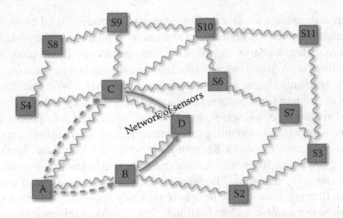

**FIGURE 11.1**  Sensor node D is not aware of data from node A that has been processed in node B as well as in node C.

measurement, that is, $\tilde{x}_B = (I - K_B)\hat{x}_B + K_B\hat{x}_A$ and $\tilde{x}_C = (I - K_C)\hat{x}_C + K_C\hat{x}_A$, where $K_B$ and $K_C$ are some locally determined gains. A node D that receives $\tilde{x}_B$ and $\tilde{x}_C$ without further information is not aware of the correlations between them due to $\hat{x}_A$. They are given by the cross-covariance term $E\left[(\tilde{x}_B - x)(\tilde{x}_C - x)^T\right] = K_B C_A K_C^T$.

## 11.3.2 COMMON PRIOR INFORMATION AND PROCESS NOISE

The second important source of dependencies is caused by parallel processing and modeling of the same noise terms. In particular, several Kalman filters that run in parallel on the same estimation problem automatically output correlated estimates. Correlations may already arise at the early initialization stage of the local Kalman filters, that is, an initialization of each estimator with the same prior estimate implies a full correlation between the local estimates. Besides possible common prior information, each Kalman filter usually employs the same state transition model Equation 11.1 and hence incorporates the same process noise in each prediction step. The problem of common process noise is especially encountered in track-to-track fusion applications [7,8] and decentralized state estimation [9]. The effect of common process noise is elucidated in the following example.

### Example 11.2

At time $k$, two local Kalman filter implementations provide the estimates $\hat{x}_A^e$ and $\hat{x}_B^e$ on $x_k$ that are supposed to have independent estimation errors, that is, $C_{AB}^e = E\left[\left(\hat{x}_A^e - x_k\right)\left(\hat{x}_B^e - x_k\right)^T\right] = 0$. The local prediction steps, according to Equation 11.3, yield $\hat{x}_A^p = A_k\hat{x}_A^e$ and $\hat{x}_B^p = A_k\hat{x}_B^e$ (no input $\hat{u}_k$). The predicted estimates now have correlated errors, and the cross-covariance matrix yields

$$
\begin{aligned}
C_{AB}^p &= E\left[\left(\hat{x}_A^p - x_k\right)\left(\hat{x}_B^p - x_k\right)^T\right] \\
&= E\left[\left(A_k\hat{x}_A^e - (A_k x_k + w_k)\right)\left(A_k\hat{x}_B^e - (A_k x_k + w_k)^T\right)\right] \\
&= A_k C_{AB}^e A_k^T + C_k^w = C_k^w.
\end{aligned}
$$

The prediction step hence causes dependencies, which appear in form of the process noise covariance matrix.

## 11.3.3 STRATEGIES FOR NETWORKED STATE ESTIMATION

To appropriately treat dependencies arising from in-network processing of estimates, the design of the Kalman filter algorithm heavily depends on and has to be adapted to the underlying and desired network architecture. Often, a coarse classification into centralized, distributed, and decentralized estimation architectures is proposed, which takes into consideration where sensor data is processed, when data are to be transmitted, and which knowledge nodes share. With this classification, different basic multisensor Kalman filter concepts can accordingly be categorized [5,10].

In a centralized architecture, a single node is charged with the task of processing all the sensor data recorded in the network and computing an estimate. Hence, all measurements have to be transmitted to a central node, which feeds a Kalman filter with the entire data. Section 11.4 discusses multisensor filtering principles. To reduce the amount of data to be communicated and to relieve the workload of the center node, distributed implementations of the Kalman filter can be pursued. For distributed Kalman filtering, cooperative nodes are typically required, and often still a central processing unit is present to assemble a global estimate from the local processing results. Section 11.5 points out that distributed estimation involves a tradeoff between an optimal estimation quality and a high robustness to node and communication failures. A decentralized architecture differs from a distributed one in that the estimation problem is solved locally on each sensor node. The nodes are usually designed to operate independently of each other and can share their information to solve the estimation problem on a global level. Although the nodes optimize the estimation quality on a local scale, a decentralized network generally cannot reach the estimation quality achieved by centralized or distributed architectures, because dependencies between the local estimates often cannot be reconstructed and exploited. Decentralized Kalman filtering is discussed in Section 11.6.

## 11.4   CENTRALIZED MULTISENSOR STATE ESTIMATION

A centralized processing of the entire sensor data offers the advantage that the conditional independence of the measurements can still be exploited and, in particular, the standard Kalman filter formulas can be used. Therefore, the problem of common process noise cannot occur in this setup, and double counting of data is avoided, for example, by labeling data. By contrast, each node must transmit its sensor data to the central processing unit, which can lead to high traffic, particularly in multihop networks. The first part of this section discusses the processing of multisensor data within the Kalman filter. In the second part, the information form is used to preprocess sensor data efficiently during transmission to reduce the amount of data to be transmitted.

### 11.4.1   Multisensor Kalman Filtering

For the processing of multiple sensor data, two possibilities can essentially be named. The set $\mathcal{Z}_k = \left\{ \hat{\underline{z}}_k^1, \ldots, \hat{\underline{z}}_k^N \right\}$ of measurements that are received at a time step $k$ can be processed either sequentially or en bloc [11]. Each measurement is related to the state through the corresponding model Equation 11.2. For a sequential processing as illustrated in Figure 11.2, the Kalman filter (Equations 11.5 through 11.7) are applied recursively to each measurement, that is,

$$\tilde{\underline{x}}_k^{e,1} = \left( \mathbf{I} - \mathbf{K}_k^1 \mathbf{H}_k^1 \right) \hat{\underline{x}}_k^p + \mathbf{K}_k^1 \hat{\underline{z}}_k^1, \qquad \tilde{\mathbf{C}}_k^{e,1} = \mathbf{C}_k^p - \mathbf{K}_k^1 \mathbf{H}_k^1 \mathbf{C}_k^p,$$

$$\tilde{\underline{x}}_k^{e,2} = \left( \mathbf{I} - \mathbf{K}_k^2 \mathbf{H}_k^2 \right) \tilde{\underline{x}}_k^{e,1} + \mathbf{K}_k^2 \hat{\underline{z}}_k^2, \qquad \tilde{\mathbf{C}}_k^{e,2} = \tilde{\mathbf{C}}_k^{e,1} - \mathbf{K}_k^2 \mathbf{H}_k^2 \tilde{\mathbf{C}}_k^{e,1},$$

$$\vdots \qquad\qquad\qquad\qquad \vdots$$

$$\hat{\underline{x}}_k^e = \left( \mathbf{I} - \mathbf{K}_k^N \mathbf{H}_k^N \right) \tilde{\underline{x}}_k^{e,N-1} + \mathbf{K}_k^N \hat{\underline{z}}_k^N, \qquad \mathbf{C}_k^e = \tilde{\mathbf{C}}_k^{e,N-1} - \mathbf{K}_k^N \mathbf{H}_k^N \tilde{\mathbf{C}}_k^{e,N-1}$$

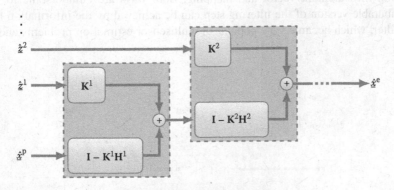

**FIGURE 11.2**   Sequential multisensor Kalman filtering.

with

$$\mathbf{K}_k^i = \tilde{\mathbf{C}}_k^{e,i-1} \left(\mathbf{H}_k^i\right)^{\mathrm{T}} \left(\mathbf{C}_k^{z,i} + \mathbf{H}_k^i \tilde{\mathbf{C}}_k^{e,i-1} \left(\mathbf{H}_k^i\right)^{\mathrm{T}}\right)^{-1}.$$

The last equation yields the filtering result, which is independent of the order of processing the measurements.

Instead of performing $N$ filtering steps, the measurement update can also be carried out in a single step. For this, the measurements are concatenated into a single vector $\hat{\underline{z}}_k^{\mathrm{total}} := \left[\left(\hat{\underline{z}}_k^1\right)^{\mathrm{T}}, \ldots, \left(\hat{\underline{z}}_k^N\right)^{\mathrm{T}}\right]^{\mathrm{T}}$. Similarly, the matrices $\mathbf{H}_k^{\mathrm{total}}$ and $\mathbf{C}_k^{z,\mathrm{total}}$ have to be compiled from local measurement and covariance matrices to compute the en-bloc filtering result

$$\hat{\underline{x}}_k^e = \left(\mathbf{I} - \mathbf{K}_k^{\mathrm{total}} \mathbf{H}_k^{\mathrm{total}}\right) \hat{\underline{x}}_k^p + \mathbf{K}_k^{\mathrm{total}} \hat{\underline{z}}_k^{\mathrm{total}},$$
$$\mathbf{C}_k^e = \mathbf{C}_k^p - \mathbf{K}_k^{\mathrm{total}} \mathbf{H}_k^{\mathrm{total}} \mathbf{C}_k^p,$$

where the gain is given by

$$\mathbf{K}_k^{\mathrm{total}} = \mathbf{C}_k^p \left(\mathbf{H}_k^{\mathrm{total}}\right)^{\mathrm{T}} \left(\mathbf{C}_k^{z,\mathrm{total}} + \mathbf{H}_k^{\mathrm{total}} \mathbf{C}_k^p \left(\mathbf{H}_k^{\mathrm{total}}\right)^{\mathrm{T}}\right)^{-1}.$$

This scheme is depicted in Figure 11.3.

Both processing schemes require that all measurements taken at time step $k$ are available at the center node. With the help of state augmentations [3], also delayed measurements can be processed and also less frequent data transfers can be established. However, the amount of data to be transmitted is not reduced. A further problem in multihop networks is that nodes without a direct link to the center node have to pass on their data to other nodes. Hence, nodes in proximity to the center node might suffer from a high load of transfers. The information form presented in the following section allows one to condense data efficiently along any communication path.

## 11.4.2 INFORMATION FILTERING

The processing of multiple measurements consists of either nested Kalman filtering steps or constructing a joint measurement vector and mapping. Both ways are cumbersome for the central node. A distributable version of the filtering step can be achieved by the information form [12] of the Kalman filter, which became very popular in multisensor estimation problems and essentially

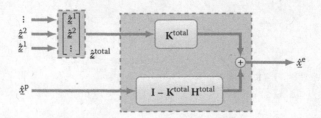

**FIGURE 11.3** Blockwise multisensor Kalman filtering.

encompasses an algebraic reformulation. In place of the mean and covariance matrix, the information vector

$$\hat{\underline{y}}_k := \mathbf{C}_k^{-1}\hat{\underline{x}}_k$$

and the information matrix

$$\mathbf{Y}_k := \mathbf{C}_k^{-1}$$

are considered. The measurements are also transformed according to

$$\underline{i}_k^i = \left(\mathbf{H}_k^i\right)^{\mathrm{T}}\left(\mathbf{C}_k^{z,i}\right)^{-1}\hat{\underline{z}}_k^i \quad \text{and} \quad \mathbf{I}_k^i = \left(\mathbf{H}_k^i\right)^{\mathrm{T}}\left(\mathbf{C}_k^{z,i}\right)^{-1}\mathbf{H}_k^i.$$

The filtering step then only requires the computation of the simple sums

$$\hat{\underline{y}}_k^e = \hat{\underline{y}}_k^p + \sum_{i=1}^{N}\underline{i}_k^i = \left(\mathbf{C}_k^p\right)^{-1}\hat{\underline{x}}_k^p + \sum_{i=1}^{N}\left(\mathbf{H}_k^i\right)^{\mathrm{T}}\left(\mathbf{C}_k^{z,i}\right)^{-1}\hat{\underline{z}}_k^i \tag{11.9}$$

and

$$\mathbf{Y}_k^e = \mathbf{Y}_k^p + \sum_{i=1}^{N}\mathbf{I}_k^i = \left(\mathbf{C}_k^p\right)^{-1} + \sum_{i=1}^{N}\left(\mathbf{H}_k^i\right)^{\mathrm{T}}\left(\mathbf{C}_k^{z,i}\right)^{-1}\mathbf{H}_k^i \tag{11.10}$$

for the information vector and information matrix, respectively. As illustrated in Figure 11.4, this reformulation offers the significant advantage that parts of the sum can be carried out along each communication path, that is, each node can collect received information vectors, combine them with its own information vector, and just needs to forward a single information vector to the next node on the path.

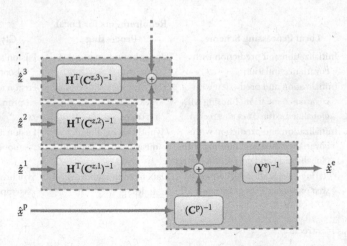

**FIGURE 11.4**  Information filtering with preprocessing of sensor data.

The prediction step can also be formulated in terms of the information parameters $\hat{\underline{y}}_k^e$ and $\mathbf{Y}_k^e$. With $\mathbf{L}_{k+1} := \mathbf{A}_k\left(\mathbf{Y}_k^e\right)^{-1}$, the predicted information matrix becomes

$$\mathbf{Y}_{k+1}^p = \left(\mathbf{L}_{k+1}\mathbf{Y}_k^e\mathbf{L}_{k+1}^T + \mathbf{B}_k\mathbf{C}_k^u\mathbf{B}_k^T\right)^{-1},$$

and the predicted information vector yields

$$\hat{\underline{y}}_{k+1}^p = \mathbf{Y}_{k+1}^p\left(\mathbf{L}_{k+1}\,\hat{\underline{y}}_k^e + \mathbf{B}_k\,\hat{\underline{u}}_k\right).$$

The corresponding state estimate is obtained through the simple reverse transformations $\hat{\underline{x}}_k = \mathbf{Y}_k^{-1}\,\hat{\underline{y}}_k$ and $\mathbf{C}_k = \mathbf{Y}_k^{-1}$ for mean and covariance matrix, respectively.

## 11.5  COOPERATIVE DISTRIBUTED STATE ESTIMATION

The previous section has pointed out that the filtering step already allows for a distributable reformulation by employing the inverse covariance form of the Kalman filter. However, the central node must access the measurement data of each node and at each time step, which can turn out to be inefficient in terms of communication and computation load. Distributed implementations of the Kalman filter algorithm aspire to distribute the workload among the participating nodes. In general, a central node is still present that finally puts the nodes' data together into an estimate of the state. In contrast to the centralized multisensor Kalman filtering scheme, this is necessary only when an estimate is requested and thereby significantly reduces the communication rate. The main challenges that have to be addressed by the concepts in this section refer to dependencies that can be traced back to the initialization step and the problem of common process noise.

Distributed Kalman filtering, in general, requires a tradeoff between estimation quality and robustness to node and communication failures. Table 11.1 provides an overview of the concepts to be discussed in the following. In particular, it highlights which knowledge must be accessible by the

**TABLE 11.1**

**Overview of Distributed Kalman Filtering Schemes**

| Method | Local Processing Scheme | Requirements for Local Processing | Global Estimate |
|---|---|---|---|
| Federated Kalman filter | Initialization and prediction with covariance inflation | Upper bound on number of nodes | Fusion at center node, suboptimal[a] (conservative) |
| Distributed Kalman filter | Initialization and prediction with covariance inflation, filtering with globalized covariance matrix | Total number of nodes, all measurement and noise matrices | Fusion at center node, optimal |
| Hypothesizing Kalman filter | Initialization and prediction with covariance inflation, filtering with globalization and correction matrix | Hypothesis on global measurement capacity | Fusion at center node, suboptimal[b] |
| Consensus Kalman filter | Filtering with additional consensus step on sensor data or estimates | Data from neighboring nodes | Synchronization, asymptotically optimal |

[a]  One-step optimal.

[b]  Optimal if hypothesis is correct.

nodes, which modifications to the local Kalman-filter-like processing schemes are necessary, and where a global estimate is computed.

## 11.5.1 Federated Kalman Filtering

The federated Kalman filter [13,14] is based on the idea that each node runs its own Kalman filter, which computes an estimate based on the locally available sensor data. For a global estimate, the local Kalman filter results then have to be fused by the central node. As stated in Section 11.3.2, the local Kalman filters can neither be initialized with the same prior estimates nor employ the standard process model without causing strong correlations between their reported estimation errors. The federated Kalman filter therefore alters the local initialization and prediction steps such that the local estimates have independent error variances.

To avoid dependencies between the local processing results, the joint estimation error

$$
\tilde{\underline{x}}_k = \begin{bmatrix} \hat{\underline{x}}_k^{e,1} - \underline{x}_k \\ \vdots \\ \hat{\underline{x}}_k^{e,N} - \underline{x}_k \end{bmatrix} = \begin{bmatrix} \hat{\underline{x}}_k^{e,1} \\ \vdots \\ \hat{\underline{x}}_k^{e,N} \end{bmatrix} - \begin{bmatrix} \mathbf{I} \\ \vdots \\ \mathbf{I} \end{bmatrix} \underline{x}_k
$$

is considered. If the local estimator of each node $i \in \{1, \ldots, N\}$ is initialized with the same prior estimate $\left(\hat{\underline{x}}_0^{p,i}, \mathbf{C}_0^{p,i}\right) := \left(\hat{\underline{x}}_0^{p}, \mathbf{C}_0^{p}\right)$, each cross-covariance matrix becomes $\mathrm{E}\left[\left(\hat{\underline{x}}_0^{p,i} - \underline{x}_k\right)\left(\hat{\underline{x}}_0^{p,j} - \underline{x}_k\right)^{\mathrm{T}}\right] = \mathbf{C}_0^{p}, i,j \in \{1, \ldots, N\}$. This implies that the joint error covariance matrix $\tilde{\mathbf{C}}_0 = \mathrm{E}\left[\tilde{\underline{x}}_0 \, \tilde{\underline{x}}_0^{\mathrm{T}}\right]$ is fully occupied. The federated Kalman filter replaces this joint covariance matrix by the inflated joint covariance matrix

$$
\begin{bmatrix} \frac{1}{\omega_1}\mathbf{C}_0^{p} & 0 & \cdots & 0 \\ 0 & \frac{1}{\omega_2}\mathbf{C}_0^{p} & \ddots & \vdots \\ \vdots & \ddots & \ddots & 0 \\ 0 & \cdots & 0 & \frac{1}{\omega_N}\mathbf{C}_0^{p} \end{bmatrix} \geq \begin{bmatrix} \mathbf{C}_0^{p} & \mathbf{C}_0^{p} & \cdots & \mathbf{C}_0^{p} \\ \mathbf{C}_0^{p} & \mathbf{C}_0^{p} & \ddots & \vdots \\ \vdots & \ddots & \ddots & \mathbf{C}_0^{p} \\ \mathbf{C}_0^{p} & \cdots & \mathbf{C}_0^{p} & \mathbf{C}_0^{p} \end{bmatrix} = \tilde{\mathbf{C}}_0 \qquad (11.11)
$$

with $\sum_{i=1}^{N} \omega_i = 1$ and $\omega_i \geq 0$, which then implies that each sensor node is initialized with $\left(\hat{\underline{x}}_0^{p,i}, \mathbf{C}_0^{p,i}\right) := \left(\hat{\underline{x}}_0^{p}, \frac{1}{\omega_i}\mathbf{C}_0^{p}\right)$. With the initial error covariance matrix $\mathbf{C}_0^{p}$ multiplied by $1/\omega_i$, the initial estimates can then be deemed as independent. This technique is known as covariance inflation [15,16] and is mainly used for the treatment of unknown correlations between estimation errors (see Section 11.6). The inflation parameters $\omega_i$ depend on the number $N$ of participating nodes.

The federated Kalman filter primarily aims at counteracting the problem of common process noise. As explained in Section 11.3.2, dependencies in form of the process noise covariance matrix $\mathbf{C}_k^{w}$ are caused by the prediction step, that is, $\mathrm{E}\left[\left(\mathbf{A}_k \, \hat{\underline{x}}_k^{e,i} - \underline{x}_{k+1}\right)\left(\mathbf{A}_k \, \hat{\underline{x}}_k^{e,j} - \underline{x}_{k+1}\right)^{\mathrm{T}}\right] = \mathbf{C}_k^{w}, i \neq j$ is the cross-covariance matrix between two local, independent estimates $\hat{\underline{x}}_k^{e,i}$ and $\hat{\underline{x}}_k^{e,j}$ after prediction. For this reason, again an inflated version

$$
\begin{bmatrix}
\dfrac{1}{\omega_1}\mathbf{C}_k^w & 0 & \cdots & 0 \\
0 & \dfrac{1}{\omega_2}\mathbf{C}_k^w & \ddots & \vdots \\
\vdots & \ddots & \ddots & 0 \\
0 & \cdots & 0 & \dfrac{1}{\omega_N}\mathbf{C}_k^w
\end{bmatrix}
\geq
\begin{bmatrix}
\mathbf{C}_k^w & \mathbf{C}_k^w & \cdots & \mathbf{C}_k^w \\
\mathbf{C}_k^w & \mathbf{C}_k^w & \ddots & \vdots \\
\vdots & \ddots & \ddots & \mathbf{C}_k^w \\
\mathbf{C}_k^w & \cdots & \mathbf{C}_k^w & \mathbf{C}_k^w
\end{bmatrix}
\tag{11.12}
$$

of the joint process noise matrix with $\sum_{i=1}^{N} \omega_i = 1$ and $\omega_i \geq 0$ is used. The prediction of each local estimate $\left(\hat{\underline{x}}_k^{e,i}, \mathbf{C}_k^{e,i}\right)$ at time $k$ is then computed according to

$$
\hat{\underline{x}}_{k+1}^{p,i} = \mathbf{A}_k\, \hat{\underline{x}}_k^{e,i}
$$

and

$$
\mathbf{C}_{k+1}^{p,i} = \mathbf{A}_k \mathbf{C}_k^{e,i} \mathbf{A}_k^{T} + \frac{1}{\omega_i}\mathbf{C}_k^w.
\tag{11.13}
$$

Compared to the standard prediction formulas (Equations 11.3 and 11.4), the only difference lies in the inflated process noise matrix $\dfrac{1}{\omega_i}\mathbf{C}_k^w$. However, this modification allows one to construct the cross-covariance terms

$$
\mathrm{E}\left[\left(\mathbf{A}_k\, \hat{\underline{x}}_k^{e,i} - \underline{x}_{k+1}\right)\left(\mathbf{A}_k\, \hat{\underline{x}}_k^{e,j} - \underline{x}_{k+1}\right)^{T}\right] =
\begin{cases}
\mathbf{A}_k \mathbf{C}_k^{e,i} \mathbf{A}_k^{T} + \dfrac{1}{\omega_i}\mathbf{C}_k^w, & i = j, \\[2mm]
\mathbf{0}, & i \neq j,
\end{cases}
$$

that is, the errors of the predicted estimates are regarded as being uncorrelated. More precisely, by using the left-hand sides of the inequalities in Equations 11.11 and 11.12, the local estimation errors can be deemed to be independent. Hence, there is no need to store and reconstruct any dependencies between the local estimates.

The filtering of local measurements can simply be carried out by applying the standard Kalman filter formulas (Equations 11.5 and 11.6) locally. In the central node, a global estimate can be formed by combining the local estimates according to

$$
\hat{\underline{x}}_k^{fus} = \mathbf{C}_k^{fus} \sum_{i=1}^{N} \left(\mathbf{C}_k^{e,i}\right)^{-1} \hat{\underline{x}}_k^{e,i}
\tag{11.14}
$$

and

$$
\mathbf{C}_k^{fus} = \left(\sum_{i=1}^{N} \left(\mathbf{C}_k^{e,i}\right)^{-1}\right)^{-1}.
\tag{11.15}
$$

This combination corresponds to the optimal fusion formulas discussed in Section 11.6.1 in the case of independent estimates. The fusion result is optimal in terms of the mean-squared error when only a single prediction and filtering step takes place before fusion. After several processing steps, the federated Kalman filter generally does not reach the quality of the schemes from Section 11.4, that is, the federated Kalman filter is optimal when a fusion and reinitialization takes place after each time step and is otherwise suboptimal. With the bounds in Equations 11.11 and 11.12, conservative estimates are computed so that nodes may also fail and still a valid fusion result is attainable.

## 11.5.2 Optimal Distributed Kalman Filtering

The distributed Kalman filter [17,18] can be characterized as a further development of the federated Kalman filter and provides the optimal estimation result after any number of time steps whereas the federated formulation is optimal only in special cases. Given the prior knowledge $\left(\hat{\underline{x}}_0^p, \mathbf{C}_0^p\right)$, each local node can be initialized with

$$\left(\overline{\underline{x}}_0^{p,i}, \overline{\mathbf{C}}_0^p\right) := \left(\hat{\underline{x}}_0^p, N \cdot \mathbf{C}_0^p\right), \tag{11.16}$$

which resembles the inflation technique (Equation 11.11) for the federated Kalman filter when each factor is set to $\omega_i = \dfrac{1}{N}$, and $N$ is the number of nodes in the network. The node index $i$ in $\overline{\mathbf{C}}_0^p$ is omitted as this matrix is the same for each node.

The prediction step performed by each node is also similar to the federated Kalman filter. For the system model in Equation 11.1 without input, each node processes its parameter set $\left(\overline{\underline{x}}_k^{e,i}, \overline{\mathbf{C}}_k^e\right)$ according to

$$\overline{\underline{x}}_{k+1}^{p,i} = \mathbf{A}_k\, \overline{\underline{x}}_k^{e,i} \tag{11.17}$$

and

$$\overline{\mathbf{C}}_{k+1}^p = \mathbf{A}_k \overline{\mathbf{C}}_k^e \mathbf{A}_k + N \cdot \mathbf{C}_k^w, \tag{11.18}$$

where $\overline{\mathbf{C}}_{k+1}^p$ is the same for each node as long as they have the same $\overline{\mathbf{C}}_k^e$. Furthermore, Equation 11.18 is identical to the federated prediction step (Equation 11.13) for $\omega_i = \dfrac{1}{N}$. If the relationship

$$\hat{\underline{x}}_k^e = \frac{1}{N} \sum_{i=1}^{N} \overline{\underline{x}}_k^{e,i} \quad \text{and} \quad \mathbf{C}_k^e = \frac{1}{N} \overline{\mathbf{C}}_k^e \tag{11.19}$$

holds for the local estimates $\left(\overline{\underline{x}}_k^{e,i}, \overline{\mathbf{C}}_k^e\right)$, that is, Equation 11.19 gives the centrally optimal estimate, then the same holds for the predicted results $\left(\overline{\underline{x}}_{k+1}^{p,i}, \overline{\mathbf{C}}_{k+1}^p\right)$.

The filtering step significantly differs from the federated Kalman filter. A locally available measurement $\hat{\underline{z}}_k^i$ at time $k$ is combined with $\overline{\underline{x}}_{k+1}^{p,i}$ through

$$\overline{\underline{x}}_k^{e,i} = \overline{\mathbf{C}}_k^e \left( \left(\overline{\mathbf{C}}_k^p\right)^{-1} \overline{\underline{x}}_k^{p,i} + \left(\mathbf{H}_k^i\right)^{\mathsf{T}} \left(\mathbf{C}_k^{z,i}\right)^{-1} \hat{\underline{z}}_k^i \right), \tag{11.20}$$

which is similar to the standard update (Equation 11.5) with the inverse covariance gain (Equation 11.8). However, the difference lies in the matrix $\bar{\mathbf{C}}_k^{\mathrm{e}}$, which is computed by

$$\bar{\mathbf{C}}_k^{\mathrm{e}} = \left( \left( \bar{\mathbf{C}}_k^{\mathrm{p}} \right)^{-1} + \left( \bar{\mathbf{C}}_k^z \right)^{-1} \right)^{-1}, \tag{11.21}$$

where the globalized measurement covariance matrix

$$\left( \bar{\mathbf{C}}_k^z \right)^{-1} = \frac{1}{N} \sum_{j=1}^N \left( \mathbf{H}_k^j \right)^{\mathrm{T}} \left( \mathbf{C}_k^{z,j} \right)^{-1} \mathbf{H}_k^j \tag{11.22}$$

instead of only $\left( \mathbf{H}_k^i \right)^{\mathrm{T}} \left( \mathbf{C}_k^{z,i} \right)^{-1} \mathbf{H}_k^i$ is used. More precisely, the matrix (Equation 11.22) incorporates the sensor parameters of all nodes in the network, that is, all measurement and noise matrices. With this globalized covariance matrix, the local updates $\left( \bar{x}_k^{\mathrm{e},i}, \bar{\mathbf{C}}_k^{\mathrm{e}} \right)$ again share the same parameter $\bar{\mathbf{C}}_k^{\mathrm{e}}$. Of course, the vectors $\bar{x}_k^{\mathrm{e},i}$ are different from each other due to the local measurement outcomes in Equation 11.20. It is important to notice that none of the sets $\left( \bar{x}_k^{\mathrm{e},i}, \bar{\mathbf{C}}_k^{\mathrm{e}} \right)$ represents a valid state estimate, because both $\bar{x}_k^{\mathrm{e},i}$ and $\bar{\mathbf{C}}_k^{\mathrm{e}}$ have been computed with the aid of the globalized matrix (Equation 11.21) instead of the matrices $\left( \mathbf{H}_k^i \right)^{\mathrm{T}} \left( \mathbf{C}_k^{z,i} \right)^{-1} \mathbf{H}_k^i$, which refer only to the local measurement model.

The reason for the globalization becomes apparent when the fusion result at the center node is scrutinized. The fusion formulas are given by the invariant relation in Equation 11.19, which yields

$$\begin{aligned} \left( \mathbf{C}_k^{\mathrm{fus}} \right)^{-1} &= \left( \frac{1}{N} \bar{\mathbf{C}}_k^{\mathrm{e}} \right)^{-1} = N \left( \bar{\mathbf{C}}_k^{\mathrm{p}} \right)^{-1} + N \left( \bar{\mathbf{C}}_k^z \right)^{-1} \\ &= \left( \mathbf{C}_k^{\mathrm{p}} \right)^{-1} + \sum_{j=1}^N \left( \mathbf{H}_k^j \right)^{\mathrm{T}} \left( \mathbf{C}_k^{z,j} \right)^{-1} \mathbf{H}_k^j, \end{aligned} \tag{11.23}$$

for the fused covariance matrix and

$$\begin{aligned} \hat{\underline{x}}_k^{\mathrm{fus}} &= \frac{1}{N} \sum_{i=1}^N \bar{x}_k^{\mathrm{e},i} = \frac{1}{N} \sum_{i=1}^N \bar{\mathbf{C}}_k^{\mathrm{e}} \left( \left( \bar{\mathbf{C}}_k^{\mathrm{p}} \right)^{-1} \bar{x}_k^{\mathrm{p},i} + \left( \mathbf{H}_k^i \right)^{\mathrm{T}} \left( \mathbf{C}_k^{z,i} \right)^{-1} \hat{\underline{z}}_k^i \right) \\ &= \mathbf{C}_k^{\mathrm{fus}} \sum_{i=1}^N \left( \bar{\mathbf{C}}_k^{\mathrm{p}} \right)^{-1} \bar{x}_k^{\mathrm{p},i} + \mathbf{C}_k^{\mathrm{fus}} \sum_{i=1}^N \left( \mathbf{H}_k^i \right)^{\mathrm{T}} \left( \mathbf{C}_k^{z,i} \right)^{-1} \hat{\underline{z}}_k^i \\ &= \mathbf{C}_k^{\mathrm{fus}} \left( \mathbf{C}_k^{\mathrm{p}} \right)^{-1} \hat{\underline{x}}_k^{\mathrm{p}} + \mathbf{C}_k^{\mathrm{fus}} \sum_{i=1}^N \left( \mathbf{H}_k^i \right)^{\mathrm{T}} \left( \mathbf{C}_k^{z,i} \right)^{-1} \hat{\underline{z}}_k^i \end{aligned} \tag{11.24}$$

for the fused estimate, where the invariant relation Equation 11.19 has also been employed for $\bar{x}_k^{\mathrm{p},i}$ and $\bar{\mathbf{C}}_k^{\mathrm{p}}$. These fusion formulas resemble Equations 11.9 and 11.10 of the information filter. Hence, the optimal estimate is obtained as if all sensor data have been processed by a centralized Kalman filter. In contrast to the centralized concepts in Section 11.4, measurements can here be processed locally and need only to be transferred to the center unit when an estimate is requested. However, the globalization (Equation 11.21) clearly demonstrates that this approach may suffer from strict requirements regarding the local availability of knowledge about utilized models. Each sensor node must be in the position to compute the globalized covariance matrix (Equation 11.22) and therefore must be aware of the other nodes' sensor models, that is, of the measurement matrices $\mathbf{H}_k^i$ and noise matrices $\mathbf{C}_k^{z,i}$ of each other node. In particular, this means that the optimal distributed Kalman filter is not robust to communication and node failures. The local parameters $\left( \overline{x}_k^{e,i}, \overline{\mathbf{C}}_k^e \right)$ itself provide no information about the state, and they must be available to the center node in their entirety before an estimate of the state can be computed. This stands in contrast to the federated Kalman filter, which provides valid local estimates and is more robust.

### 11.5.3  HYPOTHESIZING KALMAN FILTERING

The distributed Kalman filter suffers from the following severe problem: any failure prevents the computation of an unbiased estimate. If any failure occurs and is not reported to every node, the correct globalized matrix (Equation 11.21) cannot be derived. As a further consequence, each incorrect globalized matrix affects subsequent processing and fusion steps, and a valid estimate cannot be obtained anymore. This problem can be addressed by correction matrices $\Delta_k^{e,i}$ that are computed in line with the local parameters $\left( \overline{x}_k^{e,i}, \overline{\mathbf{C}}_k^e \right)$ [19,20]. With the aid of the corresponding correction matrix, each parameter $\overline{x}_k^{e,i}$ can be transformed into an unbiased estimate

$$\tilde{x}_k^e = \left( \Delta_k^{e,i} \right)^{-1} \overline{x}_k^{e,i} \tag{11.25}$$

with $\mathrm{E}\left[ \tilde{x}_k^e - x_k \right] = \underline{0}$, that is, the matrix $\Delta_k^{e,i}$ effects a debiasing of $\overline{x}_k^{e,i}$.

If the initialization of the local parameters is carried out by means of Equation 11.16, the prior correction matrix at each node $i$ is the identity $\Delta_0^{p,i} = \mathbf{I}$. At a time step $k$, the prediction formulas (Equations 11.17 and 11.18) are accompanied with the update

$$\Delta_{k+1}^{p,i} = \mathbf{A}_k \Delta_k^{e,i} \mathbf{A}_k^{-1}$$

of the current correction matrix $\Delta_k^{e,i}$, which gives the unbiased estimate

$$\tilde{x}_{k+1}^p = \left( \Delta_{k+1}^{p,i} \right)^{-1} \overline{x}_{k+1}^{p,i} = \mathbf{A}_k \left( \Delta_k^{e,i} \right)^{-1} \overline{x}_k^{e,i} = \mathbf{A}_k \tilde{x}_k^e.$$

This result complies with the prediction of the unbiased estimate (Equation 11.25) and, hence, is still unbiased.

In the filtering step, the corresponding correction matrix is given by

$$\Delta_k^{e,i} = \overline{\mathbf{C}}_k^e \left( \left( \overline{\mathbf{C}}_k^p \right)^{-1} \Delta_k^{p,i} + \left( \mathbf{H}_k^i \right)^{\mathrm{T}} \left( \mathbf{C}_k^{z,i} \right)^{-1} \mathbf{H}_k^i \right),$$

where $\overline{\mathbf{C}}_k^p$ is the predicted globalized matrix (Equation 11.18), and $\overline{\mathbf{C}}_k^e$ is the updated globalized matrix (Equation 11.21). With these matrices, any local parameter $\overline{x}_k^{e,i}$ can be turned into an unbiased estimate of the state, irrespective of whether the globalized matrix (Equation 11.22) has been computed correctly or failures have happened.

The correction matrices are of particular use when a fusion result is to be computed, but errors and node failures may have occurred during the local processing. Even missing data at the central node can be treated. Let $\mathcal{M} \subseteq \{1, \ldots, N\}$ contain the indices of those nodes whose data are available to the center node. There, the fused parameter

$$\overline{x}_k^{\text{fus}} = \frac{1}{N} \sum_{i \in \mathcal{M}} \overline{x}_k^{e,i}$$

can be computed according to Equation 11.24. Owing to node failures or missing data, $\overline{x}_k^{\text{fus}}$ is potentially biased, but with the aid of the fused correction matrix

$$\Delta_k^{\text{fus}} = \frac{1}{N} \sum_{i \in \mathcal{M}} \Delta_k^{e,i},$$

the unbiased state estimate

$$\tilde{x}_k^{\text{fus}} = (\Delta^{\text{fus}})^{-1} \overline{x}_k^{\text{fus}}$$

is obtained.

The presented technique can not only be used to make the distributed Kalman filter robust to node and communication failures, but can also be employed to replace the globalized covariance matrix (Equation 11.22) by a hypothesis on the global measurement capacity such that each node does not need to have access to each other node's sensor models. More precisely, the hypothesizing Kalman filtering scheme can be employed to assess the global measurement performance of the entire network without the need to assess the measurement quality of each node separately. Compared to the centralized Kalman filter from Section 11.4, the hypothesizing Kalman filter provides suboptimal estimates. If the hypothesis coincides with the actual globalized covariance matrix, the fusion result is even optimal.

### 11.5.4 CONSENSUS KALMAN FILTERING

Consensus Kalman filtering [21,22] has become a well-known concept for distributed state estimation and pursues a different strategy compared to the previously considered filtering schemes. A global estimate on the state is not computed at a central unit, but the nodes agree on a consensus estimate. This idea is justified by the perception that each sensor node provides an estimate that refers to the same state.

Essentially, two different possibilities have been studied to arrive at a consensus. Either a consensus on sensor data is aspired [21], or a consensus on estimates is computed [22]. It is interesting to note that the local processing that is required for a consensus on sensor data bears considerable resemblance to the optimal distributed Kalman filter from Section 11.5.2. At the initialization step, each node is also provided with an inflated covariance matrix, that is,

$$\left( \overline{x}_0^{p,i}, \overline{\mathbf{C}}_0^{p,i} \right) := \left( \hat{x}_0^p, N \cdot \mathbf{C}_0^p \right),$$

where $N$ is the number of nodes in the network. For every local estimate $\left( \overline{x}_k^{e,i}, \overline{\mathbf{C}}_k^{e,i} \right)$ at time $k$, the same number is used in prediction step

$$\overline{x}_{k+1}^{p,i} = \mathbf{A}_k \overline{x}_k^{e,i}$$

and

$$\overline{\mathbf{C}}_{k+1}^{p,i} = \mathbf{A}_k \overline{\mathbf{C}}_k^e \mathbf{A}_k + N \cdot \mathbf{C}_k^w$$

to inflate the process noise matrix.

In the filtering step, the local prior information $\left(\overline{x}_k^{p,i}, \overline{\mathbf{C}}_k^{p,i}\right)$ is updated according to

$$\overline{x}_k^{e,i} = \overline{\mathbf{C}}_k^e \left( \left(\overline{\mathbf{C}}_k^p\right)^{-1} \overline{x}_k^{p,i} + \overline{z}_k \right), \tag{11.26}$$

and

$$\overline{\mathbf{C}}_k^e = \left( \left(\overline{\mathbf{C}}_k^p\right)^{-1} + \left(\overline{\mathbf{C}}_k^z\right)^{-1} \right)^{-1}. \tag{11.27}$$

If the parameters $\overline{z}_k$ and $\overline{\mathbf{C}}_k^z$ would be given by

$$\overline{z}_k = \frac{1}{N} \sum_{i=j}^{N} \left(\mathbf{H}_k^j\right)^{\mathrm{T}} \left(\mathbf{C}_k^{z,j}\right)^{-1} \hat{z}_k^j \tag{11.28}$$

and

$$\left(\overline{\mathbf{C}}_k^z\right)^{-1} = \frac{1}{N} \sum_{j=1}^{N} \left(\mathbf{H}_k^j\right)^{\mathrm{T}} \left(\mathbf{C}_k^{z,j}\right)^{-1} \mathbf{H}_k^j, \tag{11.29}$$

then Equation 11.26 corresponds to the multisensor filtering formula 11.9 while Equation 11.29 overestimates the true error covariance matrix (Equation 11.10) by factor $N$. Hence, each local $\overline{x}_k^{e,i}$ would yield the globally optimal estimate $\hat{x}_k^e$. Note that Equation 11.29 complies with Equation 11.22, but Equation 11.28 differs from the measurement used in Equation 11.20.

As the global parameters given in Equations 11.28 and 11.29 are not available to each node, a consensus algorithm is used to approximate these parameters. Each node sends its sensor data to its neighboring nodes $\mathcal{N}(i) \subseteq \{1, \ldots, N\}$, where each local filtering step is accompanied by the consensus steps

$$\overline{z}_k = \mathbf{W}_{ii} \left(\mathbf{H}_k^i\right)^{\mathrm{T}} \left(\mathbf{C}_k^{z,i}\right)^{-1} \hat{z}_k^i + \sum_{j \in \mathcal{N}(i)} \mathbf{W}_{ij} \left(\mathbf{H}_k^j\right)^{\mathrm{T}} \left(\mathbf{C}_k^{z,j}\right)^{-1} \hat{z}_k^j$$

and

$$\left(\overline{\mathbf{C}}_k^z\right)^{-1} = \mathbf{W}_{ii}^* \left(\mathbf{H}_k^i\right)^{\mathrm{T}} \left(\mathbf{C}_k^{z,i}\right)^{-1} \mathbf{H}_k^i + \sum_{j \in \mathcal{N}(i)} \mathbf{W}_{ij}^* \left(\mathbf{H}_k^j\right)^{\mathrm{T}} \left(\mathbf{C}_k^{z,j}\right)^{-1} \mathbf{H}_k^j.$$

These parameters then enter into Equations 11.26 and 11.27. Many possibilities can be named to determine the weighting matrices $\mathbf{W}_{ij}$ and $\mathbf{W}_{ij}^*$. Examples are nearest neighboring or Metropolis weights [23].

Consensus Kalman filtering offers the advantage that no central processing unit is required, and an estimate can be retrieved from every node. In this regard, consensus filtering can also be viewed as a decentralized processing scheme. However, the focus lies on a cooperative solution of the estimation problem in the form of synchronized estimates. The local performance is not optimized, and a potential disadvantage is the problem that, in general, the error covariance matrix (Equation 11.27) neither represents the actual error nor is a valid bound.

## 11.6 DECENTRALIZED STATE ESTIMATION AND DATA FUSION

Distributed Kalman filtering usually relies on nodes that closely cooperate with each other. Even a single failure may bring down the entire system such that no estimate can be provided, as in the optimal distributed scheme from Section 11.5.2. This stands in stark contrast to a decentralized processing. In decentralized estimation architectures, sensor nodes are essentially intended to operate independently of each other without being reliant on a central processing unit. They can share information with each other for the purpose of solving the estimation problem on a higher, cooperative level, but the primary goal is to optimize the local estimation performance.

In this section, it is assumed that each node is in the position to optimally process local sensor data according to the standard Kalman filter formulas from Section 11.2.1 and to provide an estimate on the state. In contrast to distributed schemes, local prediction and filtering steps generally do not need to be altered but are accompanied by data fusion as a third processing step. Estimates can be exchanged between nodes, and data fusion contributes to significantly improving the local estimation quality. Owing to the independent local processing, dependencies between the estimates to be fused typically remain concealed. Even if dependencies are known and can be exploited, optimal fusion does not provide the same results as the centralized processing of measurements in Section 11.4.

The data fusion concepts discussed in the following subsections differ from one another in the amount of information that can be exploited. As discussed in Section 11.3, double counting of sensor data and common process noise cause dependencies and have to be addressed. Either the dependency structure can be reconstructed or has to be bounded conservatively. Ignored dependencies or the erroneous assumption of independence may lead to biased fusion results. For each of the following concepts, two estimates $(\hat{x}_A, C_A)$ and $(\hat{x}_B, C_B)$ provided by sensor nodes A, B $\in \{1, ..., N\}$ are considered which are to be fused at node A. Table 11.2 gives an overview of the considered fusion strategies.

### 11.6.1 OPTIMAL FUSION

The Bar-Shalom-Campo formulas [8] represent the optimal solution to the fusion problem and are well-known in multisensor tracking applications. These formulas can be applied when the cross-covariance matrix $C_{AB}$ is known or can be reconstructed. By means of the gain

$$K^{fus} = (C_A - C_{AB})(C_A + C_B - C_{AB} - C_{BA})^{-1},$$

## TABLE 11.2
## Overview of Data Fusion Techniques

| Method | Knowledge about Dependencies | Fusion Result |
|---|---|---|
| Bar-Shalom-Campo fusion | Dependencies are entirely known | Optimal |
| Ellipsoidal intersection | Unknown common sensor data | Conservative |
| Covariance intersection | Dependencies are unknown | Conservative |

the fused estimate

$$\underline{\hat{x}}^{\text{fus}} = (\mathbf{I} - \mathbf{K}^{\text{fus}})\underline{\hat{x}}_A + \mathbf{K}^{\text{fus}}\underline{\hat{x}}_B \tag{11.30}$$

with error covariance matrix

$$\mathbf{C}^{\text{fus}} = \mathbf{C}_A - \mathbf{K}^{\text{fus}}(\mathbf{C}_A - \mathbf{C}_{BA}) \tag{11.31}$$

can be computed. As mentioned previously, the fusion result $(\underline{\hat{x}}^{\text{fus}}, \mathbf{C}^{\text{fus}})$ does not represent the optimal Kalman filter estimate given all the measurements that have been used to compute both estimates $\hat{x}_A$ and $\hat{x}_B$. The fusion result is optimal in a maximum-likelihood sense [24].

The Bar-Shalom-Campo formulas can be generalized to Millman's formulas [25] that allow for a simultaneous fusion of more than two estimates. The prerequisite of full knowledge of the cross-covariance terms constitutes a significant drawback of the optimal fusion technique because book-keeping of the joint cross-covariance matrix is expensive in terms of both storage and processing requirements.

## 11.6.2 ELLIPSOIDAL INTERSECTION

In the situation that the estimates share common data and no other source of dependent informa-tion is present, a conservative fusion result can be achieved by means of the ellipsoidal intersection technique [23,26]. In particular, the algorithm can be employed when no local prediction steps take place or the process noise is negligible such that the problem of common process noise does not occur.

The underlying idea is to remove the "maximum" possible common information from the fusion result. For this, the information form of the estimates is considered, that is, $\hat{y}_A = \mathbf{Y}_A\underline{\hat{x}}_A$ and $\hat{y}_B = \mathbf{Y}_B\underline{\hat{x}}_B$ with $\mathbf{Y}_A = \mathbf{C}_A^{-1}$ and $\mathbf{Y}_B = \mathbf{C}_B^{-1}$, respectively. The assumption that only common data cause dependencies implies that each estimate can be decomposed into

$$\underline{\hat{y}}_A = \underline{\hat{y}}_A^I + \underline{\bar{y}} \quad \text{and} \quad \mathbf{Y}_A = \mathbf{Y}_A^I + \bar{\mathbf{Y}} \tag{11.32}$$

and

$$\underline{\hat{y}}_B = \underline{\hat{y}}_B^I + \underline{\bar{y}} \quad \text{and} \quad \mathbf{Y}_B = \mathbf{Y}_B^I + \bar{\mathbf{Y}} \tag{11.33}$$

where $\underline{\bar{y}}$ denotes the common information vector with information matrix $\bar{\mathbf{Y}}$. $\hat{y}_A^I$ and $\hat{y}_B^I$ represent information that is independent of each other. The optimal fusion result would then be obtained through

$$\underline{\hat{y}}^{\text{opt}} = \underline{\hat{y}}_A^I + \underline{\hat{y}}_B^I + \underline{\bar{y}} = \underline{\hat{y}}_A + \underline{\hat{y}}_B - \underline{\bar{y}}$$

and

$$\mathbf{Y}^{\text{opt}} = \mathbf{Y}_A^I + \mathbf{Y}_B^I + \bar{\mathbf{Y}} = \mathbf{Y}_A + \mathbf{Y}_B - \bar{\mathbf{Y}},$$

where on the right-hand side the parameters $\bar{y}$ and $\bar{\mathbf{Y}}$ have to be removed so as to prevent double counting. As the common information $\bar{y}$ and $\bar{\mathbf{Y}}$ is unknown, these terms have to be "maximized." From Equations 11.32 and 11.33, the inequalities

$$\mathbf{Y}_A \geq \bar{\mathbf{Y}} \quad \text{and} \quad \mathbf{Y}_B \geq \bar{\mathbf{Y}}$$

can be deduced, which correspond to the inequalities

$$\mathbf{C}_A \leq \bar{\mathbf{C}}, \quad \mathbf{C}_B \leq \bar{\mathbf{C}},$$

for the covariance matrices. $\bar{\mathbf{C}}$ is computed as the smallest covariance ellipsoid circumscribing the covariance ellipsoids of $\mathbf{C}_A$ and $\mathbf{C}_B$, which is the Löwner–John ellipsoid. With the diagonalizations

$$\mathbf{C}_A = \mathbf{Q}_A \mathbf{D}_A \mathbf{Q}_A^{-1} \text{ and } \mathbf{D}_A^{-\frac{1}{2}} \mathbf{Q}_A^{-1} \mathbf{C}_B \mathbf{Q}_A \mathbf{D}_A^{-\frac{1}{2}} = \mathbf{Q}_B \mathbf{D}_B \mathbf{Q}_B^{-1},$$

the smallest upper bound $\bar{\mathbf{C}}$ yields

$$\bar{\mathbf{C}} = \mathbf{Q}_A \mathbf{D}_A^{\frac{1}{2}} \mathbf{Q}_B \bar{\mathbf{D}} \mathbf{Q}_B^{-1} \mathbf{D}_A^{\frac{1}{2}} \mathbf{Q}_A^{-1},$$

where the diagonal matrix $\bar{\mathbf{D}}$ has the entries $(\bar{\mathbf{D}})_{ii} = \max\{1, (\mathbf{D}_B)_{ii}\}, i = 1, \ldots, n_x$. The resulting matrix $\bar{\mathbf{C}}$ can then be used to compute the conservative fusion result

$$\underline{\hat{x}}_{EI} = \mathbf{C}_{EI} \left( \mathbf{C}_A^{-1} \underline{\hat{x}}_A + \mathbf{C}_B^{-1} \underline{\hat{x}}_B - \bar{\mathbf{C}}^{-1} \underline{\bar{x}} \right)$$

and

$$\mathbf{C}_{EI} = \left( \mathbf{C}_A^{-1} + \mathbf{C}_B^{-1} - \bar{\mathbf{C}}^{-1} \right)^{-1}$$

with

$$\underline{\bar{x}} = \left( \mathbf{C}_A^{-1} + \mathbf{C}_B^{-1} - 2\bar{\mathbf{C}}^{-1} \right)^{-1} \left( \left( \mathbf{C}_B^{-1} - \bar{\mathbf{C}}^{-1} \right) \underline{\hat{x}}_A + \left( \mathbf{C}_A^{-1} - \bar{\mathbf{C}}^{-1} \right) \underline{\hat{x}}_B \right).$$

The last equation is prone to numerical instabilities as $\bar{\mathbf{C}}$ is a tight approximation. With a small $\gamma > 0$, the terms $\mathbf{C}_A^{-1} - \bar{\mathbf{C}}^{-1} + \gamma \mathbf{I}$ and $\mathbf{C}_B^{-1} - \bar{\mathbf{C}}^{-1} + \gamma \mathbf{I}$ become strictly positive definite.

### 11.6.3   COVARIANCE INTERSECTION

The covariance intersection algorithm [9,27] has developed into the most important concept for decentralized data fusion. Whereas the Bar-Shalom-Campo fusion formulas are reliant on a complete knowledge about the underlying dependencies, covariance intersection pursues the opposite direction and does not require any knowledge about the dependency structure. However, this advantage is paid with less informative estimation results. The fusion of the two estimates $(\underline{\hat{x}}_A, \mathbf{C}_A)$ and $(\hat{x}_B, \mathbf{C}_B)$ is performed according to

$$\underline{\hat{x}}_{CI} = \mathbf{C}_{CI} \left( \omega \mathbf{C}_A^{-1} \underline{\hat{x}}_A + (1 - \omega) \mathbf{C}_B^{-1} \underline{\hat{x}}_B \right) \tag{11.34}$$

for the fused estimate and

$$\mathbf{C}_{\text{CI}} = \left( \omega \mathbf{C}_A^{-1} + (1-\omega) \mathbf{C}_B^{-1} \right)^{-1} \tag{11.35}$$

for the corresponding error covariance matrix. The weighting parameter $\omega \in [0, 1]$ is usually determined to minimize the trace or determinant of the covariance matrix (Equation 11.35), which is a convex optimization problem. Also, approximate closed-form [28], information-theoretic [29], and set-theoretic [30] solutions have been proposed. The covariance intersection algorithm yields covariance-consistent estimates with

$$\mathbf{C}_{\text{CI}} \geq \mathrm{E}\left[ (\hat{\underline{x}}_{\text{CI}} - \underline{x}_k)(\hat{\underline{x}}_{\text{CI}} - \underline{x}_k)^{\mathrm{T}} \right],$$

irrespective of the actual cross-covariance matrix $\mathbf{C}_{AB} = \mathrm{E}\left[ (\hat{\underline{x}}_A - \underline{x}_k)(\hat{\underline{x}}_B - \underline{x}_k)^{\mathrm{T}} \right]$ and choice of $\omega$, provided that $(\hat{\underline{x}}_A, \mathbf{C}_A)$ and $(\hat{\underline{x}}_B, \mathbf{C}_B)$ are consistent estimates.

### 11.6.3.1 Covariance Bounds

The covariance intersection algorithm can also be derived in terms of upper bounds on the joint error covariance matrix. For the purpose of fusing two estimates $(\hat{\underline{x}}_A, \mathbf{C}_A)$ and $(\hat{\underline{x}}_B, \mathbf{C}_B)$, the covariance bounds algorithm [15,16] provides the conservative approximation

$$\begin{bmatrix} \dfrac{1}{\omega} \mathbf{C}_A & \mathbf{0} \\[2mm] \mathbf{0} & \dfrac{1}{1-\omega} \mathbf{C}_B \end{bmatrix} \geq \begin{bmatrix} \mathbf{C}_A & \mathbf{C}_{AB} \\ \mathbf{C}_{BA} & \mathbf{C}_B \end{bmatrix} = \mathrm{E}\left[ \left( \begin{bmatrix} \hat{\underline{x}}_A \\ \hat{\underline{x}}_B \end{bmatrix} - \begin{bmatrix} \mathbf{I} \\ \mathbf{I} \end{bmatrix} \underline{x} \right) \left( \begin{bmatrix} \hat{\underline{x}}_A \\ \hat{\underline{x}}_B \end{bmatrix} - \begin{bmatrix} \mathbf{I} \\ \mathbf{I} \end{bmatrix} \underline{x} \right)^{\mathrm{T}} \right]$$

of the joint error covariance matrix with $\omega \in (0,1)$. This bound again holds for every possible cross-covariance matrix $\mathbf{C}_{AB}$. By employing the bound as the current joint mean squared error matrix, the estimates are deemed to be uncorrelated can be fused as if they are independent. More precisely, the estimates are replaced by $\left( \hat{\underline{x}}_A, \dfrac{1}{\omega} \mathbf{C}_A \right)$ and $\left( \hat{\underline{x}}_B, \dfrac{1}{1-\omega} \mathbf{C}_B \right)$ with inflated covariance matrices. Their fusion result directly yields Equations 11.34 and 11.35.

### 11.6.3.2 Split Covariance Intersection

Partial knowledge about independent information can be incorporated into the covariance intersection algorithm. With split covariance intersection [9], independent parts of the local estimates can be exploited. For the local error covariance matrices

$$\mathbf{C}_A = \mathbf{C}_A^I + \mathbf{C}_A^D$$

and

$$\mathbf{C}_B = \mathbf{C}_B^I + \mathbf{C}_B^D,$$

where $\mathbf{C}_A^I$ and $\mathbf{C}_B^I$ refer to parts that are known to be independent. The fusion result can then be computed according to

$$\hat{\underline{x}}_{\text{sCI}} = \mathbf{C}_{\text{sCI}} \left( \omega \left( \omega \mathbf{C}_A^I + \mathbf{C}_A^D \right)^{-1} \hat{\underline{x}}_A + (1-\omega) \left( (1-\omega) \mathbf{C}_B^I + \mathbf{C}_B^D \right)^{-1} \hat{\underline{x}}_B \right)$$

and

$$\mathbf{C}_{\mathrm{sCI}}^{-1} = \omega\big(\omega\mathbf{C}_{A}^{I} + \mathbf{C}_{A}^{D}\big)^{-1} + (1-\omega)\big((1-\omega)\mathbf{C}_{B}^{I} + \mathbf{C}_{B}^{D}\big)^{-1}$$

$$= \left(\mathbf{C}_{A}^{I} + \frac{1}{\omega}\mathbf{C}_{A}^{D}\right)^{-1} + \left(\mathbf{C}_{B}^{I} + \frac{1}{1-\omega}\mathbf{C}_{B}^{D}\right)^{-1},$$

where $\omega \in [0, 1]$ is again chosen to fulfill some optimality criterion. To apply split covariance intersection, it is apparently not necessary to also split the according estimates $\hat{\underline{x}}_{A}$ and $\hat{\underline{x}}_{B}$ into dependent and independent parts. In terms of covariance bounds, split covariance intersection complies
with the outer approximation

$$\begin{bmatrix} \mathbf{C}_{A}^{I} & \mathbf{0} \\ \mathbf{0} & \mathbf{C}_{B}^{I} \end{bmatrix} + \begin{bmatrix} \dfrac{1}{\omega}\mathbf{C}_{A}^{D} & \mathbf{0} \\ \mathbf{0} & \dfrac{1}{1-\omega}\mathbf{C}_{B}^{D} \end{bmatrix} \geq \begin{bmatrix} \mathbf{C}_{A}^{I} & \mathbf{0} \\ \mathbf{0} & \mathbf{C}_{B}^{I} \end{bmatrix} + \begin{bmatrix} \mathbf{C}_{A}^{D} & \mathbf{C}_{AB}^{D} \\ \mathbf{C}_{BA}^{D} & \mathbf{C}_{B}^{D} \end{bmatrix}$$

of the joint error covariance matrix, which is then utilized to fuse the estimates.

Covariance bounds also provide the means to exploit partial information [31] about cross-covariance
terms $\mathbf{C}_{AB}^{C}$, that is,

$$\begin{bmatrix} \mathbf{C}_{A}^{C} & \mathbf{C}_{AB}^{C} \\ \mathbf{C}_{BA}^{C} & \mathbf{C}_{B}^{C} \end{bmatrix} + \begin{bmatrix} \dfrac{1}{\omega}\mathbf{C}_{A}^{D} & \mathbf{0} \\ \mathbf{0} & \dfrac{1}{1-\omega}\mathbf{C}_{B}^{D} \end{bmatrix} \geq \begin{bmatrix} \mathbf{C}_{A}^{C} & \mathbf{C}_{AB}^{C} \\ \mathbf{C}_{BA}^{C} & \mathbf{C}_{B}^{C} \end{bmatrix} + \begin{bmatrix} \mathbf{C}_{A}^{D} & \mathbf{C}_{AB}^{D} \\ \mathbf{C}_{BA}^{D} & \mathbf{C}_{B}^{D} \end{bmatrix},$$

where the superscript D again represents parts with unknown dependencies. To exploit the left-hand

side for fusion, the estimates $\left(\hat{\underline{x}}_{A}, \mathbf{C}_{A}^{C} + \dfrac{1}{\omega}\mathbf{C}_{A}^{D}\right)$ and $\left(\hat{\underline{x}}_{B}, \mathbf{C}_{B}^{C} + \dfrac{1}{\omega}\mathbf{C}_{B}^{D}\right)$ are fused by means of the Bar-
Shalom-Campo formulas (Equations 11.30 and 11.31) with the known cross-covariance term $\mathbf{C}_{AB} = \mathbf{C}_{AB}^{C}$.

## 11.7  CONCLUSIONS

Kalman filtering in networked systems of sensor nodes is a multifaceted problem. The design of
state estimation concepts cannot be detached from the technical and operational aspects of a network but can be abstracted from concrete technical realizations of a network. The problems faced
by designers of state estimation architectures can be boiled down to the proper treatment of dependent information shared by different nodes. Dependencies mainly arise from double counting of
sensor data and local state transition models that share the same process noise term, and their treatment is oriented toward the underlying estimation architecture.

Centralized, distributed, and decentralized Kalman filtering schemes have been studied. In a
centralized estimation system, a single Kalman filter has access to all sensor data. The high transmission rates are costly in terms of communication and computation requirements, but offer the
advantage that the conditional independence of measurement data can still be exploited. In this
regard, the information form of the Kalman filter appears to be most appropriate. Distributed implementations pursue the goal to distribute the workload among the sensor nodes and to reduce the
frequency of data transfers. A close-to-optimal estimation quality and robustness to failures are two

objectives that cannot be achieved simultaneously, and different concepts to reach a compromise between these objectives have been discussed. Decentralized state estimation is concerned with the question of how to fuse Kalman filter estimates. The optimal choice of a fusion strategy depends on the knowledge about the underlying dependency structure. Known dependencies between the estimates to be fused can be exploited while unknown dependencies have to be treated conservatively so as to avoid erroneous fusion results. The more knowledge is available, the more informative the fusion result will be.

# REFERENCES

1. Rudolf E. Kalman. A New Approach to Linear Filtering and Prediction Problems. *Transactions of the ASME—Journal of Basic Engineering*, (82):35–45, 1960.
2. Dan Simon. *Optimal State Estimation: Kalman, H Infinity, and Nonlinear Approaches*. John Wiley & Sons, Hoboken, NJ, 2006.
3. Wolfgang Koch. On Accumulated State Densities with Applications to Out-of-Sequence Measurement Processing. *IEEE Transactions on Aerospace and Electronic Systems*, 47(4):2766–2778, 2011.
4. Mu Hua, Tim Bailey, Paul Thompson, and Hugh Durrant-Whyte. Decentralised Solutions to the Cooperative Multi-Platform Navigation Problem. *IEEE Transactions on Aerospace and Electronic Systems*, 47:1433–1449, 2011.
5. Benjamin Noack. *State Estimation for Distributed Systems with Stochastic and Set-membership Uncertainties*. Karlsruhe Series on Intelligent Sensor-Actuator-Systems 14. KIT Scientific Publishing, Karlsruhe, Germany, 2013.
6. Dietrich Fränken and Andreas Hüpper. Improved Fast Covariance Intersection for Distributed Data Fusion. In *Proceedings of the 8th International Conference on Information Fusion (Fuşion 2005)*, Philadelphia, PA, July 2005.
7. Yaakov Bar-Shalom and Leon Campo. On the Track-to-Track Correlation Problem. *IEEE Transactions on Automatic Control*, 26(2):571–572, 1981.
8. Yaakov Bar-Shalom and Leon Campo. The Effect of the Common Process Noise on the Two-Sensor Fused-Track Covariance. *IEEE Transactions on Aerospace and Electronic Systems*, 22(6):803–805, 1986.
9. Simon J. Julier and Jeffrey K. Uhlmann. General Decentralized Data Fusion with Covariance Intersection. In Martin E. Liggins II, David L. Hall, and James Llinas, editors. *Handbook of Multisensor Data Fusion: Theory and Practice*, 2nd ed., pp. 319–343. CRC Press, Boca Raton, FL, 2009.
10. Martin E. Liggins II, David L. Hall, and James Llinas, editors. *Handbook of Multisensor Data Fusion: Theory and Practice*, 2nd ed. The Electrical Engineering and Applied Signal Processing Series. CRC Press, Boca Raton, FL, 2009.
11. Yaakov Bar-Shalom and Xiao-Rong Li. *Multitarget Multisensor Tracking: Principles and Techniques*. YBS Publishing, Storrs, CT, 1995.
12. Arthur G. O. Mutambara. *Decentralized Estimation and Control for Multisensor Systems*. CRC Press, Boca Raton, FL, 1998.
13. Neal A. Carlson. Federated Filter for Fault-tolerant Integrated Navigation Systems. In *Proceedings of the IEEE Position Location and Navigation Symposium (PLANS'88)*, pp. 110–119, Orlando, FL, 1988. Record 'Navigation into the 21st Century' (IEEE Cat. No. 88CH2675-7).
14. Neal A. Carlson. Federated Square Root Filter for Decentralized Parallel Processors. *IEEE Transactions on Aerospace and Electronic Systems*, 26:517–525, 1990.
15. Uwe D. Hanebeck, Kai Briechle, and Joachim Horn. A Tight Bound for the Joint Covariance of Two Random Vectors with Unknown but Constrained Cross-Correlation. In *Proceedings of the 2001 IEEE International Conference on Multisensor Fusion and Integration for Intelligent Systems (MFI2001)*, pp. 85–90, Baden-Baden, Germany, August 2001.
16. Steven Reece and Stephen Roberts. Robust, Low-Bandwidth, Multi-Vehicle Mapping. In *Proceedings of the 8th International Conference on Information Fusion (Fusion 2005)*, vol. 2, Philadelphia, PA, July 2005.
17. Hamid R. Hashemipour, Sumit Roy, and Alan J. Laub. Decentralized Structures for Parallel Kalman Filtering. *IEEE Transactions on Automatic Control*, 33(1):88–94, 1988.
18. Wolfgang Koch. On Optimal Distributed Kalman Filtering and Retrodiction at Arbitrary Communication Rates for Maneuvering Targets. In *Proceedings of the 2008 IEEE International Conference on Multisensor Fusion and Integration for Intelligent Systems (MFI 2008)*, Seoul, Republic of Korea, August 2008.

19. Marc Reinhardt, Benjamin Noack, and Uwe D. Hanebeck. On Optimal Distributed Kalman Filtering in Non-ideal Situations. In *Proceedings of the 15th International Conference on Information Fusion (Fusion 2012)*, Singapore, July 2012.

20. Marc Reinhardt, Benjamin Noack, and Uwe D. Hanebeck. The Hypothesizing Distributed Kalman Filter. In *Proceedings of the 2012 IEEE International Conference on Multisensor Fusion and Integration for Intelligent Systems (MFI 2012)*, Hamburg, Germany, September 2012.

21. Reza Olfati-Saber. Distributed Kalman Filter with Embedded Consensus Filters. In *Proceedings of the 44th IEEE Conference on Decision and Control and European Control Conference (CDC-ECC 2005)*, pp. 8179–8184, Sevilla, Spain, December 2005.

22. Reza Olfati-Saber. Distributed Kalman Filtering for Sensor Networks. In *Proceedings of the 46th IEEE Conference on Decision and Control (CDC 2007)*, pp. 5492–5498, New Orleans, LA, December 2007.

23. Joris Sijs. *State Estimation in Networked Systems*. PhD thesis, Technische Universiteit Eindhoven, Eindhoven, April 2012.

24. Kuo-Chu Chang, Rajat K. Saha, and Yaakov Bar-Shalom. On Optimal Track-to-Track Fusion. *IEEE Transactions on Aerospace and Electronic Systems*, 33(4):1271–1276, 1997.

25. Vladimir Shin, Younghee Lee, and Tae-Sun Choi. Generalized Millman's Formula and Its Application for Estimation Problems. *Signal Processing*, 86(2):257–266, 2006.

26. Joris Sijs, Mircea Lazar, and Paul P. J. van den Bosch. State-Fusion with Unknown Correlation: Ellipsoidal Intersection. In *Proceedings of the 2010 American Control Conference (ACC 2010)*, Baltimore, June 2010.

27. Simon J. Julier and Jeffrey K. Uhlmann. A Non-divergent Estimation Algorithm in the Presence of Unknown Correlations. In *Proceedings of the IEEE American Control Conference (ACC 1997)*, vol. 4, pp. 2369–2373, Albuquerque, NM, June 1997.

28. Wolfgang Niehsen. Information Fusion based on Fast Covariance Intersection Filtering. In *Proceedings of the 5th International Conference on Information Fusion (Fusion 2002)*, Annapolis, MD, July 2002.

29. Michael B. Hurley. An Information Theoretic Justification for Covariance Intersection and its Generalization. In *Proceedings of the 5th International Conference on Information Fusion (Fusion 2002)*, Annapolis, MD, July 2002.

30. Benjamin Noack, Marcus Baum, and Uwe D. Hanebeck. Automatic Exploitation of Independencies for Covariance Bounding in Fully Decentralized Estimation. In *Proceedings of the 18th IFAC World Congress (IFAC 2011)*, Milan, Italy, August 2011.

31. Marc Reinhardt, Benjamin Noack, Marcus Baum, and Uwe D. Hanebeck. Analysis of Set-theoretic and Stochastic Models for Fusion under Unknown Correlations. In *Proceedings of the 14th International Conference on Information Fusion (Fusion 2011)*, Chicago, July 2011.

# 12 Cubature Information Filters
## *Theory and Applications to Multisensor Fusion*

*Ienkaran Arasaratnam and Kumar Pakki Bharani Chandra*

## CONTENTS

## 12.1 INTRODUCTION

Sensor fusion is generally defined as the use of techniques that combine data from multiple sensors such that the resulting information is more accurate and more reliable than that from a single sensor. Sensor fusion techniques are widely used in many applications such as mobile robot navigation, surveillance, air traffic control, and intelligent vehicle operations. For multiple sensor fusion in a linear Gaussian environment, the information filter, which can be considered as the dual of the Kalman filter, has been a viable solution [1–3]. Although the Kalman filter and the information filter are algebraically equivalent, the Kalman filter propagates a state vector and its error covariance whereas the information filter uses an information vector and an information matrix. This difference makes the information filter superior to the Kalman filter in fusion problems because computations are straightforward and simple. Moreover, no prior information about the system state is required.

For nonlinear Kalman filter fusion problems however, it is difficult to obtain an optimal solution. In the past, researchers closely followed the linear fusion theory to obtain a suboptimal solution for nonlinear fusion problems. Recently, Vercauteren et al. derived the sigma-point information filter using the statistical linear regression theory and the unscented transformation [4], whereas Kim et al. derived a similar set of steps using a minimum mean square error criterion [5]. When we are confronted with an issue of striking the appropriate balance or trade-off between accuracy and computational complexity, the cubature Kalman filter (CKF) is considered to be the logical choice [6]. The CKF is a more accurate and stable estimation algorithm than the unscented/sigma-point filter. However, because the sigma-point filter and the CKF share a number of common characteristics, the derivation of the cubature information filter (CIF) is straightforward and hence trivial. In this chapter, we focus on deriving the square-root version of the CIF (SCIF) for improved numerical stability. Unlike the CIF, the SCIF avoids numerically sensitive matrix operations such as matrix square rooting and inversion.

The rest of the chapter is organized as follows: The next section reviews information filtering in general. We then derive the SCIF using the linear fusion theory and matrix algebra. To validate the formulation and reliability of the SCIF, it is applied to a couple of multisensor fusion problems. The chapter concludes with some remarks.

## 12.2   INFORMATION FILTERING: A BRIEF REVIEW

The information filter is a modified version of the Kalman filter. The state estimates and their corresponding covariances in the Kalman filter are replaced by the information vectors and information matrices (inverse covariances), respectively, in the information filter. The updated covariance and the updated state take the information form, as shown by

$$\mathbf{Y}_{k|k} = \mathbf{P}_{k|k}^{-1} \tag{12.1}$$

$$\hat{\mathbf{y}}_{k|k} = \mathbf{P}_{k|k}^{-1}\hat{\mathbf{x}}_{k|k} \tag{12.2}$$

Similarly, the predicted covariance and the predicted state have equivalent information forms:

$$\mathbf{Y}_{k+1|k} = \mathbf{P}_{k+1|k}^{-1} \tag{12.3}$$

$$\hat{\mathbf{y}}_{k+1|k} = \mathbf{P}_{k+1|k}^{-1}\hat{\mathbf{x}}_{k+1|k} \tag{12.4}$$

At the heart of any information filter lies the information update, which now becomes a trivial sum:

$$\mathbf{Y}_{k+1|k+1} = \mathbf{Y}_{k+1|k} + \mathcal{I}_{k+1} \tag{12.5}$$

$$\hat{\mathbf{y}}_{k+1|k+1} = \hat{\mathbf{y}}_{k+1|k} + \mathbf{i}_{k+1} \tag{12.6}$$

Here, the information contribution matrix and the information contribution vector are defined as follows, respectively [4,5]:

$$\mathcal{I}_{k+1} = \left(\mathbf{Y}_{k+1|k}\mathbf{P}_{xz,k+1|k}\right)\mathbf{R}_{k+1}^{-1}\left(\mathbf{Y}_{k+1|k}\mathbf{P}_{xz,k+1|k}\right)^{\mathrm{T}} \tag{12.7}$$

$$\mathbf{i}_{k+1} = \left(\mathbf{Y}_{k+1|k}\mathbf{P}_{xz,k|k}\right)\mathbf{R}_{k+1}^{-1}\left(\mathbf{z}_{k+1} - \hat{\mathbf{z}}_{k+1|k} + \mathbf{P}_{xz,k+1|k}^{\mathrm{T}}\hat{\mathbf{y}}_{k+1|k}\right) \tag{12.8}$$

### 12.2.1   INFORMATION FUSION

In the sensor fusion literature, there are a number of sensor networks with their own virtues and limitations [2]. In this chapter, we specifically consider a distributed configuration with feedback [3]. As shown in Figure 12.1, in this network, each local sensor has its own information processor. The locally processed results are then transmitted to the fusion center for computing a global estimate. The global estimate is broadcast so that all the local sensors utilize the global estimate for the purpose of processing the next measurement. The advantage of using the information filters within

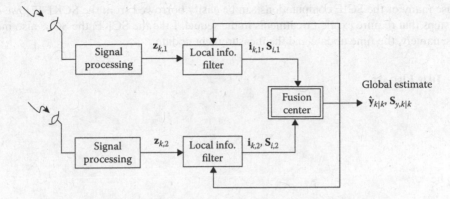

**FIGURE 12.1**   Information flow in a distributed sensor configuration with feedback.

the local sensors is that the global estimate in the fusion center can be computed from $n_s$ sensor measurements at each time step by simply summing the local information vectors and matrices:

$$\hat{\mathbf{y}}_{k|k} = \hat{\mathbf{y}}_{k|k-1} + \sum_{s=1}^{n_s} \mathbf{i}_{k,s} \tag{12.9}$$

$$\mathbf{Y}_{k|k} = \mathbf{Y}_{k|k-1} + \sum_{s=1}^{n_s} \mathcal{I}_{k,s} \tag{12.10}$$

Note that the computations outlined in this section hold under the following assumptions. (1) The tracking problem at hand is described by a linear Gaussian system. (2) The sensor measurements are uncorrelated to each other. (3) There is no measurement origin of ambiguity. (4) The sensors are synchronized. (5) There is no receipt of out-of-sequence measurements. (6) There is no communication loss among sensors. When one or more of these conditions are violated, various techniques have been proposed to get around them in the literature [2].

## 12.3   SQUARE-ROOT CUBATURE INFORMATION FILTERING

In each recursion cycle, it is important that we preserve the two properties of an information matrix, namely, its positive definitiveness and symmetry. Unfortunately, when the CIF is committed to an embedded system with limited word-length, numerical errors may lead to a loss of these properties. The accumulation of numerical errors may cause the information filter to diverge or otherwise crash. The CIF involves numerically sensitive operations such as matrix square rooting and matrix inversion, which may combine to destroy the fundamental properties of an information matrix. The logical procedure to preserve both properties of the information matrix and to improve the numerical stability is to design a square-root version of the CIF. Although the SCIF is reformulated to propagate the square roots of the information matrices, both the CIF and the SCIF are algebraically equivalent.

Before deriving the SCIF, for convenience, we introduce the following notations:

- We denote a general triangularization algorithm (e.g., QR decomposition) as $\mathbf{S} = \mathbf{Tria}(\mathbf{A})$, where $\mathbf{S}$ is a lower triangular matrix. The matrices $\mathbf{A}$ and $\mathbf{S}$ are related to each other as follows: Let $\mathbf{R}$ be an upper triangular matrix obtained from the QR decomposition on $\mathbf{A}^T$; then $\mathbf{S} = \mathbf{R}^T$.
- We use $\mathbf{S}_{Q,k}$ and $\mathbf{S}_{R,k}$ to denote the square roots of $\mathbf{Q}_k$ and $\mathbf{R}_k$, respectively. That is, $\mathbf{Q}_k = \mathbf{S}_{Q,k}\mathbf{S}_{Q,k}^T$ and $\mathbf{R}_k = \mathbf{S}_{R,k}\mathbf{S}_{R,k}^T$.

Because many of the SCIF computations can be easily borrowed from the SCKF [7], we derive only the steps that require explicit treatments in the sequel. Like the SCKF, the SCIF also includes two steps, namely, the time update and the measurement update.

### 12.3.1 TIME UPDATE

Let $\mathbf{Y}_{k|k}$ be

$$\mathbf{Y}_{k|k} = \mathbf{S}_{y,k|k}\mathbf{S}_{y,k|k}^{\mathrm{T}} \tag{12.11}$$

and $\mathbf{Y}_{k+1|k}$ be

$$\mathbf{Y}_{k+1|k} = \mathbf{S}_{y,k+1|k}\mathbf{S}_{y,k+1|k}^{\mathrm{T}} \tag{12.12}$$

As depicted in Figure 12.2, the time update of the SCIF connects $\left(\hat{\mathbf{y}}_{k|k},\mathbf{S}_{y,k|k}\right)$ to $\left(\hat{\mathbf{y}}_{k+1|k},\mathbf{S}_{y,k+1|k}\right)$ via three paths. Let us first consider how to derive Path 1, in which the information space is projected onto the state space. Taking inverse on both sides of Equation 12.1, we get

$$\mathbf{P}_{k|k} = \mathbf{Y}_{k|k}^{-1} \tag{12.13}$$

Substituting the square-root factors on both side of Equation 12.13 yields

$$\begin{aligned} \mathbf{S}_{k|k}\mathbf{S}_{k|k}^{\mathrm{T}} &= \left(\mathbf{S}_{y,k|k}\mathbf{S}_{y,k|k}^{\mathrm{T}}\right)^{-1} \\ &= \mathbf{S}_{y,k|k}^{-\mathrm{T}}\mathbf{S}_{y,k|k}^{-1} \end{aligned} \tag{12.14}$$

We may therefore write the square root of the error covariance matrix

$$\mathbf{S}_{k|k} = \mathbf{S}_{y,k|k}^{-\mathrm{T}} \tag{12.15}$$

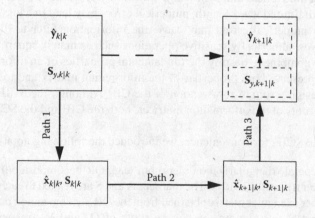

FIGURE 12.2  Time update of the SCIF.

We summarize the following important result from Equations 12.13 and 12.15 as follows:

$$\boxed{\mathbf{P}_{k|k} = \mathbf{Y}_{k|k}^{-1}} \Rightarrow \boxed{\mathbf{S}_{k|k} = \mathbf{S}_{y,k|k}^{-T}} \tag{12.16}$$

Because $\mathbf{S}_{y,k|k}$ is a triangular matrix, the least-squares method can be used to avoid computing its inversion explicitly [8]. From Equation 12.2, we write the updated state estimate

$$\hat{\mathbf{x}}_{k|k} = \mathbf{P}_{k|k}\hat{\mathbf{y}}_{k|k}$$
$$= \mathbf{S}_{k|k}\mathbf{S}_{k|k}^{T}\hat{\mathbf{y}}_{k|k} \tag{12.17}$$

which completes Path 1.

As shown in Figure 12.2, Path 3 is an inverse projection (from the state space to the information space) of Path 1. By closely following this idea and using Equation 12.3 to Equation 12.4, the state space quantities can be projected back onto the information space to obtain Path 3. Because Path 2 is identical to the time update of the SCKF, readers may refer to Section VII of Ref. [7] for a detailed derivation.

### 12.3.2 MEASUREMENT UPDATE

In the measurement update step, we will see how to fuse a new measurement with the predicted information to obtain the updated information. As depicted in Figure 12.3, the measurement update step also includes three paths. Consider Path 1. Given $\hat{\mathbf{x}}_{k+1|k}$ and $\mathbf{S}_{k+1|k}$, the predicted measurement $\hat{\mathbf{z}}_{k+1|k}$ and the submatrices of the transformation matrix $\mathbf{T}$, namely, $\mathbf{T}_{11}$ and $\mathbf{T}_{21}$ can be obtained as described in Section VII of Ref. [7]. Fortunately, $\hat{\mathbf{x}}_{k+1|k}$ and $\mathbf{S}_{k+1|k}$ are available as the by-products of the time update of the SCIF, specifically, from Path 2 of the time update. For this reason, we do not describe the derivation of Path 1 in this chapter.

Let us derive Path 2 now. The end products of Path 2 are the information contribution vector ($\mathbf{i}_{k+1}$), and the square-root information contribution matrix ($\mathbf{S}_{i,k+1}$). First, we will derive $\mathbf{S}_{i,k+1}$. By closely following Equation 12.16, we may write the inverse of the measurement noise covariance matrix

$$\overline{\mathbf{S}}_{R,k} = \mathbf{S}_{R,k}^{-T} \tag{12.18}$$

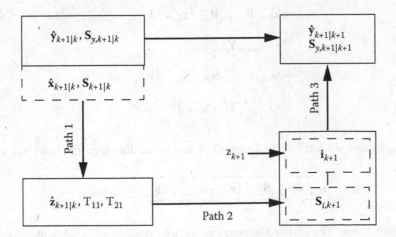

**FIGURE 12.3** Measurement update of the SCIF.

Substituting Equation 12.18 into Equation 12.7 and rearranging the right-hand side, we get

$$\mathcal{I}_{k+1} = \left(\mathbf{Y}_{k+1|k}\mathbf{P}_{xz,k+1|k}\overline{\mathbf{S}}_{R,k+1}\right)\left(\mathbf{Y}_{k+1|k}\mathbf{P}_{xz,k+1|k}\overline{\mathbf{S}}_{R,k+1}\right)^{\mathrm{T}} \tag{12.19}$$

Therefore, from Equation 12.19, we may write $\mathcal{I}_{k+1}$ in a factored form:

$$\mathcal{I}_{k+1} = \mathbf{S}_{i,k+1}\mathbf{S}_{i,k+1}^{\mathrm{T}} \tag{12.20}$$

where the square-root information matrix of dimension $(n \times m)$ is defined as

$$\mathbf{S}_{i,k+1} = \mathbf{Y}_{k+1|k}\mathbf{P}_{xz,k+1|k}\overline{\mathbf{S}}_{R,k+1} \tag{12.21}$$

$$= \mathbf{S}_{y,k+1|k}\mathbf{S}_{y,k+1|k}^{\mathrm{T}}\mathbf{P}_{xz,k+1|k}\overline{\mathbf{S}}_{R,k+1} \tag{12.22}$$

Because $\mathbf{P}_{xz,k+1|k} = \mathbf{T}_{21}\mathbf{T}_{11}^{\mathrm{T}}$ (see Equation 37 of Ref. [7]), we finally write Equation 12.22 as

$$\boxed{\mathbf{S}_{i,k+1} = \mathbf{S}_{y,k+1|k}\mathbf{S}_{y,k+1|k}^{\mathrm{T}}\mathbf{T}_{21}\mathbf{T}_{11}^{\mathrm{T}}\overline{\mathbf{S}}_{R,k+1}} \tag{12.23}$$

Moving on to determining $\mathbf{i}_{k+1}$, we rewrite Equation 12.4 as

$$\hat{\mathbf{y}}_{k+1|k} = \mathbf{Y}_{k+1|k}\hat{\mathbf{x}}_{k+1|k} \tag{12.24}$$

Because $\mathbf{Y}_{k+1|k}$ is a symmetric matrix, we may also write Equation 12.24 as

$$\hat{\mathbf{y}}_{k+1|k} = \mathbf{Y}_{k+1|k}^{\mathrm{T}}\hat{\mathbf{x}}_{k+1|k} \tag{12.25}$$

Substituting Equation 12.25 into Equation 12.8 yields

$$\begin{aligned}
\mathbf{i}_{k+1} &= \mathbf{Y}_{k+1|k}\mathbf{P}_{xz,k+1|k}\mathbf{R}_{k+1}^{-1}\left(\mathbf{z}_{k+1} - \hat{\mathbf{z}}_{k+1|k}\right. \\
&\quad + \left.\mathbf{P}_{xz,k+1|k}^{\mathrm{T}}\mathbf{Y}_{k+1|k}^{\mathrm{T}}\hat{\mathbf{x}}_{k+1|k}\right) \\
&= \mathbf{Y}_{k+1|k}\mathbf{P}_{xz,k+1|k}\overline{\mathbf{S}}_{R,k+1}\overline{\mathbf{S}}_{R,k+1}^{\mathrm{T}}\left(\mathbf{z}_{k+1} - \hat{\mathbf{z}}_{k+1|k}\right. \\
&\quad + \left.\mathbf{P}_{xz,k+1|k}^{\mathrm{T}}\mathbf{Y}_{k+1|k}^{\mathrm{T}}\hat{\mathbf{x}}_{k+1|k}\right)
\end{aligned} \tag{12.26}$$

Substituting Equation 12.21 into Equation 12.26 and expanding the right-hand side yields

$$\boxed{\mathbf{i}_{k+1} = \mathbf{S}_{i,k+1}\overline{\mathbf{S}}_{R,k+1}\left(\mathbf{z}_{k+1} - \hat{\mathbf{z}}_{k+1|k}\right) + \mathbf{S}_{i,k+1}\mathbf{S}_{i,k+1}^{\mathrm{T}}\hat{\mathbf{x}}_{k+1|k}} \tag{12.27}$$

Consider Path 3 now. The updated information vector $\hat{\mathbf{y}}_{k+1|k+1}$ can be computed by substituting Equation 12.27 into Equation 12.6. To obtain the updated information matrix $\mathbf{Y}_{k+1|k+1}$, we replace the right-hand side of with square roots and write

$$\mathbf{Y}_{k+1|k+1} = \mathbf{S}_{y,k+1|k}\mathbf{S}_{y,k+1|k}^{\mathrm{T}} + \mathbf{S}_{i,k+1}\mathbf{S}_{i,k+1}^{\mathrm{T}}$$

$$= \begin{bmatrix} \mathbf{S}_{y,k+1|k} & \mathbf{S}_{i,k+1} \end{bmatrix} \begin{bmatrix} \mathbf{S}_{y,k+1|k} & \mathbf{S}_{i,k+1} \end{bmatrix}^{\mathrm{T}}$$

Hence, the square root of the updated information matrix is given by

$$\boxed{\mathbf{S}_{y,k+1|k+1} = \mathbf{Tria}\left(\begin{bmatrix} \mathbf{S}_{y,k+1|k} & \mathbf{S}_{i,k+1} \end{bmatrix}\right)} \tag{12.28}$$

**Note:** When the local sensors employ the square-root information filtering algorithm, they are required to send the square-root information matrices to the fusion center (see Figure 12.1). For a distributed sensor network with $n_s$ sensors, we may obtain $\mathbf{S}_{y,k+1|k+1}$ by augmenting the arguments on the right-hand side of Equation 12.28 with $n_s$ square-root information contribution matrices coming from $n_s$ sensors.

$$\mathbf{S}_{y,k+1|k+1} = \mathbf{Tria}\left(\begin{bmatrix} \mathbf{S}_{y,k+1|k} & \mathbf{S}_{i,k+1}^{(1)}\cdots\mathbf{S}_{i,k+1}^{(n_s)} \end{bmatrix}\right) \tag{12.29}$$

The list below summarizes the steps involved in the SCIF algorithm.

SCIF: Time update
1. Assume that at time $k$ that $\left(\hat{\mathbf{y}}_{k|k}, \mathbf{S}_{y,k|k}\right)$ is known. Compute the square-root covariance matrix

$$\mathbf{S}_{k|k} = \mathbf{S}_{y,k|k}^{-\mathrm{T}} \tag{12.30}$$

2. Compute the state estimate

$$\hat{\mathbf{x}}_{k|k} = \mathbf{S}_{k|k}\mathbf{S}_{k|k}^{\mathrm{T}}\hat{\mathbf{y}}_{k|k} \tag{12.31}$$

3. Use the time-update of the SCKF to compute $\left(\hat{\mathbf{x}}_{k+1|k}, \mathbf{S}_{k+1|k}\right)$ from $\left(\hat{\mathbf{x}}_{k|k}, \mathbf{S}_{k|k}\right)$
4. Compute the square-root of the predicted information matrix

$$\mathbf{S}_{Y,K+1|K} = \mathbf{S}_{K+1|K}^{-\mathrm{T}} \tag{12.32}$$

5. Compute the predicted information vector

$$\hat{\mathbf{y}}_{k+1|k} = \mathbf{S}_{y,k+1|k}\mathbf{S}_{y,k+1|k}^{\mathrm{T}}\hat{\mathbf{x}}_{k+1|k} \tag{12.33}$$

SCIF: Measurement update
1. Use the measurement-update of the SCKF to compute $\mathbf{T}_{11}$ and $\mathbf{T}_{21}$ from $\left(\hat{\mathbf{x}}_{k+1|k}, \mathbf{S}_{k+1|k}\right)$
2. Compute the square root of the information contribution matrix

$$\mathbf{S}_{i,k+1} = \mathbf{S}_{y,k+1|k}\mathbf{S}_{y,k+1|k}^{\mathrm{T}}\mathbf{T}_{21}\mathbf{T}_{11}^{\mathrm{T}}\overline{\mathbf{S}}_{R,k+1} \tag{12.34}$$

where the inverse of the measurement noise covariance matrix

$$\overline{\mathbf{S}}_{R,k+1} = \mathbf{S}_{R,k+1}^{-\mathrm{T}} \tag{12.35}$$

3. Compute the information contribution vector

$$\mathbf{i}_{k+1} = \mathbf{S}_{i,k+1}\overline{\mathbf{S}}_{R,k+1}\left(\mathbf{z}_{k+1} - \hat{\mathbf{z}}_{k+1|k}\right) + \mathbf{S}_{i,k+1}\mathbf{S}_{i,k+1}^{\mathrm{T}}\hat{\mathbf{x}}_{k+1|k} \tag{12.36}$$

4. Compute the updated information vector

$$\hat{\mathbf{y}}_{k+1|k+1} = \hat{\mathbf{y}}_{k+1|k} + \mathbf{i}_{k+1} \tag{12.37}$$

5. Compute the square root of the updated information matrix

$$\mathbf{S}_{y,k+1|k+1} = \mathbf{Tria}\left(\begin{bmatrix} \mathbf{S}_{y,k+1|k} & \mathbf{S}_{i,k+1} \end{bmatrix}\right) \tag{12.38}$$

## 12.4  APPLICATIONS

### 12.4.1  Maneuvering Target Tracking in a Distributed Sensor Network with Feedback

Consider an air traffic control scenario, where an aircraft executes a maneuvering turn in a horizontal plane at a constant, but unknown turn rate $\Omega$. Figure 12.4 shows a representative trajectory of the aircraft. The kinematics of the turning motion can be modeled by the following nonlinear process equation [1]:

$$
\mathbf{x}_k = \begin{pmatrix}
1 & \dfrac{\sin\Omega T}{\Omega} & 0 & -\left(\dfrac{1-\cos\Omega T}{\Omega}\right) & 0 \\
0 & \cos\Omega T & 0 & -\sin\Omega T & 0 \\
0 & \dfrac{1-\cos\Omega T}{\Omega} & 1 & \dfrac{\sin\Omega T}{\Omega} & 0 \\
0 & \sin\Omega T & 0 & \cos\Omega T & 0 \\
0 & 0 & 0 & 0 & 1
\end{pmatrix}\mathbf{x}_{k-1} + \mathbf{v}_{k-1}
$$

where the state of the aircraft $\mathbf{x} = [\mathbf{x}[1]\ \mathbf{x}[2]\ \mathbf{x}[3]\ \mathbf{x}[4]\ \mathbf{x}[5] = \Omega]^{\mathrm{T}}$; $\mathbf{x}[1]$ and $\mathbf{x}[3]$ denote positions, and $\mathbf{x}[2]$ and $\mathbf{x}[4]$ denote velocities in the $x$ and $y$ directions, respectively; $T$ is the time interval between two consecutive mea-surements; the process noise $\mathbf{v}_{k-1} \sim \mathcal{N}(\mathbf{0},\mathbf{Q})$ with a nonsingular covariance $\mathbf{Q} = \mathrm{diag}[q_1\mathbf{M}\ q_1\mathbf{M}\ q_2 T]$, where

$$
\mathbf{M} = \begin{pmatrix}
\dfrac{T^3}{3} & \dfrac{T^2}{2} \\
\dfrac{T^2}{2} & T
\end{pmatrix}
$$

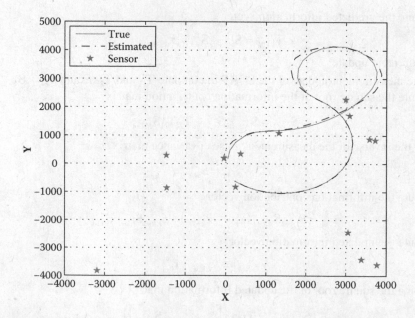

**FIGURE 12.4**  True aircraft trajectory—solid line; SCIF estimate—dotted line; radar locations—$\star$.

The scalar parameters $q_1$ and $q_2$ are related to process noise intensities. The radars were assumed to measure range and range rate. For a radar located at $\mathbf{x}_s = [\mathbf{x}_s[1], \mathbf{x}_s[2]]^T$, the measurement equation is given by

$$
\mathbf{z}_k = \left( \begin{array}{c} \sqrt{\left(\mathbf{x}_k[1] - \mathbf{x}_s[1]\right)^2 + \left(\mathbf{x}_k[3] - \mathbf{x}_s[2]\right)^2} \\ \dfrac{\left(\mathbf{x}_k[1] - \mathbf{x}_s[1]\right)\mathbf{x}_k[2] + \left(\mathbf{x}_k[3] - \mathbf{x}_s[2]\right)\mathbf{x}_k[4]}{\sqrt{\left(\mathbf{x}_k[1] - \mathbf{x}_s[1]\right)^2 + \left(\mathbf{x}_k[3] - \mathbf{x}_s[2]\right)^2}} \end{array} \right) + \mathbf{w}_k \tag{12.39}
$$

where the measurement noise covariance $\mathbf{R}$ is given by

$$
\mathbf{R} = \mathrm{cov}[\mathbf{w}_k] = \mathrm{diag}\left( \begin{bmatrix} \sigma_r^2 & \sigma_{\dot{r}}^2 \end{bmatrix} \right)
$$

To make this nonlinear tracking problem highly difficult, the target trajectory was made up of four segments, in each of which $\Omega$ was set to be $5°\ \mathrm{s}^{-1}$, $-9°\ \mathrm{s}^{-1}$, $-3°\ \mathrm{s}^{-1}$ and $9°\ \mathrm{s}^{-1}$ for the duration of 0–40 s 40–70 s, 70–90 s and 90–100 s, respectively. We used the following parameters for simulation:

$$
T = 2\mathrm{s}
$$

$$
q_1 = 0.1\ \mathrm{m}^2\mathrm{s}^{-3}
$$

$$
q_2 = 10^{-6}\ \mathrm{s}^{-3}
$$

$$
\sigma_r = 10\ \mathrm{m}
$$

$$
\sigma_{\dot{r}} = 10\ \mathrm{m/s}
$$

The true initial state was assumed to be at

$$
\mathbf{x}_0 = [0\ \mathrm{m}\ 100\ \mathrm{ms}^{-1}\ -400\ \mathrm{m}\ 120\ \mathrm{ms}^{-1}\ 2°\ \mathrm{s}^{-1}]^T
$$

The initial state estimate $\hat{\mathbf{x}}_{0/0}$ was chosen randomly from $\mathcal{N}\left(\mathbf{x}_0, \mathbf{P}_{0/0}\right)$ in each run, where

$$
\mathbf{P}_{0/0} = \mathrm{diag}[100\ \mathrm{m}^2\ 25\ \mathrm{m}^2\mathrm{s}^{-2}\ 25\ \mathrm{m}^2\ 25\ \mathrm{ms}^{-2}\ (1.7\ \mathrm{mrads}^{-1})^2]
$$

For a fair evaluation, we made a total of $N = 100$ independent Monte Carlo runs (readers may refer to http://haranarasaratnam.com/software.html for a set of MATLAB® code used to generate the results.) The radars were randomly placed in a square-shaped surveillance region with the opposite vertices at (–4000, –4000) and (4000, 4000). In this experiment, the number of radars, $n_s$, was varied from 1 to 15. We employed the following information filters for tracking the aircraft:

- Extended Information Filter (EIF)
- Cubature Information Filter (CIF)
- Square-root cubature information filter (SCIF)

Note that the unscented information filter boils down to the CIF when the free parameter $\kappa$ is forced to take 0. For this reason, the unscented information filter was excluded from comparison.

For performance comparison, we computed the accumulative root mean-squared error (ARMSE) in position and velocity. The ARMSE yields a combined measure of the bias and variance of a filter estimate. We define the ARMSE in position

$$\text{ARMSE [pos]} = \sqrt{\frac{1}{N}\sum_{n=1}^{N}\frac{1}{K}\sum_{k=1}^{K}\left(\mathbf{x}_{k,n}[1]-\hat{\mathbf{x}}_{k|k,n}[1]\right)^2+\left(\mathbf{x}_{k,n}[3]-\hat{\mathbf{x}}_{k|k,n}[3]\right)^2}$$

where $(\mathbf{x}_{k,n}[1], \mathbf{x}_{k,n}[3])$ and $\left(\hat{\mathbf{x}}_{k|k,n}[1], \hat{\mathbf{x}}_{k|k,n}[3]\right)$ are the true and globally estimated positions at the $n$th Monte Carlo run. Similarly to the ARMSE in position, we may also write formula of the ARMSE in velocity. To check the numerical robustness of an information filter, the filter divergence rate was introduced. The filter was declared to diverge when $\sqrt{\text{MSE[pos]}}$ of a specific Monte Carlo run exceeded 100 m. Subsequently, those diverged runs were excluded from the final calculations of ARMSE [pos] and ARMSE [vel].

Figure 12.5 shows the ARMSE in position and velocity, respectively, for the SCIF and EIF. As expected, as the number of sensors $n_s$ increases, the ARMSEs in position and velocity decrease. However, the performance gain quickly diminishes; as $n_s > 5$, the performance gain seems to be

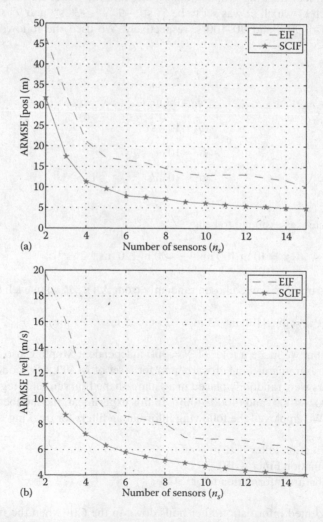

**FIGURE 12.5** Accumulative root mean square error in (a) position and (b) velocity.

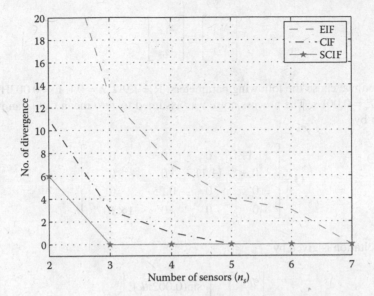

**FIGURE 12.6** Number of divergence experienced in 100 independent Monte Carlo runs.

trivial. Moreover, the SCIF consistently outperforms the EIF irrespective of $n_s$, although the performance deviation between the two reduces with $n_s$. The CIF is not included in Figure 12.5 because its performance was almost identical to the SCIF.

Figure 12.6 shows how many times each filter experienced divergence out of $N = 100$ independent Monte Carlo runs. As can be seen from Figure 12.6, the EIF diverges the most. However, its divergence rate decreases as $n_s$ increases. The CIF diverges a few times especially when $n_s$ is less. The SCIF diverges a few times only when $n_s = 2$. Unlike the CIF, the SCIF does not diverge when $n_s = 3$ and 4. This observation indicates that the SCIF is superior to other filters in terms of numerical robustness. None of the filters diverged when $n_s \geq 7$.

### 12.4.2 SPEED AND ROTOR POSITION ESTIMATION OF A TWO-PHASE PERMANENT MAGNET SYNCHRONOUS MOTOR

In this subsection, we consider the speed and rotor position estimation of a two-phase permanent magnet synchronous motor (PMSM). The kinematics of the PMSM can be expressed by the following process equation [9,10]:

$$
\begin{bmatrix} x_{1,k+1} \\ x_{2,k+1} \\ x_{3,k+1} \\ x_{4,k+1} \end{bmatrix} = \begin{bmatrix} x_{1,k} + T_s\left(-\dfrac{R}{L}x_{1,k} + \dfrac{\omega\lambda}{L}\sin x_{4,k} + \dfrac{1}{L}u_{a,k}\right) \\ x_{1,k2} + T_s\left(-\dfrac{R}{L}x_{2,k} + \dfrac{\omega\lambda}{L}\sin x_{4,k} + \dfrac{1}{L}u_{a,k}\right) \\ x_{3,k} + T_s\left(-\dfrac{3\lambda}{2J}x_{1,k}\sin x_{4,k} + \dfrac{3\lambda}{2J}x_{2,k}\cos x_{4,k} - \dfrac{Fx_{3,k}}{J}\right) \\ x_{4,k} + T_s x_{3,k} \end{bmatrix}
$$

The first two states are currents through the two windings, the third state is speed, and the fourth state is rotor angular position. The objective is to estimate the rotor angular position and speed of PMSM using the two winding currents. The measurement equation is therefore given as follows:

$$\begin{bmatrix} y_{1,k} \\ y_{2,k} \end{bmatrix} = \begin{bmatrix} x_{1,k} \\ x_{2,k} \end{bmatrix}$$

For simulation, we chose the following parameters: $R = 1.9\Omega$, $\lambda = 0.1$, $L = 0.003H$, $J = 0.00018$, $F = 0.001$, and $T_s = 0.001$ s. The process noise was assumed to zero mean Gaussian with the covariance as shown by

$$\mathbf{Q} = \begin{bmatrix} 11.11 & 0 & 0 & 0 \\ 0 & 11.11 & 0 & 0 \\ 0 & 0 & 0.25 & 0 \\ 0 & 0 & 0 & 1 \times 10^{-6} \end{bmatrix}$$

The input excitation is given by

$$\begin{bmatrix} u_{a,k} \\ u_{b,k} \end{bmatrix} = \begin{bmatrix} \sin(0.002\pi k) \\ \cos(0.002\pi k) \end{bmatrix}$$

Five current sensors were used to measure the output currents. They were assumed to provide measurements perturbed by zero mean Gaussian noise with covariances $\mathbf{R}_1 = 2.457 \times 10^{-5}$, $\mathbf{R}_2 = 6.097 \times 10^{-5}$, $\mathbf{R}_3 = 2.415 \times 10^{-5}$, $\mathbf{R}_4 = 0.1532 \times 10^{-5}$ and $\mathbf{R}_5 = 7.599 \times 10^{-5}$. The initial conditions of all the states were set to be $\mathbf{0}$ and the initial estimated state vector is selected from $\mathcal{N}\left(\begin{bmatrix} 0 & 0 & 0 & 0 \end{bmatrix}^T, 0.3162\mathbf{I}_4\right)$ for estimation. It was assumed that the first set of current sensors failed to provide measurements during the interval of 0.5 to 0.7 s and the outputs were assumed to be zero in this time interval. Fifty Monte Carlo runs were performed to evaluate the performance of the SCIF.

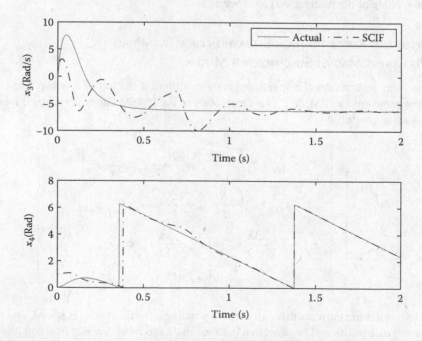

**FIGURE 12.7**   Actual and estimated speed and rotor positions using a single sensor.

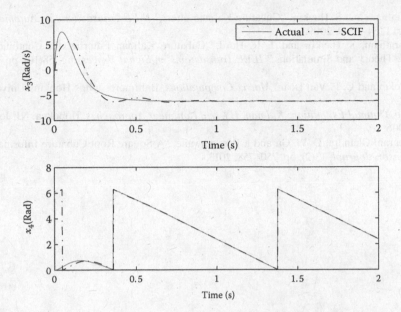

**FIGURE 12.8** Actual and estimated speed and rotor positions using multisensors.

Figures 12.7 and 12.8 show the results for the single and multisensor state estimation, respectively. It can be seen that the estimate of speed and rotor position using single sensor seems worse the multisensor estimate. The state estimation using the single sensor is not capable of providing a reliable estimate during the failure of a current sensor, whereas the multisensor-based estimator is able to provide a reasonably accurate estimate. The estimated speed using single sensor takes approximately 1.5 s to reach the actual speed, whereas in the multisensor case, the estimated speed quickly converges to the actual speed within 0.5 s. The superior performance of the multisensor state estimation can be attributed to the fact that the final fused estimate is calculated by assigning more weights to "more" accurate sensor measurements.

## 12.5 CONCLUDING REMARKS

This chapter presented the derivation of the square-root cubature information filter (SCIF) algorithm for multisensor fusion in a nonlinear Gaussian environment. Because the SCIF propagates the square-root covariance matrices, it avoids to compute numerically sensitive matrix calculations. Finally, computer experiments were performed to demonstrate that the proposed SCIF algorithm is numerically accurate and robust.

## REFERENCES

1. Y. Bar Shalom, X. R. Li and T. Kirubarajan, *Estimation with Applications to Tracking and Navigation*, Hoboken, NJ: John Wiley & Sons, 2001.
2. Y. Bar-Shalom and X. R. Li, *Multitarget-Multisensor Tracking: Principles and Techniques*, Storrs, CT: YBS, 1995.
3. Y. Zhu, Z. You, J. Zhao, K. Zhang and X. Li, "The Optimality for the Distributed Kalman Filtering Fusion," *Automatica*, 37(9), pp. 1489–1493, 2001.
4. T. Vercauteren and X. Wang, "Decentralized Sigma-Point Information Filters for Target Tracking in Collaborative Sensor Networks," *IEEE Transactions on Signal Processing*, 53(8), pp. 2997–3009, 2005.
5. Y. Kim, J. Lee, H. Do, B. Kim, T. Tanikawa, K. Ohba, G. Lee and J. Yun, "Unscented Information Filtering Method for Reducing Multiple Sensor Registration Error," *Proceedings of the IEEE International Conference on Multisensor Fusion and Integration for Intelligent Systems*, pp. 326–331, 2008.

6. I. Arasaratnam and S. Haykin, "Cubature Kalman filters," *IEEE Transactions on Automatic Control*, 54(6), pp. 1254–1269, 2009.

7. I. Arasaratnam, S. Haykin and T. R. Hurd, "Cubature Kalman Filtering for Continuous-Discrete Systems: Theory and Simulations," *IEEE Transactions on Signal Processing*, 58(10), pp. 4977–4993, 2010.

8. G. H. Golub and C. F. Van Loan, *Matrix Computations*, Baltimore: Johns Hopkins University Press, 1996.

9. D. Simon, *Optimal Estimation: Kalman, $H_\infty$, and Nonlinear Approaches*, Hoboken, NJ: John Wiley & Sons, 2006.

10. K. P. Bharani Chandra, D.-W. Gu and I. Postlethwaite, "A Square Root Cubature Information Filter," *IEEE Sensors Journal*, 13(2), pp. 750–758, 2013.

# 13 Estimation Fusion for Linear Equality Constrained Systems

*Zhansheng Duan and X. Rong Li*

## CONTENTS

## 13.1 INTRODUCTION

The state of many dynamic systems evolves subject to some equality constraints. For example, in ground target tracking [1–3], if we treat roads as curves without width, the road networks can be described by equality constraints. In an airport, an aircraft moves on the runways or taxiways. In the quaternion-based attitude estimation problem, the attitude vector must have a unit norm [4]. In a compartmental model with zero net inflow [5], mass is conserved. In undamped mechanical systems, such as one with Hamiltonian dynamics, the energy conservation law holds. Likewise, in electric circuit analysis, Kirchhoff's current and voltage laws hold.

For linear equality constrained (LEC) state estimation, numerous results are available. For example, the *dimensionality reduction method* equivalently converts a constrained state estimation problem to a reduced dimensional unconstrained one. The equivalence can be achieved by representing part of the state vector as a linear function of the remaining making use of the deterministic relationships [6], imposed by the linear equality constraint, among components of the state vector. It can also be achieved through null space decomposition as in Ref. [7]. With this method, the complexity of the dynamic system to be estimated can be reduced. However, this reduction is not significant because the computational load of the state estimator is determined mainly by the dimension of the measurement, which is unaltered in the dimensionality reduction method. Another popular method, the *projection method* [8–10], projects the unconstrained estimate onto the constraint subset by applying classical constrained optimization techniques. Specifically in Ref. [8], after the unconstrained estimate has been obtained, the problem is formulated as one of weighted least-squares estimation, in which the unconstrained estimate is treated as data and the inverse of its error covariance matrix is used as the weighting matrix. The *pseudo measurement method* has also been applied to equality constrained state estimation. It treats the equality constraint as a pseudo measurement. Thus the LEC state estimation problem is converted into a regular filtering problem with two types of measurements. However, because the pseudo measurements are noise free, the augmented measurement noise has a singular covariance. Then numerical problems may occur when the Kalman filter is applied. Moreover, the increase in the dimension of the augmented measurement will increase the computational complexity of the state estimator. That is probably why the pseudo measurement method is not popular in LEC state estimation. To avoid possible numerical problems caused by the singular covariance of the measurement noise [11] if the matrix inverse is used, the Moore–Penrose (MP) pseudoinverse was used in Refs. [12,13]. Also, to gain insight, analyze the problem, and mitigate the MP inverse of a higher dimension with a demanding computational load in batch form, two equivalent sequential forms were obtained in Refs. [12,13] by following the recursibility of linear minimum mean square error (LMMSE) estimation of Ref. [14]. It was found that under certain conditions, although equality constraints are indispensable for the evolution of the state, updating by them is redundant for filtering. If model mismatch exists, however, updating by them is helpful.

Estimation fusion, or data fusion for estimation, is the problem of how to best utilize useful information contained in multiple sets of data for the purpose of estimating a quantity—a parameter or a process [15]. It has wide application because of potentially improved estimation accuracy, enhanced reliability and survivability, extended coverage and observability, and so forth. Estimation fusion has two basic architectures: centralized fusion and distributed fusion. In centralized fusion, all raw measurements are sent to the fusion center, while in distributed fusion, each sensor sends in only processed data. They have pros and cons in terms of performance, communication requirements, reliability, survivability, information sharing, and so forth. Distributed fusion can also be split into two classes: standard and nonstandard. In standard distributed fusion, also called track fusion, local sensors send in their local tracks (estimates). All distributed fusion methods other than standard distributed fusion belong to nonstandard distributed fusion.

Estimation fusion for an unconstrained dynamic system has been well studied and results are abundant. For constrained dynamic systems, however, existing work still focuses on developing state estimators for the single-sensor case although the resultant estimators can be extended to multisensor centralized fusion easily, as shown later. For the single-sensor case, owing to the introduction of the linear equality constraint, the state estimator is different from the one for the unconstrained system. So there should be some difference when we extend fusion results for unconstrained systems to constrained systems. In this chapter, we consider the multisensor constrained estimation fusion problem. Its centralized fusion is easily tackled by applying measurement augmentation to the existing constrained state estimators for the single-sensor case. However, multisensor constrained distributed fusion is involved owing to singularity of the local estimation MSE matrices. To circumvent this singularity issue and save communication from local sensors to the fusion center, we propose to use nonstandard distributed fusion and send in dimension reduced local estimates.

This chapter is organized as follows. Section 13.2 formulates the problem. Section 13.3 presents constrained centralized fusion. Section 13.4 presents constrained nonstandard distributed fusion. Section 13.5 provides two numerical examples to illustrate the effectiveness and efficiency of the proposed fusers. Section 13.6 gives conclusions.

## 13.2   PROBLEM FORMULATION

Consider the following typical form of a linear stochastic dynamic system:

$$x_{k+1} = F_k x_k + G_k u_k + w_k \tag{13.1}$$

where $x_k \in \mathbb{R}^n$, $u_k \in \mathbb{R}^{n_u}$, $E[x_0] = \bar{x}_0$, cov $(x_0) = P_0$, $\langle w_k \rangle$ is a zero-mean white noise sequence with cov $(w_k) = Q_k$ and independent of $x_0$.

It is known in advance that the state of this system satisfies the following linear equality constraint:

$$C_k x_k = d_k \tag{13.2}$$

where $C_k \in \mathbb{R}^{m \times n}$ and $d_k \in \mathbb{R}^m$ are both known deterministically, $C_k$ is of full row rank, and $m < n$.

Assume that altogether $N$ sensors are used to observe the above constrained dynamic system simultaneously as

$$z_k^{(i)} = H_k^{(i)} x_k + v_k^{(i)}, i = 1, 2, \cdots, N$$

where $z_k^{(i)} \in \mathbb{R}^{m_i}$, $\langle v_k^{(i)} \rangle$ is a zero-mean white measurement noise sequence with $\mathrm{cov}\left(v_k^{(i)}\right) = R_k^{(i)}$, independent of $x_0$ and $\langle w_k \rangle$, and independent across sensors.

The problem of interest is to obtain centralized and distributed fused estimates of the system state $x_k$ given the measurement $\left\{z_j^{(i)}\right\}_{i=1,\cdots,N}^{j=1,\cdots,k}$ from multiple sensors.

## 13.3  CONSTRAINED CENTRALIZED FUSION

As stated in Ref. [15], multisensor centralized fusion is nothing but estimation with distributed data. And the most natural way to handle centralized fusion is to stack together the distributed measurements from multiple sensors and then apply existing state estimators. So in the following we discuss the applicability of existing single-sensor-based constrained state estimators to the multisensor case through measurement stacking first.

Let

$$z_k = \left[ \left(z_k^{(1)}\right)' \ \left(z_k^{(2)}\right)' \ \cdots \ \left(z_k^{(N)}\right)' \right]'$$

$$H_k = \left[ \left(H_k^{(1)}\right)' \ \left(H_k^{(2)}\right)' \ \cdots \ \left(H_k^{(N)}\right)' \right]'$$

$$v_k = \left[ \left(v_k^{(1)}\right)' \ \left(v_k^{(2)}\right)' \ \cdots \ \left(v_k^{(N)}\right)' \right]'$$

Then equations for the distributed measurements from multiple sensors can be written in the following compact form:

$$z_k = H_k x_k + v_k$$

where $\langle v_k \rangle$ is still a zero-mean white measurement noise sequence with $\mathrm{cov}\,(v_k) = R_k$ and

$$R_k = \mathrm{diag}\left\{R_k^{(1)}, R_k^{(2)}, \cdots, R_k^{(N)}\right\}$$

Now the existing single sensor based constrained state estimators can all be easily applied for multisensor constrained centralized fusion as follows.

### 13.3.1  PSEUDO MEASUREMENT METHOD

The idea of the pseudo measurement method [8,12,13] is to treat the linear equality constraint as an extra noise-free pseudo measurement of the system state as

$$d_k = C_k x_k$$

where $d_k$ is the measurement and $C_k$ is the measurement matrix. Then the traditional linear state estimator can be applied.

One cycle of this method can be summarized as follows.

- *Prediction:*

$$\hat{x}_{k|k-1} = F_{k-1}\hat{x}_{k-1|k-1} + G_{k-1}u_{k-1}$$

$$P_{k|k-1} = F_{k-1}P_{k-1|k-1}F'_{k-1} + Q_{k-1}$$

- *Update:*

$$\hat{x}_{k|k} = \hat{x}_{k|k-1} + P_{k|k-1}\bar{H}'_k\left(\bar{H}_k P_{k|k-1}\bar{H}'_k + \bar{R}_k\right)^+ \left(\bar{z}_k - \bar{H}_k\hat{x}_{k|k-1}\right)$$

$$P_{k|k} = P_{k|k-1} - P_{k|k-1}\bar{H}'_k\left(\bar{H}_k P_{k|k-1}\bar{H}'_k + \bar{R}_k\right)^+ \bar{H}_k P_{k|k-1}$$

where $A^+$ stands for the MP inverse of $A$ and

$$\bar{z}_k = \begin{bmatrix} z_k \\ d_k \end{bmatrix}, \bar{H}_k = \begin{bmatrix} H_k \\ C_k \end{bmatrix}, \bar{R}_k = \begin{bmatrix} R_k & 0 \\ 0 & 0 \end{bmatrix}$$

To reduce the computational complexity of the above constrained state estimator further, two equivalent sequential forms for the update step were proposed in Refs. [12,13] as follows.

Form 1

- Update by the physical measurement:

$$\hat{x}^1_{k|k} = \hat{x}_{k|k-1} + P_{k|k-1}H'_k\left(H_k P_{k|k-1}H'_k + R_k\right)^{-1}\left(z_k - H_k\hat{x}_{k|k-1}\right)$$

$$P^1_{k|k} = P_{k|k-1} - P_{k|k-1}H'_k\left(H_k P_{k|k-1}H'_k + R_k\right)^{-1} H_k P_{k|k-1}$$

- Update by the pseudo measurement:

$$\hat{x}_{k|k} = \hat{x}^1_{k|k} + P^1_{k|k}C'_k\left(C_k P^1_{k|k}C'_k\right)^+ \left(d_k - C_k\hat{x}^1_{k|k}\right)$$

$$P_{k|k} = P^1_{k|k} - P^1_{k|k}C'_k\left(C_k P^1_{k|k}C'_k\right)^+ C_k P^1_{k|k}$$

Form 2

- Update by the pseudo measurement:

$$\hat{x}^2_{k|k} = \hat{x}_{k|k-1} + P_{k|k-1}C'_k\left(C_k P_{k|k-1}C'_k\right)^+ \left(d_k - C_k\hat{x}_{k|k-1}\right)$$

$$P^2_{k|k} = P_{k|k-1} - P_{k|k-1}C'_k\left(C_k P_{k|k-1}C'_k\right)^+ C_k P_{k|k-1}$$

- Update by the physical measurement:

$$\hat{x}_{k|k} = \hat{x}_{k|k}^2 + P_{k|k}^2 H_k' \left( H_k P_{k|k}^2 H_k' + R_k \right)^{-1} \left( z_k - H_k \hat{x}_{k|k}^2 \right)$$

$$P_{k|k} = P_{k|k}^2 - P_{k|k}^2 H_k' \left( H_k P_{k|k}^2 H_k' + R_k \right)^{-1} H_k P_{k|k}^2$$

### 13.3.2 PROJECTION METHOD

The projection method [8] projects the unconstrained state estimate onto the constraint subset through the following constrained quadratic programming

$$\hat{x}_{k|k} = \arg \min_{x_k} \left( x_k - \hat{x}_{k|k}^u \right)' \left( P_{k|k}^u \right)^{-1} \left( x_k - \hat{x}_{k|k}^u \right)$$

$$\text{s.t. } C_k x_k = d_k$$

where $\hat{x}_{k|k}^u$ is the unconstrained updated estimate and $P_{k|k}^u$ is its associated MSE matrix.

One cycle of this method can be summarized as follows.

- *Prediction:* Same as in the pseudo measurement method
- *Unconstrained update:*

$$\hat{x}_{k|k}^u = \hat{x}_{k|k-1} + P_{k|k-1} H_k' \left( H_k P_{k|k-1} H_k' + R_k \right)^{-1} \left( z_k - H_k \hat{x}_{k|k-1} \right)$$

$$P_{k|k}^u = P_{k|k-1} - P_{k|k-1} H_k' \left( H_k P_{k|k-1} H_k' + R_k \right)^{-1} H_k P_{k|k-1}$$

- *Project the unconstrained estimate onto the constraint subspace:*

$$\hat{x}_{k|k} = \hat{x}_{k|k}^u + P_{k|k}^u C_k' \left( C_k P_{k|k}^u C_k' \right)^+ \left( d_k - C_k \hat{x}_{k|k}^u \right)$$

$$P_{k|k} = P_{k|k}^u - P_{k|k}^u C_k' \left( C_k P_{k|k}^u C_k' \right)^+ C_k P_{k|k}^u$$

### 13.3.3 NULL SPACE METHOD

The idea of the null space method [7] is to represent the general solution of the constraint equation as a summation of a deterministic special solution and a stochastic zero solution through null space decomposition. In this way, the original constrained state estimation is reduced to unconstrained state estimation about the zero solution part.

One specific way to fulfill the above goal is to have a QR factorization of $C_k'$ as

$$C_k' = \begin{bmatrix} Q_k^1 & Q_k^2 \end{bmatrix} \begin{bmatrix} R_k^{11} \\ 0 \end{bmatrix}$$

where $Q_k^1 \in \mathbb{R}^{n \times m}$ and $Q_k^2 \in \mathbb{R}^{n \times (n-m)}$ are orthogonal matrices, and $R_k^{11} \in \mathbb{R}^{m \times m}$ is nonsingular and upper triangular.

Then we have the following dimension reduced unconstrained system:

$$r_{k+1} = \left(Q_{k+1}^2\right)' F_k Q_k^2 r_k + \left(Q_{k+1}^2\right)' \left(F_k x_k^d + G_k u_k\right) + \left(Q_{k+1}^2\right)' w_k$$

$$z_k = H_k Q_k^2 r_k + H_k x_k^d + v_k$$

and

$$x_k = x_k^d + Q_k^2 r_k$$

where

$$r_k \in \mathbb{R}^{n-m}$$

and

$$x_k^d = Q_k^1 \left(R_k^{11}\right)^{-T} d_k$$

is a deterministic special solution of the constraint equation.

Given that

$$\hat{x}_{0|0} = \bar{x}_0, P_{0|0} = P_0$$

we have

$$\hat{r}_{0|0} = \left(Q_k^2\right)' \left(\hat{x}_{0|0} - x_0^d\right)$$

$$P_{0|0}^r = \left(Q_k^2\right)' P_{0|0} Q_k^2$$

One cycle of this method can be summarized as follows.

• *Prediction:*

$$\hat{r}_{k|k-1} = \left(Q_k^2\right)' F_{k-1} Q_{k-1}^2 \hat{r}_{k-1|k-1} + \left(Q_k^2\right)' \left(F_{k-1} x_{k-1}^d + G_{k-1} u_{k-1}\right) \tag{13.3}$$

$$P_{k|k-1}^r = \left(Q_k^2\right)' F_{k-1} Q_{k-1}^2 P_{k-1|k-1}^r \left(Q_{k-1}^2\right)' F_{k-1}' Q_k^2 + \left(Q_k^2\right)' Q_{k-1} Q_k^2 \tag{13.4}$$

$$\hat{x}_{k|k-1} = x_k^d + Q_k^2 \hat{r}_{k|k-1} \tag{13.5}$$

$$P_{k|k-1} = Q_k^2 P_{k|k-1}^r \left(Q_k^2\right)' \tag{13.6}$$

• *Update:*

$$\hat{r}_{k|k} = \hat{r}_{k|k-1} + P_{k|k-1}^r \left(H_k Q_k^2\right)' \left(H_k Q_k^2 P_{k|k-1}^r \left(H_k Q_k^2\right)' + R_k\right)^{-1}$$
$$\times \left(z_k - H_k Q_k^2 \hat{r}_{k|k-1} - H_k x_k^d\right)$$

$$P_{k|k}^r = P_{k|k-1}^r - P_{k|k-1}^r \left(H_k Q_k^2\right)' \left(H_k Q_k^2 P_{k|k-1}^r \left(H_k Q_k^2\right)' + R_k\right)^{-1}$$
$$\times H_k Q_k^2 P_{k|k-1}^r$$

$$\hat{x}_{k|k} = x_k^d + Q_k^2 \hat{r}_{k|k} \tag{13.7}$$

$$P_{k|k} = Q_k^2 P_{k|k}^r \left( Q_k^2 \right)' \tag{13.8}$$

### 13.3.4  DIRECT ELIMINATION METHOD

The direct elimination method [6] represents part of the state vector (dependent part) as a linear function of the remaining part of the state vector (independent part). As such, the dependent part can be eliminated from the state transition equation and we only need to estimate the unconstrained independent part.

Without loss of generality, assume that the components of $x_k$ has been reshuffled such that $C_k^1$ is nonsingular for

$$C_k = \left[ \begin{array}{cc} C_k^1 & C_k^2 \end{array} \right]$$

Then the state transition equation, the constraint equation and the measurement matrix can be partitioned accordingly as

$$\left[ \begin{array}{c} x_{k+1}^1 \\ x_{k+1}^2 \end{array} \right] = \left[ \begin{array}{cc} F_k^{11} & F_k^{12} \\ F_k^{21} & F_k^{22} \end{array} \right] \left[ \begin{array}{c} x_k^1 \\ x_k^2 \end{array} \right] + \left[ \begin{array}{c} G_k^1 \\ G_k^2 \end{array} \right] u_k + \left[ \begin{array}{c} w_k^1 \\ w_k^2 \end{array} \right]$$

$$C_k \left[ \begin{array}{c} x_k^1 \\ x_k^2 \end{array} \right] = d_k$$

$$H_k = \left[ \begin{array}{cc} H_k^1 & H_k^2 \end{array} \right]$$

This leads to the following unconstrained dynamic subsystem

$$x_{k+1}^2 = \left( F_k^{22} - F_k^{21} A_k \right) x_k^2 + F_k^{21} \left( C_k^1 \right)^{-1} d_k + G_k^2 u_k + w_k^2$$

$$z_k = \left( H_k^2 - H_k^1 A_k \right) x_k^2 + H_k^1 \left( C_k^1 \right)^{-1} d_k + v_k$$

and

$$x_k^1 = \left( C_k^1 \right)^{-1} d_k - A_k x_k^2$$

where

$$A_k = \left( C_k^1 \right)^{-1} C_k^2$$

One cycle of this method can be summarized as follows.

- *Prediction:*

$$\hat{x}^2_{k|k-1} = \left(F^{22}_{k-1} - F^{21}_{k-1} A_{k-1}\right)\hat{x}^2_{k-1|k-1} + F^{21}_{k-1}\left(C^1_{k-1}\right)^{-1} d_{k-1} + G^2_{k-1} u_{k-1}$$ (13.9)

$$P^2_{k|k-1} = \left(F^{22}_{k-1} - F^{21}_{k-1} A_{k-1}\right)P^2_{k-1|k-1}\left(F^{22}_{k-1} - F^{21}_{k-1} A_{k-1}\right)' + Q^2_{k-1}$$ (13.10)

$$\hat{x}^1_{k|k-1} = \left(C^1_k\right)^{-1} d_k - A_k \hat{x}^2_{k|k-1}$$ (13.11)

$$P^1_{k|k-1} = A_k P^2_{k|k-1} A'_k$$ (13.12)

$$\hat{x}_{k|k-1} = \left[ \left(\hat{x}^1_{k|k-1}\right)' \quad \left(\hat{x}^2_{k|k-1}\right)' \right]'$$ (13.13)

$$P_{k|k-1} = \begin{bmatrix} A_k P^2_{k|k-1} A'_k & -A_k P^2_{k|k-1} \\ -P^2_{k|k-1} A'_k & P^2_{k|k-1} \end{bmatrix}$$ (13.14)

where

$$Q^2_k = \text{cov}\left(w^2_k\right)$$

- *Update:*

$$\hat{x}^2_{k|k} = \hat{x}^2_{k|k-1} + P^2_{k|k-1}\left(H^2_k - H^1_k A_k\right)'$$
$$\times \left(\left(H^2_k - H^1_k A_k\right)P^2_{k|k-1}\left(H^2_k - H^1_k A_k\right)' + R_k\right)^{-1}$$
$$\times \left(z_k - \left(H^2_k - H^1_k A_k\right)\hat{x}^2_{k|k-1} - H^1_k\left(C^1_k\right)^{-1} d_k\right)$$
$$P^2_{k|k} = P^2_{k|k-1} - P^2_{k|k-1}\left(H^2_k - H^1_k A_k\right)'$$
$$\times \left(\left(H^2_k - H^1_k A_k\right)P^2_{k|k-1}\left(H^2_k - H^1_k A_k\right)' + R_k\right)^{-1}$$
$$\times \left(H^2_k - H^1_k A_k\right)P^2_{k|k-1}$$

$$\hat{x}^1_{k|k} = \left(C^1_k\right)^{-1} d_k - A_k \hat{x}^2_{k|k}$$ (13.15)

$$P^1_{k|k} = A_k P^2_{k|k} A'_k$$ (13.16)

$$\hat{x}_{k|k} = \left[ \left(\hat{x}^1_{k|k}\right)' \quad \left(\hat{x}^2_{k|k}\right)' \right]'$$ (13.17)

$$P_{k|k} = \begin{bmatrix} A_k P_{k|k}^2 A_k' & -A_k P_{k|k}^2 \\ -P_{k|k}^2 A_k' & P_{k|k}^2 \end{bmatrix} \quad (13.18)$$

Now let us summarize the properties of the above four constrained centralized fusion methods.

- The four methods have the same fusion performance although the underlying fusion rules are different, which lead to slight difference in computational complexity.
- Both the predicted and updated fused estimates satisfy the given constraint. That is,

$$C_k \hat{x}_{k|k-1} = d_k, \, C_k \hat{x}_{k|k} = d_k$$

- For the pseudo measurement method, update by the pseudo measurement is redundant. For the projection method, projection of the unconstrained estimate onto the constraint subspace is also redundant. That is,

$$\hat{x}_{k|k}^1 = \hat{x}_{k|k}, \, \hat{x}_{k|k}^2 = \hat{x}_{k|k-1}, \, \hat{x}_{k|k}^u = \hat{x}_{k|k}$$

$$P_{k|k}^1 = P_{k|k}, \, P_{k|k}^2 = P_{k|k-1}, \, P_{k|k}^u = P_{k|k}$$

- We have that

$$C_k P_{k|k-1} C_k' = 0, \, C_k P_{k|k} C_k' = 0$$

This is because from the direct elimination method, we have

$$C_k P_{k|k-1} C_k'$$

$$= \begin{bmatrix} C_k^1 & C_k^2 \end{bmatrix} \begin{bmatrix} A_k P_{k|k-1}^2 A_k' & -A_k P_{k|k-1}^2 \\ -P_{k|k-1}^2 A_k' & P_{k|k-1}^2 \end{bmatrix} \begin{bmatrix} (C_k^1)' \\ (C_k^2)' \end{bmatrix}$$

$$= C_k^1 A_k P_{k|k-1}^2 A_k' (C_k^1)' - C_k^1 A_k P_{k|k-1}^2 (C_k^2)'$$

$$\quad - C_k^2 P_{k|k-1}^2 A_k' (C_k^1)' + C_k^2 P_{k|k-1}^2 (C_k^2)'$$

$$= C_k^1 (C_k^1)^{-1} C_k^2 P_{k|k-1}^2 (C_k^2)' ((C_k^1)')^{-1} (C_k^1)' - C_k^1 (C_k^1)^{-1} C_k^2 P_{k|k-1}^2 (C_k^2)'$$

$$\quad - C_k^2 P_{k|k-1}^2 (C_k^2)' ((C_k^1)')^{-1} (C_k^1)' + C_k^2 P_{k|k-1}^2 (C_k^2)'$$

$$= 0$$

which means that the innovation associated with the pseudo measurement does not bring any new information to the constrained state estimation. Similarly we can also prove $C_k P_{k|k} C_k' = 0$.

- It also follows that both $P_{k|k-1}$ and $P_{k|k}$ are singular. This is because from the direct elimination method, we have

$$\tilde{x}_{k|k-1}^1 = -A_k \tilde{x}_{k|k-1}^2, \, \tilde{x}_{k|k}^1 = -A_k \tilde{x}_{k|k}^2$$

## 13.4  CONSTRAINED DISTRIBUTED FUSION

For multisensor constrained distributed fusion, we need to figure out what kind of local constrained estimator to use first. From the preceding discussion, we have at least four choices for the local estimators based on pseudo measurement, projection, null space, and direct elimination, respectively. For constrained centralized fusion, all of them can be applied. They have the same fusion performance except some subtle difference in computational complexity. For constrained distributed fusion, although we can still use them all as local estimators, their use will lead to different difficulty in the design of the fusion rule at the fusion center, as explained in the text that follows. Here, the design of the fusion rule and the selection of local estimators should be considered jointly.

From the preceding, clearly it does not matter which method is used as local estimators, the predicted MSE matrix $P_{k|k-1}^{(i)}$, and the updated MSE matrix $P_{k|k}^{(i)}$ are always singular at any local sensor $i$. The singularity prevents the use of most existing distributed fusion rules at the fusion center, for example, the simple convex combination method [16,17] and the information matrix fusion rule [18]. This does not mean that for constrained distributed fusion, we have to resort to some computationally intensive fusion rules, for example, MP inverse to handle singularity. This is because the singularity here is well structured. By appropriately using the structure governing the singularity of the predicted and updated MSE matrices, we can still use the existing distributed fusion rules, for example, the simple convex combination method and the information matrix fusion rule, with some minor changes as shown in the text that follows.

The preceding discussion still follows the standard distributed fusion in which the local estimates are sent to the fusion center. The singularity of the predicted and updated MSE matrices imposes a major difficulty for the design of the fusion rule. Then how about nonstandard distributed fusion, for example, local sensors sending a linear transformation of the local measurement data to the fusion center? According to Ref. [11], some nonstandard distributed fusion rules not only have the optimal performance, but also help save communication and sometimes even increase numerical robustness. In the following, we consider mainly nonstandard distributed fusion rules for the distributed fusion problem.

If we use the null space method at local sensors, other than the predicted estimate $\hat{x}_{k|k-1}^{(i)}, P_{k|k-1}^{(i)}$ and the updated estimate $\hat{x}_{k|k}^{(i)}, P_{k|k}^{(i)}$ about the full state $x_k$, we also have the dimension reduced predicted estimate $\hat{r}_{k|k-1}^{(i)}, P_{k|k-1}^{r,(i)}$ and the updated estimate $\hat{r}_{k|k}^{(i)}, P_{k|k}^{r,(i)}$ about the dimension reduced state $r_k$. Then a natural nonstandard distributed fusion is to send in $\hat{r}_{k|k-1}^{(i)}, P_{k|k-1}^{r,(i)}$ and $\hat{r}_{k|k}^{(i)}, P_{k|k}^{r,(i)}$. There are two advantages of doing so compared with the standard distributed fusion. First, $P_{k|k-1}^{r,(i)}$ and $P_{k|k}^{r,(i)}$ are usually nonsingular. This can help get rid of the singularity issue discussed above so that most existing distributed fusion rules can be applied. Second, sending in $\hat{r}_{k|k-1}^{(i)}, P_{k|k-1}^{r,(i)}$ and $\hat{r}_{k|k}^{(i)}, P_{k|k}^{r,(i)}$ instead of $\hat{x}_{k|k-1}^{(i)}, P_{k|k-1}^{(i)}$ and $\hat{x}_{k|k}^{(i)}, P_{k|k}^{(i)}$ can help save communication as $r_k$ has a dimension of $n-m$, lower than the dimension $n$ of the state $x_k$.

For the illustration, here we only introduce how to apply the information matrix fusion rule for the distributed fusion of the dimension reduced state $r_k$. The other existing distributed fusion rules, for example, the simple convex combination rules, can be applied similarly. One cycle of the information matrix fusion rule is summarized as follows.

- Fused prediction of $r_k$: Use Equations 13.3 and 13.4
- Fused prediction of $x_k$: Use Equations 13.5 and 13.6
- Fused update of $r_k$

$$\left(P_{k|k}^r\right)^{-1} = \left(P_{k|k-1}^r\right)^{-1} + \sum_{i=1}^{N}\left(\left(P_{k|k}^{r,(i)}\right)^{-1} - \left(P_{k|k-1}^{r,(i)}\right)^{-1}\right)$$

$$\left(P_{k|k}^r\right)^{-1}\hat{r}_{k|k} = \left(P_{k|k-1}^r\right)^{-1}\hat{r}_{k|k-1} + \sum_{i=1}^{N}\left(\left(P_{k|k}^{r,(i)}\right)^{-1}\hat{r}_{k|k}^{(i)} - \left(P_{k|k-1}^{r,(i)}\right)^{-1}\hat{r}_{k|k-1}^{(i)}\right)$$

- Fused update of $x_k$: Use Equations 13.7 and 13.8

Similarly, we can also use the direct elimination method at local sensors. Other than sending in the predicted estimate $\hat{x}_{k|k-1}^{(i)}, P_{k|k-1}^{(i)}$ and the updated estimate $\hat{x}_{k|k}^{(i)}, P_{k|k}^{(i)}$ about the full state $x_k$, we can just send in the dimension reduced predicted estimate $\hat{x}_{k|k-1}^{2,(i)}, P_{k|k-1}^{2,(i)}$ and the updated estimate $\hat{x}_{k|k}^{2,(i)}, P_{k|k}^{2,(i)}$ about the substate vector $x_k^2$. The two advantages of doing so compared with the standard distributed fusion are the same as the use of the null space method at local sensors. First, $P_{k|k-1}^{2,(i)}$ and $P_{k|k}^{2,(i)}$ are usually nonsingular. Second, sending in $\hat{x}_{k|k-1}^{2,(i)}, P_{k|k-1}^{2,(i)}$ and $\hat{x}_{k|k}^{2,(i)}, P_{k|k}^{2,(i)}$ instead of $\hat{x}_{k|k-1}^{(i)}, P_{k|k-1}^{(i)}$ and $\hat{x}_{k|k}^{(i)}, P_{k|k}^{(i)}$ can help save communication since $x_k^2$ has a dimension of $n - m$, lower than the dimension $n$ of the state $x_k$. One cycle of the information matrix fusion rule is summarized as follows.

- Fused prediction of $x_k^2$: Use Equations 13.9 and 13.10
- Fused prediction of $x_k$: Use Equations 13.11–13.14
- Fused update of $x_k^2$

$$\left(P_{k|k}^2\right)^{-1} = \left(P_{k|k-1}^2\right)^{-1} + \sum_{i=1}^{N}\left(\left(P_{k|k}^{2,(i)}\right)^{-1} - \left(P_{k|k-1}^{2,(i)}\right)^{-1}\right)$$

$$\left(P_{k|k}^2\right)^{-1}\hat{x}_{k|k}^2 = \left(P_{k|k-1}^2\right)^{-1}\hat{x}_{k|k-1}^2$$
$$+ \sum_{i=1}^{N}\left(\left(P_{k|k}^{2,(i)}\right)^{-1}\hat{x}_{k|k}^{2,(i)} - \left(P_{k|k-1}^{2,(i)}\right)^{-1}\hat{x}_{k|k-1}^{2,(i)}\right)$$

- Fused update of $x_k$: Use Equations 13.15–13.18

## 13.5 EXAMPLES AND DISCUSSION

Consider the following dynamic system, which describes the nearly constant velocity motion of a vehicle in a two-dimensional space:

$$x_{k+1} = F_k x_k + G_k w_k$$

where

$$x_k = \begin{bmatrix} x_k & \dot{x}_k & y_k & \dot{y}_k \end{bmatrix}$$

$$x_0 \sim \mathcal{N}(\bar{x}_0, P_0), w_k \sim \mathcal{N}(0, Q_k)$$

$$F_k = \begin{bmatrix} 1 & T & 0 & 0 \\ 0 & 1 & 0 & 0 \\ 0 & 0 & 1 & T \\ 0 & 0 & 0 & 1 \end{bmatrix}, G_k = \begin{bmatrix} T^2/2 & 0 \\ T & 0 \\ 0 & T^2/2 \\ 0 & T \end{bmatrix}, T = 2$$

It is expected to use the discrete white noise acceleration (DWNA) model to describe the nearly constant velocity motion of a vehicle along a straight line road in a two dimensional space. It is known that the slope of the straight line road is $a = \tan\left(\dfrac{\pi}{3}\right)$ and its $y$-intercept is $b = 10$. Then the desired linear equality constraint can be described as

$$C_k x_k = d_k$$

where

$$C_k = \begin{bmatrix} a & 0 & -1 & 0 \\ 0 & a & 0 & -1 \end{bmatrix}, d_k = \begin{bmatrix} -b \\ 0 \end{bmatrix}$$

To guarantee that the system state of the DWNA model satisfies this desired linear equality constraint, as discussed in Refs. [6,18], $x_0$, $w_k$ and the system parameters $\bar{x}_0$, $P_0$, $Q_k$ can not be arbitrarily designed as for an unconstrained dynamic system and they should take some structure governed by the above linear equality constraint. As discussed in Ref. [19], if we specify them as

$$x_0 = \left[ \frac{1}{a}\left( \begin{bmatrix} y_0 & \dot{y}_0 \end{bmatrix} + d_k' \right) \begin{bmatrix} y_0 & \dot{y}_0 \end{bmatrix} \right]'$$

$$[\, y_0 \ \dot{y}_0 \,]' \sim \mathcal{N}([\, 260 \ 60 \,]', \text{diag}\{625,100\})$$

$$\bar{x}_0 = \left[ \frac{1}{a}\left( \begin{bmatrix} 260 & 60 \end{bmatrix} + d_k' \right) \begin{bmatrix} 260 & 60 \end{bmatrix} \right]'$$

$$P_0 = \begin{bmatrix} \dfrac{1}{a^2}\text{diag}\{625,100\} & \dfrac{1}{a}\text{diag}\{625,100\} \\ \dfrac{1}{a}\text{diag}\{625,100\} & \text{diag}\{625,100\} \end{bmatrix}$$

$$w_k = \left[ \frac{1}{a}w_k^{a_y} \quad w_k^{a_y} \right]', w_k^{a_y} \sim \mathcal{N}(0, 0.2^2)$$

$$Q_k = \begin{bmatrix} \dfrac{1}{a^2}0.2^2 & \dfrac{1}{a}0.2^2 \\ \dfrac{1}{a}0.2^2 & 0.2^2 \end{bmatrix}$$

the desired linear equality constraint is always valid.

Assuming that two sensors are used to observe the constrained motion of the vehicle on the road as

$$z_k^{(i)} = H_k^{(i)}x_k + v_k^{(i)}, i = 1, 2$$

where

$$H_k^{(1)} = H_k^{(2)} = \begin{bmatrix} 1 & 0 & 0 & 0 \\ 0 & 0 & 1 & 0 \end{bmatrix}$$

$$R_k^{(1)} = \text{diag}\{400, 400\}, R_k^{(2)} = \text{diag}\{900, 900\}$$

## Example 13.1

In this example, we compare the estimation performance of the constrained centralized fusers using pseudo measurement (pseudo), projection (projection), null space (null), and direct elimination (direct) discussed previously in terms of RMS position and velocity errors over 200 Monte Carlo runs. Figures 13.1 and 13.2 show the estimation performance of all four methods.

It can be easily seen that the error curves for all four centralized fusers overlap with each other. This means that they have exactly the same estimation performance, as discussed previously.

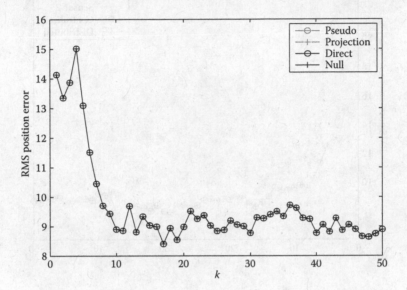

**FIGURE 13.1** RMS position error of case 1.

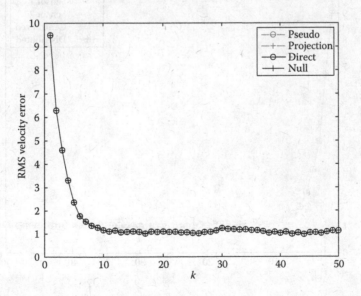

**FIGURE 13.2** RMS velocity error of case 1.

## Example 13.2

In this example, we compare the estimation performance of the proposed constrained distributed fuser with that of the constrained centralized fuser and local estimators. For illustration, only direct elimination is used by all fusers and local estimators. Figures 13.3 and 13.4 show the estimation performance of all fusers and local estimators.

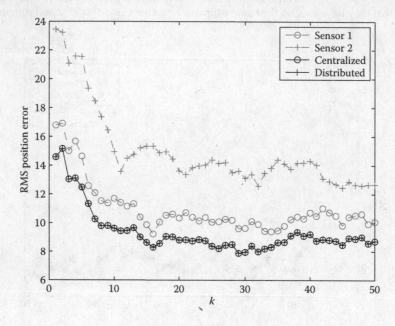

**FIGURE 13.3**   RMS position error of case 2.

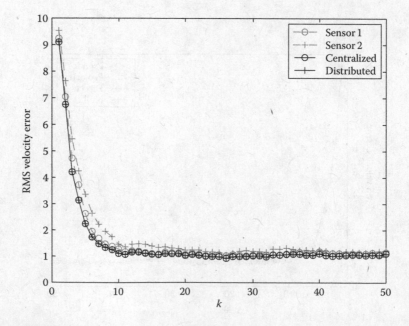

**FIGURE 13.4**   RMS velocity error of case 2.

It can be easily seen that the constrained distributed fuser has exactly the same performance as that of the constrained centralized fusion. And the fused estimation performance is better than that of any local estimator, which demonstrates the advantage of fusion. It can be also seen that sensor 2 has worse performance than sensor 1, as expected. This is because sensor 2 has worse measurement than sensor 1. In view of all these and considering that the direct elimination and null space methods have reduced communication and can help circumvent the singularity issue in distributed fusion, it seems that they should be promoted for constrained estimation fusion.

## 13.6 CONCLUSIONS

Using different techniques, state estimators for linear equality constrained dynamic systems have been well developed, including the pseudo measurement, projection, null space, and direct elimination methods. In this work, we consider extension of all these estimators to multisensor constrained estimation fusion and analyze their effectiveness and efficiency. Their extensions to centralized fusion are simple and straightforward. Their direct extension to distributed fusion is involved because the local estimation MSE matrices are singular. To circumvent this issue and save communication, we propose to send in dimension reduced local estimates to the fusion center, which are directly available in the null space and direct elimination methods.

## REFERENCES

1. T. Kirubarajan, Y. Bar-Shalom, K. R. Pattipati, and I. Kadar. Ground target tracking with variable structure IMM estimator. *IEEE Transactions on Aerospace and Electronic Systems*, 36(1):26–46, 2000.
2. C. Yang, M. Bakich, and E. Blasch. Nonlinear constrained tracking of targets on road. In *Proceedings of 2005 International Conference on Information Fusion*, pp. 235–242, Philadelphia, PA, July 25–29, 2005.
3. C. Yang, and E. Blasch. Kalman filtering with nonlinear state constraints. *IEEE Transactions on Aerospace and Electronic Systems*, 45(1):70–84, 2009.
4. D. S. Bernstein, and D. C. Hyland. Compartmental modelling and second-moment analysis of state space systems. *SIAM Journal on Matrix Analysis and Applications*, 14(3):880–901, 1993.
5. J. L. Crassidis, and F. L. Markley. Unscented filtering for spacecraft attitude estimation. *AIAA Journal of Guidance, Control, and Dynamics*, 26(4):536–542, 2003.
6. Z. S. Duan, X. R. Li, and J. F. Ru. Design and analysis of linear equality constrained dynamic systems. In *Proceedings of the 15th International Conference on Information Fusion*, pp. 2537–2544, Singapore, July 2012.
7. R. J. Hewett, M. T. Heath, M. D. Butala, and F. Kamalabadi. A robust null space method for linear equality constrained state estimation. *IEEE Transactions on Signal Processing*, 58(8):3961–3971, 2010.
8. D. Simon, and T. L. Chia. Kalman filtering with state equality constraints. *IEEE Transactions on Aerospace and Electronic Systems*, 38(1):128–136, 2002.
9. N. Gupta. Kalman filtering in the presence of state space equality constraints. In *Proceedings of the 26th Chinese Control Conference*, pp. 107–113, Zhangjiajie, Hunan, China, July 2007.
10. S. Ko, and R. R. Bitmead. State estimation for linear systems with state equality constraints. *Automatica*, 43(8):1363–1368, 2007.
11. Z. S. Duan, and X. R. Li. Lossless linear transformation of sensor data for distributed estimation fusion. *IEEE Transactions on Signal Processing*, 59(1):362–372, 2011.
12. Z. S. Duan, and X. R. Li. Best linear unbiased state estimation with noisy and noise-free measurements. In *Proceedings of the 12th International Conference on Information Fusion*, pp. 2193–2200, Seattle, WA, July 2009.
13. Z. S. Duan, and X. R. Li. The role of pseudo measurements in equality-constrained state estimation. *IEEE Transactions on Aerospace and Electronic Systems*, 49(3):1654–1666, 2013.
14. X. R. Li. Recursibility and optimal linear estimation and filtering. In *Proceedings of the 43rd IEEE Conference on Decision and Control*, pp. 1761–1766, Atlantis, Paradise Island, Bahamas, December 2004.
15. X. R. Li, Y. M. Zhu, J. Wang, and C. Z. Han. Optimal linear estimation fusion—Part I: Unified fusion rules. *IEEE Transactions on Information Theory*, 49(9):2192–2208, 2003.

16. C. Y. Chong, S. Mori, W. H. Barker, and K. C. Chang. Architectures and algorithms for track association and fusion. *IEEE Aerospace and Electronic Systems Magazine*, 15(1):5–13, 2000.

17. S. Mori, W. H. Barker, C. Y. Chong, and K. C. Chang. Track association and track fusion with non-deterministic target dynamics. *IEEE Transactions on Aerospace and Electronic Systems*, 38(2):659–688, 2002.

18. K. C. Chang, Z. Tian, and R. K. Saha. Performance evaluation of track fusion with information matrix filter. *IEEE Transactions on Aerospace and Electronic Systems*, 38(2):455–466, 2002.

19. Z. S. Duan, and X. R. Li. Constrained target motion modeling—Part I: Straight line track. In *Proceedings of the 16th International Conference on Information Fusion*, pp. 2153–2160, Istanbul, Turkey, July 2013.

# 14 Nonlinear Information Fusion Algorithm of an Asynchronous Multisensor Based on the Cubature Kalman Filter

*Wei Gao, Ya Zhang, and Qian Sun*

## CONTENTS

## 14.1 INTRODUCTION

Multisensor information fusion has been widely used in the military, defense, and high-tech fields. In multisensor information fusion theory, redundant and complementary information of multiple sensor measurements is used to improve system accuracy [1–4]. In practice, however, asynchronous information fusion problem are often encountered; for example, the sampling frequency and the inherent delay of sensors are different, and communication delays between sensors exist [5–8]. Improving system accuracy will be very difficult if the asynchronous problem cannot be solved effectively.

For asynchronous multisensor information fusion, a novel algorithm for multiple sensors with both space and time bias errors is proposed in Ref. [6], where a two-stage Kalman filtering (TSKF) and feedback system is applied to improve the entire estimation accuracy. But in this algorithm it is assumed that the sensors are independent, so the results are suboptimal. A new sequential asynchronous fusion algorithm is proposed in Ref. [8] by using the idea of sequential discretization of the sampling points on a continuous and distributed multisensor linear dynamic system. The algorithm avoids complicated calculation by using a first-come first serve criterion, but it requires information on all of the sensors to process, so it has a poor on-time performance. In consideration of the nonlinear model generated by the motion model, measurement model, and distributed measurement space transformation, the nonlinear filter must be selected when estimating the states of the multisensor information fusion system [4,9–12]. For the information fusion of a nonlinear asynchronous multisensor system, an optimal fusion framework of nonlinear multisensor data is proposed in Ref. [13]; the model is described more exactly to reduce errors. A nonlinear Stein-based estimation is proposed in Ref. [14] for wavelet de-noising of multichannel data, but the estimation results obtained with this method are suboptimal. A multisensor information fusion algorithm based on the extend Kalman filter (EKF) is proposed in Ref. [4], but the EKF solution approximates only to one order of the Taylor expansion, and the EKF needs to calculate the fuzzy Jacobian matrix, which increases the computational complexity. A multisensor asynchronous information fusion algorithm based on the unscented Kalman filter (UKF) is proposed in Ref. [11]. Compared with EKF, the filtering accuracy of UKF is higher, but the computational complexity is still a problem. In 2009, a more accurate nonlinear filtering solution based on the cubature transform named the cubature Kalman filter (CKF) was proposed by Arasaratnam and Haykin [15–19]. The CKF solution uses cubature points to approximate the mean and variance of the nonlinear system so third-order accuracy of the system can be achieved. This method has a higher accuracy and requires less calculation.

In this chapter, a nonlinear fusion algorithm for asynchronous multisensor data is presented. The nonlinear characteristics are considered while the asynchronous multisensor fusion problem is solved according to Ref. [2]. Simulation results confirm the effectiveness of the proposed algorithm.

## 14.2 BACKGROUND

### 14.2.1 INFORMATION FUSION OF A MULTISENSOR

Because the distributed processing can extend the flexibility of multisensor measuring system parameter estimation and also enhance the viability of the system [1,20], it is used in this chapter. Consider the following state equation of a discrete-time dynamic system:

$$\hat{x}(k) = \Phi(k, k-1)\hat{x}(k-1) + w(k-1) \tag{14.1}$$

where $k$ indicates the the discrete time; $\hat{x}(k)$ is the systemic state vector at time $k$; $\Phi(k, k-1)$ is the systemic state transfer matrix; $w(k-1)$ is a zero-mean Gaussian white noise process; and $w \sim (0, Q)$.

The measurement equation constituted by $N$ asynchronous sensors is

$$z_i(k) = H_i\hat{x}(k) + v_i(k) \tag{14.2}$$

where $i = 1, 2, \ldots, N$ is the $i$th measurement at time $k$; $H_i$ is the $i$th measurement matrix; $v_i(k)$ is the $i$th zero-mean Gaussian white noise; and $v \sim (0, R_i)$.

If information from $i$ sensors is measured simultaneously, each sensor obtains the optimal estimation locally utilizing its filter first, and then the fusion center weights all estimated results of the sensors. The framework of multisensor information fusion is shown in Figure 14.1.

Assume that the estimated value of the state at time $k$ can be represented by a linear combination of each sensor's state estimation. Then we have

$$\hat{x}(k) = D_1\hat{x}_1(k_1) + D_2\hat{x}_2(k_2) + \ldots + D_i\hat{x}_i(k_i) \tag{14.3}$$

where $D_1, D_2, \ldots, D_i$ are the corresponding weights of sensors.

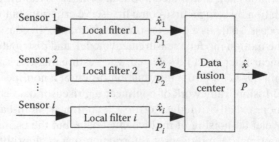

**FIGURE 14.1** The framework of multisensor information fusion.

Taking the expectation of Equation 14.3, we have

$$E[\hat{x}(k)] = D_1\Phi^{-1}(k, k_1)E[x(k)] + \dots + D_i\Phi^{-1}(k, k_i)E[x(k)]$$

If we define

$$S_1 = D_1\Phi^{-1}(k, k_1), \dots, S_i = D_i\Phi^{-1}(k, k_i) \tag{14.4}$$

then

$$P(k) = E\{[S_1\tilde{x}_1(k|k_1) + \dots + S_i\tilde{x}_i(k|k_i)][S_1\tilde{x}_1(k|k_1) + \dots + S_i\tilde{x}_i(k|k_i)]^T\}$$

According to Ref. [6], $\hat{x}(k)$ is an unbiased estimation, when

$$D_1\Phi^{-1}(k, k_1) + D_2\Phi^{-1}(k, k_2) + \dots + D_2\Phi^{-1}(k, k_2) = I$$

We define

$$A_1 = P_1(k|k_1) + P_1^T(k|k_1)$$

$$\vdots$$

$$A_i = P_i(k|k_i) + P_i^T(k|k_i)$$

When the sensors are independent, the filtering results are locally optimal when

$$S_1 = \left( I + \sum_{i=2}^{N-1}\left(A_1 A_i^- 1\right) + A_1 A_N^- 1 \right)^{-1}$$

$$S_2 = S_1 A_1 A_2^{-1} \tag{14.5}$$

$$\vdots$$

$$S_i = S_i A_1 A_i^{-1}$$

We can obtain each $D_1, D_2, \dots, D_i$ by Equations 14.4 and 14.5, and then the optimal state estimation of multisensor information fusion can also be obtained.

## 14.2.2  CKF Filter Solution

Consider the following discrete-time nonlinear state-space model

$$\begin{cases} x(k) = f(x(k-1)) + w(k-1) \\ z(k) = h(x(k)) + v(k) \end{cases} \tag{14.6}$$

where $x(k)$ is the state of the system at time $k$; $z(k)$ is the measurement at time $k$; $f(\cdot)$ and $h(\cdot)$ are some known nonlinear functions; and $w(k-1)$ and $v(k)$ are noise samples from two independent zero-mean Gaussian processes with covariance $Q(k-1)$ and $R(k)$, respectively.

CKF is proposed to solve the nonlinear filtering problem based on the spherical-radial cubature criterion. CKF first approximates the mean and variance of probability distribution through cubature points with the same weight, propagates the cubature points shown earlier by the nonlinear function, and calculates the mean and variance of the current approximation Gaussian distribution by the propagated cubature points.

The set of $2n$ cubature points are given by $[\xi_i, \omega_i]$, where $\xi_i$ is the $i$th cubature point and $\omega_i$ is the corresponding weight.

$$\begin{cases} \xi_i = \sqrt{n}\,[1]_i \\ \omega_i = \dfrac{1}{2n} \end{cases} \tag{14.7}$$

where $i = 1, 2, \ldots, 2n$ and $n$ is the dimension of the nonlinear system.

The steps involved in the time update and the measurement update of CKF are summarized as follows [18]. Assuming that at time $k - 1$ the posterior density is known.

$$p(x(k-1)) = N(\hat{x}(k-1|k-1), P(k-1|k-1)) \tag{14.8}$$

1. CKF: Time update

$$P(k-1|k-1) = S(k-1|k-1)S(k-1|k-1)^T \tag{14.9}$$

$$\xi_i(k-1|k-1) = S(k-1|k-1)\xi_i + \hat{x}(k-1|k-1) \tag{14.10}$$

$$\gamma_i(k|k-1) = f(\xi_i(k-1|k-1)) \tag{14.11}$$

$$\hat{x}(k|k-1) = \frac{1}{2n}\sum_{i=1}^{2n}\gamma_i(k|k-1) \tag{14.12}$$

$$P(k|k-1) = \frac{1}{2n}\sum_{i=1}^{2n}\gamma_i(k|k-1)\gamma_i(k|k-1)^T - \hat{x}(k|k-1)\hat{x}(k|k-1)^T + Q(k-1) \tag{14.13}$$

2. CKF: Measurement update

$$P(k|k-1) = S(k|k-1)S(k|k-1)^T \tag{14.14}$$

$$\xi_i(k|k-1) = S(k|k-1)\xi_i + \hat{x}(k|k-1) \tag{14.15}$$

$$\chi_i(k|k-1) = h(\xi_i(k|k-1)) \tag{14.16}$$

$$\hat{z}(k|k-1) = \frac{1}{2n}\sum_{i=1}^{2n}\chi_i(k|k-1) \tag{14.17}$$

$$P_{zz}(k|k-1) = \frac{1}{2n}\sum_{i=1}^{2n}\chi_i(k|k-1)\chi_i(k|k-1)^T - \hat{z}(k|k-1)\hat{z}(k|k-1)^T + R(k) \tag{14.18}$$

$$P_{xz}(k|k-1) = \frac{1}{2n}\sum_{i=1}^{2n}\xi_i(k|k-1)\chi_i(k|k-1)^T - \hat{x}(k|k-1)\hat{z}(k|k-1)^T \tag{14.19}$$

So the filter gain is

$$K(k) = P_{xz}(k|k-1)(P_{zz}(k|k-1))^{-1} \tag{14.20}$$

The filter state and the state error covariance are

$$\hat{x}(k|k) = \hat{x}(k|k-1) + K(k)(z(k) - \hat{z}(k|k-1)) \tag{14.21}$$

$$P(k|k) = P(k|k-1) - K(k)P_{zz}(k|k-1)K(k)^T \tag{14.22}$$

The CKF used a third-degree cubature rule to numerically compute the mean and variance of the probability distribution with cubature points, so the estimation accuracy can achieve third order or higher. Furthermore, this filtering solution no longer needs to calculate Jacobians and Hessians, and thus the computational complexity and consumption will decrease significantly. In a nutshell, the CKF is a new and improved algorithmic addition to the toolkit for nonlinear filtering.

## 14.3 NONLINEAR INFORMATION FUSION ALGORITHM OF AN ASYNCHRONOUS MULTISENSOR BASED ON CKF

In the information fusion problem of an asynchronous multisensor, considering the discrete-time nonlinear state-space model shown by Equation 14.6, we use CKF in data fusion. For asynchronous multisensor information fusion, we combine the multisensor data into something like single-sensor data with the time subdivision method. The sampling interval of the $i$th sensor is $T$, the corresponding discrete time is marked $k_i$, and the sampling interval of the information fusion center is denoted as $T$. The principle of multisensor asynchronous sampling is shown in Figure 14.2.

At some time intervals, data of only one sensor are measured, and the measurement data of the sensor will be directly used in the filtering process. At some time intervals, data of more than one sensor are measured, so we estimate every sensor state respectively, and then weight the filtering results. At some time intervals, there are no sensor data, and we only time update the estimation results of the previous time as the estimation results at this time. A flowchart of the proposed non-linear information fusion algorithm based on CKF is shown in Figure 14.3.

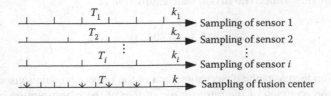

**FIGURE 14.2** The principle of multisensor asynchronous sampling.

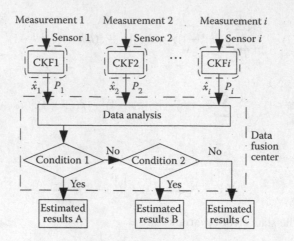

**FIGURE 14.3** Flowchart of a nonlinear asynchronous multisensor information fusion algorithm based on CKF.

In Figure 14.3, *Condition 1* means that at the current time the data of only one sensor can be measured; *Condition 2* means that at the current time the data of multiple sensors can be measured; *Estimated results A* refers to the estimated results of the sensor; *Estimated results B* refers to the weight of the estimated results of all sensors; *Estimated results C* is the time update of the estimated results from the previous moment.

The algorithm is summarized as follows.

1. At some time intervals, data of only one sensor are measured, and the measurement data of the sensor will be directly filtered based on the CKF. The fusion state estimation is the optimal estimated state values.

$$\hat{x}(k) = \hat{x}_i(k_i) \tag{14.23}$$

2. At some time intervals, data of more than one sensor are measured, so we estimate every sensor state respectively, and then weight the filtering results. The fusion state estimation is shown by

$$\hat{x}(k) = D_1\hat{x}_1(k_1) + D_2\hat{x}_2(k_2) + \ldots + D_i\hat{x}_i(k_i) \tag{14.24}$$

where $D_1, D_2, \ldots, D_i$ can be calculated according to Section 14.2.1.

3. At some time intervals, there are no sensor data and we only time update the estimation results from previous time. The fusion state estimation is

$$\hat{x}(k) = \hat{x}(k|k-1) \tag{14.25}$$

This new algorithm increases the observation data of the target by using sensor observation, and thereby the estimation accuracy of the multisensor system of the measurement parameters can be improved. In addition, the total computational complexity is moderated utilizing CKF. Thus, this algorithm can handle the nonlinear information fusion problems of an asynchronous multisensor system, and the effectiveness and viability of the system can be enhanced with this algorithm.

## 14.4  SIMULATION RESULTS

To illustrate the performance of the proposed fusion algorithm, numerical simulation examples are given in this section. Consider the discrete-time dynamic state equation of the target track:

$$x(k) = \begin{bmatrix} 1 & \dfrac{\sin\Omega\Delta}{\Omega} & 0 & \dfrac{\cos\Omega\Delta - 1}{\Omega} \\ 0 & \cos\Omega\Delta & 0 & -\sin\Omega\Delta \\ 0 & \dfrac{1-\cos\Omega\Delta}{\Omega} & 1 & \dfrac{\sin\Omega\Delta}{\Omega} \\ 0 & \sin\Omega\Delta & 0 & \cos\Omega\Delta \end{bmatrix} x(k-1) + w(k-1) \qquad (14.26)$$

where the systemic state vector is $x = [x\ \dot{x}\ y\ \dot{y}]^T$, $x$ and $y$ denote the positions of the target, and $\dot{x}$, $\dot{y}$ denote the velocities of the target; $\Delta$ is the time step between measurements; $\Omega$ is the angular rate of the target and $\Omega = 3°/s$; and the process noise $w(k) \sim N(0, Q)$ with a covariance $Q = \mathrm{diag}[qM\quad qM]$, where $q = 0.1$ and $M = \Delta^3/3\Delta^2/2\Delta^2/2\Delta$.

We consider two asynchronous sensors to observe the target. These two sensors are fixed and equipped to measure the range and bearing. Hence, we write the measurement equation

$$\begin{bmatrix} r_i(k) \\ \theta_i(k) \end{bmatrix} = \begin{bmatrix} \sqrt{x_i(k)^2 + y_i(k)^2} \\ \arctan(y_i(k)/x_i(k)) \end{bmatrix} + v_i(k)$$

where $i = 1, 2$ and the corresponding measurement noise is $v_i(k) \sim N(0, R_i)$ with $R_1 = \mathrm{diag}[r_1\sigma_\gamma\quad r_1\sigma_\theta]$ and $R_2 = \mathrm{diag}[r_2\sigma_\gamma\quad r_2\sigma_\theta]$, where $\sigma_\gamma = 10$ m, $\sigma_\theta = \sqrt{10}$ mrad, $r_1 = 0.5$, $r_2 = 1$.

The initial state and the associated covariance are

$$x_0 = [100\text{ m}\quad 10\text{ m/s}\quad 100\text{ m}\quad 10\text{ m/s}]$$

$$P_0 = \mathrm{diag}[100\text{ m}^2\quad 10\text{ m}^2/\text{s}^2\quad 100\text{ m}^2\quad 10\text{m}^2/\text{s}^2]$$

The initial state estimate $\hat{x}_0$ is chosen randomly from $N(x_0, P_0)$ in each run; the sampling interval of sensor 1 is $T_1 = 2$ s and the sampling interval of sensor 2 is $T_2 = 3$ s, so the sampling interval of the fusion center is $T_0 = T_1 - T_2 = 1$ s. The total time of each run is 100 s.

To track the maneuvering aircraft, comparing its performance against the CKF using only the data of sensor 1 and the CKF using only the data of sensor 2, we use the novel algorithm of nonlinear asynchronous data fusion based on the CKF. For a fair comparison, we make 50 independent Monte Carlo runs. All nonlinear filters are initialized with the same condition in each run.

To compare various performances, we use the root mean square error (RMSE) of the position and velocity. The RMSE yields a combined measure of the bias and variance of filter estimation. We define the RMSE in position and velocity at time $k$ as

$$\text{PMSE}_{pos}(k) = \sqrt{\frac{1}{N}\sum_{n=1}^{N}((x_n(k)-x_n(k|k))^2+(y_n(k)-y_n(k|k))^2)}$$

$$\text{PMSE}_{vel}(k) = \sqrt{\frac{1}{N}\sum_{n=1}^{N}((\dot{x}_n(k)-\dot{x}_n(k|k))^2+(\dot{y}_n(k)-\dot{y}_n(k|k))^2)}$$

where $x_n(k)$, $y_n(k)$ and $x_n(k|k)$, $y_n(k|k)$ are the true and estimated positions at the $n$th Monte Carlo run. Similarly, $\dot{x}_n(k)$, $\dot{y}_n(k)$ and $\dot{x}_n(k|k)$, $\dot{y}_n(k|k)$ are the true and estimated velocities at the $n$th Monte Carlo run. $N$ is the total number of Monte Carlo runs.

Figures 14.4 and 14.5 show the estimated RMSE of position and velocity, respectively for CKF using only the data of sensor 1, CKF using only the data of sensor 2, and the proposed fusion algorithm with two sensors. As can be seen from Figures 14.4 and 14.5, the proposed new fusion algorithm is superior to the CKF using only a single sensor in the target track model.

We use the novel algorithm based on the CKF (algorithm 1) for its numerical stability and compare it with the nonlinear fusion algorithm of the asynchronous sensors based on UKF (algorithm 2) introduced in Ref. [11]. We also make 50 independent Monte Carlo runs and the nonlinear filters are initialized with the same condition in each run. Figures 14.6 and 14.7 show the estimated RMSE in position and velocity, respectively for the proposed nonlinear fusion algorithm based on CKF (algorithm 1) and the algorithm based on UKF (algorithm 2).

The simulation results show that the accuracy of the proposed new algorithm based on the CKF is higher than that of the existing algorithm based on UKF. The main reason is that UKF gets the sigma points and corresponding weights by unscented transformation (UT), and the weights are often negative in a high-dimension system, which will introduce high-order truncation error items, reducing the accuracy of the algorithm. CKF obtains the cubature points and propagates them via the nonlinear equations; thus the weights are always positive and the errors are decreased significantly. So the numerical stability and the filtering accuracy of the novel algorithm are all better than those of the existing algorithm.

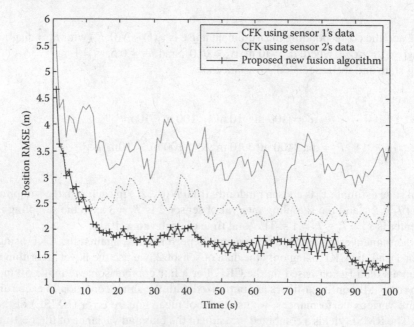

**FIGURE 14.4** Position RMSEs across 50 Monte Carlo runs.

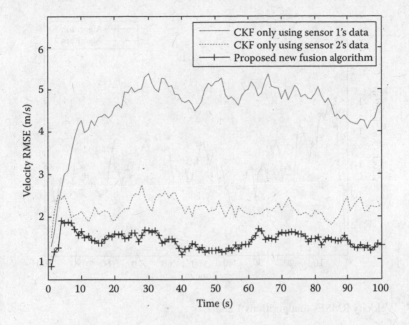

**FIGURE 14.5**   Velocity RMSEs across 50 Monte Carlo runs.

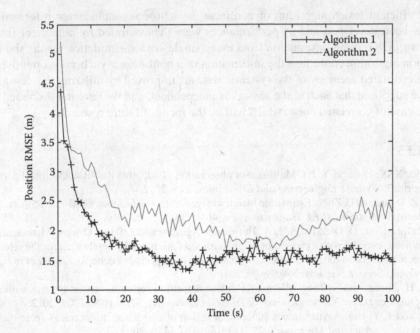

**FIGURE 14.6**   Position RMSEs of algorithms 1 and 2.

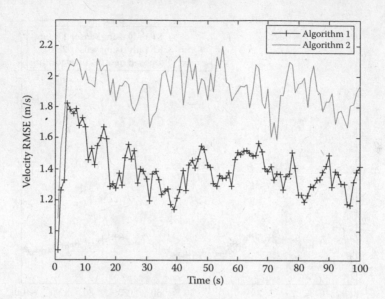

**FIGURE 14.7**   Velocity RMSEs of algorithms 1 and 2.

## 14.5   CONCLUSIONS

A new and efficient fusion algorithm of nonlinear asynchronous multisensor information based on CKF has been developed and its performances were demonstrated by numerical simulations. Compared with CKF using information from every single sensor, simulation results show that the derived fusion algorithm could fuse the information of a nonlinear asynchronous multisensor efficiently. The estimated accuracy of the systemic states is improved by utilizing the new algorithm. However, we supposed that each of the sensors is independent, and we have not discussed the algorithm in the case of correlated noise, which will be the focus of future research.

## REFERENCES

1. Z.-T. Hu, X.-X. Liu, and Y. Jin. Multi-sensor observation of adaptive Rao-Blackwellised particle filtering algorithm. *Systems Engineering and Electronics*, 2:16, 2012.
2. Y. Hu, Z. Duan, and D. Zhou. Estimation fusion with general asynchronous multi-rate sensors. *Aerospace and Electronic Systems, IEEE Transactions on*, 46(4):2090–2102, 2010.
3. J. C. Zavala-Diaz, O. Diaz-Parra, J. A. Hernandez-Aguilar, and J. Perez Ortega. Mathematical linear multi-objective model with a process of neighbourhood search and its application for the selection of an investment portfolio in the Mexican stock exchange during a period of debacle. *Advances in Information Sciences and Service Sciences*, 3(4):89–99, 2011.
4. X. Xu, H. Zhang, and X. Tang. Micro-EKF fusion algorithm for multi-sensor systems with correlated noise. *Qinghua Daxue Xuebao/Journal of Tsinghua University*, 52(9):1199–1204, 2012.
5. X. Yu, and J. Yi-Hui. Asynchronous suboptimal fusion of correlated target tracks from multisensors. *Systems Engineering and Electronics*, 25(11):1318–1320, 1440, 2003.
6. Y. Ma, and P. Niu. Based on multi-sensor information fusion circulating fluidized bed boiler combustion process clustering control research. *Advances in Information Sciences & Service Sciences*, 4(3):35–43, 2012.
7. M. Lv, W. Hou, and X. Tian. The method of virtual driving behavior in active safety based on information fusion. *Advances in Information Sciences and Service Sciences*, 4(11):337–343, 2012.
8. S. Feng, and S. Wei. Research on calibration of IFOG based on two-axis indexing. *Control and Decision*, 26(3):346–350, 2011.

9. U. Rashid, H. D. Tuan, P. Apkarian, and H. H. Kha. Globally optimized power allocation in multiple sensor fusion for linear and nonlinear networks. *IEEE Transactions on Signal Processing*, 60(2):903–915, 2012.

10. S. Gundimada, and V. K. Asari. Facial recognition using multisensor images based on localized kernel eigen spaces. *IEEE Transactions on Image Processing*, 18(6):1314–1325, 2009.

11. M. Li, Y.-F. He, and F.-Z. Nian. An improved particle filter algorithm and its performance analysis. *Advances in Information Sciences and Service Sciences*, 4(23):547–555, 2012.

12. W. Gao, Y. Zhang, and J. Wang. A strapdown interial navigation system/beidou/doppler velocity log integrated navigation algorithm based on a cubature Kalman filter. *Sensors*, 14(1):1511–1527, 2014.

13. S. Suranthiran, and S. Jayasuriya. Optimal fusion of multiple nonlinear sensor data. *IEEE Sensors Journal*, 4(5):651–663, 2004.

14. C. Chaux, L. Duval, A. Benazza-Benyahia, and J.-C. Pesquet. A nonlinear stein-based estimator for multichannel image denoising. *IEEE Transactions on Signal Processing*, 56(8 II):3855–3870, 2008.

15. I. Arasaratnam, and S. Haykin. Cubature Kalman filters. *IEEE Transactions on Automatic Control*, 54(6):1254–1269, 2009.

16. I. Arasaratnam, S. Haykin, and T. R. Hurd. Cubature Kalman filtering for continuous-discrete systems: Theory and simulations. *IEEE Transactions on Signal Processing*, 58(10):4977–4993, 2010.

17. K. Pakki, B. Chandra, D.-W. Gu, and I. Postlethwaite. Cubature information filter and its applications. pp. 3609–3614, In *Proceedings of the 2011 American Control Conference, (ACC 2011)*. June 29th–July 1st, 2011, San Francisco, 2011.

18. I. Arasaratnam, and S. Haykin. Cubature Kalman smoothers. *Automatica*, 47(10):2245–2250, 2011.

19. F. Yu, and Q. Sun. Angular rate optimal design for the rotary strap-down inertial navigation system. *Sensors*, 14(4):7156–7180, 2014.

20. Z. Hua, Z. Youguang, and L. Guoyan. Particle-filtering-based approach to undetermined blind separation. *Advances in Information Sciences and Service Sciences*, 4(6):305–313, 2012.

# 15 The Analytic Implementation of the Multisensor Probability Hypothesis Density Filter

*Fangming Huang, Kun Wang, Jian Xu, and Zhiliang Huang*

## CONTENTS

## 15.1 INTRODUCTION

Multisensor multiobject tracking has received much attention. In the multisensor tracking system, the objective of multitarget tracking is to estimate the states of targets at each time step from the sequence of noisy and cluttered observation sets obtained by each sensor [1–3]. However, the multisensor multiobject tracking problem presents the challenges of a varying number of objects and the complexity in data association between observations and objects.

Some data fusion approaches have been developed for multiobject tracking, such as the interaction multiple model [4], joint probability data association [5], and multiple hypothesis tracking [6]. Some people employed the sensor-level fusion approach for tracking [7–9]. The second approach is feature-level fusion. However, up to now, methods that are based on these approaches are computationally intensive because they have to solve the data association problem [1].

Recently, random set approaches gave a new direction for multisensor multiobject tracking. In 2003 Mahler [10] employed the random set framework to propose a probability hypothesis density (PHD) filter. This method can avoid the data association between observations and objects. Then the sequential Monte Carlo (SMC) method is used to implement the PHD filter. In particular, the implementation in Ref. [11] has the convergence proof, and it is called the particle PHD filter. Vo [12] proposed a closed-form solution for a PHD filter with assumptions on the linear Gaussian system, that is, Gaussian mixture PHD (GM-PHD) filter. The method reduced a great deal of computation compared with the particle PHD filter. Since then much research work has been conducted to improve and apply the PHD filter, such as Refs. [13–16]. However, these "core" PHD filters are all single-sensor filters.

For multisensor multiobject tracking, some methods were proposed to fuse data from multisensors in random set approaches, such as multiplication likelihood function from sensors [17] or sequential sensor updating [1,10,18]. Mahler pointed out that "Theoretically rigorous formulas for the multisensor PHD filter can be derived using the Finite-set Statistics (FISST) calculus, but are computationally intractable" [19]. For this reason, Mahler has proposed a reasonable approximation: During each measurement collection cycle, iterate the PHD filter corrector equation once for each sensor [10, p. 1169]. It is the so-called sequential PHD filtering. "This iterated-corrector approximation has a peculiarity: changing the order of the sensors produces different measurement-updated PHDs" [19]. In 2009 Mahler [19] introduced formulas for the actual "general multisensor intensity filter," that is, the exact MS-PHD filter. However, because it is computationally intractable, the exact MS-PHD theory had not been implemented since it was proposed. In 2010, Mahler derived the approximate multisensor CPHD and PHD filters, that is, product multisensor CPHD (PM-CPHD) and product multisensor PHD (PM-PHD) filters [20]. But Ouyang pointed out that there is a scale imbalance problem in PM-PHD sequential Monte Carlo implementation in 2011 [21]. In 2013, Li applied the GM-PHD filter to address the problem of PM-PHD multitarget tracking with registration errors [22]. Up to now, an analytic implementation has not been given for the exact MS-PHD filter [19].

In this chapter, the main objective is to solve the problem of analytic implementation of the exact MS-PHD filter [19]. First, we achieve the analytic implementation of exact MS-PHD formulas by using Gaussian mixture under the linear Gaussian system assumptions. Then, we propose a heuristic partition method to reduce the computational complexity of the analytic implementation of the exact MS-PHD corrector. Combining the analytic implementation of exact MS-PHD formulas with the heuristic partition method, a simple multisensor multitarget tracking algorithm is given. In addition, the proposed algorithm is applied to an example of two-sensor multitarget tracking with spontaneous birth and spawned targets to verify the performance.

## 15.2   PROBLEM FORMULATION

In this section, we adopt the random finite set (RFS) framework of Vo's [12] to model the multisensor multiobject tracking problem.

Let $\mathcal{X} \subseteq \mathcal{R}^{n_x}$ be the single object state space. Then the multiple object state at time $k$ is represented by

$$X_k = \left\{ x_{k,1}, x_{k,2}, \ldots x_{k,N_k} \right\} \in \mathcal{F}(\mathcal{X}), \tag{15.1}$$

where $\mathcal{F}(\mathcal{X})$ denotes the collection of all finite subsets of the space $\mathcal{X}$. The state is assumed to follow a Markov process on the state space $\mathcal{X}$

$$f_{k|k-1}(x_k|x_{k-1}). \tag{15.2}$$

For a multiobject state $X_{k-1}$ at time $k-1$, each $x_{k-1} \in X_{k-1}$ can die at time $k$ with probability $(1 - p_{S,k})$ or continue to exist at time $k$ with probability $p_{S,k}$. If the object $x_{k-1}$ is survived, let

$$S_{k|k-1}(x_{k-1})$$

denote the object that is transformed from $x_{k-1}$ at time $k$. Let

$$B_{k|k-1}(x_{k-1})$$

be objects spawned at time $k$ from an object $x_{k-1}$. Let $\Gamma_k$ be RFS of spontaneous births at time $k$ and can be determined by using the assumption of spontaneous birth models. Given a multiobject state $X_{k-1}$ at time $k-1$, the multiobject state $X_k$ at time $k$ is given by union of the surviving objects and new objects

$$X_k = \left[\cup_{\zeta \in X_{k-1}} S_{k|k-1}(\zeta)\right] \cup \left[\cup_{\zeta \in X_{k-1}} B_{k|k-1}(\zeta)\right] \cup \Gamma_k. \tag{15.3}$$

The RFS measurement model is described as follows, which accounts for detection uncertainty and clutter. This Markov process is partially observed in the observation space $\mathcal{Z} \subseteq \mathcal{R}^{n_z}$. Given a state $x_k$ at time $k$, the likelihood function $g_k(\cdot|\cdot)$, that is, the probability density of receiving the observation $z_k \in \mathcal{Z}$, is

$$g_k(z_k|x_k). \tag{15.4}$$

Assuming that we have $\mathcal{I}$ sensors, let $\mathcal{Z}^i$ be the measurement space of a single object at sensor $i(i = 1,\ldots,\mathcal{I})$, then measurements collected at sensor $i$ is $Z_k^i \in \mathcal{F}(\mathcal{Z}^i)$ at time $k$. For the sensor $i$, a given target $x_k \in X_k$ is either detected with probability $P_{D,k}^i$ or missed with probability $1 - P_{D,k}^i$. Consequently, at time $k$, each state generates an RFS

$$\Theta_k^i(x_k)$$

that can take on either $\{z_k^i\}$ when the target is detected or $\varnothing$ when the target is not detected. In addition to the target originated measurements, the sensor also receives a set $K_k^i$ of false measurements, or clutter. Thus, given a multitarget state $X_k$ at time $k$, the multitarget measurement $Z_k^i$ received at the sensor $i$ is formed by the union of target-generated measurements and clutter, that is,

$$Z_k^i = \left[\cup_{x \in X_k} \Theta_k^i(x)\right] \cup K_k^i,$$

The RFS of measurements at time $k$ is modeled by

$$Z_k^i = \left\{z_{k,1}^i, z_{k,2}^i, \ldots, z_{k,M_k^i}^i\right\} \in \mathcal{F}(\mathcal{Z}^i), i = 1, 2, \ldots, \mathcal{I}, \tag{15.5}$$

$$Z_k = \left\{Z_k^1, Z_k^2, \ldots, Z_k^{\mathcal{I}}\right\} \tag{15.6}$$

where $M_k^i$ is the number of measurements of the sensor $i$ at time $k$. The RFS $Z_k$ encapsulates all sensor characteristics such as measurement noise, sensor field of view, and clutter.

The multisensor multiobject tracking can be posed as follows: Given set of measurement

$$Z_{1:k} = \{Z_1, Z_2, \ldots, Z_k\}$$

collected from sensors up to time $k$, the problem is to find the multiobject state estimate $\widehat{X}_k$.

## 15.3 THE MS-PHD CORRECTOR

It is worth noting that multisensor multitarget filtering can be achieved by the two-sensor multitarget filtering (see Figure 15.1). Hence, this chapter mainly discusses the two-sensor PHD filter.

For self-contained systems, we repeat the exact formulas of the two-sensor PHD filter corrector [19] in this section. The two-sensor PHD filter predictor is exactly the same as the GM-PHD filter predictor in Ref. [12, Proposition 15.1]. Hence, it is omitted here. Suppose that there are two sensors with the following models [19]:

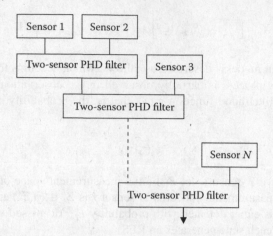

**FIGURE 15.1**   Multiple sensor fusion framework in series.

- *First sensor*: Measurement vectors $z^1 \in \mathcal{Z}^1$ and measurement-sets $Z^1 \subseteq \mathcal{Z}^1$; probability of detection $p_D^1(x)$; likelihood function $L_{z^1} = g_k^1\left(z^1|x\right)$; Poisson false alarms $\lambda_k^1, c_k^1(z^1)$.
- *Second sensor*: Measurement-vectors $z^2 \in \mathcal{Z}^2$ and measurement sets $Z^2 \subseteq \mathcal{Z}^2$; probability of detection $p_D^2(x)$; likelihood function $L_{z^2} = g_k^2\left(z^2|x\right)$; Poisson false alarms $\lambda_k^2, c_k^2(z^2)$.

Suppose that the two sensors deliver their measurement-sets $Z_k^1$ and $Z_k^2$ at the same time, and let

$$Z_k = Z_z^1 \bigcup Z_k^2 \tag{15.7}$$

be the joint measurement set. To accomplish the two-sensor measurement update step

$$v_{k|k-1}\left(x|Z_{1:k-1}\right) \to v_k\left(x|Z_{1:k-1}, Z_k^1, Z_k^2\right), \tag{15.8}$$

R. Mahler established the following result:

## Lemma 15.1 [19]

The corrector equation for the two-sensor PHD filter is

$$v_k\left(x|Z_{1:k}\right) \cong L_{Z_k}\left(x|Z_{1:k-1}\right) \cdot v_{k|k-1}\left(x|Z_{1:k-1}\right) \tag{15.9}$$

where the PHD two-sensor "pseudo-likelihood function" is

$$L_{Z_k}\left(x|Z_{1:k-1}\right) = \left(1 - p_D^1(x)\right)\left(1 - p_D^2(x)\right) \tag{15.10}$$

if $Z_k = \varnothing$ and, if otherwise,

$$L_{Z_k}\left(x|Z_{1:k-1}\right) = \left(1 - p_D^1(x)\right)\left(1 - p_D^2(x)\right)$$
$$+ \sum_{\mathcal{P} \sqsubset_2 Zk} \omega_{\mathcal{P}} \sum_{W \in \mathcal{P}} \rho_W(x) \tag{15.11}$$

where the summation is taken over all binary partitions $\mathcal{P}$ of $Z_k = Z_k^1 \cup Z_k^2$. Here

$$
\rho_w^{\cdot}(x) = \begin{cases} \dfrac{p_D^1(x) \cdot l_{z^1}^1(x) \cdot \left(1 - p_D^2(x)\right)}{1 + v_{k|k-1}\left[p_D^1 l_{z^1}^1 \left(1 - p_D^2\right)\right]}, & \text{if } W = \{z^1\} \\[4ex] \dfrac{(1 - p_D^1(x)) \cdot p_D^2(x) \cdot l_{z^2}^2(x)}{1 + v_{k|k-1}[(1 - p_D^1)p_D^2 l_{z^2}^2]}, & \text{if } W = \{z^2\}, \\[4ex] \dfrac{p_D^1(x) \cdot l_{z^1}^1(x) \cdot p_D^2(x) \cdot l_{z^2}^2(x)}{v_{k|k-1}\left[p_D^1 l_{z^1}^1 p_D^2 l_{z^2}^2\right]}, & \text{if } W = \{z^1, z^2\} \end{cases} \tag{15.12}
$$

$$
l_{z^1}^1(x) = \frac{L_{z^1}^1(x)}{\lambda_k^1 c_k^1(z^1)}, \tag{15.13}
$$

$$
l_{z^2}^2(x) = \frac{L_{z^2}^2(x)}{\lambda_k^2 c_k^2(z^2)}, \tag{15.14}
$$

$$
\omega_{\mathcal{P}} = \frac{\prod_{W \in \mathcal{P}} d_W}{\sum_{Q \sqsubset_2 Z_k} \prod_{W \in Q} d_W}, \tag{15.15}
$$

$$
d_W = \begin{cases} 1 + v_{k|k-1}\left[p_D^1 l_{z^1}^1 \left(1 - p_D^2\right)\right], & \text{if } W = \{z^1\} \\[2ex] 1 + v_{k|k-1}\left(\left(1 - p_D^1\right) p_D^2 l_{z^2}^2\right), & \text{if } W = \{z^2\}, \\[2ex] v_{k|k-1}\left[p_D^1 l_{z^1}^1 p_D^2 l_{z^2}^2\right], & \text{if } W = \{z^1, z^2\} \end{cases} \tag{15.16}
$$

The binary partition of $Z_k = Z_k^1 \cup Z_k^2$ is defined as follows [19]. An arbitrary partition of $Z_k$ is a subset $\mathcal{P}$ of the class of all subsets of $Z_k$, excluding the null set, such that the union of all of the elements of $\mathcal{P}$ is $Z_k : \bigcup_{W \in \mathcal{P}} W = Z_k$. Because every $W \in \mathcal{P}$ must have one of the following three forms:

$$
W = \{z^1\}, \ W = \{z^2\}, \ W = \{z^1, z^2\}, \tag{15.17}
$$

the $\mathcal{P}$ is named "binary." The notation $\mathcal{P} \sqsubset_2 Z_k$ is shorthand for "$\mathcal{P}$ partitions $Z_k$ into binary cells $W$" [17]. For example, suppose that $Z_k = \left\{z_1^1, z_1^2\right\} \cup \left\{z_2^1, z_2^2\right\}$. Then the binary partitions of $Z_k$ are

$$
\mathcal{P}_1 = \left\{\left\{z_1^1\right\}, \left\{z_2^1\right\}, \left\{z_1^2\right\}, \left\{z_2^2\right\}\right\}, \tag{15.18}
$$

$$
\mathcal{P}_2 = \left\{\left\{z_1^1, z_1^2\right\}, \left\{z_2^1\right\}, \left\{z_2^2\right\}\right\}, \tag{15.19}
$$

$$\mathcal{P}_3 = \left\{ \left\{ z_2^1, z_2^2 \right\}, \left\{ z_1^1 \right\}, \left\{ z_1^2 \right\} \right\},$$ (15.20)

$$\mathcal{P}_4 = \left\{ \left\{ z_1^1, z_2^2 \right\}, \left\{ z_2^1 \right\}, \left\{ z_1^2 \right\} \right\},$$ (15.21)

$$\mathcal{P}_5 = \left\{ \left\{ z_2^1, z_1^2 \right\}, \left\{ z_1^1 \right\}, \left\{ z_2^2 \right\} \right\},$$ (15.22)

$$\mathcal{P}_6 = \left\{ \left\{ z_1^1, z_1^2 \right\}, \left\{ z_2^1, z_2^2 \right\} \right\},$$ (15.23)

$$\mathcal{P}_7 = \left\{ \left\{ z_1^1, z_2^2 \right\}, \left\{ z_2^1, z_1^2 \right\} \right\}.$$ (15.24)

Mahler has discussed the computational complexity of MS-PHD as follows [19]. "If the two sensors are identical then the computational complexity of the two-sensor PHD filter is no less than $O(m_0! \cdot n)$, where $m_0$ is half of the current number of measurements and $n$ is the current number of targets" [19]. For example, if there are $n = 2$ targets present, and the half of mean number of measurements collected each time step be $m_0 = 100$, then consider those binary partitions that consist of 100 nonsingleton cells. They have the form

$$\mathcal{P}_\sigma = \left\{ \left\{ z_1^1, z_{\sigma_1}^2 \right\}, \ldots, \left\{ z_{100}^1, z_{\sigma_{100}}^2 \right\} \right\}$$ (15.25)

where $\sigma$ is any permutation on $1,\ldots, 100$. There are $100! \cong 9.3326 \times 10^{157}$ such permutations. Then the computational complexity of the two-sensor PHD filter (Lemma 15.1) is no less than $O(1.8665 \times 10^{158} n)$. Hence, the theoretically rigorous formula for the two-sensor PHD filter corrector equation (Lemma 15.1 or Proposition 15.2) is too complex for computer implement.

From the foregoing discussion, we can find that to implement the MS-PHD formulas (Equations 15.9 to 15.16), there are two problems that must be solved.

- *First problem*: How to analytically implement the MS-PHD formulas (Equations 15.9 to 15.16), especially those integral in formulas (Equations 15.12 and 15.16)?
- *Second problem*: How to implement the "binary partition," and reduce the computational complexity?

In the next two sections we will solve these problems one by one.

## 15.4 THE GAUSSIAN MIXTURE MS-PHD CORRECTOR FOR LINEAR GAUSSIAN MODELS

### 15.4.1 PRELIMINARY ASSUMPTIONS

To derive the measurement update of the GM-PHD filter, six assumptions were made in Ref. [12], which are repeated here for the sake of completeness.

A.1. Each target evolves and generates observations independently of another.
A.2. Clutter is Poisson and independent of target-originated measurements.
A.3. The predicted multitarget RFS governed by $v_{k+1|k}$ is Poisson.

A.4. Each target follows a linear Gaussian dynamical model and each sensor has a linear Gaussian measurement model, that is

$$f_{k|k-1}\left(x_k|x_{k-1}\right) = \mathcal{N}(x; F_{k-1}x_{k-1}, Q_{k-1}),\tag{15.26}$$

$$g_k^1\left(z^1|x_k\right) = \mathcal{N}\left(z^1; H_k^1 x_k, R_k^1\right),\tag{15.27}$$

$$g_k^2\left(z^2|x_k\right) = \mathcal{N}\left(z^2; H_k^2 x_k, R_k^2\right),\tag{15.28}$$

where $\mathcal{N}(; m; P)$ denotes a Gaussian density with mean $m$ and covariance $P$. $F_k$ is the state transition matrix, $Q_k$ is the process noise covariance, $H_k^i$ $(i = 1, 2)$ is the observation matrix, and $R_k^i$ $(i = 1, 2)$ is the observation noise covariance respectively.

A.5. The survival and detection probabilities are state independent, that is, $p_S(x) = p_S$, $p_S(x) = p_S, p_D^1(x) = p_D^1$ and $p_D^2(x) = p_D^2$.

A.6. The intensities of the birth and spawn RFS are Gaussian mixtures.

In this chapter, we adopt all of the above assumptions and the following additional assumption.

A.7. Each sensor's observations are independent of each other.

### 15.4.2 Summary of the Main Results

First, we solve the calculation problem of Equations 15.12 and 15.16 under the preceding assumptions.

**Proposition 15.1**

Suppose Assumptions A.4 to A.7 hold and that the predicted intensity for time $k$ is a Gaussian mixture of the form

$$v_{k|k-1}(x) = \sum_{i=1}^{J_{k|k-1}} w_{k|k-1}^{(i)} \cdot \mathcal{N}\left(x; m_{k|k-1}^{(i)}, P_{k|k-1}^{(i)}\right).\tag{15.29}$$

∎

Then, the $d_W$ and $\rho_W(x)$ can be given as the following.

When $W = \{z^1\}$,

$$d_W = 1 + \frac{p_D^1 \cdot \left(1 - p_D^2\right)}{\lambda_k^1 \cdot c_k^1(z^1)} \sum_{i=1}^{J_{k|k-1}} w_{k|k-1}^{(i)} \cdot \phi_{z^1}^{(i)},\tag{15.30}$$

$$\rho_W^{(i)} = \frac{p_D^1 \cdot \left(1 - p_D^2\right) \cdot \phi_{z^1}^{(i)}}{\lambda_k^1 \cdot c_k^1(z^1) + p_D^1\left(1 - p_D^2\right) \sum_{j=1}^{J_{k|k-1}} w_{k|k-1}^{(j)} \cdot \phi_{z^1}^{(j)}},\tag{15.31}$$

$$\phi_{z^1}^{(i)} = \mathcal{N}\left(z^1; H_k^1 m_{k|k-1}^{(i)}, R_k^1 + H_k^1 P_{k|k-1}^{(i)}\left(H_k^1\right)^T\right),\tag{15.32}$$

When $W = \{z^2\}$,

$$d_W = 1 + \frac{\left(1 - p_D^1\right) \cdot p_D^2}{\lambda_k^2 \cdot c_k^2(z^2)} \sum_{i=1}^{J_{k|k-1}} w_{k|k-1}^{(i)} \cdot \phi_{z^2}^{(i)}, \tag{15.33}$$

$$\rho_W^{(i)} = \frac{\left(1 - p_D^1\right) \cdot p_D^2 \cdot \phi_{z^2}^{(i)}}{\lambda_k^2 \cdot c_k^2(z^1) + \left(1 - p_D^1\right) p_D^2 \sum_{j=1}^{J_{k|k-1}} w_{k|k-1}^{(j)} \cdot \phi_{z^2}^{(j)}}, \tag{15.34}$$

$$\phi_{z^2}^{(i)} = \mathcal{N}\left(z^2; H_k^2 m_{k|k-1}^{(i)}, R_k^2 + H_k^2 P_{k|k-1}^{(i)} \left(H_k^2\right)^T\right), \tag{15.35}$$

When $W = \{z^1, z^2\}$,

$$d_W = \frac{p_D^1 \cdot p_D^2}{\lambda_k^1 \cdot c_k^1(z^1) \cdot \lambda_k^2 \cdot c_k^2(z^2)} \sum_{i=1}^{J_{k|k-1}} w_{k|k-1}^{(i)} \phi_{z^1 z^2}^{(i)}, \tag{15.36}$$

$$\rho_W^{(i)} = \frac{p_D^1 \cdot p_D^2 \cdot \phi_{z^1 z^2}^{(i)}}{p_D^1 p_D^2 \sum_{j=1}^{J_{k|k-1}} w_{k|k-1}^{(j)} \cdot \phi_{z^1 z^2}^{(j)}}, \tag{15.37}$$

$$\phi_{z^1 z^2}^{(i)} = \mathcal{N}\left(\tilde{z}; \tilde{H}_k m_{k|k-1}^{(i)}, \tilde{R}_k + \tilde{H}_k P_{k|k-1}^{(i)} (\tilde{H}_k)^T\right), \tag{15.38}$$

where

$$\tilde{z} = \left[(z^1)^T, (z^2)^T\right]^T, \tag{15.39}$$

$$\tilde{H}_k = \left[\left(H_k^1\right)^T, \left(H_k^2\right)^T\right]^T, \tag{15.40}$$

$$\tilde{R}_k = \mathrm{diag}\left(R_k^1, R_k^2\right). \tag{15.41}$$

## Remark 15.1

When $W = \{z^1\}$ or $W = \{z^2\}$, the derivation of Equations 15.30 through 15.35 is similar as the derivation of Proposition 15.2 in Vo [12]. For the sake of space, the derivation is omitted here. When $W = \{z^1, z^2\}$, we adopt the approximate formula

$$\begin{aligned} &v_{k|k-1}\left[p_D^1 l_{z^1}^1 p_D^2 l_{z^2}^2\right] \\ &:= \int p_D^1 \cdot l_{z^1}^1(x) \cdot p_D^2 \cdot l_{z^2}^2(x) \cdot v_{k|k-1}\left(x \mid Z_{1:k-1}\right) dx \end{aligned} \tag{15.42}$$

$$= p_D^1 \cdot p_D^2 \cdot \int l_{z^1}^1(x) \cdot l_{z^2}^2(x) \cdot v_{k|k-1}\left(x \middle| Z_{1:k-1}\right) dx \qquad (15.43)$$

$$\cong p_D^1 \cdot p_D^2 \cdot \int l_{\tilde{z}}^{1,2}(x) \cdot v_{k|k-1}\left(x \middle| Z_{1:k-1}\right) dx \qquad (15.44)$$

where

$$l_{\tilde{z}}^{1,2}(x) = \frac{L_{\tilde{z}}^{1,2}(x)}{\lambda_k^1 c_k^1(z^1) \cdot \lambda_k^2 c_k^2(z^2)} \qquad (15.45)$$

$$L_{\tilde{z}}^{1,2}(x) = \mathcal{N}(\tilde{z}; \tilde{H}_k x, \tilde{R}_k). \qquad (15.46)$$

∎

The ":=" stands for "definition." The "$\cong$" stands for "approximately equal." The approximation is based on the centralized enlarged dimensional fusion. As the rest of the derivation of Equations 15.36 to 15.41 is similar to the derivation of Proposition 15.2 in Vo [12], it is omitted here. From the approximation (Equation 15.44), it can be seen that the proposed algorithm is not equivalent to successive application of the single sensor GM-PHD. The performance comparison between sequential PHD and the proposed algorithm will be presented in future work.

Now, for the linear Gaussian multitarget model, the following proposition presents an analytic solution to the MS-PHD corrector (Equations 15.9 to 15.11). More concisely, the proposition shows how the Gaussian components of the predicted intensity are analytically propagated to the posterior intensity.

## Proposition 15.2

Suppose that Assumptions A.4 to A.7 hold and that the predicted intensity for time $k$ is a Gaussian mixture of the form

$$v_{k|k-1}(x) = \sum_{i=1}^{J_{k|k-1}} w_{k|k-1}^{(i)} \cdot N\left(x; m_{k|k-1}^{(i)}, P_{k|k-1}^{(i)}\right). \qquad (15.47)$$

Then, the posterior intensity at time $k$ is also a Gaussian mixture, and is given by

$$v_k(x) = v_k^{ND}(x) + \sum_{\mathcal{P} \sqsubset_2 Z_k} \sum_{W \in \mathcal{P}} v_k^D(x, W). \qquad (15.48)$$

The Gaussian mixture $v_k^{ND}(x)$, handling the no detection cases, is given by

$$v_k^{ND}(x) = \sum_{i=1}^{J_{k|k-1}} w_{k|k}^{(i)} \cdot N\left(x; m_{k|k}^{(i)}, P_{k|k}^{(i)}\right), \qquad (15.49)$$

$$w_{k|k}^{(i)} = \left(1 - p_D^1\right)\left(1 - p_D^2\right) \cdot w_{k|k-1}^{(i)}, \qquad (15.50)$$

$$m_{k|k}^{(i)} = m_{k|k-1}^{(i)}, \tag{15.51}$$

$$P_{k|k}^{(i)} = P_{k|k-1}^{(i)}. \tag{15.52}$$

The Gaussian mixture $v_k^D(x, W)$, handling the detected target cases, is given by

$$v_k^D(x, W) = \sum_{i=1}^{J_{k|k-1}} w_{k|k}^{(i)} \cdot N\left(x; m_{k|k}^{(i)}, P_{k|k}^{(i)}\right), \tag{15.53}$$

$$w_{k|k}^{(i)} = \omega_{\mathcal{P}} \cdot \rho_W^{(i)} \cdot w_{k|k-1}^{(i)}, \tag{15.54}$$

$$\omega_{\mathcal{P}} = \frac{\prod_{W \in \mathcal{P}} d_W}{\sum_{\mathcal{Q} \sqsubset_2 Z_{k+1}} \prod_{W \in \mathcal{Q}} d_W}. \tag{15.55}$$

where, the $d_W$ and $\rho_W^{(i)}$ can be taken as in Proposition 15.1, and when $W = \{z^1\}$,

$$m_{k|k}^{(i)} = m_{k|k}^{(i)} + K_k^{(i)}\left(z^1 - H_k^1 m_{k|k}^{(i)}\right), \tag{15.56}$$

$$P_{k|k}^{(i)} = \left[I - K_k^{(i)} H_k^1\right] P_{k|k-1}^{(i)}, \tag{15.57}$$

$$K_k^{(i)} = P_{k|k-1}^{(i)} \left(H_k^1\right)^T \left(R_k^V + H_k^1 P_{k|k-1}^{(i)} \left(H_k^1\right)^T\right)^{-1}. \tag{15.58}$$

When $W = \{z^2\}$,

$$m_{k|k}^{(i)} = m_{k|k}^{(i)} + K_k^{(i)}\left(z^2 - H_k^2 m_{k|k}^{(i)}\right), \tag{15.59}$$

$$P_{k|k}^{(i)} = \left[I - K_k^{(i)} H_k^2\right] P_{k|k-1}^{(i)}, \tag{15.60}$$

$$K_k^{(i)} = P_{k|k-1}^{(i)} \left(H_k^2\right)^T \left(R_k^2 + H_k^2 P_{k|k-1}^{(i)} \left(H_k^2\right)^T\right)^{-1}. \tag{15.61}$$

When $W = \{z^1, z^2\}$,

$$m_{k|k}^{(i)} = m_{k|k}^{(i)} + K_k^{(i)}\left(\tilde{z} - \tilde{H}_k m_{k|k}^{(i)}\right), \tag{15.62}$$

$$P_{k|k}^{(i)} = \left[I - K_k^{(i)} \tilde{H}_k\right] P_{k|k-1}^{(i)}, \tag{15.63}$$

$$K_k^{(i)} = P_{k|k-1}^{(i)} \tilde{H}_k \left( \tilde{R}_k + \tilde{H}_k P_{k|k-1}^{(i)} (\tilde{H}_k)^T \right)^{-1} \tag{15.64}$$

where $I$ is the corresponding dimension unit matrix, and

$$\tilde{z} = \left[ (z^1)^T, (z^2)^T \right]^T, \tag{15.65}$$

$$\tilde{H}_k = \left[ \left( H_k^1 \right)^T, \left( H_k^2 \right)^T \right]^T, \tag{15.66}$$

$$\tilde{R}_k = \mathrm{diag} \left( R_k^1, R_k^2 \right). \tag{15.67}$$

## 15.5   EFFECTIVE BINARY PARTITION

In this section, we discuss how to implement and simplify the "binary partition" of $Z_k$ in Section 15.3, and how can we reduce the computational complexity.

It is worth noting that the binary partition $\mathcal{P} \sqsubset_2 Z_k$ gives all the possible partitions of measure sets $Z_k = Z_k^1 \cup Z_k^2$. However, in these partitions, most contain the impossible measurement pair $W = \left\{ z_k^1, z_k^2 \right\}$ with $\omega_{\mathcal{P}} \cong 0$. Although to take all of these partition into consideration can guarantee the theoretical rigor of the MS-PHD filter, it is not necessary in practical application. On the contrary, if we only consider those binary partitions which contain the "effective" measurement pair, the computational complexity of Proposition 15.2 would be dramatically reduced. Specifically, we structure on the effective partition of $Z_k$ for each predict target $m_{k|k-1}^{(i)}$. To be more specific, we only take a few measurements of $Z_k^1$ and $Z_k^2$ to partition respectively, and the remaining measurements of $Z_k^1$ and $Z_k^2$ are disposed as single-skeleton set $\left\{ z_k^1 \right\}$ or $\left\{ z_k^2 \right\}$. Thus, as long as the predicted target number is controlled, the number of partitions of $Z_k$ can be controlled in a certain range. It provides an executable direction to structure binary partition.

Now, the first problem we have to deal with is how to select the numbered measurements for each predicted target. The second problem is how to structure the partition.

We can handle the first problem by the "effective" principle, that is, the nearest-neighbor principle. Hence, the partition method in this chapter is named "effective binary partition (EBP)." First of all, we introduce distance measure into the measurement space $\mathcal{Z}^1$ and $\mathcal{Z}^2$ respectively, such as Mahalanobis distance. For example, for $\left\{ z_{k,1}^1, z_{k,2}^1 \right\} \subseteq \mathcal{Z}^1$,

$$d_M \left( z_{k,1}^1, z_{k,2}^1 \right) = \sqrt{ \left( z_{k,1}^1 - z_{k,2}^1 \right)^T \left( S_k^1 \right)^{-1} \left( z_{k,1}^1 - z_{k,2}^1 \right) } \tag{15.68}$$

where

$$S_k^1 = R_k^1 + H_k^1 P_{k|k-1}^{(i)} \left( H_k^1 \right)^T \tag{15.69}$$

is the predicted measurement covariance of sensor 1 at time step $k$. It is similar for measurement space $\mathcal{Z}^2$. Thus, for each predicted target $m_{k|k-1}^{(i)}$, we can choose a few measurements (such as $l_0 = 1,2,3$) from $Z_k^1$ and $Z_k^2$ which are closest to the predict measurement $\eta_{k|k-1}^{(1,i)} = H_k^1 m_{k|k-1}^{(i)}$ and $\eta_{k|k-1}^{(2,i)} = H_k^2 m_{k|k-1}^{(i)}$ respectively. For convenience, we name these selected measurements "effective measurement." In addition, given a threshold $d_l$, we can further cut down the number of partitions by getting rid of the effective measurement which is outside of the threshold.

For the second problem, an example is given to illustrate how to construct the "effective binary partition." Suppose there are $J_{k|k-1}$ predicted targets

$$\left\{ w_{k|k-1}^{(i)}, m_{k|k-1}^{(i)}, P_{k|k-1}^{(i)} \right\}_{i=1}^{J_{k|k-1}} \tag{15.70}$$

and measurement set

$$Z_k^1 = \left\{ z_{k,1}^1, z_{k,2}^1, \cdots z_{k,M_k^1}^1 \right\}, \tag{15.71}$$

$$Z_k^2 = \left\{ z_{k,1}^2, z_{k,2}^2, \cdots z_{k,M_k^2}^2 \right\}. \tag{15.72}$$

For the predicted measurement of predicted target $m_{k|k-1}^{(i)}$, suppose the nearest-neighbor measurement set, that is, effective measurement set $Z_k^A \in Z_k^1$ and $Z_k^B \in Z_k^2$ are

$$Z_k^A = \{a, b, c\}, \tag{15.73}$$

$$Z_k^B = \{d, e, f\}. \tag{15.74}$$

Then we can list the all possible binary partition of $Z_k^A \cup Z_k^B$ as described in Figure 15.2. It can be seen from Figure 15.2 that the partition number of $Z_k^A \cup Z_k^B$ is $9 + 18 + 6 = 33$. Through combining these partition respectively with the single-skeleton cell partition of $\left( Z_k^1 / Z_k^A \right) \cup \left( Z_k^2 / Z_k^B \right)$, we can construct the effective binary partition of $Z_k^1 \cup Z_k^2$.

Once the number $l_0$ of selected measurement is given, the element number of $Z_k^A$ and $Z_k^B$ are also determined, and then the up-boundary of the binary partition number is fixed. For example, $l_0 = 3$, there are $\tau(l_0) = 9 + 18 + 6 = 33$ effective binary partition for predicted target $m_{k|k-1}^{(i)}$. When $l_0 = 2$, $\tau(l_0) = 6$; $l_0 = 1$, $\tau(l_0) = 1$. Thus, the total number of effective binary partitions for $\left\{ w_{k|k-1}^{(i)}, m_{k|k-1}^{(i)}, P_{k|k-1}^{(i)} \right\}_{i=1}^{J_{k|k-1}}$ is at most $\tau(l_0) \cdot J_{k|k-1}$. Besides the single-skeleton cell partition of $Z_k^1 \cup Z_k^2$, there are at most $\text{Num}^{\mathcal{P}} = \tau(l_0) \cdot J_{k|k-1} + 1$ effective binary partitions for $Z_k^1 \cup Z_k^2$. For example, if we select at most $l_0 = 3$ measurements from $Z_k^1$ and $Z_k^2$ about the predicted state $m_{k|k-1}^{(i)}$ respectively, then there are at most 33 partitions (see Figure 15.2). Thus, if the number of predicted targets is $J_{k|k-1}$, then there are at most $33 J_{k|k-1}$ binary partitions. Altogether, we take into account the partition which contains only single-skeleton subset of $Z_k$, at most $33 J_{k|k-1} + 1$ partitions should be considered. Consequently, the number of the binary partitions is related only to the number of predicted targets, and the relationship is linear.

The computational complexity of the EBP of the two-sensor PHD filter is $O(\tau(l_0) \cdot J_{k|k-1} + 1)$, where $\tau(l_0)$ is a constant corresponding to the effective measurement number $l_0$, $J_{k|k-1}$ is the number of predicted targets.

## Remark 15.2

For GM-PHD, there are several different implementations in the literature. Hence, it is necessary to distinguish GM-MSPHD from the previous existing GM-PHD filters, [12,23] for instance. The essential difference between GM-PHD [12,23] and GM-MSPHD is that GM-PHD filter is a single

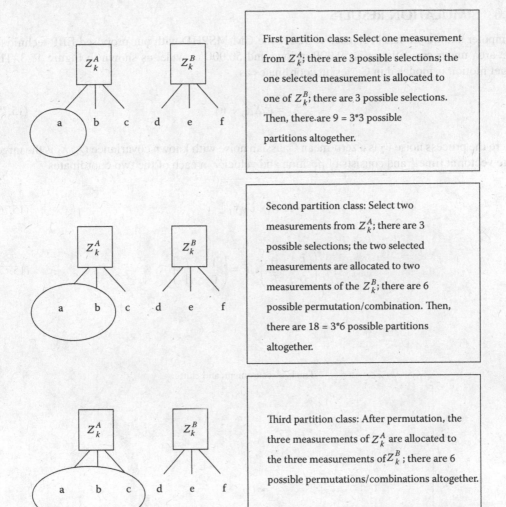

First partition class: Select one measurement from $Z_k^A$; there are 3 possible selections; the one selected measurement is allocated to one of $Z_k^B$; there are 3 possible selections. Then, there are 9 = 3*3 possible partitions altogether.

Second partition class: Select two measurements from $Z_k^A$; there are 3 possible selections; the two selected measurements are allocated to two measurements of the $Z_k^B$; there are 6 possible permutation/combination. Then, there are 18 = 3*6 possible partitions altogether.

Third partition class: After permutation, the three measurements of $Z_k^A$ are allocated to the three measurements of $Z_k^B$; there are 6 possible permutations/combinations altogether.

**FIGURE 15.2** Partition.

sensor filtering, and GM-MSPHD is a multisensor filtering. Although the Mahalanobis distance is used in the partition of GM-MSPHD filtering, the partition in GM-MSPHD filtering is obviously different from the partition in Ref. [24]. The partition in Ref. [24] is based on the distance between measurements, but the partition in this chapter is based on the distance between predicted measurement and measurement value of each sensor.                                                    ■

### Remark 15.3

Similarly to Refs. [12,22], the pruning scheme is required after the updated step because the number of Gaussian components increases without bound as time progresses. A simple pruning procedure has been provided by truncating components that have weak weights to mitigate this problem.

In addition, the track extracting from the intensity function is similar to the method in Ref. [12]. Interested readers are referred to Ref. [12] for the details.                                                    ■

## 15.6   SIMULATION RESULTS

Computer simulation has been used to verify the GM-MSPHD with our proposed EBP technique. The area under surveillance is 50,000 m long and 50,000 m wide, as shown in Figure 15.3. The target motion is modeled in Cartesian coordinates as

$$X_k = F_k X_{k-1} + w_k \tag{15.75}$$

where the process noise $w_k$ is a zero-mean Gaussian noise with known covariance $Q_k$, $X_k$ is the target state vector at time $k$ and consists of position and velocity in each of the two coordinates

$$X = [x\ \dot{x}\ y\ \dot{y}]^T; \tag{15.76}$$

$$F_k = \begin{pmatrix} \tilde{F} & 0 \\ 0 & \tilde{F} \end{pmatrix}, \tilde{F} = \begin{pmatrix} 1 & T \\ 0 & 1 \end{pmatrix} \tag{15.77}$$

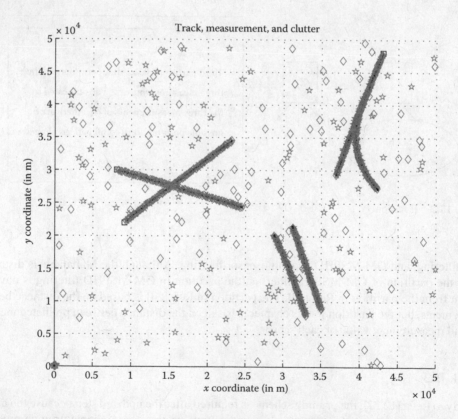

**FIGURE 15.3**   Target trajectories, measurements, and clutters. Only the last time step clutters are given. ◊ are measurements of sensor 1 with the last time step clutters. ☆ are measurements of sensor 2 with the last time step clutters. The points near the origin are the missing measurements, which are forced to be zero value.

$$Q_k = \begin{pmatrix} \tilde{Q} & 0 \\ 0 & \tilde{Q} \end{pmatrix}, \; \tilde{Q} = \sigma_w^2 \begin{pmatrix} \dfrac{T^4}{4} & \dfrac{T^3}{2} \\ \dfrac{T^3}{2} & T^2 \end{pmatrix} \tag{15.78}$$

where $\sigma_w = 4(m/s^2)$, $T = 1s$ is the sample period.

Targets can appear from five possible locations as well as spawned from other targets. Specifically, a Poisson RFS with intensity

$$\gamma_k(x) \sum_{i=1}^{5} w_i \mathcal{N}\left(x; \tilde{m}_i, \tilde{P}_{\gamma,i}\right) \tag{15.79}$$

where $w_i = 0.01$,

$$\tilde{m}_1 = [9130; 200; 22000; 190]; \tag{15.80}$$

$$\tilde{m}_2 = [8130; 240; 30000; -80]; \tag{15.81}$$

$$\tilde{m}_3 = [33000; -50; 8000; 180]; \tag{15.82}$$

$$\tilde{m}_4 = [35000; -50; 9000; 180]; \tag{15.83}$$

$$\tilde{m}_5 = [43000; -80; 48000; -230]; \tag{15.84}$$

and

$$\tilde{P}_{\gamma,i} = \begin{pmatrix} 400 & 0 & 0 & 0 \\ 0 & 25 & 0 & 0 \\ 0 & 0 & 400 & 0 \\ 0 & 0 & 0 & 25 \end{pmatrix}, i = 1, \ldots, 5. \tag{15.85}$$

In addition, the RFS $B_{k|k-1}(\zeta)$ of targets spawned from a target with previous state $\zeta$ is Poisson with intensity

$$\beta_k(x) = 0.2 \mathcal{N}(x; \zeta, Q_\beta), \tag{15.86}$$

$$Q_\beta = \text{diag}(400, 400, 400, 400). \tag{15.87}$$

Each target is detected with probability $p_{D,k} = 0.98$.

Target generated measurements $z_k$ contain the $(x, y)$ target coordinates spoiled by noise $v_k^1$ and $v_k^2$, that is,

$$z_k^1 = H_k^1 X_k + v_k^1, \tag{15.88}$$

$$z_k^2 = H_k^2 X_k + v_k^2 \tag{15.89}$$

where

$$H_k^1 = H_k^2 = \begin{pmatrix} 1 & 0 & 0 & 0 \\ 0 & 0 & 1 & 0 \end{pmatrix} \tag{15.90}$$

the measurement noise $v_k^1$ and $v_k^2$ are the zero-mean white Gaussian noise with the known covariance

$$R_k^1 = \sigma_{v^1}^2 \begin{pmatrix} 1 & 0 \\ 0 & 1 \end{pmatrix}, \; R_k^2 = \sigma_{v^2}^2 \begin{pmatrix} 1 & 0 \\ 0 & 1 \end{pmatrix}, \tag{15.91}$$

where $\sigma_{v^1} = 50 \; m$ and $\sigma_{v^2} = 100 \; m$ respectively. The detected measurements are immersed in clutter that can be modeled as a Poisson RFS $\mathcal{K}_k$ with intensity

$$\mathcal{K}_k^1 = \mathcal{K}_k^2 = \lambda_k V u_k(z). \tag{15.92}$$

where $u_k(z)$ is the uniform density over the surveillance region, $V = 2.5 \times 10^9 \; m^2$ is the "volume" of the surveillance region, and $\lambda_k = 4 \times 10^{-8} \; m^{-2}$ is the average number of clutter returns per unit volume (i.e., $M^1 = M^2 = 100$ clutter returns over the surveillance region of each sensor). In this case, for each predicted target, we choose $l_0 = 3$ effective measurements. If we control the number of predicted targets by at most $J_{k|k-1} = 40$, then the corresponding computational complexity of proposed GM-MSPHD filter would be $O((\tau(l_0) \cdot J_{k|k-1} + 1) = O(13,21)$. Through the threshold technology,

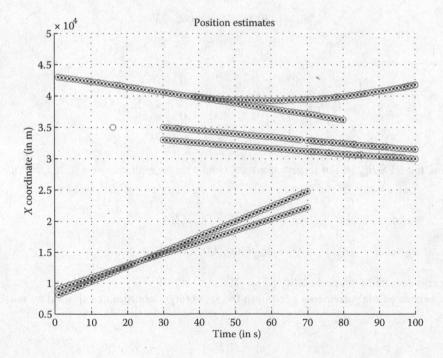

**FIGURE 15.4** Position estimates of X coordinate. O, filter estimates. •, true tracks.

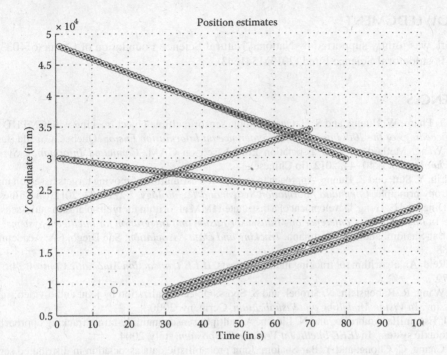

**FIGURE 15.5**   Position estimates of $Y$ coordinate. O, GM-MS-PHD filter estimates. •, true tracks.

the actual computational complexity in simulation is further reduced. Because lots of predicted Gaussian components are false, corresponding effective measurements do not exist. They also won't produce effective partition.

Figure 15.3 shows the true target trajectories, and measurements. To illustrate intensity of clutters, the clutters on the last time step are also given in Figure 15.3. The points near the origin are the missing measurements, which are forced to be zero value. At the beginning time, three targets are born at three different locations. At $k = 30s$, two targets are born and one target is spawned. At $k = 70s$, two targets die, and at $k = 80s$, another one target disappears.

Figures 15.4 and 15.5 show the position estimates of $X$ and $Y$ coordinates respectively. From the position estimates shown in Figures 15.4 and 15.5, it can be seen that the Gaussian mixture MS-PHD filter provides accurate tracking performance. The filter not only successfully detects and tracks the new born targets, but also manages to detect and track the spawned target. When the target is dead, the corresponding track will terminate immediately. The filter does generate anomalous estimates occasionally, but these false estimates die out very quickly.

## 15.7   CONCLUSION

This chapter describes an analytic suboptimal solution of the exact MS-PHD corrector for linear Gaussian multisensor multitarget models. A multisensor multitarget filter is proposed by combining the analytic suboptimal solution with a simple EBP procedure to reduce the computational complexity. The method makes the practical application of exact MS-PHD filter become possible. Simulations have demonstrated that the proposed approach is an attractive alternative to multisensor multitarget tracking with unknown time-varying number of targets.

## ACKNOWLEDGMENT

This work was jointly supported by National Natural Science Foundation of China (61403352) and China's Postdoctoral Science Fund (2013M541643).

## REFERENCES

1. N. T. Pham, W. Huang, and S. H. Ong. Multiple sensor multiple object tracking with GMPHD filter. In *Proceedings of the 10th International Conference on Information Fusion*, Quebec, Canada, July 2007.
2. Y. Wang. Method for data association ESM based on GAM. *Command Information System and Technology*, 3(3):40–44, 2012. (in Chinese).
3. T. Liu, Y. Sun, and Z. Huang. Approach to state fusion estimation of moving targets based on maximum entropy principle. *Command Information System and Technology*, 2(2):19–22, 2011. (in Chinese).
4. Z. Ding, and L. Hong. Development of a distributed IMM algorithm for multi-platform multi-sensor tracking. In *International Conference on Multisensor Fusion and Integration for Intelligent Systems*, 1996.
5. Y. Bar-Shalom, and T. E. Fortmann. *Tracking and Data Association*. San Diego, CA: Academic Press, 1988.
6. D. Reid. An algorithm for tracking multiple targets. *IEEE Transaction Automatic Control*, 24(6):84–90, 1979.
7. G. Wang, R. Rabenstein, N. Strobel, and S. Spors. Object localization by joint audio-video signal processing. In *Vision Modelling and Visualization*, Germany, 2000.
8. P. J. Escamilla-Ambrosio, and N. Lieven. A multiple-sensor multiple-target tracking approach for the autotaxic system. In *IEEE Intelligent Vehicles Symposium*, Italy, 2004.
9. K. Chang, C. Chong, and Y. Bar-Shalom. Joint probabilistic data association in distributed sensor networks. *IEEE Transactions on Automatic Control*, 31(10):889–897, 1986.
10. R. Mahler. Multi-target Bayes filtering via first-order multi-target moments. *IEEE Transactions on Aerospace and Electronic Systems*, 39(4):1152–1178, 2003.
11. B. N. Vo, S. Singh, and A. Doucet. Sequential Monte Carlo methods for Bayesian multi-target filtering with random finite sets. *IEEE Transactions on Aerospace and Electronic Systems*, 41(4):1224–1245, 2005.
12. B. N. Vo, and W. K. Ma. The Gaussian mixture probability hypothesis density filter. *IEEE Transaction Signal Processing*, 54(11):4091–4104, 2006.
13. R. Mahler. A theory of PHD filters of higher order in target number. *Proceedings of SPIE*, 6235, p. 62350K, 2006.
14. H. Zhang, Z. Jing, and S. Hu. Localization of multiple emitters based on the sequential PHD filter. *Signal Processing*, 90:34–43, 2010.
15. R. Mahler, B. T. Vo, and B. N. Vo. CPHD filtering with unknown clutter rate and detection profile. *IEEE Transactions on Signal Processing*, 59(8):3497–3513, 2011.
16. B. Ristic, and D. Clark. Particle filter for joint estimation of multi-object dynamic state and multi-sensor bias. In *IEEE International Conference on Acoustics, Speech and Signal Processing (ICASSP)*, pp. 3877–3880, 2012.
17. W. K. Ma, B. Vo, S. Singh, and A. Baddeley. Tracking an unknown time-varying number of speakers using TDOA measurements: A random finite set approach. *IEEE Transactions on Signal Processing*, 54(9):3291–3304, 2006.
18. B. N. Vo, S. Singh, and W. K. Ma. Tracking multiple speakers with random sets. In *IEEE International Conference on Acoustics, Speech and Signal Processing (ICASSP)*, pp. 357–360, Montreal, Canada, 2004.
19. R. Mahler. The multisensor PHD filter, I: General solution via multi-target calculus. *Proceedings of SPIE*, 7336, 2009.
20. R. Mahler. Approximate multisensor CPHD and PHD filters. In *The 13th International Conference on Information Fusion*. IEEE, 2010.
21. C. Ouyang, and H. Ji. Scale unbalance problem in product multisensor PHD filter. *Electronics Letters*, 47(22):1247–1249, 2011.
22. W. Li, Y. Jia, J. Du, and F. Yu. Gaussian mixture PHD filter for multi-sensor multi-target tracking with registration errors. *Signal Processing*, 93(1):86–99, 2013.
23. D. Clark, and B.-N. Vo. Convergence analysis of the Gaussian mixture PHD filter. *IEEE Transactions on Signal Processing*, 55(4):1204–1212, 2007.
24. K. Granstrom, C. Lundquist, and U. Orguner. A Gaussian mixture PHD filter for extended target tracking. In *Proceedings of the International Conference on Information Fusion*, Edinburgh, UK, July 2010.

# 16 Information Fusion Estimation for Multisensor Multirate Systems with Multiplicative Noises

## Shuli Sun, Jing Ma, and Fangfang Peng

## CONTENTS

## 16.1 INTRODUCTION

In many practical systems, there often exist various uncertainties owing to the completely or partially unknown parameters and environmental disturbances. The uncertainties can be approximated mathematically by an additive noise or a multiplicative noise [1–6]. These systems are widely used in target tracking, detection, signal processing, and other areas. Thus, the research on systems with additive and multiplicative noises has an important practical significance. For single-sensor systems, many results have been reported, including the least mean square optimal linear filters with uncertain observations of multiplicative noises [1], the least mean square optimal linear filter with stochastic parameters [2], and the nonlinear polynomial filters [3–5] with high computational cost. In networked systems, the phenomena of random delays and packet dropouts can be depicted by stochastic variables. Such models can be transformed into the stochastic parameterized systems with multiplicative noises; for the work in this aspect, see Refs. [6–11]. However, the abovementioned literature does not take multiple sensors into account.

As the sensor technology is widely used in military, civilian, scientific research, and many other fields, a single sensor has failed to meet the performance requirements in many practical applications. Moreover, multiple sensors can provide more information than any single sensor in time and space. Hence, multisensor information fusion has received considerable research attention in recent years [12]. For systems with a single sampling rate, the optimal matrix-weighting fusion filter in the linear minimum variance sense [13] and the self-tuning fusion filter with unknown noise variances [14] have been presented. Recently, the multirate multisensor asynchronous fusion algorithms have been studied in Refs. [15–17]. Yan et al. [18] and Xiao et al. [19] adopt the lift state augmentation approach, yielding estimators with a high computational cost. Although Yan et al. [20,21] adopt the nonaugmented approach to design the filters, a modeling error is made by ignoring the process noise. By incorporating the process noise into the model to eliminate the modeling error, an optimal filter is presented to improve the estimation accuracy [22]. Further, the missing observations are also taken into account in Ref. [23]. In Ref. [24], a two-sensor multirate distributed fusion estimator is proposed for two-sensor systems with one-step cross-covariance noises. In Ref. [25], a distributed fusion filter (DFF) is presented by pseudo-observation method. However, most of the abovementioned literatures do not take into account the parameter uncertainties of multiplicative noises. Peng and Sun [26] study the distributed fusion filtering problem for multisensor multirate systems with observation uncertainties of multiplicative noises. However, state uncertainty is not considered. In sensor networks, there often exist various sensors with different sampling rates and stochastic uncertainties of multiplicative noises. It is significant to use the nonaugmented approach to deal with the multirate multisensor systems.

In this chapter, the information fusion estimation problem is investigated for a class of multisensor multirate systems with observation uncertainties of multiplicative noises. State is sampled uniformly at the fastest rate. Different sensors have different sampling periods that are integer multiples of the state update period. A centralized fusion filter (CFF) in the linear minimum variance sense is designed by using the observations received from different sensors at each time. It involves a time-varying Kalman filter with time-varying observation dimensions. It has the best accuracy when all sensors work healthily. However, a faulty sensor can lead to the failure of the CFF. To improve the reliability, a DFF is also designed. First, the pseudo-observations are introduced by employing a group of Bernoulli distributed stochastic variables. Local estimators at the state sampling points are designed based on the pseudo-observations. Then, the estimation error cross-covariance matrices between any two local estimators are derived. At last, a distributed optimal fusion estimator is obtained by using the distributed matrix-weighting fusion estimation algorithm in the linear minimum variance sense [13]. It has a good reliability since it has a distributed parallel structure.

## 16.2   PROBLEM FORMULATION

Consider the following multisensor multirate systems with multiplicative noises:

$$x(th_0 + h_0) = (\Phi + \lambda(th_0)\widehat{\Phi})x(th_0) + \Gamma w(th_0) \tag{16.1}$$

$$z_i(th_i) = (H_i + \beta_i(th_i)\widehat{H}_i)x(th_i) + v_i(th_i), \ i = 1,2,\cdots,L, \tag{16.2}$$

where $x(th_0) \in \mathbb{R}^n$ is the state; $z_i(th_i) \in \mathbb{R}^{n_i}$, $i = 1,2, \cdots, L$, are the measured outputs; $\Phi$, $\widehat{\Phi}$, $\Gamma$, $H_i$, and $\widehat{H}_i, i = 1,2,\cdots,L$, are constant matrices with suitable dimensions; $\lambda(th_0) \in \mathbb{R}$, $w(th_0) \in \mathbb{R}^q$, $v_i(th_i) \in \mathbb{R}^{n_i}$, and $\beta_i(th_i) \in \mathbb{R}$, $i = 1, 2, \cdots, L$, are white noises. The state is updated at the fastest rate with a period $h_0$, and the observation is sampled at a lower rate with a period $h_i = m_i h_0$, where $m_i$ is a positive integer. The subscript $i$ denotes the $i$th sensor and $L$ is the number of sensors.

### Assumption 1

The terms $w(th_0)$ and $v_i(th_i)$ are uncorrelated white noises with zero means and covariance matrices $\mathrm{E}[w(th_0)w^\mathrm{T}(th_0)] = Q_w$ and $\mathrm{E}\left[v_i(th_i)v_i^\mathrm{T}(th_i)\right] = Q_{v_i}, i = 1, 2, \cdots, L$, where E denotes the mathematical expectation and the superscript T denotes the transpose.

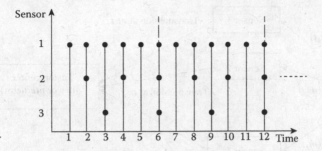

**FIGURE 16.1**    Sampling case of sensors.

## Assumption 2

Multiplicative noises $\lambda(th_0)$ and $\beta_i(th_i)$, $i = 1, 2, \cdots, L$, are mutually uncorrelated scalar white noise sequences with zero means and variances $Q_\lambda$ and $Q_{\beta_i}$ and are uncorrelated with other random variables.

## Assumption 3

The initial state $x(0)$ is uncorrelated with $w(th_0)$ and $v_i(th_i)$, $i = 1, 2, \cdots, L$, and satisfies that $E\{x(0)\} = \mu$, $E[(x(0) - \mu)(x(0) - \mu)^T] = P_0$.

## Remark 1

The sampling case of multisensor multirate systems can be described in Figure 16.1. The horizontal axis denotes time whereas the vertical axis denotes different sensors. Three sensors are shown in Figure 16.1. Solid black circles represent the sampling time of different sensors. The sampling rate goes from the highest (sensor 1) to the lowest (sensor 3). As shown in Figure 16.1, three sensors all sample uniformly. The first sensor has the same sampling rate as the state update rate; that is, the sampling period is $h_0$. The sampling period of the second sensor is $2h_0$ and the third is $3h_0$. It is clear that the least common multiple of three sample periods is $6h_0$. This means the samplings of different sensors are asynchronous in each data block of the length $6h_0$.

The objective of this chapter is to find the following:

a. The centralized fusion estimator $\hat{x}_c(t|t)$ based on the observations received from different sensors at each time
b. The distributed fusion estimator $\hat{x}_o(t|t)$ based on the local estimators from all sensors at each time

For the sake of easy manipulation, we use $t$ to denote the $th_0$ time moment in the following text.

## 16.3    CENTRALIZED FUSION FILTER

The CFF has the structure shown in Figure 16.2.

Each sensor samples at its own rate and transmits its observations to the fusion center for estimate update. In the fusion center, combining the observations from the sensors that have samples at $th_0$ moment, we have the augmented observation as follows:

$$z_c(t) = (H_c(t) + \beta_c(t)\widehat{H}_c(t))x(t) + v_c(t)$$    (16.3)

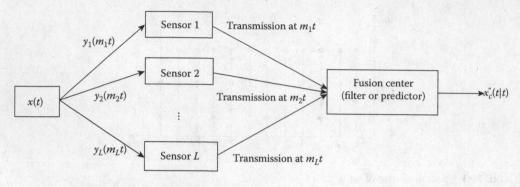

**FIGURE 16.2** Centralized fusion estimation scheme.

where the augmented observation $z_c(t)$, observation matrices $H_c(t)$, $\hat{H}_c(t)$, observation noise $v_c(t)$, and stochastic matrix $\beta_c(t)$ are defined as follows:

$$z_c(t) = \left[ z_{p_1(t)}^{\mathrm{T}}(t), z_{p_2(t)}^{\mathrm{T}}(t), \cdots, z_{p_k(t)}^{\mathrm{T}}(t) \right]^{\mathrm{T}},$$

$$H_c(t) = \left[ H_{p_1(t)}^{\mathrm{T}}, H_{p_2(t)}^{\mathrm{T}}, \cdots, H_{p_{k(t)}}^{\mathrm{T}} \right]^{\mathrm{T}},$$

$$\hat{H}_c(t) = \left[ \hat{H}_{p_1(t)}^{\mathrm{T}}, \hat{H}_{p_2(t)}^{\mathrm{T}}, \cdots, \hat{H}_{p_k(t)}^{\mathrm{T}} \right]^{\mathrm{T}},$$

$$v_c(t) = \left[ v_{p_1(t)}^{\mathrm{T}}(t), v_{p_2(t)}^{\mathrm{T}}(t), \cdots, v_{p_k(t)}^{\mathrm{T}}(t) \right]^{\mathrm{T}},$$

$$\beta_c(t) = \mathrm{diag}\left( \beta_{p_1(t)}(t) I_{n_{p_1(t)}}, \beta_{p_2(t)}(t) I_{n_{p_2(t)}}, \cdots, \beta_{p_k(t)}(t) I_{n_{p_k(t)}} \right),$$

where $p_k(t)$ denotes the $p_k(t)$th sensor, $1 \le p_1(t) < p_2(t) < \cdots < p_k(t) \le L$. Diag() denotes a diagonal matrix.

The variance matrix of the augmented observation noise $v_c(t)$ is given as follows:

$$R_c(t) = \mathrm{diag}\left( Q_{v_{p_1(t)}}, Q_{v_{p_2(t)}}, \cdots, Q_{v_{p_k(t)}} \right)$$

The augmented observation equation (Equation 16.3) is time varying in the dimensions and coefficients. For Equations 16.1 and 16.3, we can derive the CFF. The following theorem gives the result.

**Theorem 1**

For Equations 16.1 and 16.3, the CFF is computed by

$$\hat{x}_c(t|t) = \hat{x}_c(t|t-1) + K_c(t)\varepsilon_c(t) \tag{16.4}$$

The centralized fusion predictor is computed by

$$\hat{x}_c(t|t-1) = \Phi \hat{x}_c(t-1|t-1)$$ (16.5)

The innovation $\varepsilon_c(t)$ and its covariance matrix $Q_{\varepsilon_c}(t)$ are computed by

$$\varepsilon_c(t) = z_c(t) - H_c(t)\hat{x}_c(t|t-1)$$ (16.6)

$$Q_{\varepsilon_c}(t) = H_c(t)P_c(t|t-1)H_c^T(t) + Q_{\beta_c}(t) \odot \left( \hat{H}_c(t)q(t)\hat{H}_c^T(t) \right) + R_c(t)$$ (16.7)

where $\odot$ is the Hadamard product, $Q_{\beta_c}(t) = \text{diag}\left( Q_{\beta_{p_1(t)}} 1_{n_{p_1(t)}}, Q_{\beta_{p_2(t)}} 1_{n_{p_2(t)}}, \cdots, Q_{\beta_{p_k(t)}} 1_{n_{p_k(t)}} \right)$, where $1_{n_{p_k(t)}}$ denotes an $n_{p_k(t)}$ by $n_{p_k(t)}$ matrix of all ones.

The filtering gain matrix is computed by

$$K_c(t) = P_c(t|t-1)H_c^T(t)Q_{\varepsilon_c}^{-1}(t)$$ (16.8)

The prediction and filtering error covariance matrices are computed by

$$P_c(t|t-1) = \Phi P_c(t-1|t-1)\Phi^T + Q_\lambda \hat{\Phi} q(t-1)\hat{\Phi}^T + \Gamma Q_w \Gamma^T$$ (16.9)

$$P_c(t|t) = (I_n - K_c(t)H_c(t))P_c(t|t-1).$$ (16.10)

The state second-order moment matrix $q(t)$ is computed by

$$q(t+1) = \Phi q(t)\Phi^T + Q_\lambda \hat{\Phi} q(t)\hat{\Phi}^T + \Gamma Q_w \Gamma^T.$$ (16.11)

The initial values are $\hat{x}_c(0|0) = \mu$, $P_c(0|0) = P_0$ and $q(0) = P_0 + \mu\mu^T$. ∎

**Proof**

From projection theory, we readily have Equations 16.4 through 16.6. Substituting Equation 16.3 into Equation 16.6, we can rewrite Equation 16.6 as follows:

$$\varepsilon_c(t) = H_c(t)\tilde{x}_c(t|t-1) + \beta_c(t)\hat{H}_c(t)x(t) + v_c(t),$$ (16.12)

where the prediction error $\tilde{x}_c(t|t-1) = x(t) - \hat{x}_c(t|t-1)$. Then, we have the covariance matrix of innovation sequence as

$$Q_{\varepsilon_c}(t) = E\left\{ \varepsilon_c(t)\varepsilon_c^T(t) \right\} = H_c(t)P_c(t|t-1)H_c^T(t) + E\left\{ \beta_c(t)\hat{H}_c(t)q(t)\hat{H}_c^T(t)\beta_c(t) \right\} + R_c(t)$$ (16.13)

Noting that $E\left\{ \beta_c(t)\hat{H}_c(t)q(t)\hat{H}_c^T(t)\beta_c(t) \right\} = Q_{\beta_c}(t) \odot \left( \hat{H}_c(t)q(t)\hat{H}_c^T(t) \right)$, we have Equation 16.7 to hold.

From Equations 16.4 and 16.5, we have the filtering and prediction error equations as follows:

$$\tilde{x}_c(t|t) = \tilde{x}_c(t|t-1) - K_c(t)\varepsilon_c(t), \tilde{x}_c(t|t-1) = \Phi\tilde{x}_c(t-1|t-1) + \lambda(t)\hat{\Phi}x(t-1) + \Gamma w(t). \quad (16.14)$$

Substituting Equation 16.14 into $P_c(t|t) = \mathrm{E}\left\{\tilde{x}_c(t|t)\tilde{x}_c^{\mathrm{T}}(t|t)\right\}$ and $P_c(t|t-1) = \mathrm{E}\left\{\tilde{x}_c(t|t-1)\right.$ $\left.\tilde{x}_c^{\mathrm{T}}(t|t-1)\right\}$ yields Equations 16.9 and 16.10. Substituting Equation 16.1 into $q(t) = \mathrm{E}\{x(t)x^{\mathrm{T}}(t)\}$ yields Equation 16.11.

### Remark 2

In Theorem 1, we assume that there is at least one observation arriving at the fusion center. If there are no observations arriving at the fusion center, a predictor is implemented based on the latest fusion filter.

### Remark 3

The proposed CFF provides the linear optimal estimate in the linear minimum variance sense when all sensors work healthily. However, it does not have the reliability. A faulty sensor can lead to the failure of the CFF. It is not convenient for the detection and isolation of the faults. To improve the reliability, the DFF will be discussed in Section 16.4.

## 16.4  DISTRIBUTED FUSION FILTER

The DFF has the structure shown in Figure 16.3.

Individual sensors provide their local estimators including filters and predictors based on their own observations. Then, local estimators are transmitted to the fusion center for fusion estimate at each time. A DFF has better robustness and flexibility since it has a parallel structure, which is convenient for the detection and isolation of the faults. To obtain the distributed fusion estimator by using the optimal weighted fusion estimation algorithm in the linear minimum variance sense [13], we need to compute the local estimators and variance matrices from each sensor and the cross-covariance matrices between any two local estimators. Later, we will give the computation of local estimators and cross-covariance matrices.

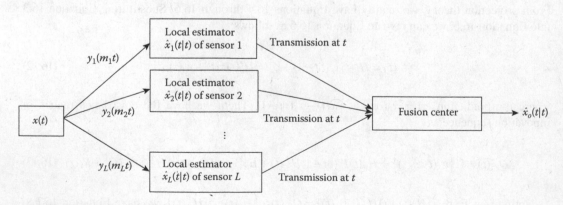

**FIGURE 16.3**  Distributed fusion estimation scheme.

## 16.4.1  System Transformation

Next, we transform the multirate fusion estimation problem into a single-rate fusion estimation problem. Similar to Ref. [25], we introduce white Bernoulli distributed variables $\gamma_i(t)$ as follows:

$$\gamma_i(t) = \begin{cases} 1, & t = km_i, k = 1, 2, \cdots \\ 0, & \text{else} \end{cases}$$

Based on $\gamma_i(t)$, we define the following variables:

$$y_i(t) = \begin{cases} z_i(t), & \gamma_i(t) = 1 \\ 0, & \gamma_i(t) = 0, \end{cases} \quad V_i(t) = \begin{cases} v_i(t), & \gamma_i(t) = 1 \\ 0, & \gamma_i(t) = 0, \end{cases} \quad \xi_i(t) = \begin{cases} \beta_i(t), & \gamma_i(t) = 1 \\ 0, & \gamma_i(t) = 0, \end{cases} \quad (16.15)$$

From the above definition, we see that $\gamma_i(t) = 1$ denotes $y_i(t) = z_i(t)$ and $\gamma_i(t) = 0$ denotes $y_i(t) = 0$. Then, multirate Equations 16.1 and 16.2 with multiplicative noises can be transformed into the following single-rate multisensor system:

$$x(t+1) = (\Phi + \lambda(t)\hat{\Phi})x(t) + \Gamma w(t) \tag{16.16}$$

$$y_i(t) = \gamma_i(t)(H_i + \xi_i(t)\hat{H}_i)x(t) + V_i(t), i = 1, 2, \cdots, L. \tag{16.17}$$

Further, we have the following noise statistical information:

$$Q_{V_i}(t) = E\left[V_i(t)V_i^{\mathrm{T}}(t)\right] = \begin{cases} Q_{v_i}, \gamma_i(t) = 1 \\ \sigma I, \gamma_i(t) = 0, \end{cases} \quad Q_{\xi_i}(t) = E\left[\xi_i(t)\xi_i^{\mathrm{T}}(t)\right] = \begin{cases} Q_{\beta_i}, \gamma_i(t) = 1 \\ 0, \gamma_i(t) = 0, \end{cases} \tag{16.18}$$

where $\sigma$ is a sufficiently large positive number.

Next, we shall give local filters (LFs) based on the random variables $(\gamma_i(t), \gamma_i(t-1), \cdots, \gamma_i(0))$ and pseudo-observations $(y_i(t), y_i(t-1), \cdots, y_i(0))$, $i = 1, 2, \cdots, L$, of Equation 16.17.

## 16.4.2  Local State Estimators

Similarly to the derivation of Theorem 1, we have the LF at the state update points of each sensor based on the above model (Equations 16.16 and 16.17). The detailed proof is omitted here.

**Theorem 2**

Under Assumptions 1 through 3, the LF of the $i$th sensor subsystem of Equations 16.16 and 16.17 is computed by

$$\hat{x}_i(t|t) = \hat{x}_i(t|t-1) + K_i(t)\varepsilon_i(t) \tag{16.19}$$

$$\hat{x}_i(t|t-1) = \Phi\hat{x}_i(t-1|t-1) \tag{16.20}$$

$$\varepsilon_i(t) = y_i(t) - \gamma_i(t)H_i\hat{x}_i(t|t-1) \tag{16.21}$$

$$K_i(t) = \gamma_i(t)P_i(t|t-1)H_i^{\mathrm{T}}Q_{\varepsilon_i}^{-1}(t) \tag{16.22}$$

$$Q_{\varepsilon_i}(t) = \gamma_i(t)H_iP_i(t|t-1)H_i^{\mathrm{T}} + \gamma_i(t)Q_{\xi_i}(t)\widehat{H}_iq(t)\widehat{H}_i^{\mathrm{T}} + Q_{V_i}(t) \tag{16.23}$$

$$P_i(t|t) = (I - \gamma_i(t)K_i(t)H_i)P_i(t|t-1) \tag{16.24}$$

$$P_i(t|t-1) = \Phi P_i(t-1|t-1)\Phi^{\mathrm{T}} + Q_\lambda\widehat{\Phi}q(t-1)\widehat{\Phi}^{\mathrm{T}} + \Gamma Q_w\Gamma^{\mathrm{T}}, \tag{16.25}$$

where $\varepsilon_i(t)$ is the innovation process with variance $Q_{\varepsilon_i}(t)$, $K_i(t)$ is the filtering gain matrix, and $P_i(t|t)$ and $P_i(t|t-1)$ are the filtering and prediction error variance matrices. The initial values are $\hat{x}_i(0|0) = \mu$ and $P_i(0|0) = P_0$. The state second-order moment $q(t)$ is computed by Equation 16.11. ∎

## Remark 4

From Theorem 2, it is clear that the filter is implemented when $\gamma_i(t) = 1$ and the predictor is implemented when $\gamma_i(t) = 0$. Thus, the local estimators at the state update points have been obtained by filtering and prediction. They are simple and have a good real-time property.

Now, we have obtained the local estimators at the state update points based on the observations of each sensor. Next, we compute the cross-covariance matrices between any two local estimators.

### 16.4.3 Computation of Cross-Covariance Matrices

#### Theorem 3

Under Assumptions 1 through 3, the cross-covariance matrix of filtering errors between the $i$th and the $j$th sensor subsystems of Equations 16.16 and 16.17 is computed by

$$\begin{aligned} P_{ij}(t|t) &= P_{ij}(t|t-1) - \gamma_j(t)P_{ij}(t|t-1)H_j^{\mathrm{T}}K_j^{\mathrm{T}}(t) \\ &\quad - \gamma_i(t)K_i(t)H_iP_{ij}(t|t-1) + K_i(t)Q_{\varepsilon_{ij}}(t)K_j^{\mathrm{T}}(t) \end{aligned} \tag{16.26}$$

$$P_{ij}(t|t-1) = \Phi P_{ij}(t-1|t-1)\Phi^{\mathrm{T}} + Q_\lambda\widehat{\Phi}q(t-1)\widehat{\Phi}^{\mathrm{T}} + \Gamma Q_w\Gamma^{\mathrm{T}} \tag{16.27}$$

$$Q_{\varepsilon_{ij}}(t) = \gamma_i(t)\gamma_j(t)H_iP_{ij}(t|t-1)H_j^{\mathrm{T}}, \tag{16.28}$$

where $Q_{\varepsilon_{ij}}(t)$ is the innovation cross-covariance matrix between the $i$th and the $j$th sensor subsystems. The initial value is $P_{ij}(0|0) = P_0$. ∎

## Proof

From Equations 16.17 and 16.21, we can rewrite the innovation equation as follows:

$$\varepsilon_i(t) = \gamma_i(t)H_i\tilde{x}_i(t|t-1) + \gamma_i(t)\xi_i(t)\hat{H}_i x(t) + V_i(t). \tag{16.29}$$

Using $\gamma_i^2(t) = \gamma_i(t)$, $\tilde{x}_i(t|t-1) \perp V_i(t)$, and $\mathrm{E}[\xi_i(t)] = 0$, the innovation cross-covariance (Equation 16.28) is obtained from $Q_{\varepsilon_{ij}}(t) = \mathrm{E}\left[\varepsilon_i(t)\varepsilon_j^{\mathrm{T}}(t)\right]$.

Subtracting Equation 16.20 from Equation 16.16, we have the one-step prediction error equation

$$\tilde{x}_i(t|t-1) = \Phi\tilde{x}_i(t-1|t-1) + \lambda_i(t-1)\hat{\Phi}x(t-1) + \Gamma w(t-1) \tag{16.30}$$

Using $\tilde{x}_i(t|t) \perp w(t)$ and $\mathrm{E}[\lambda(t)] = 0$, Equation 16.26 follows from $P_{ij}(t|t-1) = \mathrm{E}\left[\tilde{x}_i(t|t-1)\tilde{x}_j^{\mathrm{T}}(t|t-1)\right]$.

Subtracting Equation 16.19 from $x(t)$, we have the filtering error equation

$$\tilde{x}_i(t|t) = \tilde{x}_i(t|t-1) - K_i(t)\varepsilon_i(t). \tag{16.31}$$

Hence, we have the filtering error cross-covariance matrix

$$
\begin{aligned}
P_{ij}(t|t) &= \mathrm{E}\left[\tilde{x}_i(t|t)\tilde{x}_j^{\mathrm{T}}(t|t)\right] \\
&= P_{ij}(t|t-1) - \mathrm{E}[\tilde{x}_i(t|t-1)\varepsilon_j^{\mathrm{T}}(t)]K_j^{\mathrm{T}}(t) - K_i(t)\mathrm{E}[\varepsilon_i(t)\tilde{x}_j^{\mathrm{T}}(t|t-1)] + K_i(t)Q_{\varepsilon_{ij}}(t)K_j^{\mathrm{T}}(t),
\end{aligned}
\tag{16.32}
$$

where

$$\mathrm{E}\left[\tilde{x}_i(t|t-1)\varepsilon_j^{\mathrm{T}}(t)\right] = \gamma_j(t)H_j P_{ij}(t|t-1)K_j^{\mathrm{T}}(t). \tag{16.33}$$

Substituting Equation 16.33 into Equation 16.32, we have Equation 16.26.

## Remark 5

The computation formulas of cross-covariance matrices in Theorem 3 have the simple form, which is different from Ref. [26] where cross-covariance matrices are recursively computed by three cases.

### 16.4.4 DISTRIBUTED FUSION ESTIMATOR WEIGHTED BY MATRICES

In the preceding sections, we have obtained the local estimators at the state update points and their covariance matrices. Applying the distributed matrix-weighting optimal fusion estimation algorithm in the linear minimum variance sense [13], we can obtain the distributed fusion estimator as follows:

$$\hat{x}_o(t|t) = \sum_{i=1}^{L} A_i(t)\hat{x}_i(t|t). \tag{16.34}$$

The optimal weighted matrices are computed by

$$[A_1(t),\cdots, A_L(t)] = (e^{\mathrm{T}}\Omega^{-1}(t)e)^{-1}e^{\mathrm{T}}\Omega^{-1}(t), \tag{16.35}$$

where $\Omega(t) = (P_{ij}(t|t))$ is an $nL \times nL$ matrix whose the $(i,j)$ block is $P_{ij}(t|t)$ and $e = [I_n, \cdots, I_n]^{\mathrm{T}}$ is an $nL \times n$ matrix. Then, the optimal fusion estimation error variance matrix is computed by

$$P_o(t|t) = (e^{\mathrm{T}}\Omega^{-1}(t)e)^{-1}. \tag{16.36}$$

Furthermore, we have $P_o(t|t) \le P_i(t|t)$.

### Remark 6

Compared with the centralized fusion estimator in Section 16.3, the distributed fusion estimator has flexibility and robustness since it has the distributed parallel structure [13].

## 16.5  SIMULATION

Consider the following tracking system with three sensors of different sampling rates:

$$x(th_0 + h_0) = \left(\begin{bmatrix} 0.95 & h_0 \\ 0 & 0.95 \end{bmatrix} + \lambda(th_0)\begin{bmatrix} 0.01 & 0 \\ 0 & 0.01 \end{bmatrix}\right)x(th_0) + \begin{bmatrix} 0.8 \\ 0.6 \end{bmatrix}w(th_0) \tag{16.37}$$

$$z_i(th_i) = (H_i + \xi_i(th_i)\widehat{H}_i)x(th_i) + v_i(th_i), i = 1,2,3, \tag{16.38}$$

where the observation noises $v_i(th_i)$, $h_i = m_i h_0$, $i = 1, 2, 3$, are uncorrelated with the white noise $w(th_0)$ with zero mean and variance $Q_w$, and $\lambda(th_0)$ and $\zeta_i(th_i)$ are independent Gaussian noises with zero means and variances $Q_\lambda$ and $Q_{\zeta_i}$. Our aim is to find the CFF $\hat{x}_c(th_0|th_0)$ and DFF $\hat{x}_o(th_0|th_0)$ weighted by matrices.

In the simulation, we set $h_0 = 1$, $m_1 = 1$, $m_2 = 3$, $m_3 = 2$, $H_1 = [0.5 \quad 1]$, $H_2 = [1 \quad 1]$, $H_3 = [1 \quad 0.5]$, $\widehat{H}_1 = [0.1 \quad 0.1]$, $\widehat{H}_2 = [0.1 \quad 0.05]$, $\widehat{H}_3 = [0.2 \quad 0.1]$, $Q_w = 1$, $Q_{v_1} = 10$, $Q_{v_2} = 8$, $Q_{v_3} = 6$, $Q_\lambda = 0.1$, $Q_{\beta_1} = 0.1$, $Q_{\beta_2} = 0.2$, $Q_{\beta_3} = 0.05$, $x(0) = [0 \quad 0]^{\mathrm{T}}$, and $P_0 = 0.1I_2$. We take 120 sampling data.

Figure 16.4 gives the tracking performances of the DFF, where the solid curves denote the true values and the dashed ones denote the filters. Figure 16.5 gives the comparison curves of variances for all the LFs, CFF, and DFF. From Figure 16.5, we see that the proposed fusion filters outperform

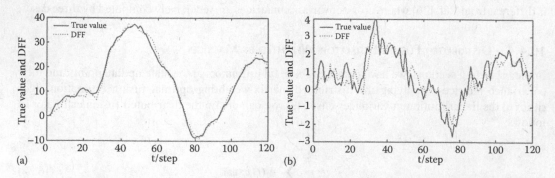

**FIGURE 16.4**  Distributed fusion filter: (a) the first state component and (b) the second state component.

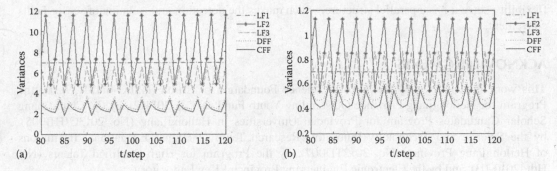

**FIGURE 16.5**  Comparison of variances of DFF, CFF, and LFs: (a) the first state component and (b) the second state component.

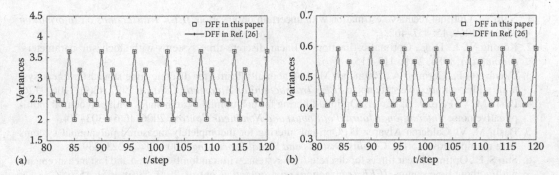

**FIGURE 16.6**  Comparison of variances of DFFs in this chapter and Ref. [26]: (a) the first state component and (b) the second state component.

the LFs. Compared with CFF, DFF has a small accuracy loss. However, it is significant that the DFF has better reliability than CFF since it has a parallel structure that makes it convenient to detect and isolate the faults of sensors.

Figure 16.6 shows the comparison curves of variances in the interval [80, 120] for DFFs in this chapter and Ref. [26] under the condition of $\lambda(th_0) = 0$, which means no uncertainty in state equation. From Figure 16.6, we see that they have the same accuracy. Hence, the DFF in this chapter generalizes the results in Ref. [26].

## 16.6  CONCLUSION

Centralized and distributed fusion estimators have been designed for systems with multiple sensors and multiplicative noises. Each sensor samples uniformly, and different sensors have different sampling rates, whose sampling periods are the integer multiples of the state update period. By using the innovation analysis method, a CFF is designed in the linear minimum variance sense. The centralized fusion estimator has the best accuracy when all sensors work healthily. However, it has poor reliability. To improve the reliability, a DFF is designed. First, the pseudo-observations at the state update point are introduced by employing a group of Bernoulli distributed variables. LFs dependent on the stochastic variables are designed at the state update points. They are equivalent to the filter at sampling points and predictors at no sampling points. They avoid the state augmentation by lift method and have a good real-time property. Then, cross-covariance matrices between any two LFs are derived. At last, the DFF is obtained by using the matrix-weighting fusion estimation algorithm in the linear minimum variance sense. The given distributed fusion estimator has a small accuracy loss compared with the centralized fusion estimator. However, it has better robustness and flexibility since it has a parallel structure, which makes the detection and isolation of sensor faults convenient.

## ACKNOWLEDGMENTS

This work is supported by the Natural Science Foundation of China (NSFC-61174139), by the Program for Heilongjiang Province Outstanding Youth Fund (No. JC201412), by the Chang Jiang Scholar Candidates Program for Provincial Universities in Heilongjiang (No. 2013CJHB005), by the Science and Technology Innovative Research Team in Higher Educational Institutions of Heilongjiang Province (No. 2012TD007), by the Program for High-Qualified Talents (No. Hdtd2010-03), and by the Electronic Engineering Provincial Key Laboratory.

## REFERENCES

1. Nahi N., Optimal recursive estimation with uncertain observation. *IEEE Transactions on Information Theory*, 1969, 15: 457–462.
2. Koning W. L. D. E., Optimal estimation of linear discrete-time systems with stochastic parameters. *Automatica*, 1984, 20(1): 113–115.
3. Carravetta F., Germani A., Raimondi M., Polynomial filtering for discrete-time stochastic linear systems with multiplicative state noise. *IEEE Transactions on Automatic Control*, 1997, 42(9): 1106–1126.
4. Basin M. V., Pérez J., Skliar M., Optimal filtering for polynomial system states with polynomial multiplicative noise. *International Journal of Robust and Nonlinear Control*, 2006, 16(6): 303–314.
5. Basin M. V., Calderón Álvarez D., Optimal filtering for incompletely measured polynomial systems with multiplicative noises. *Circuits, Systems and Signal Processing*, 2009, 28(2): 223–239.
6. Sun S. L., Optimal linear filters for discrete-time systems with randomly delayed and lost measurements with/without time stamps. *IEEE Transactions on Automatic Control*, 2013, 58(6): 1551–1556.
7. Sun S. L., Ma J., Linear estimation for networked control systems with random transmission delays and packet dropouts. *Information Sciences*, 2014, 269: 349–365.
8. Sun S. L., Xiao W. D., Optimal linear estimators for systems with multiple random measurement delays and packet dropouts. *International Journal of Systems Science*, 2013, 44(2): 358–370.
9. Sun S. L., Linear optimal state and input estimators for networked control systems with multiple packet dropouts. *International Journal of Innovative, Computing, Information and Control*, 2012, 8(10B): 7289–7305.
10. Ma J., Sun S., Optimal linear estimation for systems with multiplicative noise uncertainties and multiple packet dropouts. *IET Signal Processing*, 2012, 6(9): 839–848.
11. Ma J., Sun S. L., Optimal linear estimators for systems with random sensor delays, multiple packet dropouts and uncertain observations. *IEEE Transactions on Signal Processing*, 2011, 59(11): 5181–5192.
12. Deng Z. L., *Information Fusion Filtering Theory with Applications*. Harbin: Harbin Institute of Technology Press, 2007.
13. Sun S. L., Deng Z. L., Multi-sensor optimal information fusion Kalman filter. *Automatica*, 2004, 40(6): 1017–1023.
14. Deng Z. L., Gao Y., Li C. B., Hao G., Self-tuning decoupled information fusion Wiener state component filters and their convergence. *Automatica*, 2008, 44(3): 685–695.
15. Wen C. L., Lu B., Ge Q. B., A data fusion algorithm based on filtering step by step. *Acta Electronica Sinica*, 2004, 32(8): 1264–1267.
16. Ge Q. B., Wang G. A., Tang T. H. et al. The research on asynchronous date algorithm based on sampling of rational number times. *Electronic Journal*, 2006, 34(3): 543–548.
17. Wen C. L., Chen Z. G., Yan L. P. et al. The multiscale recursive fusion estimation based on dynamic systems of multirate sensors. *Journal of Electronics and Information Technology*, 2003, 25(3): 306–312.
18. Yan L. P., Liu B. S., Zhou D. H., The modeling and estimation of asynchronous multirate multisensory dynamic systems. *Aerospace Science and Technology*, 2006, 10(1): 63–71.
19. Xiao C. Y., Ma J., Sun S. L., Design of information fusion filter for a class of multi-sensor asynchronous sampling systems. *Chinese Control and Decision Conference*, 2011: 1081–1084.
20. Yan L. P., Liu B. S., Zhou D. H., Asynchronous multirate multisensor information fusion algorithm. *IEEE Transactions on Aerospace and Electronic Systems*, 2007, 43(3): 1135–1146.

21. Yan L. P., Zhou D. H., Fu M. Y., Xia Y. Q., State estimation for asynchronous multirate multisensory dynamic systems with missing measurements. *IET Signal Processing*, 2010, 4(6): 728–739.
22. Lin H. L., Ma J., Sun S. L., Optimal state filters for a class of non-uniform sampling systems. *Journal of Systems Science and Mathematics*, 2012, 32(6): 768–779.
23. Deng Z. H., Yan L. P., Fu M. Y., Multirate multisensor date fusion based on missing measurements. *Systems Engineering and Electronics*, 2010, 32(5): 886–890, 958.
24. Liu Y. L., Yan L. P., Xia Y. Q., Multirate multisensor distributed data fusion algorithm for state estimation with cross-correlated noises. Technical Committee on Control Theory, Chinese Association of Automation. *Proceedings of the 32nd Chinese Control Conference*, Xi'an, China, July 26–28, 2013.
25. J. Ma, Sun S. L., Distributed fusion filter for multi-rate multi-sensor systems with packet dropouts. *Proceedings of the 10th World Congress on Intelligent Control and Automation*, 4502–4506, Beijing, China, 2012.
26. F. F. Peng, Sun S. L., Distributed fusion estimation for multi-sensor multi-rate systems with stochastic observation multiplicative noises. *Mathematical Problems in Engineering*, 2014, 373270: 1–8.

# 17 Optimal Distributed Kalman Filtering Fusion with Singular Covariances of Filtering Errors and Measurement Noises

*Enbin Song*

## CONTENTS

## 17.1 INTRODUCTION

Multiple sensors estimation fusion has been used pervasively from civil to military fields, for example, target tracking and localization, fault diagnosis, surveillance and monitoring, air traffic control, and so forth and showed some merits for estimation fusion, for example, increasing reliability and survivability, improving estimation accuracy, reducing communication burden, and so on.

Generally speaking, there are two basic fusion architectures [1]: *centralized* and *decentralized/distributed* (with respect to *measurement fusion* and *track fusion*, respectively) depending on whether the raw measurements are sent to the fusion center or not. More precisely, centralized fusion means that the fusion center receives all raw measurements whereas distributed fusion implies that each sensor sends preprocessed measurements to the fusion center. It is well known that Kalman filtering is one of the most popular recursive least mean square error (LMSE) algorithms to optimally estimate the unknown state or process of a dynamic system. Centralized *Kalman filtering fusion* means that the fusion center can use all raw measurements from the local sensors in time to obtain globally optimal state estimates in the sense of LMSE. However, sending raw measurements needs more communication bandwidth, computation, and power consumption. Consequently, centralized fusion has a poor survivability of the system (in particular, in a war situation). For the case of distributed fusion, every local sensor first implements a Kalman filter based on its own observations for local requirements, and then sends the processed data–local state estimate to a fusion center. The fusion center fuses all received local estimates to yield an optimal state estimate in terms of LMSE. This distributed processing seems more preferred for many practical issues.

When the estimation error and sensor noise covariance matrices are invertible, an optimal Kalman filtering fusion formula (Equation 17.23) in Section 17.2, has been proposed in Refs. [2–6]. More importantly, it was proved to be globally optimal in the sense that the fused Kalman filtering is equivalent to the centralized Kalman filtering using all sensor measurements. Moreover, a Kalman filtering fusion with feedback was also proposed there. Moreover, a rigorous performance analysis for Kalman filtering fusion with feedback was provided in Refs. [6,7].

In this chapter, we consider the distributed Kalman filtering fusion for the case where covariances of estimation errors and measurement noises are singular, which is motivated by the following reasons. First, the existing fusion algorithms proposed in Refs. [2–6] strongly depend on invertible estimation error covariance matrices (see Equations 17.22 and 17.23). However, this condition is not always guaranteed to hold. Hence, the proposed fusion formula here could be applied to more general and extensive cases than the results there. Specific practical applications in Refs. [8–14] are Kalman filtering for linear dynamic systems with state equality constraints. In particular, in Refs. [9,10,12], the authors provided good reasons why one should not use a reduced state space for treating the constrained system in some practical problems. Specifically, Ref. [9] considers the biped locomotion problem and reduced it to a state estimation with equality constraint. Also, Refs. [10] and [12] consider a linear system and measurements describing movement of a land-based vehicle. Furthermore, the vehicle moving on a road is described by a linear equation. Similarly, there are many practical dynamic systems with constrained movement trajectories in space and air vehicle systems. The preceding examples imply that the state belongs to a subspace of whole space with respect its original dimension space, which leads to the deduction that the state is a degenerated random vector and its covariance must be singular. Second, our proposed fused state estimate is still equivalent to the centralized Kalman filtering using all sensor measurements, which means that the centralized Kalman filtering can also be obtained by fusing local estimates even if covariances of estimation errors and measurement noises are singular. In addition, our result is not same as the result in Ref. [15], where the authors investigated the problem of estimation fusion based on local transformation of raw measurements (not on the local estimate) under the condition that the filtering error covariances are singular. Furthermore, because our globally optimal distributed fusion is of the form of convex linear combination of one-step prediction of centralized Kalman filtering and all the local Kalman filtering and its corresponding one-step prediction, this obviously provides a theoretical support to global optimality of the convex combination fusion algorithm in Ref. [16]. In Ref. [16], the global optimality of algorithm was only supported by numerical examples without rigorous theoretical analysis. As stated in the last paragraph of Section III on page 67 Ref. [16], "this opens a hope that it is possible to derive a globally optimal distributed Kalman filtering fusion. Of course, a rigorous analysis for the equivalence is worth studying in the future." A more detailed explanation for the theoretical significance of our result here is provided in Remark 17.7 in Section 17.5.

Although the the convex combination fusion algorithm in Ref. [16] can be theoretically proved via our result here to have the same performance as *centralized Kalman filtering*, numerical examples here show that our proposed algorithm could save computation significantly compared with the fusion algorithm in Ref. [16]. For more details, see Table 17.1 in Section 17.6.

To derive globally optimal distributed Kalman filtering fusion equivalent to the centralized Kalman filtering, as in Refs. [2–6], a key skill is to find sufficient statistics of all sensor observations that can be equivalently expressed in terms of two-step sensor estimates. If this is done well, the centralized Kalman filtering can be easily rewritten as the corresponding distributed Kalman filtering fusion. To the best of our knowledge, so far, without the assumption of matrix invertibility there has not been any work similar to that done 20 years ago in Refs. [2–5]. In fact, what we technically are doing in this chapter is finding such sufficient statistics even without invertible covariances of filtering errors and measurement noises.

In addition, when there is a *feedback*, we obtain performance analysis results similar to those given in Refs. [6,7] for the case in which the *estimation error covariance matrices* are singular, that is, the corresponding fusion formula with feedback is, like the fusion without feedback, exactly identical to the corresponding centralized Kalman filtering fusion formula using all sensor measurements. Moreover, the various $P$ matrices in the feedback Kalman filtering at both local filters and the fusion center are still the covariance matrices of tracking errors. Furthermore, the feedback can reduce the covariance of each local tracking error.

The rest of the chapter is organized as follows. Section 17.2 describes the system model of Kalman filtering and presents the existing fusion formula. Section 17.3 presents a new distributed Kalman filtering fusion formula with singular estimation error covariance matrices and we prove the optimality of the proposed fusion formula. In Section 17.4, we also prove the optimality of the proposed fusion formula for Kalman filtering fusion with feedback when covariances of filtering errors are singular. We consider optimal Kalman filtering fusion with both singular covariances of filtering errors and measurement noises in Section 17.5. In Section 17.6, we demonstrate through several examples that our proposed fusion formula not only is globally optimal, but also consumes less computation than the algorithm in Ref. [16]. Finally, Section 17.7 contains some concluding remarks.

Throughout this chapter, we adopt the following notations. $I$ denotes the identity matrix of appropriate dimension, $(\cdot)'$ denotes the transpose of the corresponding matrix, and $(\cdot)^*$ denotes the Hermitian conjugate of the corresponding matrix.

## 17.2  PROBLEM FORMULATION

Assume the $l$-sensor distributed linear dynamic system is given by

$$\mathbf{x}_{k+1} = \Phi_k \mathbf{x}_k + v_k, \quad k = 0,1,\ldots, \tag{17.1}$$

$$\mathbf{y}_k^i = H_k^i \mathbf{x}_k + w_k^i, \quad i = 1,\ldots,l, \tag{17.2}$$

where $\Phi_k$ is a matrix of order $(r \times r)$, $\mathbf{x}_k, v_k \in \mathbb{R}^r$, $H_k^i \in \mathbb{R}^{N_i \times r}$, $\mathbf{y}_k^i, w_k^i \in \mathbb{R}^{N_i}$. The process noise $v_k$ and measurement noise $w_k^i$ are both zero-mean random variables independent of each other temporally and are not cross correlated.

The stacked measurement equation can be expressed as

$$\mathbf{y}_k = H_k \mathbf{x}_k + w_k, \tag{17.3}$$

where

$$\mathbf{y}_k = \left(\mathbf{y}_k^{1'},\ldots,\mathbf{y}_k^{l'}\right)',$$

$$H_k = \left(H_k^{1'},\ldots,H_k^{l'}\right)', \tag{17.4}$$

$$w_k = \left(w_k^{1'},\ldots,w_k^{l'}\right)'.$$

Moreover, the covariances of sensor noises are given by

$$R_k^i = \mathrm{Cov}\left(w_k^i\right), \quad i = 1,\ldots,l, \tag{17.5}$$

and

$$R_k = \text{Cov}(w_k) = \text{diag}\left(R_k^1, \ldots, R_k^l\right), \tag{17.6}$$

where $R_k$ and $R_k^i$ are both invertible for all $i$, that is, $R_k$ could be any positive definite matrix with on-diagonal blocks $R_k^i$ of full rank.

Due to results in Kalman filtering Refs. [17–19], it is well known that the centralized Kalman filtering by using all sensor observations is expressed in the following form:

$$\begin{aligned}
\mathbf{x}_{k/k} &= \mathbf{x}_{k/k-1} + K_k(\mathbf{y}_k - H_k\mathbf{x}_{k/k-1}) \\
&= (I - K_k H_k)\mathbf{x}_{k/k-1} + K_k\mathbf{y}_k,
\end{aligned} \tag{17.7}$$

$$\begin{aligned}
K_k &= P_{k/k}H_k'R_k^{-1} \\
&= P_{k/k-1}H_k'\left(H_k P_{k/k-1}H_k' + R_k\right)^{-1}
\end{aligned} \tag{17.8}$$

with covariance of filtering error given by

$$\begin{aligned}
P_{k/k} &= (I - K_k H_k)P_{k/k-1} \\
&= P_{k/k-1} - P_{k/k-1}H_k'\left(H_k P_{k/k-1}H_k' + R_k\right)^{-1} H_k P_{k/k-1}
\end{aligned} \tag{17.9}$$

or

$$P_{k/k}^\dagger = P_{k/k-1}^\dagger + P_{k/k-1}^\dagger P_{k/k-1}H_k'R_k^{-1}H_k P_{k/k-1}P_{k/k-1}^\dagger. \tag{17.10}$$

Equation 17.10 comes from Lemma 17.4 in Appendix I, where

$$\mathbf{x}_{k/k-1} = \Phi_k \mathbf{x}_{k-1/k-1} \tag{17.11}$$

$$P_{k/k} = E[(\mathbf{x}_{k/k} - \mathbf{x}_k)(\mathbf{x}_{k/k} - \mathbf{x}_k)' \mid \mathbf{y}_0, \ldots, \mathbf{y}_k] \tag{17.12}$$

$$P_{k/k-1} = E[(\mathbf{x}_{k/k-1} - \mathbf{x}_k)(\mathbf{x}_{k/k-1} - \mathbf{x}_k)' \mid \mathbf{y}_0, \ldots, \mathbf{y}_{k-1}] \tag{17.13}$$

**Remark 17.1**

Equation 17.10 shows that even if $P_{k/k}$ may be singular, similarly, the pseudo-inverse of $P_{k/k}$ still can be expressed in terms of the pseudo-inverse of $P_{k/k-1}$. The result could be regarded as an extension of

$$P_{k/k}^{-1} = P_{k/k-1}^{-1} + H_k'R_k^{-1}H_k, \tag{17.14}$$

when $P_{k/k}$ and $P_{k/k-1}$ are both invertible.                                              ∎

**Remark 17.2**

As stated in Section 17.1, Kalman filtering for linear dynamic system with state equality constraints in Refs. [11,12] certainly leads to the deduction that $\mathrm{Cov}(v_k)$ and $\mathrm{Cov}(\mathbf{x}_k)$ must be singular, which means that $P_{k/k}$ and $P_{k/k-1}$ may be singular. To the best of our knowledge, Equation 17.10 is new and will play an important role in our results. For details, see Equation 17.24.  ∎

Likewise, the local Kalman filtering at the $i$th sensor is

$$
\begin{aligned}
\mathbf{x}_{k/k}^i &= \mathbf{x}_{k/k-1}^i + K_k^i \left( \mathbf{y}_k^i - H_k^i \mathbf{x}_{k/k-1}^i \right) \\
&= \left( I - K_k^i H_k^i \right) \mathbf{x}_{k/k-1}^i + K_k^i \mathbf{y}_k^i,
\end{aligned}
\tag{17.15}
$$

$$
\begin{aligned}
K_k^i &= P_{k/k}^i H_k^{i'} R_k^{i-1} \\
&= P_{k/k-1}^i H_k^{i'} \left( H_k^i P_{k/k-1}^i H_k^{i'} + R_k^i \right)^{-1}
\end{aligned}
\tag{17.16}
$$

with covariance of filtering error given by

$$
\begin{aligned}
P_{k/k}^i &= \left( I - K_k^i H_k^i \right) P_{k/k-1}^i \\
&= P_{k/k-1}^i - P_{k/k-1}^i H_k^{i'} \left( H_k^i P_{k/k-1}^i H_k^{i'} + R_k^i \right)^{-1} H_k^i P_{k/k-1}^i
\end{aligned}
\tag{17.17}
$$

or

$$
P_{k/k}^{i\dagger} = P_{k/k-1}^{i\dagger} + P_{k/k-1}^{i\dagger} P_{k/k-1}^i H_k^{i'} R_k^{i-1} H_k^i P_{k/k-1}^i P_{k/k-1}^{i\dagger}
\tag{17.18}
$$

where

$$
\mathbf{x}_{k/k-1}^i = \Phi_k \mathbf{x}_{k-1/k-1}^i
\tag{17.19}
$$

$$
P_{k/k}^i = E\left[ \left( \mathbf{x}_{k/k}^i - \mathbf{x}_k \right) \left( \mathbf{x}_{k/k}^i - \mathbf{x}_k \right)' \Big| \mathbf{y}_0^i, \dots, \mathbf{y}_k^i \right]
\tag{17.20}
$$

$$
P_{k/k-1}^i = E\left[ \left( \mathbf{x}_{k/k-1}^i - \mathbf{x}_k \right) \left( \mathbf{x}_{k/k-1}^i - \mathbf{x}_k \right)' \Big| \mathbf{y}_0^i, \dots, \mathbf{y}_{k-1}^i \right]
\tag{17.21}
$$

We first recall some known results about Kalman filtering fusion. In Refs. [2–6], if covariances of filtering errors are invertible, the centralized filtering and error matrix can be expressed by the local filtering and error matrices as follows:

$$
P_{k/k}^{-1} = P_{k/k-1}^{-1} + \sum_{i=1}^{l} \left( P_{k/k}^{i-1} - P_{k/k-1}^{i-1} \right)
\tag{17.22}
$$

and

$$P_{k/k}^{-1}\mathbf{x}_{k/k} = P_{k/k-1}^{-1}\mathbf{x}_{k/k-1} + \sum_{i=1}^{l}\left(P_{k/k}^{i-1}\mathbf{x}_{k/k}^{i} - P_{k/k-1}^{i-1}\mathbf{x}_{k/k-1}^{i}\right) \tag{17.23}$$

Moreover, Ref. [7] provided a rigorous performance analysis for the case in which there is feedback from the fusion center to sensors, that is, the one-step predictions $\mathbf{x}_{k/k-1}^{i}$ and $P_{k/k-1}^{i}$ at every local sensor in Equations 17.22 and 17.23 are replaced by the feedback $\mathbf{x}_{k/k-1}$ and $P_{k/k-1}$, respectively. Zhu et al. [7] have shown that although the feedback cannot improve the performance at the fusion center, the feedback can reduce the covariance of each local tracking error. Obviously, for Equations 17.22 and 17.23 to be true, invertible error covariance matrices are necessary.

From then on, the assumption of invertibility of estimation error covariance matrices has been a restrictive condition for the preceding results.

Thus, when covariances of filtering errors are not invertible,

- Can we find sufficient statistics of sensor observations similar to

$$H_k' R_k^{-1}\mathbf{y}_k = \sum_{i=1}^{l} H_k^{i} R_k^{i-1}\mathbf{y}_k^{i}$$

so that the centralized Kalman filtering can still be expressed in terms of the local filtering?
- Can we yet obtain a similar performance analysis for the distributed Kalman filtering with feedback?

The positive answers to the preceding questions are presented below.

## 17.3   DISTRIBUTED KALMAN FILTERING FUSION WITH SINGULAR COVARIANCES OF FILTERING ERRORS

First, we focus on how to express the centralized filtering as a combination of local filtering when only covariances of filtering errors are singular.

Equations 17.10 and 17.18 and Lemma 17.7 in Appendix I lead to the deduction that the estimation error covariance of the centralized Kalman filtering can be expressed in terms of the estimation error covariances of all local filters as follows:

$$P_{k/k}^{\dagger} = P_{k/k-1}^{\dagger} + \sum_{i=1}^{l} P_{k/k-1}^{\dagger} P_{k/k-1}\left(P_{k/k}^{i\dagger} - P_{k/k-1}^{i\dagger}\right)P_{k/k-1}P_{k/k-1}^{\dagger} \tag{17.24}$$

**Remark 17.3**

To the best of our knowledge, the updated formula (Equation 17.24) is new and it could deal with the case in which the filtering error covariance is singular. It is worth noting that in this case, the error covariance of centralized Kalman filtering can still be expressed in terms of the estimation error covariances of all local filters. Compared with Equation 17.22, we not only replaced $P_{k/k}^{-1}$ and $P_{k/k}^{i-1}$

by $P_{k/k}^\dagger$ and $P_{k/k}^{i\dagger}$, respectively, but also multiplied an orthogonal projection operator $P_{k/k-1}^\dagger P_{k/k-1}$ at the both sides of $\left(P_{k/k}^{i\dagger} - P_{k/k-1}^{i\dagger}\right)$. Of course, the new updated formula includes Equation 17.22 as a special case when error covariances are invertible. The orthogonal projection operator $P_{k/k-1}^\dagger P_{k/k-1}$ means the information matrices (error covariances) are concentrated in some subspace (in fact, the space is the column space of $P_{k/k-1}^\dagger P_{k/k-1}$) instead of the whole space because the centralized and local information matrices (error covariances) of the previous step are not invertible. ∎

To derive globally optimal distributed Kalman filtering fusion equivalent to the centralized Kalman filtering, as done in Refs. [2–6], a key skill is to find sufficient statistics of all sensor observations $\{\mathbf{y}_k^1,\ldots,\mathbf{y}_k^l\}$ that can be equivalently expressed in terms of two-step sensor estimates $\{\mathbf{x}_{k/k}^1,\mathbf{x}_{k/k-1}^1;\ldots;\mathbf{x}_{k/k}^l,\mathbf{x}_{k/k-1}^l\}$. If this is done well, the centralized Kalman filtering can be easily rewritten as the corresponding distributed Kalman filtering fusion, which naturally is globally optimal. When the covariances of filtering errors and sensor noises are both invertible, such sufficient statistics for the centralized Kalman is simply $H_k' R_k^{-1} \mathbf{y}_k = \sum_{i=1}^{l} H_k^{i'} R_k^{i-1} \mathbf{y}_k^i$ (for details, see $K_k \mathbf{y}_k$ in Equation 17.7), where $H_k^{i'} R_k^{i-1} \mathbf{y}_k^i = P_{k/k}^{i-1} \mathbf{x}_{k/k}^i - P_{k/k-1}^{i-1} \mathbf{x}_{k/k-1}^i$ (see Equation 17.23).

## Remark 17.4

If the covariances of filtering errors and sensor noises are not invertible, it is difficult to find the aforementioned sufficient statistics. Hence, to the best of our knowledge, so far there has not been any work without the assumption of matrix invertibility as done 20 years ago in Refs. [2–5]. In fact, what we are doing technically in this chapter is just to find such sufficient statistic even without invertible covariances of filtering errors and measurement noises. Specifically, we carefully apply the matrix analysis technique (it equivalently analyzes information matrices corresponding to the state of the system) to obtain the globally optimal distributed fusion formula. ∎

Equations 17.4 through 17.7 imply that

$$
\begin{aligned}
K_k \mathbf{y}_k &= P_{k/k} H_k' R_k^{-1} \mathbf{y}_k \\
&= \sum_{i=1}^{l} P_{k/k} H_k^{i'} R_k^{i-1} \mathbf{y}_k^i \\
&= \sum_{i=1}^{l} P_{k/k} P_{k/k}^{i\dagger} P_{k/k}^i H_k^{i'} R_k^{i-1} \mathbf{y}_k^i \\
&= \sum_{i=1}^{l} P_{k/k} P_{k/k}^{i\dagger} K_k^i \mathbf{y}_k^i,
\end{aligned}
\tag{17.25}
$$

where the third equality is due to Lemma 17.6 in Appendix I, and the fourth equality uses Equation 17.16.

## Remark 17.5

From Equations 17.7 and 17.15, it is easy to see that Equation 17.25 plays a very important role in the key technique for the distributed Kalman filtering fusion equivalent to the centralized one because $K_k^i \mathbf{y}_k^i$ can be expressed simply in terms of $\mathbf{x}_{k/k-1}^i$ and $\mathbf{x}_{k/k}^i$ in following Equation 17.26. ∎

To express the centralized filtering $\mathbf{x}_{k/k}$ in terms of the local filtering, we use Equations 17.15 and 17.25 to eliminate $\mathbf{y}_k$ from 17.7. Note that Equation 17.15 means that

$$K_k^i \mathbf{y}_k^i = \mathbf{x}_{k/k}^i - \left(I - K_k^i H_k^i\right)\mathbf{x}_{k/k-1}^i. \tag{17.26}$$

Thus, substituting Equations 17.25 and 17.26 into 17.7 yields

$$
\begin{aligned}
\mathbf{x}_{k/k} &= (I - K_k H_k)\mathbf{x}_{k/k-1} + \sum_{i=1}^{l} P_{k/k} P_{k/k}^{i\dagger} \left[ \mathbf{x}_{k/k}^i - \left(I - K_k^i H_k^i\right)\mathbf{x}_{k/k-1}^i \right] \\
&= (I - K_k H_k)\mathbf{x}_{k/k-1} + \sum_{i=1}^{l} P_{k/k} P_{k/k}^{i\dagger} \mathbf{x}_{k/k}^i - \sum_{i=1}^{l} P_{k/k} P_{k/k}^{i\dagger}\left(I - K_k^i H_k^i\right)\mathbf{x}_{k/k-1}^i \\
&= \left(I - P_{k/k} H_{k'} R_k^{-1} H_k\right)\mathbf{x}_{k/k-1} + \sum_{i=1}^{l} P_{k/k} P_{k/k}^{i\dagger} \mathbf{x}_{k/k}^i + \sum_{i=1}^{l}\left[-P_{k/k} P_{k/k}^{i\dagger}\left(I - P_{k/k}^i H_k^{i'} R_k^{-1} H_k^i\right)\right]\mathbf{x}_{k/k-1}^i \\
&= \left(I - P_{k/k} H_{k'} R_k^{-1} H_k\right)\mathbf{x}_{k/k-1} + \sum_{i=1}^{l} P_{k/k} P_{k/k}^{i\dagger} \mathbf{x}_{k/k}^i + \sum_{i=1}^{l}\left[-P_{k/k}\left(P_{k/k}^{i\dagger} - H_k^{i'} R_k^{i-1} H_k^i\right)\right]\mathbf{x}_{k/k-1}^i,
\end{aligned}
\tag{17.27}
$$

where the third equality uses Equation 17.16, and the last equality is again thanks to Lemma 17.6 in Appendix I.

Equations 17.24 and 17.27 mean that the centralized filtering and its error matrix are explicitly expressed in terms of the filtering and error matrices of the local filtering, respectively. Consequently, the distributed Kalman filtering fusion given in Equations 17.24 and 17.27 has the same performance as that of the centralized Kalman filtering.

In addition, when error matrices of the fusion center and local sensor, $P_{k/k}$ and $P_{k/k}^i$, are invertible, we can easily see that Equations 17.24 and 17.27 now reduce to 17.22 and 17.23.

## 17.4  OPTIMALITY OF KALMAN FILTERING FUSION WITH FEEDBACK WHEN COVARIANCES OF FILTERING ERRORS ARE SINGULAR

In this section, we consider the optimality of Kalman filtering fusion with feedback under the condition that covariances of filtering errors are singular. When there is feedback, all the local sensors receive the latest estimate of the fusion center. Thus, similarly to the results in Ref. [4], the following local and global one-stage predictions were modified naturally as

$$\hat{\mathbf{x}}_{k/k-1}^i = \Phi_k \hat{\mathbf{x}}_{k-1/k-1} = \hat{\mathbf{x}}_{k/k-1} \tag{17.28}$$

$$\hat{P}_{k/k-1}^i = \hat{P}_{k/k-1}, \quad \forall i \tag{17.29}$$

where $\widehat{(\cdot)}$ denotes an estimated vector or covariance matrix with feedback after this modification. Therefore, similarly, for Equations 17.24 and 17.27, we consider the following filtering fusion with feedback,

$$\hat{P}_{k/k}^{\dagger} = \hat{P}_{k/k-1}^{\dagger} + \sum_{i=1}^{l} \hat{P}_{k/k-1}^{\dagger} \hat{P}_{k/k-1}\left(\hat{P}_{k/k}^{i\dagger} - \hat{P}_{k/k-1}^{i\dagger}\right)\hat{P}_{k/k-1}\hat{P}_{k/k-1}^{\dagger} \tag{17.30}$$

$$\hat{\mathbf{x}}_{k/k} = \left(I - \hat{P}_{k/k}H'_kR_k^{-1}H_k\right)\hat{\mathbf{x}}_{k/k-1} + \sum_{i=1}^{l}\hat{P}_{k/k}\hat{P}_{k/k}^{i\dagger}\hat{\mathbf{x}}_{k/k}^{i}$$

$$+ \sum_{i=1}^{l}\left[-\hat{P}_{k/k}\left(\hat{P}_{k/k}^{i\dagger} - H_k^{i'}R_k^{i-1}H_k^{i}\right)\right]\hat{\mathbf{x}}_{k/k-1}^{i}. \tag{17.31}$$

Because all the results and the mathematical deductions with feedback are similar to those in Ref. [7], we directly provide the results without detailed proof as follows:

1. The filtering fusion Equations 17.28 to 17.31 with feedback also can achieve the same performance as that of the centralized filtering fusion, that is,

$$\hat{\mathbf{x}}_{k/k} = \mathbf{x}_{k/k}, \qquad \hat{P}_{k/k} = P_{k/k} \tag{17.32}$$

2. The matrices $\hat{P}_{k/k}$ and $\hat{P}_{k/k}^{i}$ are still the covariances of the global and local filtering errors respectively, that is to say,

$$\hat{P}_{k/k} = E[(\hat{\mathbf{x}}_{k/k} - \mathbf{x}_k)(\hat{\mathbf{x}}_{k/k} - \mathbf{x}_k)'|\mathbf{y}_0,\ldots,\mathbf{y}_k] \tag{17.33}$$

and

$$\hat{P}_{k/k}^{i} = E[(\hat{\mathbf{x}}_{k/k}^{i} - \mathbf{x}_k)(\hat{\mathbf{x}}_{k/k}^{i} - \mathbf{x}_k)'|\mathbf{y}_0^{i},\ldots,\mathbf{y}_k^{i}] \tag{17.34}$$

3. The feedback can benefit the local filters in the sense of reducing the covariance of each local tracking error, that is,

$$\hat{P}_{k/k}^{i} \preceq P_{k/k}^{i}, \quad i = 1,2,\ldots,l \tag{17.35}$$

## 17.5 OPTIMAL KALMAN FILTERING FUSION WITH SINGULAR COVARIANCES OF FILTERING ERRORS AND MEASUREMENT NOISES

In this section, we consider the case where the covariances of both filtering errors and measurement noise are singular. Obviously, it is more challenging than the previous case where the covariances of filtering errors are singular and the covariances of the measurement noise are invertible. Similarly, according to the standard results in Kalman filtering Refs. [17–19],

$$K_k = P_{k/k}H'_kR_k^{-1} \tag{17.36}$$

and

$$K_k^{i} = P_{k/k}^{i}H_k^{i'}R_k^{i-1} \tag{17.37}$$

do not hold in general because $R_k$ and $R_k^i$ may be singular. Fortunately,

$$K_k = P_{k/k-1}H_k'\left(H_k P_{k/k-1}H_k' + R_k\right)^\dagger \tag{17.38}$$

and

$$K_k^i = P_{k/k-1}^i H_k^{i'}\left(H_k^i P_{k/k-1}^i H_k^{i'} + R_k^i\right)^\dagger \tag{17.39}$$

still hold true. The preceding analysis implies that we cannot use the technique as Equations 17.25 to 17.27.

In what follows, we deal with the problem by using another trick.

We first give the limit formula for the Moore–Penrose inverse (see e.g., Ref. [20]). Assume that $A \in \mathbb{C}^{m \times n}$, then

$$\lim_{\epsilon \to 0^+}(A^*A + \epsilon I)^{-1}A^* = A^\dagger, \tag{17.40}$$

which plays a key role in the following analysis. From Remark 17.5 and Equation 17.25, we know that the key is to express the centralized sufficient statistics $K_k \mathbf{y}_k$ by linear combination of sensor sufficient statistics $K_k^i \mathbf{y}_k^i, i \le l$. Therefore, we give the following result.

## Proposition 17.1

For system 17.1, assume $\mathbf{y}_k^i$ and $\mathbf{y}_k$ are defined by Equations 17.2 and 17.3 respectively. Moreover, under the assumptions that both covariances of filtering errors and measurement noises are singular, $K_k$ and $K_k^i$ are defined by Equations 17.38 and 17.39, respectively, for all $k = 0,1,\ldots,$

$$K_k \mathbf{y}_k = \sum_{i=1}^l K_k(i) K_k^{i\dagger} K_k^i \mathbf{y}_k^i, \tag{17.41}$$

where

$$K_k = \begin{pmatrix} K_k(1) & K_k(2) & \ldots & K_k(l) \end{pmatrix} \tag{17.42}$$

is an appropriate partition of matrix $K_k$ such that $K_k \mathbf{y}_k = \sum_{i=1}^l K_k(i)\mathbf{y}_k^i$.  ∎

## Proof 17.1

The proof is lengthy and has been relegated to Appendix II.  ∎

In the following, we give the optimal Kalman filtering fusion formula with singular covariances of filtering errors and measurement noises.

## Proposition 17.2

Under the assumptions of Proposition 17.1, for all $k = 0, 1, \ldots,$

$$
\begin{aligned}
\mathbf{x}_{k/k} &= \mathbf{x}_{k/k-1} + K_k(\mathbf{y}_k - H_k \mathbf{x}_{k/k-1}) \\
&= (I - K_k H_k)\mathbf{x}_{k/k-1} + K_k \mathbf{y}_k \\
&= (I - K_k H_k)\mathbf{x}_{k/k-1} + \sum_{i=1}^{l} K_k(i) K_k^{i^\dagger} K_k^i \mathbf{y}_k^i \\
&= (I - K_k H_k)\mathbf{x}_{k/k-1} \\
&\quad + \sum_{i=1}^{l} K_k(i) K_k^{i^\dagger} \left( \mathbf{x}_{k/k}^i - \left( I - K_k^i H_k^i \right) \mathbf{x}_{k/k-1}^i \right),
\end{aligned}
\tag{17.43}
$$

where $K_k(i)$ defined by Equation 17.42. ∎

## Proof 17.2

Clearly, Equation 17.7 implies that the second equality holds. It is easily seen that the third equality holds because of Proposition 17.1. Moreover, it follows from Equation 17.26 that the last equality is true. ∎

## Remark 17.6

Equation 17.43 shows that centralized Kalman filtering can still be expressed as a linear combination in terms of local Kalman filtering even if both covariance matrices of filter errors and measurement noises are singular. Clearly, it is a generalization of fusion formula 17.23. However, so far, we have not derived a simple distributed recursive formula for $P_{k/k}$ similar to Equation 17.24, and it is not certain if such a recursive formula really exists. ∎

## Remark 17.7

Furthermore, because our globally optimal distributed fusion is of the form of convex linear combination of $\left\{ \mathbf{x}_{k/k-1}, \mathbf{x}_{k/k}^1, \mathbf{x}_{k/k-1}^1; \ldots; \mathbf{x}_{k/k}^l, \mathbf{x}_{k/k-1}^l \right\}$, this obviously provides a theoretical support to global optimality of the convex combination fusion algorithm of $\left\{ \tilde{\mathbf{x}}_{k/k-1}, \mathbf{x}_{k/k}^1, \mathbf{x}_{k/k-1}^1; \ldots; \mathbf{x}_{k/k}^l, \mathbf{x}_{k/k-1}^l \right\}$ in Ref. [16], where $\tilde{\mathbf{x}}_{k/k-1}$ is the prediction of convex linear combination fusion algorithm. Our result implies that the algorithm in Ref. [16] is equivalent to the centralized Kalman filtering provided that $\tilde{\mathbf{x}}_{k/k-1}$ there is identical to the prediction $\mathbf{x}_{k/k-1}$ of the centralized Kalman filtering. This can be easily implemented while one takes the same initial values for the two recursive algorithms. ∎

## 17.6 NUMERICAL EXAMPLES

As we have emphasized in the introduction, the three recursive algorithms—centralized Kalman filtering fusion (CKF), convex linear minimum mean square error (LMMSE) combination optimal distributed Kalman filtering fusion (CLDKF) in Ref. [16], and globally optimal distributed Kalman filtering fusion (GODKF) (Equation 17.27)—are theoretically the same as globally optimal LMSE estimation provided that they have the same initial values. In this section, we provide several numerical examples that demonstrate that our GODKF not only provides theoretical support to the CLDKF, but also uses less computation quantity than the CLDKF.

In what follows, we present two types of dynamic systems modeled as two objects moving on a circle and a straight line with process noise and measurement noise, respectively. Moreover, similar to the models proposed in Ref. [10], the two dynamical systems subject to state equality constraints are the models with singular estimation error covariance and measurement noise covariance matrices. All codes are written by MATLAB 7.8® and performed on a laptop computer with Intel CPU 2.00GHZ processor and 2.98GB memory.

All the following examples under the assumptions that the original object dynamics and measurement equations are modeled as follows:

$$\mathbf{x}_{k+1} = \Phi_k \mathbf{x}_k + v_k, \tag{17.44}$$

$$\mathbf{y}_k^i = H_k^i \mathbf{x}_k + w_k^i, \quad i = 1, 2, \tag{17.45}$$

where the state $\mathbf{x}_k$ is known to actually be constrained in the null space of $D_k$:

$$\mathcal{N}(D_k) \triangleq \{\mathbf{x} : D_k \mathbf{x} = \mathbf{0}\} \tag{17.46}$$

For example, the objects are actually moving on piecewise straight road in the following examples. Besides, $v_k, w_k^i, k = 0, 1, 2, \ldots$, satisfy the assumptions of standard Kalman filtering.

### Example 17.1: Circle Moving Model with Singular Estimation Error Covariance Only

In this example, for program simplicity, $D_k$ is assumed to be constant matrix $D = \begin{pmatrix} 1 & 0 \\ 1 & 0 \end{pmatrix}$.

More specifically, the system (Equations 17.44 and 17.45) subject to constraint (Equation 17.46) can be converted into the following form:

$$\begin{aligned} \mathbf{x}_{k+1} &= P\Phi_k \mathbf{x}_k + P v_k, \\ \mathbf{y}_k^i &= H_k^i \mathbf{x}_k + w_k^i, \quad i = 1, 2, \end{aligned} \tag{17.47}$$

where

$$P = I - D^\dagger D = \begin{pmatrix} 0 & 0 \\ 0 & 1 \end{pmatrix},$$

$$P\Phi_k = P \begin{pmatrix} \cos(2\pi/300) & \sin(2\pi/300) \\ -\sin(2\pi/300) & \cos(2\pi/300) \end{pmatrix}$$

and

$$R_{Pv_k} = \begin{pmatrix} 0 & 0 \\ 0 & 15 \end{pmatrix}$$

are all constant matrices.

The two measurement matrices and measurement noise covariance matrices are given by

$$H_k^1 = \begin{pmatrix} 1 & 0 \\ 1 & -1 \end{pmatrix}, \quad H_k^2 = \begin{pmatrix} 1 & 0 \\ 1 & -2 \end{pmatrix},$$

and

$$R_k^i = R_{w_k^i} = \begin{pmatrix} 15 & 0 \\ 0 & 15 \end{pmatrix}, \quad i = 1, 2$$

The stacked measurement equation is written as

$$\mathbf{y}_k = H_k \mathbf{x}_k + w_k,$$

where

$$\mathbf{y}_k = \begin{pmatrix} \mathbf{y}_k^1 \\ \mathbf{y}_k^2 \end{pmatrix}, \quad H_k = \begin{pmatrix} H_k^1 \\ H_k^2 \end{pmatrix}, \quad w_k = \begin{pmatrix} w_k^1 \\ w_k^2 \end{pmatrix},$$

and the covariance of the noise $w_k$ is given by

$$R_k = \mathrm{diag}\left(R_k^1, R_k^2\right)$$

The initial values are given as follows: $Ex_0 = (50,0)'$, $x_{0|0} = (50,0)'$, $P_{0|0}^i = \mathrm{Var}(x_0) = \begin{pmatrix} 0 & 0 \\ 0 & 15 \end{pmatrix}$.

Using a Monte Carlo method of 1000 runs, time index $k = 1, 2, \ldots, 300$, that is, we assume that Kalman filtering implements 300 steps. We evaluate tracking performance of an algorithm by

$$E_k = \frac{1}{1000} \sum_{j=1}^{1000} \left\| \mathbf{x}_{k/k}^{(j)} - \mathbf{x}_k^{(j)} \right\|^2, \quad k = 1, 2, \ldots, 300$$

The numerical results of the three algorithms are given in Figure 17.1.

## Example 17.2: Circle Moving Model with Singular Estimation Error and Measurement Noise Covariances

In this example, we only change the two measurement matrices and singular measurement noise covariance matrices in Example 17.1 to

$$R_k^i = R_{Pw_k^i} = \begin{pmatrix} 0 & 0 \\ 0 & 15 \end{pmatrix}, \quad i = 1, 2, \quad H_k^1 = \begin{pmatrix} 2.5 & 0 \\ 2.5 & -2.5 \end{pmatrix},$$

and

$$H_k^2 = \begin{pmatrix} 1 & 0 \\ 1 & -2 \end{pmatrix}.$$

The numerical results are given in Figure 17.2.

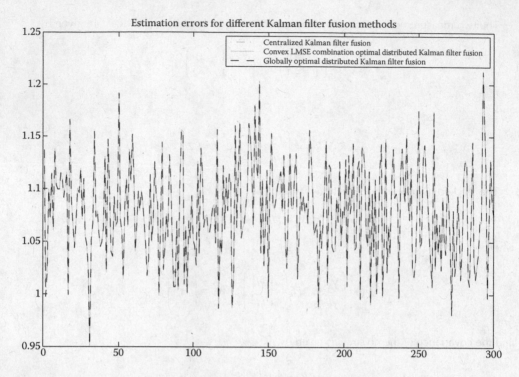

**FIGURE 17.1**  The average tracking error of three algorithms in Example 17.1.

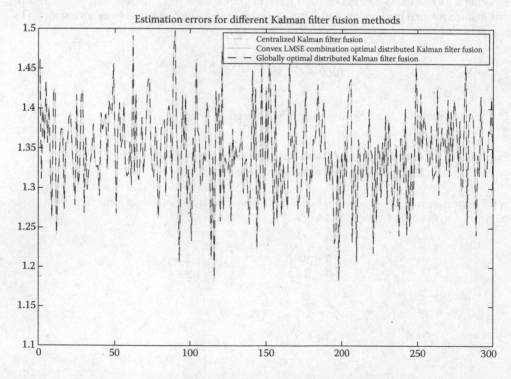

**FIGURE 17.2**  The average tracking errors of three fusion algorithms in Example 17.2.

## Example 17.3:  Straight Line Moving Model with Singular Estimation Error and Measurement Noise Covariances

In this example, we let

$$D = \begin{pmatrix} 0 & 1 \\ 0 & 1 \end{pmatrix}, \quad \text{and} \quad P = I - D^\dagger D = \begin{pmatrix} 1 & 0 \\ 0 & 0 \end{pmatrix}.$$

Consequently,

$$P\Phi_k = P \begin{pmatrix} 1 & 1 \\ 0 & -1 \end{pmatrix} \quad \text{and} \quad R_{Pv_k} = \begin{pmatrix} 15 & 0 \\ 0 & 0 \end{pmatrix}.$$

The two measurement matrices and singular measurement noise covariance matrices are given by:

$$R_k^i = R_{Pw_k} = \begin{pmatrix} 15 & 0 \\ 0 & 0 \end{pmatrix}, \quad i = 1,2, \quad H_k^1 = \begin{pmatrix} 2.5 & 2.5 \\ 0 & -2.5 \end{pmatrix}$$

and

$$H_k^2 = \begin{pmatrix} 3 & 6 \\ 0 & -6 \end{pmatrix}.$$

The numerical results are shown in Figure 17.3.

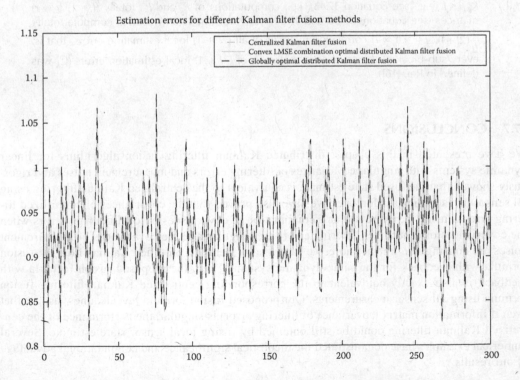

**FIGURE 17.3**   The average tracking errors of the three fusion algorithms in Example 17.3.

**TABLE 17.1**

**Comparison of Time Consumption of the Two Distributed Fusion Algorithms**

|  | Example 17.1 | Example 17.2 | Example 17.3 |
|---|---|---|---|
| GODKF | 133.0454(s) | 129.3726(s) | 130.4649(s) |
| CLDKF | 216.0183(s) | 217.1235(s) | 218.1664(s) |
| $\dfrac{\text{GODKF}}{\text{CLDKF}}$ | 61.59% | 59.58% | 59.80% |

Although Figures 17.1 to 17.3 show that the performances of both GODKF and CLDKF are the same as that of CKF, their computational burdens are not the same. To show this difference, we provide computation times of the two distributed fusion algorithms as follows.

From Table 17.1 and Figures 17.1 to 17.3 in our numerical results, we have the following three observations

1. Figures 17.1, 17.2, and 17.3 show that the numerical results of GODKF and CLDKF are both generally the same as those of CKF, which is consistent to the theoretical analysis in this paper.
2. Although the two distributed fusion algorithms work well in all of the preceding examples, Table 17.1 shows that, compared with CLDK in Ref. [6], our proposed GODKF algorithm could save computation significantly.
3. Roughly speaking, at each iteration, the main difference of computation burdens of the fusion coefficients of GODKF and CLDKF is that the former need to compute $K_k$ and $K_k^i, i \leq l$ (e.g., see Equation 17.43), i.e., computations of $P_k$ and $P_k^i$, totally $(l + 1)$ $(r \times r)$ matrices (see Equations 17.38 and 17.39); however, the latter needs to compute totally $\frac{1}{2}(2l + 1)(2l + 2)$ $(r \times r)$ correlation matrices of all $(2l + 1)$ local estimation errors, that is, every sub-block of the covariance matrix $C_k$ of $(2l + 1)$ local estimation errors ($C_k$ was defined in Ref. [16]).

## 17.7   CONCLUSIONS

We have presented in this chapter distributed Kalman filtering fusion algorithms for linear dynamic systems with singular covariances of filtering errors and measurement noises and rigorously showed that the fused state estimate is equivalent to the centralized Kalman filtering using all sensor measurements. Moreover, we have also proved that the covariances of centralized filtering error can still be expressed as a combination of covariances of local filtering errors when the covariances of centralized filtering errors are singular and the covariances of measurement noises are nonsingular. Furthermore, we have also presented a modified Kalman filtering fusion formula with feedback for the same dynamic system. Then, the proposed fusion formula with feedback is also exactly equivalent to the corresponding centralized Kalman filtering fusion formula using all sensor measurements. Our proposed fusion formula reveals a new aspect that even if information matrix (covariance of filtering error) is singular, the performance of the centralized Kalman filtering could be still obtained by fusing local sensor state estimate. Several numerical examples have demonstrated the theoretical significance and computational advantage of our results.

## APPENDIX I

### Lemma 17.1

For $A \in \mathbb{C}_{\succeq}^m$, $B \in \mathbb{C}_{\succeq}^m$, if $A \succeq B$, then,

$$\mathcal{R}(A) \supseteq \mathcal{R}(B). \tag{17.48}$$

∎

### Proof 17.3

For an arbitrary $x \in \mathcal{R}(A)^\perp = \mathcal{N}(A)$, it follows from $A \succeq B$ that

$$0 = x^* A x \geq x^* B x \geq 0,$$

which further implies that

$$0 = x^* B x = \left( B^{\frac{1}{2}} x \right)^* B^{\frac{1}{2}} x$$

That is to say, $x \in \mathcal{N}(B) = \mathcal{R}(B)^\perp$. Note that $\mathcal{R}(A)^\perp \subseteq \mathcal{R}(B)^\perp$ amounts to $\mathcal{R}(A) \supseteq \mathcal{R}(B)$, the lemma has been verified.  ∎

### Lemma 17.2

For $A \in \mathbb{C}_{\succeq}^m$, $B \in \mathbb{C}_{\succeq}^m$, if $A \succeq B$, then,

$$AA^\dagger B = B = BA^\dagger A. \tag{17.49}$$

∎

### Proof 17.4

Due to Lemma 17.1, there must exist a matrix $Y$ such that

$$AY = B, \tag{17.50}$$

which means that

$$AA^\dagger B = AA^\dagger AY = B \tag{17.51}$$

and

$$\begin{aligned}
BA^\dagger A &= B^* A^\dagger A = (AY)^* A^\dagger A \\
&= Y^* A^* A^\dagger A = Y^* AA^\dagger A \\
&= Y^* A = Y^* A^* = (AY)^* \\
&= B^* = B.
\end{aligned} \tag{17.52}$$

The following lemma is attributable to Radoslaw Kala and Krzysztof Klaczyński [21].  ∎

**Lemma 17.3**

For $A \in \mathbb{C}_{\succeq}^m$, $D \in \mathbb{C}_{\succeq}^r$ and $B \in \mathbb{C}^{m \times r}$ define

$$\tilde{A} = A + BDB^* \tag{17.53}$$

and

$$V = B^* \, GB + D^\dagger. \tag{17.54}$$

Assume

$$\mathcal{R}(B^*) \subseteq \mathcal{R}(D) \tag{17.55}$$

and

$$\mathcal{R}(B) \subseteq \mathcal{R}(A) \tag{17.56}$$

are satisfied. Then

$$\tilde{A}^\dagger = A^\dagger - A^\dagger BV^\dagger B^* A^\dagger \tag{17.57}$$

∎

**Lemma 17.4**

Suppose that $H_k^i \in \mathbb{C}^{N_i \times r}$, $R_k^i \in \mathbb{C}_{\succeq}^{N_i}$ and $P_{k/k-1}^i \in \mathbb{C}_{\succeq}^r$, $i = 1, 2, \cdots, l$. Moreover, let $H_k = \left( H_k^{1*}, H_k^{2*}, \cdots, H_k^{l*} \right)^*$, $R_k = \text{diag}\left( R_k^1, R_k^2, \cdots, R_k^l \right)$,

$$P_{k/k} = P_{k/k-1} - P_{k/k-1} H_k' \left( H_k P_{k/k-1} H_k' + R_k \right)^\dagger H_k P_{k/k-1} \tag{17.58}$$

and

$$P_{k/k}^i = P_{k/k-1}^i - P_{k/k-1}^i H_k^{i'} \left( H_k^i P_{k/k-1}^i H_k^{i'} + R_k^i \right)^\dagger H_k^i P_{k/k-1}^i \tag{17.59}$$

Then

$$P_{k/k}^\dagger = P_{k/k-1}^\dagger + P_{k/k-1}^\dagger P_{k/k-1} H_{k'} R_k^{-1} H_k P_{k/k-1} P_{k/k-1}^\dagger \tag{17.60}$$

and

$$P_{k/k}^{i^\dagger} = P_{k/k-1}^{i^\dagger} + P_{k/k-1}^{i^\dagger} P_{k/k-1}^i H_k^{i'} R_k^{i-1} H_k^i P_{k/k-1}^i P_{k/k-1}^{i^\dagger} \tag{17.61}$$

∎

## Proof 17.5

It is obvious that

$$\mathcal{R}\left(P_{k/k-1}^\dagger P_{k/k-1} H_k'\right) \subseteq \mathcal{R}\left(P_{k/k-1}^\dagger\right) \tag{17.62}$$

Note that $R_k \in \mathbb{R}^{\left(\sum_{i=1}^l N_i\right) \times \left(\sum_{i=1}^l N_i\right)}$ and is invertible, which leads to $\mathcal{R}\left(R_k^{-1}\right)$ is the entire linear space $\mathbb{R}^{\sum_{i=1}^l N_i}$. Thus, together with the relation $H_k \in \mathbb{R}^{\left(\sum_{i=1}^l N_i\right) \times r}$, it is obvious that

$$\mathcal{R}\left(\left(P_{k/k-1}^\dagger P_{k/k-1} H_{k'}\right)^*\right) = \mathcal{R}\left(\left(P_{k/k-1}^\dagger P_{k/k-1} H_{k'}\right)'\right)$$

$$= \mathcal{R}\left(H_k \left(P_{k/k-1}^\dagger P_{k/k-1}\right)'\right) \subseteq \mathcal{R}(H_k) \subseteq \mathbb{R}^{\sum_{i=1}^l N_i} \tag{17.63}$$

$$= \mathcal{R}\left(R_k^{-1}\right)$$

Furthermore, according to property of Moore–Penrose generalized inverse of matrix, we have

$$P_{k/k-1}\left(P_{k/k-1}^\dagger P_{k/k-1} H_k'\right) = P_{k/k-1} H_k' \tag{17.64}$$

and

$$\left(P_{k/k-1}^\dagger P_{k/k-1} H_k'\right)^* P_{k/k-1} \left(P_{k/k-1}^\dagger P_{k/k-1} H_k'\right)$$

$$= \left(H_k P_{k/k-1} P_{k/k-1}^\dagger\right) P_{k/k-1} \left(P_{k/k-1}^\dagger P_{k/k-1} H_k'\right) \tag{17.65}$$

$$= H_k P_{k/k-1} H_k'.$$

Note that

$$P_{k/k} = P_{k/k-1} - P_{k/k-1} H_k' \left(H_k P_{k/k-1} H_k' + R_k\right)^\dagger H_k P_{k/k-1}, \tag{17.66}$$

combined with Equations 17.62 through 17.65 and Lemma 17.3, we know Equation 17.60 holds. Consequently, Equation 17.61 also holds.                                                                                                      ∎

**Lemma 17.5**

For any $i$, $1 \leq i \leq l$, and $k$, $k = 2, 3, \ldots$, we have

$$P_{k/k-1} \preceq P^i_{k/k-1} \tag{17.67}$$

and

$$P_{k/k} \preceq P^i_{k/k} \tag{17.68}$$

where $P_{k/k-1}$, $P_{k/k}$, $P^i_{k/k-1}$ and $P^i_{k/k}$ are defined as Equations 17.12, 17.13, 17.20, and 17.21, respectively. ∎

**Proof 17.6**

The local filtering $\mathbf{x}^i_{k/k}$ can be regarded as a particular multisensor filtering fusion that only uses measurements $\left\{ \mathbf{y}^i_1, \ldots, \mathbf{y}^i_k \right\}$ and the combination coefficients of all other sensor measurements are zeros. However, the centralized filtering $\mathbf{x}_{k/k}$ is optimal by using all the measurements $\left\{ \mathbf{y}^1_1, \ldots, \mathbf{y}^l_1; \ldots; \mathbf{y}^1_k, \ldots, \mathbf{y}^l_k \right\}$. Moreover, it implies that $\mathbf{x}_{k/k}$ is the best one in the sense that its estimation error matrix $P_{k/k}$ is the smallest with respect to matrix Löwner partial order (matrix positive semi-definite order). Consequently, the lemma is true. ∎

**Lemma 17.6**

For any $i$, $1 \leq i \leq 1$, we have

$$P_{k/k-1} P^{i\dagger}_{k/k-1} P^i_{k/k-1} = P_{k/k-1} \tag{17.69}$$

and

$$P_{k/k} P^{i\dagger}_{k/k} P^i_{k/k} = P_{k/k}, \tag{17.70}$$

where $P_{k/k-1}$, $P_{k/k}$, $P^i_{k/k-1}$ and $P^i_{k/k}$ are defined as Equations 17.12, 17.13, 17.20, and 17.21, respectively. ∎

**Proof 17.7**

It is obvious because Lemmas 17.2 and 17.5 hold. ∎

**Lemma 17.7**

Under the assumptions of Lemma 17.4, we have

$$P^\dagger_{k/k} = P^\dagger_{k/k-1} + \sum_{i=1}^{l} P^\dagger_{k/k-1} P_{k/k-1} \left( P^{i\dagger}_{k/k} - P^{i\dagger}_{k/k-1} \right) P_{k/k-1} P^\dagger_{k/k-1} \tag{17.71}$$

∎

## Proof 17.8

According to Lemma 17.4, we have

$$
\begin{aligned}
P_{k/k}^{\dagger} &= P_{k/k-1}^{\dagger} + P_{k/k-1}^{\dagger} P_{k/k-1} H_k' R_k^{-1} H_k P_{k/k-1} P_{k/k-1}^{\dagger} \\
&= P_{k/k-1}^{\dagger} + \sum_{i=1}^{l} P_{k/k-1}^{\dagger} P_{k/k-1} H_k^{i'} R_k^{i-1} H_k^i P_{k/k-1} P_{k/k-1}^{\dagger} \\
&= P_{k/k-1}^{\dagger} + \sum_{i=1}^{l} \left\{ \left[ P_{k/k-1}^{\dagger} P_{k/k-1} \left( P_{k/k-1}^{i\dagger} P_{k/k-1}^i \right) \right] H_k^{i'} R_k^{-1} H_k^i \right. \\
&\qquad \left. \times \left[ \left( P_{k/k-1}^i P_{k/k-1}^{i\dagger} \right) P_{k/k-1} P_{k/k-1}^{\dagger} \right] \right\} \\
&= P_{k/k-1}^{\dagger} + \sum_{i=1}^{l} \left\{ P_{k/k-1}^{\dagger} P_{k/k-1} \left[ \left( P_{k/k-1}^{i\dagger} P_{k/k-1}^i \right) H_k^{i'} R_k^{-1} H_k^i \right. \right. \\
&\qquad \left. \left. \times \left( P_{k/k-1}^i P_{k/k-1}^{i\dagger} \right) \right] P_{k/k-1} P_{k/k-1}^{\dagger} \right\} \\
&= P_{k/k-1}^{\dagger} + \sum_{i=1}^{l} P_{k/k-1}^{\dagger} P_{k/k-1} \left( P_{k/k}^{i\dagger} - P_{k/k-1}^{i\dagger} \right) P_{k/k-1} P_{k/k-1}^{\dagger},
\end{aligned}
\tag{17.72}
$$

where the third equality uses the fact that

$$
P_{k/k-1}^{\dagger} P_{k/k-1} \left( P_{k/k-1}^{i\dagger} P_{k/k-1}^i \right) = \left( P_{k/k-1}^i P_{k/k-1}^{i\dagger} \right) P_{k/k-1} P_{k/k-1}^{\dagger}
\tag{17.73}
$$

$$
= P_{k/k-1}^{\dagger} P_{k/k-1},
$$

thanks to Lemmas 17.2 and 17.5, and the last equality uses Lemma 17.4 again. ∎

## Lemma 17.8

Suppose that $P \in \mathbb{C}_{\succeq}^r$, $H \in \mathbb{C}^{N \times r}$, $R \in \mathbb{C}_{\succeq}^N$, then

$$
\lim_{\epsilon \to 0^+} PH^*(HPH^* + R + \epsilon I_N)^{-1} = PH^*(HPH^* + R)^{\dagger}
\tag{17.74}
$$

∎

## Proof 17.9

Let $A = \begin{pmatrix} P^{\frac{1}{2}} H^* \\ R^{\frac{1}{2}} \end{pmatrix}$, where $P^{\frac{1}{2}} \succeq 0$ and $R^{\frac{1}{2}} \succeq 0$ denote the square roots of $P$ and $R$ respectively,

because $P \succeq 0$ and $R \succeq 0$. Therefore,

$$
A^* A = HPH^* + R.
\tag{17.75}
$$

Equation 17.40 implies that

$$
\lim_{\epsilon \to 0^+} A(A^*A + \epsilon I_N)^{-1} = (A^*)^{\dagger} = A(A^*A)^{\dagger}
\tag{17.76}
$$

Moreover, Equations 17.75 and 17.76 lead to

$$\lim_{\epsilon \to 0^+} \begin{pmatrix} P^{\frac{1}{2}} H^*(HPH^* + R + \epsilon I_N)^{-1} \\ R^{\frac{1}{2}} (HPH^* + R + \epsilon I_N)^{-1} \end{pmatrix} = \lim_{\epsilon \to 0^+} \begin{pmatrix} P^{\frac{1}{2}} H^* \\ R^{\frac{1}{2}} \end{pmatrix} (HPH^* + R + \epsilon I_N)^{-1}$$

$$= \begin{pmatrix} P^{\frac{1}{2}} H^* \\ R^{\frac{1}{2}} \end{pmatrix} (HPH^* + R)^\dagger$$

$$= \begin{pmatrix} P^{\frac{1}{2}} H^*(HPH^* + R)^\dagger \\ R^{\frac{1}{2}} (HPH^* + R)^\dagger \end{pmatrix}. \tag{17.77}$$

It immediately follows that

$$\lim_{\epsilon \to 0^+} P^{\frac{1}{2}} H^*(HPH^* + R + \epsilon I_N)^{-1} = P^{\frac{1}{2}} H^*(HPH^* + R)^\dagger \tag{17.78}$$

Note that

$$PH^*(HPH^* + R + \epsilon I_N)^{-1} = P^{\frac{1}{2}} P^{\frac{1}{2}} H^*(HPH^* + R + \epsilon I_N)^{-1} \tag{17.79}$$

and the proof is complete.                                                                    ∎

## Lemma 17.9

Assume that $(\Omega, \mathcal{F}, P)$ is a probability space, $\mathbf{y} : \Omega \to \mathcal{R}^n$ is a random vector with zero mean and invertible covariance $R$. If there exists a matrix $A \in \mathcal{R}^{m \times n}$ such that

$$AY(\omega) = \mathbf{0}, \ \forall \ \omega \in \Omega, \tag{17.80}$$

Then,

$$A = \mathbf{0} \tag{17.81}$$

∎

## Proof 17.10

Note $E(Y) = \mathbf{0}$ and by Equation 17.80, we have

$$\mathbf{0} = \text{Cov}(\mathbf{0}) = \text{Cov}(AY) = ARA'. \tag{17.82}$$

This implies that $A = \mathbf{0}$ because $R$ is invertible.                                ∎

**Lemma 17.10**

Suppose that $P \in \mathbb{C}^r_\succeq$, $H \in \mathbb{C}^{N \times r}$, $R \in \mathbb{C}^N_\succeq$; then $\forall \epsilon \geq 0$,

$$\text{rank}\left(PH^*(HPH^* + R + \epsilon I_N)^\dagger\right) = \text{rank}(PH^*) \qquad (17.83)$$

∎

**Proof 17.11**

For $\forall \epsilon \geq 0$, it is obviously that

$$HPH^* + R + \epsilon I_N \succeq HPH^* \qquad (17.84)$$

It follows from Equation 17.84 and Lemma 17.1 that

$$\mathcal{R}(HPH^* + R + \epsilon I_N) \supseteq \mathcal{R}(HPH^*), \qquad (17.85)$$

which further implies that

$$HPH^*(HPH^* + R + \epsilon I_N)^\dagger(HPH^* + R + \epsilon I_N) = HPH^*. \qquad (17.86)$$

On the one hand, we have

$$\text{rank}\left(PH^*(HPH^* + R + \epsilon I_N)^\dagger(HPH^* + R + \epsilon I_N)\right)$$
$$\geq \text{rank}\left(HPH^*(HPH^* + R + \epsilon I_N)^\dagger(HPH^* + R + \epsilon I_N)\right)$$
$$= \text{rank}(HPH^*) = \text{rank}\left(HP^{\frac{1}{2}}P^{\frac{1}{2}}H^*\right) \qquad (17.87)$$
$$= \text{rank}\left(\left(P^{\frac{1}{2}}H^*\right)^* P^{\frac{1}{2}}H^*\right) = \text{rank}\left(P^{\frac{1}{2}}H^*\right)$$
$$\geq \text{rank}\left(P^{\frac{1}{2}}P^{\frac{1}{2}}H^*\right) = \text{rank}(PH^*),$$

where the first equality holds because of Equation 17.86, and the second equality is due to $P^{\frac{1}{2}} \succeq 0$ denotes the square root of $P$ because $P \succeq 0$.

On the other hand, it is well known that

$$\text{rank}\left(PH^*(HPH^* + R + \epsilon I_N)^\dagger(HPH^* + R + \epsilon I_N)\right)$$
$$\leq \text{rank}(PH^*). \qquad (17.88)$$

Equations 17.87 and 17.88 mean that Lemma 17.10 holds.

The following lemma is due to Ref. [22].

∎

**Lemma 17.11**

Assume that $A, B \in \mathbb{C}^{m \times n}$, where $B$ is regarded as a variable. Then

$$\lim_{B \to A} B^\dagger = A^\dagger$$

is equivalent to

$$\text{rank}\,(B) = \text{rank}\,(A)$$

when $B$ is approaching sufficiently close to $A$.                                                    ■

## APPENDIX II

### Proof of Proposition 17.1

### Proof 17.12

For an arbitrary given integer $t \geq 0$ and $\varepsilon > 0$, we consider the l-sensor distributed linear dynamic system as follows:

$$\mathbf{x}_{k+1} = \Phi_k \mathbf{x}_k + v_k, \quad k = 0,1,\ldots, \tag{17.89}$$

$$\mathbf{y}_k^i = H_k^i \mathbf{x}_k + w_k^i, \quad i = 1,\ldots,l, \quad k = 0,1,\ldots t-1, \tag{17.90}$$

$$\mathbf{y}_k^i(\epsilon) = H_k^i \mathbf{x}_k + w_k^i + \sqrt{\epsilon}\breve{w}_k^i, \quad i = 1,\ldots,l, \quad k = t, \tag{17.91}$$

where $\Phi_k$ is a matrix of order $(r \times r)$, $\mathbf{x}_k, v_k \in \mathbb{R}^r$, $H_k^i \in \mathbb{R}^{N_i \times r}$, $\mathbf{y}_k^i, w_k^i \in \mathbb{R}^{N_i}$. The process noise $v_k$ and measurement noise $w_k^i$ ($w_k^i + \sqrt{\epsilon}\breve{w}_k^i$ when $k = t$) are both zero-mean random variables independent of each other temporally and are not cross correlated. Compared with system (Equations 17.1 and 17.2), here we have replaced $\mathbf{y}_t^i$ and $w_t^i$ by $\mathbf{y}_t^i(\epsilon)$ and $w_t^i + \sqrt{\epsilon}\breve{w}_t^i$ when $k = t$, respectively; moreover, $w_t^i$ and $\sqrt{\epsilon}\breve{w}_t^i$ are independent random variables with zero mean and covariance

$$\text{Cov}\left(w_t^i\right) = R_t^i \tag{17.92}$$

and

$$\text{Cov}\left(\breve{w}_t^i\right) = I_{N_i} \tag{17.93}$$

Consequently, similar to the definition of Equations 17.5 and 17.6, we let

$$R_t^i(\epsilon) \triangleq \text{Cov}\left(w_t^i + \sqrt{\epsilon}\breve{w}_t^i\right) = \text{Cov}\left(w_t^i\right) + \epsilon\,\text{Cov}\left(\breve{w}_t^i\right)$$
$$= R_t^i + \epsilon I_{N_i}, \quad i = 1,\ldots,l. \tag{17.94}$$

and

$$R_t(\epsilon) \triangleq \mathrm{Cov}\left(w_t + \sqrt{\epsilon}\breve{w}_t\right)$$

$$= \mathrm{Cov}\left(\left(w_k^{1'},\ldots,w_k^{l'}\right)' + \sqrt{\epsilon}\left(\breve{w}_k^{1'},\ldots,\breve{w}_k^{l'}\right)'\right) \tag{17.95}$$

$$= \mathrm{diag}\left(R_t^1,\ldots,R_t^l\right)' + \mathrm{diag}\left(\epsilon I_{N_1},\ldots,\epsilon I_{N_l}\right)'$$

$$= R_t + \epsilon I_N$$

where

$$N = \sum_{i=1}^{l} N_i. \tag{17.96}$$

Denote

$$\mathbf{y}_t(\epsilon) \triangleq \left(\mathbf{y}_t^{1'}(\epsilon),\ldots,\mathbf{y}_t^{l'}(\epsilon)\right)' \tag{17.97}$$

Consequently, according to Kalman filtering theory,

$$K_t(\epsilon) = P_{t/t-1}H_t'\left(H_tP_{t/t-1}H_t' + R_t + \epsilon I_N\right)^{-1}$$

$$= P_{t/t}(\epsilon)H_t'(R_t + \epsilon I_N)^{-1} \tag{17.98}$$

and

$$K_t^i(\epsilon) = P_{t/t-1}^i H_t^{i'}\left(H_t^i P_{t/t-1}^i H_t^{i'} + R_k^i + \epsilon I_N\right)^{-1}$$

$$= P_{t/t}^i(\epsilon)H_t^{i'}\left(R_t^i + \epsilon I_N\right)^{-1}, \tag{17.99}$$

where

$$P_{t/t}(\epsilon) = (I - K_t(\epsilon)H_t)P_{t/t-1} \tag{17.100}$$

and

$$P_{t/t}^i(\epsilon) = \left(I - K_t^i(\epsilon)H_t^i\right)P_{t/t-1}^i \tag{17.101}$$

Lemma 17.8 in Appendix I implies that

$$\lim_{\epsilon \to 0^+} K_t(\epsilon) = P_{t/t-1}H_t'\left(H_tP_{t/t-1}H_t' + R_t\right)^\dagger = K_t \tag{17.102}$$

and

$$\lim_{\epsilon \to 0^+} K_t^i(\epsilon) = P_{t/t-1}^i H_t^{i'} \left( H_t^i P_{t/t-1}^i H_t^{i'} + R_t^i \right)^\dagger = K_t^i \tag{17.103}$$

also hold.

Let

$$K_t(\epsilon) = \left( K_t(\epsilon,1) \quad K_t(\epsilon,2) \quad \dots \quad K_t(\epsilon,l) \right) \tag{17.104}$$

be an appropriate partition of matrix $K_t(\varepsilon)$ such that $K_t(\epsilon)\mathbf{y}_t(\epsilon) = \sum_{i=1}^{l} K_t(\epsilon,i)\mathbf{y}_t^i(\epsilon)$.

Because $\varepsilon > 0$, Equation 17.94 means we are dealing with the case where measurement noises possess invertible covariances. Thus, by the similar analysis in Equation 17.25, it follows that

$$\begin{aligned}
K_t(\epsilon)\mathbf{y}_t(\epsilon) &= \left( K_t(\epsilon,1)K_t(\epsilon,2)\dots K_t(\epsilon,l) \right)\mathbf{y}_t(\epsilon) \\
&= \sum_{i=1}^{l} K_t(\epsilon,i)\mathbf{y}_t^i(\epsilon) = \sum_{i=1}^{l} P_{t/t}(\epsilon)P_{t/t}^{i\dagger}(\epsilon)K_t^i(\epsilon)\mathbf{y}_t^i(\epsilon) \\
&= \left( P_{t/t}(\epsilon)P_{t/t}^{1\dagger}(\epsilon)K_t^1(\epsilon)\cdots P_{t/t}(\epsilon)P_{t/t}^{l\dagger}(\epsilon)K_t^l(\epsilon) \right)\mathbf{y}_t(\epsilon).
\end{aligned} \tag{17.105}$$

Because $\mathbf{y}_t(\epsilon)$ denotes a random variable with an invertible covariance $H_t\text{Cov}(\mathbf{x}_t)H_t' + R_t + \epsilon I_N$, combined with Lemma 17.9 in Appendix I, Equations 17.104 and 17.105 lead to the deduction that

$$K_t(\epsilon,i) = P_{t/t}(\epsilon)P_{t/t}^{i\dagger}(\epsilon)K_t^i(\epsilon), \quad i = 1,2,\dots,l \tag{17.106}$$

Furthermore, according to the property of linear matrix equation, it follows from Equation 17.106 that the row space of $K_t(\varepsilon, i)$ is contained in the row space of $K_t^i(\epsilon)$. Consequently,

$$K_t(\epsilon,i) = K_t(\epsilon,i)K_t^{i\dagger}(\epsilon)K_t^i(\epsilon), \quad i = 1,2,\dots,l. \tag{17.107}$$

It follows from Lemma 17.10 in Appendix I that $\forall \epsilon \geq 0$,

$$\begin{aligned}
\text{rank}\left( K_t^i(\epsilon) \right) &= \text{rank}\left( P_{t/t-1}^i H_t^{i'} \left( H_t^i P_{t/t-1}^i H_t^{i'} + R_t^i + \epsilon I_{N_i} \right)^\dagger \right) \\
&= \text{rank}\left( P_{t/t-1}^i H_t^{i'} \right).
\end{aligned} \tag{17.108}$$

Because Moore–Penrose generalized inverse is continuous when rank is unchanged (see Lemma 17.11 in Appendix I), Equations 17.103 and 17.108 mean that

$$\lim_{\epsilon \to 0^+} \left( K_t^i(\epsilon) \right)^\dagger = \left( \lim_{\epsilon \to 0^+} K_t^i(\epsilon) \right)^\dagger = K_t^{i\dagger} \tag{17.109}$$

Note that $K_t(i)$ has been defined in Equation 17.42, taking the limits for both hands of Equation 17.106, and combining Equations 17.102–17.104, and 17.109, we achieve

$$
\begin{aligned}
K_t(i) &= \lim_{\epsilon \to 0^+} K_t(\epsilon, i) = \lim_{\epsilon \to 0^+} \left[ K_t(\epsilon, i) K_t^{i^\dagger}(\epsilon) K_t^i(\epsilon) \right] \\
&= \left[ \lim_{\epsilon \to 0^+} K_t(\epsilon, i) \right] \times \left[ \lim_{\epsilon \to 0^+} K_t^{i^\dagger}(\epsilon) \right] \times \left[ \lim_{\epsilon \to 0^+} K_t^i(\epsilon) \right] \\
&= K_t(i) K_t^{i^\dagger} K_t^i, \quad i = 1, 2, \ldots, l.
\end{aligned}
\tag{17.110}
$$

Thus we have finished the proof [8,13,14]. ∎

# REFERENCES

1. X. R. Li, Y. M. Zhu, J. Wang, and Han C. Z. Optimal linear estimation fusion—Part I: Unified fusion rules. *IEEE Transactions on Information Theory*, 49(9):2192–2208, 2003.
2. Y. Bar-Shalom. *Multitarget-Multisensor Tracking: Advanced Applications*, vol. 1. Artech House, Norwood, MA, 1990.
3. C. Y. Chong, K. C. Chang, and S. Mori. Distributed tracking in distributed sensor networks. In *Proceedings of 1987 American Control Conference*, Seattle, WA, pp. 1863–1868, June 1986.
4. C. Y. Chong, S. Mori, and K. C. Chang. Chapter 8 *Distributed multitarget multisensor tracking*. In *Multitarget in Multisensor Tracking: Advanced Applications*, vol. 1, Y. Bar-Shalom, ed. Artech House, Norwood, MA, 1990.
5. H. R. Hashmipour, S. Roy, and A. J. Laub. Decentralized structures for parallel Kalman filterings. *IEEE Transactions on Automatic Control*, 33(1):88–93, 1998.
6. Y. M. Zhu, J. Zhou, X. J. Shen, E. B. Song, and Y. T. Luo. *Networked Multisensor Decision and Estimation Fusion based on Advanced Mathematical Methods*. Kluwer Academic Publishers, Dordrecht, the Netherlands, 2012.
7. Y. M. Zhu, Z. S. You, J. Zhao, K. S. Zhang, and X. R. Li. The optimality for the distributed Kalman filter with feedback. *Automatica*, 37(9):1489–1493, 2001.
8. S. T. Chen. Comments on "State estimation for linear systems with state equality constraints." [*Automatica*, 43(2007):1363–1368, 2007]. *Automatica*, 46(11):1929–1932, 2010.
9. H. Hemami, and B. F. Wyman. Modeling and control of constrained dynamic systems with application to biped locomotion in the frontal plane. *IEEE Transactions on Automatic Control*, 24(4):526–535, 1979.
10. S. Ko, and R. R. Bitmead. State estimation for linear systems with state equality constraints. *Automatica*, 43(8):1363–1368, 2007.
11. D. Simon. Kalman filtering with state constraints: a survey of linear and nonlinear algorithms. *IET Control Theory Applications*, 4(8):1303–1318, 2010.
12. D. Simon, and T. L. Chia. Kalman filtering with state equality constraints. *IEEE Transactions on Aerospace Electronic Systems*, 38(1):128–136, 2002.
13. E. B. Song, Y. M. Zhu, J. Zhou, and Z. S. You. Optimal Kalman filtering fusion with cross-correlated sensor noises. *Automatica*, 43(8):1450–1456, 2007.
14. B. Teixeira, J. Chandrasekar, L. Torres, L. Aguirre, and D. Bernstein. State estimation for linear and non-linear equality constrained systems. *International Journal of Control*, 82(5):918–936, 2009.
15. Z. S. Duan, and X. R. Li. Lossless linear transformation of sensor data for distributed estimation fusion. *IEEE Transactions on Signal Processing*, 59(1):362–372, 2011.
16. J. Xu, E. B. Song, Y. T. Luo, and Y. M. Zhu. Optimal distributed Kalman filtering fusion algorithm without invertibility of estimation error and sensor noise covariances. *IEEE Signal Processing Letters*, 19(1):55–58, 2012.
17. G. C. Goodwin, and R. Payne. *Dynamic System Identification: Experimental Design and Data Analysis*. Academic Press, New York, 1977.
18. S. Haykin. *Adaptive Filter Theory*. Prentice-Hall, Englewood Cliffs, NJ, 1996.
19. L. Ljung. *System Identification: Theory for the User*. Prentice-Hall, Englewood Cliffs, NJ, 1987.

20. C. G. den Broeder Jr., and A. Charnes. *Contributions to the Theory of Generalized Inverses for Matrices*. Department of Mathemathics, Purdue University, Lafayette, IN, 1957 (reprinted as ONR Res. Memo. No. 39, Northwestern University, Evanston, IL, 1962).
21. R. Kala, and K. Klaczyński. Generalized inverses of a sum of matrices. *Sankhyā Series A*, 56:458–464, 1994.
22. G. W. Stewart. On the continuity of the generalized inverse. *SIAM Journal on Applied Mathematics*, 17(1):33–45, 1969.

# 18 Accumulated State Densities and Their Applications in Object Tracking

*Wolfgang Koch*

## CONTENTS

## 18.1 INTRODUCTION

In tracking and sensor data fusion applications, the full information on kinematic object properties accumulated over a certain discrete time window up to the present time is contained in the conditional joint probability density function of the kinematic state vectors referring to each time step in this window. This density is conditioned by the time series of all sensor data collected up to the present time and has accordingly been called an accumulated state density (ASD) [1]. ASDs provide a unified treatment of filtering and retrodiction insofar as by marginalizing them appropriately, the standard filtering and retrodiction densities are obtained. In addition, ASDs fully describe the posterior correlations between the states at different instants of time.

Here, provide an introduction into the notion of ASDs, derive closed formulae for calculating them and discuss their relevance for problem solving in target tracking and distributed fusion applications such as exact processing of out-of-sequence (OoS) measurements in case of non-ideal communications, solving data association problems within the framework of probabilistic multiple hypothesis tracking (PMHT), and exact track-to-track fusion in distributed sensor netwoks.

## 18.2   ON THE PROBLEM OF TRACKING OBJECTS

To a degree never known before, decision makers in a net-centric world have access to vast amounts of data. For effective use of this information potential in real-world applications, however, the data streams must not overwhelm the decision makers involved. On the contrary, the data must be fused in such a way that high-quality information for situation pictures results, the basis for decision making.

Situation pictures are produced by spatio-temporally processing various pieces of sensor information that in themselves often have only limited value for understanding the underlying situation. In this context, object "tracks" are of particular importance [2–4]. Tracking faces an omnipresent aspect in real-world application insofar as it is dealing with fusion of data produced at different instants of time; that is, tracking is important in all applications where particular emphasis is placed on sensor data given by time series.

In most tracking algorithms, the characteristics of conditional probability densities $p(\mathbf{x}_l|Z^k)$ of (joint) object states $\mathbf{x}_l$ are calculated, which describe the available knowledge of the object properties at a certain instant of time $t_l$, given a time series $Z^k$ of imperfect sensor data accumulated up to time $t_k$. In certain applications, however, the kinematic object states $\mathbf{x}_k, ..., \mathbf{x}_n, n \le k$, accumulated over a certain time window from a past instant of time $t_m$ up to the present time $t_k$ is of interest. The statistical properties of the accumulated state vectors are completely described by their joint probability density function, $p(\mathbf{x}_k, ..., \mathbf{x}_n|Z^k)$, which is conditioned by the time series $Z^k$. These densities may be called accumulated state densities (ASDs). By marginalizing them, the standard filtering and retrodiction densities directly result; in other words, ASDs provide a unified description of filtering and retrodiction. In addition, ASDs fully describe the correlations between the state estimates at different instants of time. In Ref. [5], for example, ASDs are considered to provide a more comprehensive treatment of issues in particle filtering.

To some extent, the notion of ASDs might be considered as a step backward insofar as in the old days of object tracking it was known that one could express a linear-Gaussian estimation problem in a joint, that is, "accumulated," fashion, while Kalman's approach was a way to find a recursive solution. Nevertheless, a chief contribution of the discussion here is a *recursive* algorithm to find the parameters of an ASD.

ASDs are useful in tracking applications, where out-of-sequence (OoS) measurements are to be processed, that is, when the sensor data do not arrive in the temporal order, in which they have been produced. The OoS problem is unavoidable in any real-world multiple sensor tracking application. To the authors' knowledge, Y. Bar-Shalom was the first who picked up the problem and provided a solution in the case of Kalman filtering [6]. For the subsequent development and generalizations see, for example, Refs. [7–11]. To avoid storing and reprocessing the entire time series of sensor data as well as to avoid the temporal delay connected to it, OoS measurements have to be inserted into the running tracking process in a particular way.

In any real-world tracking application, moreover, data association conflicts have to be taken into account, that is, ambiguities in the origin of the sensor data that are caused by unwanted targets and false returns. In this case, an exhaustive and mutually exclusive enumeration of association hypotheses results in multiple model tracker (MHT)- and probabilistic data association (PDA)-type approaches to target tracking as discussed in Refs. [2–4], for example. Tracking in the presence of data association conflicts, however, can also be considered as an "incomplete data problem" [12], which can be solved by the expectation maximization (EM) methodology [13,14]. Under linear Gaussian assumptions regarding the sensor and object evolution models, ASDs play a crucial role here as well.

Finally, we discuss why ASDs provide an exact solution to the track-to-track fusion (T2TF) problem. Recently an exact solution for T2TF has been published [15], the distributed Kalman filter (DKF). However, the DKF is exact only if global knowledge in terms of the measurement models for all sensors is available at a local processor. An exact solution for T2TF can also be achieved as a convex combination of local ASDs generated at each node in a distributed sensor system. This method crucially differs from the DKF in that an exact solution is achieved without each processing platform being required to have knowledge of the global information.

This chapter is organized as follows. Section 18.3 summarizes basic facts on the Bayesian tracking paradigm. In Section 18.4, the notion of an ASD is introduced along with a discussion of closed-formulae for the parameters of the ASD in the case, where Kalman filtering can be applied to tracking. The concept of ASDs is generalized to multiple hypothesis and interaction multiple model trackers (MHT, IMM). In Section 18.5, ASDs are applied to the out-of-sequence tracking problem. Section 18.6 discusses the role of ADS within the PMHT framework, while the use of ASDs for solving the T2TF problem is the topic of Section 18.7. Simulated examples illustrate the discussions. For the sake of notational simplicity we confine the discussion to the case of synchronous sensors. The generalization of the methodological framework to asynchronous sensors is straightforward.

## 18.3  BAYESIAN TRACKING PARADIGM

A Bayesian tracking algorithm is an iterative updating scheme for calculating conditional probability density functions $p(\mathbf{x}_l|Z^k)$ that represent all available knowledge on the object states $\mathbf{x}_l$ at discrete instants of time $t_l$. The densities are explicitly conditioned by the sensor data $Z^k$ accumulated up to some time $t_k$, typically the present time. Implicitly, they are also determined by all available context knowledge on the sensor characteristics, the dynamical object properties, the environment of the objects, topographical maps, or tactical rules governing the objects' overall behavior.

With respect on the instant of time $t_l$ at which estimates of the object states $\mathbf{x}_l$ are required, the related density iteration process is referred to as *prediction* $(t_l > t_k)$, *filtering* $(t_l = t_k)$, or *retrodiction* $(t_l < t_k)$. The propagation of the probability densities involved is given by three basic update equations.

*Prediction.* The prediction density $p(\mathbf{x}_k|Z^{k-1})$ is obtained by combining the evolution model $p(\mathbf{x}_k|\mathbf{x}_{k-1})$ with the previous filtering density $p(\mathbf{x}_{k-1}|Z^{k-1})$:

$$p(\mathbf{x}_{k-1}|Z^{k-1}) \xrightarrow[\text{constraints}]{\text{evolution model}} p(\mathbf{x}_k|Z^{k-1})$$

$$p(\mathbf{x}_k|Z^{k-1}) = \int d\mathbf{x}_{k-1}\, \underbrace{p(\mathbf{x}_k|\mathbf{x}_{k-1})}_{\text{evolution model}}\, \underbrace{p(\mathbf{x}_{k-1}|Z^{k-1})}_{\text{previous filtering}}. \tag{18.1}$$

*Filtering.* The filtering density $p(\mathbf{x}_k|Z^k)$ is obtained by combining the sensor model $p(Z_k|\mathbf{x}_k)$, also called the "likelihood function," with the prediction density $p(\mathbf{x}_k|Z^{k-1})$ according to

$$p(\mathbf{x}_k|Z^{k-1}) \xrightarrow[\text{sensor model}]{\text{current sensor data}} p(\mathbf{x}_k|Z^k)$$

$$p(\mathbf{x}_k|Z^k) = \frac{p(Z_k, m_k|\mathbf{x}_k)\, p(\mathbf{x}_k|Z^{k-1})}{\int d\mathbf{x}_k\, \underbrace{p(Z_k, m_k|\mathbf{x}_k)}_{\text{sensor model}}\, \underbrace{p(\mathbf{x}_k|Z^{k-1})}_{\text{prediction}}}. \tag{18.2}$$

*Retrodiction.* The retrodiction density $p(\mathbf{x}_l|Z^k)$ is obtained by combining the object evolution model $p(\mathbf{x}_{l+1}|\mathbf{x}_l)$ with the previous prediction and filtering densities according to

$$p(\mathbf{x}_{l-1}|Z^k) \xleftarrow[\text{evolution model}]{\text{filtering, prediction}} p(\mathbf{x}_l|Z^k)$$

$$p(\mathbf{x}_l|Z^k) = \int d\mathbf{x}_{l+1} \underbrace{\frac{\overbrace{p(\mathbf{x}_{l+1}|\mathbf{x}_l)}^{\text{evolution}} \overbrace{p(\mathbf{x}_l|Z^l)}^{\text{prev. filtering}}}{\underbrace{p(\mathbf{x}_{l+1}|Z^l)}_{\text{prev. prediction}}} \underbrace{p(\mathbf{x}_{l+1}|Z^k)}_{\text{prev. retrodiction}}.$$ (18.3)

Being the natural antonym of "prediction," the technical term "retrodiction" was introduced by Oliver Drummond in a series of papers [16]. Adopting the standard terminology [17], we could speak of *fixed-interval* retrodiction.

According to this paradigm, an *object track* represents all relevant knowledge on a time-varying object state of interest, including its history and measures that describe the quality of this knowledge. As a technical term, "track" is therefore either a synonym for the collection of densities $p(\mathbf{x}_l|Z^k)$, $l = 1,...,k,...$, or of suitably chosen parameters characterizing them, such as estimates related to appropriate risk functions and the corresponding estimation error covariance matrices.

## 18.4   NOTION OF ACCUMULATED STATE DENSITIES

All information on the object states accumulated over a time window $t_k, t_{k-1}, ..., t_n$ of length $k - n + 1$,

$$\mathbf{x}_{k:n} = (\mathbf{x}_k,...,\mathbf{x}_n)$$ (18.4)

that can be extracted from the time series of accumulated sensor data $Z^k$ up to and including time $t_k$ is contained in a joint density function $p(\mathbf{x}_{k:n}|Z^k)$, which may be called accumulated state density (ASD). Via marginalizing over $\mathbf{x}_k, ..., \mathbf{x}_{l+1}, \mathbf{x}_{l-1}, ..., \mathbf{x}_n,$

$$p(\mathbf{x}_l|Z^k) = \int d\mathbf{x}_k,...,d\mathbf{x}_{l+1},d\mathbf{x}_{l-1},...,d\mathbf{x}_n\, p(\mathbf{x}_k,...,\mathbf{x}_n|Z^k),$$ (18.5)

the filtering density $p(\mathbf{x}_k|Z^k)$ for $l = k$ and the retrodiction densities $p(\mathbf{x}_l|Z^k)$ for $l < k$ result from the ASD. ASDs thus in a way unify the notions of filtering and retrodiction. In addition, ASDs also contain all mutual correlations between the individual object states at different instants of time. Bayes' theorem provides a recursion formula for updating ASDs:

$$p(\mathbf{x}_{k:n}|Z^k) = \frac{p(Z_k, m_k|\mathbf{x}_k)p(\mathbf{x}_k|\mathbf{x}_{k-1})p(\mathbf{x}_{k-1:n}|Z^{k-1})}{\int d\mathbf{x}_{k:n} p(Z_k, m_k|\mathbf{x}_k)p(\mathbf{x}_k|\mathbf{x}_{k-1})p(\mathbf{x}_{k-1:n}|Z^{k-1})}.$$ (18.6)

The sensor data $Z_k$ consisting of $m_k$ measurements explicitly appear in this representation. A little formalistically speaking, "sensor data processing" means nothing other than to achieve by certain reformulations that the sensor data are no longer explicitly present.

### 18.4.1  CLOSED-FORM REPRESENTATION FOR ASDs

Under conditions in which Kalman filtering is applicable (perfect data sensor-data-to-track association, linear Gaussian sensor, and evolution models), a closed-form representation of $p(\mathbf{x}_{k:n}|Z^k)$ can be derived. In this case, let the likelihood function be given by

$$p(Z_k, m_k|\mathbf{x}_k) = \mathcal{N}(\mathbf{z}_k; \mathbf{H}_k\mathbf{x}_k, \mathbf{R}_k),\tag{18.7}$$

where $Z_k = \mathbf{z}_k$ denotes the vector of sensor measurements at time $t_k$, $\mathbf{x}_k = \mathbf{x}_k$ the kinematic state vector of the object, $\mathbf{H}_k$ the measurement matrix, and $\mathbf{R}_k$ the measurement error covariance matrix, while the Markovian evolution model of the object is represented by

$$p(\mathbf{x}_k|\mathbf{x}_{k-1}) = \mathcal{N}(\mathbf{x}_k; \mathbf{F}_{k|k-1}\mathbf{x}_{k-1}, \mathbf{D}_{k|k-1})\tag{18.8}$$

with an evolution matrix $\mathbf{F}_{k|k-1}$ and a corresponding evolution covariance matrix $\mathbf{D}_{k|k-1}$.

By repeated use of Bayes theorem and the Markov property we get from Equation 18.6

$$
\begin{aligned}
p(\mathbf{x}_{k:n}|Z^k) &= \frac{p(\mathbf{z}_k|\mathbf{x}_k)p(\mathbf{x}_k|\mathbf{x}_{k-1})p(\mathbf{x}_{k-1:n}|Z^{k-1})}{\int d\mathbf{x}_{k:n}p(\mathbf{z}_k|\mathbf{x}_k)p(\mathbf{x}_k|\mathbf{x}_{k-1})p(\mathbf{x}_{k-1:n}|Z^{k-1})} \\[2mm]
&= \frac{\mathcal{N}(\mathbf{z}_k; \mathbf{H}_k\mathbf{x}_k, \mathbf{R}_k)\mathcal{N}(\mathbf{x}_k; \mathbf{F}_{k|k-1}\mathbf{x}_{k-1}, \mathbf{D}_{k|k-1})p(\mathbf{x}_{k-1:n}|Z^{k-1})}{\int d\mathbf{x}_{k:n}\mathcal{N}(\mathbf{z}_k; \mathbf{H}_k\mathbf{x}_k, \mathbf{R}_k)\mathcal{N}(\mathbf{x}_k; \mathbf{F}_{k|k-1}\mathbf{x}_{k-1}, \mathbf{D}_{k|k-1})p(\mathbf{x}_{k-1:n}|Z^{k-1})} \\[2mm]
&= \frac{\prod_{l=n+1}^{k} \mathcal{N}(\mathbf{z}_l; \mathbf{H}_l\mathbf{x}_l, \mathbf{R}_l)\mathcal{N}(\mathbf{x}_l; \mathbf{F}_{l|l-1}\mathbf{x}_{l-1}, \mathbf{D}_{l|l-1})\mathcal{N}(\mathbf{x}_n; \mathbf{x}_{n|n}, \mathbf{P}_{n|n})}{\int d\mathbf{x}_{k:n}\prod_{l=n+1}^{k} \mathcal{N}(\mathbf{z}_l; \mathbf{H}_l\mathbf{x}_l, \mathbf{R}_l)\mathcal{N}(\mathbf{x}_l; \mathbf{F}_{l|l-1}\mathbf{x}_{l-1}, \mathbf{D}_{l|l-1})\mathcal{N}(\mathbf{x}_n; \mathbf{x}_{n|n}\mathbf{P}_{n|n})}.
\end{aligned}\tag{18.9}
$$

A successive use of a well-known product formula for Gaussians (see Appendix, Equation 18.150) now directly yields a product representation of the augmented state density:

$$p(\mathbf{x}_{k:n}|Z^k) = \mathcal{N}(\mathbf{x}_k; \mathbf{x}_{k|k}, \mathbf{P}_{k|k})\prod_{l=n}^{k-1}\mathcal{N}(\mathbf{x}_l; \mathbf{h}_{l|l+1}(\mathbf{x}_{l+1}), \mathbf{R}_{l|l+1}),\tag{18.10}$$

where the auxiliary quantities $\mathbf{h}_{l|l+1}$, $\mathbf{R}_{l|l+1}$, $l \le k$, are defined by

$$\mathbf{h}_{l|l+1}(\mathbf{x}_{l+1}) = \mathbf{x}_{l|l} + \mathbf{W}_{l|l+1}(\mathbf{x}_{l+1} - \mathbf{x}_{l+1|l})\tag{18.11}$$

$$\mathbf{R}_{l|l+1} = \mathbf{P}_{l|l} - \mathbf{W}_{l|l+1}\mathbf{P}_{l|l+1}\mathbf{W}_{l|l+1}^{\top}\tag{18.12}$$

and a "retrodiction gain" matrix

$$\mathbf{W}_{l|l+1} = \mathbf{P}_{l|l}\mathbf{F}_{l+1|l}^{\top}\mathbf{P}_{l+1|l}^{-1}.\tag{18.13}$$

Note that $\mathcal{N}(\mathbf{x}_l; \mathbf{h}_{l|l+1}(\mathbf{x}_{l+1}), \mathbf{R}_{l|l+1})$ can be interpreted in analogy to a Gaussian likelihood function with a linear measurement function $\mathbf{h}_{l|l+1}(\mathbf{x}_{l+1})$. $\mathbf{h}_{l|l+1}$, $\mathbf{R}_{l|l+1}$ are defined by the parameters of $p(\mathbf{x}_l|Z^l) = \mathcal{N}(\mathbf{x}_l; \mathbf{x}_{l|l}, \mathbf{P}_{l|l})$,

$$\mathbf{x}_{l|l} = \begin{cases} \mathbf{x}_{l|l-1} + \mathbf{W}_{l|l-1}(\mathbf{z}_l - \mathbf{H}_l\mathbf{x}_{l|l-1}) \\ \mathbf{P}_{l|l}\left(\mathbf{P}_{l|l-1}^{-1}\mathbf{x}_{l|l-1} + \mathbf{H}_l^{\top}\mathbf{R}_l^{-1}\mathbf{z}_l\right) \end{cases} \tag{18.14}$$

$$\mathbf{P}_{l|l} = \begin{cases} \mathbf{P}_{l|l-1} - \mathbf{W}_{l|l-1}\mathbf{S}_{l|l-1}\mathbf{W}_{l|l-1}^{\top} \\ \left(\mathbf{P}_{l|l-1}^{-1} + \mathbf{H}_l^{\top}\mathbf{R}_l^{-1}\mathbf{H}_l\right)^{-1} \end{cases}. \tag{18.15}$$

There exist two equivalent formulations of the Kalman update formulae according to the two versions of the product formula (Equation 18.150). The innovation covariance matrix $\mathbf{S}_{l|l-1}$ and the Kalman gain matrix are given by

$$\mathbf{S}_{l|l-1} = \mathbf{H}_l\mathbf{P}_{l|l-1}\mathbf{H}_l^{\top} + \mathbf{R}_l \tag{18.16}$$

$$\mathbf{W}_{l|l-1} = \mathbf{P}_{l|l-1}\mathbf{H}_l^{\top}\mathbf{S}_{l|l-1}^{-1}. \tag{18.17}$$

Also the parameters of the prediction density $p(\mathbf{x}_{l+1}|Z^l) = \mathcal{N}(\mathbf{x}_{l+1}; \mathbf{x}_{l+1|l}, \mathbf{P}_{l+1|l})$,

$$\mathbf{x}_{l|l-1} = \mathbf{F}_{l|l-1}\mathbf{x}_{l-1|l-1} \tag{18.18}$$

$$\mathbf{P}_{l|l-1} = \mathbf{F}_{l|l-1}\mathbf{P}_{l-1|l-1}\mathbf{F}_{l|l-1}^{\top} + \mathbf{D}_{l-1}, \tag{18.19}$$

enter into the product representation in Equation 18.10. With $\mathbf{x}_{l|k}$, and $\mathbf{P}_{l|k}$ known from the Rauch–Tung–Striebel recursion,

$$\mathbf{x}_{l|k} = \mathbf{x}_{l|l} + \mathbf{W}_{l|l+1}(\mathbf{x}_{l+1|k} - \mathbf{x}_{l+1|l}) \tag{18.20}$$

$$\mathbf{P}_{l|k} = \mathbf{P}_{l|l} + \mathbf{W}_{l|l+1}(\mathbf{P}_{l+1|k} - \mathbf{P}_{l+1|l})\mathbf{W}_{l|l+1}^{\top}, \tag{18.21}$$

we can rewrite $p(\mathbf{x}_{k:n}|Z^k)$ by the following product:

$$p(\mathbf{x}_{k:n}|Z^k) = \mathcal{N}(x_k; \mathbf{x}_{k|k}, \mathbf{P}_{k|k})\prod_{l=n}^{k-1}\mathcal{N}(x_l - \mathbf{W}_{l|l+1}x_{l+1}; \mathbf{x}_{l|k} - \mathbf{W}_{l|l+1}x_{l+1|k}\mathbf{Q}_{l|k}), \tag{18.22}$$

where we used the abbreviation

$$\mathbf{Q}_{l|k} = \mathbf{P}_{l|k} - \mathbf{W}_{l|l+1}\mathbf{P}_{l+1|k}\mathbf{W}_{l|l+1}^{\top}. \tag{18.23}$$

From elementary matrix algebra manipulations it can be shown that this product can be represented by a single Gaussian,

$$p(\mathbf{x}_{k:n}|Z^k) = \mathcal{N}(\mathbf{x}_{k:n}; \mathbf{x}_{k:n|k}, \mathbf{P}_{k:n|k}), \tag{18.24}$$

with a joint expectation vector $\mathbf{x}_{k:n|k}$ defined by

$$\mathbf{x}_{k:n|k} = \left(\mathbf{x}_{k|k}^\top, \mathbf{x}_{k-1|k}^\top, \ldots, \mathbf{x}_{n|k}^\top\right)^\top, \tag{18.25}$$

while the corresponding joint covariance matrix $\mathbf{P}_{k:n|k}$ can be written as an inverse of a tridiagonal block matrix:

$$\mathbf{P}_{k:n|k} =$$

$$\begin{pmatrix}
\mathbf{T}_{k|k} & -\mathbf{W}_{k-1|k}^\top \mathbf{Q}_{k-1|k}^{-1} & \mathbf{O} & \cdots & \mathbf{O} \\
-\mathbf{Q}_{k-1|k}^{-1}\mathbf{W}_{k-1|k} & \mathbf{T}_{k-1|k} & -\mathbf{W}_{k-2|k}^\top \mathbf{Q}_{k-2|k}^{-1} & \ddots & \vdots \\
\mathbf{O} & -\mathbf{Q}_{k-2|k}^{-1}\mathbf{W}_{k-2|k} & \ddots & \ddots & \mathbf{O} \\
\vdots & \ddots & \ddots & \mathbf{T}_{n+1|k} & -\mathbf{W}_{n|k}^\top \mathbf{Q}_{n|k} \\
\mathbf{O} & \cdots & \mathbf{O} & -\mathbf{Q}_{n|k}\mathbf{W}_{n|k} & \mathbf{T}_{n|k}
\end{pmatrix}^{-1}. \tag{18.26}$$

This can be seen by considering projectors $\Pi_l$ defined by

$$\Pi_l \mathbf{x}_{k:n} = \begin{cases} (\mathbf{1}, \mathbf{O}, \ldots, \mathbf{O})\mathbf{x}_{k:n}, & l = k \\ (\mathbf{O}, \ldots, -\mathbf{W}_{l|l+1}, \mathbf{1}, \ldots, \mathbf{O})\mathbf{x}_{k:n}, & n \leq l < k \end{cases}$$

$$= \begin{cases} \mathbf{x}_k, & l = k \\ \mathbf{x}_l - \mathbf{W}_{l|l+1}\mathbf{x}_{l+1}, & n \leq l < k. \end{cases} \tag{18.27}$$

Using $\Pi_l$ and $\mathbf{Q}_{l|k}$, $l = 1,\ldots,k$, the ASD can be rewritten:

$$p(\mathbf{x}_{k:n|k}|Z^k) = \prod_{l=n}^{k} \mathcal{N}\left(\Pi_l \mathbf{x}_{k:n}; \Pi_l \mathbf{x}_{k:n}^k, \mathbf{Q}_{l|k}\right) \tag{18.28}$$

$$\propto \prod_{l=n}^{k} e^{-\frac{1}{2}(\Pi_l \mathbf{x}_{k:n} - \Pi_l \mathbf{x}_{k:n|k})^\top \mathbf{Q}_{l|k}^{-1}(\Pi_l \mathbf{x}_{k:n} - \Pi_l \mathbf{x}_{k:n|k})} \tag{18.29}$$

$$= e^{-\frac{1}{2}(\mathbf{x}_{k:n} - \mathbf{x}_{k:n|k})^\top \left(\sum_{l=n}^{k} \Pi_l^\top \mathbf{Q}_{l|k}^{-1} \Pi_l\right)(\mathbf{x}_{k:n} - \mathbf{x}_{k:n|k})} \tag{18.30}$$

$$= \mathcal{N}(\mathbf{x}_{k:n}; \mathbf{x}_{k:n|k}, \mathbf{P}_{k:n|k}) \tag{18.31}$$

with a covariance matrix $\mathbf{P}_{k:n|k}$, which is given by a harmonic mean:

$$\mathbf{P}_{k:n|k} = \left( \sum_{l=n}^{k} \mathbf{\Pi}_l^\top \mathbf{Q}_{l|k}^{-1} \mathbf{\Pi}_l \right)^{-1}. \tag{18.32}$$

The summation of the matrices $\mathbf{\Pi}_l^\top \mathbf{Q}_{l|k}^{-1} \mathbf{\Pi}_l$ directly yields the inverse ASD covariance matrix as a tridiagonal block matrix (Equation 18.26). In this matrix, the following auxiliary quantities $\mathbf{T}_{l|k}$, $m \le l \le k$ are defined by

$$\mathbf{T}_{l|k} = \begin{cases} \mathbf{P}_{k|k}^{-1} + \mathbf{W}_{l-1|l}^\top \mathbf{Q}_{l-1|k}^{-1} \mathbf{W}_{l-1|l} & \text{for } l = k \\ \mathbf{Q}_{l|k}^{-1} + \mathbf{W}_{l-1|l}^\top \mathbf{Q}_{l-1|k}^{-1} \mathbf{W}_{l-1|l} & \text{for } n < l < k. \\ \mathbf{Q}_{n|k}^{-1} & \text{for } l = n \end{cases} \tag{18.33}$$

The tridiagonal structure is a consequence of the Markov property of the underlying evolution model. This representation of the inverse of $\mathbf{P}_{k:n|k}$ is useful in practical calculations.

By repeated use of the matrix inversion lemma (see Appendix) and an induction argument, the inverse of this tridiagonal block matrix can be calculated. The resulting block matrix is given by

$$\mathbf{P}_{k:n|k} =$$

$$\begin{pmatrix} \mathbf{P}_{k|k} & \mathbf{P}_{k|k}\mathbf{W}_{k-1|k}^\top & \mathbf{P}_{k|k}\mathbf{W}_{k-2|k}^\top & \cdots & \mathbf{P}_{k|k}\mathbf{W}_{n|k}^\top \\ \mathbf{W}_{k-1|k}\mathbf{P}_{k|k} & \mathbf{P}_{k-1|k} & \mathbf{P}_{k-1|k}\mathbf{W}_{k-2|k-1}^\top & * & \mathbf{P}_{k-1|k}\mathbf{W}_{n|k-1}^\top \\ \mathbf{W}_{k-2|k}\mathbf{P}_{k|k} & \mathbf{W}_{k-2|k-1}\mathbf{P}_{k-1|k} & \mathbf{P}_{k-2|k} & * & \vdots \\ \vdots & * & * & * & \mathbf{P}_{n+1|k}\mathbf{W}_{n|n+1}^\top \\ \mathbf{W}_{n|k}\mathbf{P}_{k|k} & \mathbf{W}_{n|k-1}\mathbf{P}_{k-1|k} & \cdots & \mathbf{W}_{n|n+1}\mathbf{P}_{n+1|k} & \mathbf{P}_{n|k} \end{pmatrix}, \tag{18.34}$$

where the following abbreviations are used:

$$\mathbf{W}_{l|k} = \prod_{\lambda=l}^{k-1} \mathbf{W}_{\lambda|\lambda+1} = \prod_{\lambda=l}^{k-1} \mathbf{P}_{\lambda|\lambda} \mathbf{F}_{\lambda+1|\lambda}^\top \mathbf{P}_{\lambda+1|\lambda}^{-1}. \tag{18.35}$$

The densities $\left\{ \mathcal{N}(\mathbf{x}_l; \mathbf{x}_{l|k}, \mathbf{P}_{l|k}) \right\}_{l=n}^{k}$ are directly obtained via marginalizing, as the covariance matrices $\mathbf{P}_{l|k}$, $n \le l \le k$ appear on the diagonal of this block matrix. Note that the ASD is completely defined by the results of prediction, filtering, and retrodiction obtained for the time window $t_k, \ldots, t_n$, that is, it is a byproduct of Kalman filtering and Rauch–Tung–Striebel smoothing.

## 18.4.2 ASDs for MHT/IMM Filtering

These considerations can be generalized to the case of ambiguity with respect to the origin of the sensor data or with respect to the evolution model, which is currently in effect, that is, to multiple hypothesis tracking (MHT) and interacting multiple model filters (IMMs).

A sensor output at time $t_k$, consisting of $m_k$ measurements collected in the set $Z_k$, can be ambiguous, that is, the origin of the sensor data has to be explained by a set of data interpretations, which are assumed to be exhaustive and mutually exclusive. As an example, let us consider measurements $Z_k = \left\{ \mathbf{z}_k^j \right\}_{j=1}^{m_k}$ possibly related to the kinematic state $\mathbf{x}_k$ of well-separated objects. "Well-separated" here means that measurements potentially originated by one object could not have been originated by another. Even in this simplified situation, ambiguity can arise from imperfect detection and due to false measurements, often referred to as clutter, or of measurements from unwanted objects. Let the detection properties of the sensor be summarized by its detection probability $P_D$ and the clutter background by the spatial false return density $\rho_F$.

Let $j_k = 0$ denote the data interpretation hypothesis that the object has not been detected at all by the sensor at time $t_k$, that is, all sensor data have to be considered as false measurements, while $1 \leq j_k \leq m_k$ represents the hypothesis that the object has been detected, $\mathbf{z}_k^{j_k} \in Z_k$ being the corresponding measurement of the object properties, and the remaining sensor data being false. Evidently, $\{0, \ldots, m_k\}$ denotes a set of mutually exclusive and exhaustive data interpretations. Standard reasoning yields for this simple example a likelihood function for ambiguous data given by a weighted sum of Gaussians and a constant (see, e.g., Ref. [18]):

$$p(Z_k, m_k | \mathbf{x}_k) = \sum_{j_k=0}^{m_k} p(Z_k, m_k | j_k, \mathbf{x}_k) p(j_k) \tag{18.36}$$

$$\propto (1 - P_D)\rho_F + P_D \sum_{j_k=0}^{m_k} \mathcal{N}\left(\mathbf{z}_{j_k}; \mathbf{H}_k \mathbf{x}_k, \mathbf{R}_k\right). \tag{18.37}$$

Data interpretation hypotheses are the basis for MHT techniques (see, e.g., Ref. [19]). In such situations, the origin of a time series $Z^k = \{Z_k, m_k, Z^{k-1}\}$ of sensor data accumulated up to the time $t_k$ can be interpreted by interpretation histories,

$$\mathbf{j}_k = (j_k, \ldots, j_1), \quad \text{where } 0 \leq j_l \leq m_l, \tag{18.38}$$

that assume at each data collection time $t_l$, $1 \leq l \leq k$, a certain data interpretation $j_l$ to be true. Via marginalizing, the previous filtering density $p(\mathbf{x}_{k-1} | Z^{k-1})$ can be written as a mixture over such interpretation histories $\mathbf{j}_{k-1}$:

$$p(\mathbf{x}_{k-1} | Z^{k-1}) = \sum_{\mathbf{j}_{k-1}} p(\mathbf{j}_{k-1} | Z^{k-1}) \mathcal{N}\left(\mathbf{x}_{k-1}; \mathbf{x}_{k-1|k-1}^{\mathbf{j}_{k-1}}, \mathbf{P}_{k-1|k-1}^{\mathbf{j}_{k-1}}\right). \tag{18.39}$$

By making use of likelihood functions for uncertain data, we directly obtain

$$p(\mathbf{x}_k | Z^k) = \sum_{\mathbf{j}_k} p_{k|k}^{\mathbf{j}_k} \mathcal{N}\left(\mathbf{x}_k; \mathbf{x}_{k|k}^{\mathbf{j}_k}, \mathbf{P}_{k|k}^{\mathbf{j}_k}\right). \tag{18.40}$$

Mixture reduction techniques, such as described in Ref. [20], keep the number of mixture components involved manageable.

According to these preliminary considerations and using the data interpretation histories $\mathbf{j}_k$, the accumulated state density is given by

$$p(\mathbf{x}_{k:n} | Z^k) = \sum_{\mathbf{j}_k} p(\mathbf{j}_k | Z^k) \, p(\mathbf{x}_{k:n} | \mathbf{j}_k, Z^k) \tag{18.41}$$

$$= \sum_{\mathbf{j}_k} p_{k|k}^{\mathbf{j}_k} \mathcal{N}\left(\mathbf{x}_{k:n}; \mathbf{x}_{k:n|k}^{\mathbf{j}_k}, \mathbf{P}_{k:n|k}^{\mathbf{j}_k}\right), \tag{18.42}$$

That is, the ASD for MHT applications is simply a weighted sum of individual ASDs, which are completely defined by the results of prediction, filtering, and retrodiction along a certain branch of the hypothesis tree defined by a particular interpretation history $\mathbf{j}_k$. The corresponding weighting factor is given by the probability of $\mathbf{j}_k$ being true at time $t_k$ given the data: $p(\mathbf{j}_k|Z^k) = p_{k|k}^{\mathbf{j}_k}$. The ASD for MHT application is thus a byproduct of MHT tracking and retrodiction. Because PDA filtering [3] can be considered as a second-order approximation to MHT, an approximate representation of the corresponding ASD by a single Gaussian can be derived.

In practical applications, it may be uncertain which evolution model out of a set of $r$ possible alternatives is currently in effect (different flight phases such as, e.g., no turn, slight maneuver, high-$g$ turn). The maneuvering class $1 \le i_k \le r$ an object belongs to at time $t_k$ can thus be considered as a part of its state. Markovian IMM evolution models (see Ref. [21] and the literature cited therein) for object states $\mathbf{x}_k = (\mathbf{x}_k, i_k)$ have the form

$$p(x_k, i_k | x_{k-1}, i_{k-1}) = p_{i_k i_{k-1}} \mathcal{N}\left(\mathbf{x}_k; \mathbf{F}_{k|k-1}^{i_k} \mathbf{x}_{k-1}, \mathbf{D}_{k|k-1}^{i_k}\right). \tag{18.43}$$

IMM models are thus characterized by $r$ kinematic linear Gaussian transition densities $p(\mathbf{x}_k|\mathbf{x}_{k-1}, i_k)$ and class transition probabilities $p_{i_k i_{k-1}} = p(i_k|i_{k-1})$ that are to be specified and part of the modeling assumptions.

By making use of the total probability theorem, the IMM approach can easily be combined with Kalman or MHT filtering. In the probability density

$$p(\mathbf{x}_k|Z^k) = \sum_{\mathbf{j}_k} p(\mathbf{x}_k, \mathbf{j}_k|Z^k) \tag{18.44}$$

at each step $k$ of the filtering loop (Equation 18.2), the individual terms of the sum become mixture densities themselves,

$$p(\mathbf{x}_k, \mathbf{j}_k|Z^k) = \sum_{i_k, \dots i_1} p(i_k, \dots i_1, \mathbf{j}_k|Z^k) p(\mathbf{x}_k|i_k, \dots i_1, \mathbf{j}_k, Z^k). \tag{18.45}$$

Hence, in the optimal approach to IMM filtering the conditional densities $p(\mathbf{x}_k, \mathbf{j}_k|Z^k)$ of the kinematic state $\mathbf{x}_k$ of the object are sums over every possible sequence of dynamics models $i_k, \dots i_1$ from the initial observation through the most recent measurement at scan $k$ ("dynamics histories"). As the number of terms in the sum (Equation 18.45) exponentially increases with increasing $k$, various techniques have been developed that approximately represent the densities (Equation 18.45) by mixtures with a *constant* or fluctuating but small number of components.

Let us denote the dynamics histories "$m$ scans back" by $\mathbf{i}_k$, an $m$-tuple of indices,

$$\mathbf{i}_k = (i_k, i_{k-1}, \dots i_{k-m+1}). \tag{18.46}$$

In particular, we are looking for approximations by Gaussian mixtures,

$$p(x_k, \mathbf{j}_k|Z^k) \approx \sum_{\mathbf{i}_k} p_{k|k}^{\mathbf{i}_k, \mathbf{j}_k} \mathcal{N}\left(\mathbf{x}_k; \mathbf{x}_{k|k}^{\mathbf{i}_k, \mathbf{j}_k}, \mathbf{P}_{k|k}^{\mathbf{i}_k, \mathbf{j}_k}\right). \tag{18.47}$$

The weighting factors of the data association history $\mathbf{j}_k$, $p_{k|k}^{\mathbf{j}_k} = p(\mathbf{j}_k|Z^k)$ are given by $p_{k|k}^{\mathbf{j}_k} = \sum_{\mathbf{i}_k} p_{k|k}^{\mathbf{i}_k,\mathbf{j}_k}$. Due to Bayes' theorem, the expectation vectors $\mathbf{x}_{k|k}^{\mathbf{i}_k,\mathbf{j}_k}$ and covariance matrices $\mathbf{P}_{k|k}^{\mathbf{i}_k,\mathbf{j}_k}$ of the mixtures are iteratively obtained by formulae that are essentially based on Kalman filtering. Also the weighting factors $p_{k|k}^{\mathbf{i}_k,\mathbf{j}_k}$ obey simple update formulae. For $m = 1$ and assuming well-separated objects, the density $p(\mathbf{x}_k, \mathbf{j}_k|Z^k)$ is approximated by a mixture with $r$ components according to the $r$ dynamics models used. GPB and IMM algorithms are possible realizations of this scheme [21]. The IMM approach may easily be adopted to fixed-interval retrodiction. In direct analogy to Equation 18.47, the densities $p(\mathbf{x}_l|Z^k) = \sum_{\mathbf{j}_k} p(\mathbf{x}_l, \mathbf{j}_k|Z^k)$ in the retrodiction loop (Equation 18.3) are approximately represented by the same class of functions previously used in the filtering loop:

$$p(\mathbf{x}_l, \mathbf{j}_k|Z^k) \approx \sum_{\mathbf{i}_l} p_{l|k}^{\mathbf{i}_l,\mathbf{j}_k} \mathcal{N}\left(\mathbf{x}_l; \mathbf{x}_{l|k}^{\mathbf{i}_l,\mathbf{j}_k}, \mathbf{P}_{l|k}^{\mathbf{i}_l,\mathbf{j}_k}\right) \tag{18.48}$$

with $l < k$. The backward iteration is initialized by the filtering result at the present scan $k$: $p_{k|k}^{\mathbf{i}_k,\mathbf{j}_k}$, $\mathbf{x}_{k|k}^{\mathbf{i}_k,\mathbf{j}_k}$, $\mathbf{P}_{k|k}^{\mathbf{i}_k,\mathbf{j}_k}$. According to Ref. [22], approximate update formulae for the parameters defining $p(\mathbf{x}_l, \mathbf{j}_k|Z^k)$ can be derived. Therefore, and for the same reasons as before, the accumulated state density is as mixture of individual ASDs for each data interpretation and model history:

$$p(\mathbf{x}_{k:n}|Z^k) = \sum_{\mathbf{i}_k \mathbf{j}_k} p_{k|k}^{\mathbf{i}_k \mathbf{j}_k} \mathcal{N}\left(\mathbf{x}_{k:n}; \mathbf{x}_{k:n|k}^{\mathbf{i}_k \mathbf{j}_k}, \mathbf{P}_{k:n|k}^{\mathbf{i}_k \mathbf{j}_k}\right). \tag{18.49}$$

Each ASD component is defined by the results of prediction, filtering, and retrodiction given these histories. The corresponding weighting factor results from the filtering step. In the case of standard IMM filtering with $r$ evolution models, the ASD is approximately given by marginalizing the joint ASD:

$$p(\mathbf{x}_{k:n}, i_k, \ldots, i_n|Z^k) \approx \prod_{l=n}^{k} p_{l|k}^{i_l} \mathcal{N}\left(\mathbf{x}_{k:n}; \mathbf{x}_{k:n|k}^{i_k,\ldots,i_n}, \mathbf{P}_{k:n|k}^{i_k,\ldots,i_n}\right), \tag{18.50}$$

where $\mathbf{x}_{k:n|k}^{i_k,\ldots,i_n}$ is given by the expectation vectors resulting from standard IMM filtering and retrodiction,

$$\mathbf{x}_{k:n|k}^{i_k,\ldots,i_n} = \left(\mathbf{x}_{k|k}^{i_k,\top}, \ldots, \mathbf{x}_{n|k}^{i_n,\top}\right)^{\top}, \tag{18.51}$$

while $\mathbf{P}_{k:n|k}^{i_k,\ldots,i_n}$ essentially has the same structure as the Kalman ASD excepting the covariances of IMM filtering and retrodiction are used. The weighing factors $p_{l|k}^{i_l}$ are the same as in standard IMM filtering and retrodiction.

## 18.5 ASDS AND OUT-OF-SEQUENCE MEASUREMENTS

In any real-world application of sensor data fusion, we have to be aware of out-of-sequence measurements. Owing to latencies in the underlying communication infrastructure, for example, such measurements arrive at a processing node in a distributed data fusion system "too late," that is, after sensor data with a time stamp newer than the time stamp of an out-of-sequence measurement have already been processed. Accumulated object state densities are useful means for dealing with this type of sensor data, which may provide valuable information on an object state of interest in spite of their latency, especially if the sensor involved is of a high quality.

Under conditions, where Kalman filtering is applicable, let us consider a measurement $\mathbf{z}_m$ produced at time $t_m$ with $n \leq m$, that is, possibly before the "present" time $t_k$, where the time series $Z^k$ is available and has been exploited. We wish to understand the impact this new, but late sensor information has on the present and the past object states $\mathbf{x}_l$, $n \leq l$, that is, on the accumulated object state $\mathbf{x}_{k:n}$. Let $\mathbf{z}_m$ be a measurement of the object state $\mathbf{x}_m$ at time $t_m$ characterized by a Gaussian likelihood function, which is defined by a measurement matrix $\mathbf{H}_m$ and a corresponding measurement error covariance matrix $\mathbf{R}_m$. We further renumber the object states $\mathbf{x}_k, \ldots, \mathbf{x}_n$ such that $\mathbf{x}_k, \ldots, \mathbf{x}_m, \ldots, \mathbf{x}_n =: \mathbf{x}_{k:m:n}$ are consistent with their time stamps $(t_l)_{l=k,\ldots,m,\ldots,n}$.

To process the measurement $\mathbf{z}_m$, we need the extended ASD $p(\mathbf{x}_{k:m:n})$. This new joint probability density function can be calculated as above by successively using the Bayes theorem and the product formula (Equation 18.150). To this end, let us first assume $m < k$. We obtain

$$
\begin{aligned}
p(\mathbf{x}_{k:m:n}|Z^k) \quad & \propto p(\mathbf{z}_k|\mathbf{x}_k)p(\mathbf{x}_k|\mathbf{x}_{k-1})\ldots p(\mathbf{x}_{m+1}|\mathbf{x}_m)p(\mathbf{x}_m|\mathbf{x}_{m-1})p(\mathbf{z}_{m-1}|\mathbf{x}_{m-1})\cdots p(\mathbf{x}_n|Z^n) \\
& = \prod_{l \neq m} \mathcal{N}(\mathbf{z}_l; \mathbf{H}_l\mathbf{x}_l, \mathbf{R}_l)\mathcal{N}(\mathbf{x}_l; \mathbf{F}_{l|l-1}\mathbf{x}_{l-1}, \mathbf{D}_{l|l-1}) \\
& \quad \times \mathcal{N}(\mathbf{x}_m; \mathbf{F}_{m|m-1}\mathbf{x}_{m-1}, \mathbf{D}_{m|m-1})\mathcal{N}(\mathbf{x}_n; \mathbf{x}_{n|n}, \mathbf{P}_{n|n}).
\end{aligned}
$$

Applying the product formula to all terms from $n$ up to $m-1$, we can proceed as in Equation 18.9. From there on, we observe

$$
\begin{aligned}
& \mathcal{N}(\mathbf{x}_{m-1}; \mathbf{x}_{m-1|m-1}, \mathbf{P}_{m-1|m-1})\mathcal{N}(\mathbf{x}_m; \mathbf{F}_{m|m-1}\mathbf{x}_{m-1}, \mathbf{D}_{m|m-1})\mathcal{N}(\mathbf{x}_{m+1}; \mathbf{F}_{m+1|m}\mathbf{x}_m, \mathbf{D}_{m+1|m}) \\
& = \mathcal{N}(\mathbf{x}_{m-1}; \mathbf{h}_{m-1|m}(\mathbf{x}_m), \mathbf{R}_{m-1|m})\mathcal{N}(\mathbf{x}_m; \mathbf{x}_{m|m-1}, \mathbf{P}_{m|m-1})\mathcal{N}(\mathbf{x}_{m+1}; \mathbf{F}_{m+1|m}\mathbf{x}_m, \mathbf{D}_{m+1|m}) \\
& = \mathcal{N}(\mathbf{x}_{m-1}; \mathbf{h}_{m-1|m}(\mathbf{x}_m), \mathbf{R}_{m-1|m})\mathcal{N}(\mathbf{x}_m; \mathbf{h}_{m|m+1}(\mathbf{x}_{m+1}), \mathbf{R}_{m|m+1}) \\
& \quad \times \mathcal{N}(\mathbf{x}_{m+1}; \mathbf{x}_{m+1|m-1}, \mathbf{P}_{m+1|m-1}).
\end{aligned}
$$

This is due to the fact that $\mathbf{F}$ and $\mathbf{D}$ describe a linear flow:

$$
\mathbf{F}_{m+1|m-1} = \mathbf{F}_{m+1|m}\mathbf{F}_{m|m-1} \tag{18.52}
$$

$$
\mathbf{D}_{m+1|m-1} = \mathbf{F}_{m+1|m}\mathbf{D}_{m|m-1}\mathbf{F}_{m+1|m} + \mathbf{D}_{m+1|m}. \tag{18.53}
$$

Particularly, the function $\mathbf{h}_{m|m+1}$ and the covariance matrix $\mathbf{R}_{m|m+1}$ are given by

$$
\mathbf{h}_{m|m+1}(\mathbf{x}_{m+1}) = \mathbf{x}_{m|m-1} + \mathbf{W}_{m|m+1}(\mathbf{x}_{m+1} - \mathbf{x}_{m+1|m-1}) \tag{18.54}
$$

$$
\mathbf{W}_{m|m+1} = \mathbf{P}_{m|m-1}\mathbf{F}_{m+1|m}\mathbf{P}_{m+1|m-1}^{-1} \tag{18.55}
$$

$$
\mathbf{R}_{m|m+1} = \mathbf{P}_{m|m-1} - \mathbf{W}_{m|m+1}\mathbf{P}_{m+1|m-1}\mathbf{W}_{m|m+1}^{\top}. \tag{18.56}
$$

This type of reasoning is also known as continuous time retrodiction; see, for example, Ref. [23] for a more detailed discussion. The resulting terms fit well into the product and we obtain by a continued use of the product formula:

$$
p(\mathbf{x}_{k:m:n}|Z^k) = \mathcal{N}(\mathbf{x}_k; \mathbf{x}_{k|k}, \mathbf{P}_{k|k})\prod_{l=n}^{k-1} \mathcal{N}(\mathbf{x}_l; \mathbf{h}_{l|l+1}(\mathbf{x}_{l+1}), \mathbf{R}_{l|l+1}).
$$

For the case of $m = k + 1$, that is, $t_m > t_l$ for all times $t_l$ of $Z^k$, we directly obtain a prediction factor as it is known from the standard Kalman equations. The predicted joint probability density function then is:

$$p(\mathbf{x}_{k:m:n}|Z^k) = \mathcal{N}(\mathbf{x}_m; \mathbf{x}_{m|m-1}, \mathbf{P}_{m|m-1}) \prod_{l=n}^{k} \mathcal{N}(\mathbf{x}_l; \mathbf{h}_{l|l+1}(\mathbf{x}_{l+1}), \mathbf{R}_{l|l+1}). \tag{18.57}$$

In both cases, the same reasoning as above now leads to the extended ASD as a single Gaussian density function:

$$p(\mathbf{x}_{k:m:n}|Z^k) = \mathcal{N}(\mathbf{x}_{k:m:n}; \mathbf{x}_{k:m:n|k}, \mathbf{P}_{k:m:n|k}). \tag{18.58}$$

### 18.5.1 Modified Kalman Update Step

We furthermore introduce a projection matrix $\Pi_m$, defined by $\Pi_m \mathbf{x}_{k:m:n} = \mathbf{x}_m$, which extracts the object state $\mathbf{x}_m$ from the accumulated state vector $\mathbf{x}_{k:m:n}$. The likelihood function of the out-of-sequence measurement with respect to the accumulated object state is thus given by

$$p(\mathbf{z}_m|\mathbf{x}_{k:m:n}) = \mathcal{N}(\mathbf{z}_m; \mathbf{H}_m \Pi_m \mathbf{x}_{k:n}, \mathbf{R}_m). \tag{18.59}$$

Standard Bayesian reasoning and the product formula for Gaussians (Equation 18.150) directly yields the accumulated state density:

$$p(\mathbf{x}_{k:m:n}|\mathbf{z}_m, Z^k) = \frac{p(\mathbf{z}_m|\mathbf{x}_{k:m:n})p(\mathbf{x}_{k:m:n}|Z^k)}{\int d\mathbf{x}_{k:m:n} p(\mathbf{z}_m|\mathbf{x}_{k:m:n})p(\mathbf{x}_{k:m:n}|Z^k)} \tag{18.60}$$

$$= \mathcal{N}(\mathbf{x}_{k:m:n}; \mathbf{x}_{k:m:n|k,m}, \mathbf{P}_{k:m:n|k,m}) \tag{18.61}$$

with parameters obtained by a version of the Kalman update equations:

$$\mathbf{x}_{k:m:n|k,m} = \mathbf{x}_{k:m:n|k} + \mathbf{W}_{k:m:n}(\mathbf{z}_m - \mathbf{H}_m \Pi_m \mathbf{x}_{k:m:n|k}) \tag{18.62}$$

$$\mathbf{P}_{k:m:n|k,m} = \mathbf{P}_{k:m:n|k} - \mathbf{W}_{k:m:n} \mathbf{S}_{k:m:n} \mathbf{W}_{k:m:n}^\top, \tag{18.63}$$

where the corresponding Kalman gain and innovation matrices are given by

$$\mathbf{S}_{k:m:n} = \mathbf{H}_m \Pi_m \mathbf{P}_{k:m:n|k} \Pi_m^\top \mathbf{H}_m^\top + \mathbf{R}_m \tag{18.64}$$

$$\mathbf{W}_{k:m:n} = \mathbf{P}_{k:m:n|k} \Pi_m^\top \mathbf{H}_m^\top \mathbf{S}_{k:m:n}^{-1}. \tag{18.65}$$

Note that the matrix $\mathbf{S}_{k:m:n}$ to be inverted when calculating the Kalman gain matrix has the same dimension as the measurement vector $\mathbf{z}_m$, that is, is low-dimensional matrix, just as in standard Kalman filtering. Nevertheless, the processing of an out-of-sequence measurement $\mathbf{z}_m$ has impact on all state estimates and the related error covariance matrices in the time window considered. The strongest impact is observed for the time $t_m$, where the measurement has actually been produced, while it declines the further we proceed to the present time $t_k \geq t_l > t_m$ or deeper into the past $t_m > t_l \geq t_n$.

Accumulated state densities are therefore well suited to quantitatively discuss the question to what extent an OoS measurement is still useful or not, a phenomenon that is sometimes called "information aging." If we are interested in the updated state estimates for the time $t_l$, $n \leq l \leq k$, we simply have to consider the density:

$$p(\mathbf{x}_l | \mathbf{z}_m, Z^k) = \mathcal{N}\left(\mathbf{x}_l; \mathbf{\Pi}_l \mathbf{x}_{k:m:n|k,m}, \mathbf{\Pi}_l \mathbf{P}_{k:m:n|k,m} \mathbf{\Pi}_l^\top\right) \tag{18.66}$$

which results from applying the projection matrix $\mathbf{\Pi}_l$ previously introduced. In a practical application, we will usually be interested in the effect of out-of-sequence measurements, which were produced not too long ago, on the present time and the most recent past. It is therefore sufficient to consider accumulated state densities $p(\mathbf{x}_{k:n} | Z^k)$ characterized by lower dimensional parameters $\mathbf{x}_{l|k}$, $\mathbf{P}_{l|k}$. To determine the actual size of $n$ to be taken into account is an important task in designing sensor networks.

By using the accumulated state densities for MHT and IMM filtering, these considerations can directly be generalized to the case of ambiguous sensor data. Their impact on practical implementations for the general case is given in the subsequent section.

### 18.5.2  ON COMPUTATIONAL COSTS

In cases where the state vector $\mathbf{x}_l$ has a large dimension and a long time window $k : n$ is chosen, the accumulated expectation vector and the corresponding block covariance matrix gets large as well. Nevertheless, the block covariance matrix has a particular structure that can explicitly be exploited in the calculations involved. In this section, we consider the difference in costs of the presented accumulated formula to the standard Kalman filtering and Rauch–Tung–Striebel recursion. The latter is given by

$$\mathbf{x}_{k-1|k} = \mathbf{x}_{k-1|k-1} + \mathbf{W}_{k-1|k}(\mathbf{x}_{k|k} - \mathbf{x}_{k|k-1}) \tag{18.67}$$

$$\mathbf{x}_{k-2|k-1} = \mathbf{x}_{k-2|k-2} + \mathbf{W}_{k-2|k-1}(\mathbf{x}_{k-1|k-1} - \mathbf{x}_{k-1|k-2}) \tag{18.68}$$

$$\mathbf{x}_{k-2|k} = \mathbf{x}_{k-2|k-2} + \mathbf{W}_{k-2|k-1}(\mathbf{x}_{k-1|k} - \mathbf{x}_{k-1|k-2}) \tag{18.69}$$

Combining these equations and an induction argument directly yields the retrodicted state $\mathbf{x}_{l|k}$ for $n \leq l \leq k - 1$ as derived in the accumulated expectation value (see Equation 18.174 in the Appendix):

$$\mathbf{x}_{l|k} = \mathbf{x}_{l|k-1} + \mathbf{W}_{l|k}(\mathbf{x}_{k|k} - \mathbf{x}_{k|k-1}). \tag{18.70}$$

The same reasoning enables us to derive the retrodicted covariance matrix $\mathbf{P}_{l|k}$:

$$\mathbf{P}_{l|k} = \mathbf{P}_{l|k-1} + \mathbf{W}_{l|k}(\mathbf{P}_{k|k} - \mathbf{P}_{k|k-1})\mathbf{W}_{l|k}^\top, \tag{18.71}$$

which is the block diagonal entry version of Equation 18.171 in the Appendix. This not only shows the formal equivalence of the accumulated knowledge to a $k - n$ step retrodiction, but also shows that the same matrices have to be calculated. Even further, when processing a new measurement $\mathbf{z}_m$, the preceding section enables us to obtain a closed formula for filtering and retrodiction combined.

To this end, we state the Gaussian density $p(\mathbf{x}_l|Z^k, \mathbf{z}_m) = \mathcal{N}(\mathbf{x}_l; \mathbf{x}_{l|k,m}, \mathbf{P}_{l|k,m})$ by calculating the $l$th block element of Equations 18.62 and 18.63:

$$\mathbf{x}_{l|k,m} = \mathbf{x}_{l|k} + \mathbf{W}_{l|k,m}(\mathbf{z}_m - \mathbf{x}_{m|k}) \tag{18.72}$$

$$\mathbf{P}_{l|k,m} = \mathbf{P}_{l|k} - \mathbf{W}_{l|k,m}\mathbf{S}_{m|m-1}\mathbf{W}_{l|k,m}^\top, \tag{18.73}$$

where the accumulated gain matrix $\mathbf{W}_{l|k,m}$ is given by

$$\mathbf{W}_{l|k,m} = \mathbf{W}_{l|m}\mathbf{P}_{max\{l,m\}|k}\mathbf{W}_{m|l}^\top\mathbf{H}_m^\top\mathbf{S}_{m|m-1}^{-1}. \tag{18.74}$$

Using this notation, we have to define $\mathbf{W}_{r|s} = \mathbf{1}$ for all $r \geq s$ to keep the formalism complete. Note that due to the symmetry of the accumulated covariance $\mathbf{P}_{k:n|k}$, it is sufficient to calculate the $m$th column of it. This gives us an algorithm of filtering and retrodiction in combination done with minimal effort.

### 18.5.3 Discussion of a Simulated Example

Figure 18.1 shows a simulated trajectory of a maneuvering air object defined by $x(t) = \dfrac{v^2}{q}\sin\left(\dfrac{q}{2v}t\right)$, $y(t) = 2x(t)$, $v = 300$ m/s, $q = 4g$. It is observed by two typical mid-range radars located at $(-50, 0)$km and $(0, 50)$km (scan period: $T = 5$ s, measurement error standard deviations: $\sigma_r = 20$ m (range), $\sigma_\varphi = 0.2°$ (azimuth). Let us consider measurement fusion, where the fusion center is receiving measurements from sensor one without communication delay, while the measurements of sensor two are arriving "out-of-sequence," that is with a temporal delay of zero, one, two, and five scan periods.

Figure 18.2 shows numerical results based on 1000 Monte Carlo runs (mean error of the expectation vector in position of the filtering steps and the mean trace of corresponding filtering error

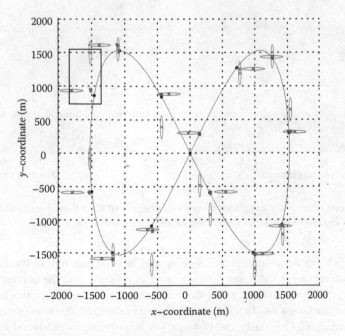

**FIGURE 18.1** Simulated trajectory of a highly maneuvering object.

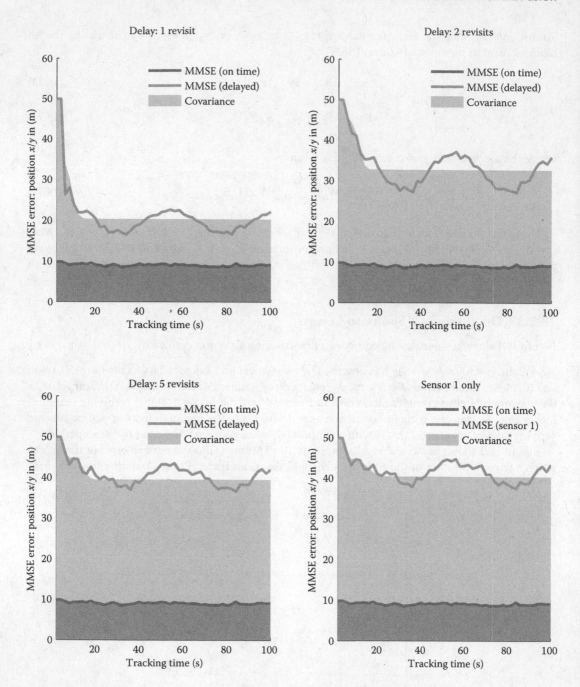

**FIGURE 18.2** Mean error of the expectation vector in position of the filtering steps and the mean trace of corresponding filtering error covariance matrix for various communication delays.

covariances matrix). For no communication delay and in each sub-figure, the dark lines and the dark-shaded region represent the mean filtering error in position and the corresponding variance, respectively. The gray lines and the regions shaded in gray show these quantities for different delays as a function of the tracking time. In the right sub-figure in the second row no measurements of sensor 2 are processed at all. Obviously, out-of-sequence measurements produced by sensor 2, which arrive at the fusion center with a delay of five scan periods or more, are nearly useless and do not significantly improve the filtering result at the present time. There may be a significant improvement for retrodicted estimates at the time at which this OoS measurement has actually been produced.

## 18.6  ASDS AND DATA AUGMENTATION METHODS

More generally speaking and in full accordance to the Bayesian approach previously discussed, data augmentation methods, such as expectation maximization (EM), intend to make statements about a quantity $X$ given that measurements $Z$ of $X$ are available. Because of missing information, for example, lacking knowledge of the correct measurement-to-target associations, the calculation of the related conditional probability density $p(X|Z)$ may be difficult. If additional information $A$ were known, however, for example, the data associations, the augmented density $p(X|A, Z)$, could more easily be calculated.

According to this strategy, data augmentation algorithms use the augmented conditional probability density $p(X|A, Z)$ to calculate at least certain characteristics of the original probability density $p(X|Z)$. This concept is guided by the general observation that $p(X|Z)$ can be expressed by probability densities involving the additional information $A$ as a direct consequence of $p(A|X, Z) = p(X|A, Z)$ $p(A|Z)/p(X|Z)$, that is, of Bayes' rule:

$$p(X|Z) = \frac{p(X|A, Z)p(A|Z)}{p(A|X, Z)}. \tag{18.75}$$

The EM algorithm is a particular realization of this more general concept. Let us assume that $X^i$ is a certain preliminary estimate of $X$. By exploiting the augmented density $p(X|A, Z)$, we wish to calculate a "better" estimate $X^{i+1}$ in the sense that

$$p(X^{i+1}|Z) > p(X^i|Z). \tag{18.76}$$

In other words, we are looking for an iterative algorithm to localize the "posterior mode" of the conditional density $p(X|Z)$. Assuming that $p(A|X_i, Z)$ is available as well, we are looking for an $X$ for which the following is true:

$$\log p(X|Z) > \log p(X^i|Z) \Leftrightarrow \tag{18.77}$$

$$\int dA \log p(X|Z)p(A|X^i, Z) > \int dA \log p(X^i|Z)p(A|X^i, Z). \tag{18.78}$$

Using as an abbreviation a function $Q$ defined by

$$Q(X; X^i) = \int dA \log p(X|A, Z)p(A|X^i, Z), \tag{18.79}$$

we obtain from Equation 18.78 by exploiting Equation 18.75 and Jensen's inequality:

$$Q(X; X_i) - Q(X^i; X^i) > \int dA \log \frac{p(A|X, Z)}{p(A|X^i, Z)} p(A|X^i, Z) \tag{18.80}$$

$$\geq \log \int dA \frac{p(A|X, Z)}{p(A|X^i, Z)} p(A|X^i, Z) \tag{18.81}$$

$$= 0. \tag{18.82}$$

According to these considerations, an EM algorithm essentially consists of two consecutive steps:

1. *Expectation*: With $X^i$ denoting the current estimate according to $p(X|Z)$ (initialization required!), compute the function $Q(X; X^i)$ defined previously.
2. *Maximization*: Find the next update by maximizing $Q(X; X^i)$,

$$X^{i+1} = \text{argmax}_X Q(X; X^i), \tag{18.83}$$

and repeat the EM step until the condition $|Q(X^{i+1}; X^i) - Q(X^i; X^i)| < \varepsilon$ holds. A theorem guarantees convergence [12].

In multiple target tracking, auxiliary information on which measurement has been originated by which target is particularly useful. Let

$$a_l^{j \to s_j}, \quad l = k,\dots,n, \quad j = 1,\dots,m_l, \quad s_j = 1,\dots,S, \tag{18.84}$$

denote the hypothesis that at time $t_l$ the measurement $\mathbf{z}_l^j$ is to be associated with a target indexed by $s_j$. Let a set of all feasible associations that map the measurements in the time window $t_{k:n}$ to each target be denoted by $a_{k:n}$. According to the more general discussion, the function $Q$ is in this case obtained by calculating an expectation over data association probabilities. By using the notion of accumulated state vectors as in the previous section, $Q$ is given by

$$Q\left(\mathbf{x}_{k:n}; \mathbf{x}_{k:n}^i\right) = \sum_{a_{k:n}} \log\left(p(\mathbf{x}_{k:n}|a_{k:n}, Z^k)\right) p\left(a_{k:n}|\mathbf{x}_{k:n}^i, Z^k\right). \tag{18.85}$$

As will be shown in Section 18.6.2, under linear Gaussian assumptions regarding the sensor likelihood functions and the transition densities describing the targets' evolution, these functions prove to be given by Gaussian ASDs as discussed in the previous section (up to a constant irrelevant to maximization). The maximization step in the EM loop is thus trivial and given by the corresponding accumulated vector of expectation vectors obtained by Kalman filtering and retrodiction while processing suitably chosen "synthetic" data.

### 18.6.1 Expectation and Maximization Steps

By applying Bayes' rule to the argument of the logarithm in the previous function $Q$,

$$p(\mathbf{x}_{k:n}|a_{k:n}, Z^k) \propto p(Z_{k:n}|a_{k:n}, \mathbf{x}_{k:n}) p(\mathbf{x}_{k:n}|Z^{n-1}), \tag{18.86}$$

we obtain for $Q$ an expression, which is given by (up to a constant independent of the state variables):

$$Q\left(\mathbf{x}_{k:n}; \mathbf{x}_{k:n}^i\right) = \log p(\mathbf{x}_{k:n}|Z^{n-1})$$

$$+ \sum_{a_{k:n}} \log\left(p(Z_{k:n}|a_{k:n}, \mathbf{x}_{k:n})\right) p\left(a_{k:n}|\mathbf{x}_{k:n}^i, Z^k\right) + \text{const.} \tag{18.87}$$

$Q$ can be expressed more explicitly by using the known transition densities, sensor likelihood functions, prior information on the object states at time $t_{n-1}$, and the posterior association probabilities yielding

$$Q\left(\mathbf{x}_{k:n}; \mathbf{x}_{k:n}^i\right) = \sum_{s=1}^{S} \log p\left(\mathbf{x}_{n-1}^s \Big| Z^{n-1}\right) + \sum_{l=n}^{k} \sum_{s=1}^{S} \log p\left(\mathbf{x}_l^s \Big| \mathbf{x}_{l-1}^s\right)$$

$$+ \sum_{l=n}^{k} \sum_{a_l} \log\left(p(Z_l | a_l, \mathbf{x}_l)\right) p\left(a_l \Big| \mathbf{x}_l^i, Z_l\right) + \text{const.},$$

(18.88)

where we assumed $p\left(a_l \big| \mathbf{x}_{k:n}^i, Z_k\right) = p\left(a_l \big| \mathbf{x}_l^i, Z_l\right)$. To proceed, let us first calculate the posterior association probabilities with the additional assumption

$$p\left(a_l \Big| \mathbf{x}_l^i, Z_l\right) = p\left(a_l^{1 \to s_1}, \ldots, a_l^{m_l \to s_{m_l}} \Big| \mathbf{x}_l^i, Z_l\right)$$

(18.89)

$$= \prod_{j=1}^{m_l} p\left(a_l^{j \to s_j} \Big| \mathbf{x}_l^i, Z_l\right).$$

(18.90)

According to Bayes' rule we obtain

$$p\left(a_l \Big| \mathbf{x}_l^i, Z_l, m_l\right) = \frac{p\left(Z_l \big| a_l, m_l, \mathbf{x}_l^i\right) \pi_l^{j \to s_j}}{\sum_{a_l} p\left(Z_l \big| a_l, m_l, \mathbf{x}_l^i\right) \pi_l^{j \to s_j}}$$

(18.91)

with $\pi_l^{j \to s_j}$ denoting the prior probability for associating at time $t_l$ the measurement $\mathbf{z}_l^j$ to the object indexed by $s_j$:

$$\pi_l^{j \to s_j} = p\left(a_l^{j \to s_j} \Big| \mathbf{x}_l^i, m_l\right).$$

(18.92)

These quantities will be discussed in Section 18.6.2 in greater detail. The nominator can be rewritten by using an induction argument:

$$\sum_{a_l} p\left(Z_l \big| a_l, m_l, \mathbf{x}_l^i\right) p\left(a_l \big| \mathbf{x}_l^i, m_l\right)$$

(18.93)

$$= \sum_{s_1=1}^{S} \cdots \sum_{s_{m_l}=1}^{S} \prod_{j=1}^{m_l} p\left(\mathbf{z}_l^j \Big| \mathbf{x}_l^{i;s_j}\right) \pi_l^{j \to s_j}$$

(18.94)

$$= \sum_{s_1=1}^{S} \cdots \sum_{s_{m_l-1}=1}^{S} \prod_{j=1}^{m_l-1} p\left(\mathbf{z}_l^j \middle| \mathbf{x}_l^{i;s_j}\right) \pi_l^{j \to s_j}$$

$$\times \left\{ \sum_{s=1}^{S} p\left(\mathbf{z}_l^{m_l} \middle| \mathbf{x}_l^{i;s}\right) \pi_l^{j \to s_j} \right\} \tag{18.95}$$

$$= \prod_{j=1}^{m_l} \sum_{s_j=1}^{S} p\left(\mathbf{z}_l^j \middle| \mathbf{x}_l^{i;s_j}\right) \pi_l^{j \to s_j}. \tag{18.96}$$

The posterior association probabilities are thus given by:

$$p\left(a_l \middle| \mathbf{x}_l^i, Z_l, m_l\right) = \prod_{j=1}^{m_l} \frac{\mathcal{N}\left(\mathbf{z}_l^j; \mathbf{H}_l \mathbf{x}_l^{i;s_j}, \mathbf{R}_l^j\right) \pi_l^{j \to s_j}}{\sum_{s=1}^{S} \left(\mathbf{z}_l^j; \mathbf{H}_l \mathbf{x}_l^{i;s}, \mathbf{R}_l^j\right) \pi_l^{j \to s_j}} \tag{18.97}$$

$$= \prod_{j=1}^{m_l} w_l^{i;j \to s_j}, \tag{18.98}$$

where we use as an abbreviation individual weighting factors $w_l^{i;j \to s}$ denoting the posterior association probability for associating in the iteration step $i$ at time $t_l$ the measurement $\mathbf{z}_l^j$ to the object indexed by $s$. With this result and using

$$\log\left(p(Z_l|a_l, \mathbf{x}_l)\right) p\left(a_l \middle| \mathbf{x}_l^i, Z^k\right) = \log\left\{ \prod_{j=1}^{m_l} \mathcal{N}\left(\mathbf{z}_l^j; \mathbf{H}_l \mathbf{x}_l^{s_j}, \mathbf{R}_l^j \middle| w_l^{i;j \to s_j}\right) \right\}, \tag{18.99}$$

we can finally calculate the expectation according to the previous considerations:

$$\sum_{a_l} \log\left(p(Z_l \mid a_l, \mathbf{x}_l)\right) p\left(a_l \middle| \mathbf{x}_l^i, Z_l\right) \tag{18.100}$$

$$= \sum_{s_1=1}^{S} \cdots \sum_{s_{m_l}=1}^{S} \sum_{j=1}^{m_l} \log \mathcal{N}\left(\mathbf{z}_l^j; \mathbf{H}_l \mathbf{x}_l^{s_j}, \mathbf{R}_l^j \middle| w_l^{i;j \to s_j}\right) \tag{18.101}$$

$$= \sum_{s_1=1}^{S} \cdots \sum_{s_{m_l-1}=1}^{S} \sum_{j=1}^{m_l-1} \sum_{s=1}^{S} \log \mathcal{N}\left(\mathbf{z}_l^{m_l}; \mathbf{H}_l \mathbf{x}_l^s, \mathbf{R}_l^{m_l} \middle| w_l^{i;j \to s}\right) \tag{18.102}$$

$$= \sum_{j=1}^{m_l} \sum_{s=1}^{S} \log \mathcal{N}\left(\mathbf{z}_l^j; \mathbf{H}_l \mathbf{x}_l^{s_j}, \mathbf{R}_l^j \middle| w_l^{i;j \to s}\right) \tag{18.103}$$

Because a sum of logarithms is the logarithm of a product, the resulting product of linear Gaussian likelihood functions is equivalent to a likelihood function characterized by a "fused" measurement as previously discussed,

$$p\left(\mathbf{z}_l^{*s} \middle| \mathbf{x}_l^s\right) \propto \prod_{j=1}^{m_l} \mathcal{N}\left(\mathbf{z}_l^j; \mathbf{H}_l \mathbf{x}_l^s, \mathbf{R}_l^j \middle| w_l^{i;j \to s}\right) \tag{18.104}$$

$$\propto \mathcal{N}\left(\mathbf{z}_l^{*s}; \mathbf{H}_l \mathbf{x}_l^s, \mathbf{R}_l^{*s}\right) \tag{18.105}$$

For each object $s$, the "synthetic" measurement $\mathbf{z}_l^{*s}$ obtained by combining all measurements at the corresponding time frame $t_l$ according to weighting factors specific for each object $s$ is given by

$$\mathbf{z}_l^{*s} = \sum_{j=1}^{m_l} w_l^{i;j \to s} \mathbf{R}_l^{j^{-1}} \mathbf{z}_l^j \tag{18.106}$$

$$\mathbf{R}_l^{*s^{-1}} = \sum_{j=1}^{m_l} w_l^{i;j \to s} \mathbf{R}_l^{j^{-1}}. \tag{18.107}$$

## 18.6.2 ASDs and Q Functions

Inserting these results into the function $Q$ and taking the logarithm of $Q$, it is proportional to a double product, which can be represented by a product of $S$ ASDs, according to the discussion in the previous section:

$$\log Q\left(\mathbf{x}_{k:n}; \mathbf{x}_{k:n}^i\right) \propto \prod_{s=1}^{S} \prod_{l=n}^{k} p\left(\mathbf{z}_l^{*s} \middle| \mathbf{x}_l^s\right) p\left(\mathbf{x}_l^s \middle| \mathbf{x}_{l-1}^s\right) p\left(\mathbf{x}_{n-1}^s \middle| Z^{n-1}\right) \tag{18.108}$$

$$\propto \prod_{s=1}^{S} \mathcal{N}\left(\mathbf{x}_{k:n}^s; \mathbf{x}_{k:n}^{*s}, \mathbf{P}_{k:n}^{*s}\right). \tag{18.109}$$

The maximization of the function $Q$ is thus obtained by running $S$ independent Kalman filters on the synthetic measurements followed by Kalman retrodiction. The resulting expectation vectors $\mathbf{x}_{k:n}^* = \mathbf{x}_{k:n}^{i+1}$ are input for calculating the $Q\left(\mathbf{x}_{k:n}; \mathbf{x}_{k:n}^{i+1}\right)$ to be maximized in the next step. The EM loop is repeated until convergence. Instead of enumeration association hypotheses followed by pruning and merging, the EM philosophy thus tries to solve the association problem by an iteration.

Linearity in the number of targets and measurements is the main motivation for a further development and extension of this methodology, which is often called probabilistic multiple hypothesis tracking (PMHT).

Unfortunately, compared with alternatives such as the probabilistic data association filter (PDAF), PMHT has not yet shown its superiority in terms of track-lost statistics. Furthermore, the problem of track extraction and deletion is apparently not yet satisfactorily solved within this framework. Four properties of PMHT are responsible for its problems in track maintenance: nonadaptivity, hospitality, narcissism, and local maxima [24]. Approaches toward a solution for each of these phenomena and derivations of improved PMHT trackers have been proposed such as in Ref. [25]. Moreover, a sequential likelihood ratio (LR) test for track extraction has been developed, which in the context of the PMHT methodology has the potential for track extraction [26]. As PMHT provides all required ingredients for a sequential LR calculation, the LR is thus a byproduct of the PMHT iteration process.

Here, we have considered point-source objects. A generalization to extended objects is straightforward and is discussed in Ref. [27]. For introducing missing and false measurements into the framework, let $P_D^s$ denote a detection probability related to object $s$, which produces at most *one* measurement at time $t_l$ per object and sensor. Moreover, $\mathbf{z}_l^0$ indicates a noninformative measurement representing a missing detection, that is, a measurement with a very large measurement error. The probability of having $m_l$ false measurements is $p_F(m_l)$ given by a Poisson distribution characterized by a spatial clutter density as before. Let us introduce a fictitious target $s = 0$ producing $m_l \geq 0$ false measurements at time $t_l$, which are equally distributed in space. The detection probability related to this fictitious target $s = 0$ is given by $P_D^0 = 1 - p_F(0)$. As before, we assume that the data associations are independent of each other, which is a critical one because it can be justified as an approximation only.

For performance improvements, much depends on the proper formulation of *prior* probabilities $\pi_l^{j \to s_j}$ under the hypothesis $S$ targets are existent. See Ref. [25] for a detailed discussion.

## 18.7  DISTRIBUTED ASD FUSION

It is assumed that a single target is observed by $S$ synchronized sensors, and that all time-varying target properties of interest are collected by a state vector $\mathbf{x}_k$. The knowledge on the state $\mathbf{x}_k$ can be expressed by a posterior probability density function, which is conditioned on all measurements up to the current time $t_k$. If Kalman filter conditions apply, this density is given by a Gaussian $\mathcal{N}(\mathbf{x}_k; \mathbf{x}_{k|k}, \mathbf{P}_{k|k})$ with an expectation vector $\mathbf{x}_{k|k}$ and covariance matrix $\mathbf{P}_{k|k}$ [3]. These conditions include a linear Gaussian Markovian transition density of the state which is given by $p(\mathbf{x}_k | \mathbf{x}_{k-1}) = \mathcal{N}(\mathbf{x}_k; \mathbf{F}_{k|k-1} \mathbf{x}_{k-1}, \mathbf{Q}_{k|k-1})$. It is assumed that each of the $S$ sensors produces measurements at the same instants of time $t_l$, $l = 1, \ldots, k$ denoted by $Z_l^s$ where $s = 1, \ldots, S$. The accumulation of the sensor data $Z_l$ up to and including the time $t_k$ is a time series recursively defined by $\mathcal{Z}^k = \left\{ Z_k^1, \ldots, Z_k^S, \mathcal{Z}^{k-1} \right\}$. The time series produced by the measurements of an individual sensor $s \in \{1, \ldots, S\}$ only is denoted by $\mathcal{Z}_s^k$. The statistical properties of an individual sensor measurement $\mathbf{z}_l^s$ is given by a probability density function $p\left(\mathbf{z}_l^s | \mathbf{x}_l\right)$, also called the sensor likelihood function, which needs to be known up to a constant factor only: $p\left(\mathbf{z}_l^s | \mathbf{x}_l\right) \propto \ell_l^s\left(\mathbf{x}_l; \mathbf{z}_l^s\right)$.

For the DKF, the local sensor data of each node in the network is processed to a local state variable $\mathbf{x}_{k|k}^s$:

$$\mathbf{x}_{k|k}^s \leftarrow \mathcal{Z}_k^s. \tag{18.110}$$

If the local parameters are obtained by a fusion center, they can be combined to the global estimate $\mathbf{x}_{k|k}$:

$$\mathbf{x}_{k|k} \leftarrow \left\{ \mathbf{x}_{k|k}^s \right\}_{s=1,\ldots,S}. \tag{18.111}$$

The fusion rules for both the local data processing and the central parameter fusion can vary for different DKF implementations and may involve additional weighting and debiasing matrices. However, it should be noted that in most cases the local parameters are different to local optimal Kalman filter results because some methodologies of decorrelation are applied. Furthermore, a DKF scheme is called *exact* if its resulting fused track is identical to a central Kalman filter processing all measurements from every sensor. As Gaussian linear models are assumed, this implies that an exact T2TF algorithm is equivalent to a minimum mean squared error estimator [3] for the global state.

This chapter focuses on decorrelated tracking in multisensor applications whereby a product representation for the posterior density of a state $\mathbf{x}_k$ at time $t_l \le t_k$ is achieved. As stated in Ref. [15], for the decorrelated, independent parameters $\left\{\mathbf{x}_{k|k}^s\right\}_{s=1,\ldots,S}$ the fused posterior pdf can be expressed as

$$p(\mathbf{x}_k|\mathcal{Z}^k) = c_{k|k} \prod_{s=1}^{S} \mathcal{N}\left(\mathbf{x}_k; \mathbf{x}_{k|k}^s, \mathbf{P}_{k|k}^s\right). \tag{18.112}$$

The normalizing factor $c_{k|k}$ does not depend on the state $\mathbf{x}_k$ and therefore will not be calculated in practical applications. Then, by an application of a well-known product formula for Gaussians, the resulting Gaussian posterior density obtained by the fusion centre is given by $p(\mathbf{x}_k|\mathcal{Z}^k) = \mathcal{N}(\mathbf{x}_k; \mathbf{x}_{k|k}, \mathbf{P}_{k|k})$ where

$$\mathbf{P}_{k|k}^{-1} = \sum_{s=1}^{S} \mathbf{P}_{k|k}^{s-1}, \tag{18.113}$$

$$\mathbf{x}_{k|k} = \mathbf{P}_{k|k}\left(\sum_{s=1}^{S} \mathbf{P}_{k|k}^{s-1} \mathbf{x}_{k|k}^s\right). \tag{18.114}$$

Note that this fusion scheme is independent of previous transmissions. As a consequence, it is possible to obtain all parameters of the exact posterior density if communication of all $S$ local tracks for time $t_k$ is possible.

In the multisensor case, the posterior density can be decomposed as follows. An application of Bayes' theorem yields

$$p(\mathbf{x}_{k:n}|\mathcal{Z}^k) =$$

$$\frac{p\left(Z_k^1, \ldots, Z_k^S \middle| \mathbf{x}_k\right) \cdot p(\mathbf{x}_{k:n}|\mathcal{Z}^{k-1})}{\int d\mathbf{x}_{k:n} p\left(Z_k^1, \ldots, Z_k^S \middle| \mathbf{x}_k\right) \cdot p(\mathbf{x}_{k:n}|\mathcal{Z}^{k-1})} \tag{18.115}$$

$$= \frac{\prod_{s=1}^{S} p\left(Z_k^s \middle| \mathbf{x}_k\right) \cdot p(\mathbf{x}_{k:n}|\mathcal{Z}^{k-1})}{\int d\mathbf{x}_{k:n} \prod_{s=1}^{S} p\left(Z_k^s \middle| \mathbf{x}_k\right) \cdot p(\mathbf{x}_{k:n}|\mathcal{Z}^{k-1})} \tag{18.116}$$

$$\propto \prod_{s=1}^{S} p\left(Z_k^s \middle| \mathbf{x}_k\right) \cdot p(\mathbf{x}_k|\mathbf{x}_{k-1:n}) p(\mathbf{x}_{k-1:n}|\mathcal{Z}^{k-1}). \tag{18.117}$$

After applying the Gaussian assumptions for the sensor and the dynamics model as well as for the posterior of the previous time step, the following terms are obtained:

$$p(\mathbf{x}_{k:n}|\mathcal{Z}^k) \propto$$

$$\prod_{s=1}^{S} \mathcal{N}\left(\mathbf{z}_k^s; \mathbf{H}_k^s \mathbf{x}_k, \mathbf{R}_k^s\right) \cdot \mathcal{N}(\mathbf{x}_k; \mathbf{F}_{k|k-1}\mathbf{x}_{k-1}, \mathbf{Q}_{k|k-1}) \tag{18.118}$$

$$\cdot \mathcal{N}(\mathbf{x}_{k-1:n}; \mathbf{x}_{k-1:n|k-1}, \mathbf{P}_{k-1:n|k-1}).$$

One of many implications of the product formula, given in the Appendix, is the following:

$$\mathcal{N}(\mathbf{x}_k; \mathbf{F}_{k|k-1}\mathbf{x}_{k-1}, \mathbf{Q}_{k|k-1})$$

$$\propto \mathcal{N}(\mathbf{x}_k; \mathbf{F}_{k|k-1}\mathbf{x}_{k-1}, S\mathbf{Q}_{k|k-1})^S. \tag{18.119}$$

The Gaussian density for the dynamics model with a covariance which is "blown up" by a factor $S$ is also called the relaxed evolution model. This can be used to transform the above posterior density into

$$p(\mathbf{x}_{k:n}|\mathcal{Z}^k) \propto \prod_{s=1}^{S} \left\{ \mathcal{N}\left(\mathbf{z}_k^s; \mathbf{H}_k^s \mathbf{x}_k, \mathbf{R}_k^s\right) \right.$$

$$\left. \mathcal{N}(\mathbf{x}_k; \mathbf{F}_{k|k-1}\mathbf{x}_{k-1}, S\mathbf{Q}_{k|k-1}) \right\} \tag{18.120}$$

$$\cdot \mathcal{N}(\mathbf{x}_{k-1:n}; \mathbf{x}_{k-1:n|k-1}, \mathbf{P}_{k-1:n|k-1}).$$

A recursion argument on the posterior at time $t_{k-1}$ yields

$$p(\mathbf{x}_{k:n}|\mathcal{Z}^k) \propto \prod_{l=n+1}^{k} \left[ \prod_{s=1}^{S} \left\{ \mathcal{N}\left(\mathbf{z}_l^s; \mathbf{H}_l^s \mathbf{x}_l, \mathbf{R}_l^s\right) \right. \right.$$

$$\left. \left. \mathcal{N}(\mathbf{x}_l; \mathbf{F}_{l|l-1}\mathbf{x}_{l-1}, S\mathbf{Q}_{l|l-1}) \right\} \right] \cdot \mathcal{N}(\mathbf{x}_n; \mathbf{x}_{n|n}, \mathbf{P}_{n|n}), \tag{18.121}$$

where $\mathbf{x}_{n|n}$ and $\mathbf{P}_{n|n}$ refer to the result of a standard Kalman filter after processing all data up to time $t_n$. It can also be an initial estimate density as it represents "the base" of the ASD.

Following the calculation of a posterior ASD, it can be seen from Equation 18.121 that the global estimate in a multisensor system is equivalent to the convex combination of local ASDs where a relaxed evolution model is used:

$$p(\mathbf{x}_{k:n}|\mathcal{Z}^k) \propto \prod_{s=1}^{S} \mathcal{N}\left(\mathbf{x}_{k:n}; \mathbf{x}_{k:n|k}^s, \mathbf{P}_{k:n|k}^s\right), \tag{18.122}$$

and the fused ASD estimate and covariance are given by

$$\mathbf{x}_{k:n|k} = \mathbf{P}_{k:n|k} \sum_{s=1}^{S} \mathbf{P}_{k:n|k}^{s-1} \mathbf{x}_{k:n|k}^{s} \qquad (18.123)$$

$$\mathbf{P}_{k:n|k} = \left( \sum_{s=1}^{S} \mathbf{P}_{k:n|k}^{s-1} \right)^{-1}. \qquad (18.124)$$

### 18.7.1  DASD Filter Implementation

The following sections show how to iteratively calculate the local parameters $\mathbf{x}_{k:n|k}^{s}$ and $\mathbf{P}_{k:n|k}^{s}$. Assume, the local posterior ASD at time $t_{k-1}$ for sensor $s$ is given in terms of $\mathbf{x}_{k-1:n|k-1}^{s}$ and $\mathbf{P}_{k-1:n|k-1}^{s}$. The prediction of the state is straightforward due to the Markov proposition:

$$\mathbf{x}_{k:n|k-1}^{s} = \begin{pmatrix} \mathbf{x}_{k|k-1}^{s} \\ \mathbf{x}_{k-1|k-1}^{s} \\ \vdots \\ \mathbf{x}_{n|k-1}^{s} \end{pmatrix}, \qquad (18.125)$$

where $\mathbf{x}_{k|k-1}^{s} = \mathbf{F}_{k|k-1} \mathbf{x}_{k-1|k-1}^{s}$ is equivalent to a local Kalman filter prediction. For the ASD covariance prediction, a recursive formulation according to the previous discussion is used:

$$\mathbf{P}_{k:n|k-1}^{s} = \begin{pmatrix} \mathbf{P}_{k|k-1}^{s} & \mathbf{P}_{k|k-1}^{s} \mathbf{W}_{k-1:n}^{s\top} \\ \mathbf{W}_{k-1:n}^{s} \mathbf{P}_{k|k-1}^{s} & \mathbf{P}_{k-1:n|k-1}^{s} \end{pmatrix}, \qquad (18.126)$$

where

$$\mathbf{P}_{k|k-1}^{s} = \mathbf{F}_{k|k-1} \mathbf{P}_{k-1|k-1}^{s} \mathbf{F}_{k|k-1}^{\top} + S \mathbf{Q}_{k|k-1}, \qquad (18.127)$$

$$\mathbf{W}_{k-1:n}^{s} = \begin{pmatrix} \mathbf{W}_{k-1|k}^{s} \\ \mathbf{W}_{k-2:n}^{s} \mathbf{W}_{k-1|k}^{s} \end{pmatrix}, \qquad (18.128)$$

$$\mathbf{W}_{l|l+1}^{s} = \mathbf{P}_{l|l}^{s} \mathbf{F}_{l+1|l}^{\top} \left( \mathbf{F}_{l+1|l} \mathbf{P}_{l|l}^{s} \mathbf{F}_{l+1|l}^{\top} + S \mathbf{Q}_{l+1|l} \right)^{-1}. \qquad (18.129)$$

The preceding notations are given by

$$\mathbb{C}\mathbf{x}_{l}, \mathbf{x}_{k} | \mathcal{Z}^{k-1} = \mathbf{W}_{l|k} \mathbf{P}_{k|k-1}, \qquad (18.130)$$

where

$$\mathbf{W}_{llk} = \prod_{i=l}^{k-1} \mathbf{W}_{ili+1} . \tag{18.131}$$

For the filtering step, it is assumed that the prior parameters $\mathbf{x}_{k:n|k-1}^s$ and $\mathbf{P}_{k:n|k-1}^s$ for sensor $s$ are given. As the measurement error is assumed to be independent from the past, the sensor likelihood function can be expressed by an application of projections $\Pi_k$ onto the current state:

$$p\left(\mathbf{z}_k^s \middle| \mathbf{x}_k\right) = p\left(\mathbf{z}_k^s \middle| \Pi_k \mathbf{x}_{k:n}\right) \tag{18.132}$$

$$= \mathcal{N}\left(\mathbf{z}_k^s ; \mathbf{H}_k^s \Pi_k \mathbf{x}_{k:n}, \mathbf{R}_k^s\right), \tag{18.133}$$

where

$$\Pi_k = (\mathbf{1}_d, \mathbf{O}, \ldots, \mathbf{O}). \tag{18.134}$$

In the above notation, $\mathbf{1}_d$ is an identity matrix in the dimension of the state and $\mathbf{O}$ is the corresponding zero matrix. Then, the posterior parameters are obtained by the multiplication of the local prior density and the likelihood function. An application of the product formula in the Appendix yields

$$\mathbf{x}_{k:n|k}^s = \mathbf{x}_{k:n|k-1}^s + \mathbf{W}_{k:n|k-1}^s \left(\mathbf{z}_k^s - \mathbf{H}_k^s \Pi_k \mathbf{x}_{k:n|k-1}^s\right), \tag{18.135}$$

$$\mathbf{W}_{k:n|k-1}^s = \mathbf{P}_{k:n|k-1}^s \Pi_k^\top \mathbf{H}_k^{s\top} \mathbf{S}_k^{s-1}, \tag{18.136}$$

$$\mathbf{S}_k^s = \mathbf{H}_k^s \Pi_k \mathbf{P}_{k:n|k-1}^s \Pi_k^\top \mathbf{H}_k^{s\top} + \mathbf{R}_k^s, \tag{18.137}$$

$$\mathbf{P}_{k:n|k}^s = \mathbf{P}_{k:n|k-1}^s - \mathbf{W}_{k:n|k-1}^s \mathbf{S}_k^s \mathbf{W}_{k:n|k-1}^{s\top} . \tag{18.138}$$

### 18.7.2 Sliding Window Mechanism

After the filtering step of the multisensor ASD, the local parameters have the following structure:

$$\mathbf{x}_{k:n|k}^s = \begin{pmatrix} \mathbf{x}_{k|k}^s \\ \mathbf{x}_{k-1|k}^s \\ \vdots \\ \mathbf{x}_{n|k}^s \end{pmatrix}, \tag{18.139}$$

$$\mathbf{P}^s_{k:n|k} = \begin{pmatrix} \mathbf{P}^s_{k|k} & \mathbf{P}^s_{k|k} \mathbf{W}^{s\top}_{k-1:n} \\ \mathbf{W}^s_{k-1:n} \mathbf{P}^s_{k|k} & \mathbf{P}^s_{k-1:n|k} \end{pmatrix}. \tag{18.140}$$

If $\dim_x$ denotes the dimension of the state $\mathbf{x}_k$, both, the ASD parameter $\mathbf{x}^s_{k:n|k}$ and the square matrix $\mathbf{P}^s_{k:n|k}$ have the dimension $(k - n + 1)\dim_x$. To keep the dimension fixed during the prediction-filtering iterations, the block row entry which refers to time $t_n$ can be skipped because it has the least impact on the estimate of the current time $t_k$. This can be seen by the off-diagonal entries of the posterior ASD covariance matrix because theses submatrices describe the correlations of the estimates. As given in Equation 18.130), the cross covariance of $\mathbf{x}_k$ and $\mathbf{x}_l$ for some $l < k$ is given by

$$\mathbb{C}\mathbf{x}_n, \mathbf{x}_k | \mathcal{Z}^k = \mathbf{W}_{n|k} \cdot \mathbf{P}_{k|k} \tag{18.141}$$

$$= \prod_{l=n}^{k-1} \mathbf{W}_{l|l+1} \cdot \mathbf{P}_{k|k} \tag{18.142}$$

Using the definition of $\mathbf{W}_{l|l+1}$ and the fact that $\det(\mathbf{F}_{l+1|l}) = 1$ for most tracking applications, one can see that

$$\det(\mathbf{W}_{l|l+1}) = \frac{\det(\mathbf{P}_{l|l})}{\det(\mathbf{P}_{l|l} + \mathbf{Q}_{l+1|l})} < 1. \tag{18.143}$$

This implies the following relation:

$$\det(\mathbf{W}_{n|k}) < \det(\mathbf{W}_{l|k}) < \det(\mathbf{W}_{k-1|k}) \text{ for } n < l < k, \tag{18.144}$$

which clearly shows that the estimate for time $t_n$ has the least impact on time $t_k$. As a result, the corresponding block entries are omitted and one obtains the fixed-size parameters:

$$\mathbf{x}^s_{k:n+1|k} = \begin{pmatrix} \mathbf{x}^s_{k|k} \\ \mathbf{x}^s_{k-1|k} \\ \vdots \\ \mathbf{x}^s_{n+1|k} \end{pmatrix}, \tag{18.145}$$

$$\mathbf{P}^s_{k:n+1|k} = \begin{pmatrix} \mathbf{P}^s_{k|k} & \mathbf{P}^s_{k|k} \mathbf{W}^{s\top}_{k-1:n+1} \\ \mathbf{W}^s_{k-1:n+1} \mathbf{P}^s_{k|k} & \mathbf{P}^s_{k-1:n+1|k} \end{pmatrix}. \tag{18.146}$$

It should be noted that the fused estimate density is an approximation to the exact result which is obtained by transmitting the full ASD length. However, it can be seen in the numerical evaluation following in the next section that already for ASDs with only few states no difference to the optimal centralized solution can be observed for the estimate of the current time $t_k$.

### 18.7.3  Numerical Evaluation

This section presents a numerical evaluation of a simulated scenario with a single target and four sensors. The target starts at the origin of the Cartesian reference system with an initial velocity of 400 ms. The dynamics of the target are simulated according to a continuous white noise acceleration (CWNA) model and a process noise with a power spectral density $q = 10.0$ for 100 time steps.

Each sensor measures the position of the target with a Gaussian distributed noise term. The measurement error covariances are given by

$$\mathbf{R}_k^s := \mathrm{diag}(100 \cdot s, 100 \cdot s) \text{ for } s = 1, \ldots, 4. \tag{18.147}$$

For a given time $t_k$, the local ASD of each sensor is extended to its estimate $\mathbf{x}_{k:n|k}$ and covariance $\mathbf{P}_{k:n|k}$ where the ASD length $k - n + 1$ is fixed.

Figure 18.3 shows the *normalized root mean square error* (RMSE) in position with respect to the RMSE of a centralized Kalman filter (CKF) averaged over 100 Monte Carlo runs for different ASD lengths. It can be seen that in this scenario an ASD length of 10 states is already sufficient to provide a result that perfectly matches the estimate of the CKF. The limiting case of ASD length 1 is equivalent to a Federated Kalman filter which is also known as the "relaxed evolution model" filter.

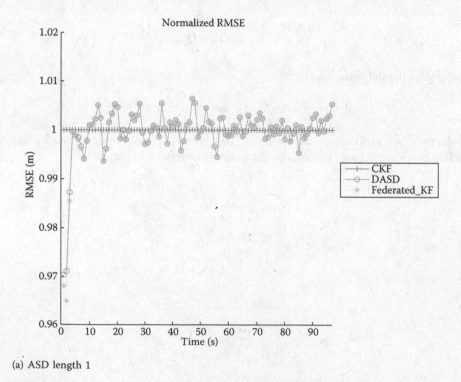

(a) ASD length 1

**FIGURE 18.3**  Normalized RMSE in position for a centralized CKF and the DASD and a federated Kalman filter. The RMSE values are relative to the CKF results and normalized over Monte Carlo simulations. The ASD length was fixed to include 1 (a), 5 (b), and 10 (c) states, respectively.

*(Continued)*

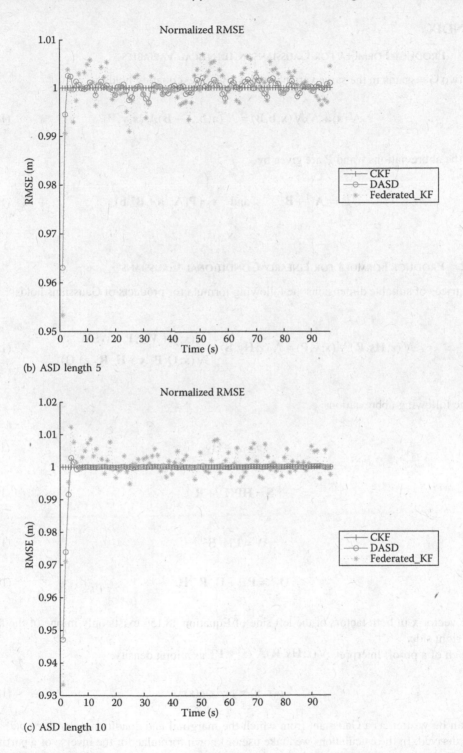

(b) ASD length 5

(c) ASD length 10

**FIGURE 18.3 (CONTINUED)** Normalized RMSE in position for a centralized CKF and the DASD and a federated Kalman filter. The RMSE values are relative to the CKF results and normalized over Monte Carlo simulations. The ASD length was fixed to include 1 (a), 5 (b), and 10 (c) states, respectively.

## APPENDIX

### 18A.1　Product Formula for Gaussians in Identical Variables

Given two Gaussians in the same variable $\mathbf{x}$, the following equation holds:

$$\mathcal{N}(\mathbf{x};\mathbf{a},\mathbf{A})\mathcal{N}(\mathbf{x};\mathbf{b},\mathbf{B}) = \mathcal{N}(\mathbf{a};\mathbf{b},\mathbf{A}+\mathbf{B})\mathcal{N}(\mathbf{x};\mathbf{y},\mathbf{P}), \tag{18.148}$$

where the abbreviations $\mathbf{y}$ and $\mathbf{P}$ are given by

$$\mathbf{P} = (\mathbf{A}^{-1}+\mathbf{B}^{-1})^{-1} \quad \text{and} \quad \mathbf{y} = \mathbf{P}(\mathbf{A}^{-1}\mathbf{a}+\mathbf{B}^{-1}\mathbf{b}). \tag{18.149}$$

### 18A.2　Product Formula for Linearly Conditional Gaussians

For matrices of suitable dimensions the following formula for products of Gaussians holds:

$$\mathcal{N}(\mathbf{z};\mathbf{Hx},\mathbf{R})\mathcal{N}(\mathbf{x};\mathbf{y},\mathbf{P}) = \mathcal{N}(\mathbf{z};\mathbf{Hy},\mathbf{S}) \begin{cases} \mathcal{N}(\mathbf{x};\mathbf{y}+\mathbf{W}\nu,\mathbf{P}-\mathbf{WSW}^{\top}) \\ \mathcal{N}(\mathbf{x};\mathbf{Q}(\mathbf{P}^{-1}\mathbf{y}+\mathbf{H}^{\top}\mathbf{R}^{-1}\mathbf{z}),\mathbf{Q}) \end{cases} \tag{18.150}$$

with the following abbreviations:

$$\nu = \mathbf{z}-\mathbf{Hy} \tag{18.151}$$

$$\mathbf{S} = \mathbf{HPH}^{\top}+\mathbf{R} \tag{18.152}$$

$$\mathbf{W} = \mathbf{PH}^{\top}\mathbf{S}^{-1} \tag{18.153}$$

$$\mathbf{Q}^{-1} = \mathbf{P}^{-1}+\mathbf{H}^{\top}\mathbf{R}^{-1}\mathbf{H}. \tag{18.154}$$

The vector $\mathbf{x}$ in both factors of the left side of Equation 18.150 exists only in one of the factors on the right side.

Sketch of a proof: Interpret $\mathcal{N}(\mathbf{z};\mathbf{Hx},\mathbf{R})\mathcal{N}(\mathbf{x};\mathbf{y},\mathbf{P})$ as a joint density:

$$p(\mathbf{z},\mathbf{x}) = p(\mathbf{z}|\mathbf{x})p(\mathbf{x}) \tag{18.155}$$

It can be written as a Gaussian, from which the marginal and conditional densities $p(\mathbf{z})$, $p(\mathbf{x}|\mathbf{z})$ can be derived. In the calculations we make use of known formulae for the inverse of a partitioned matrix. From

$$p(\mathbf{z},\mathbf{x}) = p(\mathbf{x}|\mathbf{z})p(\mathbf{z}) \tag{18.156}$$

the formula results.

## 18A.3  INVERSE OF BLOCK MATRICES

The inversion of the inverse ASD covariance matrix makes use of well-known results on block matrices [28]. The inverse of a partitioned symmetric matrix is given by

$$\begin{pmatrix} \mathbf{A} & \mathbf{C} \\ \mathbf{C}^\top & \mathbf{B} \end{pmatrix}^{-1} = \begin{pmatrix} \mathbf{T}^{-1} & -\mathbf{T}^{-1}\mathbf{C}\mathbf{B}^{-1} \\ -\mathbf{B}^{-1}\mathbf{C}^\top\mathbf{T}^{-1} & \mathbf{B}^{-1} + \mathbf{B}^{-1}\mathbf{C}^\top\mathbf{T}^{-1}\mathbf{C}\mathbf{B}^{-1} \end{pmatrix} \tag{18.157}$$

where $\mathbf{T} = \mathbf{A} - \mathbf{C}\mathbf{B}^{-1}\mathbf{C}^\top$ is called the Schur complement of the matrix $\mathbf{B}$.

## 18A.4  GAUSSIAN ASDs: DETAILS OF THE PROOF

While the expectation vector $\mathbf{x}_{k:n|k}$ of the accumulated object states $\mathbf{x}_{k:n}$ is defined by

$$\mathbf{x}_{k:n|k} = \left( \mathbf{x}_{k|k}^\top, \mathbf{x}_{k-1|k}^\top, \ldots, \mathbf{x}_{n+1|k}^\top, \mathbf{x}_{n|k}^\top \right)^\top,$$

the corresponding covariance matrix $\mathbf{P}_{k:n|k}$ can recursively be written as

$$\mathbf{P}_{l:n|k} = \begin{pmatrix} \mathbf{P}_{l|k} & \mathbf{P}_{l|k}\mathbf{W}_{l-1:n}^\top \\ \mathbf{W}_{l-1:n}\mathbf{P}_{l|k} & \mathbf{P}_{l-1:n|k} \end{pmatrix},$$

$n + 1 \le l \le k$, with $\mathbf{P}_{n:n|k} = \mathbf{P}_{n|k}$ and $\mathbf{W}_{l:n}$ given by

$$\mathbf{W}_{l:n} = \begin{pmatrix} \mathbf{W}_{l|l+1} \\ \mathbf{W}_{l-1:n}\mathbf{W}_{l|l+1} \end{pmatrix}$$

with $\mathbf{W}_{n:n} = \mathbf{W}_{n|n+1}$ and the retrodiction gain matrices $\mathbf{W}_{l|l+1}$ given by

$$\mathbf{W}_{l|l+1} = \mathbf{P}_{l|l}\mathbf{F}_{l+1|l}^\top \mathbf{P}_{l+1|l}^{-1}$$

$$= \mathbf{P}_{l|l}\mathbf{F}_{l+1|l}^\top \left( \mathbf{F}_{l+1|l}\mathbf{P}_{l|l}\mathbf{F}_{l+1|l}^\top + \mathbf{D}_{l+1|l} \right)^{-1}.$$

This statement directly follows from a straightforward induction argument, though the necessary calculations are perhaps somewhat tedious. The proposition holds for $k = n$. Let us assume that it is true at time $t_k$. Owing to standard assumptions, the ASD at time $t_{k+1}$ can be represented by

$$p(\mathbf{x}_{k+1:n}|Z^{k+1}) = \frac{p(\mathbf{z}_{k+1}|\mathbf{x}_{k+1})p(\mathbf{x}_{k+1}|\mathbf{x}_k)p(\mathbf{x}_{k:n}|Z^k)}{\int d\mathbf{x}_{k+1:n}p(\mathbf{z}_{k+1}|\mathbf{x}_{k+1})p(\mathbf{x}_{k+1}|\mathbf{x}_k)p(\mathbf{x}_{k:n}|Z^k)}. \tag{18.158}$$

Using the projection matrices $\Pi_k = (\mathbf{1}, \mathbf{O}, \ldots, \mathbf{O})$ defined by

$$\Pi_k \mathbf{x}_{k:n} = \mathbf{x}_k \tag{18.159}$$

and $\Pi_{k:n} = (-\mathbf{W}_{k:n}, \mathbf{1})$ defined by

$$\Pi_{k:n} \left( \mathbf{x}_{k+1}^\top, \mathbf{x}_{k:n}^\top \right)^\top = -\mathbf{W}_{k:n} \mathbf{x}_{k+1} + \mathbf{x}_{k:n}, \tag{18.160}$$

a repeated use of the product formula 18.150 yields

$$p(\mathbf{z}_{k+1}|\mathbf{x}_{k+1}) p(\mathbf{x}_{k+1}|\mathbf{x}_k) p(\mathbf{x}_{k:n}|Z^k) = \mathcal{N}(\mathbf{z}_{k+1}; \mathbf{H}_{k+1}\mathbf{x}_{k+1}, \mathbf{R}_{k+1}) \mathcal{N}(\mathbf{x}_{k+1}; \mathbf{x}_{k+1|k}, \mathbf{P}_{k+1|k})$$
$$\times \mathcal{N}(\mathbf{x}_{k:n}; \mathbf{x}_{k:n|k} + \mathbf{W}_{k:n}(\mathbf{x}_{k+1} - \mathbf{x}_{k+1|k}), \mathbf{R}_{k:n}) \tag{18.161}$$

$$= \mathcal{N}(\mathbf{z}_{k+1}; \mathbf{H}_{k+1}\mathbf{x}_{k+1|k}, \mathbf{S}_{k+1|k}) \mathcal{N}(\mathbf{x}_{k+1}; \mathbf{x}_{k+1|k+1}, \mathbf{P}_{k+1|k+1})$$
$$\times \mathcal{N} \left( \Pi_{k:n} \mathbf{x}_{k+1:n}; \Pi_{k:n} \left( \mathbf{x}_{k+1|k}^\top, \mathbf{x}_{k:n|k}^\top \right)^\top, \mathbf{R}_{k:n} \right) \tag{18.162}$$

with $\mathbf{W}_{k:n}$ and $\mathbf{R}_{k:n}$ given by

$$\mathbf{W}_{k:n} = \mathbf{P}_{k:n|k} \Pi_k^\top \mathbf{F}_{k+1|k}^\top \mathbf{P}_{k+1|k}^{-1}$$
$$= \begin{pmatrix} \mathbf{W}_{k|k+1} \\ \mathbf{W}_{k-1:n} \mathbf{W}_{k|k+1} \end{pmatrix}$$
$$\mathbf{R}_{k:n} = \mathbf{P}_{k:n|k} - \mathbf{W}_{k:n} \mathbf{P}_{k+1|k} \mathbf{W}_{k:n}^\top.$$

In particular

$$\Pi_{k:n}^\top \mathbf{R}_{k:n}^{-1} \Pi_{k:n} = \begin{pmatrix} \mathbf{W}_{k:n}^\top \mathbf{R}_{k:n}^{-1} \mathbf{W}_{k:n} & -\mathbf{W}_{k:n}^\top \mathbf{R}_{k:n}^{-1} \\ -\mathbf{R}_{k:n}^{-1} \mathbf{W}_{k:n} & \mathbf{R}_{k:n}^{-1} \end{pmatrix}.$$

By a second use of the product formula for Gaussians in identical variables (Equation 18.148), we thus obtain up to a constant independent of the state vectors:

$$p(\mathbf{z}_{k+1}|\mathbf{x}_{k+1}) p(\mathbf{x}_{k+1}|\mathbf{x}_k) p(\mathbf{x}_{k:n}|Z^k) \propto \mathcal{N}(\mathbf{z}_{k+1}; \mathbf{H}_{k+1}\mathbf{x}_{k+1|k}, \mathbf{S}_{k+1|k}) \mathcal{N}\left( \mathbf{x}_{k+1:n}; \mathbf{x}_{k+1:}^{k+1}, \mathbf{P}_{k+1:n}^{k+1} \right),$$

where the covariance matrix $\mathbf{P}_{k+1:n|k+1}$ is given by

$$\mathbf{P}_{k+1:n|k+1} = \left( \Pi_{k+1}^\top \mathbf{P}_{k+1|k+1}^{-1} \Pi_{k+1} + \Pi_{k:n}^\top \mathbf{R}_{k:n}^{-1} \Pi_{k:n} \right)^{-1} \tag{18.163}$$

$$= \begin{pmatrix} \mathbf{P}_{k+1|k+1}^{-1} + \mathbf{W}_{k:n}^\top \mathbf{R}_{k:n}^{-1} \mathbf{W}_{k:n} & -\mathbf{W}_{k:n}^\top \mathbf{R}_{k:n}^{-1} \\ -\mathbf{R}_{k:n}^{-1} \mathbf{W}_{k:n} & \mathbf{R}_{k:n}^{-1} \end{pmatrix}^{-1}. \tag{18.164}$$

This block matrix can directly be inverted by using Equation 18.157. The corresponding Schur complement is particularly simple and given by

$$\mathbf{T} = \mathbf{P}_{k+1|k+1}^{-1} + \mathbf{W}_{k:n}^{\top}\mathbf{R}_{k:n}^{-1}\mathbf{W}_{k:n} - \mathbf{W}_{k:n}^{\top}\mathbf{R}_{k:n}^{-1}\mathbf{R}_{k:n}\mathbf{R}_{k:n}^{-1}\mathbf{W}_{k:n} = \mathbf{P}_{k+1|k+1}^{-1} . \tag{18.165}$$

We thus obtain for the covariance matrix of the ASD:

$$\mathbf{P}_{k+1:n|k+1} = \begin{pmatrix} \mathbf{P}_{k+1|k+1} & \mathbf{P}_{k+1|k+1}\mathbf{W}_{k:n}^{\top} \\ \mathbf{W}_{k:n}\mathbf{P}_{k+1|k+1} & \mathbf{R}_{k:n} + \mathbf{W}_{k:n}\mathbf{P}_{k+1|k+1}\mathbf{W}_{k:n}^{\top} \end{pmatrix} \tag{18.166}$$

$$= \begin{pmatrix} \mathbf{P}_{k+1|k+1} & \mathbf{P}_{k+1|k+1}\mathbf{W}_{k:n}^{\top} \\ \mathbf{W}_{k:n}\mathbf{P}_{k+1|k+1} & \mathbf{P}_{k:n}^{k} + \mathbf{W}_{k:n}(\mathbf{P}_{k+1|k+1} - \mathbf{P}_{k+1|k})\mathbf{W}_{k:n}^{\top} \end{pmatrix}. \tag{18.167}$$

Using the identity

$$\mathbf{W}_{k|k+1}(\mathbf{P}_{k+1|k+1} - \mathbf{P}_{k+1|k})\mathbf{W}_{k|k+1}^{\top} = \mathbf{P}_{k|k+1} - \mathbf{P}_{k|k} \tag{18.168}$$

resulting from the Rauch–Tung–Striebel equations, the matrix $\mathbf{W}_{k:n}(\mathbf{P}_{k+1|k+1} - \mathbf{P}_{k+1|k})\mathbf{W}_{k:n}^{\top}$ can be transformed, yielding:

$$\mathbf{W}_{k:n}(\mathbf{P}_{k+1|k+1} - \mathbf{P}_{k+1|k})\mathbf{W}_{k:n}^{\top}$$
$$= \begin{pmatrix} \mathbf{P}_{k|k+1} - \mathbf{P}_{k|k} & (\mathbf{P}_{k|k+1} - \mathbf{P}_{k|k})\mathbf{W}_{k-1:n}^{\top} \\ \mathbf{W}_{k-1:n}(\mathbf{P}_{k|k+1} - \mathbf{P}_{k|k}) & \mathbf{W}_{k-1:n}(\mathbf{P}_{k|k+1} - \mathbf{P}_{k|k})\mathbf{W}_{k-1:n}^{\top} \end{pmatrix}. \tag{18.169}$$

With this result, the block matrix $\mathbf{P}_{k:n}^{k} + \mathbf{W}_{k:n}(\mathbf{P}_{k+1|k+1} - \mathbf{P}_{k+1|k})\mathbf{W}_{k:n}^{\top}$ on the right-lower corner on the right side of Equation 18.167 is given by

$$\mathbf{P}_{k:n|k} + \mathbf{W}_{k:n}(\mathbf{P}_{k+1|k+1} - \mathbf{P}_{k+1|k})\mathbf{W}_{k:n}^{\top}$$
$$= \begin{pmatrix} \mathbf{P}_{k|k+1} & \mathbf{P}_{k|k+1}\mathbf{W}_{k-1:n}^{\top} \\ \mathbf{W}_{k-1:n}\mathbf{P}_{k|k+1} & \mathbf{P}_{k-1:n|k} + \mathbf{W}_{k-1:n}(\mathbf{P}_{k|k+1} - \mathbf{P}_{k|k})\mathbf{W}_{k-1:n}^{\top} \end{pmatrix}. \tag{18.170}$$

An induction argument for the block matrix on the right-lower corner directly yields

$$\mathbf{P}_{k:n|k} + \mathbf{W}_{k:n}(\mathbf{P}_{k+1|k+1} - \mathbf{P}_{k+1|k})\mathbf{W}_{k:n}^{\top} = \mathbf{P}_{k:n|k+1} . \tag{18.171}$$

According to the product formula Equation 18.150, $\mathbf{x}_{k+1:n|k+1}$ is the sum of the following vectors:

$$\mathbf{P}_{k+1:n|k+1}\mathbf{\Pi}_{k:n}^{\top}\mathbf{R}_{k:n}^{-1}\mathbf{\Pi}_{k:n}\left(\mathbf{x}_{k+1|k}^{\top},\mathbf{x}_{k:n|k}^{\top}\right)^{\top} = \begin{pmatrix} \mathbf{O} \\ -\mathbf{W}_{k:n}\mathbf{x}_{k+1|k} + \mathbf{x}_{k:n|k} \end{pmatrix} \tag{18.172}$$

$$\mathbf{P}_{k+1:n|k+1}\mathbf{\Pi}_{k+1}^{\top}\mathbf{P}_{k+1|k+1}^{-1}\mathbf{\Pi}_{k+1}\mathbf{x}_{k+1|k+1} = \begin{pmatrix} \mathbf{x}_{k+1|k+1} \\ \mathbf{W}_{k:n}\mathbf{x}_{k+1|k+1} \end{pmatrix}. \tag{18.173}$$

By using an induction argument, we thus obtain

$$\mathbf{x}_{k+1:n|k+1} = \begin{pmatrix} \mathbf{x}_{k+1|k+1} \\ \mathbf{x}_{k|k+1} \\ \mathbf{x}_{k-1:n|k} + \mathbf{W}_{k-1:n}(\mathbf{x}_{k|k+1} - \mathbf{x}_{k|k}) \end{pmatrix}. \tag{18.174}$$

An induction argument concludes the proof.

## REFERENCES

1. W. Koch, F. Govaers. On accumulated state densities with applications to out-of-sequence measurement processing. *IEEE Transactions on Aerospace and Electronic Systems*, Vol. 47, No. 4, pp. 2766–2778, 2011.
2. W. Koch. *Tracking and Sensor Data Fusion. Methodological Framework and Selected Applications.* Springer Mathematical Engineering Series. Springer, New York, 2014.
3. Y. Bar-Shalom, P.K. Willett, X. Tian. *Tracking and Data Fusion. A Handbook of Algorithms.* YBS Publishing, Storrs, CT, 2011.
4. S. Blackman, R. Populi. *Design and Analysis of Modern Tracking Systems.* Artech House, Norwood, MA, 1999.
5. M.S. Arulampalam, S. Maskell, N. Gordon, T. Clapp. A tutorial on particle filters for online nonlinear/non-Gaussian Bayesian tracking. *IEEE Transactions on Signal Processing*, Vol. 50, No. 2, pp. 174–188, 2002.
6. Y. Bar Shalom. Update with out-of-sequence measurements in tracking: Exact solution. *IEEE Transactions on Aerospace and Electronic Systems*, Vol. 38, No. 3, pp. 769–777, 2002.
7. M. Mallick. Out-of-sequence track filtering using the decorrelated pseudo measurement approach. *SPIE Signal and Data Processing of Small Targets*, Vol. 5428, pp. 154–166, 2004.
8. Y. Bar-Shalom, H. Chen. IMM estimator with out-of-sequence measurements. *IEEE Transactions on Aerospace and Electronic Systems*, Vol. 41, No. 1, pp. 90–98, 2005.
9. M. Orton, A. Marrs. Particle filters for tracking with out-of-sequence measurements. *IEEE Transactions on Aerospace and Electronic Systems*, Vol. 41, No. 2, pp. 693–702, 2005.
10. R.G. Everitt, S.R. Maskell, R. Wright, M. Briers. Multi-target out-of-sequence data association: Tracking using graphical models. *Information Fusion*, Vol. 7, No. 4, pp. 434–447, 2007.
11. J. Zhen, A. Balasuriya, S. Challa. Sensor fusion-based visual target tracking for autonomous vehicles with the out-of-sequence measurements solution. *Robotics and Autonomous Systems*, Vol. 56, No. 2, pp. 157–176, 2008.
12. M.A. Tanner. *Tools for Statistical Inference.* Springer, New York, 1996.
13. R.L. Streit (ed.). *Studies in Probabilistic Multi-Hypothesis Tracking and Related Topics.* Scientific and Engineering Studies. Vol. SES-98-01. Naval Undersea Warfare Center Division, Newport, RI, 1998.
14. H. Gauvrit, J.-P. LeCadre, C. Jauffet. A formulation of multitarget tracking as an incomplete data problem. *IEEE Transactions on Aerospace and Electronic Systems*, Vol. AES-33, No. 4, pp. 1242–1257, 1997.

15. F. Govaers, W. Koch. An exact solution to track-to-track-fusion at arbitrary communication rates. *IEEE Transactions on Aerospace and Electronic Systems*, Vol. 48, No. 3, pp. 2718–2729, 2012.
16. O.E. Drummond. Target tracking with retrodicted discrete probabilities. *SPIE Signal and Data Processing of Small Targets*, Vol. 3163, pp. 249–268, 1997.
17. A. Gelb (ed.). *Applied Optimal Estimation*. MIT Press, Cambridge, MA, 1974.
18. W. Koch. Target tracking, Chapter 8. In S. Stergiopoulos (ed.), *Signal Processing Handbook*. CRC Press, Boca Raton, FL, 2001.
19. S. Blackman. Multiple hypothesis methods. *IEEE Aerospace and Electronic Systems Society Magazine*, Vol. 19, No. 1, pp. 41–52, 2004.
20. D. Salmond. Mixture reduction algorithms for target tracking in clutter. *SPIE Signal and Data Processing of Small Targets*, Vol. 1305, pp. 434–445, 1990.
21. X.R. Li, V.P. Jilkov. Survey of maneuvering target tracking. Part V. Multiple model methods. *IEEE Transactions on Aerospace and Electronic Systems*, Vol. 41, No. 4, pp. 1255–1321, 2005.
22. W. Koch. Fixed-interval retrodiction approach to Bayesian IMM-MHT for maneuvering multiple targets. *IEEE Transactions on Aerospace and Electronic Systems*, Vol. 36, No. 1, pp. 2–14, 2002.
23. W. Koch, J. Koller, M. Ulmke. Ground target tracking and road map extraction. *ISPRS Journal of Photogrammetry & Remote Sensing*, Vol. 61, pp. 197–208, 2006.
24. P. Willett, R. Yanhua, R. Streit. PMHT: Problems and some solutions. *IEEE Transactions on Aerospace and Electronic Systems*, Vol. 38, No. 3, p. 738, 2002.
25. M. Wieneke, W. Koch. The PMHT: Solutions for some of its problems. *Proceedings of SPIE Vol. 6699, Signal and Data Processing of Small Targets 2007*, p. 669917, 2007. doi:10.1117/12.734388.
26. M. Wieneke, W. Koch. On sequential track extraction within the PMHT framework. *EURASIP Journal on Advances in Signal Processing*, Vol. 2008, Article ID 276914, 2008.
27. M. Wieneke, W. Koch. A PMHT approach for extended objects and object groups. *IEEE Transactions on Aerospace and Electronic Systems*, Vol. 48, No. 3, pp. 2349–2370, 2012.
28. D.A. Harville. *Matrix Algebra from a Statistician's Perspective*. Springer, New York, 1997.

# 19 Belief Function Based Multisensor Multitarget Classification Solution

*Samir Hachour, François Delmotte, and David Mercier*

## CONTENTS

## 19.1 INTRODUCTION

This chapter proposes a multisensor multitarget classification architecture. Each sensor is supposed to track locally a set of randomly appearing and disappearing targets using interacting multiple models (IMM) algorithms [1,2]. Based on estimated kinematic data for targets, a local classification step is performed that aims to recognize the classes of targets (e.g., go-fast boat, military boat, cargo, etc.) based on their behaviors. A special case of classification is considered in which classes are nested: A go-fast boat can behave as a military boat or cargo, a military boat can evolve as slowly as a cargo but cannot go as fast as a go-fast boat, and finally a cargo can evolve neither as fast as a military boat nor as fast as a go-fast boat. In such a classification, belief function based classifiers are shown to perform better than classical Bayesian classifiers, namely in Ref. [3] for a single target classification and in Ref. [4] for multiple target classification. Bayesian and credal classifiers locally processed at each sensor level and are considered in this chapter. Classification results are shown to deteriorate when it comes to managing high sensor uncertainties. To enhance the classification results of the targets, a multisensor approach is needed [5,6]. The question is, How can one fuse data coming from multiple sensors?

This chapter focuses on the multisensor fusion center and provides two main contributions. The first concerns a parameterless track-to-track assignment solution where already existing solutions are mostly parameter dependent [2,7,8]. The track-to-track solution aims to obtain a consensus on the tracked targets; it is shown in Ref. [9] that the proposed solution has the advantage of getting over parameter training. The track-to-track solution receives target estimates provided by the sensors' IMMs at the entry and matches them to obtain a consensus on the commonly tracked targets. Local classification results of the commonly tracked targets are then fused to obtain a better classification performance.

The second contribution of this chapter concerns the way local classification results are fused to obtain the best classification results. Accordingly, some credal and Bayesian fusion rules are tested. It is shown, for the considered classification case (nested classes), that the credal disjunctive rule of combination provides the best results. Notice that in the Bayesian framework, no equivalence to the credal disjunctive rule of combination exists; some other Bayesian rules of combination, provided in Ref. [10], are instead tested. The comparison result was presented in Ref. [11] and extended in this chapter.

Section 19.2 provides some basics on belief function theory. Section 19.3 gives some information on the adopted tracking solution. Section 19.4 describes the Bayesian and the credal classifiers that are locally processed at each sensor level. Section 19.5 highlights the two main contributions of this chapter: the parameterless track-to-track solution and the motivation of using the disjunctive credal rule of combination, which are processed at the fusion center. Finally, a nested classes simulation example is provided in Section 19.6 that describes a piracy surveillance where the proposed fusion solution is shown to be efficient, especially in the case of high sensor uncertainties.

## 19.2 BASICS OF BELIEF FUNCTION THEORY

Belief functions are often referred to as Dempster–Shafer theory, named for Arthur Dempster and Glenn Shafer, who introduced the formalism [12,13]. The theory is also known as credal formalism. Belief functions in the works of Smets are gathered in a framework called the Transferable Belief Model (TBM) [14]. Some basic notions on belief function theory are provided in this section.

### 19.2.1 KNOWLEDGE REPRESENTATION

In credal theory, knowledge is expressed on a finite set of mutually exclusive and exhaustive hypotheses $H = \{h_1, h_2, \ldots, h_N\}$ that is referred to as the *frame of discernment*. Belief can be given to the hypotheses as singletons or sets of hypotheses $A \subseteq H$. This is the major advantage over the probability measures, which require a precise knowledge about the singletons only. The uncertainty is represented by a mass function (basic belief assignment) $m$ expressed on $2^H$ that satisfies

$$\sum_{A \subseteq H} m(A) = 1. \tag{19.1}$$

Subsets $A \subseteq H$ with $m(A) > 0$ are referred to as *focal elements*.

The plausibility function $Pl$ can also be used to model knowledge. It is in one-to-one correspondence with $m$, given as follows:

$$Pl(A) \sum_{A \cap B \neq \varnothing} m(B). \tag{19.2}$$

Some other functions are used to express knowledge and can be found in Ref. [15].

## 19.2.2 Knowledge Combination

Information is often propagated in the form of mass functions. When more than one mass function is expressed on the same frame of discernment, they can be combined. Several combination rules are presented in Ref. [15]. In this chapter, the conjunctive and disjunctive rules of combination are considered.

### 19.2.2.1 Conjunctive Rule of Combination

The conjunctive rule of combination assumes that the masses to combine are provided from independent and reliable sources. For two mass functions $m_{s_1}$ and $m_{s_2}$ provided by sources $s_1$ and $s_2$, the combination rule is given by

$$m_{s_1} \cap s_2(A) = \left( m_{s_1} \cap m_{s_2} \right)(A) = \sum_{A_1 \cap A_2 = A} m_{s_1}(A_1) m_{s_2}(A_2). \tag{19.3}$$

A normalized version of the formula in Equation 19.3 exists; it is referred to as Dempster's rule of combination and is defined by

$$m_{s_1} \cap s_2(A) = \frac{\displaystyle\sum_{A_1 \cap A_2 = A} m_{s_1}(A_1) m_{s_2}(A_2)}{\displaystyle\sum_{A_1 \cap A_2 = \varnothing} m_{s_1}(A_1) m_{s_2}(A_2)}. \tag{19.4}$$

### 19.2.2.2 Disjunctive Rule of Combination

The disjunctive rule of combination supposes that at least one of the sources is reliable. Disjunctive combination of two mass functions $m_{s_1}$ and $m_{s_2}$ that are provided by sources $s_1$ and $s_2$ is defined as follows:

$$m_{s_1} \cup s_2(A) = \sum_{A_1, A_2 | A_1 \cup A_2 = A} m_{s_1}(A_1) m_{s_2}(A_2). \tag{19.5}$$

## 19.2.3 Decision Making

When necessary, probability measures can be calculated from mass functions. This refers to the pignistic transformation [16]. The pignistic probability $BetP$ over hypothesis $h_i \in H$ is defined by

$$BetP(\{h_i\}) = \sum_{h_i \in A} \frac{m(A)}{|A|(1 - m(\varnothing))}. \tag{19.6}$$

### 19.2.4 Generalized Bayes Theorem

Based on the idea of the well-known Bayesian theorem, Smets introduced the generalized Bayesian theorem (GBT) [14], which requires two steps. Based on likelihoods $l(h_i|z) = Pl(h_i|z)$, a conditional mass function $m(A|z)$ for each $A \subseteq H$ is firstly calculated as follows:

$$m(A|z) = \prod_{h_i \in A} l\left(h_i|z\right) \prod_{h_i \in \bar{A}} \left(1 - l\left(h_i|z\right)\right). \tag{19.7}$$

The second step of the GBT consists in updating the result with priors, as with the classical Bayes theorem. This is done using the conjunctive rule (Equation 19.4). If no prior knowledge is available, an a priori vacuous belief function is considered ($m(H) = 1$).

## 19.3 LOCAL TRACKING SOLUTION

This section explains the idea behind multitarget tracking problem and provides some information about the adopted solution to estimate trajectories of targets.

### 19.3.1 Evolution of Targets Model

A jump Markov chain model (JMCM) is often used to represent the evolution of a maneuvering target. An example of a JMCM is given by Equation 19.8, illustrating the evolution of a given target $t$ being in evolution model $ml$, with $l = \{1, ..., r\}$ and $r$ being the number of possible known maneuvering models.

$$x_k^t = Fx_{k-1}^t + Gu_k^t(m_l) + w_k^t, \tag{19.8}$$

where $x_k^t \in \mathbb{R}^p$ represents the $t$th target state vector at time $k$, with $F$ being the $(p \times p)$ state matrix; and $u_k^t$ represents the $t$th target deterministic input, which represents simply a known acceleration mode $m_l$. The parameter $w_k^t$ represents the state Gaussian noise with covariance matrix $Q$. The input matrix is denoted $G$. Detailed information about the adopted models can be found in Refs. [17–20]. For simplicity, the measurements are taken according to a linear model given as follows:

$$z_k^t = Hx_k + v_k, \tag{19.9}$$

where $z_k^j \in \mathbb{R}^q$ is the $j$th received observation at time $k$, with $j \in \{1, 2, ..., m\}$. The observation matrix of dimension $(q \times p)$ is denoted $H$ and $v_k$ represents the measurement error; it is considered as a Gaussian noise with zero mean value and covariance matrix $R$. The set of measurements taken by the sensor $i$ at time $k$ is denoted $Z_i = \left\{z_k^1, z_k^2, ..., z_k^{m_i}\right\}$, with $i \in \{1, 2, ..., S\}$ where $S$ is the number of sensors. Estimation of the state vector in Equation 19.8 involves calculation of the following probability density function:

$$p\left(x_k^t|u_k^t(m_{1,...,r}), z_{1,...,k}^t\right), \quad t = 1, ..., n, \tag{19.10}$$

where $z_{1,...,k}^t$ represents the cumulative measurement for the target $t$ until time step $k$.

### 19.3.2 Local Tracking Algorithm

In default of being able to analytically calculate the complex multimodal density in Equation 19.10, it is estimated using interacting multiple model (IMM) algorithms. A set of IMMs is performed by each sensor, and each IMM tracks a given target. Estimation is performed through two main steps: prediction and update steps. The local tracking algorithm processed at each sensor level is summarized in Algorithm 19.1.

**Algorithm 19.1 Local tracking algorithm**

**Requires:** Measurements: $z^j$, $j = \{1, 2, ..., m\}$
**Ensures:** State estimates: $\hat{x}^t$, $t = \{1, 2, ..., n\}$

1. State vectors $\bar{x}^t$ and measurements $\bar{z}^t$, $t = \{1, 2, ..., n\}$ prediction using IMM prediction step
2. Reception of the real measurements $z^j$, $j = \{1, 2, ..., m\}$
3. Assignment of the real measurement $z^j$, $j = \{1, 2, ..., m\}$ to predicted ones using the matching algorithm described in Section 19.5.1
   a. Measurements are used to update the estimations if they are assigned to known targets.
   b. Measurements that are not assigned to any known target are used to initiate new targets.
   c. Known targets that do not receive any measurement are considered nondetected.
4. State estimates $\hat{x}^t$, $t = \{1, 2, ..., n\}$ update, using IMM update step

At this level each sensor would have estimated the state of $n_i$, $i \in \{1, 2, ..., S\}$ targets and each IMM among the $n_i$ ones would have calculated $r$ evolution mode probabilities $\mu_l$ and likelihoods $\lambda_l$ with $l = \{1, 2, ..., r\}$. These data are used in the classification step.

## 19.4 LOCAL CLASSIFICATION

It is considered that IMM algorithms contain an exhaustive list of all the possible evolution models of the targets. The list of models is given by:

$$M = [m_1, m_2, ..., m_r], \tag{19.11}$$

where $r$ represents the total number of models.

A priori knowledge of target behaviors allows the clustering of the $r$ different models in $M$ to be made. The set of possible behaviors can be defined by $B = [b_1, b_2, ..., b_{nb}]$, where $nb$ is the number of behaviors. The set of models belonging to the behavior $b_i$, for example, is defined by $M_{b_i} \subseteq M$, with $i = 1, ..., nb$. The number of models in $M_{b_i}$ is noted by $r_{b_i}$. Once defined, behavior likelihoods $l_{b_i}$ are calculated from the different model likelihoods $\lambda_j$ and probabilities $\mu_j$, with $j = 1, ..., r$ which are provided by the IMM update steps.

### 19.4.1 BEHAVIOR LIKELIHOODS CALCULATION

Calculation of the behavior likelihoods is performed as follows:

$$l(b_i) = \sum_{j:m_j \in M_{b_i}} \mu'_j \lambda_j, \qquad i = 1, ..., nb, \tag{19.12}$$

with:

$$\mu'_j = \frac{\mu_j}{\sum\limits_{j:m_j \in M_{b_i}} \mu_j}, \qquad j = 1, ..., r_{b_i}. \tag{19.13}$$

To obtain the classes probabilities or pignistic probabilities, the calculated behavior likelihoods can be processed by either Bayesian or credal classifiers, respectively.

### 19.4.2  BAYESIAN CLASSIFIER

Algorithm 19.2 describes the principal steps processed by the Bayesian classifier.

**Algorithm 19.2 Bayesian classifier**

**Requires:** Class a priori probabilities $P(c_i|z_{1,...,k-1})$ and behavior likelihoods $l_{b_j}$, where $i = 1, ..., nc$ and $j = 1, ..., nb$
**Ensures:** Class a posteriori probabilities $P(c_i|z_{1,...,k-1})$

1. Behaviors to class likelihoods

$$l(C) = M \times l(B), \tag{19.14}$$

where $M$ is a matrix representing the relations between behaviors and classes, $l(B)$ represents likelihoods of the behaviors, and $l(C)$ represents likelihoods of the classes.
2. Class a posteriori probabilities calculation

$$P\left(c_i|z_{1,...,k}\right) = \frac{l(c_i)}{\sum_{j=1}^{S} l(c_i)P\left(c_j|z_{1,...,k-1}\right)} P\left(c_i|z_{1,...,k-1}\right), \tag{19.15}$$

At each time step, each sensor $i$ returns a set $\Phi_i = \left\{\phi_k^1, \phi_k^2, ..., \phi_k^{n_i}\right\}$ of probabilities, where $i = 1, ..., S$ with $S$ representing the number of sensors and $\phi_k^t$ the probability distribution concerning the classification of target $t$.

### 19.4.3  CREDAL CLASSIFIER

The credal classifier is executed in two main steps. First, the mass function of behaviors is computed, and then it is transformed into the mass function of classes and a decision is made using the pignistic transformation.

**Algorithm 19.3 Credal classifier**

**Requires:** Class a priori mass function $m_k^C$ and behavior likelihoods $l(B)$
**Ensures:** Class a posteriori pignistic probabilities $BetP^C(c_i)$

1. Calculation of the mass function of behaviors using the GBT:

$$m_k(D) = \prod_{b_i \in D} l(b_i) \prod_{b_i \in \bar{D}} \left(1 - l(b_i)\right), \tag{19.16}$$

where $D \subseteq B$.

2. Projection of the mass function into the class space $C = \{c_1, c_1, ..., c_{nc}\}$ with $nc$ representing the number of classes

$$m_k^C = \bar{M} \times m_k^B,$$
(19.17)

where $\bar{M}$ is a matrix expressing the relations between behaviors and classes and contains the conditional masses $m(A|D)$, with $A \subseteq C$ and $D \subseteq B$.

3. Recursive mass functions combination: $m_k$ and $m_{k-1}$ are combined using the conjunctive rule [14]:

$$m_k^C(D) = \sum_{D_1, D_2 | D_1 \cap D_2 = D} m_k^C(D_1) m_{k-1}^C(D_2),$$
(19.18)

where the initial belief $m_0^C$ is a vacuous mass function [21].

4. To make a decision, the class mass function $m_k^C$ is transformed into pignistic probabilities using Equation 19.6.

For the purposes of a global credal classification, the local decision making (transformation into pignistic probabilities) can be avoided and the local mass functions are used to perform a global classification. The set of mass functions given by a sensor $i$ at time $k$ is denoted $M_i = \left\{m_k^1, m_k^2, ..., m_k^{n_i}\right\}$, where $i = 1, ..., S$ and $S$ is the number of sensors.

## 19.5 GLOBAL CLASSIFICATION

Figure 19.1 illustrates the complete multisensor multitarget algorithm. Other multisensor fusion architectures can be found in Ref. [6]. In the proposed solution, each sensor $i$ among the $S$ used ones makes a set of estimated state vectors $\hat{X}_i = \left\{\hat{x}_k^1, \hat{x}_k^1, ..., \hat{x}_k^{n_i}\right\}$ based on the set of taken measurements $Z_i$ at time $k$ by performing the described local tracking algorithm. Using the local classification algorithm, the sensor provides a set of mass functions $M_i = \left\{m_k^1, m_k^2, ..., m_k^{n_i}\right\}$ for the credal classification and a set of probability distributions $\Phi_i = \left\{\phi_k^1, \phi_k^2, ..., \phi_k^{n_i}\right\}$ provided by the Bayesian classifiers.

The idea behind the global classification is first, to obtain a consensus about the commonly tracked targets through a track-to-track assignment (association) solution and second, to fuse their local classifications.

### 19.5.1 TRACK-TO-TRACK ASSOCIATION

The box "track-to-track association" in Figure 19.1 aims to establish a matching between sensors estimates sets $\hat{X}_{1,...,S}$ elements to recognize in which order the local classification results have to be combined. The matching step is ensured by a newly proposed parameterless assignment method.

1. Plausibility calculation: Let $r_{t,\ell} \in \{0, 1\}$ be the relation that $\hat{x}_i^t$ is associated or not with $\hat{x}_j^\ell$ ($r_{t,l} = 1$ means that target $t$ estimated by sensor $i$ corresponds to target $\ell$ estimated by sensor $j$, $r_{t,\ell} = 0$ otherwise).

For each estimated target $t \in \{1, ..., n_i\}$ given by a sensor $i$, a plausibility function $Pl_t$ is built on the set $\hat{X}_j^{*t} = \left\{\hat{x}_j^1, ..., \hat{x}_j^{n_j}, *_j\right\}$ of sensor $j$'s known targets. Element $*t$ represents the hypothesis that target $t$ is not known by sensor $j$. The plausibility $Pl_t$ is defined by

$$Pl_t(\{\ell\}) = G_{\ell,t}, \quad \forall \ell \in \{1, ..., n_j\},$$
(19.19)

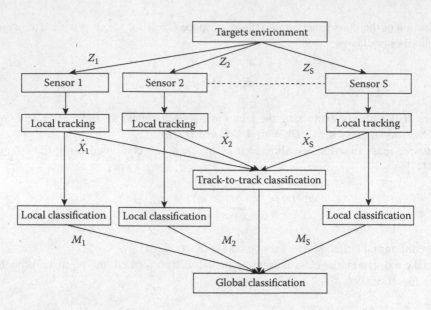

**FIGURE 19.1** Multisensor classification approach flowchart.

where $G_{\ell,t}$ is a likelihood measure calculated as in Ref. [17].

Plausibility $Pl_t(\{\ell\})$ represents the amount of belief supporting the association of $\hat{x}_i^t$ and $\hat{x}_j^\ell$.

The maximum plausibility that target with state $\hat{x}_i^t$ will be associated to one of the $n_j$ already known targets by sensors $j$ corresponds to $\max_{\ell=1,..,n_j} \left( Pl_t(\{\ell\}) \right) \leq 1$. This maximum can be lower than 1, in particular if the frame of discernment formed by the set of known targets is not exhaustive. Indeed, a target with state $\hat{x}_i^t$ can correspond to a new object (*$t$) for sensor $j$. The plausibility of this event is thus defined by

$$Pl_t\left(\{{}^*_t\}\right) = 1 - \max_{\ell=1,..,n_j} \left( Pl_t\left(\{\ell\}\right) \right). \tag{19.20}$$

2. Mass functions calculation: Once a complete plausibility function $Pl_t$ is calculated on a an exhaustive closed-world $\hat{X}_j \cup \{*t\}$, a corresponding mass function denoted by $m_t$ is obtained by a direct application of the GBT [14,22], recalled here for convenience:

$$m_t(A) = \prod_{\ell \in A} Pl_t\left(\{\ell\}\right) \prod_{\ell \in \bar{A}} \left(1 - Pl_t\left(\{\ell\}\right)\right), \quad \forall A \subseteq \hat{X}_j^{*t}. \tag{19.21}$$

3. Assignment decision making: First, mass functions $m_t$ are transformed into pignistic probabilities $BetP_t$ and then the best assignment relation is chosen as the one maximizing the following criterion:

$$\max \sum_{\ell,t} BetP_t\left(\{\ell\}\right) r_{\ell,t}, \quad \ell = \{1, \ldots, n_i + n_j\}, \quad t = \{1, \ldots, n_j\}. \tag{19.22}$$

with respect to the following constraints:

$$\sum_{\ell}^{n_i+n_j} r_{\ell,t} \leq 1, \quad \sum_{t}^{n_j} r_{\ell,t} = 1, \tag{19.23}$$

$$r_{\ell,t} \in \{0,1\} \; \forall \ell \in \{1, \ldots, n_i + n_j\}, \quad \forall t \in \{1, \ldots, n_j\} \tag{19.24}$$

The constraint at the left in Equation 19.23 means that sensor $i$'s estimation of a given target can be assigned to sensor $j$'s known target; if not, it is considered as a new target for sensor $j$. The constraint at right in Equation 19.23 means that a target known by sensor $j$ can be matched with a target of sensor $i$. If the target is not known by sensor $i$, it is assigned to the extraneous element ($*j$).

As in the Denœux et al. [8] and GNN approaches, this problem can be solved using the Hungarian or Munkres algorithms [23].

Once a consensus is reached between sensors about the commonly tracked targets, their local classifications are fused using different Bayesian and credal rules of combination.

### 19.5.2   Local Classification Fusion

The following Bayesian and credal fusion rules are used to fuse local classification probabilities and mass functions respectively.

#### 19.5.2.1   Bayesian Rules of Combination

Some Bayesian rules of combination are tested to obtain the best global classification result:

- Conjunctive rule of combination:

$$P_{s_1} \cap s_2(c_j) = \frac{P_{s_1}(c_j) P_{s_2}(c_j)}{\sum_{c_i \in C} P_{s_1}(c_i) P_{s_2}(c_i)}, \tag{19.25}$$

- Other Bayesian combinations:

$$P_{s_1,s_2}(c_j) = \Psi\left(P_{s_1}(c_j), P_{s_2}(c_j)\right), \tag{19.26}$$

Notice that no disjunctive rule of combination exists in the Bayesian framework; therefore, other combination rules are tested. As reported in Ref. [10], the operator $\Psi$ can represent the average, min, max, and so forth.

#### 19.5.2.2   Credal Rules of Combination

Local credal classification results are fused using two different rules:

- Conjunctive rule of combination given by Equation 19.4
- Disjunctive rule of combination given by Equation 19.5

The presented theoretical knowledge is highlighted through the following simulation example.

## 19.6   MARITIME TARGETS SIMULATION EXAMPLE

### 19.6.1   DESCRIPTION

A maritime targets classification scenario is considered. The simulation aims to identify target types (cargo, military boat, go-fast boat, etc.) where the classification is based on the complexity of the performed maneuvers. A set $B = \{b_1, b_2, b_3\}$ of three behaviors is defined based on the targets' maneuvering capacities:

- Behavior 1 ($b_1$): Behavior of targets having low maneuvering capacities (e.g., cargo)
- Behavior 2 ($b_2$): Behavior of targets having medium maneuvering capacities (e.g., military boat)
- Behavior 3 ($b_3$): Behavior of targets having high maneuvering capacities (e.g., go-fast boat)

where the set of targets possible classes $C = \{c_1, c_2, c_3\}$ corresponds to $C = \{$cargo, military boat, go-fast boat$\}$.

The state vector of a given target is represented by $x = [x \quad \dot{x} \quad y \quad \dot{y}]$, which represents the position and the velocity on $(x, y)$ directions. The state vector of each target evolves following the model in Equation 19.8, with a state matrix $F$ given by

$$F = \begin{bmatrix} 1 & \Delta T & 0 & 0 \\ 0 & 1 & 0 & 0 \\ 0 & 0 & 1 & \Delta T \\ 0 & 0 & 0 & 1 \end{bmatrix}$$

where $\Delta T$ is the sampling time.

The deterministic input vector $u(m_l) = [a_x \, a_y]^T$ in Equation 19.8 represents different acceleration modes of the targets. For example $u(m_1) = [0 \; 0]^T$ represents a constant velocity mode. The differences in the acceleration capabilities allow the classification to be made.

In the performed simulation, each target's IMM is composed of 13 evolution models according to the different maneuvers that can be made in $x$ and $y$ directions (more details can be found in Ref. [24]). The different evolution models are distributed over the three possible target behaviors, as follows:

$M_{b_1} = [m_1]$: models belonging to the behavior $b_1$.

$M_{b_2} = [m_1, \ldots, m_5]$: models belonging to the behavior $b_2$.

$M_{b_3} = [m_1, \ldots, m_{13}]$: models belonging to the behavior $b_3$.

Once behaviors are defined, their likelihoods are calculated as illustrated by Equation 19.12. Knowledge on likelihoods space $B$ is then projected to classes space $C$ according to the following relations.

- Relation 1: Target in behavior 1 can correspond to cargo, military, or go-fast boats. All of them can evolve with a constant velocity. This relation can be written as $b_1 = \{c_1, c_2, c_3\}$.
- Relation 2: Target in behavior 2, which has performed a medium maneuver, may correspond to a military or go-fast boat only. Cargos are supposed unable to perform any maneuver. This relation can be written as $b_2 = \{c_2, c_3\}$.
- Relation 3: Target in behavior 3, which has performed a sharp maneuver, can only be a go-fast boat, because cargos and military boats cannot perform sharp maneuvers. This relation can be written as $b_3 = \{c_3\}$.

Behavior knowledge is transferred to class knowledge using Equation 19.14 for the Bayesian classifier, where the matrix $M$ is given by

$$M = \begin{bmatrix} 1/3 & 0 & 0 \\ 1/3 & 1/2 & 0 \\ 1/3 & 1/2 & 1 \end{bmatrix},$$

which corresponds to the following conditions:

$$\text{If } P(b_1) = 1 \Rightarrow P(c_1|b_1) = \frac{1}{3}, \quad P(c_2|b_1) = \frac{1}{3}, \quad P(c_3|b_1) = \frac{1}{3}.$$

$$\text{If } P(b_2) = 1 \Rightarrow P(c_1|b_2) = 0, \quad P(c_2|b_2) = \frac{1}{2}, \quad P(c_3|b_2) = \frac{1}{2}.$$

$$\text{If } P(b_3) = 1 \Rightarrow P(c_1|b_3) = 0, \quad P(c_2|b_3) = 0, \quad P(c_3|b_3) = 1.$$

Behavior belief is more precisely transferred into the class space using Equation 19.17 according to the relations described previously. Knowledge transfer is performed from the space $2^B$ into space $2^C$ which is illustrated by $(2^3 = 8) \times (2^3 = 8)$ transfer matrix $\bar{M}$ given by

$$\bar{M} = \begin{bmatrix} 1 & 0 & 0 & 0 & 0 & 0 & 0 & 0 \\ 0 & 0 & 0 & 0 & 0 & 0 & 0 & 0 \\ 0 & 0 & 0 & 0 & 0 & 0 & 0 & 0 \\ 0 & 0 & 0 & 0 & 0 & 0 & 0 & 0 \\ 0 & 0 & 0 & 0 & 1 & 0 & 0 & 0 \\ 0 & 0 & 0 & 0 & 0 & 0 & 0 & 0 \\ 0 & 0 & 1 & 0 & 0 & 0 & 1 & 0 \\ 0 & 1 & 0 & 1 & 0 & 1 & 0 & 1 \end{bmatrix}.$$

according to the following knowledge conditioning:

- $m_k(\{c_1, c_2, c_3\}|b_1) = 1$ (cf. Relation 1)
- $m_k(\{c_2, c_3\}|b_2) = 1$ (cf. Relation 2)
- $m_k(\{c_1, c_2, c_3\}|\{b_1, b_2\}) = 1$ (cf. Relations 1 and 2)
- $m_k(c_3|b_3) = 1$ (cf. Relation 3)
- $m_k(\{c_1, c_2, c_3\}|\{b_1, b_3\}) = 1$ (cf. Relations 1 and 3)
- $m_k(\{c_2, c_3\}|\{b_2, b_3\}) = 1$ (cf. Relations 2 and 3)
- $m_k(\{c_1, c_2, c_3\}|\{b_1, b_2, b_3\}) = 1$ (cf. Relations 1, 2, and 3)

This matrix enables computation of the mass functions on the classes space $C$, to take a decision on the targets' classifications.

## 19.6.2 SIMULATION AND RESULTS

Some simulation results are given in this section, performed using MATLAB®. The first one concerns the track-to-track assignment step where the parameterless proposed solution is compared with the widely used global nearest neighbor (GNN) algorithm [17], which is a parameter-dependent solution. The second result recalls a local classifications comparison performed by a reliable sensor. The result highlights the outperforming of the credal classifier over the Bayesian one. Finally, the last results consider sensors with high uncertainties such that local classifications are deteriorated, and the results of their fusion are evaluated. The global classification is shown to be the best when using the disjunctive credal rule of combination.

### 19.6.3 Assignment Simulation

A scenario of two nearby targets is considered in this section. It represents a conflicting situation illustrated by Figure 19.2a. The simulation aims to calculate the rate of false decisions made by the parameterless assignment solution compared to the GNN algorithm. Notice that GNN is a parameter-dependent algorithm; it depends on parameters $\lambda$ that can be seen as a detection distance. False decision rates in Figure 19.2b are given for different values of $\lambda$.

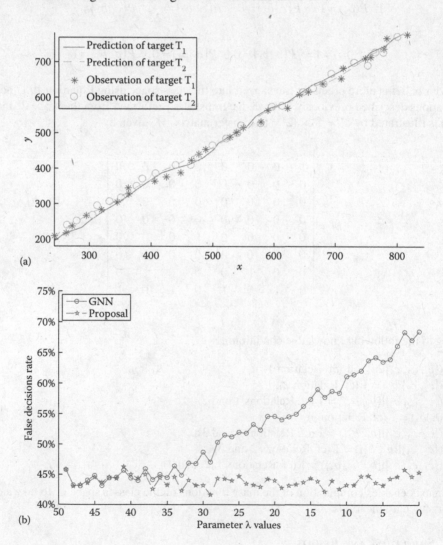

**FIGURE 19.2** Track-to-track assignment test. (a) Conflictual scenario of nearby targets and (b) false decisions rates with detection distance variation.

The simulation presented in this section shows that the value chosen for the parameter λ influences the performance of the corresponding algorithm. In fact, the parameter needs to be trained to ensure an optimal performance. A similar comparison is made in Ref. [9] including another parameter-dependent solution [8]. The comparison highlights the advantage of using the proposed parameterless solution.

### 19.6.4   LOCAL CLASSIFICATION RESULTS

In this section, the scenario given in Figure 19.3 is considered. First, local Bayesian and credal classifications of a reliable sensor are given. Classification results concern only target 2 in Figure 19.3. Knowing that target 2 evolves most of the time with a constant velocity, it performs two maneuvers: a first medium maneuver at time interval [62, 66] and a sharp maneuver at time interval [80, 86]. Normally, target 2 is expected to be in doubt between the three classes during the first constant velocity evolution step; to be in doubt between the second and the third classes after its medium maneuver, because targets of class 1 cannot perform any maneuver; and finally, to be classified in class 3 after its sharp maneuver.

The Bayesian and the credal classification results are, respectively, depicted in Figure 19.4a and b. It can be seen that the credal classifier provides almost the expected classification, namely, perfect doubt between all classes in the first step of movement, perfect doubt between the second and third class in the second step of movement, and the final classification in the third class after the sharp maneuver. At the final step, the Bayesian classifier succeeds in classifying the target, but it fails to manage the imprecision of the first and second steps of the movement. More details and explanations about this result can be found in Ref. [4].

The local classification results for unreliable sensors can be given by Figure 19.5a and b, respectively, for the Bayesian and the credal classifiers.

The objective of the proposed multisensor approach is to enhance the local classifications by fusing them using the described Bayesian and credal rules of combination.

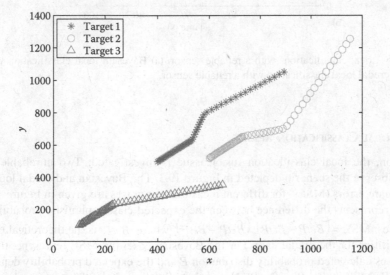

**FIGURE 19.3**   Multitarget scenario.

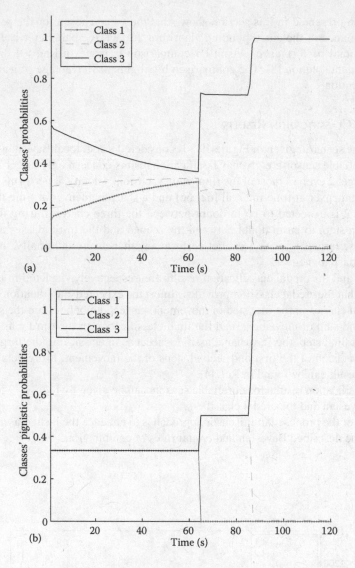

**FIGURE 19.4** Local classifications with a reliable sensor. (a) Bayesian local classification with a reliable sensor and (b) credal local classification with a reliable sensor.

### 19.6.5 GLOBAL CLASSIFICATION RESULTS

In this section, the local classification fusion issue is investigated. Two unreliable sensors are designed to observe the scenario depicted in Figure 19.3. The Bayesian and credal local classification mean square errors (MSEs) for different measurement noises are given in Figure 19.6.

The MSE represents the difference between the expected classes pignistic probabilities and the calculated ones: $\text{MSE} = \left(\widehat{BetP} - BetP\right)'\left(\widehat{BetP} - BetP\right)$, where $\widehat{BetP}$ is the theoretical expected pignistic probabilities for the credal case. For the Bayesian case, the MSE represents the difference between classes calculated probability distribution $P$ and the expected probability density distribution denoted $\hat{P}$: $\text{MSE} = (\hat{P} - P')'(\hat{P} - P)$. Notice that all the results of this section are averaged over 30 Monte Carlo simulations.

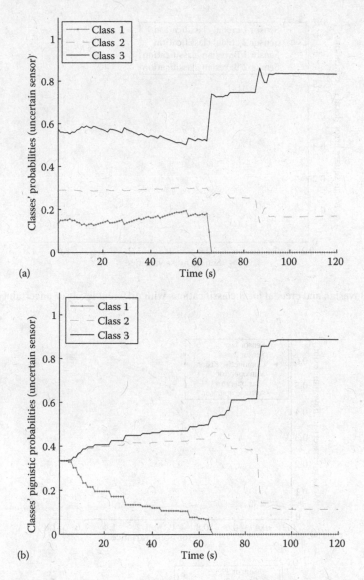

**FIGURE 19.5** Local classifications with an unreliable sensor. (a) Bayesian local classification with an unreliable sensor and (b) credal local classification with an unreliable sensor.

The results presented in Figure 19.6 confirm the result depicted in Figure 19.4 about the outperforming of the credal classifier over the Bayesian classifier. Figure 19.6 shows also the increasing aspect of the classification errors when sensor uncertainty increases. The multisensor approach aims to enhance the classification results presented in Figure 19.6. Bayesian and credal global classification results are given in Figure 19.7a and b respectively.

Local and global Bayesian classification results are given in Figure 19.7a. As can be seen, the best classification result is given by max operator. Concerning the credal classification results in Figure 19.7a, the best classification is given by the disjunctive rule of combination, and worst is given by the conjunctive combination rule. This is due to the nature of the conjunctive combination, which favors singletons or specific subsets over the doubt. In fact, for example, if sensor 1's belief is given

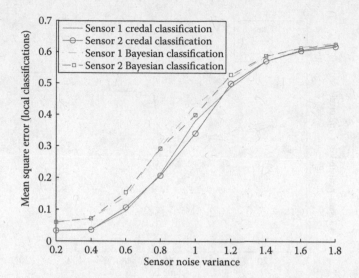

**FIGURE 19.6**   Bayesian and creedal local classifications with different levels of uncertainty.

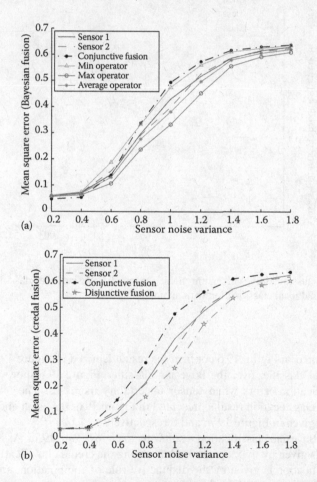

**FIGURE 19.7**   Global classification results using different rules of combination. (a) Global classification results using Bayesian fusion rules and (b) global classification results using credal fusion rules.

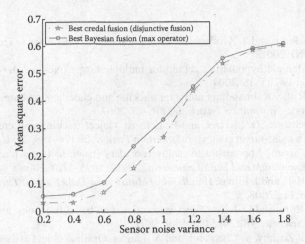

**FIGURE 19.8** Multitarget scenario.

by $m_1(\{c_2, c_3\}) = 0.4$ and $m_1(\{c_1, c_2, c_3\}) = 0.6$, and sensor 2's belief is given by $m_2(\{c_2, c_3\}) = 0.5$ and $m_2(\{c_1, c_2, c_3\}) = 0.5$, the conjunctive combination gives $m_{1 \cap 2}(\{c_2, c_3\}) = 0.7$ and $m_{1 \cap 2}(\{c_1, c_2, c_3\}) = 0.3$. This explains the accentuation of the classification divergence caused by sensor noises during the first and second steps of movement where doubt is preferred. On the other hand, the disjunctive combination of the same mass functions gives $m_{1 \cup 2}(\{c_2, c_3\}) = 0.2$ and $m_{1 \cup 2}(\{c_1, c_2, c_3\}) = 0.8$. This illustrates the cautious nature of the disjunctive combination, which helps one to obtain the best global classification result for the considered problem.

Figure 19.8 compares the best Bayesian classification result with the best credal classification result. It can be seen the credal global classification is more robust than the Bayesian one.

## 19.7 CONCLUSION

This chapter recalls some previous results of kinematic data-based classification [3,4]. In a case of nested classes, the credal classifier was shown to outperform the classical Bayesian classifier. It is shown in this chapter that classification results can become deteriorated when it comes to manage high sensor uncertainties. Therefore, a centralized multisensor approach was proposed that aims to enhance the classification results by fusing local classifications of sensors.

Multiple Bayesian and credal fusion rules are tested. It was shown that the best classification result is given by the credal disjunctive rule of combination. It is a cautious fusion rule that seems to be adapted to highly imprecise classification scenarios such as the case here for a nested classes situation.

Notice that before fusing the local classifications, a consensus on the tracked targets needs to be reached. This is done here using a new parameterless track-to-track matching method. In contrast to the existing methods, such as GNN algorithm, for example, the proposed method does not need any parameter training, which is appreciable for random tracking environment applications.

Through the results of this chapter, it can be seen that the application of the credal formalism in multiple target tracking and classification contexts is a challenging and promising task. It is here applied to the association and the classification steps. In the future, filtering methods with belief function will be considered.

# REFERENCES

1. Y. Bar-Shalom, X.R. Li, and T. Kirubarajan. *Estimation with Applications to Tracking and Navigation*. Wiley, Hoboken, NJ, 2001.
2. S.S. Blackman. Multiple hypothesis tracking for multiple target tracking. *Aerospace and Electronic Systems Magazine*, 19(1):5–18, 2004.
3. P. Smets, and B. Ristic. Kalman filter and joint tracking and classification based on belief functions in the TBM framework. *Information Fusion*, 8(1):16–27, 2007.
4. S. Hachour, F. Delmotte, D. Mercier, and E. Lefèvre. Object tracking and credal classification with kinematic data in a multi-target context. *Information Fusion*, 20:174–188, 2014.
5. D. Smith, and S. Singh. Approaches to multisensor data fusion in target tracking: A survey. *IEEE Transactions on Knowledge and Data Engineering*, 18(12):1696–1710, 2006.
6. M. Liggins II, D. Hall, and J. Llinas. *Handbook of Multisensor Data Fusion: Theory and Practice*. CRC Press, Boca Raton, FL, 2008.
7. D. Mercier, É. Lefèvre, and D. Jolly. Object association with belief functions, an application with vehicles. *Information Sciences*, 181(24):5485–5500, 2011.
8. T. Denœux, N. El Zoghby, V. Cherfaoui, and A. Jouglet. Optimal object association in the Dempster-Shafer framework. *IEEE Transactions on Cybernetics*, 44(12):2521–2531, 2014.
9. S. Hachour, F. Delmotte, and D. Mercier. A new parameterless credal method to track-to-track assignment problem. In *3rd International Conference on Belief Functions*, pp. 403–411, Oxford, UK, September 2014.
10. I. Bloch. Information combination operators for data fusion: A comparative review with classification. *IEEE Transactions on Systems, Man, and Cybernetics*, 26(1):52–67, 1996.
11. S. Hachour, F. Delmotte, E. Lefèvre, and D. Mercier. Multi-sensor multi-target tracking with robust kinematic data based credal classification. In *8th Workshop, Sensor Data Fusion: Trends, Solutions, Application, SDF 2013*, paper Thu1130, Bonn, Germany, October 9–11, 2013.
12. A.P. Dempster. A generalization of Bayesian inference. *Journal of the Royal Statistical Society. Series B*, 30(2):205–247, 1968.
13. G. Shafer. *A Mathematical Theory of Evidence*. Princeton University Press, Princeton, NJ, 1976.
14. P. Smets. Belief functions: The disjunctive rule of combination and the generalized Bayesian theorem. *International Journal of Approximate Reasoning*, 9(1):1–35, 1993.
15. P. Smets. Belief functions on real numbers. *International Journal of Approximate Reasoning*, 40(3):181–223, 2005.
16. P. Smets. Decision making in the TBM: The necessity of the pignistic transformation. *International Journal of Approximate Reasoning*, 38(2):133–147, 2005.
17. S.S. Blackman, and R. Popoli. *Design and Analysis of Modern Tracking Systems*. Artech House, Norwood, MA, 1999.
18. G.A. Watson, and W.D. Blair. IMM algorithm for tracking targets that maneuver through coordinated turns. *Proceedings of Signal and Data Processing of Small Targets*, 1698:236–247, 1992.
19. R.L. Moose, H.F. Vanlandingham, and D.-H. McCabe. Modeling and estimation for tracking maneuvering targets. *IEEE Transactions on Aerospace and Electronic Systems*, 15(3):448–456, 1979.
20. R.A. Singer. Estimating optimal tracking filter performance for manned maneuvering targets. *IEEE Transactions on Aerospace and Electronic Systems*, 6(4):473–483, 1970.
21. P. Smets, and R. Kennes. The transferable belief model. *Artificial Intelligence*, 66(2):191–234, 1994.
22. F. Delmotte, and P. Smets. Target identification based on the transferable belief model interpretation of Dempster-Shafer model. *IEEE Transactions on Systems, Man and Cybernetics, Part A: Systems and Humans*, 34(4):457–471, 2004.
23. F. Bourgeois, and J.-C. Lassalle. An extension of the Munkres algorithm for the assignment problem to rectangular matrices. *Communications of the ACM*, 14(12):802–804, 1971.
24. B. Ristic, N. Gordon, and A. Bessell. On target classification using kinematic data. *Information Fusion*, 5(1):15–21, 2004.

# 20 Decision Fusion in Cognitive Wireless Sensor Networks

*Andrea Abrardo, Marco Martalò, and Gianluigi Ferrari*

## CONTENTS

## 20.1 INTRODUCTION

With the development of wireless communications in the last few years, most of the available spectrum has been fully allocated. On the other hand, recent investigations on the actual spectrum utilization have shown that a portion of the licensed spectrum is largely underutilized [1]. As a matter of fact, the so-called spectrum scarcity problem is due mostly to an inefficient spectrum allocation policy rather than to actual physical spectrum shortage. Accordingly, to "chase" the explosion of wireless communications, novel solutions should be envisaged.

Dynamic spectrum access (DSA) has been considered to achieve a more efficient radio spectrum utilization [2,3]. In DSA, part of the spectrum can be allocated to one or more users, which are called primary users (PUs). Such a spectrum is not exclusively dedicated to PUs, although they have higher priority than other users, which are referred to as secondary users (SUs). In particular, SUs can access the same spectrum as long as the PUs are not temporally using it or can share the spectrum with the PUs as long as the PUs can be properly protected. By doing so, the radio spectrum can be reused in an opportunistic manner or shared all the time; thus the spectrum utilization efficiency can be improved significantly.

To support DSA, SUs are required to capture or sense the radio environment, and an SU with such a capability is also called a cognitive radio (CR) [4,5]. One of the main tasks of CR is represented by spectrum sensing (SS), defined as the task of finding spectrum holes [6], that is, portions of the spectrum allocated (licensed) to some primary users but left unused for a certain time. On the other hand, SS from a single node does not always guarantee satisfactory performance because of noise uncertainty, the intrinsic random nature of the nodes' positions, and unpredictable channel fluctuations. For example, a CR user cannot detect the signal from a primary transmitter behind a high building and may decide to access the licensed channel, thus interfering with the primary receiver.

On the other hand, collaboration of multiple users in SS may highly improve SS performance by introducing a form of spatial diversity [7,8]. In cooperative SS, CR users first send the collected data to a combining user or fusion center (FC). Alternatively, each user may independently perform local decisions and then report binary decisions to the FC. Finally, the FC takes a decision on the presence or absence of the licensed signal based on the received information.

In this chapter, we focus on a cognitive WSN, where a primary wireless sensor network (PWSN) is co-located with a cognitive (or secondary) sensor network (CWSN). In particular, the nodes of the CWSN reach their associated access point (AP) directly (single hop). The frequency spectrum that is shared by PWSN and CWSN is divided into subchannels, which can be assigned by the PWSN, whereas the nodes of the CWSN cooperate to sense the frequency spectrum and estimate the free subchannels that can be used to transmit their data. The secondary nodes transmit the packets containing the observations on the channels' statuses to their FC, embedded in the secondary AP, which makes a final decision about the status (free or busy) of each subchannel and broadcasts this information to all secondary nodes. The correlation, in the sensing operation, among the secondary nodes is taken into account. In particular, we provide a simple analytical model to characterize the local sensing performance per subchannel, in terms of probabilities of missed detection (MD) and false alarm (FA). Moreover, we propose a joint source channel coding (JSCC) scheme, in the CWSN, to exploit the source correlation to improve the reliability of the final decision taken by the secondary FC. By relying on recent results on the design of practical decoding and fusion rules for multiple access schemes with correlated sources [9], we derive an effective fusion rule and an associated iterative decoding strategy for joint channel decoding (JCD) at the secondary FC.

The rest of the chapter is structured as follows. In Section 20.2, we present the reference scenario. In Section 20.3, we analyze the performance from a single node perspective. Then, in Section 20.4 we analyze the performance from a network point of view, distinguishing between uncoded and coded scenarios. In Section 20.5, both theoretical and simulation results are provided. Finally, concluding remarks are given in Section 20.6.

## 20.2  SYSTEM MODEL

The scenario of interest is shown in Figure 20.1. The FC is placed at the center of the cell, while secondary and primary nodes are independent and identically distributed (i.i.d.) according to a uniform distribution in a circular cell with a given radius $R$. A logical description of the scenario is given in Figure 20.2. Let us denote as cognitive user equipment (CUE) and primary user equipment (PUE) the secondary and primary nodes, respectively.* The number of PUEs and CUEs is equal to $P$ and $N$, respectively. Each PUE is assigned one channel among a set of $N_{ch}$ orthogonal subchannels, for example, a set of nonoverlapping $N_{ch}$ frequency bands. Moreover, each subchannel is assigned to at most one PUE, which transmits its own data (when available) with fixed power $P_T$ over the assigned subchannel. The status of the $i$th ($i \in \{1,\ldots,N_{ch}\}$) subchannel $S_i$ is assumed binary, namely:

$$S_i = \begin{cases} S_0 & \text{with probability } P(S_0) \\ S_1 & \text{with probability } P(S_1) = 1 - P(S_0). \end{cases}$$

Data transmissions follow a classical model for cellular environments, where the path loss is thoroughly characterized by two parameters: (1) the distance attenuation factor $\alpha$ (adimensional, in the range 2–4) and (2) the standard deviation $\sigma$ (in decibels) of the log-normal shadowing [10].

Each CUE scans all $N_{ch}$ channels to detect the presence of a primary signal transmission. In other words, the CUEs perform a binary hypothesis test on the presence of primary signals in each

---

* This is a slight abuse of notation, as CUE and PUE are notations typically adopted in cellular systems. However, as smartphones are today advanced sensing systems, we keep this notation also in the realm of WSNs.

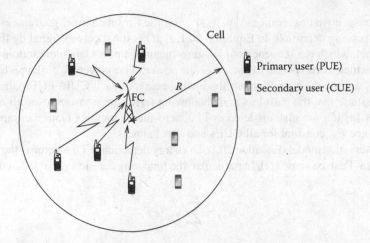

**FIGURE 20.1** Cognitive scenario of interest.

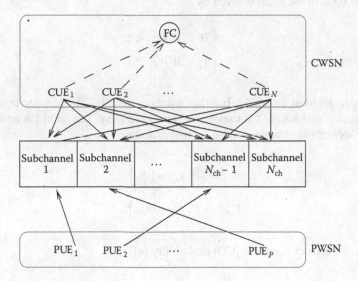

**FIGURE 20.2** Logical description of the scenario of interest.

subchannel: A subchannel is idle under hypothesis $S_0$ and busy under hypothesis $S_1$. As for the signal model, we assume that the primary signal can be modeled as a zero-mean stationary white Gaussian process: This is a reasonable assumption when the CWSN has no a priori knowledge about the possible modulation and pulse shaping formats adopted by the PWSN.

On the basis of the preceding model assumptions, the $k$th CUE ($k = 1, ..., N$) has to distinguish, for the $i$th subchannel ($i = 1, ..., N_{ch}$), between two independent Gaussian sequences:

$$v_{k,i}(\ell) = \begin{cases} n_{k,i}(\ell) & \text{if } S_0 \\ s_{k,i}(\ell) + n_{k,i}(\ell) & \text{if } S_1 \end{cases} \qquad \ell = 1, ..., m \qquad (20.1)$$

where $m$ is the number of observed samples—for simplicity, the time index $\ell$ refers to a single block of observations. Note that an implicit assumption in Equation 20.1 is that the phenomenon status ($S_0$ or $S_1$) does not change over $m$ consecutive observations: This is realistic given that $m$ is typically much shorter than the duration of a primary signal transmission. Moreover, the fact that the $k$th node takes $m$ consecutive observations $\left\{ v_{k,i}(\ell) \right\}_{\ell=1}^{m}$ on the $i$th subchannel guarantees a sort of time diversity in the sensing operation. In Equation 20.1, $s_{k,i}(\ell)$ is the received signal by the $k$th CUE on the $i$th subchannel, which is a sequence of i.i.d. zero-mean complex Gaussian random variables with variance $P_R^{(k,i)}$, which corresponds to the received power. The power $P_R^{(k,i)}$ depends on the transmitted power $P_T$ and on the path loss and shadowing terms of each CUE–PUE pair. Note that it is reasonable to assume that the path loss and shadowing terms are constant over all $m$ acquisitions. The noise terms $\{n_{k,i}(\ell)\}$ are also modeled as i.i.d. zero-mean complex Gaussian random variables with fixed variance $P_N$, constant for all CUEs and subchannels.

Under the observation model (Equation 20.1), an energy detection (ED) scheme is the optimal detector in the Neyman–Pearson sense [11]. In particular, the following decision variable has to be evaluated:

$$W_{k,i} = \sum_{\ell=1}^{m} \left| v_{k,i}(\ell) \right|^2$$

and the binary decision of the CUE is given by

$$x_{k,i} = \begin{cases} 0 & \text{if } W_{k,i} < \tau \\ 1 & \text{if } W_{k,i} \geq \tau \end{cases} \qquad (20.2)$$

where $\tau$ is a proper decision threshold. In other words, each CUE decides for 0 if the channel is sensed idle, whereas it decides for 1 if the channel is sensed busy. The local FA and MD probabilities, under the proposed ED scheme, can be defined as follows:

$$P_{FA}^{(k,i)} \triangleq P\left( x_{k,i} = 1 \middle| S_0 \right)$$

$$P_{MD}^{(k,i)} \triangleq P\left( x_{k,i} = 0 \middle| S_1 \right).$$

Consequently, the correct detection (CD) probability is

$$P_{CD}^{(k,i)} = 1 - P_{MD}^{(k,i)}.$$

The $k$th CUE then generates the decision vector $x_k = \left( x_{k,1}, \ldots, x_{k,N_{ch}} \right)$ $(k = 1,\ldots,N)$, where $x_{k,i} \in \{0, 1\}$ $(i = 1,\ldots,N_{ch})$ corresponds to its local decision on the absence (0) or presence (1) of a primary signal in the $i$th subchannel. This vector is then transmitted to the secondary FC, which, on receiving decision vectors from all nodes of the CWSN, applies a proper fusion strategy to make a final decision on the status (free or busy) of each subchannel. The FC can thus broadcast this information to all secondary nodes, possibly with the assignment of the free subchannels to a subset of them, to avoid multiple access interference.

## 20.3   SINGLE-NODE PERSPECTIVE

We now analyze the performance from a single-node perspective. As the subchannels are independent, without loss of generality we focus on a single subchannel and derive the corresponding FA and MD probabilities—for notational simplicity, we drop the superscript/subscript referring to the subchannel. Because each subchannel is assigned to at most one PUE, without loss of generality we also focus on a single PUE.

With reference to Figure 20.1, assume that CUEs and PUEs are uniformly distributed within the cell with radius $R$ and that the FC is positioned at the cell center. Denote as $X_k = x_k e^{j2\pi\theta_k}$ and $X_p = y e^{j2\pi\phi}$ the positions of the $k$th CUE and the PUE, respectively, where $0 \le x_k, y \le R$ and $0 \le \theta_k, \phi \le 2\pi$. The distance $d$ between the two nodes is

$$d_k = |X_p - X_k| = \sqrt{x_k^2 + y^2 - 2x_k y \cos(\theta_k - \phi)}.$$

Assuming a fixed transmit power $P_T$ for primary nodes, the power $P_R^{(k)}$ received by the CUEs can be expressed as

$$P_R^{(k)} = \frac{K}{d_k^\alpha} h_k P_T$$

where $K$ is the gain at 1 meter from the emitter; $h_k$ is the log-normal shadowing coefficient of the link between the PUE and the $k$th CUE; and $d_k$ is the distance between the PUE and the $k$th CUE. Therefore, the sensing signal-to-noise ratio (SNR) experienced by the $k$th CUE, with respect to the PUE, can be expressed as follows:

$$\gamma_k(d_k, h) = \frac{P_R^{(k)}}{P_N} = \frac{K h_k P_T}{P_N d_k^\alpha}.$$

The local FA and MD probabilities for ED can then be evaluated as [11]

$$P_{FA}^{(k)} = \Gamma_u(m\tau_N, m)$$

$$P_{MD}^{(k)} = 1 - \Gamma_u\left(\frac{m\tau_N}{1 + \gamma_k(d_k, h)}, m\right)$$

where $\tau_N = P_N \tau$ is the normalized threshold (with $\tau$ introduced in Equation 20.2) and $\Gamma_u$ is the upper incomplete gamma function:

$$\Gamma_u(a, n) \triangleq \int_a^\infty x^{n-1} e^{-x} \, dx.$$

Note that the FA probability is the same for all CUEs and does not depend on their distances from the PUE, that is, $P_{FA}^{(k)} = P_{FA}$. The MD probability, instead, depends on the distance $d_k$ and on the shadowing term $h_k$. Averaging with respect to the statistical distribution of the shadowing term, the following expression for the average MD probability at distance $d_k$ is obtained:

$$\bar{P}_{MD}^{(k)} = 1 - \frac{1}{\sqrt{2\pi\sigma^2}} \int_{-\infty}^\infty \Gamma_u\left(\frac{m\tau_N}{1 + \gamma_k(d_k, 10^{S/10})}, m\right) e^{-\frac{S^2}{2\sigma^2}} \, dS. \qquad (20.3)$$

Even though the integral in Equation 20.3 has no closed-form solution, it can be numerically evaluated. Finally, the average MD probability for any CUE, denoted as $\bar{P}_{MD}$, can be obtained from Equation 20.3 by averaging over the distance $d_k$, thus obtaining:

$$\bar{P}_{MD} = \int_\rho \bar{P}_{MD}^{(k)}(\rho) f_D(\rho) \, d\rho \qquad (20.4)$$

where $f_D(\rho)$ is the probability density function (PDF) of the CUE–PUE distance. The average CD probability for CUE, denoted as $\bar{P}_{CD}$, can be straightforwardly expressed as

$$\bar{P}_{CD} = 1 - \bar{P}_{MD}.$$

Expression 20.4 for $\bar{P}_{MD}$ is an average measure for all possible CUEs. The PDF $f_D(\rho)$ is associated with the distance between two randomly chosen points in a circle and, according to the Crofton fixed points theorem [12], can be given by the following expression:

$$f_D(\rho) = 2\frac{\rho}{R^2}\left[\frac{2}{\pi}\arccos\left(\frac{\rho}{2R}\right) - \frac{\rho}{\pi R}\sqrt{1 - \frac{\rho^2}{4R^2}}\right] \quad 0 \le \rho \le 2R.$$

The choice of a proper value for the threshold $\tau_N$ is crucial to maximize the ultimate system performance. Indeed, a small value of $\tau_N$ yields frequent FAs, whereas a high value of $\tau_N$ entails a high MD probability. The optimized value of $\tau_N$ obviously depends on the sensing SNR experienced by each CUE, which, in turn, depends on several other uncontrollable characteristics, such as the positions of CUEs/PUEs and shadowing. We preliminarily observe that the selection of a different and optimized threshold for each CUE would involve a huge amount of message exchange between the FC and the CUEs. Therefore, we make the reasonable assumption that the threshold is a predefined system parameter to be optimized off-line (e.g., in a training phase) through statistical considerations. More generally, we make the assumption that the FC has no knowledge about the actual positions of CUEs and PUEs inside the cell.

## 20.4  NETWORK PERSPECTIVE

We now analyze the performance from a network perspective, considering the use of the local decisions coming from all the CUEs, considering possible data fusion rules. In particular, in Section 20.4.1 we first focus on an uncoded scenario with error-free communication links. Then, in Section 20.4.2 we extend our analysis to the presence of communication noise and channel coding.

### 20.4.1  UNCODED SCENARIO WITH NO COMMUNICATION NOISE

We first consider the case where data (i.e., local decisions on the statuses of the subchannels) are transmitted as uncoded by each CUE to the secondary FC, using a set of orthogonal error-free communication channels. This makes it possible to derive the ultimate performance limits that can be achieved in this scenario. In reality, noisy communications from CUEs to the secondary FC are likely to degrade the system performance, as shown by means of simulations in Section 20.5.

To derive the network-wide performance in the absence of channel coding, we now consider all the $N_{ch}$ subchannels, where the characteristics of the local decisions by the CUEs, in terms of FA and MD probabilities, have been derived in Section 20.3. Using the considered observation model, it is possible to write the a priori joint (among all CUEs) probability mass function (PMF) of the decisions for the $i$-th subchannel ($i \in \{1,\dots,N_{ch}\}$) as [13]

$$P(x_i) = P(x_i|S_i = S_0)P(S_0) + P(x_i|S_i = S_1)P(S_1)$$
$$= P(S_0)\prod_{k\in\mathcal{X}_0}(1 - P_{FA})^k \prod_{k\in\mathcal{X}_1}(P_{FA})^{N-k} + P(S_1)\prod_{k\in\mathcal{X}_0}(1 - \bar{P}_{CD})^k \prod_{k\in\mathcal{X}_1}(\bar{P}_{CD})^{N-k} \quad (20.5)$$

where

$$\mathcal{X}_0 = \{h \in \{1,\ldots,N\} : x_{k,h} = 0\}$$

$$\mathcal{X}_1 = \{j \in \{1,\ldots,N\} : x_{k,j} = 1\}.$$

Obviously, $\mathcal{X}_0 \cup \mathcal{X}_1 = \{1,\ldots,N\}$.

As the communication channels between the CUEs and the secondary FC are error free, the transmitted data are received correctly and the fusion rule at the FC can be written as

$$\Lambda_i = \sum_{k=1}^{N} x_{k,i} \underset{S_0}{\overset{S_1}{\gtrless}} T \tag{20.6}$$

where $T$ is a "global" decision threshold (i.e., equal for all subchannels) to be optimized. In other words, the hypothesis testing problem turns out be a counting problem: the number of local decisions, at the CUEs, in favor of $S_1$ is first counted and then compared with the threshold $T$. The fusion rule (Equation 20.6) is shown to be optimal under the assumption, as in this chapter, of blind detection, that is, the FC has no knowledge about the actual sensing accuracy (in terms of local FA and MD probabilities) of each CUE [14].

Let us now evaluate the performance at the FC by computing the final CD and FA probabilities, denoted as $P_{CD,f}$ and $P_{FA,f}$, respectively—note that these probabilities do not depend on the subchannel. To this end, we approximate the performance of the system following an approach similar to that in Ref. [14]. In particular, we assume that all users are characterized by the same $\bar{P}_{MD}$ given by Equation 20.4, regardless of the position of the PUE. However, unlike Ref. [14], we do not consider a Gaussian approximation for $\Lambda$, as the number of CUEs may not necessarily be large. As the observations $\{x_k\}$ are i.i.d. with a Bernoulli distribution, one can write

$$P_{CD,f} = P(\Lambda > T | S_1)$$
$$\simeq \sum_{k=T}^{N} \binom{N}{k} (\bar{P}_{CD})^k (1 - \bar{P}_{CD})^{N-k}. \tag{20.7}$$

As the local FA probability does not depend on the position of the PUE, the FA probability at the FC has exactly the following expression:

$$P_{FA,f} = P(\Lambda > T | S_0)$$
$$= \sum_{k=T}^{N} \binom{N}{k} (P_{FA})^k (1 - P_{FA})^{N-k}. \tag{20.8}$$

At this point, one can observe that the probabilities in Equations 20.7 and 20.8 depend on the local and global thresholds, that is, one can write:

$$P_{CD,f} = \mathcal{G}(\tau_N, T)$$

$$P_{FA,f} = \mathcal{F}(\tau_N, T)$$

where $\mathcal{G}(\cdot,\cdot)$ and $\mathcal{F}(\cdot,\cdot)$ are proper functions. The optimized values of the local ($\tau_N$) and global ($T$) thresholds can be determined by observing that protecting primary users against secondary interference is, in general, more important than giving opportunities to secondary users. Therefore, we consider a Neyman–Pearson detector at the FC where we fix the requested CD probability, that is, we impose $P_{CD,f} = P_{CD,tgt}$. For each possible value of $\tau_N$, the optimal value of the threshold $T$, denoted as $T^*$, can be determined by finding the maximum value (if existing) of $T$ that allows to achieve $P_{CD,tgt}$, that is,

$$T^*(\tau_N) = \max\Big\{T \in \mathbb{N} : \mathcal{G}(\tau_N, T) = P_{CD,tgt}\Big\}. \qquad (20.9)$$

In Equation 20.9, the maximum value is considered because $\mathcal{G}(\tau_N, T)$ is, as shown later, a decreasing function of $T$ for fixed $\tau_N$. The optimal value of $\tau_N$, denoted as $\tau_N^*$, is then selected as the value that allows to achieve the minimum $P_{FA,f}$:

$$\tau_N^* = \arg\min_{\tau_N} \mathcal{F}[\tau_N, T^*(\tau_N)].$$

We now show an illustrative example of the general scenario of interest shown in Figure 20.1. In particular, the main system parameters are the following: $R = 1$ km, $P_T = 30$ mW, $\alpha = 4$, $\sigma = 5$ dB, $P_N = -110$ dBm, and $m = 10$. In Figure 20.3, $P_{CD,f}$ is shown, as a function of $T$, for $N = 15$ and various values of $\tau_N$. In all cases, the target CD probability $P_{CD,tgt}$ is set to 0.95. One can observe that the optimal value of $T$ depends on $\tau_N$, for instance, for $\tau_N = 1$ the optimal threshold is $T = 6$, whereas $T = 1$ for $\tau_N = 1.6$. These values are then used to determine the optimal values of $\tau_N$, as shown in Figure 20.4, where $P_{FA,f}$ is shown, as a function of $\tau_N$, for various values of $T$. The points highlighted with circles correspond to those obtained from Figure 20.3. Because the optimized thresholds correspond to the minimum FA, in the considered settings the optimized thresholds are $\left(T^*, \tau_N^*\right) = (1, 1.8)$.

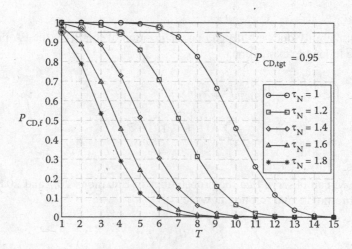

**FIGURE 20.3** $P_{CD,f}$ as a function of $T$, for $N = 15$ and various values of $\tau_N$. In all cases, the target CD probability $P_{CD,tgt}$ is set to 0.95.

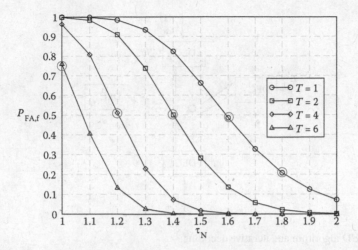

**FIGURE 20.4** $P_{\mathrm{FA,f}}$, as a function of $\tau_{\mathrm{N}}$, for various values of $T$. The points highlighted with circles correspond to those obtained from Figure 20.3.

### 20.4.2 Coded Scenario with Communication Noise

Consider now a scenario where the data transmitted by the CUEs to the FC are corrupted by communication noise. For simplicity, assume that the communication links are orthogonal and affected by additive white Gaussian noise (AWGN), so that the observable model at the FC is

$$r_{k,i} = \tilde{x}_{k,i} + w_{k,i} \qquad k = 1,\ldots,N \qquad i = 1,\ldots,N_{\mathrm{ch}} \tag{20.10}$$

where $\left\{\tilde{x}_{k,i}\right\}_{i=1}^{N_{\mathrm{ch}}}$ is the binary data sequence obtained by channel encoding (with coding rate $r_{\mathrm{cod}}$) the decision sequence $\{x_{k,i}\}_{i=1}^{N_{\mathrm{ch}}}$. The communication channel is characterized by an SNR denoted as $\gamma_{\mathrm{ch}}^{(k)}$.* In Section 20.5, a common average value of the communication SNR, denoted as $\gamma_{\mathrm{ch}}$, will be considered. The choice of proper channel coding strategies is an interesting open issue, due to the fact that the decision packet length is typically small (in fact, it corresponds to the number $N_{\mathrm{ch}}$ of subchannels). However, this problem goes beyond the scope of this chapter—in Section 20.5, a regular low-density parity-check (LDPC) code will be considered.

Because the transmitted data are inherently correlated according to the PMF in Equation 20.5, inspired by the work in Ref. [9], we use JCD at the receiver to improve the detection/decoding performance. In this case, $N$ subdecoders, one per CUE, are present at the AP. Each subdecoder works on the basis of its channel LLRs and the a priori soft information obtained from the soft-output information generated by the other subdecoders (associated with the remaining CUEs), properly combined taking into account the source correlation. The corresponding scheme of the JCD receiver is shown in Figure 20.5. This procedure is then iterated to refine at each step the decisions of each decoder.

The log-likelihood ratio (LLR) relative to the $i$th observable at the input of the $k$th subdecoder can be written as follows [15]:

$$\mathcal{L}_{i,\mathrm{in}}^{(k)} = \begin{cases} \mathcal{L}_{i,\mathrm{ch}}^{(k)} + \mathcal{L}_{i,\mathrm{ap}}^{(k)} & i = 1,\ldots,N_{\mathrm{ch}} \\ \mathcal{L}_{i,\mathrm{ch}}^{(k)} & i = N_{\mathrm{ch}} + 1,\ldots,N_{\mathrm{ch}}/r_{\mathrm{cod}} \end{cases}$$

---

* We are making the reasonable assumption that the SNR is constant over the transmission of an entire packet $\mathbf{x}_k = \left(x_{k,1},\ldots,x_{k,N_{\mathrm{ch}}}\right)$ and changes independently from CUE to CUE.

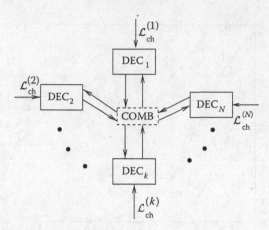

**FIGURE 20.5**   JCD algorithm and iterative decoding.

where the channel LLR can be written as

$$\mathcal{L}_{i,\mathrm{ch}}^{(k)} = \frac{2r_{i,k}}{\sigma_{\mathrm{N}}^2} \qquad k = 1,\ldots,N$$

with $\sigma_{\mathrm{N}}^2$ being the variance of the AWGN. The a priori component of the LLR can be instead written as [9]

$$P(x_{i,k}) = \sum_{x_i'} \underbrace{P\left(x_{i,k}=0\,\middle|\,x_i'\right)}_{\text{a priori source correl.}} \prod_{\substack{\ell=1 \\ \ell \neq k}}^{N} \underbrace{\hat{P}(x_{i,\ell})}_{\text{from decoder } \ell} .$$

Because each decoder outputs the LLRs on the bits of the information sequence, the following soft-input fusion (SF) rule can be used:

$$\hat{S}_i = \operatorname*{argmax}_{S_i=0,1} \beta_i \sum_{\{x_i\}} \underbrace{\frac{P(x_i|S_i)P(S_i)}{\sum_{S^*=0,1} P(x_i|S^*)P(S^*)}}_{\text{from the correlation model}} \underbrace{\prod_{j=1}^{n} P\left(x_{i,j}\,\middle|\,\mathcal{L}_i^{(j)}\right)}_{\text{from decoder LLRs}} \tag{20.11}$$

where

$$\beta_i \triangleq \beta(1 - S_i) + (1 - \beta)S_i \qquad i \in \{0,1\}$$

with $\beta \in (0,1)$. In other words, the fusion rule (Equation 20.11), derived in Ref. [9] from a maximum a posteriori probability (MAP) criterion, can still be used in this scenario using proper coefficients $\beta_0 = \beta$ and $\beta_1 = 1 - \beta$ to obtain proper FA and CD probabilities. These coefficients need to be optimized according to a given criterion: In particular, in this chapter we consider the same procedure used in Section 20.4.1 to optimize $T$. In other words, for each value of $\tau_{\mathrm{N}}$, the optimal value of the threshold $\beta$, denoted as $\beta^*$, can be determined by finding the maximum value (if it exists) of $\beta$ that allows to achieve $P_{\mathrm{CD,tgt}}$. The optimal value of $\tau_{\mathrm{N}}$, denoted as $\tau_{\mathrm{N}}^*$, is then selected as the value which allows to achieve, for $\beta = \beta^*$, the minimum $P_{\mathrm{FA,f}}$. Note also that these optimal values obviously depend on the sensing SNR.

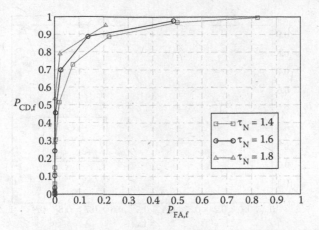

**FIGURE 20.6** Uncoded ROC for $N = 15$ and various values of $\tau_{N}$.

Moreover, a hard-input fusion (HF) rule can be considered if the LLRs at the output of the channel decoders are quantized, thus obtaining an estimate of the local decisions $\hat{x}_i (i = 1,\ldots,N_{ch})$. From a fusion point of view, the estimated decisions $\{\hat{x}_i\}_{i=1}^{N_{ch}}$ are uncoded and, therefore, the same fusion rule as in Equation 20.6 can be applied:

$$\Lambda_{i,cod} = \sum_{k=1}^{N} \hat{x}_{k,i} \underset{S_0}{\overset{S_1}{\gtrless}} T_{cod}$$

where $T_{cod}$ is a proper global threshold. The same optimization procedure for local $\tau_N$ and global $T_{cod}$ thresholds described at the end of Section 20.4.1 can be applied in this case as well.

## 20.5 NUMERICAL RESULTS

In this section, we present simulation results considering the same scenario as in Section 20.4.1. We assume that the frequency band is divided into $N_{ch} = 100$ subchannels and, in the presence of channel coding, a half-rate (i.e., $r_{cod} = 1/2$) (3, 6) regular LDPC code, with pseudo-random parity-check matrix generation, is considered. The JCD algorithm summarized in Section 20.4.2 is performed with five external iterations, while each LDPC decoder performs internal iterations until a valid codeword is obtained or a maximum number of 50 internal iterations is reached. To eliminate statistical fluctuations of the communication noise, the results in the presence of channel coding are averaged over 100 different trials. In both uncoded and coded scenarios, results are presented in terms of the receiver operating characteristic (ROC) curve, for the binary hypothesis testing problem of interest, showing the CD probability as a function of the FA probability [16].

### 20.5.1 UNCODED SCENARIO

In Figure 20.6, the uncoded ROC is shown for $N = 15$ and various values of $\tau_N$. Each point of the curves corresponds to a different value of $T \in \{1, 2, \ldots, N\}$; in particular, $T = 1$ corresponds to the point in the upper right corner of the figure.* Note that for increasing values of $T$, a smaller FA probability can be obtained at the price of a lower CD probability. Moreover, increasing $\tau_N$ for a

---

* Simulation results for the uncoded case, not shown here for brevity, are in agreement with the theoretical results obtained from our analytical framework in Section 20.4.1.

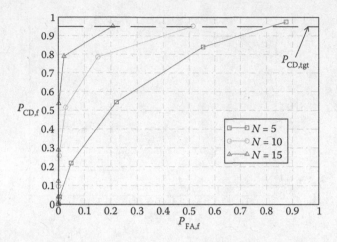

**FIGURE 20.7** Uncoded ROC for various values of $N$, considering the optimal values of $T$ and $\tau_N$ obtained with $P_{CD,tgt} = 0.95$.

fixed value of $T$ allows to reduce both FA and CD probabilities, thus showing the inherent tradeoff between these two quantities.

In Figure 20.7, the uncoded ROC is shown for various values of $N$, considering the optimal values of $T$ and $\tau_N$ obtained with $P_{CD,tgt} = 0.95$. In particular, for $N = 5$ the optimal point is ($T^* = 1$, $\tau_N = 1.1$), for $N = 10$ ($T^* = 1$, $\tau_N = 1.5$), and for $N = 15$ ($T^* = 1$, $\tau_N = 1.8$), as detailed in Table 20.1. Note that in all presented cases $T = 1$, that is, the optimal decision rule is the OR decision rule. This means that, if CUE observations are available without errors are the FC, it is sufficient to have at least one of them in favor of $S_1$ to decide for this status. Moreover, for increasing values of $N$ both $\tau_N$ and $T$ increase as well and the minimum possible $P_{FA,f}$ can be noticeably reduced from 0.88 ($N = 5$) to 0.52 ($N = 10$) and 0.21 ($N = 15$). Obviously, these results depend on the considered target CD probability, as summarized in Table 20.1. In particular, one can observe that with $N = 10$ and $P_{CD,tgt}$ the minimum FA probability is achieved by fusing two decisions instead of one.

### 20.5.2 Coded Scenario

In Figure 20.8, we show the ROC curves obtained considering $N = 10$, $P_{CD,tgt} = 0.9$, and different values of $\gamma_{ch}$. The performance of SF is compared with that of HF. One can observe that the better the communication channels (i.e., the higher the channel SNR), the closer the performance to the theoretical limit given by the uncoded system: in particular, the performance with $\gamma_{ch} = 10$ dB overlaps with the theoretical limit given by the uncoded system. Moreover, in the presence of a bad communication link quality, the use of SF allows one to obtain better performance than HF because the likelihood information coming from the decoder is properly exploited. The ROC curves are

**TABLE 20.1**

**Optimal Working Points for the Uncoded Case and Various Values of $N$ and $P_{CD,tgt}$**

| $P_{CD,tgt}$ | $N = 5$ | $N = 10$ | $N = 15$ |
|---|---|---|---|
| 0.9 | $T^* = 1$, $\tau_N^* = 1.2$ | $T^* = 1$, $\tau_N^* = 1.6$ | $T^* = 1$, $\tau_N^* = 2.1$ |
| 0.95 | $T^* = 1$, $\tau_N^* = 1.1$ | $T^* = 1$, $\tau_N^* = 1.5$ | $T^* = 1$, $\tau_N^* = 1.8$ |
| 0.99 | $T^* = 1$, $\tau_N^* = 1$ | $T^* = 2$, $\tau_N^* = 1.1$ | $T^* = 1$, $\tau_N^* = 1.4$ |

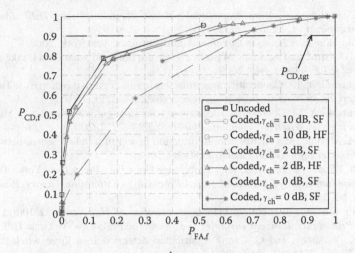

**FIGURE 20.8** ROC curves for $N = 10$, $P_{CD,tgt} = 0.9$, and different values of $\gamma_{ch}$. The performance of SF is compared with that of HF.

shown for the optimal values of thresholds, for example, $T^*$ and $\tau_N^*$ in the uncoded case (indicated as reference in Figure 20.8), $\beta^*$ and $\tau_N^*$ for the coded case and SF, and $T_{cod}^*$ and $\tau_N^*$ for the coded case and HF. In particular, for high SNR (i.e., $\gamma_{ch} = 10$ dB) the optimal value is $\tau_N^* = 1.6$ for both uncoded and coded (either HF or SF) cases. When the SNR decreases, for example, $\gamma_{ch} = 2$ dB, the optimal value is, in the coded case (either HF or SF), $\tau_N^* = 1.7$. Finally, for very bad communication link quality, for example, $\gamma_{ch} = 0$ dB, $\tau_N^* = 1.3$ with HF and 2 with SF. The fact that the optimized threshold differs in the coded case for at least medium–low communication SNRs, leads to the conjecture that the choice of $\tau_N$ changes the correlation model, which should be properly taken into account. Therefore, the dimensionality of the optimization problem increases. This issue will be the subject of our future work.

## 20.6 CONCLUDING REMARKS

In this chapter, we have analyzed a cognitive WSN, where CUEs sense the frequency spectrum and cooperate to estimate the free subchannels that can be used to transmit their data. First, a simple analytical framework for characterizing the local sensing performance per subchannel, in terms of MD and FA probabilities, has been derived for a scenario with uncoded transmissions over error-free communication channels. This allows one to determine the ultimate achievable performance in such a scenario. Then, we have considered a realistic scenario with communication noise and the use of channel coding. In this context, the performance of JSCC schemes with JCD at the receiver has been investigated. Our results have shown the beneficial impact of JCD at the receiver (with improved quality of the subchannel status estimation), thus making it possible to achieve performance close to the theoretical limit for sufficiently large values of the SNR.

## REFERENCES

1. D. Datla, A. Wyglinski, and G. Minden. A spectrum surveying framework for dynamic spectrum access networks. *IEEE Transactions on Vehicular Technology*, 58(2):4158–4168, 2009.
2. J. Mitola, and G. Q. Maguire. Cognitive radio: Making software radios more personal. *IEEE Personal Communications*, 6(4):13–18, 1999.
3. J. Mitola. *Cognitive Radio: An Integrated Agent Architecture for Software-Defined radio*. PhD dissertation, Royal Institute of Technology, Stockholm, 2000.

4. S. Heykin. Cognitive radio: Brain-empowered wireless communications. *IEEE Journal on Selected Areas in Communication*, 23:201–220, 2005.
5. S. Heykin. *Fundamentals Issues in Cognitive Radio*. Springer, New York, 2007.
6. R. Tandra, M. Mishra, and A. Sahai. What is a spectrum hole and what does it take to recognize one? *Proceedings of IEEE*, 97:824–848, 2009.
7. G. Ganesan, and G. Y. Li. Cooperative spectrum sensing in cognitive radiopart I: Two user networks. *IEEE Transactions on Wireless Communications*, 6(6):2204–2213, 2007.
8. G. Ganesan, and G. Y. Li. Cooperative spectrum sensing in cognitive radiopart II: Multiuser networks. *IEEE Transactions on Wireless Communication*, 6(6):2214–2222, 2007.
9. G. Ferrari, M. Martalò, and A. Abrardo. Information fusion in wireless sensor networks with source correlation. *Elsevier Information Fusion*, 15:80–89, 2014.
10. A. Goldsmith. *Wireless Communications*. Cambridge University Press, New York, 2005.
11. S. M. Kay. *Fundamentals of Statistical Signal Processing: Estimation Theory*. Prentice-Hall, Upper Saddle River, NJ, 1993.
12. D. Moltchanov. Distance distributions in random networks. *Ad Hoc Networks*, 10(6):1146–1166, 2012.
13. A. Papoulis. *Probability, Random Variables and Stochastic Processes*. McGraw-Hill, New York, 1991.
14. R. Niu, P. K. Varshney, and Q. Cheng. Distributed detection in a large wireless sensor network. *Information Fusion*, 7(4):380–394, 2006.
15. A. Abrardo, G. Ferrari, M. Martalò, and F. Perna. Feedback power control strategies in wireless sensor networks with joint channel decoding. *MDPI Sensors*, 9(11):8776–8809, 2009.
16. T. Fawcett. An introduction to ROC analysis. *Pattern Recognition*, 27(8):861–874, 2006.

# 21 Dynamics of Consensus Formation among Agent Opinions

*Thanuka Wickramarathne, Kamal Premaratne,*
*and Manohar Murthi*

## CONTENTS

## 21.1 INTRODUCTION

In a typical consensus scenario, a group of adaptive agents (i.e., sensors, fusion nodes, people, etc.) iteratively interact with their neighbors and update their (current) *states* or *opinions* toward collectively estimating some phenomenon of interest without any global coordination. Understanding the dynamics of consensus formation in such settings, where the collective behavior is governed solely by the *local interactions*, is a topic of overlapping interest for researchers in multiple domains [1–19]. With the emergence of social networks as a dominant force in modern society, the importance of understanding the mathematical underpinnings that may lead to a consensus in such complex fusion environments [20] is becoming more and more important. However, in addition to agent interactions being ad hoc and/or dynamic with possibly asynchronous communications (i.e., delays), sensing in such environments are often distributed and involves a large numbers of heterogeneous sources that include a mix of both soft (i.e., human or human-based) and hard (i.e., conventional physics-based) sensors. Therefore, convergence analysis in such complex fusion networks is a challenging task due mainly to the unstructured environment and the difficulties associated with adequately modeling the agent opinions itself.

Much of the previous work in consensus has focused on applications in distributed control, estimation, or fusion [4–8,10–12,14,16–18]. But most (if not all) of these previous work deals with an agent state that is modeled as a real-valued vector, that is, the consensus state is a real-valued vector. The work in Ref. [19] constitutes a slight divergence, where agents observe signal data and use estimation theory (e.g., frequentist approaches) to collectively obtain a consensus probability mass function (p.m.f.). But the agent state remains a real-valued vector. In contrast, in our work, the agent state is a Dempster–Shafer theoretic (DST) body of evidence (BoE) (which subsumes a p.m.f.), not a real-valued vector. In contrast to other models that are perhaps better suited for control and estimation purposes, such an agent state model is ideally suited for capturing the data imperfections and nuances associated with soft evidence (e.g., human-generated input/opinions, human domain expert opinions, etc., in social network settings). In addition to this difference, our work also focuses on achieving a rational consensus. This notion of a rational consensus is different than what appears in existing literature, where a rational consensus is typically taken as a weighted average [1]. The work in Ref. [21] on nonlinear asynchronous paracontracting operators provides the mathematical framework for convergence analysis of asynchronous iterations involving both finite and infinite pools of nonlinear operators. Fang et al. [22] exploit this work to develop asynchronous consensus protocols for finite pools of convex combination operators [23] by viewing consensus protocols as asynchronous iterations.

Even though most of the previous work [4–18] explores consensus in unstructured network environments [4–6,10–12,14,16–18], simple models of agent states cannot adequately well capture the types of uncertainties and the nuances characteristic of soft evidence in complex fusion networks. Imprecise probabilistic formalisms, such as Dempster–Shafer (DS) theory [24], are better suited for handling these data uncertainties (see Table 21.1 [25]), but convergence properties of such formalisms is often difficult to establish under complex fusion settings.

In this chapter, we explore properties of consensus formation in complex fusion environments via a certain DST fusion operator [26–30], as demonstrated in Ref. [20]. This belief revision strategy in fact generates a rational consensus, which we have defined as an agreement that is consistent with an estimate of ground truth (when available). As the work in Ref. [31] demonstrates, a rational consensus can be utilized to estimate agent credibility even when the ground truth is known only partially (or even unknown). The DST framework bears a close relationship to the Bayesian framework, and the attraction of this approach to researchers working primarily in probabilistic fusion environments is immediately apparent: the DST approach engenders the use of belief functions that generalize p.m.f.s, but provide the capability for capturing uncertainty regarding the actual probabilities through belief and plausibility uncertainty bounds, and through the ability to allocate a mass to propositions that do not consist solely of singletons. This allows for a smoother transition between DST notions and Bayesian notions, which after all are quite effective and hence well entrenched in work related to hard sensors/sensor networks. Our previous work in establishes the theoretical convergence properties of the DST consensus protocol in complex fusion environments. This work constitutes the first instance where convergence of DST data fusion schemes is studied in complex fusion networks and an infinite number of fusion agents. For detailed proofs and convergence properties in complex fusion environments, we refer interested readers to Ref. [20]. Let us proceed by providing a brief summary of preliminary notions utilized throughout the chapter.

**TABLE 21.1**

**DST Models for Various Types of Data Imperfections**

| Type of Imperfection | $x_i[k]$ | | Remarks |
| | Focal Set | $m_i(\cdot)[k]$ | |
| --- | --- | --- | --- |
| Hard evidence | $\theta_1$ | 1.0 | One singleton focal element |
| Probabilistic | $\theta_1$ | 0.1 | Singleton focal elements only |
| | $\theta_2$ | 0.7 | |
| | $\theta_3$ | 0.2 | |
| Possibilistic | $\theta_1$ | 0.7 | Consonant focal elements only |
| | $(\theta_1,\theta_3)$ | 0.2 | |
| | $\Theta$ | 0.1 | |
| Ambiguity | $(\theta_1,\theta_2)$ | 1.0 | Inability to discern |
| Vacuous | $\Theta$ | 1.0 | Missing/unknown entry |
| Dempster–Shafer | $\sum_{B \subseteq \Theta} m_i(B)[k] = 1.0$ | | Encapsulates all the preceding and more general forms |

*Source:* P. Vannoorenberghe, *Information Fusion*, 5(3):179–188, 2004.

## 21.2 PRELIMINARIES

In our consensus analyses, notions from graph theory and DS belief theory [24] are utilized for modeling agent interactions and complex agent opinions that consist of numerous uncertainties, respectively. Convergence properties of the DST belief revision process under these conditions are then established utilizing the properties of paracontracting operators [20,26]. Let us now provide a brief summary of the essentials of these notions.

### 21.2.1 GRAPH THEORETIC NOTIONS

**Definition 21.1**

(Basic Notions [10,32]): A *graph* $G = (V, E)$ consists of a *vertex set* $V \subset \mathbb{N}$, an *edge set* $E \subseteq V \times V$, and a relation that associates each edge with two vertices. A *directed graph* or *digraph* $G = (V, E)$ is a graph where $E$ is a set of ordered pairs of vertices. An ordered pair $(u, v) \in E$, or $u \to v$, denotes a directional edge from $u \in V$ to $v \in V$. For $u \to v$, $u$ and $v$ are the *predecessor* of $v$ and *successor* of $u$, respectively. A vertex $v \in V$ is a *root* of $G$, if $\exists\, u \to v, \forall\, u \in V, u \neq v$. A *rooted graph* has at least one root. ∎

**Definition 21.2**

(Graph Composition [10]): Let $G[k] \equiv (V[k], E[k]) \in \mathbb{G}$ be arbitrary graphs with $V[k] \subset \mathbb{N}$, $k = 1,2$, where $\mathbb{G}$ is the set of all graphs with vertex set $\subset \mathbb{N}$. The *composition* $G[2] \circ G[1]$ is the graph s.t. the edge set contains edges $(u, w)$, if $(u, v) \in E[1]$ and $(v, w) \in E[2]$.

A finite sequence of directed graphs $(G[k] \in \mathbb{G})_{k=1}^{\ell}, \ell \in \mathbb{N}$ is *jointly rooted* if $G[\ell] \circ G[\ell - 1] \circ \cdots \circ G[1]$ is a rooted graph. An infinite sequence of graphs $(G[k] \in \mathbb{G})_{k=1}^{\infty}$ is a *repeatedly jointly rooted* if $\exists\, \ell \in \mathbb{N}$ for which each finite sequence $(G[i])_{i=\ell(k-1)+1}^{k\ell}, k \geq 1$ is jointly rooted. ∎

### 21.2.2 DS Framework

In the DS framework, the total set of mutually exclusive and exhaustive propositions of interest (i.e., the "sample space" or "scope of expertise") is referred to as the frame of discernment (FoD) $\Theta = \{\theta_1,\ldots,\theta_n\}$ [23]. A singleton proposition $\theta_i \in \Theta$ represents the lowest level of discernible information. Elements in the power set of $\Theta$, denoted via $2^\Theta$, form all the propositions of interest. We use $A\backslash B$ to denote all singletons in $A$ that are not in $B$; $\bar{A}$ denotes $\Theta\backslash A$.

**Definition 21.3**

Consider the FoD $\Theta$ and $B \subseteq \Theta$. The mapping $m(\cdot) : 2^\Theta \mapsto [0, 1]$ is a basic probability assignment (BPA) or mass assignment if $\sum_{B\subseteq\Theta} m(B) = 1$ with $m(\varnothing) = 0$. The belief and plausibility of $B$ are $\mathrm{Bl}(B) = \sum_{C\subseteq B} m(C)$ and $\mathrm{Pl}(B) = 1 - \mathrm{Bl}(\bar{B})$, respectively. A proposition $B \subseteq \Theta$ that possesses nonzero mass is a focal element. The set of focal elements is the core $\mathfrak{F}$; the triplet $\mathcal{E} \equiv \{\Theta, \mathfrak{F}, m(\cdot)\}$ is the corresponding body of evidence (BoE). The set of all BoEs defined on $\Theta$ is denoted by $\mathfrak{E}_\Theta = \left\{ \mathcal{E} \middle| \mathcal{E} = \{\Theta, \mathfrak{F}, m(\cdot)\} \right\}$. ∎

DS theory supports modeling of ignorance by allowing partial probability specifications. Nonzero support can be assigned to composite propositions (i.e., nonsingleton propositions) $B \subseteq \Theta$. While $m(B)$ measures the support that is directly assigned to proposition $B$ only, the belief $\mathrm{Bl}(B)$ represents the total support that can move into $B$ without any ambiguity; $\mathrm{Pl}(B)$ represents the extent to which one finds $B$ plausible. These DST belief and plausibility measures are closely related to the inner and outer measures of a nonmeasurable event $B \subseteq \Theta$ w.r.t. p.m.f.s defined on $\Theta$. Furthermore, when focal elements are constituted of singletons only (i.e., $\sum_{B\in\Theta} m(B) = 1$), the mass, belief and plausibility all reduce to a p.m.f.

**Theorem 21.1**

(Fagin–Halpern [FH] Conditionals [33]): For any $B \subseteq \Theta$ and a conditioning event $A \subseteq \Theta$ s.t. $\mathrm{Bl}(A) > 0$, conditional belief $\mathrm{Bl}(B|A) : 2^\Theta \mapsto [0, 1]$ and conditional plausibility $\mathrm{Pl}(B|A) : 2^\Theta \mapsto [0, 1]$ are given by $\mathrm{Bl}(B|A) = \mathrm{Bl}(A \cap B)/[\mathrm{Bl}(A \cap B) + \mathrm{Pl}(A \cap \bar{B})]$ and $\mathrm{Pl}(B|A) = \mathrm{Pl}(A \cap B)/[\mathrm{Pl}(A \cap B) + \mathrm{Bl}(A \cap \bar{B})]$, respectively. ∎

The Fagin–Halpern (FH) conditionals extend the notions of inner and outer measures of non-measurable events to conditioning in DST. A proposition with positive mass after conditioning is referred to as a conditional focal element. The collection of conditional focal elements that are generated with respect to the conditioning event $A$ is referred to as the conditional core and denoted by $\mathfrak{F}_{|A}$. Thus, $\mathfrak{F}_{|A} = \left\{ B \subseteq \Theta \middle| m(B|A) > 0 \right\}$ and $m(\cdot|A) : 2^\Theta \mapsto [0, 1]$ is the corresponding conditional BPA related to $\mathrm{Bl}(\cdot|A)$ via the Möbius transformation [24]

$$m(B|A) = \sum_{C\subseteq B} (-1)^{|B-C|} \mathrm{Bl}(C|A), \quad \forall B \subseteq \Theta. \tag{21.1}$$

The conditional computations may take a significant amount of time and computational resources, especially when the FoDs are large. An important recent result that can be used to directly identify the FH conditional focal elements is

**Theorem 21.2**

(Conditional Core Theorem [CCT] [30]): Given $Bl(A) > 0$ in the BoE $\mathcal{E} \equiv \{\Theta, \mathfrak{F}, m(\cdot)\}$, $m(B|A) > 0$ iff $B$ can be expressed as $B = X \cup Y$, for some $X \in \text{in}(A)$ and $Y \in \text{OUT}(A) \cup \{\varnothing\}$. Here, $\text{in}(A) = \{B \subseteq A | B \in \mathfrak{F}\}$ and $\text{OUT}(A) = \{B \subseteq A | B = \bigcup_{i \subseteq \mathcal{I}} C_i, C_i \in \text{out}(A)\}$, where $\text{out}(A) = \{B \subseteq A | B \cup C \in \mathfrak{F}, \varnothing \neq B, \varnothing \neq C \subseteq \overline{A}\}$. ∎

The DST fusion operator that we utilize employs FH conditional notions for belief revision, and hence CCT indeed helps one to better understand the fusion process. For a comprehensive discussion on CCT, we refer interested readers to Ref. [30]. The following example Ref. [34] illustrates the application of the CCT.

**Example 21.1**

Consider the BoE, $\mathcal{E} \equiv \{\Theta, \mathfrak{F}, m(\cdot)\}$ with $\Theta = \{a, b, c, d, e, f, g, h, i\}$, $m = \{a, b, h, df, beg, \Theta\}$ and $m(B) = \{0.1, 0.1, 0.1, 0.2, 0.2, 0.3\}$, for $B \in \mathfrak{F}$ (in the same order given in $\mathfrak{F}$). Then, for $A = (abcde)$,

$$\text{in}(A) = \{a, b\}; \qquad \text{out}(A) = \{d, be, abcde\};$$
$$\text{IN}(A) = \{a, b, ab\}; \qquad \text{OUT}(A) = \{d, be, bde, abcde\}.$$

Note that $\mathcal{B} = \{ad, bd, be, abe, bde, abde, abcde\}$ are the only propositions that can be expressed as $B = X \cup Y$, for some $X \in \text{in}(A)$ and $Y \in \text{OUT}(A)$. So, according to the CCT, the nine elements of $\mathcal{B}$ and $\text{in}(A)$ are the only propositions that will belong to the conditional core (w.r.t. $A = (abcde)$).

### 21.2.3 PARACONTRACTING OPERATORS

We interpret consensus in terms of notions of paracontracting operators, where fusion operators are established as paracontractions and their fixed points are viewed as consensus states.

Let $\mathbb{D} \subset \mathfrak{R}^{|2^\Theta|}$ be the domain of interest (in our case, all probabilistic assignments on the sample space $\Theta$). A vector $\xi \in \mathbb{D}$ is referred to as a *fixed point* of an operator $F : \mathbb{D}^m \mapsto \mathbb{D}$ if $F(\xi, \ldots, \xi) = \xi$, where $m \in \mathbb{N} \equiv \{1, 2, \ldots\}$. Further, the set of all fixed points of operator $F$ is denoted by $\text{fix}(F) \equiv \{\xi \in \mathbb{D} | F(\xi, \ldots, \xi) = \xi\}$. A vector $\zeta \in \mathbb{D}$ is a *common fixed point* if $\zeta$ is a fixed point of each operator $F \in \mathcal{F}$, that is, $\zeta \in \text{fix}(F), \forall F \in \mathcal{F}$. Let $\mathbb{I} \subset \mathbb{N}$ be a set of indices and $m \in \mathbb{N}$ a fixed number. Henceforth, we deal with the *pool of operators* $\mathcal{F} \equiv \{F^i : \mathbb{D}^{m_i} \mapsto \mathbb{D} | m \geq m_i \in \mathbb{N}, \forall i \in \mathbb{I}\}$, where $\mathbb{D}$ is closed. Let $\mathbf{X} \equiv [\mathbf{x}_1, \ldots, \mathbf{x}_{m_i}] \in \mathbb{D}^{m_i}$ and $\mathbf{Y} \equiv [\mathbf{y}_1, \ldots, \mathbf{y}_{m_i}] \in \mathbb{D}^{m_i}$ be arbitrary vectors on $\mathbb{D}^{m_i}$. Also, let $\mathbb{N}_0 \equiv \{0, 1, \ldots\}$ and $\|\cdot\|$ be a vector norm defined on $\mathbb{D}$.

**Definition 21.4**

(Paracontracting Operators [21]):

i. The pool $\mathcal{F}$ is *contractive* on $\mathbb{D}$, if $\|F^i(\mathbf{X}) - F^i(\mathbf{Y})\| \leq \omega \cdot \max_j \|\mathbf{x}_j - \mathbf{y}_j\|, \forall i \in \mathbb{I}, \mathbf{X}, \mathbf{Y} \in \mathbb{D}^{m_i}$, for some $0 \leq \omega < 1$.

ii. An operator $F^i \in \mathcal{F}$ is *paracontracting* on $\mathbb{D}$ if $\|F^i(\mathbf{X}) - \xi\| < \max_j \|\mathbf{x}_j - \xi\|, \forall \mathbf{X} \in \mathbb{D}^{m_i}$ and any $\xi \in \text{fix}(F^i)$, unless $\mathbf{X} \in \text{fix}(F^i)$.

iii. Let $F^i$ be continuous on $\mathbb{D}^{m_i}$ for all $i \in \mathbb{I}$. Then, the pool $\mathcal{F}$ is said to be *paracontracting* on $\mathbb{D}$, if $\|F^i(\mathbf{X}) - \xi\| < \max_j \|\mathbf{x}_j - \xi\|$ for any $\xi \in \text{fix}(F^i)$, unless $\mathbf{X} \in \text{fix}(F^i)$. ∎

**Definition 21.5 [21]**

An infinite pool of operators $\mathcal{G} \equiv \left\{ G^k : \mathbb{D}^{\tilde{m}_k} \mapsto \mathbb{D} \middle| \tilde{m}_k \in \{1,\ldots,m\}, \forall k \in \mathbb{N} \right\}$ is said to *approximate* the pool $\mathcal{F}$, if there exists $i_k \in \mathbb{I}, \forall k \in \mathbb{N}_0$, s.t. $\tilde{m}_k = m_{i_k}$,

$$\lim_{j \to \infty} \left\| G^j(\cdot) - F^{i_j}(\cdot) \right\| = 0, \forall j \in \mathbb{N}_0,$$

uniformly for all $\mathbf{X} \in \mathbb{D}^m$. Here, $\|\cdot\|$ is some norm on $\mathbb{D}$.                    ∎

Let us now review asynchronous iterations, which we will utilize to model iterative belief revision processes.

**Definition 21.6**

(Asynchronous Iteration [21]): Let $\mathcal{X}_0 = \left\{ \mathbf{x}[-\ell] \in \mathbb{D} \middle| \ell = 1,\ldots,M \right\}$ be a set of initial conditions, where $M \in \mathbb{N}_0$. Let $\mathcal{S}$ denote the sequence of $m_i$-tuples of the form $(s^1[k],\ldots,s^{m_{I[k]}}[k])$, where $s^\ell[k] \in \mathbb{N}_0 \cup \{-1,\ldots,-M\}$ s.t. $s^\ell[k] \le k$ for all $\ell \in \{1, \ldots, m_{I[k]}\}$, where $I[k] \in \mathbb{I}, \forall k \in \mathbb{N}_0$. Then, for sequences $\mathcal{I} \equiv \left\{ I[k] \in \mathbb{I} \middle| k \in \mathbb{N}_0 \right\}$ and $\mathcal{S}$, the sequence $\mathbf{x}[k+1] = F^{I[k]} \left( \mathbf{x}\left[ s^1[k] \right],\ldots,\mathbf{x}\left[ s^{m_{I[k]}}[k] \right] \right)$ is referred to as an *asynchronous iteration* and is denoted by $(\mathcal{F},\mathcal{X}_0,\mathcal{I},\mathcal{S})$.                    ∎

On operator coupling in asynchronous iterations, we have

**Definition 21.7 [21]**

Let $(\mathcal{F},\mathcal{X}_0,\mathcal{I},\mathcal{S})$ be an asynchronous iteration.

  i. $\mathcal{S}$ is *admissible*, if $s^\ell[k] \to \infty$, $\forall \ell = 1,\ldots,m$, whenever $k \to \infty$;
 ii. $\mathcal{S}$ is *regulated*, if $s \equiv \max_{k,\ell}(k - s^\ell[k])$ exists;
iii. $\mathcal{I}$ is *admissible*, if $I[k] \cup I[k+1] \cup \cdots = \mathbb{I}, \forall k \in \mathbb{N}_0$;
 iv. $\mathcal{I}$ is an *index-regulated sequence* if, for all $i \in \mathbb{I}$, $\exists c_i \in \mathbb{N}_0$, s.t. $i \in \left\{ \{I[k]\} \cup \{I[k+1]\} \cup \cdots \cup \{I[k+c_i]\} \right\}, \forall k \in \mathbb{N}_0$;
  v. $\mathcal{I}$ is *regulated* if $\exists c \in \mathbb{N}_0$ s.t. $I[k] \cup I[k+1] \cup \cdots \cup I[k+c] = \mathbb{I}, k \in \mathbb{N}_0$.

                    ∎

Operator coupling for convergence analysis can be studied by associating a directed graph with the asynchronous iterations [21]: every iteration including initial conditions is assigned a vertex; a pair $(j, k)$ is an edge of the graph iff the $j$th iteration is used for the $k$th iteration.

**Definition 21.8**

(Confluent Iteration [21]): Let $(\mathcal{F},\mathcal{X}_0,\mathcal{I},\mathcal{S})$ be an asynchronous iteration. Then the *graph* of $(\mathcal{F},\mathcal{X}_0,\mathcal{I},\mathcal{S})$ is the directed graph $G \equiv (V, E)$, where the vertex set $V$ is given by $\mathbb{N}_0 \cup \{-1,\ldots,-M\}$ and the edge set $E$ is given by $(k, k_0) \in E$ iff $\exists 1 \le \ell \le m_{I[k_0-1]}$ s.t. $s^\ell[k_0 - 1] = k$.

Then, $(\mathcal{F},\mathcal{X}_0,\mathcal{I},\mathcal{S})$ is called *confluent* if there are numbers $n_0, b \in \mathbb{N}$ and a sequence $\left\{ b_k \in \mathbb{N} \middle| k = n_0, n_0 + 1, \ldots \text{ s.t. } k \ge n_0 \right\}$ s.t. the following are true:

   i. for every vertex $k_0 \geq k$, $\exists b_k \to k_0$ in $G$;
  ii. $k - b_k \leq b$;
 iii. $S$ is regulated; and
 iv. for every $i \in \mathcal{I}$, there is a $c_i \in \mathbb{N}$, so that for all $k \geq n_0$, there is a vertex $w_k^i \in \mathcal{V}$, which is a
     successor of $b_k$ and a predecessor of $b_{k+c_i}$, for which $I(w_k^i - 1) = i$.

                                                                      ■

A confluent iteration establishes the necessary conditions for adequate coupling among agents to be able to reach a consensus. The main convergence result in Ref. [21] on the convergence of confluent asynchronous iterations is stated in Ref. [22] as follows.

## Theorem 21.3 [21,22]

Let $\mathcal{F}$ be a finite paracontracting pool on $\mathbb{D}$ and assume that $\mathcal{F}$ has common fixed points. Then, any confluent asynchronous iteration $(\mathcal{F}, \mathcal{X}_0, \mathcal{I}, \mathcal{S})$ converges to some fixed point $\xi \in \mathrm{fix}(\mathcal{F})$.    ■

## 21.3 AGENT OPINIONS AND THEIR INTERACTIONS

A consensus scenario of a group of adaptive agents is usually characterized by individual agents having access only to imperfect (i.e., incomplete, uncertain, etc.) information, absence of global control, decentralized evidence, communication impairments, and so forth [3]. We can formally define the process of agent interactions and belief revision toward reaching a consensus as follows.

## Definition 21.9

Consider the FoD (i.e., sample space of interest) $\Theta \equiv \{\theta_1, \ldots, \theta_n\}$. Let $\mathbb{D}$ denote the domain of all probabilistic (including partial) assignments on $\Theta$. Now, consider a multiagent system $\mathcal{N} \equiv \{\mathcal{A}_1, \ldots, \mathcal{A}_m\}$ consisting of a fixed number of $m \in \mathbb{N}$ adaptive agents who interact toward collectively estimating some phenomenon $\mathcal{X}$, whose "truth" lies in the discrete set $\Theta$ of alternatives. The estimate maintained by $\mathcal{A}_i \in \mathcal{N}$ at time $t_k$ is referred to as the *"state" of $\mathcal{A}_i$ at $t_k$* and we denote it as $\mathbf{x}_i[k] \in \mathbb{D}$; $t_k, k \in \mathbb{N}_0$ denote the *discrete event-based time* [22]. Let $\mathcal{X}_0 \equiv \left\{ \mathbf{x}_i[0] \in \mathbb{D} \mid i = 1, \ldots, m \right\}$ denote the set of initial conditions of $\mathcal{N}$, where $\mathbf{x}_i[0]$ denote the initial estimate of $\mathcal{X}$ generated by $\mathcal{A}_i \in \mathcal{N}$. The agents in $\mathcal{N}$ iteratively exchange information with their neighbors (which is a subset of $\mathcal{N}$) and update their states (or revise their beliefs) via some fusion operator $F \in \mathcal{F}_\lhd \equiv \left\{ F^i : \mathbb{D}^{m_i} \mapsto \mathbb{D} \mid m \geq m_i \in \mathbb{N}, i \in \mathbb{I} \right\}$. Here, $\mathbb{I} \subset \mathbb{N}$ is an index set and $\mathcal{F}_\lhd$ denotes the pool of fusion operators used by all agents throughout the entire fusion process. Furthermore, an operator $F \in \mathcal{F}_\lhd$ uniquely identifies the agent $\mathcal{A}_i \in \mathcal{N}$ and its neighbors that are being used for belief revision via $F$.    ■

## 21.3.1 AGENT STATE UPDATES

Notice that the only time the system undergoes a change is when some agent $\mathcal{A}_i \in \mathcal{N}$ updates its state. Accordingly, $t_k, k \in \mathbb{N}_0$ refer to an event-based time, not necessarily to an absolute time reference. Therefore, without loss of generality, we assume that only one agent updates its state at any given $t_k$, and the sequence $\{t_k\}_{k=0}^{\infty}$ satisfies $t_k < t_{k+1}, k \in \mathbb{N}_0$ (see Figure 21.1). With this in place, let us now formalize the iterative belief revision process that takes place in a multiagent system.

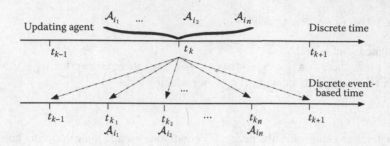

**FIGURE 21.1** Discrete event-based time refers to event-based time, not necessarily to an absolute time reference. In a situation where $n$ agents $\{\mathcal{A}_{i_1},...,\mathcal{A}_{i_n}\} \subseteq \mathcal{N}$ update their states at "absolute" time $t_k$, a discrete event-based time sequence $\cdots < t_{k-1} < t_{k_1} < t_{k_2} < \cdots \leq t_{k_n} < t_{k+1} \cdots$ can still be extracted s.t. only agent $\mathcal{A}_{i_j}$ updates its state at time $t_{k_j}$ s.t. $k_j \in \mathbb{N}_0, j = 1,...,n$.

## Definition 21.10

Let $\mathcal{I} = \left\{ I[k] \in \mathbb{I} \right\}_{k=0}^{\infty}$ denote the sequence of indices identifying the fusion operator $F^{I[k]} \in \mathcal{F}_{\lhd}$ that is "active" at $t_k, k \in \mathbb{N}_0$. Let $i \leftarrow I[k]$ denote the mapping $\mathcal{F} \mapsto \mathcal{N}$ identifying the agent $\mathcal{A}_i \in \mathcal{N}$ that is being updated via $F^{I[k]}$ at $t_k, k \in \mathbb{N}_0$. Then, the belief revision process that takes place at $t_k$ can be expressed as:

$$\mathbf{x}_j[k+1] = \begin{cases} F^{I[k]}\left( \mathbf{x}_{i_1}\left[ s^1[k] \right],...,\mathbf{x}_{i_{m_{I[k]}}}\left[ s^{m_{I[k]}}[k] \right] \right), & j = i \leftarrow I[k]; \\ \mathbf{x}_j[k], & \forall j \neq i. \end{cases}$$

Here, $\mathcal{S} = \left\{ S[k] \right\}_{k=0}^{\infty}$ denotes a sequence of $m_i$-tuples $S[k] \equiv \left( s^1[k],...,s^{m_{I[k]}}[k] \right)$ with $s^\ell[k] \in \mathbb{N}_0 \cup \{-1,...,-m\}$ s.t. $s^\ell[k] \leq k, \forall \ell \in \{1,...,m_{I[k]}\}, m_i \leq m, \forall i \in \mathbb{I}$.  ∎

As we have elaborated, notice that at a given time $t_k$, only one agent state gets updated and states of all other agents remain the same. $I[k]$ identifies the operator $F^{I[k]} \in \mathcal{F}$ performing the state update at $t_k$, which in turn uniquely identifies the agent $\mathcal{A}_i, i \leftarrow I[k]$ that is getting updated with the information received from its set of neighbors $\mathcal{N}_{I[k]} \subseteq \mathcal{N}$.

## Example 21.2

Consider the multiagent system $\mathcal{N} = \{\mathcal{A}_1,...,\mathcal{A}_5\}$ [20]. Suppose the agents update their states via the pool of operators $\mathcal{F} \equiv \left\{ F^i \middle| i = 1,...,6 \right\}$, where $F^1 : \mathbb{D}^2 \mapsto \mathbb{D}$, $F^2 : \mathbb{D}^2 \mapsto \mathbb{D}$, $F^3 : \mathbb{D}^4 \mapsto \mathbb{D}$, $F^4 : \mathbb{D}^3 \mapsto \mathbb{D}$, $F^5 : \mathbb{D}^2 \mapsto \mathbb{D}$ and $F^6 : \mathbb{D}^3 \mapsto \mathbb{D}$. Take the initial conditions of $\mathcal{N}$ as $\mathcal{X}_0 = \left\{ \mathbf{x}_i[0] \middle| i = 1,...,5 \right\}$. Consider the following sequence of state updates:

$$\begin{aligned} \mathbf{x}_1[1] &= F^1(\mathbf{x}_1[0], \mathbf{x}_3[0]), & k &= 0; \\ \mathbf{x}_2[2] &= F^2(\mathbf{x}_2[1], \mathbf{x}_3[0]), & k &= 1; \\ \mathbf{x}_4[3] &= F^4(\mathbf{x}_3[0], \mathbf{x}_4[0], \mathbf{x}_5[0]), & k &= 2; \\ \mathbf{x}_5[4] &= F^5(\mathbf{x}_4[3], \mathbf{x}_5[3]), & k &= 3; \\ \mathbf{x}_3[5] &= F^3(\mathbf{x}_1[1], \mathbf{x}_2[2], \mathbf{x}_3[4], \mathbf{x}_4[3]), & k &= 4; \\ \mathbf{x}_5[6] &= F^6(\mathbf{x}_1[1], \mathbf{x}_4[3], \mathbf{x}_5[4]), & k &= 5; \\ \mathbf{x}_1[7] &= F^1(\mathbf{x}_1[1], \mathbf{x}_3[5]), & k &= 6, \text{etc.} \end{aligned}$$

We have omitted the cases for which $\mathbf{x}_j[k]$ is not updated, for which $\mathbf{x}_j[k+1] = \mathbf{x}_j[k]$, $k \geq 0$. Define the sequence of $m_i$-tuples $\mathcal{S} \equiv \{(s^1[0] = 0, s^2[0] = 0), (s^1[1] = 0, s^2[1] = 0), (s^1[2] = 0, s^2[2] = 0, s^3[2] = 0), \ldots\}$ and the sequence $\mathcal{I} \equiv \{1, 2, 4, 5, 3, 6, 1, \ldots\}$ of elements of the index set $\mathbb{I} \equiv \{1, \ldots, 6\}$ of $\mathcal{F}$. Now, with $\mathcal{F}$ and the two sequences $\mathcal{I}$ and $\mathcal{S}$, the belief revision process can be expressed as given in Definition 21.10.

Belief revision process described in Definition 21.10 is very general. It can be modeled via $\mathcal{F}_\triangleleft$ and the two sequences $\mathcal{I}$ and $\mathcal{S}$. In fact, our work in Ref. [20] is based on devising a pool of fusion operators that is applicable for soft/hard fusion environments and characterizing $\mathcal{I}$ and $\mathcal{S}$ under which a belief revision process of the type in Definition 21.10 converges.

### 21.3.2 Modeling Opinions and Belief Revision

To be able to model complex agent opinions, we utilize DST notions for modeling agent states and the belief revision process.

#### 21.3.2.1 State Models

Let us model $\mathbf{x}_i[k]$ via a DST BoE $\mathcal{E}_i[k] \equiv \{\Theta, \mathfrak{F}_i[k], m_i(\cdot)[k]\}, k \in \mathbb{N}_0$, for $i = 1, \ldots, m$. Thus, $\mathbf{x}_i[k]$ can be expressed via any valid mass assignment on $2^\Theta$. This allows for a convenient representation of a wide variety of data imperfections, including probabilistic uncertainties (see Table 21.1). Further, because the DST model is completely specified by mass, belief, or plausibility values, depending on the application, $\mathbf{x}_i[k]$ can be equivalently represented via $m_i(\cdot)[k]$, $\mathrm{Bl}_i(\cdot)[k]$, or $\mathrm{Pl}_i(\cdot)[k]$.

#### 21.3.2.2 Belief Revision

To be able to study a wider array of application domains, we model belief revision via the conditional update equation (CUE) proposed in Ref. [27], which provides a versatile belief revision mechanism that is applicable to multiple fusion scenarios.

**Definition 21.11**

(Conditional Update Equation [CUE]): The CUE that updates $\mathcal{E}_i[k] \in \mathfrak{E}_\Theta$ with the evidence in $\mathcal{E}_j[k] \in \mathfrak{E}_\Theta, j \in \{1, \ldots, m\}\backslash\{i\}$, is denoted as $\mathcal{E}_i[k] \triangleleft_{j=1}^m \mathcal{E}_j[k]$, and the corresponding belief function is given by

$$\mathrm{Bl}_i(B)[k+1] = \alpha_i[k]\mathrm{Bl}_i(B)[k] + \sum_{j \neq i} \sum_{A \in \mathfrak{F}_j[k]} \beta_{ij}(A)[k]\mathrm{Bl}_j(B|A)[k],$$

where the CUE parameters $\alpha_i[k]$, $\beta_j(A)[k] \in \mathfrak{R}_{+0}$ satisfy

$$\alpha_i[k] + \sum_{j \neq i} \sum_{A \in \mathfrak{F}_j[k]} \beta_{ij}(A)[k] = 1, \forall k \in \mathbb{N}_0.$$

Here, $\sum_{j \neq i}$ denotes the sum over $j \in \{1, \ldots, m\}\backslash\{i\}$. ∎

CUE, which is a convex sum formulation of conditionals, provides a robust mechanism for handling contradictory evidence, vacuous evidence, evidence sources possessing nonidentical scopes of expertise, and so forth; uncertainty of the fused result given by $Un(\cdot)[k] = \mathrm{Pl}_i(\cdot)[k] - \mathrm{Bl}_i(\cdot)[k]$ provides valuable information regarding the underlying uncertainties of the fused result [27,29,35]. Moreover, the convex sum formulation is also consistent with the most widely used weighted average view of consensus analysis [1].

### 21.3.2.3 Characteristics of Agent Interactions

The CUE parameters $\alpha_i[k]$ and $\beta_{ij}(A)[k]$ provide the flexibility to model different belief revision scenarios. As we illustrate in the sections to follow, convergence of multiple fusion scenarios can be analyzed in a unified setting via an appropriate parameterization. We now provide some insight into CUE parameter selection from a consensus perspective and refer interested readers to Refs. [27,35] for a detailed discussion.

a. **Selection of $\alpha_i[k]$**: Parameter $\alpha_i[k] \in [0, 1]$ can capture the flexibility of $\mathcal{E}_i[k]$ toward changes, perhaps depending on the credibility of the incoming evidence and/or "inertia" of the available evidence. For instance, a lower $\alpha_i[k]$ is suitable in the initial phase of evidence collection, when $\mathcal{E}_i[k]$ has little or no credible knowledge base to begin with. Another strategy is *inertia-based selection* [35], where both existing and incoming evidence are considered to be equally important.

b. **Selection of $\beta_{ij}(\cdot)[k]$**: Parameters $\beta_{ij}(\cdot)[k]$ are used for both "refining" and "weighing" incoming evidence. We see two immediate choices [27].

   i. **Receptive CUE (rCUE)**: Choose $\beta_{ij}(A)[k] = C_{ij}[k]m_j(A)[k]$, $C_{ij}[k] \in \mathfrak{R}_+$ for all $A \in \mathfrak{F}_j[k]$ s.t. the condition in Definition 21.11 is satisfied. The rCUE strategy weighs incoming evidence based on the support that already exists in the source providing the evidence for each focal element, which can be interpreted as the source receiving evidence being receptive to incoming evidence. Not surprisingly, rCUE has an interesting Bayesian interpretation: It yields a weighted average of the corresponding p.m.f.s (see Section 21.3.2.4).

   ii. **Cautious CUE (cCUE)**: Choose $\beta_{ij}(A)[k] = C_{ij}[k]m_i(A)[k]$, $C_{ij}[k] \in \mathfrak{R}_+$ for all $A \in \mathfrak{F}_j[k]$ s.t. the condition in Definition 21.11 is satisfied. The cCUE strategy weighs incoming evidence based on the support that already exists in the source receiving the evidence for each focal element, which can be interpreted as the source receiving evidence being cautious to incoming evidence. Therefore, the focal elements of the updated BoE are restricted to within itself and only refinements are allowed, that is, a new focal element $B \in \mathfrak{F}_i[k+1]$ is created only if there exists a $C \in \mathfrak{F}_i[k]$ s.t. $B \subset C \subseteq \Theta$. This is characteristic of an agent who is being cautious to incoming evidence owing to high credibility of its existing knowledge. For instance, as we discuss in the following section, an opinion leader, or a highly reliable estimate of the ground truth (*GT*), can be modeled as an agent employing a cCUE strategy who allows refinement only of the evidence it already has. Unlike previous work on consensus analysis, the work in this chapter can accommodate a cautious agent, thus allowing us to establish the relationship between the consensus and the *GT*.

### 21.3.2.4 Probabilistic Case

A probabilistic assignment $P_i[k]: \Theta \mapsto [0, 1]$ that is consistent with the evidence in $\mathcal{E}_i[k]$ satisfies $\mathrm{Bl}_i(B)[k] \leq P_i(B)[k] \leq \mathrm{Pl}_i(B)[k], \forall B \subseteq \Theta$. Therefore, for probabilistic modeling (which allows probabilities to be assigned to singletons of $\Theta$, and not partial probability specifications), we have $P_i(B)[k] = m_i(B)[k] = \mathrm{Bl}_i(B)[k] = \mathrm{Pl}_i(B)[k]$. Thus, DST state modeling encapsulates probabilistic state models for which the CUE-based updates reduce to

$$P_i(B)[k+1] = \begin{cases} \alpha_i[k]P_i(B)[k] + \displaystyle\sum_{j \neq i} C_{ij}[k]P_{ij}(B)[k], & \text{for rCUE;} \\ P_i(B)[k], & \text{for cCUE.} \end{cases} \quad (21.2)$$

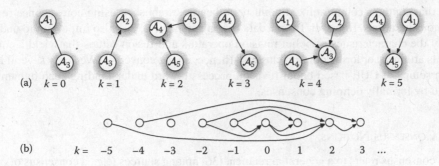

**FIGURE 21.2** (a) Spatial and (b) temporal coupling of $\mathcal{N} = \{\mathcal{A}_1, \dots, \mathcal{A}_5\}$ as given in Example 21.2. Note that (a) spatial coupling specified via $F \in \mathcal{F}$ can only capture the agents that are involved in a given iteration. In an asynchronous setup, agent coupling alone cannot guarantee convergence.

Note that rCUE reduces to a weighted average of p.m.f.s. Therefore, rCUE belief revision in conjunction with DST state models encapsulates the traditional weighted average modeling approach of consensus. However, with the cCUE, the agent state remains the same. This in fact makes perfect sense—cCUE allows refinements only to originally cast evidence, and no further refinements are possible with probabilistic modeling, as only singleton propositions possess nonzero support.

### 21.3.3 AGENT INTERACTIONS MODELING

Both spatial and temporal coupling among agents play key roles in the convergence characteristics of a multiagent system.

#### 21.3.3.1 Spatial Coupling

Spatial coupling among agents is best described via a graph. Consider the state update that takes place at time $t_k$ via $F^{I[k]}$: the agent $\mathcal{A}_i$, where $i \leftarrow I[k]$, updates its state by interacting with agents $\mathcal{A}_j \in \mathcal{N}_{I[k]}$. The rooted digraph $G[k] = (V[k], E[k])$, with the vertex set $V[k] = \{\mathcal{A}_i \cup \mathcal{N}_{I[k]}\}$ and the edge set $E[k] = \{e_{ji} | \mathcal{A}_j \in \mathcal{N}_{I[k]}\}$, represents unidirectional information flow from $\mathcal{A}_j \in \mathcal{N}_{I[k]}$ to $\mathcal{A}_i$ via the directed edge $e_{ji}$. Such a graph is referred to as an agent interaction topology (AIT) in Ref. [22]. The multiagent system is said to be fully connected if $V[k] = N$, $\forall k \in \mathbb{N}$ (i.e., all AITs used throughout the belief revision process are fully connected); otherwise the system is is said to be partially connected. Now, consider the set of all AITs used by agent $\mathcal{A}_i \in \mathcal{N}$, that is, the set of graphs $\mathbb{G}_i = \{G[k] = (\{\mathcal{A}_i \cup \mathcal{N}_{I[k]}\}, E[k]) | i \leftarrow I[k]\}$. We say that the multiagent system is static, if $|\mathbb{G}_i| = 1, \forall \mathcal{A}_i \in \mathcal{N}$ (i.e., each agent $\mathcal{A}_i$ uses a single AIT throughout the updating process); otherwise, it is said to be *dynamic*.

#### 21.3.3.2 Temporal Coupling

When the communication among agents is asynchronous (e.g., when delays exists), spatial coupling alone is insufficient for convergence analysis. As illustrated in Ref. [21], the graph of an asynchronous iteration (see Definition 21.8) can be used to analyze temporal coupling among fusion operators. Figure 21.2 illustrates the spatial and temporal coupling (of the corresponding asynchronous iteration) of agents in Example 21.2.

## 21.4 DYNAMICS OF CONSENSUS FORMATION

With the notions in Section 21.2.3 in place, one can view a consensus state $\mathbf{x}^* \in \mathbb{D}$ in a multiagent system as a common fixed point of the pool of belief revision operators used by agents. As we showed in Ref. [20], convergence of CUE-based belief revision using a pool of (para)contractive operators that contains common fixed points can then be established using convergence results on asynchronous iterations. However, from a data fusion perspective, it is also important to understand adequately the characteristics of belief revision operators and fusion setups that yield a consensus state (if it is attained) or form "opinion clusters" (hence, do not converge). We study several interesting fusion setups of CUE-based belief revision process toward understanding such dynamics. Let us proceed by formally defining consensus.

### 21.4.1 CONSENSUS NOTIONS

The word consensus refers to a general agreement [36] among sources (e.g., a consensus of opinions among a jury pool), and has been adopted in many sensor research/applications [4,6,9–11,14,22] where an "agreement" among opinions is sought. Consensus, as applicable to a complex fusion environment, can be defined as follows.

**Definition 21.12**

(Consensus [20]): Consider the multiagent system $\mathcal{N} = \{\mathcal{A}_1,\ldots,\mathcal{A}_m\}$ of adaptive agents that interact toward collectively estimating $\mathcal{X}$. The initial conditions are taken as $\mathcal{X}_0 = \left\{\mathcal{E}_i[0] \big| i = 1,\ldots,m\right\}$. Each agent $\mathcal{A}_i$ iteratively updates its state $\mathcal{E}_i[k] = \{\Theta, \mathfrak{F}_i[k], m_i(\cdot)[k]\}$ at $t_k$ via some fusion operator $F \in \mathcal{F}_{\lhd}$ as identified by the sequences $\mathcal{I}$ and $\mathcal{S}$. Then, we say a *consensus* is reached among agents in $\mathcal{N}$, if $\left\|\mathcal{E}_i[k] - \mathcal{E}_j[k]\right\| \to 0$ as $k \to \infty$ for all $\mathcal{A}_i, \mathcal{A}_j \in \mathcal{N}$, for some norm $\|\cdot\| : \mathfrak{E}_\Theta \mapsto \mathfrak{R}_{+0}$.   ∎

The seminal work in Ref. [1], which has since become popular as the Lehrer–Wagner model of rational consensus, espouses the weighted average as the basis for consensus formation. This constitutes the core of many consensus related papers. However, in complex fusion environments that are rife with data uncertainties, satisfying this requirement alone is not sufficient toward generating a rational consensus. One reassuring property in a majority of soft/hard fusion environments is the availability of ground truth $GT$ estimates. These estimates $\widehat{GT}$ are often vague but highly reliable (e.g., prevailing threat level estimates extracted via satellite imagery, initial hypotheses of opinion leaders in a social network setting). Hence, we believe a more meaningful, or "rational" consensus must satisfy the following properties:

$\mathcal{P}_1$: It is based on a weighted average of agent states [1].
$\mathcal{P}_2$: It is consistent with $\widehat{GT}$ when it is available.
$\mathcal{P}_3$: It converges to the $GT$ when it is known.

More formally, in terms of DST BoEs, these properties of a rational consensus can be expressed as

**Definition 21.13**

(Rational Consensus [20]): Let $\mathcal{E}^t = \{\Theta, \mathfrak{F}^t = \theta^t \in \Theta, m^t(\theta^t) = 1.0\}$ and $\hat{\mathcal{E}}^t = \left\{\Theta, \hat{\mathfrak{F}}^t, \hat{m}^t(\cdot)\right\}$ denote the GT and $\widehat{GT}$, respectively. Let the BoE $\mathcal{E}^* = \left\{\Theta, \mathfrak{F}^*, m^*(\cdot)\right\}$ denote a consensus state reached by the agents in $\mathcal{N}$ via some belief revision process. We say $\mathcal{E}^*$ is *rational* if

i.  $\mathcal{E}^*$ is reached as some convergence point of weighted averages of $\mathcal{E}_i[\cdot], i = 1,\ldots,m$.
ii. $\mathcal{E}^*$ is a refinement of $\hat{\mathcal{E}}^t$ when it is known (i.e., for all $B \in \mathfrak{F}^*$, there is a $C \in \hat{\mathfrak{F}}^t$ s.t. $B \subseteq C$).
iii. $\mathcal{E}^* \to \mathcal{E}^t$, whenever $\hat{\mathcal{E}}^t \to \mathcal{E}^t$ is known.   ∎

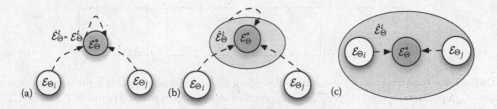

**FIGURE 21.3** Graphical illustration of rational consensus. Clearly, $\mathcal{E}^t$ must take the form s.t. $m^t(\theta^t) = 1.0$ for some $\theta^t \in \Theta$, where $\theta^t$ is the $\widehat{GT}$. Furthermore, since $\hat{\mathcal{E}}^t$ is an estimate of the $GT$, we assume that $\exists B \in \mathfrak{F}^t$ s.t. $\theta^t \in B$. (a) $GT$ is known. (b) $\widehat{GT}$ is known. (c) Unknown $GT$.

Figure 21.3 illustrates the nature of a rational consensus as defined in Definition 21.13.

As we will show presently, it turns out that a particular CUE parameterization (see Section 21.3.2.2) guarantees the formation of a rational consensus as defined in Definition 21.13. It can also be shown that, under certain mild conditions, this parameterization is also contractive on $\mathfrak{E}_\Theta$, which is a necessary condition for convergence analysis under the umbrella of asynchronous iterations [20].

## 21.4.2 Paracontracting Operators' View of Consensus

Consider the following CUE-based fusion operator.

### Definition 21.14

($F_\triangleleft$ Operator): Consider a set of BoEs $\mathcal{E}_i \in \mathfrak{E}_\Theta, i = 1,\ldots,m'$, that possibly contains $\hat{\mathcal{E}}^t$ (i.e., $\exists j \in \{1, \ldots, m'\}$, s.t. $\mathcal{E}_j \equiv \hat{\mathcal{E}}^t$, whenever $\hat{\mathcal{E}}^t$ exists). Then, the operator $F_\triangleleft^i : \mathfrak{E}_\Theta^{m'} \mapsto \mathfrak{E}_\Theta$ that updates $\mathcal{E}_i$ with the evidence in $\mathcal{E}_j, j \in \{1,\ldots,m'\}\backslash\{i\}$ is defined via the CUE as $F_\triangleleft^i(\mathcal{E}_1,\ldots,\mathcal{E}_{m'}) = \mathcal{E}_i \triangleleft_{j=1}^{m'} \mathcal{E}_j$, where the CUE parameters are given by

$$\alpha_i = \lambda_i \neq 1; \quad \beta_{ij}(A) = \begin{cases} \lambda_j m_i(A), & \text{for } \mathcal{E}_i = \hat{\mathcal{E}}^t; \\ \lambda_j m_j(A), & \text{otherwise,} \end{cases}$$

s.t. $\alpha_i + \sum_{j\neq i}\sum_{A\in\mathfrak{F}_j}\beta_{ij}(A) = 1$, for $\lambda_i \in \mathfrak{R}_{+0}$, $i = 1, \ldots, m'$. ∎

Note that the parameter choices are based on rCUE and cCUE strategies: cCUE is used for updating the $\widehat{GT}$, thus only allowing refinements on initial assignments; rCUE is used for the other agents, so that they are receptive to information coming from other agents. The definition of $F_\triangleleft$ allows the inclusion of multiple $\hat{\mathcal{E}}^t$ in the mix as long as they are "consistent" (or do not contradict) one another. We address the exact requirements for inclusion of multiple $\widehat{GT}$ after establishing the following results.

### Claim 21.1 [20]

The $F_\triangleleft$ operator in Definition 21.14 has infinitely many fixed points in $\mathfrak{E}_\Theta$. Furthermore, if $\mathcal{E}^* \in \text{fix}(F_\triangleleft)$, then for all $B \in \mathfrak{F}^*$, $\nexists C \in \mathfrak{F}^*$ s.t. $B \subset C$ or $B \supset C$. ∎

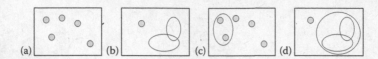

**FIGURE 21.4** The structure of consensus BoEs. Shaded and nonshaded objects represent singleton and composite propositions, respectively. $\mathcal{F}_i^*, i = 1,2$, represent two examples of fixed points of $F_\lhd$; $\mathcal{F}_i, i = 1,2$, are example that cannot be fixed points of $F_\lhd$. (a) $\mathfrak{F}_1^*$; (b) $\mathfrak{F}_2^*$; (c) $\mathfrak{F}_1$; (d) $\mathfrak{F}_2$.

Claim 21.1 establishes the fact that the operator $F_\lhd$ can in fact generate a consensus, viz., some fixed-point in $\mathfrak{E}_\Theta$. Furthermore, it also states that the consensus BoE will have a unique structure: any proposition (with nonzero support) will not contain or be contained in any other proposition (see Figure 21.4 for a set theoretic interpretation). Therefore, consensus BoEs will contain only propositions that do not allow further refinements. However, such propositions need not be singletons. Also, notice that this scenario is not important in pure probabilistic assignments, as all propositions with nonzero support are elements of $\Theta$ (i.e., $\forall B \in \mathfrak{F} \Rightarrow B \in \Theta$). However, in the case of partial probability specifications (as allowed in DST), BoEs are allowed to contain composite propositions (i.e., $B \subseteq \Theta$) to represent uncertainty. Claim 21.1 establishes that, even under this case, consensus can only contain propositions that cannot be further refined. The assurance that the generated consensus state will be the "most" refined as allowed by evidence is in fact a useful observation, especially in the case of soft/hard fusion environments, where soft sources often provide vague evidence (hence composite propositions).

### 21.4.3 Reaching a Consensus

With Claim 21.1 in place, we are now in a position to define a DST consensus protocol (i.e., a scheme that can generate a consensus via iterative belief revision) for complex fusion environments.

**Definition 21.15 [20]**

Let $\mathcal{N} \equiv \{\mathcal{A}_1, \ldots, \mathcal{A}_m\}$ be a multiagent system, where agent states and the initial conditions are given via $\mathcal{E}_i[k] = \{\Theta, \mathfrak{F}_i[k], m_i(\cdot)[k]\}, i = 1, \ldots, m$, and $\mathcal{X}_0 = \left\{ \mathcal{E}_i[0] \middle| i = 1, \ldots, m \right\}$, respectively. Then, for an operator sequence $\mathcal{I} \equiv \left( I[k] \in \mathbb{I} \right)_{k=0}^{\infty}$, a DST asynchronous consensus protocol on $\mathcal{N}$ is given via

$$\mathcal{E}_j[k+1] = \begin{cases} F_\lhd^{I[k]}\left( \mathcal{E}_{i_1}[s^1(k)], \ldots, \mathcal{E}_{i_{m_{I[k]}}}[s^{m_{I[k]}}(k)] \right), \\ \quad \text{for } j = i \leftarrow I[k]; \\ \mathcal{E}_j[k], \text{ for } j \neq i. \end{cases}$$

Here, $\mathcal{S} = \left\{ S[k] \right\}_{k=0}^{\infty}$ denotes a sequence of $m_i$-tuples $S[k] = \left( s^1[k], \ldots, s^{m_{I[k]}}[k] \right)$ with $s^\ell[k] \in \mathbb{N}_0 \cup \{-1, \ldots, -m\}$ s.t. $s^\ell[k] \leq k, \forall \ell \in \{1, \ldots, m_{I[k]}\}$, where $m_i \leq m, \forall i \in \mathbb{I}$. ■

All agents $\mathcal{A}_i \in \mathcal{N}$ start off with the initial states $\mathcal{E}_i[0]$ and update their states in the sequence specified via $\mathcal{I}$. The state updating process will take the following form:

i. The BoE of an agent which is not a $\widehat{GT}$ will receive all the focal elements from its neighbors, that is, $\mathfrak{F}_i[k+1]$ of an agent $\mathcal{A}_i, i \leftarrow I[k]$, via neighbor set $\mathcal{N}_{I[k]}$ will have $\mathfrak{F}_i[k+1] = \mathfrak{F}_i[k] \cup_{j \in \mathcal{N}_{I[k]}} \mathfrak{F}_j[k]$;

ii. The BoE of an agent having a $\widehat{GT}$ will only receive "new" focal elements that are "refinements" to its existing focal set, that is, $B \in \mathfrak{F}_i[k+1] \backslash \mathfrak{F}_i[k]$ iff $\exists C \in \mathfrak{F}_i[k]$ s.t. $B \subseteq C$. After sufficient number of initial iterations, (a) all agents who do not possess a $\widehat{GT}$ will have received focal elements of all other agents directly or indirectly (i.e., via another agent who has updated its state with other agents); and (b) all agents who possess a $\widehat{GT}$ will have received all focal elements from all other agents that are a refinement to initial state.

We now show that the iteration operators $F_\lhd^i, i \in \mathbb{I}$, of the form in Definition 21.15 are paracontractive on $\mathfrak{E}_\Theta$. This sets the foundation for CUE-based consensus analysis.

## Claim 21.2

When used in an iteration of the form Definition 21.15, the $F_\lhd$ operator in Definition 21.14 is paracontractive on $\mathfrak{E}_\Theta$ w.r.t. the $p$-norm $\|\mathcal{E}\|$, where $\|\mathcal{E}\|^p = \sum_{B \subseteq \Theta} |m(B)|^p$. ∎

Another result that is necessary for reaching a consensus state under the iterations given in Definition 21.15 is

## Claim 21.3

Let $\mathcal{F}_\lhd \equiv \left\{ F_\lhd^i : \mathfrak{E}_\Theta^{m_i} \mapsto \mathfrak{E}_\Theta \middle| m \geq m_i \in \mathbb{N}, i \in \mathbb{I} \right\}$ denote the pool of $F_\lhd$ operators enumerated via the index set $\mathbb{I}$ for updating agents in $\mathcal{N}$ with initial conditions $\mathcal{X}_0 \equiv \{\mathcal{E}_i[0]|i = 1,\ldots,m\}$. Then, the pool $\mathcal{F}_\lhd$ contains common-fixed points under the following conditions.

  i. $\mathcal{F}_\lhd$ contains only operators corresponding to at most one agent having access to a $\widehat{GT}$.

  ii. If $\mathcal{F}_\lhd$ contains operators corresponding to more than one agent having access to $\widehat{GT}$ then the initial conditions $\mathcal{E}_i[0], \mathcal{E}_j[0] \in \mathcal{X}_0$ of such agents are s.t. $\exists C \in \mathfrak{F}_j^t[0]$ for which $B \subseteq C$ or $B \supseteq C$, for all $B \in \mathfrak{F}_i^t[0]$.

∎

This claim establishes the conditions that lead to consensus state generation via the iteration in Definition 21.15 using the pool $\mathcal{F}_\lhd$ of fusion operators. The belief revision process given in Definition 21.15 employs fusion operators from $\mathcal{F}_\lhd$. Thus Claims 21.1, 21.2, and 21.3 on fixed-points and contractive properties in $\mathfrak{E}_\Theta$ are satisfied. In fact, we now have

## Claim 21.4

A consensus BoE generated via the protocol in Definition 21.15 is a rational consensus in accordance with Definition 21.13. ∎

Claim 21.4 guarantees the ability of the iteration in Definition 21.15 to generate a rational consensus in complex fusion environments. However, convergence of belief revision in such environments is governed by the properties of fusion operators as well as the characteristics of agent interactions (i.e., $\mathcal{I}$ and $\mathcal{S}$). Let us now look at several agent interactions that may lead to an agreement among agent opinions.

### 21.4.3.1  Synchronous, Fully Connected Network

This represents perhaps the simplest multiagent system, where each agent is connected to all the other agents and information is exchanged without any iteration delay (i.e., $k - s^\ell[k] = 0, k \in \mathbb{N}_0$). In this case, iterated belief revision in Definition 21.15 reduces to

$$\mathcal{E}_j[k+1] = \begin{cases} F_\triangleleft^{I[k]}(\mathcal{E}_1[k],\ldots,\mathcal{E}_m[k]), & \text{for } j = i \leftarrow I[k]; \\ \mathcal{E}_j[k], & \text{for } j \neq i. \end{cases} \tag{21.3}$$

This iteration converges, thus guaranteeing the ability of agents to reach a consensus.

### 21.4.3.2 Synchronous, Static, Partially Connected Network

For this network, agents communicate without iteration delays (i.e., $k - s^j[k] = 0$) with (at least one) partially connected AIT that does not change over time (e.g., hierarchical, static scale-free, networks). Therefore, some agents cannot communicate with others. In this case, the consensus protocol in Definition 21.15 reduces to

$$\mathcal{E}_j[k+1] = \begin{cases} F_\triangleleft^{I[k]}(\mathcal{E}_{i_1}[k],\ldots,\mathcal{E}_{i_{m_i}}[k]), & \text{for } j = i \leftarrow I[k]; \\ \mathcal{E}_j[k], & \text{for } j \neq i. \end{cases} \tag{21.4}$$

It can be shown that, if $\mathcal{I}$ is regulated, iterated belief revision in Equation 21.4 converges as long as the graph union of AITs of all agents is connected.

### 21.4.3.3 Synchronous, Dynamic, Partially Connected Network

Here, agents communicate without iteration delays (i.e., $k - s^j[k] = 0$), and with (at least one) partially connected AITs that change over time (e.g., Erdös-Rényi [37], ad hoc networks). This setup is similar to the preceding one, except that the neighbor set of agents changes over time. In this case, iterated belief revision in Definition 21.15 reduces to

$$\mathcal{E}_j[k+1] = \begin{cases} F_\triangleleft^{I[k]}(\mathcal{E}_{i_1}[k],\ldots,\mathcal{E}_{i_{m_{I[k]}}}[k]), & \text{for } j = i \leftarrow I[k]; \\ \mathcal{E}_j[k], & \text{for } j \neq i, \end{cases} \tag{21.5}$$

where $m_{I[k]} \in \{1,\ldots,m\}$. As previously, if $\mathcal{I}$ is regulated, the iteration in Equation 21.5 will converge whenever the AITs $G[k]$, $k \geq 1$, of the network are repeatedly jointly rooted.

### 21.4.3.4 Asynchronous, Fully Connected Network

For this case, each agent is connected to all the other agents, but the information exchange is asynchronous (or has delays) (i.e., $k - s^\ell[k] < 0$). In this case, the iterated belief revision in Definition 21.15 reduces to

$$\mathcal{E}_j[k+1] = \begin{cases} F_\triangleleft^{I[k]}(\mathcal{E}_1[s^1[k]],\ldots,\mathcal{E}_m[s^m[k]]), & \text{for } j = i \leftarrow I[k]; \\ \mathcal{E}_j[k], & \text{for } j \neq i. \end{cases} \tag{21.6}$$

If $\mathcal{I}$ is regulated, the iterated belief revision in Equation 21.6 converges whenever the iteration delays $k - s^\ell[k]$ are finite.

Similar results can be obtained for asynchronous, dynamic, and partially connected networks. In this case, similar to the cases we have studied in the preceding text, one needs to impose conditions to guarantee adequate coupling among agents to reach a consensus. Given that the pool $\mathcal{F}_\triangleleft$ of CUE-based operators is paracontractive on $\mathfrak{E}_\Theta$ (under the conditions on $\widehat{GT}$), the belief revision given in Definition 21.15 is capable of generating a rational consensus in any network setup as long as these coupling conditions are satisfied. In the analysis of an arbitrary network topology, one needs to make sure that it satisfies Definition 21.8; and, then the convergence is automatically guaranteed via Theorem 21.3.

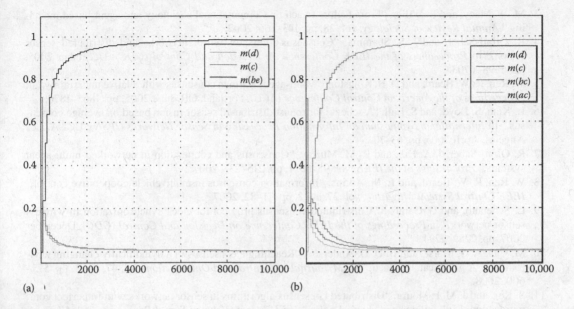

**FIGURE 21.5** Convergence behavior of an agent under an infinite pool of CUE operators. Mass $m_2(\cdot)[k]$ appears on the $y$-axis. Note the initially cast evidence for $\mathcal{A}_2$ supports only the two propositions $\{c\}$ and $\{cd\}$. (a) $\widehat{GT} = \{be\}$. (b) $\widehat{GT} = \{bc\}$.

### 21.4.3.5 Convergence with an Infinite Pool

In the absence of sensor reliability and/or credibility of evidence, an intuitive technique for belief revision involves weighting the incoming evidence based on its "distance" to the BoE that is updated, that is, in updating $\mathcal{E}_i$, (in Definition 21.14) are chosen s.t. $\lambda_j \propto \left\| \mathcal{E}_i - \mathcal{E}_j \right\|$. This would essentially lead to an infinite pool of fusion operators. However, as the iteration progresses, it is easy to see that such a pool approximates a finite pool where the $\lambda_{(\cdot)}$ of fusion operators corresponding to each agent are constant. Figure 21.5 shows the convergence behavior of this scheme under two different $\widehat{GT}$s. Notice that in each case, the convergence time has increased, but the consensus state is still consistent with the $\widehat{GT}$.

## 21.5 CONCLUDING REMARKS

We explored dynamics of consensus formation among a group of adaptive agents whose states are modeled as DST BoEs. Both asynchronous networks and situations in which the link structure are dynamic are accounted for by utilizing the framework of asynchronous iterations of paracontracting operators. The DST agent model provides significant flexibility: p.m.f. agent models can be accommodated as a special case, and the DST model allows for capturing the types of imperfections and nuances that one encounters in soft evidence. So, this work is ideally suited to study opinion and consensus dynamics in social networks. An important consequence of the proposed DST belief revision is its ability to generate a rational consensus, implying that the consensus reached can be used to estimate agent credibility (even when the ground truth is only partially known or unknown).

## REFERENCES

1. K. Lehrer, and C. Wagner, *Rational Consensus in Science and Society*, Philosophical Studies Series in Philosophy. Dordrecht: D. Reidel, 1981.
2. H. E. Stephanou, and S.-Y. Lu, "Measuring consensus effectiveness by a generalized entropy criterion," *IEEE Transactions on Pattern Analysis and Machine Intelligence*, vol. 10, no. 4, pp. 544–554, 1988.

3. M. W. Macy, and R. Willer, "From factors to actors: Computational sociology and agent-based model-ing," *Annual Review of Sociology*, vol. 28, pp. 143–166, 2002.

4. R. Olfati-Saber, and J. S. Shamma, "Consensus filters for sensor networks and distributed sensor fusion," in *Proceedings of the IEEE Conference on Decision and Control (CDC)*, December 2005, pp. 6698–6703.

5. W. Ren, R. W. Beard, and D. B. Kingston, "Multi-agent Kalman consensus with relative uncertainty," in *Proceedings of the American Control Conference (ACC)*, Portland, OR, June 2005, pp. 1865–1870.

6. L. Xiao, S. Boyd, and S. Lall, "A scheme for robust distributed sensor fusion based on average consen-sus," in *International Symposium on Information Processing in Sensor Networks (IPSN)*, UCLA, Los Angeles, April 2005, pp. 63–70.

7. R. Olfati-Saber, J. A. Fax, and R. M. Murray, "Consensus and cooperation in networked multi-agent systems," *Proceedings of the IEEE*, vol. 95, no. 1, pp. 215–233, 2007.

8. W. Ren, R. W. Beard, and E. M. Atkins, "Information consensus in multivehicle cooperative control," *IEEE Control Systems Magazine*, vol. 27, no. 2, pp. 71–82, 2007.

9. L. Schenato, and G. Gamba, "A distributed consensus protocol for clock synchronization in wireless sensor network," in *Proceedings of the IEEE Conference on Decision and Control (CDC)*, December 2007, pp. 2289–2294.

10. M. Cao, A. S. Morse, and B. D. O. Anderson, "Reaching a consensus in a dynamically changing envi-ronment: A graphical approach," *SIAM Journal of Control and Optimization*, vol. 47, no. 2, pp. 575–600, 2008.

11. S. Kar, and J. M. F. Moura, "Distributed consensus algorithms in sensor networks with imperfect com-munication: Link failures and channel noise," *IEEE Transactions on Signal Processing*, vol. 57, no. 1, pp. 355–369, 2009.

12. Y. G. Sun, and L. Wang, "Consensus of multi-agent systems in directed networks with nonuniform time-varying delays," *IEEE Transactions on Automatic Control*, vol. 54, no. 7, pp. 1607–1613, 2009.

13. R. F. Weiss, "Consensus technique for the variation of source credibility," *Psychological Reports*, vol. 20, no. 3c, pp. 1159–1162, 2011.

14. Z. Wu, H. Fang, and Y. She, "Weighted average prediction for improving consensus performance of second-order delayed multi-agent systems," *IEEE Transactions on Systems, Man, and Cybernetics, Part B: Cybernetics*, vol. 42, no. 5, pp. 1501–1508, 2012.

15. F. Borran, M. Hutle, N. Santos, and A. Schiper, "Quantitative analysis of consensus algorithms," *IEEE Transactions on Dependable and Secure Computing*, vol. 9, no. 2, pp. 236–249, 2012.

16. C. Altafani, "Consensus problems on networks with antagonistic interactions," *IEEE Transactions on Automatic Control*, vol. 58, no. 4, pp. 935–946, 2013.

17. Z. Li, W. Ren, X. Liu, and M. Fu, "Consensus of multi-agent systems with general linear and Lipschitz nonlinear dynamics," *IEEE Transactions on Automatic Control*, vol. 58, no. 7, pp. 1786–1791, 2013.

18. X. Lu, R. Lu, S. Chen, and J. Lu, "Finite-time distributed tracking control for multi-agent systems with a virtual leader," *IEEE Transactions on Circuits and Systems—I*, vol. 60, no. 2, pp. 352–362, 2013.

19. H. Terelius, D. Varagnolo, C. Baquero, and K. H. Johansson, "Fast distributed estimation of empirical mass functions over anonymous networks," in *Proceedings of the IEEE Conference on Decision and Control (CDC)*, Florence, Italy, December 2013, pp. 6771–6777.

20. T. Wickramarathne, K. Premaratne, M. Murthi, and N. Chawla, "Convergence analysis of iterated belief revision in complex fusion environments," *IEEE Journal of Selected Topics in Signal Processing*, vol. 8, no. 4, pp. 598–612, 2014.

21. M. Pott, "On the convergence of asynchronous iteration methods for nonlinear paracontractions and consistent linear systems," *Linear Algebra and Its Applications*, vol. 283, no. 13, pp. 1–33, 1998.

22. L. Fang, and P. J. Antsaklis, "Asynchronous consensus protocols using nonlinear paracontractions the-ory," *IEEE Transactions on Automatic Control*, vol. 53, no. 10, pp. 2351–2355, 2008.

23. L. Moreau, "Stability of multiagent systems with time–dependent communication links," *IEEE Transactions on Automatic Control*, vol. 50, no. 2, pp. 169–182, 2005.

24. G. Shafer, *A Mathematical Theory of Evidence*. Princeton, NJ: Princeton University Press, 1976.

25 P. Vannoorenberghe, "On aggregating belief decision trees," *Information Fusion*, vol. 5, no. 3, pp. 179–188, 2004.

26. T. L. Wickramarathne, K. Premaratne, and M. N. Murthi, "Convergence analysis of consensus belief functions within asynchronous ad-hoc fusion networks," in *Proceedings of the International Conference on Statistical Signal Processing (ICASSP)*, Vancouver, Canada, May 2013, pp. 3612–3616.

27. K. Premaratne, M. N. Murthi, J. Zhang, M. Scheutz, and P. H. Bauer, "A Dempster-Shafer theoretic conditional approach to evidence updating for fusion of hard and soft data," in *Proceedings of the 12th International Conference on Information Fusion (FUSION)*, Seattle, WA, July 2009, pp. 2122–2129.

28. T. L. Wickramarathne, K. Premaratne, M. N. Murthi, and M. Scheutz, "A Dempster-Shafer theoretic evidence updating strategy for non-identical frames of discernment," in *Proceedings of the Workshop on the Theory of Belief Functions (BELIEF)*, Brest, France, April 2010.

29. T. L. Wickramarathne, K. Premaratne, M. N. Murthi, M. Scheutz, and S. Kübler, "Belief theoretic methods for soft and hard data fusion," in *Proceedings of the International Conference on Statistical Signal Processing (ICASSP)*, Prague, Czech Republic, May 2011, pp. 2388–2391.

30. T. L. Wickramarathne, K. Premaratne, and M. N. Murthi, "Toward efficient computation of the Dempster-Shafer belief theoretic conditionals," *IEEE Transactions on Cybernetics*, vol. 43, no. 2, pp. 712–724, 2012.

31. T. L. Wickramarathne, K. Premaratne, and M. N. Murthi, "Consensus-based credibility estimation of soft evidence for robust data fusion," in *Belief Functions, Advances in Intelligent and Soft Computing*, vol. 164, T. Denoeux, and M.-H. Masson, Eds. Berlin/Heidelberg: Springer, pp. 301–309.

32. D. B. West, *Introduction to Graph Theory*, 2nd ed. Upper Saddle River, NJ: Prentice-Hall, 2001.

33. R. Fagin, and J. Y. Halpern, "A new approach to updating beliefs," in *Proceedings of the Conference on Uncertainty in Artificial Intelligence (UAI)*, P. P. Bonissone, M. Henrion, L. N. Kanal, and J. F. Lemmer, Eds. New York: Elsevier Science, 1991, pp. 347–374.

34. T. L. Wickramarathne, K. Premaratne, and M. N. Murthi, "Focal elements generated by the Dempster-Shafer theoretic conditionals: A complete characterization," in *Proceedings of the 13th International Conference on Information Fusion (FUSION)*, Scotland, UK, July 2010, pp. 1–8.

35. K. Premaratne, D. A. Dewasurendra, and P. H. Bauer, "Evidence combination in an environment with heterogeneous sources," *IEEE Transactions on Systems, Man and Cybernetics, Part A: Systems and Humans*, vol. 37, no. 3, pp. 298–309, 2007.

36. *Merriam Webster's Collegiate Dictionary*. Merriam Webster, July 2003.

37. P. Erdös, and A. Rényi, "On the evolution of random graphs," *Institute of Mathematics, Hungarian Academy of Sciences*, vol. 5, pp. 17–61, 1960.

# 22 Decentralized Bayesian Fusion in Networks with Non-Gaussian Uncertainties

*Nisar R. Ahmed, Simon J. Julier, Jonathan R. Schoenberg, and Mark E. Campbell*

## CONTENTS

## 22.1 INTRODUCTION AND MOTIVATION

A decentralized data fusion (DDF) system is a general, flexible system for fusing information in a robust, distributed, and scalable manner. A DDF network is composed of a set of connected processing centers or nodes. Each node attempts to estimate the underlying state of a system of interest. Nodes can use information that comes from two sources: local or remote. Local information is collected from nodes that are equipped with sensors. Remote information is obtained from neighboring nodes in the network. Although this information could be a set of raw observations, in most cases these are actual, fused estimates of the underlying state of interest.

Each node fuses this information together to produce estimates that are distributed back into the network, where they can be used by other nodes or by some kind of output process. This definition subsumes a wide variety of real data fusion systems, including sensor networks aboard mobile vehicles [1], cooperative target tracking unmanned aerial vehicles (UAVs) [2], formation flying spacecraft [3], autonomous robotic map-building teams [4,5], hierarchical sensor networks of heterogeneous maritime surveillance assets [6], or hierarchical wireless networks of aggregative "clusterheads" for energy-efficient data collection and communication [7].*

DDF can lead to solutions that are more scalable, modular, *and* robust than centralized solutions, in which all raw observation data are communicated to a centralized processing node. Scalability arises from the fact that nodes send fused estimates, which summarize the state given all information up to the current time, rather than raw observations, which summarize information collected at each time. Through careful exploitation of conditional independence relationships among the information sets embedded within each node's local posterior distribution, DDF obtains significant reductions in local processing and communication, thus bypassing the shortcomings of alternative decentralized fusion strategies such as consensus [8,9]. Modularity arises from the fact that different nodes can be specialized using their own kinds of sensing systems. However, because only fused estimates are distributed, it is not necessary for every node in the network to have the likelihood model of all types of observations collected by all sensors. Finally, robustness arises from the fact that the network has no central point of failure. As nodes or communication links fail, the quality of estimates in the network gradually declines. Even in the extreme case where the network becomes bisected, the individual subnetworks can still operate on their own and can combine back together at a later date [10,11].

However, these advantages in terms of flexibility and robustness come at significant theoretical cost. Because fused estimates are distributed between nodes, remote information is not simply another kind of sensor information. Rather, dependencies exist, and these must be explicitly modeled or accounted for if the system is not to become overconfident. Initial work on Bayesian DDF focused on problems characterized by linear systems or Gaussian distributions [10]. However, in practice most system models are nonlinear and most probability distributions are non-Gaussian, for example, robotic mapping [12–14] or target search and tracking [15–17].† Although the same mathematical framework applies to both Gaussian and non-Gaussian DDF, exact solutions for the latter are unfortunately analytically and computationally intractable in general.

This chapter highlights several theoretical and algorithmic advances in tackling these issues in a broad range of applications and discusses some new avenues for non-Gaussian DDF research suggested by these approaches. Section 22.2 introduces the basic theory, advantages, and limitations of Bayesian DDF, with particular emphasis on the important distinction between algorithms for exact DDF systems (where all interagent information dependencies are explicitly known and accounted for) and "rumor robust" conservative DDF systems (which maintain consistent fusion results even though exact DDF is infeasible). Approximate DDF solutions are then examined for cases where non-Gaussian common information between network nodes can be (approximately) ascertained. Implementation issues are discussed for semiparametric Gaussian mixture filters for distributed robotic terrain height mapping. Algorithms are presented for conservative DDF in cases where non-Gaussian common information cannot be ascertained. Emphasis here is placed exclusively on information-theoretical optimization techniques for weighted exponential product (WEP) fusion. Examples of non-Gaussian approximate and conservative DDF are discussed in the context of distributed robotic target search and tracking with Gaussian mixture models. Finally, some possible directions for future work are discussed.

---

* With this definition, a sensor platform that does not locally fuse data or communicate directly with other fusion nodes is not a processing node in its own right. Rather, it is associated with a "parent" processing node.

† Time delays are also significant non-Gaussian sources of estimation error. We neglect them here for simplicity. However, they can be handled in a Bayesian framework [18,19].

## 22.2  OPTIMAL BAYESIAN DISTRIBUTED DATA FUSION

It is assumed that (1) the number of active processing nodes in the network may vary; (2) each node is aware only of its neighbors; and (3) no node ever knows the complete network topology. In more general cases, it also assumed that; (4) the communication topology is arbitrary (i.e., possibly not strongly connected) and possibly dynamic/ad hoc; and (5) no node knows the receipt status of data sent to other nodes.

The state of a system is to be estimated by a decentralized network of $N_A$ processing nodes. The state of the system at time step $k$ is $x_k$.* The estimate of the state at the $i$th node is the random vector $x_k^i$.† Given a common prior $p(x_0)$, each node operates resursively, predicting the state to the current time, and then fuses information from local sensors and communicated to it from other nodes in the network. Suppose $I_{1:k}^i$ is the set of all information, including both local and remote, available to node $i$ at time step $k$. The estimate is the distribution $p\left(x_k^i \big| I_{1:k}^i\right)$. For brevity, we sometimes write this as $p\left(x_k^i\right)$.

### 22.2.1  LOCAL FUSION ONLY

First consider the case in which node $i$ only exploits the locally collected sensor data. In this case, $I_{1:k}^i = Z_{1:k}^i = \left\{z_1^i, \ldots, z_k^i\right\}$, where $z_k^i$ is the local sensor information collected at time step $k$. This is a conventional tracking problem, and the filtering equations are

$$p_i\left(x_k \big| I_{1:k}^i\right) \propto p\left(z_k^i \big| x_k\right) \cdot p_i\left(x_k \big| I_{1:k-1}^i\right), \tag{22.1}$$

where $p\left(z_k^i \big| x_k\right)$ is the observation likelihood and $p_i\left(x_k \big| z_{1:k-1}^i\right)$ is the predicted distribution computed from

$$p\left(x_k^i \big| I_{1:k-1}^i\right) = \int p\left(x_k^i \big| x_{k-1}^i\right) \cdot p\left(x_{k-1}^i \big| I_{1:k-1}^i\right) dx_{k-1}^i, \tag{22.2}$$

where $p\left(x_k \big| x_{k-1}^i\right)$ is the state transition equation. However, an important but subtle issue arises even in this case. Because each node attempts to estimate the state of the *same* system, the process noise is *identical* for each node. Therefore,

$$p\left(x_k^i, x_k^j \big| x_{k-1}^i, x_{k-1}^j\right) \neq p\left(x_k^i \big| x_{k-1}^i\right) \cdot p\left(x_k^j \big| x_{k-1}^j\right). \tag{22.3}$$

As a result, the estimates are not independent of one another.

$$p\left(x_k^i, x_k^j \big| I_{1:k-1}^i, I_{1:k-1}^j\right) \neq p\left(x_k^i \big| I_{1:k-1}^i\right) \cdot p\left(x_k^j \big| I_{1:k-1}^j\right). \tag{22.4}$$

This has a direct implication when developing a DDF system.

### 22.2.2  LOCAL AND REMOTE FUSION

Suppose each node now exploits the information that is distributed to it from the other nodes as well. In this case, $I_{1:k}^i = Z_{1:k}^i \cup R_{1:k}^i$, where $Z_{1:k}^i$ is the local sensor information as before, but $R_{1:k}^i$ is the

---

* The length of each time step does not have to be the same.

† More general representations, such as random sets, can be used as well [20]. However, we do not discuss these generalities in this chapter.

set of information received from the other nodes. Although $R_{1:k}^i$ could be the set of raw observations from the other nodes, the power and flexibility of DDF arises when these are just the estimates. The Bayesian DDF problem is for each agent $i$ to find the fused information probability density function (PDF)

$$p_{f,i}(x_k) \equiv p\left(x_k^i \big| I_{1:k}^i\right) = p_i\left(x_k \big| Z_{1:k}^i \cup R_{1:k}^i\right) \tag{22.5}$$

which is ideally equivalent to the PDF obtained by a single centralized fusion node receiving $I_{1:k}^i$. It is not possible simply to assume that the information sets are conditionally independent. As already noted, dependencies arise between nodes even if they have never communicated with one another before. If they do communicate between one another, further dependencies arise because of the propagation of common observation: Sensor observations collected at a node $i$ implicitly flow throughout the whole DDF network. Therefore, the exact formulation must account for these effects.

### 22.2.3 Exact Bayesian DDF

The exact formulation uses the following insight. The estimates at each node, $p_i(x_k)$ and $p_j(x_k)$, can be factored into

$$p_i(x_k) \propto p\left(x_k \big| I_k^i \cap I_k^j\right) p\left(x_k \big| I_k^{i|j}\right), \quad p_j(x_k) \propto p\left(x_k \big| I_k^i \cap I_k^j\right) p\left(x_k \big| I_k^{j|i}\right), \tag{22.6}$$

where $I_k^{i|j}$ is the set of information at node $i$ that $j$ does not already possess; $I_k^{j|i}$ is the set of information at $j$ that $i$ does not already possess; and $I_k^i \cap I_k^j$ is the set of information common to both $i$ and $j$; and $p_{ij}^c(x_k) \equiv p\left(x_k \big| I_k^i \cap I_k^j\right)$ is the information common to both nodes. Because these sets are disjoint,

$$p_{ij}^f(x_k) \propto p_{ij}^c(x_k) p\left(x_k \big| I_k^{i|j}\right) p\left(x_k \big| I_k^{j|i}\right). \tag{22.7}$$

Substituting Equation 22.6 into Equation 22.7, the exact Bayesian solution for DDF is

$$p_{ij}^f(x_k) \propto \frac{p_i(x_k) p_j(x_k)}{p_{ij}^c(x_k)} \tag{22.8}$$

Therefore, an optimal DDF solution requires the following. First, each node must maintain its own estimate $p_i(x_k)$. Second, each pair of nodes that communicate must store the common information between each node, $p_{ij}^c(x_k)$. If the common information is not canceled, it is treated as independent information ("double counted"). As a result the DDF network apparently has access to more information than is genuinely the case. This leads to highly overconfident fusion results. The resulting statistical inconsistencies can spread quickly and cripple network estimation performance via the "rumor propagation" effect (e.g., see Ref. [6] for a concrete example).

There are two major challenges in performing optimal DDF. The first issue lies with the required bookkeeping. The second issue is computational complexity.

For the first issue, $p_{ij}^c(x_k)$ must somehow be tracked locally. This can be achieved when information between any two nodes travels along a uniquely traceable path. Unfortunately this occurs only for two network topologies: When the network is either fully connected (all nodes talk to all other nodes, i.e., maximum redundancy) or tree connected (in which case there is no redundancy) [10,11]. Extensions to general nontree/dynamic topologies exist, but require significantly greater computational cost. For instance, in Ref. [21] Martin and Chang propose an algorithm in which nodes share local "information ancestry trees" to explicitly determine $p_{ij}^c(x_k)$ on the basis of all previous communication instances. However, the storage, communication, and processing requirements for exact tracking of $p_{ij}^c(x_k)$ can become very difficult to manage and may not scale well for practical applications.

The second challenge is that, even if the network topology or information history can be suitably constrained, the form of $p_{ij}^c(x_k)$ might not be analytically tractable. When the systems are linear and Gaussian, $p_i(x_k)$, $p_j(x_k)$ and $p_{ij}^c(x_k)$ are all Gaussian. More generally, when $p_i(x_k)$, $p_k(x_k)$ are members of the exponential family of distributions, then the update equations are the sums, differences, and linear combinations of the functions in the exponents.* However, for other choices of $p_i(x_k)$ and $p_j(x_k)$, the computation becomes extremely complicated. For example, if $p_i(x_k)$ and $p_j(x_k)$ are Gaussian mixtures (GMs), $p_{ij}^c(x_k)$ will not, in general, be a GM itself.

Although non-Gaussian $p_{ij}^c(x_k)$ cannot in general be tracked or removed exactly, these operations can be reasonably well approximated so that each node obtains useful information updates. This implies the following general "recipe" for exact DDF with arbitrary non-Gaussian distributions between nodes $i$ and $j$:

1. Approximate $p_{ij}^c(x_k)$, ideally with a representation that permits compact storage/communication.
2. Approximate and normalize the division of $p_i(x_k)p_j(x_k)$ by the approximate common information.
3. Store the result of step 2 for calculation of (the approximate) $p_{ij}^c(x_k)$ for the next fusion instance between $i$ and $j$.

Steps 1 and 2 are often the most challenging, and satisfactory solutions will be heavily application dependent. Some techniques for addressing these challenges are discussed next in the context of a terrain mapping application for a tree-connected robot networks.

## 22.3  APPROXIMATE DDF WITH NON-GAUSSIAN DISTRIBUTIONS

### 22.3.1  APPLICATION 1: DISTRIBUTED TERRAIN MAPPING

In this application, a team of mobile robots equipped with laser range sensors jointly map the height of the surrounding terrain. Each robot constructs a 2.5D probabilistic height map in a shared global East–North–Up (ENU) coordinate frame and fuses this model with other probabilistic height maps produced by neighbors in a network with a static tree topology.

The mixture-model based terrain estimation algorithm introduced by Miller and Campbell [23] translates laser scanner terrain detections into a conditional distribution over the terrain height $U$ in each cell of a planar $EN$ grid. The algorithm begins by transforming the terrain detections into an inertial coordinate system accounting for uncertainties in the sensor alignments and measurements errors. Next, the terrain detections are probabilistically associated to cells in the terrain grid. A Gaussian estimate of the conditional elevation distribution in the cell is then generated from each raw laser measurement associated to each grid cell. The conditional elevation densities in all grid cells are assumed to be independent of one another. In Ref. [23], Miller and Campbell show that the resulting total multiscan conditional height distribution in the $l$th $EN$ grid cell is a Gaussian mixture (GM),

$$p(U_l \mid r_{0:K}, q_{0:K}) \approx \frac{1}{c_l} \sum_{k=0}^{K} \sum_{m=1}^{M} p_{kml} \cdot \mathcal{N}\left(\hat{u}_{kml}, \sigma_{\hat{u}_{kml}}^2\right) \qquad (22.9)$$

---

* Tonkes and Blair, for example, developed the fusion operations for ExPoly distributions of the form $e^{-P(x)}$, where $P(x)$ is an elliptic polynomial [22]. However, although the update is extremely simple, the prediction equations can become extremely complicated and cumbersome in many cases.

where $U_l$ is the height estimate conditioned on all laser scans $r_{0:K}$ up to time $K$ for robot positions $q_{1:K}$; $p_{km}$ is cell $l$ association probability for the $m$th height estimate at time $k$ with mean $\hat{u}$ and variance $\sigma^2$; and $\sum_{k=0}^{K}\sum_{m=1}^{M} p_{km\in l}$ is a normalizing constant, equal to the expected number of measurements falling in cell $l$ up to time $K$.

Considering just the first and second moments of this GM, the $l$th cell's conditional height is summarized by the mean and covariance of the GM [24]:

$$\hat{U}_{GM_l} = \frac{1}{c_l} \sum_{k=0}^{K} \sum_{m=1}^{M} p_{km\in l} \hat{u}_{km\in l} \tag{22.10}$$

$$\sigma^2_{\hat{U}_{GM_l}} = \frac{1}{c_l} \sum_{k=0}^{K} \sum_{m=1}^{M} p_{km\in l} \left( \hat{u}^2_{km\in l} + \sigma^2_{\hat{u}_{km\in l}} \right) - \hat{U}^2_{GM_l} \tag{22.11}$$

This allows each sensor node to maintain a recursively calculated information set with sufficient statistics for each grid cell via

$$Z_l^K \triangleq \left\{ Z_p^K = \sum_{k=0}^{K} \sum_{m=1}^{M} p_{km\in l}, Z_{p\hat{u}}^K = \sum_{k=0}^{K} \sum_{m=1}^{M} p_{km\in l} \hat{u}_{km\in l}, \right.$$

$$\left. Z_{p\hat{u}^2}^K = \sum_{k=0}^{K} \sum_{m=1}^{M} p_{km\in l} \hat{u}^2_{km\in l}, Z_{p\sigma_{\hat{u}}^2}^K = \sum_{k=0}^{K} \sum_{m=1}^{M} p_{km\in l} \sigma^2_{\hat{u}_{km\in l}} \right\} \tag{22.12}$$

The first and second GM moments are thus computed as

$$\hat{U}_{GM_l} = \frac{1}{Z_p^K} Z_{p\hat{u}}^K \tag{22.13}$$

$$\sigma^2_{\hat{U}_{GM_l}} = \frac{1}{Z_p^K} \left( Z_{p\hat{u}^2}^K + Z_{p\sigma_{\hat{u}}^2}^K \right) - \left( \frac{Z_{p\hat{u}}^K}{Z_p^K} \right)^2 \tag{22.14}$$

Hence, the conditional height estimates corresponding to new laser scans for cell $l$ can be easily fused with conditional height estimates derived from previous scans by updating the cell $l$ statistics given in Equation 22.12). Although this approximation loses higher-order moment information for each cell's conditional height PDF (because the GM in Equation 22.9 is repeatedly condensed into a single Gaussian), it greatly eases the ability to accumulate additional laser data into the local terrain estimate and requires far less memory for mapping large areas.

### 22.3.1.1 Bayesian DDF via the Channel Filter

Schoenberg et al. [25] show that the Gaussian approximation of each cell's conditional height distribution also naturally lends itself to Bayesian DDF in tree-connected topologies.* In this case, because each robot's local terrain map is represented by a succinct information set in each grid cell

---

* Schoenberg further extends the mapping algorithm DDF with arbitrary topologies in Ref. [4].

(Equation 22.12), the union of the information sets at sensor nodes $i$ and $j$ can be directly tracked for each grid cell using a channel filter, where the union is given by

$$Z_i^K \bigcup Z_j^K = Z_{i\backslash j}^K + Z_{j\backslash i}^K + Z_{i\cap j}^K = Z_i^K + Z_j^K - Z_{i\cap j}^K \tag{22.15}$$

and where $Z_{i\cap j}^K$ is the common information contained in sets $Z_i^K$ and $Z_j^K$. Because the tree-connected network topology ensures all common information between nodes $i$ and $j$ comes across the $i$–$j$th communication link, the common information up to time $K$ is the union of all information previously shared by $i$ and $j$,

$$Z_{i\cap j}^K = Z_i^{K-1} \bigcup Z_j^{K-1} \tag{22.16}$$

The local information set at node $i$ up to time $k$, given all of the information sets in the neighborhood $N_i$ (sensor nodes connected to $i$) where $i \notin N_i$ can now be updated,

$$Z_i^K = Z_i^{K-1} + Z_i^{K-1,K} + \sum_{j\in N_i} \left[ \tilde{Z}_{i\leftarrow j}^k - Z_{i\cap j}^{K-1} \right] \tag{22.17}$$

where $Z_i^{K-1,K}$ is the new information accumulated at sensor $i$ locally (i.e., from new laser measurements) from $t_{k-1}$ to $t_k$ and $\tilde{Z}_{i\leftarrow j}^k$ is the information received at node $i$ from neighboring sensor node $j$ at time $t_k$. Note that the addition and subtraction operators in Equation 22.17 imply direct addition and subtraction of the corresponding moment statistics in each information set.

In Ref. [25] Schoenberg et al. formally prove that the channel filter is robust to communication failure and does not need to verify receipt of a transmitted message from one node to another to ensure consistent operation, as long as two rules are followed. First, each sensor node commits not to send back any information already received over a given channel,

$$\tilde{Z}_{i\rightarrow j}^k = Z_i^K - Z_{i\cap j}^{K-1} = Z_i^K - Z_{i\leftarrow j}^{K-1} \tag{22.18}$$

where $Z_{i\leftarrow j}^{K-1}$ indicates the information received at node $i$ from node $j$ up to $t_{k-1}$; this may not be the same as $Z_{j\leftarrow i}^{K-1}$ because of communication failure. Second, each sensor node is required to keep track of all information received from a given node. These two rules allow the estimate of the common information between nodes $i$ and $j$ at node $i$ to be a simple assignment of the received information,

$$Z_{i\cap j}^K = \begin{cases} \tilde{Z}_{i\leftarrow j}^k & \text{if} \quad \tilde{Z}_{i\leftarrow j}^k \neq \varnothing \\ Z_{i\cap j}^{K-1} & \text{if} \quad \tilde{Z}_{i\leftarrow j}^k = \varnothing \end{cases} \tag{22.19}$$

The channel filter thus always accounts for the common information and the unique information contained at each node. The consequence is that, in the presence of communication losses, the estimates of the common information at each node are not necessarily symmetric across the $i$–$j$th channel, that is,

$$Z_{i\cap j}^K \neq Z_{j\cap i}^K \tag{22.20}$$

where the first node index indicates the estimation of common information between nodes $i$ and $j$ resides on that particular node. In the case of communication failure, the local information set at node $i$ is updated via Equation 22.17 without information $\tilde{Z}_j^k$ from nodes whose communication links have failed, so that the updated information is $Z_i^K = Z_i^{K-1} + Z_i^{K-1,K}$. The channel filter for the $ij$th communication link remains unchanged while the link is down according to Equation 22.19. If communication failure persists for $\kappa$ time steps, the local sensor nodes continue to update local information via Equation 22.17). When communication is restored, all information $\tilde{Z}_j^{K+\kappa}$ from neighboring nodes is received and assimilated at node $i$ into $Z_i^{K+\kappa}$ in a single step.

### 22.3.1.2 Experimental Results

Data were collected using Pioneer P3-DX mobile robots equipped with a Hokuyo URG-04X laser scanner, which features a 240° field-of-view and angular resolution of 0.36° and maximum range of ≈ 5 meters. As shown in Figure 22.1, the laser is pitched downward 45° and scans along the ground

(a)

(b)

**FIGURE 22.1**    (a) Mobile robot setup with Hokuyo lidar and onboard computer for distributed occupancy grid map fusion experiment. (b) Laboratory environment for the occupancy grid mapping experiment is 15 × 8 meters and contains boxes of different sizes.

as the robots move forward. The test environment is $15 \times 8$ meters and is instrumented with a Vicon MX+ precision tracking system that provides 3D object pose. The features in the environment consist of boxes between 10 and 25 centimeters tall that are meant to simulate traffic cones or other similarly sized obstacles for a full-size traffic vehicle. Eight robots are run in different paths around the environment for a 120-second data collection. The robots are run sequentially to avoid sensing each other during map construction. The channel filter requires a singly connected tree topology to ensure information is not double counted. Network communication links are artificially broken to create a chain topology (each node connected to two others, with nodes 1 and 8 singly connected as leaf nodes).

Two measures of information content in the global network are used to assess the consistency of the exact DDF solution: the association probability metric (APM) and the cumulative entropy metric (CEM). For the terrain mapping problem, the APM defines the information content in the observed E–N grid cells $C_O$ for sensor node $i$ (i.e., cells that are guaranteed to have at least one registered scan from $i$ with nonzero probability) as

$$\text{APM} \triangleq \sum_{l \in C_O} z_{p_{i_l}}^K \tag{22.21}$$

The APM in node $i$ is monotonically increasing because there is no way to remove association probability from a given grid cell. The CEM is defined as

$$\text{CEM} \triangleq \frac{1}{\sum_{l \in C_O} H\left[ p(U_l \mid \underline{r}^K, q^K) \right]} \approx \frac{1}{\sum_{l \in C_O} 0.5 \log\left( 2\pi e \sigma_{\hat{U}_{GM_l}}^2 \right)} \tag{22.22}$$

The approximation in the CEM (Equation 22.22) follows from computing the entropy of a Gaussian distribution whose first and second moments match those of the original Gaussian mixture model (GMM). This is not the exact entropy of a GMM, but it is a conservative approximation [26].

For a chain topology with eight agents, the APM and CEM are shown in Figure 22.2. The APM (top subplot) shows the delay across the network because the information must propagate along the maximum diameter graph. The leaf nodes (1 and 8) are lagging in information content compared to the other nodes; this demonstrates the delayed arrival of information content at the extreme nodes of the graph. At the end of the simulation, additional communication steps are required to propagate the latest collected local information across the entire network to reach the exact centralized solution; this was not done here for the sake of clarity. The stair steps visible in the APM (top subplot) illustrate the robustness of DDF to communication failure (i.e., failure to send or receive) using the channel filter: Each robot continues to assimilate local information, until remote information is received and the information content is updated without double counting. It is clear that double counting does not occur, as the APM (top subplot) does not exceed the centralized solution at any point. The CEM (bottom subplot) shows that each node converges to a similar plateau as the centralized solution, but the solutions are not exact because of propagation delays in the network.

These results highlight the effectiveness of Gaussian approximations of non-Gaussian PDFs in certain applications. In particular, this application shows how to recursively approximate non-Gaussian common information PDFs using the zeroth, first, and second moments of Gaussian

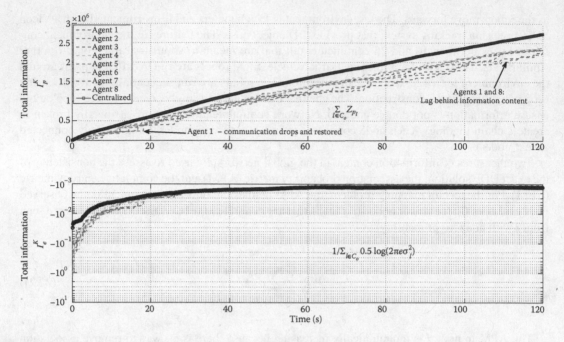

**FIGURE 22.2** APM and CEM for the channel filter chain topology at each individual node are shown along with the centralized solution. The information is delayed propagating across the network as the leaf nodes (agents 1 and 8) trail the information content of the nonleaf nodes from the APM (top subplot).

mixtures. This approximation sacrifices some higher order information, but, when combined with the channel filter, enables Bayesian DDF to be performed without significant information loss.

## 22.4 CONSERVATIVE DDF

The previous sections have shown that it is possible to develop algorithms for approximating optimal DDF solutions with non-Gaussian PDFs. However, these approximations can be used only for sensor networks with static tree communication topologies, which are overly restrictive and brittle in practice. In this section, we describe alternative conservative approximation algorithms for non-Gaussian DDF, which are suboptimal but can be used with arbitrary dynamic sensor network topologies. It will be shown that conservative approximations lead to their own set of implementation challenges for non-Gaussian DDF, but efficient algorithms can also be developed to address these.

### 22.4.1 CHARACTERIZATION OF CONSERVATIVE DDF ALGORITHMS

Conservative DDF algorithms replace optimal Bayesian solutions with conservative "rumor robust" fusion rules that closely approximate Bayesian updates, that is, that remain statistically consistent without ever explicitly computing and removing $p_{ij}^c(x_k)$. Following Ref. [27], we define a conservative fusion rule to be a method for combining $p_i(x_k)$ and $p_j(x_k)$ that does not double-count $p_{ij}^c(x_k)$. Suboptimality arises because this rule might lose some information from $p\left(x_k \middle| I_k^{i/j}\right)$ or $p\left(x_k \middle| I_k^{j/i}\right)$.

### 22.4.2 CONSERVATIVE DDF RULES

Perhaps somewhat surprisingly, two fusion rules automatically obey these properties irrespective of the network architecture, the dependency structure, and the form of the probability distributions: the arithmetic mean and the geometric mean. The arithmetic mean is

$$
\begin{aligned}
p^A(x_k) &\propto \omega p_i(x_k) + (1-\omega) p_j(x_k) \\
&= \omega p\left(x_k \big| I_k^i \cap I_k^j\right) p\left(x_k \big| I_k^{i/j}\right) + (1-\omega) p\left(x_k \big| I_k^i \cap I_k^j\right) p\left(x_k \big| I_k^{j/i}\right) \\
&= p\left(x_k \big| I_k^i \cap I_k^j\right)\left[ \omega p\left(x_k \big| I_k^{i/j}\right) + (1-\omega) p\left(x_k \big| I_k^{j/i}\right)\right] \\
&= p_{ij}^c(x_k)\left[ \omega p\left(x_k \big| I_k^{i/j}\right) + (1-\omega) p\left(x_k \big| I_k^{j/i}\right)\right].
\end{aligned}
\tag{22.23}
$$

This has the property that

$$
\min[p_i(x_k), p_j(x_k)] \le p^A(x_k) \le \max[p_i(x_k), p_j(x_k)] \ \forall x_k
\tag{22.24}
$$

The geometric mean, or weighted exponential product (WEP), is given by

$$
\begin{aligned}
p^W(x_k) &\propto [p_i(x_k)]^\omega + [p_j(x_k)]^{1-\omega} \\
&= \left[ p\left(x_k \big| I_k^i \cap I_k^j\right) p\left(x_k \big| I_k^{i/j}\right)\right]^\omega \left[ p\left(x_k \big| I_k^i \cap I_k^j\right) p\left(x_k \big| I_k^{j/i}\right)\right]^{1-\omega} \\
&= p\left(x_k \big| I_k^i \cap I_k^j\right)\left[ p\left(x_k \big| I_k^{i/j}\right)\right]^\omega \left[ p\left(x_k \big| I_k^{j/i}\right)\right]^{1-\omega} \\
&= p_{ij}^c(x_k)\left[ p\left(x_k \big| I_k^{i/j}\right)\right]^\omega \left[ p\left(x_k \big| I_k^{j/i}\right)\right]^{1-\omega}.
\end{aligned}
\tag{22.25}
$$

This has the property that

$$
p^W(x_k) \ge \min[p_i(x_k), p_j(x_k)]
\tag{22.26}
$$

Because the WEP "clumps" information, it guarantees information gains that cannot be obtained via the arithmetic mean. Other important related properties of the WEP are discussed in Ref. [27].

There are many interpretations of the WEP: It can be seen as an example associated with Chernoff information [28], as a logarithmic opinion pool, as a mixture of experts, as an application of Kullback's Principle of Minimum Discrimination Information, or as using a particular memory-less approximation of the common information PDF [29].

One important issue is that there is an essential tradeoff at the core of WEP fusion: Considerable flexibility in the network topology and choice of probability distributions is gained at the expense of conservative information loss. Several approaches for choosing the WEP fusion parameter $\omega$ have been considered to help balance this tradeoff.

### 22.4.3 OPTIMIZATION OF THE WEIGHTING FUNCTION

It is natural to consider information-theoretic cost functions for optimizing $\omega$. Motivated by the connection between WEP and covariance intersection, in Ref. [28] Hurley proposed minimization of the entropy of Equation 22.25. Despite its intuitive appeal, this cost function is not always justified from a general Bayesian perspective, as proper fusion of specific observations can sometimes *increase* the entropy of $p_{ij}^f(x_k)$ [27].* In Ref. [24] Hurley also considers minimization of the Chernoff information between $p_i(x_k)$ and $p_j(x_k)$; this equates the Kullback–Leibler divergences (KLDs) between $p_{ij}^f(x_k)$ and the individual node posteriors, ensuring the result is "half-way" between $p_i(x_k)$ and $p_j(x_k)$ in an information gain/loss sense. An alternative interpretation of the Chernoff rule is that it tries to estimate which member of the family of possible $\hat{p}_{ij}^c(x_k)$ PDFs looks "most common" to $p_i(x_k)$ and $p_j(x_k)$. Yet, Chernoff fusion ignores potential imbalances in local information gain or sensor quality between $i$ and $j$, meaning that fusion could cause $i$ or $j$ to consistently lose more information than the other gains.

In Ref. [13], Ahmed et al. proposed two alternative information-theoretic WEP rules to address these issues. The first *weighted Chernoff* fusion rule applies information weighting factors $c_i, c_j \geq 0$ to account for relative local information gains by $i$ and $j$. This biases $\omega$ toward the node $i$ or $j$ with greater information content, so that $\omega$ satisfies

$$c_i\, D_{KL}[p^W(x_k)\|p_i(x_k)] = c_j\, D_{KL}[p^W(x_k)\|p_k(x_k)], \tag{22.27}$$

where $D_{KL}[\cdot\|\cdot]$ is the KLD. The second *minimum information loss* rule selects $\omega$ to minimize $D_{KL}\left[p_{ij}^f(x_k)\,\|\,p^W(x_k)\right]$, that is, the true information lost by using Equation 22.25 instead of Equation 22.8. Because $p_{ij}^f(x_k)$ is unavailable, it is approximated by the naive Bayes fusion PDF $p_{NB}(x_k) \propto p_i(x_k)p_j(x_k)$, that is, so that

$$D_{KL}\left[p_{ij}^f(x_k)\,\|\,p^W(x_k)\right] \approx D_{KL}\left[p_{NB}(x_k)\,\|\,p^W(x_k)\right] \tag{22.28}$$

Ahmed et al. [13] and Schoenberg [4] showed that $D_{KL}[p_{NB}(x_k)\|p^W(x_k)]$ gives a worst case upper bound on $D_{KL}\left[p_{ij}^f(x_k)\,\|\,p^W(x_k)\right]$, thus yielding a "minimax" WEP criterion (also referred to as the "minimax naive Bayes" WEP rule [29]).

It is straightforward to show that the Shannon, Chernoff, weighted Chernoff, and minimax WEP fusion functionals are always convex in $\omega$.

## 22.5 CONSERVATIVE DDF WITH NON-GAUSSIAN DISTRIBUTIONS

Non-Gaussian distributions generally lead to analytically intractable WEP fusion PDFs in Equation 22.25 as well as analytically intractable integrals for the WEP fusion costs described previously. Fortunately, fast and accurate techniques for online WEP parameter optimization and fusion PDF approximation can be developed for some widely used non-Gaussian distribution models, such as GMs and particle approximations. As shown next, heuristic methods can be developed to solve the $\omega$ optimization and $p^W(x_k)$ approximation problems separately, while other more direct methods can address these problems simultaneously by exploiting certain properties of the underlying distributions being fused.

### 22.5.1 Application 2: Target Search with Ad Hoc Networks

Network connectivity and topology cannot be guaranteed in many target search and tracking applications, especially when dealing with contested communication environments or situations in which network nodes must reposition themselves to maintain target observability. Opportunistic ad hoc communication can propagate information quickly across a tracking network, although

---

\ * Bayesian updating decreases the *average* posterior entropy, that is, the conditional entropy averaged over all possible measurements [30].

non-Gaussian sources of uncertainties (data association ambiguities, non-Gaussian process or sensor noise, etc.) make the problem of maintaining statistical consistency much more challenging.

WEP fusion of particle or GM target state PDFs requires overcoming two additional major hurdles: (1) numerical optimization of an analytically intractable high-dimensional cost integral (subject to online computational constraints) and (2) approximation of the corresponding WEP-optimal fusion PDF $p^W(x_k)$ as either a particle set or GM. Two approaches for dealing with these issues are considered here. In the first approach, the intractable WEP cost integral is approximated by a heuristic cost that is easier to optimize for an assumed $p^W(x_k)$ model. In the second approach, the intractable cost integral is directly approximated by importance sampling (IS), which enables simultaneous high-fidelity optimization of the WEP fusion parameter $\omega$ and approximation of $p^W(x_k)$.

### 22.5.1.1 Heuristic Optimization

Upcroft in Ref. [31] and Julier in Ref. [16] propose different adaptations of the covariance intersection (CI) fusion rule for GM WEP fusion, which takes the full form

$$p^W(x_k) \propto \left[ \sum_{q=1}^{M_i} w^q \mathcal{N}(x_k; \mu^q; \Sigma^q) \right]^\omega \left[ \sum_{r=1}^{M_j} w^r \mathcal{N}(x_k; \mu^r; \Sigma^r) \right]^{1-\omega} \qquad (22.29)$$

In Ref. [31], CI is heuristically performed between each mixand of the GMs $p_i(x_k)$ and $p_j(x_k)$ to produce another GM; $M_i M_j$ fusion operations are required, involving separate $\omega_{ij}$ terms selected to minimize the determinant of each fused mixand's covariance. In light of the connection between CI and Chernoff fusion discussed in Ref. [28], Julier in Ref. [16] notes that this technique does not lead to a good approximation of the Chernoff fusion PDF, and proposes a "first order CI" (FOCI) approximation to address this,

$$p^W(x_k) \approx \sum_{m=1}^{M_i M_j} w^m \mathcal{N}\left(x_k; \mu^m; \Sigma^m\right) \qquad (22.30)$$

$$\Sigma^m = \left( \omega(\Sigma^q)^{-1} + (1-\omega)(\Sigma^r)^{-1} \right)^{-1} \qquad (22.31)$$

$$\mu^m = \Sigma^m \left( \omega(\Sigma^q)^{-1}\mu^q + (1-\omega)(\Sigma^r)\mu^r \right)^{-1} \qquad (22.32)$$

$$w^m = \frac{(w^q)^\omega (w^r)^{1-\omega}}{\sum_{q',r'} \left(w^{q'}\right)^\omega \left(w^{r'}\right)^{1-\omega}}, \qquad (22.33)$$

where $q, q' \in \{1, ..., M_i\}$ and $r, r' \in \{1, ..., M_j\}$ respectively index the mixands of $p_i(x_k)$ and $p_j(x_k)$. Although only one $\omega$ needs to be selected, it is still nontrivial to numerically/analytically minimize the Chernoff fusion cost for the FOCI PDF. As such, Julier in Ref. [16] heuristically selects $\omega$ to minimize the FOCI mixture covariance (calculated via the weighted FOCI mixand covariances and mixture mean). In the context of Chernoff fusion for arbitrary (possibly non-GM) PDFs, in Ref. [32] Farrell and Ganesh derive a heuristic direct estimate of the Chernoff-optimal $\omega$ using the entropies of $p_i(x_k)$ and $p_j(x_k)$, but do not specify how to actually approximate the fused PDF. In principle, this latter technique could also be used to define $\omega$ for the FOCI approximation.

For WEP fusion of particle sets, in Refs. [33,34] Ong et al. first transform $p_i(x_k)$ and $p_j(x_k)$ into GMs and then combine these using either of the CI-based WEP fusion methods described previously; the fused GM PDF is then sampled to recover a fused particle approximation. Although the compression of particle sets into finite GMs risks losing information, this also beneficially reduces the size of messages passed between network nodes. The particle-to-GM compression step can be implemented via standard statistical learning methods such as the weighted expectation-maximization algorithm (WEM, which generalizes standard EM learning by incorporating external weights for each sample) [35], or by placing Parzen kernels around each sample and applying suitable mixture condensation techniques, for example, Ref. [36].

### 22.5.1.2 Direct Optimization via Monte Carlo

In Refs. [37] and [13] Ahmed et al. present a general stochastic optimization procedure for direct minimization of convex WEP fusion costs such as Equations 22.27 and 22.28. The key idea is that the required integral cost function can be estimated as a function of $\omega$ via Monte Carlo (IS) with respect to ·some suitably chosen importance density $g_{ij}(x_k, \omega)$. Ideally, $g_{ij}(x_k, \omega)$ match the shape and support of $p^W(x_k)$ as closely as possible. Then, by drawing $N_s$ samples from $g_{ij}(x_k, \bar{\omega})$ only once for some fixed $\omega = \bar{\omega}$, cost estimates for different $\omega$ values can be obtained very quickly by recomputing and summing the modified importance weights of all samples. Fast derivative-free 1D convex optimization algorithms, such as golden section or bisection search, can then be used to obtain highly accurate estimates of the optimal $\omega$ value using a fixed number of sampled-based cost evaluations, which depends only on a desired error tolerance. In the case of GM fusion, $g_{ij}(x_k, \omega)$ is selected to be the FOCI approximation in Equation 22.30, which can be further "precompressed" into a GM with $M_g \ll M_i M_j$ components to accelerate IS PDF calculations. As the number of importance samples $N_s \to \infty$, the estimated $\omega$ minimizer converges to the true optimal solution, although tight convergence is obtained very quickly in practice.

IS optimization was compared to other approximations for GM Chernoff fusion across 200 randomly generated pairs of GMs in $\mathbb{R}^D$ for $D \in \{1, 2, 4, 6\}$. Component means in each dimension were uniformly sampled in the range $[-10, 10]$; $M_i$ and $M_j$ were uniformly sampled in the integer range $[1, 20]$, while component weights were uniformly sampled in the range $[0, 1]$ and then normalized to sum to 1. Component covariance matrices were sampled from a Wishart distribution using 12 degrees of freedom and a base covariance matrix equal to $0.25 \cdot I(n)$, where $I(n)$ is the identity matrix. IS optimization was initially used with the *uncompressed* FOCI PDF as the IS PDF, that is, so that $M^g \in [1, 400]$ GM components (denoted hereafter as no compression IS, or NCIS). Three other approximate Chernoff fusion cost optimizations were also assessed: (1) FOCI; (2) direct Monte Carlo sampling (DSA), which draws samples directly from the GM $p_i(x_k)$ to exploit the identity

$$\int p^W(x_k)\,dx_k = \int p_i(x_k)\left[\frac{p_j(x_k)}{p_i(x_k)}\right]^{1-\omega} dx_k = E\left[\left[\frac{p_j(x_k)}{p_i(x_k)}\right]^{1-\omega}\right]_{p_i(x_k)} ; \qquad (22.34)$$

and (3) unscented transform sampling (UTA), which is the same as (2) except that direct Monte Carlo is replaced by deterministic GM sampling [35]. NCIS and DSA were both applied 10 times in each case with $N_s \in \{100, 500, 1000, 2500, 5000\}$. Ground-truth cost values for estimating $\omega^*_{\text{TRUE}}$ were obtained via grid-based summation for 1D/2D cases and Monte Carlo sampling for 4D/6D cases with 100,000 samples. The 1D optimization for $\omega$ was carried out in all cases with the golden section algorithm and a convergence tolerance of 1e-03. The direct entropy-based approximation for $\omega^*$ proposed by Farrell and Ganesh in Ref. [32] was also assessed, using 100,000 direct samples from each GM to be fused to estimate its entropy precisely. The calculations for all these methods were performed in MATLAB® on a Windows 7® cluster (Intel i5 64-bit 2.66GHz quad-core processors, 12 GB RAM).

Figure 22.3a shows optimization errors $\omega^* - \hat{\omega}^*$ for $D = 4$ and $N_s = 500$, where the results have been sorted according to the true $\omega^*$ values in ascending order along the horizontal axis. NCIS clearly produces more accurate and consistent $\omega^*$ estimates compared with all other approximations. Whereas NCIS provides unbiased estimates of $\omega^*$, all other methods produce noticeably inconsistent estimates; the UTA and Farrell approximations in particular tend to under-/overestimate $\omega^*$. NCIS also obtains a much smaller sample variance for $N_s = 500$ than DSA at the same sample size. Figure 22.3b shows corresponding run time statistics for each of the various approximations. The

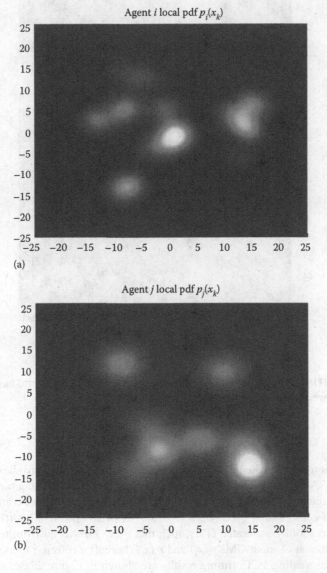

(a)

(b)

**FIGURE 22.3** Results for approximating $p^W(x_k)$ for GMs (a) $p_i(x_k)$ and (b) $p_j(x_k)$ with minimax cost metric (Equation 22.28) ($\omega = 0.55205$).

(*Continued*)

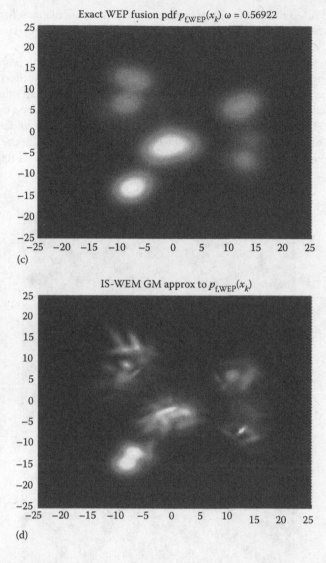

(c)

(d)

**FIGURE 22.3 (CONTINUED)** Results for approximating $p^W(x_k)$ for (c) exact grid-based fusion result and (d) GM obtained via weighted EM compression of IS particles.

*(Continued)*

Farrell approximation requires the most computation time (because the entropies are precisely estimated), whereas DSA and UTA require the least. Despite its improved accuracy, NCIS generally leads to longer computation times than FOCI if the IS PDF $g_{ij}(x_k, \omega)$ is not compressed. To investigate the effects of compressing $g_{ij}(x_k, \omega)$, a separate set of IS optimization trials was conducted on the same set of randomly generated GMs, such that $g_{ij}(x_k, \omega)$ was now limited to $M^g \in \{10, 50, 100\}$ via precompression of input GMs $p_i(x_k)$ and $p_j(x_k)$ (hereafter referred to as precompressed IS, or PCIS). The corresponding PCIS timing results are shown in Figure 22.3c, along with baseline NCIS times. PCIS leads to significantly faster computation times compared to NCIS, especially if $g_{ij}(x_k, \omega)$ is limited to fewer than a dozen components. Figure 22.3d shows that PCIS yields little to no loss in $\hat{\omega}^*$ accuracy compared to NCIS, although the variance of $\hat{\omega}^*$ tends to increase slightly (an expected effect that can be offset by a suitable increase in $N_s$).

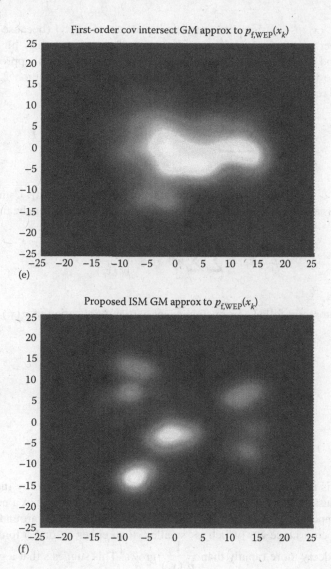

First-order cov intersect GM approx to $p_{f,\text{WEP}}(x_k)$

(e)

Proposed ISM GM approx to $p_{f,\text{WEP}}(x_k)$

(f)

**FIGURE 22.3 (CONTINUED)**   Results for approximating $p^{\text{W}}(x_k)$ for (e) FOCI approximation (Equation 22.30) and (f) component-wise IS GM approximation (Equation 22.37).

## 22.5.2   GM APPROXIMATION OF THE WEP FUSION DISTRIBUTION

To recover a GM approximation of $p^{\text{W}}(x_k)$ at the optimum $\omega$, the final weighted samples produced by the IS optimization procedure can be immediately compressed into a GM via the weighted EM algorithm. However, this approach can still converge to poor GM approximations without careful initialization. Alternatively, in Ref. [38] Ahmed et al. showed that $p^{\text{W}}(x_k)$ is equivalent to a mixture of non-Gaussian PDFs that can be closely approximated by a GM. This follows from the interesting fact that the WEP rule implicitly provides an estimate of the common information $p_{ij}^c(x_k)$, since Equation 22.25 can be rewritten in a form similar to Equation 22.8,

$$p^{\text{W}}(x_k) \propto \frac{p_i(x_k)p_j(x_k)}{\left[p_i(x_k)\right]^{1-\omega}\left[p_j(x_k)\right]^{\omega}} \propto \frac{p_i(x_k)p_j(x_k)}{\hat{p}_{ij}^c(x_k)}, \tag{22.35}$$

where $\hat{p}_{ij}^c(x_k) \propto p_i^{1-\omega}(x_k) p_j^{\omega}(x_k)$ is a "memoryless" estimate of $p_{ij}^c(x_k)$ (because $\omega$ is recomputed each time $i$ and $j$ perform fusion). Note that $\hat{p}_{ij}^c(x_k)$ and $p^W(x_k)$ are typically highly non-Gaussian PDFs when $p_i(x_k)$ and $p_j(x_k)$ are GMs, although only $p^W(x_k)$ needs to be well approximated by a GM to achieve recursive fusion updates. This follows from refactoring Equation 22.29 according to Equation 22.35 and rewriting the result as

$$p^W(x_k) \propto \sum_{q=1}^{M_i} \sum_{r=1}^{M_j} w^q w^r \frac{\mathcal{N}(x_k;\mu^q,\Sigma^q)\mathcal{N}(x_k;\mu^r,\Sigma^r)}{\hat{p}_{ij}^c(x_k)}, \tag{22.36}$$

where we note that $\hat{p}_{ij}^c(x_k)$ can be evaluated pointwise for arbitrary $\omega$ and $x_k$. Using the fact that the product of two Gaussian pdfs is another unnormalized Gaussian PDF, this can be further simplified to

$$p^W(x_k) \propto \sum_{q=1}^{M^i} \sum_{r=1}^{M^j} w^{qr} \frac{\overline{z}^{qr} \mathcal{N}(x_k;\mu^{qr},\Sigma^{qr})}{\hat{p}_{ij}^c(x_k)}, \tag{22.37}$$

where each numerator term follows from component-wise multiplication of $p_i(x_k)$ and $p_j(x_k)$,

$$\Sigma^{qr} = \left[ (\Sigma^q)^{-1} + (\Sigma^r)^{-1} \right]^{-1}$$

$$\mu^{qr} = \Sigma^{qr} \left[ (\Sigma^q)^{-1}\mu^q + (\Sigma^r)^{-1}\mu^r \right],$$

$$\tilde{w}^{qr} = w^q w^r \overline{z}^{qr},$$

$$\overline{z}^{qr} = \mathcal{N}(\mu^q;\mu^r,(\Sigma^q+\Sigma^r)).$$

Equation 22.37 is thus a mixture of non-Gaussian components formed by the ratio of a single (unnormalized) Gaussian PDF and non-Gaussian PDF $\hat{p}_{ij}^c(x_k)$. Although not a normalized closed-form pdf, each component of Equation 22.37 tends to concentrate mass around $\mu^{qr}$. As $x_k$ moves away from $\mu^{qr}$, the covariance $\Sigma^{qr}$ (which is "smaller" than either $\Sigma^q$ or $\Sigma^r$) forces each Gaussian numerator term to decay more rapidly than $\frac{1}{p_{ij}^c(x_k)}$ grows. This suggests that a good GM approximation to $p_{ij}^f(x_k)$ can be found by replacing each PDF ratio term in Equation 22.37 with a moment-matched Gaussian PDF. Although the required moments cannot be found analytically, they can be quickly estimated via IS, as detailed in Ref. [38].* In Ref. [38], IS PDFs are selected to be Gaussians centered at $\mu^{qr}$ with heurstically determined conservative spherical covariances, although more sophisticated approximations could be used instead. For instance, the Laplace approximation [39] can be used to better approximate the local shape of $p^W(x_k)$ near the maximum a posteriori (MAP) value of each mixand in Equation 22.37, while GMs could also be used to construct heavy-tailed IS PDFs that help control sample variance [40].

Figure 22.4 shows that this mixture IS approach leads to a much smoother approximation of $p^W(x_k)$ than weighted EM compression of IS particles and captures the shape of $p^W(x_k)$ more accurately than the FOCI approximation, even when $p_i(x_k)$, and $p_j(x_k)$ are highly non-Gaussian. The improved performance of the mixture IS technique stems from the fact that it is generally easier to approximate the individual mixands of $p^W(x_k)$ by separate "local" importance distributions than it is

---

* In Ref. [38] Ahmed et al. show that this same mixture approximation also applies directly to GM channel filter DDF, because the actual common information pdf $p_{ij}^c(x_k)$ simply replaces the estimate $\hat{p}_{ij}^c(x_k)$.

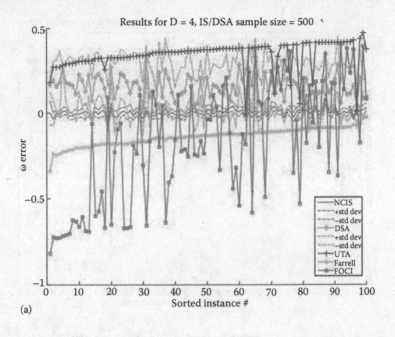

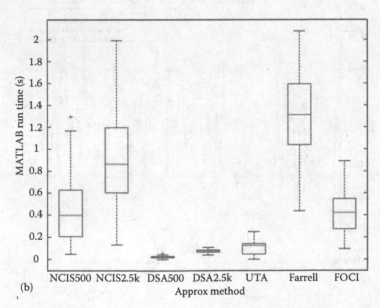

(b)

**FIGURE 22.4** Optimization results for Chernoff fusion of 200 randomly generated pairs of 4D GMs: (a) error between ground truth and estimated $\omega$, sorted by GM instance number along the horizontal axis; (b) corresponding computation times for all methods.

(*Continued*)

to approximate the entire shape of $p^W(x_k)$ by a single "global" distribution. Although mixture IS also leads to $M_i M_j$ sets of IS calculations per fusion update, these are easily parallelized and generally lead to estimated GM moments with lower sample variance than those produced by WEM, thanks to component-wise conditioning.

However, it should be noted that the mixands in Equation 22.36 can become highly multimodal or skewed in certain cases, and thus will not always be accurately approximated by moment-matched

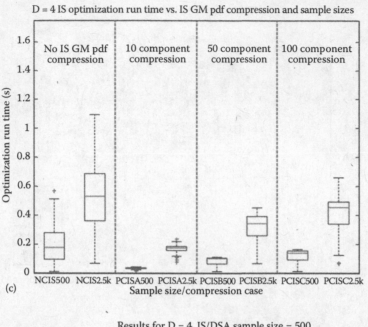

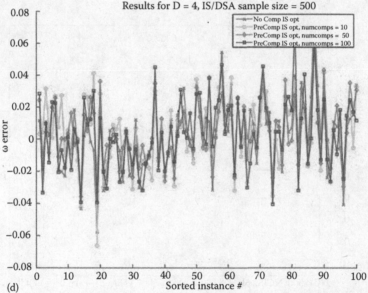

**FIGURE 22.4 (CONTINUED)** Optimization results for Chernoff fusion of 200 randomly generated pairs of 4D GMs: (c) optimization times for uncompressed and precompressed IS optimization runs for same GM pairs; (d) error between ground truth and estimated ω for uncompressed and pre-compressed IS optimization runs.

Gaussians. It remains an open problem to develop reliable methods for detecting such cases, so that Equation 22.37 can be adapted accordingly. Furthermore, when $x_k$ has large dimension, steps must be taken to limit the well-known sensitivity of IS to small differences between IS and target PDFs, which can lead to high variance parameter estimates [40]. Figure 22.5 demonstrates this sensitivity for two separate GM fusion problems using the effective sample size (ESS) for each component's IS approximation. As noted in Ref. [41], the ESS estimates the fraction of $N_s$ samples that are contributing to each component's moment estimates with nontrivial IS weights, and thus indicates the relative closeness of each IS pdf to its target PDF (i.e., low/high IS sample variance is achieved as ESS

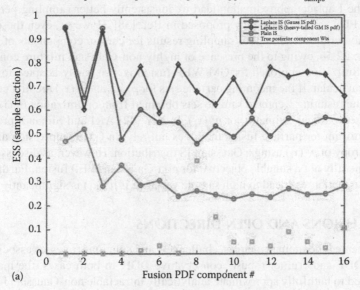

(a)

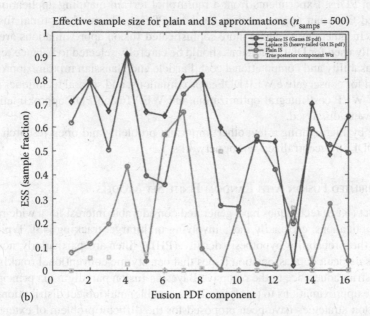

(b)

**FIGURE 22.5** Expected sample sizes (ESS) for mixture IS GM approximation on toy 2D GM fusion problems, with $M_i = M_j = 4$ initial components and 16 resultant components (fused mixture component weights shown to indicate relative importance to overall fusion PDF shape): (a) when individual components of Equation 22.37 are all approximately Gaussian, "simple" Gaussian IS PDFs work well but are outperformed by more sophisticated IS PDFs based on the Laplace approximation; (b) both simple and sophisticated unimodal IS PDFs break down for GM fusion problems where components of Equation 22.37 are significantly non-Gaussian (e.g., component 13, which has two distinct modes).

approaches 1/0). For the GM fusion problem in Figure 22.5a, sophisticated Gaussian and GM IS PDFs based on the Laplace approximation lead to consistently better sampling performance over simple Gaussian PDF sampling strategy proposed in Ref. [38]. However, even these sophisticated approximations sometimes lead to poor sampling results for certain components of the GM fusion problem in Figure 22.5b, owing to the presence of highly non-Gaussian mixture components.

The direct optimization approach for GM WEP fusion is also easily adapted to particle-based WEP fusion. In particular, if the incoming particle sets for $p_i(x_k)$ and $p_j(x_k)$ are first compressed into GM PDFs, then the resulting weighted sample sets obtained from optimization of $\omega$ can be directly interpreted as fused particle estimates for $p^W(x_k)$. In Ref. [42] Aigl and Simandl provide a similar optimization algorithm for particle fusion that does not rely on GM compression and selects $\omega$ to minimize the entropy of $p^W(x_k)$ using a Gaussian IS distribution. However, as discussed in Ref. [27], the entropy is generally not a suitable objective for non-Gaussian WEP fusion. Furthermore, use of a Gaussian IS distribution will lead to high sample variance if $p^W(x_k)$ is significantly non-Gaussian.

## 22.6  CONCLUSIONS AND OPEN DIRECTIONS

This chapter presented the main principles, challenges, and solution strategies associated with non-Gaussian exact DDF and "rumor robust" conservative DDF. In both cases, the major challenges arise from the need to faithfully approximate analytically intractable non-Gaussian fusion and common information PDFs. Experiments from a multirobot terrain mapping application showed that moment-matched Gaussian approximations to these PDFs yield nearly optimal fusion results in certain settings. In other cases, where more sophisticated fusion approximations are required, the PDF model family used for DDF recursions should be carefully selected to balance representational flexibility with usability and computational cost. Particle and Gaussian mixture approximations are especially useful for conservative WEP DDF approximations, and several techniques for addressing the problems of WEP cost integral optimization and WEP fusion PDF approximation with these representations were discussed.

We conclude by highlighting a few other important problems and open research directions for non-Gaussian DDF in decentralized sensor networks.

### 22.6.1  DISTRIBUTED FUSION WITH RANDOM FINITE SET MODELS

Random finite set (RFS) techniques have generated considerable interest for a wide range of multisensor fusion applications, especially those involving multitarget tracking [43]. Typical RFS algorithms such as the probability hypothesis density (PHD) filter are particularly notable for their ability to bypass difficult data association issues that can stymie conventional tracking techniques, as well as their strict adherence to the recursive Bayesian fusion paradigm via principled computationally tractable approximations to (analytically intractable) multiobject distributions [20].

Several solution strategies have been proposed for the difficult problem of extending Bayesian RFS fusion to a distributed setting in sensor networks [44–46]. It is interesting to note that all methods developed so far in this domain rely on either communication of raw sensor data between fusion centers or WEP-like fusion rules to avoid double-counting of common information between fused multiobject distributions. In Ref. [45] Uney et al. explicitly mention the possibility of using the RFS analog of the Bayesian channel filter to track common information exactly between multiobject distributions, but argues that this approach is brittle because of its reliance on tree-connected topologies. As such, practical implementations for exact DDF with RFS have been overlooked in the RFS literature, despite the valuable insights they could provide for many applications (especially with regard to establishing baseline performance comparisons for alternative suboptimal approximations).

## 22.6.2  Distributed Fusion with Factorized Probabilistic Belief Models

Probabilistic graphical models (PGMs) are important tools for statistical sensor fusion [47,48]. In their most general form, PGMs encode high-dimensional joint probability distributions as products of conditionally independent "factors" that are functions of (possibly overlapping) subsets of random variables. These factorized models can be represented by structures such as bipartite graphs or (un)directed trees, where nodes represent random variable factors and edges denote conditional dependencies (and hence "information flows") between different factors. This general formalism permits the development and use of highly efficient "message passing" algorithms for computing Bayesian belief updates for each factor as new data arrive, and thus greatly simplifies the calculation of posterior beliefs for high-dimensional joint distributions [39]. As such, PGMs are quite useful for data fusion in the presence of densely coupled uncertainties and help bypass computational difficulties that arise in practice even for "simple" distributions, for example, multivariate Gaussians for robotic mapping and navigation [49].

Although PGMs enjoy widespread use, DDF algorithms have not yet been adapted to take full advantage of their nice "modular" computational properties in distributed sensor networks. In Ref. [11] Makarenko et al. note the close relationship between Bayesian message-passing in PGMs and common information tracking/removal in exact DDF, but do not consider the influence of local conditional posterior belief factorizations on the fusion process. Some recent work on posegraph-based robotic navigation and mapping explicitly considers this issue [5], but restricts attention to Gaussian distributions and assumes that all agents exchange the entirety of their local joint posterior PDFs to carry out nonfactorized/"monolithic" DDF updates (i.e., where beliefs over all random variables are always updated simultaneously). Recent work in Ref. [29] shows that *partial decentralized Bayesian information fusion* (where only certain parts of the joint distribution are updated throughout the network) is possible if DDF updates are defined in terms of conditionally independent factors. However, this concept has so far been examined only with simple hierarchical models over multinomial distributions.

## 22.6.3  Distributed Fusion with Bayesian Nonparametric Models

Bayesian nonparametric (BNP) models, such as those based on hierarchical Dirichlet processes and Gaussian processes, have emerged as important tools for learning-based sensor fusion in autonomous robotics and intelligent control systems [50]. BNP models are particularly useful for unsupervised Bayesian learning problems, in which unlabeled feature data are clustered into uncertain categories or hierarchies. The formal learning procedures for BNP models yield posterior distributions over probabilistic mixture/product models that effectively contain an "infinite" number of components, where model parametrizations that best match the underlying data tend to accumulate the largest "weights." BNPs thereby circumvent the issues of model order assignment and data-to-model association encountered by conventional parametric modeling techniques. Learning agents thus remain agnostic about exact model order/parametrization and marginalize over these unknowns to perform reasoning in practice.

Although they hold great promise for the development of adaptive intelligent robotic sensor networks, BNP inference methods remain too computationally expensive for online operations aboard a single platform. Distributed BNP learning has been proposed as a possible solution in the context of large-scale machine learning [51], but methods considered so far rely only on the explicit exchange of raw observations between learning agents, rather than on posterior belief exchange as in the DDF framework. However, it is as yet unclear whether or not the data-driven nature of BNP models fundamentally precludes the possibility of adapting them to "belief-driven" DDF implementations.

# REFERENCES

1. J. K. Uhlmann, S. J. Julier, B. Kamgar-Parsi, M. O. Lanzagorta, and H.-J. S. Shyu. Nasa Mars rover: A testbed for evaluating applications of covariance intersection. *Proceedings of SPIE*, 3693:140–149, 1999.
2. W. W. Whitacre, and M. E. Campbell. Cooperative estimation using mobile sensor nodes in the presence of communication loss. *Journal of Aerospace Information Systems*, 10(3):114–130, 2013.
3. T. H. McLoughlin, and M. Campbell. Distributed estimate fusion filter for large spacecraft formations. In *Proceedings of the AIAA Guidance, Navigation and Control Conference and Exhibit*, 2008.
4. J. Schoenberg. Data fusion and distributed robotic perception. PhD thesis, Cornell University, Ithaca, NY, 2012.
5. A. Cunningham, V. Indelman, and F. Dellaert. DDF-Sam 2.0: Consistent distributed smoothing and mapping. In *Robotics and Automation (ICRA), 2013 IEEE International Conference on*, pp. 5220–5227. IEEE, 2013.
6. S. J. Julier, and R. Mittu. Distributed data fusion and maritime domain awareness for harbour protection. In *Harbour Protection through Data Fusion Technologies*, E. Shahbazian, G. Rogova, and M. J. de Weert (eds.), pp. 181–190. Springer, New York, 2009.
7. X. Yao, X. Wang, and S. Li. A new data fusion routing algorithm for wireless sensor networks. In *Computer Science and Automation Engineering (CSAE), 2011 IEEE International Conference on*, vol. 1, pp. 676–680, June 2011.
8. R. Olfati-Saber, E. Franco, E. Frazzoli, and J. S. Shamma. Belief consensus and distributed hypothesis testing in sensor networks. In *Networked Embedded Sensing and Control*, P. J. Antsaklis, and P. Tabuada (eds.), pp. 169–182. Springer, New York, 2006.
9. D. Gu. Distributed EM algorithm for Gaussian mixtures in sensor networks. *Neural Networks, IEEE Transactions on*, 19(7):1154–1166, 2008.
10. S. Grime, and H. F. Durrant-Whyte. Data fusion in decentralized sensor networks. *Control Engineering Practice*, 2(5):849–863, 1994.
11. A. Makarenko, A. Brooks, T. Kaupp, H. Durrant-Whyte, and F. Dellaert. Decentralised data fusion: A graphical model approach. In *2009 International Conference on Information Fusion (FUSION 2009)*, pp. 545–554, July 2009.
12. J. R. Schoenberg, and M. Campbell. Distributed terrain estimation using a mixture-model based algorithm. In *Information Fusion, 2009. FUSION'09. 12th International Conference on*, pp. 960–967. IEEE, 2009.
13. N. Ahmed, J. Schoenberg, and M. Campbell. Fast weighted exponential product rules for robust general multi-robot data fusion. In *Proceedings of Robotics: Science and Systems*. Sydney, Australia, July 2012.
14. R. Tse, N. Ahmed, and M. Campbell. Unified mixture-model based terrain estimation with Markov random fields. In *Multisensor Fusion and Integration for Intelligent Systems (MFI), 2012 IEEE Conference on*, pp. 238–243. IEEE, 2012.
15. F. Bourgault. Decentralized control in a Bayesian world. PhD thesis, University of Sydney, New South Wales, Australia, 2005.
16. S. J. Julier. An empirical study into the use of Chernoff information for robust, distributed fusion of Gaussian mixture models. In *2006 International Conference on Information Fusion (FUSION 2006)*, 2006.
17. T. Kaupp, B. Douillard, F. Ramos, A. Makarenko, and B. Upcroft. Shared environment representation for a human-robot team performing information fusion. *Journal of Field Robotics*, 24(11–12):911–942, 2007.
18. S. J. Julier, and J. K. Uhlmann. Fusion of time delayed measurements with uncertain time delays. In *Proceedings of the 2005 American Control Conference*, pp. 4028–4033, Portland, OR, June 2005.
19. S. Challa, R. J. Evans, and X. Wang. A Bayesian solution and its approximations to out-of-sequence measurement problems. *Information Fusion*, 4(3):185–199, 2003.
20. R. P. S. Mahler. *Statistical Multisource-Multitarget Information Fusion*, vol. 685. Artech House, Boston, 2007.
21. T. Martin, and K. C. Chang. A distributed data fusion approach for mobile ad hoc networks. In *2005 International Conference on Information Fusion (FUSION 2005)*, pp. 1062–1069, 2005.
22. B. Tonkes, and A. D. Blair. Decentralised data fusion with exponentials of polynomials. In *2007 International Conference on Intelligent Robotics and Systems (IROS 2007)*, pp. 3727–3732. San Diego, CA, 2007.

23. I. Miller, and M. E. Campbell. A mixture-model based algorithm for real-time terrain estimation. *Journal of Robotic Systems*, 23(9):755–775, 2006.
24. Y. Bar-Shalom, X. R. Li, and T. Kirubarajan. *Estimation with Applications to Tracking and Navigation: Theory, Algorithms and Software*. John Wiley & Sons, Hoboken, NJ, 2004.
25. J. R. Schoenberg, M. Campbell, and I. Miller. Posterior representation with a multi-modal likelihood using the Gaussian sum filter for localization in a known map. *Journal of Field Robotics*, 29(2):240–257, 2012.
26. M. F. Huber, T. Bailey, H. Durrant-Whyte, and U. D. Hanebeck. On entropy approximation for Gaussian mixture random vectors. In *2008 IEEE International Conference on Multisensor Fusion and Integration for Intelligent Systems*, pp. 181–188. IEEE, 2008.
27. T. Bailey, S. Julier, and G. Agamennoni. On conservative fusion of information with unknown non-Gaussian dependence. In *2012 International Conference on Information Fusion (FUSION 2012)*, pp. 1876–1883. IEEE, 2012.
28. M. B. Hurley. An information theoretic justification for covariance intersection and its generalization. In *2002 International Conference on Information Fusion (FUSION 2002)*, pp. 505–511, 2002.
29. N. Ahmed. Conditionally factorized DDF for general networked Bayesian estimation. In *2014 IEEE Conference on Multisensor Fusion and Integration for Intelligent Systems (MFI 2014)*, September 2014.
30. T. M. Cover, and J. A. Thomas. *Elements of Information Theory*. John Wiley & Sons, Hoboken, NJ, 1991.
31. B. Upcroft. Rich probabilistic representations for bearing-only decentralised data fusion. In *Proceedings of the 8th International Conference on Information Fusion (FUSION 2005)*, pp. 25–29. Philadelphia, PA, 2005.
32. W. J. Farrell, and C. Ganesh. Generalized Chernoff fusion approximation for practical distributed data fusion. In *12th International Conference on Information Fusion (FUSION 2009)*, pp. 555–562, 2009.
33. L.-L. Ong, B. Upcroft, T. Bailey, M. Ridley, S. Sukkarieh, and H. Durrant-Whyte. A decentralised particle filtering algorithm for multi-target tracking across multiple flight vehicles. In *2006 International Conference on Intelligent Robotics and Systems (IROS 2006)*, pp. 4539–4544. Beijing, China, 2006.
34. L.-L. Ong, B. Upcroft, M. Ridley, T. Bailey, S. Sukkarieh, and H. Durrant-Whyte. Consistent methods for decentralised data fusion using particle filters. In *International Conference on Multisensor Fusion and Integration for Intelligent Systems*, 2006.
35. J. Goldberger, and H. Greenspan. Simplifying mixture models using the unscented transform. *IEEE Transactions on Pattern Analysis and Machine Intelligence*, 30(8):1496–1502, 2008.
36. M. West. Approximating posterior distributions by mixtures. *Journal of the Royal Statistical Society B*, 55(2):409–422, 1993.
37. N. Ahmed, and M. Campbell. Fast consistent Chernoff fusion of Gaussian mixtures for ad hoc sensor networks. *IEEE Transactions on Signal Processing*, 60(12):6739–6745, 2012.
38. N. Ahmed, T.-L. Yang, and M. Campbell. On generalized Bayesian data fusion with complex models in large scale networks. *ArXiv e-prints*, August 2013.
39. C. M. Bishop. *Pattern Recognition and Machine Learning*, vol. 1. Springer, New York, 2006.
40. C. Robert, and G. Casella. *Monte Carlo Statistical Methods*, 2nd ed. Springer, New York, 2004.
41. J. S. Liu. *Monte Carlo Strategies in Scientific Computing*. Springer, New York, 2008.
42. J. Ajgl, and M. Simandl. Particle based probability density fusion with differential Shannon entropy criterion. In *2011 Proceedings of the 14th International Conference on Information Fusion (FUSION 2011)*, pp. 1–8. IEEE, 2011.
43. B.-N. Vo, S. Singh, and A. Doucet. Sequential Monte Carlo implementation of the PHD filter for multi-target tracking. In *Proceedings of the 6th International Conference on Information Fusion*, pp. 792–799, 2003.
44. D. Clark, S. Julier, R. Mahler, and B. Ristic. Robust multi-object sensor fusion with unknown correlations. In *Sensor Signal Processing for Defence (SSPD 2010)*, 29–30. London, Sept. 2010.
45. M. Uney, D. E. Clark, and S. J. Julier. Distributed fusion of PHD filters via exponential mixture densities. *Selected Topics in Signal Processing, IEEE Journal of*, 7(3):521–531, 2013.
46. B. K. Habtemariam, A. Aravinthan, R. Tharmarasa, K. Punithakumar, T. Lang, and T. Kirubarajan. Distributed tracking with a PHD filter using efficient measurement encoding. *Journal of Advances in Information Fusion*, 7(2):114–130, 2012.
47. M. J. Beal, H. Attias, and N. Jojic. Audio-video sensor fusion with probabilistic graphical models. In *Computer Vision ECCV 2002*, A. Heyden, G. Sparr, M. Nielsen, and P. Johansen (eds.), pp. 736–750. Springer, New York, 2002.
48. C.-Y. Chong, and S. Mori. Graphical models for nonlinear distributed estimation. In *Proceedings of the 2004 International Conference on Information Fusion (FUSION 2004)*, pp. 614–621. Stockholm, Sweden, July 2004.

49. T. Bailey, M. Bryson, H. Mu, J. Vial, L. McCalman, and H. Durrant-Whyte. Decentralised cooperative localisation for heterogeneous teams of mobile robots. In *Robotics and Automation (ICRA), 2011 IEEE International Conference on*, pp. 2859–2865. IEEE, 2011.
50. T. Campbell, S. Ponda, G. Chowdhary, and J. P. How. Planning under uncertainty using Bayesian nonparametric models. In *AIAA Guidance, Navigation, and Control Conference*, 2012.
51. S. Williamson, A. Dubey, and E. Xing. Parallel Markov chain Monte Carlo for nonparametric mixture models. In *Proceedings of the 30th International Conference on Machine Learning*, pp. 98–106, 2013.

# 23 Attack-Resilient Sensor Fusion for CPS

*Radoslav Ivanov, Miroslav Pajic, and Insup Lee*

## CONTENTS

## 23.1 INTRODUCTION

Modern cyber physical systems (CPS) are a complex mixture of cyber and physical components. Although the two aspects have been analyzed individually both in terms of performance and security, ensuring the safety of systems combining the two components is a challenging task. In addition to the cyber attacks deployed over the years (e.g., denial of service), CPS are also vulnerable to physical attacks (e.g., sensor spoofing). In particular, recent work has shown that an attacker may gain control of a system by either hijacking its sensors [1] or gaining direct access to the system's internal network [2,3].

Yet, this problem can be alleviated in modern CPS because multiple sensors can now estimate the same physical variable. In automotive CPS, for example, velocity can be estimated by a wheel encoder, GPS, camera, smart phone applications, and so forth. Even though they may have different precisions, their data can be combined to obtain an estimate that is more accurate than that of any single sensor [4]. In addition, sensor diversity increases the system's robustness to external disturbances (e.g., a tunnel that blocks GPS connection to satellites) and decreases its dependence on a particular sensor.

On the other hand, increased sensor diversity provides a malicious attacker with new opportunities to sabotage the system. In particular, while some sensors are difficult to access, others may be easy to spoof or eliminate completely. Consequently, in this chapter we focus on two aspects of the development of a resilient sensor fusion algorithm. We consider a standard CPS model (Figure 23.1), in which multiple sensors measuring the same physical variable communicate with the system over a shared bus. Communication is time triggered—each sensor transmits its measurement during its allocated time slot, according to a predefined schedule. Therefore, in the first part of the chapter we analyze the effect of the communication schedule on the impact of an attacker trying to increase the uncertainty in the system. In the second part we investigate ways of using past measurements to improve the precision of the fusion algorithm at the current time. This chapter is an overview of our previous work [5–7], and in the interest of space some proofs are omitted.

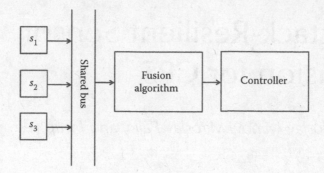

**FIGURE 23.1** A typical architecture of a modern CPS. Sensors communicate via a shared bus. Their measurements are used in the fusion algorithm, which sends its output to the controller.

## 23.2    BASIC CONCEPTS

This section describes the basic concepts used in this chapter, that is, the sensor model and the fusion algorithm.

### 23.2.1    SENSOR MODEL

When designing a fusion algorithm, the first consideration is the underlying sensor model; it affects both the purpose and the approach of the algorithm. The two most commonly used models can be broadly classified as *probabilistic* and *abstract*. In the probabilistic model, sensors provide a single measurement that is corrupted by noise, for example, $y = \theta + w$, where $y$ is the obtained measurement, $\theta$ is the true value, and $w$ is noise with a known probability distribution (e.g., Gaussian, uniform). By their nature, probabilistic models are used in claims about the expected operation of a system, that is, what is the most likely outcome of a certain experiment. They are by far the most widely used in the literature, beginning with the seminal work in the area, the Kalman filter [4]. By using the probabilistic model and by assuming that system dynamics are known, the filter produces a best linear unbiased estimator (BLUE) of the true state. Numerous extensions to the Kalman filter have been proposed, notably a distributed version with local measurement propagation [8].

However, obtaining probabilistic sensor models is challenging and often requires unrealistic assumptions about the distribution of noise. Furthermore, safety analysis reasons about the worst-case operation of CPS are usually caused by rare low-probability events. Thus, the probabilistic model is not well suited to the task at hand. The *abstract* sensor model, on the other hand, can be used to draw conclusions about rare events without taking into consideration the likelihood of their occurrence. In the abstract model a sensor provides an interval instead of a single measurement. The interval is usually computed in two stages. First the sensor's measurement is obtained, similar to the one in the probabilistic setting. Yet, this model does not assume a probability distribution on $w$ but rather an absolute bound on its magnitude, that is, $|w| \leq M$ for some constant $M$. Thus, the output of the sensor is $[y - M, y + M]$. $M$ is different for each sensor and is computed based on the sensor manufacturer's specifications and implementation guarantees [9]. Example 23.1 illustrates how to compute the interval of a standard wheel encoder. Note that the interval is computed in such a way as to guarantee that it contains the true value of the variable if the sensor is *correct*, that is, it follows its specification.

## Example 23.1: Computing the Interval of a Wheel Encoder

Consider an encoder with 192 cycles per revolution and a sampling time of 20 ms. It is mounted on a vehicle moving at 10 mph whose wheels have a radius of 6 inches.

To compute the size of the encoder's interval, we use the equation for angular velocity: $\omega = \theta/t$, where $\theta$ is angular displacement and $t$ is elapsed time. Suppose $\theta \pm d\theta$ and $t \pm dt$, where $d\theta$ and $dt$ are the encoder's cycle measurement noise and sampling jitter, respectively. One way of calculating $d\omega$, the largest measurement deviation from the true value, is through equalizing the average errors, that is, $d\omega/\omega = d\theta/\theta + dt/t$. Note that $\theta = (2\pi rK)/192$, where $K$ is the number of encoder sectors passed since the previous sampling. With a sampling time of 20 ms, $K$ is roughly 18 at 10 mph. Suppose the encoder may mismeasure $K$ by at most one sector, that is, $d\theta = (2\pi r)/192$, and that there is sampling jitter of at most 100 ns, that is, $dt = 0.0001$. Thus, one may compute $d\omega \approx 0.1$ mph, that is, the encoder's interval size is about 0.2 mph at 10 mph.

Intuitively, the abstract model is well suited for worst-case analysis for the following reason. Suppose interval $A$ is an unsafe region that the system should avoid, for example, $A = [100 \text{ mph}, \infty)$. Thus if every sensor's interval has an empty intersection with $A$ one can conclude that the system is not in state $A$.

Yet sensors do not always perform according to specifications. In particular, sometimes they may be utilized in adverse environments (e.g., a wheel encoder will provide a wrong velocity when moving on sand); they may have other transient faults; or they may be spoofed. Thus, it is possible for a sensor's interval to not contain the true value. Hence, it is necessary to incorporate this information when combining data from different sensors.

### 23.2.2 FUSION ALGORITHM

Consider a system with $n$ sensors measuring the same physical variable. Suppose it is known that at most $f$ of the obtained intervals would not contain the true value in any round. In this case the true value would lie in at least $n - f$ intervals, and any interval that contains all intersections of $n - f$ intervals is guaranteed to contain it. Because it is desired that the interval be as precise as possible, the smallest such interval is selected; it is referred to as the fusion interval in this chapter. The fusion interval for different values of $f$ is illustrated in Figure 23.2. Note that its size increases as $f$ increases.

The fusion interval was first introduced by Marzullo in his work on abstract sensors [10]. He developed an efficient algorithm, $O(n \log n)$, for computing the fusion interval and provided worst-case results about its size depending on $f$.

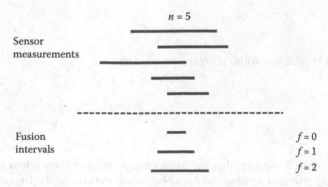

**FIGURE 23.2**  The fusion interval for three values of $f$. The dashed horizontal line separates sensor intervals from fusion intervals in all figures in this work.

**Theorem 23.1**

If $f < \lceil n/3 \rceil$, then the size of the fusion interval is at most the size of some correct interval. If $\lceil n/3 \rceil \leq f < \lceil n/2 \rceil$, then the size of the fusion interval is at most the size of some (not necessarily correct) interval. If $f \geq \lceil n/2 \rceil$, then the fusion interval may be arbitrarily large.*                    ∎

**Proof**

See Ref. [10].                    ∎

Multiple extensions to Marzullo's work have been proposed. For example, the size of the fusion interval could be reduced if a version of majority voting is performed such that points contained in more intervals are more likely to be the true value than others [11]. However, this approach may output a fusion interval that is not guaranteed to contain the true value. Alternatively, one may combine the abstract and probabilistic models by assuming a probability distribution over the interval [12].

## 23.3 SCHEDULING MEASUREMENT TRANSMISSIONS

While the fusion algorithm described in the preceding section was developed for faulty sensors, in this section we analyze the effect of a malicious attacker on the performance of the algorithm and investigate ways of limiting his impact. In particular, a smaller fusion interval is always desired when arguing about the safety of CPS because a larger interval may contain more points that are considered unsafe, for example, it may contain points belonging to region $A$ described previously.

In this chapter, we assume that sensors are not randomly faulty but are under attack. If an attacker takes control of a sensor (whether through spoofing, signal jamming, or another means), he can send any interval to the controller on the sensor's behalf. In addition, we assume the attacker has access to the shared bus (Figure 23.1) and can examine other sensors' measurements before sending his own. We do not fix a specific attack model (i.e., knowledge about the system, algorithms used) because the results provided here are worst-case results and are, in some sense, independent of the attacker's abilities.[†] However, we assume the number of attacked sensors, denoted by $f_a$, satisfy $f_a < \lceil n/2 \rceil$.

The following result puts the problem in perspective. Regardless of the attacker's model, it provides an upper bound on the size of the fusion interval. Note that, if $\mathcal{N}$ is a set of sensors, then $S_{\mathcal{N},f}$ denotes their fusion interval with $f$ as input ($f$ is set conservatively high such that $f \geq f_a$, e.g., $f = \lceil n/2 \rceil - 1$). In addition, we use $|\cdot|$ to denote the size of an interval.

**Theorem 23.2**

Let $s_{c_1}$ and $s_{c_2}$ be the two largest-width correct sensors. Then $|S_{\mathcal{N},f}| \leq |s_{c_1}| + |s_{c_2}|$.                    ∎

**Proof**

See Ref. [5].                    ∎

Although Theorem 23.2 suggests that the worst case is improved when less precise sensors are compromised, it only provides an absolute upper bound on the size of the fusion interval. We now

---

* Note that $f$ is usually not known in practice, hence it will be set conservatively high, e.g., $f = \lceil n/2 \rceil - 1$. If $f$ is larger, the result cannot be trusted because the interval may not contain the true value. Therefore, we assume a bound on the number of incorrect intervals.

† Two specific attack strategies are analyzed in our previous work [7].

investigate how this worst case can be achieved and avoided. Note that the attacker's impact in general depends on what intervals he has seen when sending his sensors' measurements. In particular, if he has to transmit without any knowledge of other measurements, he has to guess as to what would constitute a good attack; on the other hand, if he has observed all other measurements, he can make a much more informed decision. Therefore, in this section we explore how different schedules of sending measurements over a shared bus affect the uncertainty in the system, as reflected in the size of the fusion interval. Because the size of the intervals is the only information available a priori to system designers, we focus on two precision-based schedules: *Ascending*, in which most precise sensors transmit first, and *Descending*, in which least precise sensors transmit first.

We first note that neither schedule is better than the other for all possible attack strategies and all possible measurements—that is, there exist cases in which one schedule produces a larger fusion interval than the other and vice versa. For example, consider the two scenarios shown in Figure 23.3. The attack strategy used there is to send the attacked interval (denoted by sinusoidal) around the intersection of the seen intervals. Whereas in the scenario in Figure 23.3a it is better for the attacker to know the position of the larger sensors that coincide, in Figure 23.3b it is more beneficial for the attacker to see the measurements of the more precise sensors $s_1$ and $s_2$.

Because it is not known in advance which sensors are under attack and what the attacker's strategy is, it is challenging to find a useful metric to compare schedules. For example, in our previous work [5] we compare them using the expected size of the fusion interval. However, computing expectations requires additional assumptions on the distributions of sensors measurements (e.g., uniform, normal). Therefore, this chapter provides worst-case results to describe which sensors the system should protect to limit the attacker's impact.

To formulate the theorems, we use the following notation. Given a system of $n$ sensors, let $S_{na}$ be largest width fusion interval when no sensor is attacked. With $S_{\mathcal{F}}$ we denote the worst-case fusion interval for a fixed set of attacked sensors $\mathcal{F}, |\mathcal{F}| = f_a$, whereas $S_{f_a}^{wc}$ is the worst-case fusion interval for a given number of attacked sensors, $f_a$. Note that $\left|S_{f_a}^{wc}\right| \geq |S_{\mathcal{F}}| \geq |S_{na}|$ by definition.

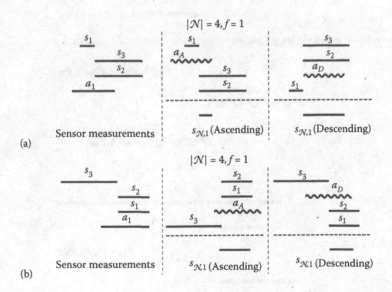

**FIGURE 23.3** Two examples that show that neither schedule is better in all situations. The first column shows the measurements by the sensors, including the attacked one. The other columns contain the intervals sent to the controller (the attacker shown by the sinusoidal), and the corresponding fusion interval. (a) An example where the "Ascending" schedule is better for the system. (b) An example where the "Descending" schedule is better for the system.

**Theorem 23.3**

If the $f_a$ largest intervals are under attack, then $|S_{na}| = |S_{\mathcal{F}}|$.                                    ∎

**Proof**

See Ref. [5].                                                                                              ∎

**Theorem 23.4**

$|S_{f_a}^{wc}| = |S_{\mathcal{F}}|$ if $\mathcal{F}$ consists of the smallest intervals in the system.                            ∎

**Proof**

The proof is similar to that of Theorem 23.3. See Ref. [5].                                              ∎

Figure 23.4a illustrates Theorem 23.3. $a_1$ and $a_2$ are both attacked and do not contain the true value (at the intersection of the other intervals). Because $a_1$ and $a_2$ are the largest intervals, they can

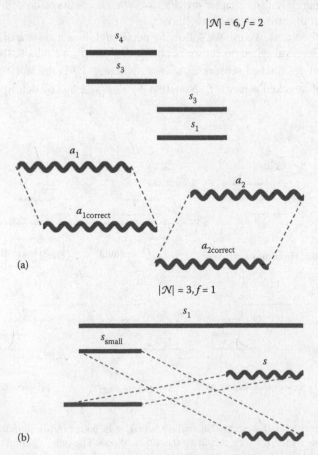

(a)

(b)

**FIGURE 23.4** Illustrations of Theorems 23.3 and 23.4. (a) Attacking the biggest intervals does not change the worst case in the system. (b) Attacking the smallest intervals can achieve the absolute worst case.

be moved and made correct while preserving the size of the fusion interval. Hence, the same worst case is achieved with correct intervals. Figure 23.4b gives an example of Theorem 23.4. The worst case for the setup can be achieved when $s$ or $s_{small}$ is attacked (but not $s_1$).

Theorems 23.3 and 23.4 suggest that the attacker can gain more power in the worst case by corrupting the precise sensors as opposed to the imprecise ones. The intuitive explanation is that the uncertainty in the system is greater in the former case—attacking imprecise sensors does not greatly affect the fusion interval because their intervals are large even when correct; attacking precise sensors, however, leaves the system with less information about the true value.

Based on these observations, one may conclude that the Ascending schedule is the more beneficial for CPS. We note that by scheduling the most precise sensors first, the system achieves two goals. First of all, if these sensors are compromised, the attacker will be forced to transmit without any information about the correct measurements; thus, although the precise sensors provide the attacker with the most power as described in the Theorems 23.3 and 23.4, the Ascending schedule would preclude this. In addition, if the attacker compromises the less precise sensors instead, he would have more information but would control the less powerful sensors. Finally, in previous work [5], we also illustrate the advantages of the Ascending schedule in the presence of an optimal attack strategy; in addition to worst-case analysis, we provide average-case results to argue that safety critical systems should adopt this schedule.

## 23.4 USING SYSTEM DYNAMICS AND MEASUREMENT HISTORY

In this section we investigate the use of system dynamics and measurement history to improve further the precision of the sensor fusion algorithm. Furthermore, we adopt a more general sensor model in this section. We note that many modern sensors provide multidimensional measurements (e.g., position in three dimensions); in addition, as a result of internal filtering techniques used (e.g., Kalman filters in GPS) the constraints may be more complex than rectangular constraints. Thus, in this section we consider a more general model of sensors providing multidimensional polyhedra instead of intervals. Formally, each sensor provides a set $P = \{x \in \mathbb{R}^d \mid Bx \le b\}$, where $B \in \mathbb{R}^{m \times d}$, $b \in \mathbb{R}^m$.

### Algorithm 23.1: Sensor Fusion Algorithm for Polyhedra

Input: An array of polyhedra $P$ of size $n$ and an upper bound on the number of attacked polyhedra $f$

```
1: C ← combinations_n_choose_n_minus_f(P)
2: R_{N,f} ← ∅
3: for each K in C do
4:     add(R_{N,f}, intersection(K))
5: end for
6: return conv(R_{N,f})
```

### 23.4.1 EXTENDED SENSOR FUSION ALGORITHM

Before analyzing how one could use history, we extend the sensor fusion algorithm to polyhedra while preserving the fact that the output is guaranteed to contain the true value. The inputs to the algorithm are similar to the one-dimensional case: $n$ polyhedra and a number $f$ that denotes an upper bound on the number of attacked sensors. The intuition of the algorithm is the same: Because the true value may be in any intersection of $n - f$ polyhedra, we consider all such intersections (denoted by $R_{N,f}$); because the final output needs to be a polyhedron as well, the convex hull of the intersections is returned from the algorithm, that is, $S_{N,f} = \text{conv}(R_{N,f})$. The algorithm is formalized in

$n = 3, f = 1$

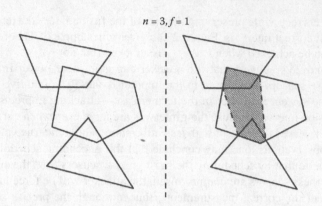

**FIGURE 23.5**  An illustration of the sensor fusion algorithm for polyhedra.

Algorithm 23.1. Unlike the one-dimensional case, the complexity of this algorithm is exponential as it is more difficult to exploit the structure of polyhedra. Figure 23.5 illustrates the algorithm. In this case, $n = 3$ and $f = 1$; hence there are at least two correct sensors. Thus, we consider all intersections of two polyhedra and output their convex hull (shaded area) as the fusion polyhedron.

### 23.4.2  INTRODUCING SYSTEM DYNAMICS

Having developed the fusion algorithm for polyhedra, we investigate ways of incorporating historical measurements to improve the accuracy of the algorithm. We assume a linear time-invariant system of the form $x(t + 1) = Ax(t) + w$, where $x \in \mathbb{R}^d$ is the state, $A \in \mathbb{R}^{d \times d}$ is the transition matrix, and $w \in \mathbb{R}^d$ is bounded noise such that $\|w\|_\infty \leq M$, with $\|\cdot\|_\infty$ being the infinity norm and $M$ a constant. To simplify notation, we introduce the map

$$m(x(t)) = \{y \in \mathbb{R}^d \mid Ax(t) + w, \|w\|_\infty \leq M\}. \tag{23.1}$$

Thus, given a polyhedron $P(t)$, $m(P(t))$ is its image under $m$, that is, it is its mapping one step in the future. We also introduce the notation $\cap_p$ (named pairwise intersection), in which each sensor's polyhedron from time $t$ is mapped to $t + 1$ and intersected with the same sensor's polyhedron at time $t + 1$, that is,*

$$m(\mathcal{N}(t)) \cap_p \mathcal{N}(t+1) = \{P_i \mid m(P_i(t)) \cap P_i(t+1), i = 1, \ldots, n\}. \tag{23.2}$$

Note that the abstract sensor model imposes restrictions on the ways of using historical measurements. For example, one cannot only map subsets of polyhedra to future time steps, as that may lead to the fusion polyhedron not including the true value. Thus, the following are all possible combinations given the assumptions in this chapter.

1. *map_n* (formally, $S_{m(\mathcal{N}(t)) \cup \mathcal{N}(t+1), 2f}$): In this approach we map all polyhedra in $\mathcal{N}(t)$ to time $t + 1$, and obtain $2n$ polyhedra in time $t + 1$. We then compute their fusion polyhedron with $2f$ as the the input to the fusion algorithm. This is illustrated in Figure 23.6a.

---

* Pairwise intersection produces a set that again contains $n$ polyhedra. The fusion polyhedron can now be computed by following Algorithm 23.1 (with an additional assumption as described later).

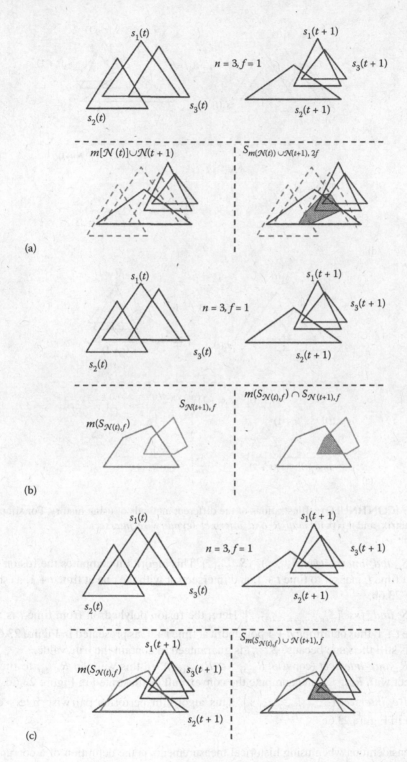

**FIGURE 23.6** Illustrations of the different methods of using history. For simplicity $A = I$, the identity matrix, and $w = 0$. (a) *map_n*. (b) *map_S_and_intersect*. (c) *map_S_and_fuse*.

*(Continued)*

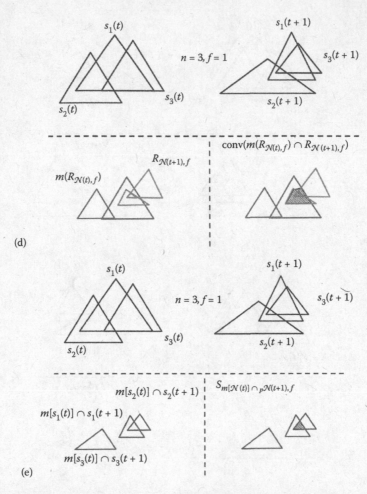

**FIGURE 23.6 (CONTINUED)** Illustrations of the different methods of using history. For simplicity $A = I$, the identity matrix, and $w = 0$. (d) *map_R_and_intersect*. (e) *pairwise_intersect*.

2. *map_S_and_intersect* $(m(S_{\mathcal{N}(t),f}) \cap S_{\mathcal{N}(t+1),f})$: This algorithm computes the fusion polyhedron at time $t$, maps it to time $t + 1$, and intersects it with the one at time $t + 1$, as shown in Figure 23.6b.

3. *map_S_and_fuse* $\left(S_{m(S_{\mathcal{N}(t),f}) \cup \mathcal{N}(t+1),f}\right)$: Here, the fusion polyhedron from time $t$ is mapped to time $t + 1$, thus obtaining $n + 1$ polyhedra at time $t + 1$, as presented in Figure 23.6c. Note that $f$ is still the same because $S_{\mathcal{N}(t),f}$ is guaranteed to contain the true value.

4. *map_R_and_intersect* $(\text{conv}(m(R_{\mathcal{N}(t),f}) \cap R_{\mathcal{N}(t+1),f}))$: In this we map $R_{\mathcal{N}(t),f}$ to time $t + 1$, intersect with $R_{\mathcal{N}(t+1),f}$, and compute the convex hull as illustrated in Figure 23.6d.

5. *pairwise_intersect* $\left(S_{m(\mathcal{N}(t)) \cap_p \mathcal{N}(t+1),f}\right)$: This algorithm performs pairwise intersection as shown in Figure 23.6e.

A new consideration when using historical measurements is the definition of a corrupted sensor. Note that with the definition used in the first part of this work, one cannot use the pairwise intersection method—if different subsets of sensors are attacked in different rounds, it is possible that not enough pairwise intersections will contain the true value. In this case, a stricter definition is necessary, that is, a sensor is correct if its polyhedron contains the true value at all time steps. If such an assumption is not realistic for a system, it could still use the other methods, which work with the weaker assumption.

One may compare the mapping algorithms through the volumes of the fusion polyhedron. The first four can be directly compared, as shown next.

## Theorem 23.5

The region obtained using *map_R_and_intersect* is a subset of the regions derived by *map_n*, *map_S_and_intersect*, and *map_S_and_fuse*.                                                                                ■

## Proof

See Ref. [6].                                                                                                       ■

Theorem 23.5 intuitively makes sense because *map_R_and_intersect* is only mapping enough uncertainty from previous measurements to guarantee that the true value is preserved. In particular, it is not computing the convex hull at time $t$ as *map_S_and_intersect* and *map_S_and_fuse* do (and potentially introduce additional points to the fused region), nor is it mapping potentially corrupted polyhedra as does *map_n*.

We note, however, that without additional assumptions about the rank of $A$, it is hard to reason about the relationship between sensor fusion polyhedra obtained by *map_R_and_intersect* and *pairwise_intersect*. Counterexamples are presented in Figure 23.7. In Figure 23.7a, $R_{\mathcal{N}(t),f}$ is a single point that is projected onto the $x$ axis. Hence *map_R_and_intersect* is a subset of *pairwise_intersect*, which produces an interval of points. Conversely, Figure 23.7b shows an example where *pairwise_intersect* is a point, and *map_R_and_intersect* is an interval containing that point. It is worth noting, however, that regardless of which of the two approaches is used, *pairwise_intersect* can be used as a preliminary step to detect attacked sensors; if the two polyhedra of a certain sensor have an empty intersection, then the sensor must be corrupted (faulty or tampered with) in one of the rounds; thus, it can be discarded from both, effectively reducing $n$ and $f$ by 1.

Finally, we note that if $A$ is a full rank matrix and $w = 0$, then *pairwise_intersect* outperforms all other methods, as shown next.                                                                                      ■

## Theorem 23.6

If $A$ is full rank and $w = 0$, the polyhedron obtained by *pairwise_intersect* is a subset of the polyhedron derived using *map_R_and_intersect*.                                                                       ■

## Proof

See Ref. [6].                                                                                                       ■

Based on these results, we argue that systems that incorporate past measurements in sensor fusion should use *pairwise_intersect*. We now show that *pairwise_intersect* guarantees the fusion polyhedron contains the true value and that it is never larger than the one produced by the no-history algorithm.

## Proposition 23.1

The fusion polyhedron computed using *pairwise_intersect* will always contain the true value and is never larger than the fusion polyhedron computed without using history.                                          ■

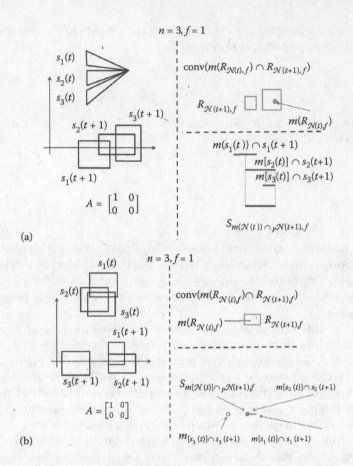

**FIGURE 23.7** Examples showing that, in general, polyhedra obtained using *map_R_and_intersect* and *pairwise_intersect* are not subsets of each other if $A$ is not full rank. (a) *map_R_and_intersect* is not a subset of *pairwise_intersect*. (b) *pairwise_intersect* is not a subset of *map_R_and_intersect*.

**Proof**

See Ref. [6].                                                                                     ∎

## 23.5 CONCLUSION

This chapter described two aspects of the development of an attack-resilient sensor fusion algorithm for CPS. It analyzed the effect of communication schedules on the impact a malicious attacker might have on sensor fusion performance and suggested that safety critical systems should implement the Ascending schedule. In addition, we improved the precision of the algorithm by incorporating system dynamics and historical measurements. Furthermore, the two methods can be combined to leverage their complementary properties [7]. Finally, we have illustrated their effectiveness on a real robot (a video is available at http://www.seas.upenn.edu/~pajic/research/CPS_security .html#videos).

# REFERENCES

1. A. H. Rutkin. "Spoofers" use fake GPS signals to knock a yacht off course. *MIT Technology Review*, August 2013.
2. K. Koscher, A. Czeskis, F. Roesner et al. Experimental security analysis of a modern automobile. In *SP'10: IEEE Symposium on Security and Privacy*, pp. 447–462, Claremont, Berkeley, CA, May 22–25, 2010.
3. S. Checkoway, D. McCoy, B. Kantor et al. Comprehensive experimental analyses of automotive attack surfaces. In *SEC'11: Proceedings of the 20th USENIX Conference on Security*, pp. 77–92, San Francisco, August 8–12, 2011.
4. R. E. Kalman. A new approach to linear filtering and prediction problems. *Transactions of the ASME–Journal of Basic Engineering*, 82(Series D):35–45, 1960.
5. R. Ivanov, M. Pajic, and I. Lee. Attack-resilient sensor fusion. In *DATE'14: Design, Automation and Test in Europe*, Dresden, Germany, March 24–28, 2014.
6. R. Ivanov, M. Pajic, and I. Lee. Resilient multidimensional sensor fusion using measurement history. In *HiCoNS'14: High Confidence Networked Systems*, pp. 1–10, Berlin, Germany, April 14–17, 2014.
7. R. Ivanov, M. Pajic, and I. Lee. Resilient sensor fusion for safety-critical cyber-pysical systems. Technical report, University of Pennsylvania, 2014, Philadelphia, PA. Available at http://www.seas .upenn.edu/~pajic/pubs/ARSF4CPS_TechReport.pdf.
8. L. Xiao, S. Boyd, and S. Lall. A scheme for robust distributed sensor fusion based on average consensus. In *IPSN'05: International Symposium on Information Processing in Sensor Networks*, pp. 63–70, Los Angeles, April 25–27, 2005.
9. M. Pajic, J. Weimer, N. Bezzo et al. Robustness of attack-resilient state estimators. In *Cyber-Physical Systems (ICCPS), 2014 ACM/IEEE International Conference on*, pp. 163–174, 2014.
10. K. Marzullo. Tolerating failures of continuous-valued sensors. *ACM Transactions on Computer Systems*, 8(4):284–304, 1990.
11. R. R. Brooks and S. S. Iyengar. Robust distributed computing and sensing algorithm. *Computer*, 29(6):53–60, 1996.
12. Y. Zhu and B. Li. Optimal interval estimation fusion based on sensor interval estimates with confidence degrees. *Automatica*, 42(1):101–108, 2006.

# Section II

---

*Multisensor Data Fusion Showcases Advancements*

# 24 Multisensor Data Fusion for Water Quality Evaluation Using Dempster–Shafer Evidence Theory

*Zhou Jian*

## CONTENTS

## 24.1 INTRODUCTION

Water quality evaluation is important in providing a reliable supply of potable water. Empirical evidence shows that water quality parameters, such as dissolved oxygen (DO), $NH_3$-N, total phosphorus (TP), and total nitrogen (TN), are sensitive indicators of contaminants. In 2005, Hall et al. demonstrated that changes in water quality parameters, which potentially indicate contamination, can be detected using sensors. Subsequently, wireless sensor networks (WSNs) have been extensively applied in monitoring and evaluating water quality.

Multisensor data fusion is a technology that enables combining information from several sources to form a unified picture. It is an important tool for improving the performance of a monitoring system when various sensors are available. Multisensor data fusion seeks to combine data from multiple sensors to perform inferences that will be more efficient and potentially more accurate than if they were achieved by means of a single sensor. The multisensor data fusion technique has received much attention recently, but it is more about applications to target identification, signal and image processing, biomedicine, and so forth. In this chapter, we apply multisensor data fusion technology in water quality evaluation.

The main problem is that data obtained from sensors have different degrees of uncertainty, which may arise for a number of reasons. For instance, the sensing error increases with the age of the sensor, and the sensor is disturbed by the environment. The quality of the wireless links is another major limiting factor. Furthermore, this uncertainty may lead to different sensors reaching conflicting conclusions. Because data obtained from the sensors are inherently incomplete, uncertain, and imprecise, it is imperative that a fusion mechanism be devised so as to minimize such imprecision and uncertainty.

Dempster–Shafer evidence theory (DS evidence theory) and Bayesian methods are commonly used to handle uncertainty. The basic strategy of Bayesian methods is that if the prior probabilities and conditional probabilities are determined in advance, then the a posteriori probabilities can be estimated using Bayes' formula. Nonetheless, effective fusion performance can be achieved only if adequate and appropriate a priori and conditional probabilities are available. In some situations, assumptions can be made with respect to a priori and conditional probabilities, but these assumptions can become unreasonable in many other situations. DS evidence theory can be regarded as an extension of classical probabilistic reasoning, which makes inferences from incomplete and uncertain data provided by different independent sources. A key advantage of DS evidence theory is its ability to deal with uncertain data without adequate a priori probabilities.

This chapter introduces a novel multisensor data fusion approach for water quality evaluation using DS evidence theory. We view each sensor measurement as a piece of evidence that reveals some uncertain information about the water quality. And we get water quality evaluation through the fusion of uncertain data from each sensor. The rest of this chapter is organized as follows. In Section 24.2, we introduce some preliminary concepts of the DS evidence theory. We present our multisensor data fusion approach for water quality evaluation using DS evidence theory in terms of mass function based on water quality parameters, reliability discounting, and decision rule using the fusion mass function values. Section 24.3 describes experiments in which we demonstrate the effectiveness of the proposed approach. Section 24.4 provides some concluding remarks.

## 24.2   WATER QUALITY EVALUATION USING DS EVIDENCE THEORY

### 24.2.1   DS Evidence Theory

DS evidence theory, originating from Dempster's work and further extended by Shafer, is a generalization of traditional probability that allows us to better quantify uncertainty. The theory is based on a number of key propositions, which are summarized as follows.

1. *Frame of discernment*: Let $\Theta$ be a finite set of elements; an element can be a hypothesis, an object, or in our case a water quality evaluation. We refer to $\Theta$ as the frame of discernment; the set consisting of all the subsets of $\Theta$ is called the power set of $\Theta$, and is denoted by $\Omega(\Theta)$.
2. *Mass function*: The mass function $m$ is also called a basic probability assignment function. It is defined as a mapping of the power set $\Omega(\Theta)$ to a number between 0 and 1:

$$m : \Omega(\Theta) \rightarrow [0, 1] \tag{24.1}$$

$$m(\Phi) = 0, \quad \sum_{A \subseteq \Omega(\Theta)} m(A) = 1 \tag{24.2}$$

where $m(A)$ is an expression of the level of confidence exactly in $A$. It does not include the confidence in any particular subset of $A$. In water quality evaluation, $m(A)$ can be considered as a degree of belief held by a certain class of water quality. If $m(A) > 0$, the subset $A$ of $\Theta$ is called a focal element. When a mass value is committed to a subset that has more than one element, it is explicitly stating that there is not enough information to distribute this belief more precisely to each individual element in the subset. In particular, the total belief is assigned to the whole frame of discernment, $m(\Theta) = 1$, when there is no evidence about $\Theta$ at all. $m(\Theta)$ is the uncertain function.

3. *Dempster's rule of combination (DS rule):* Suppose $m_1$ and $m_2$ are two mass functions formed based on data obtained from two different and independent sources in the same frame of discernment $\Theta$. We can get a new $m$ according to the DS rule as follows:

$$m(A) = m_1 \oplus m_2 = \begin{cases} 0 & A = \Phi \\ \dfrac{\displaystyle\sum_{B \cap C = A} m_1(B) m_2(C)}{1 - K} & A \neq \Phi \end{cases} \quad (24.3)$$

where

$$K = \sum_{B \cap C = \Phi} m_1(B) m_2(C) \quad (24.4)$$

$K \in (0, 1)$ is a normalization constant and can be viewed as a measure of conflict among the sources of evidence. The higher the $K$, the more conflicting are the sources. Dempster's rule of combination can blend measures of evidence from different sources.

## 24.2.2 WATER QUALITY EVALUATION

This section focuses on DS evidence theory applications in the multisensor environment and presents our implementation for water quality evaluation.

Consider a wireless sensor network with planktonic sensor node as shown in Figure 24.1, used to monitor and evaluate the water quality in a measurement area, for example, a lake or a pool. Each sensor node has several subsensors used to measure the water quality parameters. Let $F = \{f_1, f_2 \cdots f_m\}$ denote the water quality parameters measured by each sensor node; depending on the application,

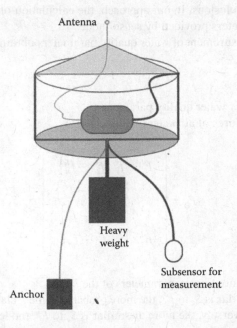

Antenna

Heavy
weight

Subsensor for
measurement

Anchor

**FIGURE 24.1** Planktonic sensor node model.

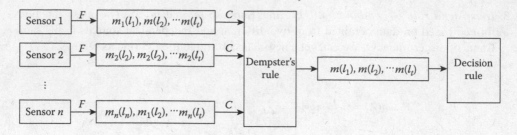

**FIGURE 24.2** Water quality evaluation using DS evidence theory.

$f$ may represent a quality parameter such as DO, $NH_3$-N, and so forth and $m$ is the number of water quality parameters. Through the water quality parameters, we can determine the class of water quality. Let $\Theta = \{l_1, l_2 \cdots l_i \cdots l_t\}$ be the frame of discernment for water quality evaluation; $l_i$ means that the water quality is classified in the $l_i$ class and $t$ is the number of water quality classes.

Assume $N$ sensor nodes, for the sake of simplicity, and suppose that all sensor nodes are independent. Each sensor node measures water quality parameters ($F$) of measurement area. When applying the DS evidence theory to multisensor data fusion, data obtained from each sensor node are the theory's evidence. Figure 24.2 shows a block diagram of our multisensor data fusion approach for water quality evaluation using DS evidence theory.

According to DS evidence theory, for each sensor node, the possibility of water quality class can be described by mass function values. $m_n(l_1)$, $m_n(l_2)$,...$m_n(l_t)$ are mass function values obtained by $F$ from sensor $n$. $m_n(l_t)$ is the confidence assigned to the $l_t$ class of water quality provided by sensor $n$. Multisensor data fusion amounts to combining several pieces of evidence to form a new comprehensive piece of evidence. For uncertainty, each sensor node is then given a reliability discounting ($C$) before combination. By using the fusion mass function values obtained by the DS rule, the class of water quality can be determined.

### 24.2.2.1 Mass Function Based on Water Quality Parameters

The derivation of mass function is the most crucial step in DS evidence theory, because it determines the reliability of conclusions. In our approach, the calculation of the mass function is based on the water quality parameters provided by sensor node.

Let $S_k$ represent the measurement of water quality parameters obtained from sensor $k$:

$$S_k = \{s_1, s_2 \cdots s_i \cdots s_m\} \tag{24.5}$$

where $s_i$ is the $i$th element of water quality parameters.

Let $F_\Theta$ represent the features of all water quality classes:

$$F_\Theta = \left\{ \begin{array}{l} F^{l_1} : \left\{ f_1^{l_1}, f_2^{l_1} \cdots f_m^{l_1} \right\} \\ F^{l_2} : \left\{ f_1^{l_2}, f_2^{l_2} \cdots f_m^{l_2} \right\} \\ F^{l_i} : \left\{ f_1^{l_i}, f_2^{l_i} \cdots f_m^{l_i} \right\} \\ F^{l_t} : \left\{ f_{11}^{l_t}, f_2^{l_t} \cdots f_m^{l_t} \right\} \end{array} \right\} \tag{24.6}$$

where $F^{l_i}$ describes the feature quality parameters of the $l_i$ class.

Intuitively, the more similar is $S_k$ to $F^{l_i}$, the more probable is the $l_i$ class of water quality, as far as sensor $k$ is concerned. Conversely, the more dissimilar is $S_k$ to $F^{l_i}$, the less probable is the $l_i$ class of water quality, again as far as sensor $k$ is concerned.

There are many measures for quantifying the distance between the measured parameters and the feature of water quality class. We propose to use the Minkowski distance measure, as follows:

$$d_k^{li} = \left( \sum_{x=1}^{m} \left( \frac{s_x - f_x^{li}}{f_{max} - f_{min}} \right)^{\alpha} \right)^{\frac{1}{\alpha}}$$

(24.7)

where $d_k^{li}$ is the distance between $S_k$ and $F^{li}$ and $\alpha$ is a constant; division by $f_{max} - f_{min}$ is for normalizing. The distances between the measurement of sensor $k$ and the features of all water quality classes can be captured:

$$D_k = \left\{ d_k^{l1}, d_k^{l2} \cdots d_k^{li} \cdots d_k^{lt} \right\}$$

(24.8)

The smaller the distance $d_k^{li}$, the more probable is the $l_i$ class of water quality. Defining

$$m_k(l_i) = \frac{\left( 1/d_k^{li} \right)}{\sum_{i=1}^{t} \left( 1/d_k^{li} \right)}$$

(24.9)

Then we have the mass function of water quality class from sensor $k$:

$$m_k = \{ m_k(l_1),\ m_k(l_2) \cdots m_k(l_i) \cdots m_k(l_t) \}$$

(24.10)

where $m_k(l_i)$ is the mass function assigned by sensor $k$ to the $l_i$ class of water quality.

### 24.2.2.2 Reliability Discounting

Some sensor nodes are more vulnerable to misreading or malfunctioning because of many factors, such as their age, type, and location. The impact of evidence is discounted to reflect the sensor node's reliability, in terms of reliability discounting $C$ ($0 < C < 1$). For uncertainty, evidence from each sensor node is then given a reliability discounting as follows:

$$C_k = \begin{cases} \beta & N_k \le N_{min} \\ \min\left( \dfrac{R_k}{N_k}, \beta \right) & N_k > N_{min} \end{cases}$$

(24.11)

where $N_k$ is the total number of water quality evaluations from sensor $k$; $R_k$ is the number of correct water quality evaluations from sensor $k$; $\beta$ and $N_{min}$ are fixed values. If the number of water quality evaluations is less than $N_{min}$, we use a fixed value $\beta$ as reliability discounting. $\beta$ and $N_{min}$ are usually selected as 0.9, 10 respectively.

If its reliability discounting ($C$) is high, this evidence will be given more weight and have a greater effect on the modified combinatorial rule. Then, the mass function $m_k$ in Equation 24.10 is updated to $m_k'$:

$$m'_k = \begin{cases} m'_k(l_i) = C_k \cdot m_k(l_i) \\ m'_k(\Theta) = 1 - \sum_{i=1}^{t} m'_k(l_i) \end{cases} \tag{24.12}$$

where $m'_k(l_i)$ is the modified mass function assigned by sensor $k$ to the $l_i$ class of water quality and $m'_k(\Theta)$ is the uncertain function (unidentified mass function) from sensor $k$.

### 24.2.2.3  Decision Rule Using the Fusion Mass Function Values

As stated earlier, all sensors are assumed to be independent. The multievidence combinatorial rule becomes:

$$m(A) = m'_1 \oplus m'_2 \oplus \cdots \oplus m'_k \cdots \oplus m'_n = \begin{cases} 0 & A = \Phi \\ \dfrac{\displaystyle\sum_{\cap_{k=1}^{t} A_k = A} m'_1(A_1) m'_2(A_2) \cdots m'_k(A_k) \cdots m'_n(A_n)}{1 - K'} & A \neq \Phi \end{cases} \tag{24.13}$$

where

$$K' = \sum_{\cap_{k=1}^{t} A_k = \Phi} m'_1(A_1) m'_2(A_2) \cdots m'_k(A_k) \cdots m'_n(A_n) \tag{24.14}$$

Using the combinatorial rule, a fusion mass function ($m(A)$) converts the confidence of each class of water quality arising from different evidence sources (sensor nodes) into a fraction in (0, 1). By using the fusion mass function values, water quality can be evaluated on the decision rule as follows:

1. The current determined class of water quality should have a maximal mass function value and should be greater than a certain value; this value should be at least greater than $1/t$, where $t$ stands for the number of water quality classes.
2. The difference of the mass function values between the current determined class of water quality and other classes should be greater than a certain gate limit value, and here it is 0.2.
3. The uncertain function value should be less than a certain gate limit value, and here it is 0.1.

If these three rules are not satisfied simultaneously, the current class of water quality is uncertain.

## 24.3  EXPERIMENTS AND RESULTS

In this section, we describe two experiments to validate the performance of the proposed approach. In our experiments, we use the senor nodes we developed to measure water quality parameters, as shown in Figure 24.3. The measured parameters include DO, $NH_3$-N, TP, TN; then $F = \{DO, NH_3$-N, $TP, TN\}$.

According to "Environmental quality standards for surface water of China" (GB 3838-2002), water quality is categorized into five classes, and the features of all water quality classes are shown in Table 24.1. Thus, in our experiments, the frame of discernment for water quality evaluation is $\Theta = \{l_1 = I, l_2 = II, l_3 = III, l_4 = IV, l_5 = V\}$, where I means that the water quality is classified in the first class. From Table 24.1, Equation 24.6 can be expressed as

**FIGURE 24.3** Sensor nodes used in experiments.

**TABLE 24.1**
**Features of All Water Quality Classes**

|     | DO  | NH$_3$-N | TP    | TN  |
|-----|-----|----------|-------|-----|
| I   | 7.5 | 0.15     | 0.01  | 0.2 |
| II  | 6.0 | 0.5      | 0.025 | 0.5 |
| III | 5.0 | 1.0      | 0.05  | 1.0 |
| IV  | 3.0 | 1.5      | 0.1   | 1.5 |
| V   | 2.0 | 2.0      | 0.2   | 2.0 |

$$F_\Theta = \begin{cases} F^{l_1} : \{7.5, & 0.15, & 0.01, & 0.2\} \\ F^{l_2} : \{6.0, & 0.5, & 0.025, & 0.5\} \\ F^{l_3} : \{5.0, & 1.0, & 0.05, & 1.0\} \\ F^{l_4} : \{3.0, & 1.5, & 0.1, & 1.5\} \\ F^{l_5} : \{2.0, & 2.0, & 0.2, & 2.0\} \end{cases}$$

In the first experiment, we use three sensor nodes ($S_1$, $S_2$, and $S_3$) to monitor the water quality of a pool. The objective is to determine the water quality class of the pool based on the three sensor nodes. Table 24.2 lists the water quality parameters measured by sensors $S_1$, $S_2$, and $S_3$.

From Equation 24.7, when $\alpha = 2$, $f_{max} = \max(f_i)$, $f_{min} = 0$, we can calculate that:

$$d_1^{l1} = \sqrt{\left(\frac{6.1-7.5}{7.5}\right)^2 + \left(\frac{2.55-0.15}{2}\right)^2 + \left(\frac{0.122-0.01}{0.2}\right)^2 + \left(\frac{3.432-0.2}{2}\right)^2} = 2.0976$$

$$d_1^{l2} = \sqrt{\left(\frac{6.1-6.0}{7.5}\right)^2 + \left(\frac{2.55-0.5}{2}\right)^2 + \left(\frac{0.122-0.025}{0.2}\right)^2 + \left(\frac{3.432-0.5}{2}\right)^2} = 1.8534$$

**TABLE 24.2**

**Water Quality Parameters Measured by Sensors $S_1$, $S_2$, and $S_3$**

|       | DO   | NH$_3$-N | TP    | TN    |
|-------|------|----------|-------|-------|
| $S_1$ | 6.10 | 2.550    | 0.122 | 3.432 |
| $S_2$ | 6.65 | 0.660    | 0.038 | 0.706 |
| $S_3$ | 7.21 | 0.467    | 0.022 | 0.690 |

$$d_1^{l3} = \sqrt{\left(\frac{6.1-5.0}{7.5}\right)^2 + \left(\frac{2.55-1.0}{2}\right)^2 + \left(\frac{0.122-0.05}{0.2}\right)^2 + \left(\frac{3.432-1.0}{2}\right)^2} = 1.4935$$

$$d_1^{l4} = \sqrt{\left(\frac{6.1-3.0}{7.5}\right)^2 + \left(\frac{2.55-1.5}{2}\right)^2 + \left(\frac{0.122-0.1}{0.2}\right)^2 + \left(\frac{3.432-1.5}{2}\right)^2} = 1.1797$$

$$d_1^{l5} = \sqrt{\left(\frac{6.1-2.0}{7.5}\right)^2 + \left(\frac{2.55-2.0}{2}\right)^2 + \left(\frac{0.122-0.2}{0.2}\right)^2 + \left(\frac{3.432-2.0}{2}\right)^2} = 1.0194$$

From Equation 24.9, we can get the mass function of water quality class from $S_1$: $m_1$ = {0.135, 0.154, 0.191, 0.241, 0.279}. Similarly, we can also calculate mass functions from $S_2$ and $S_3$, as shown in Table 24.3.

Suppose that the numbers of water quality evaluations from sensors $S_1$, $S_2$, and $S_3$ are the same, 15, and the numbers of correct evaluations are respectively 9, 12, 14. From Equations 24.11 and 24.12, we can get the modified mass functions and the uncertain functions from $S_1$, $S_2$, and $S_3$, as shown in Table 24.4.

We can see from Table 24.2 that the water quality parameters measured by $S_1$ are obviously different from those measured by $S_2$ and $S_3$. $S_1$ gives a different conclusion. For uncertainty, the data provided by $S_1$ may be incorrect. One need only consider the data of $S_1$ and $S_2$ from Table 24.4, where two pieces of evidence are in conflict. According to the multievidence combinatorial rule, which is realized by Equations 24.13 and 24.14, we can combine the evidence provided by $S_1$ and $S_2$, as shown in the first row of Table 24.5. The second class of water quality has the maximal mass function value. It is seen that our approach is effective when there is a conflict of evidence. However, according to our decision rule, the uncertain function value is great and we cannot determine the class of current water quality.

Then, considering the data of $S_3$, the fusion mass function and uncertain function combining $S_1$, $S_2$, and $S_3$ are also shown in Table 24.5. We reach the water quality evaluation result in favor of II

**TABLE 24.3**

**Mass Functions from $S_1$, $S_2$, and $S_3$**

|       | I     | II    | III   | IV    | V     |
|-------|-------|-------|-------|-------|-------|
| $m_1$ | 0.135 | 0.154 | 0.191 | 0.241 | 0.279 |
| $m_2$ | 0.185 | 0.438 | 0.232 | 0.091 | 0.054 |
| $m_3$ | 0.265 | 0.422 | 0.177 | 0.084 | 0.052 |

**TABLE 24.4**

**Modified Mass Functions and Uncertain Functions from $S_1$, $S_2$, and $S_3$**

|        | I     | II    | III   | IV    | V     | $\Theta$ |
|--------|-------|-------|-------|-------|-------|----------|
| $m_1'$ | 0.081 | 0.092 | 0.114 | 0.145 | 0.168 | 0.4      |
| $m_2'$ | 0.148 | 0.350 | 0.186 | 0.073 | 0.043 | 0.2      |
| $m_3'$ | 0.238 | 0.380 | 0.159 | 0.076 | 0.047 | 0.1      |

**TABLE 24.5**

**Fusion Mass Functions and Uncertain Functions**

|                           | I     | II    | III   | IV    | V     | $\Theta$ |
|---------------------------|-------|-------|-------|-------|-------|----------|
| $m_1' \oplus m_2'$        | 0.145 | 0.316 | 0.196 | 0.114 | 0.096 | 0.133    |
| $m_1' \oplus m_2' \oplus m_3'$ | 0.193 | 0.483 | 0.172 | 0.072 | 0.048 | 0.032    |

with a degree of belief of 0.483, and the uncertain function value is reduced to 0.032. According to the decision rule, we can determine that the water quality of the pool is in the second class. This result is the combination of the three sensor nodes and is considered to be more believable compared to that of the single-sensor node.

We can see clearly from Table 24.5 that by comparing the fusion mass function value with the single-sensor mass function value, the mass function value of the current determined water quality class is enlarged, the difference of the mass function values between the current determined class of water quality and other classes is enlarged, and at the same time, the uncertain function value is reduced.

In the second experiment, we use different numbers of sensor nodes to monitor the water quality of a small lake at different times and different locations, and use the approach of this chapter to evaluate the water quality. Here we already know that water quality of this lake is in the fourth class. The objective is to demonstrate that our approach could improve the accuracy of water quality evaluation. According to simulation, Table 24.6 shows the correct rate and uncertain rate compared between different numbers of sensor nodes used in multisensor data fusion for water quality evaluation.

From Table 24.6, it can be seen that our approach can increase the correct rate of water quality evaluation, compared to the approach using a single-sensor node. It can also be seen that fusion can reduce the uncertain rate of water quality evaluation.

**TABLE 24.6**

**Correct Rate and Uncertain Rate with Different Sensor Node Combinations**

|                   | Single-Sensor Node | Three-Sensor Nodes | Five-Sensor Nodes |
|-------------------|--------------------|--------------------|-------------------|
| Correct rate (%)  | 38                 | 84                 | 96                |
| Uncertain rate (%)| 46                 | 10                 | 2                 |

In summary, our experiments have indicated that the proposed multisensor data fusion approach for water quality evaluation using DS evidence theory has improved water quality evaluation performance greatly.

## 24.4  CONCLUSIONS

In this chapter, we have applied multisensor data fusion technology in water quality evaluation. We have introduced a novel multisensor data fusion approach for water quality evaluation using DS evidence theory. We have proposed a method of calculating mass function based on water quality parameters and a reliability discounting to reflect the sensor node's reliability. Furthermore, we have proposed the decision rule to determine the class of water quality by using the fusion mass function values. Our experiments have indicated that the proposed approach can evaluate water quality from uncertain sensor data, increase the correct rate, and reduce the uncertain rate of water quality evaluation compared to the approaches using individual sensors.

## REFERENCES

"Environmental quality standards for surface water," GB 3838-2002, Beijing: Environmental Science Press of P.R. China, 2002.

Hall, J., A.D. Zaffino, R.B. Marx, P.C. Kefauver, E.R. Krishnan, R.C. Haught, and J.G. Herrmann. On-line Water Quality Parameters as Indicators of Distribution System Contamination. Submitted to *JAWWA*, July, 2005.

# 25 A Granular Sensor-Fusion Method for Regenerative Life Support Systems

*Gregorio E. Drayer and Ayanna M. Howard*

## CONTENTS

## 25.1 INTRODUCTION

Multisensor data fusion consists of combining observations and measurements from a number of different sensors to provide a complete description of a system and its environment [1]. The main multisensor fusion methods are probabilistic in nature and derive from the application of tools in statistics, estimation, and control theory. These are (1) the Bayes rule, (2) probabilistic grids, (3) the Kalman filter, and (4) sequential Monte Carlo methods. However, shortcomings to probabilistic methods are found in their apparent inability to address unknown situations, which grows in importance for anomaly detection and management of emergent phenomena. There are four main limitations for probabilistic methods in multisensor data fusion [1]:

1. *Complexity*: This limitation in found in the large number of probabilities required to correctly apply probabilistic reasoning.
2. *Inconsistency*: This refers to the difficulty in obtaining consistent deductions about the state of a system from sets of belief that are not necessarily consistent.
3. *Precision of models*: This refers to the difficulty to obtain system representations, caused primarily by the inability to describe probabilities of quantities for which there is not enough available information.
4. *Uncertainty about uncertainty*: It is difficult to assign probabilities in the presence of unknowns and uncertainty about sources of information.

Less traditional methods, such as interval calculus, fuzzy logic [2], and evidential reasoning [3–5], provide alternative approaches that help overcome these limitations [1]. Such approaches will support current research efforts in managing large-scale/ubiquitous sensor systems and anomaly detection applications. This chapter represents a step toward a multisensor data fusion method for the development of monitoring and automation systems. The approach presented in this chapter was originally developed for applications in regenerative life support systems (RLSSs). The following two subsections provide background on RLSS and on how these systems may benefit from such sensor fusion methodology.

### 25.1.1 Background on Regenerative Life Support Systems

In the development of automated systems it may be of interest to reach a certain level of automation (LOA). Possible LOAs range from completely manual systems to fully autonomous ones. Most people, however, believe that systems can be controlled only either manually or automatically, and discard the possibility of having humans work with automation at various degrees. Such is the case in the domain of space exploration: Most people take extreme sides and consider fully robotic missions versus manned ones, and do not highlight the advantages of having humans and automation collaborate in a shared mission [6]. This polarization is most probably due to the lower costs of conducting purely robotic missions.

The cost of both robotic and manned missions is determined by requirements in mass, volume, and power [7]. Manned missions differ from robotic ones in that, in addition to science instruments, they also need to support the physiological demands and quality of life of a human crew [8,9]. The subsystems that keep the crew alive while contributing to mission success are called life support systems (LSSs). These subsystems add mass and volume to mission elements, resulting in the need for greater launch capacities, which as a consequence increase their overall cost [10]. In addition, the presence of a human crew has traditionally created the need for expensive management structures to minimize the risk of loss-of-mission (LOM) and loss-of-crew (LOC) events. For example, the space shuttle program management at the National Aeronautics and Space Administration (NASA) used to absorb 69% of the total budget allocated to generic operations and infrastructure functions [11]. Recent innovations in commercial spaceflight aim to reduce costs considerably while increasing autonomy of operations [12].

Besides mission requirements, mission duration also increases the cost of manned missions. If the LSS operates in open loop, that is, byproducts are not recycled on board the spacecraft, the total mass of consumables must be launched, stored, and consumed throughout the duration of the mission. In consequence, the mass of consumables, as well as byproducts, increase with mission duration. Although early space exploration systems were able to revitalize air, they were unable to recycle water from urine or to produce food, thus limiting the autonomy of the spacecraft to only 14 days [8].

One way to cope with this problem is to regenerate consumables by recycling byproducts. The components that provide these capabilities are called *regenerative* LSSs, or RLSSs; they include a suite of technologies based on physicochemical and biological processes aimed to transform wastes and byproducts back into consumables. RLSSs are meant to be autonomous and to help maximize crew time dedicated to mission objectives. However, their operation is not trivial: RLSS processes require considerable effort and time, and they constitute complex mass and energy transfer networks subject to the behaviors of their unit processes and to crew demands. As a consequence, they pose novel challenges for their integration and operation.

The International Space Station (ISS) is today the platform used for the development of LSS technologies to enable future exploration missions to the moon and to other destinations in the solar system. Private companies, such as Space Exploration Technologies (SpaceX) and Bigelow Aerospace, are expected to build capacities and destinations to join in these efforts.

The ISS is also the primary space-based environmental control and life support system (ECLSS) research platform. Among its purposes is to test, incorporate, and mature technologies to reduce the need for resupply missions and to enable long-term manned space flight beyond low Earth orbit [13]. Its three key components are the Water Recovery System (WRS), the Oxygen Generating System (OGS), and the Carbon Dioxide Reduction Assembly (CRA) [14]. These processes are entirely physicochemical and help to close the water and atmosphere regeneration cycles.

### 25.1.2 CHALLENGES FOR RLSSs

RLSSs offer various options to recycle metabolic byproducts, such as urine, and to achieve an incremental closure of gaseous and liquid material cycles. Such material loop closure *increases the autonomy of space habitats* and helps reduce the frequency of resupply missions and their overall cost. But as researchers continue their efforts to integrate regenerative technologies and to achieve system closure, new challenges arise from unintended interactions between chemical species in the closed-loop system. Material loop closure not only makes possible the interconnection of complex material networks, but may also promote unintended interactions between chemical species within the habitat. Such interactions may lead to the accumulation of unexpected chemical compounds that could affect individual life-support processes or crew health. They could also lead to the depletion of a vital element for the crew, such as oxygen. By now, the life support and automation communities have had the opportunity to learn that such uncertainty is to be expected and its effects to be discovered as anomalies during operation [6].

Beyond methods in robust [15] and adaptive control [16,17], paradigms in switched control [18–21] offer advantages for the management of the uncertainty caused by material loop closure in RLSS. The contribution discussed in this chapter makes use of a perception-based approach to a switched control paradigm. Switched control introduces attributes of flexibility and modularity to the control system [21]. These attributes may be used to allow for different control actions depending on the *operational condition* of the physical system and its *situation* in a given context. In other words, these changes may depend on the internal state of the physical system and on external factors defined by its environment and active goals.

### 25.1.3 SENSOR FUSION FOR RLSS

Slow-changing characteristics of controlled environmental systems and the increasing availability of data from sensors and measurements offer opportunities for the development of computational methods to enhance situation observability, decrease human workload, and support real-time decision making. The invention of methods to measure environmental variables by means of microsystems or optical devices tends to reduce the unit cost of novel sensor technology and opens opportunities for engineers to integrate ever more complex systems. Multisensor data fusion can be used in user-centered interfaces in support situation awareness and observability. Situation

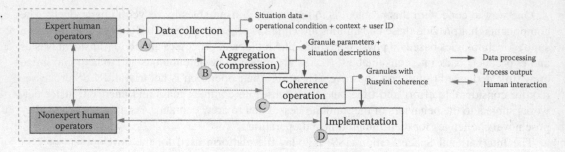

**FIGURE 25.1**  Granular multisensor fusion method.

observability may also enable humans to perceive and comprehend the state of the system at a given instant and to help human operators decide what actions to take at any given time that may affect the projection of such state into the near future [22].

Toward this purpose, this chapter makes use of the fuzzy associative memory (FAM)-based agent architecture as a framework to enable a granular approach to automation and control. Such an approach is presented in Section 25.2 and is conceived as a switching control paradigm with attributes of flexibility and modularity. The combination of sensor information is used to create a *sensing space* in which the operational conditions of the system may be found. This granular approach takes advantage of the opportunity to define perceptual elements or granules within the sensing space, in which each granule represents a specific situation. In particular, this work employs intelligent agents based on FAM made of granular structures composed of $n$-dimensional noninteractive fuzzy sets [2,23–25]. Granular structures [26–29] define the situations in which each control action governs the system, thus implementing a switched control paradigm to their automation. Situation-rich signals serve as the switching mechanism and provide observability of the operational condition and context of the system; that is, situation observability. Simulations are supported on the process of respiration in an aquatic habitat [30] acting as an RLSS.

The difficulty of manually defining fuzzy sets for each individual condition makes such techniques impractical. Therefore, the contribution of Section 25.3 takes advantage of the interaction of human experts with the system to collect situation-rich data useful to represent their situation knowledge base (SKB). The SKB is then used in the perception function of the FAM-based agents to generate the switching signals that combine control laws into an integrated control signal. Those switching signals contain information about the situation of the system and may also be used in user interfaces for human–automation coordination. This general contribution is composed of four specific steps: (1) data collection, (2) aggregation algorithm, (3) coherence operation, and (4) implementation. The steps are represented in Figure 25.1 as blocks in the diagram and described in Section 25.3.1, with a numerical example in Section 25.3.2.

## 25.2   GRANULAR APPROACH TO THE AUTOMATION

This section introduces the use of agents based on FAM [2,23,24] to develop granular structures composed of $n$-dimensional noninteractive fuzzy sets, used to define operating conditions and the control law that governs an action in each situation. The objectives are the following: (1) to implement a switched control strategy on the dynamic model of a reconfigurable aquatic habitat, introducing flexibility into the dynamics of the system, and (2) to explore how the granular structure of FAM-based agents may generate useful information to enhance situation observability and thus potentially provide human operators with resources for real-time decision making. Such exploration is oriented toward the development of methods in user-centered design that take into account situation awareness to inform better ecological interfaces [6,22]. Although results presented herein are based on simulations, hardware of the system described is used to identify model parameters.

### 25.2.1 THE FAM-BASED AGENT ARCHITECTURE

The FAM-based agent architecture has found motivation in the monitoring and automation of LSS [31] and implements a switched control approach [20] that assigns a control action to each situation in which the system may operate in the form of (Situation, Controller). The switching capability introduces flexibility in the behavior of the system and enables its development in a modular and incremental fashion. The architecture is characterized by a perception function, a set of controllers, and a correspondence function. The latter associates a controller to each situation defined in the perception function and combines them into an integrated control signal. Figure 25.2 shows a diagram of a single FAM-based agent with a user interface manipulating a single variable in a small-scale aquatic habitat. The diagram describes the components of the FAM-based agent with the following blocks: (1) Perception, (2) Control Signals, and (3) Correspondence Function. Some advantages of this approach have been shown in previous work [31].

#### 25.2.1.1 Perception as a Granular Structure

The perception function assumes the availability of $n$ measurable variables $x_i$ for $i = 1, 2, ..., n$ from sensors and their universes of discourse $X_i$ so that $x_i \in X_i \subseteq \mathfrak{R}$; the variables are nonredundant and noninteractive: $X_i \neq X_j; j = 1, 2, ..., n; i \neq j$. Each universe $X_i$ is partitioned in $k_i$ subsets, each of which is denoted as $X_i^\alpha \subset X_i$, $\alpha = 1, 2, ..., k_i$. Continuous membership functions describe each one of the subsets as $\mu_{X_i^\alpha}(x_i)$, which are normal and convex [32]. Such partitions are coherent when complying with the Ruspini condition [33]:

$$\sum_{\alpha=1}^{k_i} \mu_{X_i^\alpha}(x_i) = 1 \qquad \forall i = 1, 2, ..., n \tag{25.1}$$

This condition becomes of importance in automation systems because of its capacity to ensure that there will not be regions in $X$ for which a control policy is not assigned or with conflicts between controllers. As a result, a number of $l$ possible situations or operating conditions are defined as noninteractive fuzzy sets $\tilde{A}_j$, for $j = 1, 2, ..., l$. The $l$ situations are the Cartesian product of the combination of the subsets $X_i^\alpha$ in $X_i$. The Cartesian product is implemented with the *minimum* operator as in Equation 25.2, for $l = \prod_{i=1}^{n} = k_i = k_1 \cdot k_2 \cdots k_n$.

$$\tilde{A}_j(x_1, ..., x_n) = \min_{\substack{i=1,...,n \\ \alpha=1,2,...,k_i}} \left( \mu_{X_i^\alpha}(x_i) \right) \tag{25.2}$$

The set $\tilde{A} = \{\tilde{A}_j\}$ represents the granular structure in which each granule $\tilde{A}_j$ describes a different situation and a percept of the FAM-based agent.

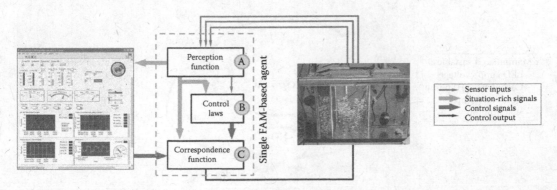

**FIGURE 25.2** Diagram of the FAM-based agent and its components.

### 25.2.1.2 Control Signals

In the same fashion, the set of control signals $U = \{u_j\}$ are obtained from up to $l$ different control laws. Controllers generate signals $u_j$ that correspond to each condition $\tilde{A}_j$. These signals may be treated modularly to form the set $U = \{u_1, u_2, ..., u_l\}$, with the maximum number of different control signals limited by $l$. The control signals can be generated by model-based methods or techniques in soft-computing and computational intelligence. The error modulation solution [24] or a similar technique is required for controllers with integral control action (poles in zero). Considerations on switched control [20,21] should be included in this component of the FAM-based agent and in the correspondence function $\Omega$ described in the next subsection.

### 25.2.1.3 Correspondence Function

With the sets $\tilde{A}$ and $U$ defined, the correspondence function $\Omega$ can be expressed as a rule-base or in pairs (Situation, Control Signal) as in Equation 25.3.

$$\Omega : \tilde{A} \rightarrow U$$

$$\Omega = \{\Omega_j\} = \left\{ \left( \tilde{A}_j(x_1,...,x_n), u_j(t) \right) \right\} \tag{25.3}$$

The resulting FAM is defuzzified with the weighted average technique to obtain an integrated control signal $u_I$. This signal drives a single actuator in the system. Thus, each actuator and its controller in a physical system may be conceived as an agent, constituting a FAM-based multiagent system. The weights used in Equation 25.4 are the membership values of each corresponding situation, and the weighted arguments are their corresponding control signals.

$$u_I(x_1,...,x_n,t) = \frac{\sum_{i=1}^{l} \mu_{\tilde{A}_i}(x_1,...,x_n) \cdot u_i(t) \sum}{\sum_{i=1}^{l} \mu_{\tilde{A}_i}(x_1,...,x_n)} \tag{25.4}$$

### 25.2.2 Application to the Model of the Habitat

This subsection presents the application of the FAM-based agent architecture to the control of the dissolved oxygen (DO) levels in the model of the aquatic habitat [30], whose regenerative cycles and variables are shown in Figure 25.3. It defines (1) the operating range of the life support variables considered; (2) the operational conditions that result from the combination of the operating ranges, and their corresponding control actions; and (3) the simulation performed on the habitat for this subsection.

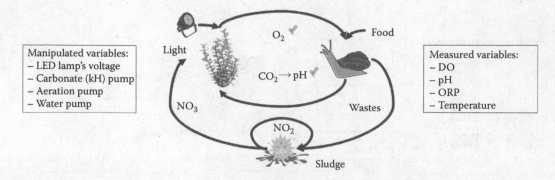

**FIGURE 25.3** Regenerative cycle of the aquatic habitat.

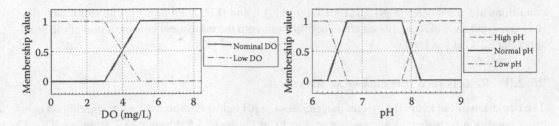

**FIGURE 25.4** Fuzzy partitions of the DO and pH variables.

### 25.2.2.1 Life Support Variables

The life support variables are the DO and pH in the fourth compartment. Their operating ranges and fuzzy membership functions are shown in Figure 25.4. These ranges are defined considering the minimum DO concentration allowed for fresh water animals (2 [mg/L]), and the pH values in which most aquatic organisms may live with low stress [34]. There are two conditions for the DO concentrations: nominal and low. The pH has three ranges: nominal, high, and low.

### 25.2.2.2 Control Laws and Operating Conditions

Two control actions are used to drive the power level of the light-emitting diode (LED) lamp: (1) power on and constant at 100% and (2) a proportional-integral (PI) controller that may dim the lamp in the 0–100% range. The PI controller is used in most of the operating conditions, with $P = 200$ and $I = 50$. The differences of the PI controllers for each operating condition is in the controlled variable and its reference. A representation of the operating conditions and the control actions for each case is shown in Table 25.1.

These operating conditions result from the combination of the operating ranges of each variable, according to Section 25.2.1.1. To ensure that the system works correctly, note that the PI controller uses the error modulation technique presented in Ref. [24]. The nominal and $DO_{min}$ controllers make use of a reference signal with a duty cycle of 18 hours for every 24. This duty cycle helps to account for the physiological requirements of the botanical elements. For the nominal condition, the reference alternates between 6.0 [mg/L] and 5.0 [mg/L], while for the $DO_{min}$ the reference switches between 4.5 [mg/L] and 4.0 [mg/L].

### 25.2.2.3 Simulation of the Habitat Model

The simulation presented in this subsection explores the transitions between operational conditions triggered by the depletion of kH and the lack of supply from the dosifier pump in the second compartment. This substance is consumed by the bacteria of the biofilter during the process of nitrification. The source of kH is inhibited until day 14; on day 14 the kH source is restored. The purpose of this simulation is to explore the operating condition transitions of the FAM-based agent and the time response of the life support variables considered. In addition, the simulation also shows the evolution of the membership values of the life support variables in each of the operating conditions, making the system observable from this perspective at any given time. The simulations are implemented with a stiff Mod. Rosenbrock numeric method with maximum step of 0.01. Initial

**TABLE 25.1**
**Control Actions for Different Operating Conditions**

|  | Low pH | Nominal pH | High pH |
|---|---|---|---|
| Nominal DO | $pH_{ref} = 6.3$ | Nominal | $pH_{ref} = 8.0$ |
| Low DO | Lamp on | $DO_{ref} = 3$ | $DO_{min}$ |

conditions are [DO] = 8.4 [mg/L], [CD] = 0.69 [mg/L], and [kH] = 20 [mg/L]. The simulation time is 21 days and its initial conditions are in equilibrium with the equivalent concentration of the atmosphere at 22°C at sea level.

### 25.2.3 RESULTS FROM THE GRANULAR APPROACH

The depletion of the kH in the system deteriorates the pH below nominal values, triggering an operating condition transition as shown during day 12 in Figures 25.5 through 25.7. Between days 12 and 15 the system continues to transition into different situations and recovers its nominal condition thereafter, when the kH supply is reenabled. These results show the dynamics of the transitions from three perspectives: (1) the evolution of life support variables, DO and pH, is shown in Figure 25.5; (2) the behavior of the LED lamp is presented in Figure 25.6; and (3) the membership value of the operational condition of the system over time is shown in Figure 25.7. The system remains "fail-op/fail-safe" within the conditions defined in Table 25.1. From day 5 to about day 11, the system shows consistent and mostly periodic temporal responses as evidenced in Figures 25.5 and 25.6.

During this period of time and, without looking at Figure 25.7, it can be said that the system remains within a single operating condition, in this case in the "nominal" condition, and seems to be performing well. However, once the first transition enters into effect at around day 12, it becomes harder to assess in which situation the system is in until it goes back into "nominal" on day 15. The lack of situation observability evident for only two or three signals in this case, is more so true for large-scale sociotechnical systems composed of many more sensors and signals. Hence, having a granular structure that allows the system to identify its mode of operation becomes helpful not only to allow for automation strategies that adapt the system to various situations, but also to generate new signals that better describe their evolution. This is what Figure 25.7 presents in the signals (a) through (f); it shows the history of the situation of the system. In these figures, membership values are shown for the conditions defined in Table 25.1 as follows: (a) nominal DO/high pH; (b) all nominal; (c) nominal DO/low pH; (d) low DO/high pH; (e) low DO/nominal pH; (f) low DO/low pH.

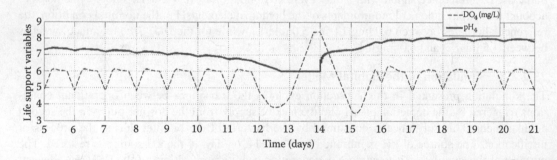

**FIGURE 25.5**  Evolution of the life support variables during the simulation.

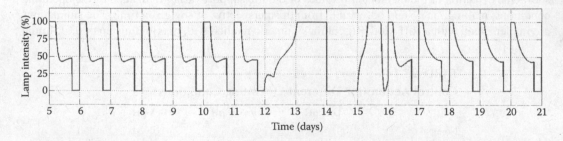

**FIGURE 25.6**  Lamp intensity for the simulation.

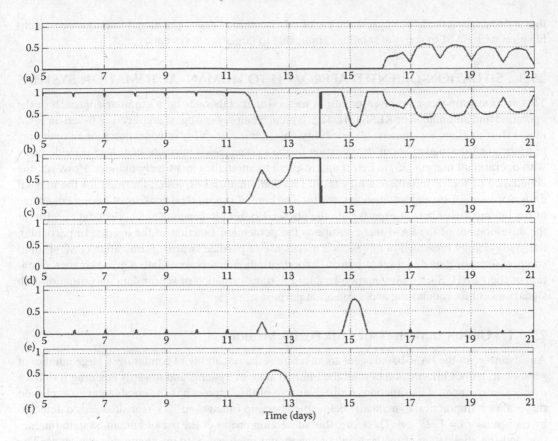

**FIGURE 25.7** Membership values of the conditions defined in Table 25.1. (a) Nominal DO/high pH, (b) all nominal, (c) nominal DO/low pH, (d) low DO/high pH, (e) low DO/nominal pH, and (f) low DO/low pH.

These signals allow us to understand not only the situation in a real-time scenario, but also to perform forensic analysis on Figures 25.5 and 25.6. For example, between days 12 and a slightly after day 15, the system transits between three different operating conditions before going back to "nominal." These conditions are (not in chronological order): nominal DO/low pH, low DO/nominal pH, and low DO/low pH. The last of these conditions to enter into effect before the system is dominantly back at "nomimal" is low DO/nominal pH. Under this condition and according to Table 25.1, the system sets the dissolved oxygen reference value to 3 [mg/L]. Thus, it can be seen in Figure 25.5 that the oxygen goes from being in saturation at 8.4 [mg/L] down to about 3.5 [mg/L], just before the last transition back to "nominal" comes into effect. Note also, that to bring down the oxygen level from 8.4 [mg/L] to about 3.5 [mg/L], the LED lamp has to be at 0% (turned off) in Figure 25.6 to prevent the plants from generating oxygen and allowing other organisms to consume it. After this transition the system alternates between "nominal" and nominal DO/high pH, which then again, according to Table 25.1, alternates the controlled variable between the pH level at 8.0 and the dissolved oxygen concentration following the LED lamp duty cycle. The forensic analysis of Figures 25.5 and 25.6 described earlier is possible to most observers because of the information provided by Figure 25.7. Such information could be displayed in ecological interfaces to support human operators in real-time decision-making tasks. The signals generated by the FAM could also be used to assess the evolution of systems in the future, looking at further applications in diagnosis and prognosis of engineering systems. Beyond RLSS, the granular decomposition of sensing spaces presented in this method is applicable to a wider range of complex sociotechnical systems. Questions then arise on how to manage high-dimensional sensing spaces and what type of methods are required to make this approach practical. We suggest the use of other methods in computational intelligence in combination with FAM to arrive at solutions applicable to larger-scale systems.

## 25.3 SITUATION-ORIENTED APPROACH TO HUMAN–AUTOMATION SYSTEMS

This section proposes a multisensor fusion method that elaborates on a granular approach to the operation and automation of RLSS [31]. The approach employs an agent architecture based on FAM in an effort to allow for situation observability, that is, *the capability* of nonexpert human operators to probe for information about the situation of the system. Such attribute may also provide users with operational margin [35] to detect and respond to anomalies in a timely manner. However, the abundance of sensor information may result in a combinatorial explosion unsuited for the manual design of monitoring and automation systems. The core of this method consists of taking advantage of the interaction of human experts with the physical system to generate and collect data useful for the development of the FAM that constitutes the perception function of the agents. In particular, the methodology presented in this section makes use of particle swarm optimization (PSO) [36] to *compress* sensor data and a set of human–expert situation assessments into a granular representation of their SKB. Such representation enables the transformation of sensor data into situation-rich signals useful for monitoring and automation purposes.

### 25.3.1 GRANULAR MULTISENSOR DATA FUSION METHOD

An advantage of the FAM-based agent architecture is the possibility of combining a large number of sensors, to gather information beyond the internal state of systems and towards enabling it to have a better assessment of its situation. A disadvantage of this approach is the combinatorial explosion that makes it impractical to manually define membership functions $\mu_{x_i^\alpha}(x_i)$ for situations $\alpha$ detected by each sensor $i = 1, 2,\ldots, n$. Therefore, this subsection proposes the use of human–system interaction and the application of methods in computational intelligence to overcome this challenge. The method shown in Figure 25.8 collects situation assessments from expert human operators, that is, system snapshots, to obtain situation-rich data sets that may be useful to generate a representation of the SKB of experts. Datasets containing a number of $N$ snapshots are aggregated (compressed) into a parametric representation. The aggregation consists of a particle swarm optimization process that adapts $\pi$-membership functions to the data contained in the data set for each sensor and each situation. The result is a granular structure useful for decision support tools and, when coherent, susceptible for adoption as the perception function of the FAM-based agent architecture. The following subsections describe each one of these steps.

#### 25.3.1.1 Data Collection

Data collection consists of taking advantage of the interaction between expert human operators and the system to obtain situation-rich data sets. These data sets include measurements of the *operating*

| | | Measurements | | | | Expert input | | |
|---|---|---|---|---|---|---|---|---|
| | Time | $x_1$ | $x_2$ | ... | $x_n$ | Situation | Confidence | User code |
| Dataset | $t_1$ | $x_{11}$ | $x_{21}$ | ... | $x_{n1}$ | $s_1$ | $c_1$ | $h_1$ |
| | $t_2$ | $x_{12}$ | $x_{22}$ | ... | $x_{n2}$ | $s_2$ | $c_2$ | $h_2$ |
| | ⋮ | ⋮ | ⋮ | | ⋮ | ⋮ | ⋮ | ⋮ |
| | $t_N$ | $x_{1N}$ | $x_{2N}$ | ... | $x_{nN}$ | $s_G$ | $c_N$ | $h_N$ |

**FIGURE 25.8** Data set description for the data collection process.

*condition* of the system (internal state), its *context* (external state), and an *identifier* of the expert. Datasets contain $N$ snapshots of the system at times $t_j$ for $j = 1, 2, ..., N$ as shown in Figure 25.8.

*Measurements* of the system state, both internal and external, include values $x_{ij}$ recorded by sensors $x_i$ for $i = 1, 2, ..., n$. If sensors are not available, values may be systematically obtained and introduced by the expert through a user interface, depending on the nature of the measurement. In addition to measurements, the data set includes *expert input* that defines to which situation $s_\gamma$ the snapshots belong in each case, for $\gamma = 1, 2, ..., G$, and with what degree of confidence $c_j \in [0, 1]$. If $c_j = 1$, the expert is fully confident that the system snapshot taken at $t_j$ belongs to situation $s_\gamma$. The number $G \geq l$ depends on the presence of hierarchical structures in the situation assessments according to the notion of *levels of resolution* in granular computing [29]; that is, *a situation* assessed as "nominal" may be subdivided in more specific situations, such as "nominal-high" and "nominal-low." This subsection does not address hierarchical granular structures, making $G = l$. Finally, the user code $h_j$ allows the data collection process to identify the expert that contributed to each snapshot of the data set, enabling approaches in crowdsourcing [37,38]. The intention of the following steps is to compress the data set into a more compact and meaningful representation.

### 25.3.1.2    Aggregation or Data Compression

The aggregation algorithm transforms (compresses) situation-rich data sets into granular structures described by an array of parameters that define membership functions $\mu_{X_i^\alpha}$ for each situation $\gamma$ susceptible to detection by sensors $i$. How situation knowledge is represented, how it is obtained from data sets, and a suggested approach to achieve coherence of the fuzzy sets are described in the following paragraphs.

#### 25.3.1.2.1    *Knowledge Representation*

Given the need to allow for flexible adaptation of a membership function $\mu_{X_i^\alpha}$ to collections of snapshots found in the data sets, the aggregation algorithm makes use of a piece-wise differentiable function defined by four parameters and known as a $\pi$-membership function, as defined in Equation 25.5:

$$\mu_{X_i^\alpha}(x_i; a, b, c, d) = \begin{cases} 0 & x_i \leq a \\ 2\left(\dfrac{x_i - a}{b - a}\right)^2 & a < x_i \leq \dfrac{a+b}{2} \\ 1 - 2\left(\dfrac{x_i - b}{b - a}\right)^2 & \dfrac{a+b}{2} < x_i \leq b \\ 1 & b < x_i \leq c \\ 1 - 2\left(\dfrac{x_i - c}{d - c}\right)^2 & c < x_i \leq \dfrac{c+d}{2} \\ 2\left(\dfrac{x_i - d}{d - c}\right)^2 & \dfrac{c+d}{2} < x_i \leq d \\ 0 & x_i \geq d \end{cases} \tag{25.5}$$

The $\pi$-membership function results in the curve shown in Figure 25.9, with parameters $P = [a, b, c, d]$ defining the "feet" and "shoulders" of the curve. Each membership function in the aggregation process represents a single situation $\gamma = 1, ..., G$ for a single sensor $x_i$. The PSO process obtains the four parameters in each case, as described next.

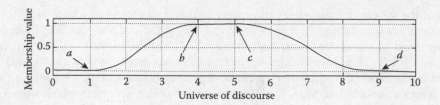

**FIGURE 25.9**   π-Membership function for $P = [a, b, c, d] = [1, 4, 5, 9]$.

### 25.3.1.2.2   Particle Swarm Optimization

PSO [36] is the process that transforms data sets into a granular structure. For each situation $\gamma$ and sensor $i$, find $P^* \in X_i$ such that the condition in Equation 25.6 is found, where $f(x_i) = \sum \left( \mu_{X_i^\alpha}(x_{ij}) - c_j \right)^2$ for $j = 1, 2, \ldots, N$ and in each case subject to the initial constraints shown in Table 25.2.

$$P^* = \arg \min_{x_i \in X_i} f(x_i) = \left\{ x_i^* \in X_i : f\left(x_i^*\right) \le f(x_i) \forall x_i \in X_i \right\} \tag{25.6}$$

The swarm is subject to random variables $\zeta_1 \in [0, 1]$ and $\zeta_2 = 1 - \zeta_1$, to parameters $W = 0.99$, $\varphi = 0.02$, and follows the steps enumerated in Table 25.3 with $p$ representing an agent (particle) in the population.

---

**TABLE 25.2**

**Initial Constraints of the Particle Swarm Optimization**

| | Constraints |
|---|---|
| 1 | $a \le b \le c \le d$ |
| 2 | $\min x_{ij} - 0.25 \mid \max x_{ij} - \min x_{ij} \mid \le a \le \min x_{ij}$ |
| 3 | $\min x_{ij} \le b \le \max x_{ij}; \min x_{ij} \le c \le \max x_{ij}$ |
| 4 | $\max x_{ij} \le d \le \max x_{ij} + 0.25 \mid \max x_{ij} - \min x_{ij} \mid$ |

---

**TABLE 25.3**

**Particle Swarm Optimization Algorithm**

| Step | Description |
|---|---|
| 1 | Randomly distribute particle swarm (or swarm of agents) in the search space. |
| 2 | Evaluate the performance of each particle according to $f(x_i)$. |
| 3 | If the current position is better than previous ones, then update with the best. |
| 4 | Determine the best particle so far according to their previous and present positions. |
| 5 | Update velocities with $v_p^{t+1} = W \cdot v_p^t + \varphi \left[ \zeta_1 \left( x_{lp}^t - x_p^t \right) + \zeta_2 \left( x_g^t - x_p^t \right) \right]$ $\le \dfrac{\left\| \max x_{ij} - \min x_{ij} \right\|}{100}.$ |
| 6 | Update positions of particles according to $x_p^{t+1} = x_p^t + v_p^{t+1}$. |
| 7 | Repeat from (2) until $f(x_i^*) < \dfrac{\left\| \max x_{ij} - \min x_{ij} \right\|}{500}$ or iterations = 2000. |

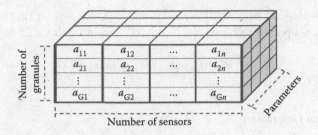

**FIGURE 25.10** Three-dimensional array containing a granular structure.

The process results in a granular structure described by an array of dimensions $G \times n \times 4$ as shown in Figure 25.10. Although the PSO may converge to a "best" result, the irregularities introduced by the data collection step make it necessary to employ a coherence operation to obtain granular structures that comply with the Ruspini condition in Equation 25.1. The advantage of using PSO is the flexibility it provides to vary the computation power invested in the aggregation process.

### 25.3.1.3 Coherence Operation

The coherence operation adjusts parameters $P = [a, b, c, d]$ of each fuzzy set $\mu_{X_i^\alpha}$ by determining their similarity or adjacency, and performing operations on these parameters in each case. It performs searches of fuzzy sets that are similar or adjacent, making use of its descriptive parameters. The coherence operation separates the search for similar or adjacent conditions by employing two suboperations: the *similarity operation* and the *adjacency operation*. Each suboperation separately performs searches in the granular structure (shown in Figure 25.10) described by parameters $P = [a, b, c, d]$ for each sensor and situation. The following paragraphs describe the suboperations.

### 25.3.1.4 Similarity Operation

The similarity operation searches for fuzzy sets in each sensor for all situations. The search identifies those sets that comply with specific similarity conditions based on the parameters illustrated in Figure 25.11.

Two sets of parameters are defined in Figure 25.11: $P_r = [a_r, b_r, c_r, d_r]$ and $P_s = [a_s, b_s, c_s, d_s]$. During the search, $P_r$ serves as the set of parameters that describes the current *reference* fuzzy set of a given situation for a particular sensor. Each fuzzy set is used as the reference set during the search of similar fuzzy sets in each sensor. The fuzzy sets described by $P_s$ are those to which $P_r$ is compared. The algorithm explores the entire partition for each sensor comparing the reference parameters $P_r$ to all other parameters $P_s$ being searched. Similar sets are identified when:

$$a_r < \langle P_s \rangle < d_r \tag{25.7}$$

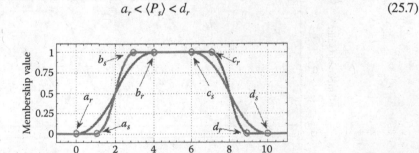

**FIGURE 25.11** Plot of two similar fuzzy sets.

with $\langle P_s \rangle$ being the average of parameters $a_s$, $b_s$, $c_s$, and $d_s$. Once similar fuzzy sets $P_{r,s}$ have been identified, their value is updated to $P'_{r,s} = \left[ a'_{r,s}, b'_{r,s}, c'_{r,s}, d'_{r,s} \right]$ as in Equation 25.8.

$$P'_{r,s} = \left[ \min\left(a_{r,s}\right), \langle b_{r,s} \rangle, \langle c_{r,s} \rangle, \max\left(d_{r,s}\right) \right] \qquad (25.8)$$

### 25.3.1.5 Adjacency Operation

The adjacency operation, in analogy to the similarity operation, searches for fuzzy sets in the partition of each sensor. In this case, however, the search will focus on the fuzzy sets adjacent to the reference fuzzy set defined by parameters $P_r = [a_r, b_r, c_r, d_r]$, shown as the central fuzzy set in Figure 25.12.

Two other fuzzy sets are shown in Figure 25.12 to the right and to the left, with parameters $P_R = [a_R, b_R, c_R, d_R]$ and $P_L = [a_L, b_L, c_L, d_L]$, respectively. The adjacency operation makes use of these parameters to search for fuzzy sets in each sensor that are adjacent to a reference fuzzy set in all situations. As with the similarity operation, here again each fuzzy set in the partition serves as the reference in one opportunity. However, in this case the condition used to identify adjacent fuzzy sets to the right and to the left are expressed as inequalities in Equations 25.9 and 25.10, respectively.

$$b_r < a_R < d_r \qquad (25.9)$$

$$a_r < d_L < c_r \qquad (25.10)$$

Once adjacent fuzzy sets have been identified, the parameters of the central fuzzy set are updated to $P'_{RL} = [a_{RL}, b_{RL}, c_{RL}, d_{RL}]$ as in Equation 25.11.

$$P'_{RL} = \left[ \langle a_r, c_L \rangle, \langle b_r, d_L \rangle, \langle c_r, a_R \rangle, \langle d_r, d_R \rangle \right] \qquad (25.11)$$

The terms $\langle a_r, c_L \rangle$, $\langle b_r, d_L \rangle$, $\langle c_r, a_R \rangle$, and $\langle d_r, b_R \rangle$ are the average of a parameter in $P_r$ and the corresponding parameters of all adjacent fuzzy sets found to the right or to the left. Such correspondence is illustrated in Figure 25.12. The purpose of the adjacency operation is to obtain fuzzy partitions that comply with the Ruspini condition in Equation 25.1.

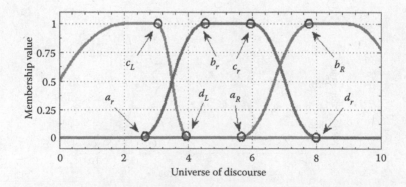

**FIGURE 25.12**  Plot of three adjacent fuzzy sets.

## 25.3.2 NUMERICAL EXAMPLE ON THE AQUATIC HABITAT MODEL

The model of the aquatic habitat [30] was used to perform simulations of anomalies that exhibit transitions between various operating conditions. The purpose was to operate under all possible situations so that data could be collected. This example makes use of two sensors: DO and pH. Possible levels of pH are high, good, or low levels, while DO levels are good or low, resulting in six possible situations. Expert human operators were modeled as a prototype granular structure to collect data for confidence values greater than 0.1. They read a different situation every 5 minutes throughout 21 days, allowing for each situation to be monitored every 30 minutes. The data obtained are processed with steps A, B, C, and D of Figure 25.1.

### 25.3.2.1 Results and Discussion

Figure 25.13 shows three 3-D graphs comparing results obtained from the sensor fusion algorithm with the prototype granular structure. Each situation is defined by a different color. Figure 25.13a provides a spatial distribution of the confidence values $c_j$. The number of data points collected in each situation is not uniform. The algorithm obtains granules independently of the number of data points. Figure 25.13b shows the resulting granules. The output of step (b) is processed with a coherence operation based on similarity and proximity, resulting in Figure 25.13c.

The lack of uniformity in the distribution of data points collected by expert human operators poses a challenge to the application of tools in computational intelligence for the development of decision aids and automation systems. Special attention should be given to how experts collect data and on the number of data points needed to guarantee coherence of granular structures. With better data sets, the particle swarm optimization should arrive at solutions without excessive overlaps or holes, as those shown in Figure 25.13b. However, the result also exhibits regularity in the distribution of the granules, even if some situations register a few number of data inputs. This regularity can be observed when comparing Figure 25.13b to the prototype granular structure used to model the SKB of expert human operators. Another question related to quality of data sets is how parameters of the particle swarm may compensate for the lack of human inputs. One advantage of making use of PSO is the flexibility it provides to increase computing power to arrive at solutions to the optimization problem. This supports and suggests the need for research that may help to characterize the performance of the particle swarm aggregation algorithm working under different search parameters, particle population sizes, and sizes of data sets. A final observation can be made on the borders of the output granular structures as compared to the prototype. Because the granules obtained are the product of the data sets used, they are not able to define situations beyond those values. In other words, those areas not covered by the granules represent *unknown* situations. This implies that under such conditions a nonexpert human operator should request assistance from experts, either to record new assessments in the data sets or to evaluate the need for intervention.

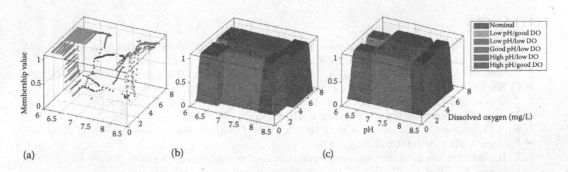

**FIGURE 25.13**    Steps a, b, and c and prototype granular structure.

### 25.3.2.2  Remarks from the Numerical Example

This subsection presented a numerical example of the granular multisensor data fusion method. It collected assessments from simulated expert human operators to generate a granular structure suitable for decision support tools and automation systems. The methodology presented offers an approach to overcome the combinatorial explosion of merging information from a large number of sensors. It makes use of human–system interaction to generate data sets that are processed with tools in computational intelligence. Expert assessments define the operational condition of the system with a subjective assessment of its situation. An algorithm based on particle swarm optimization obtains a representation of the SKB of human experts.

## 25.4  SUMMARY AND FUTURE DIRECTIONS

Issues in the integration of humans and automation pose challenges that will continue to evolve with innovations in sensing, processing, and actuation. Situation awareness research in particular will play an important role in preventing accidents and promoting the safety and reliability of systems in which humans and automation are meant to interact. This research presented an effort that aims to contribute to situation-oriented and user-centered design approaches. It does so by incorporating human expert assessments in the development of the perception function of agents developed using the FAM-based architecture.

This chapter introduced a granular approach to the automation of a small-scale RLSS that enhances situation observability [31]. It made use of the FAM-based agent architecture as an approach that enables the interaction of humans and automation. Although the work presented here focused on applications in regenerative life support systems, the granular approach and observations on the perception function of the FAM-based agent architecture may also be relevant in fields of application beyond life support systems. One of such observations is that the noninteractiveness of the FAM makes the results obtained in this work scalable to multiple sensors.

The granular approach to automation has the advantage of introducing flexibility in automation design in a way that allows for coordination with tasks performed by human operators. One limitation of this particular approach is found in its dependence on data sets containing situation assessments collected by human experts with the objective of obtaining a representation of their knowledge.

Future research directions focus on increasing the degree of automation for the development agents based on the FAM-based agent architecture. In particular, it looks at translating the approach presented in this work into tools that facilitate the steps of (1) collecting and processing data from a number of human experts; (2) generating crowdsourced SKBs as inputs to the FAM-based agent architecture; (3) developing the controllers and correspondence function; and (4) implementing the approach in compliance with industry certification standards for code generation, design verification and validation, and deployment.

## REFERENCES

1. H. Durrant-Whyte, and T. C. Henderson. *Springer Handbook of Robotics*, Chapter 25, pp. 585–614. Springer-Verlag, Berlin, Heidelberg, 2008.
2. T. Ross. *Fuzzy Logic with Engineering Applications*, 3rd edition. John Wiley & Sons, Hoboken, NJ, 2010.
3. G. Shafer. *A Mathematical Theory of Evidence*. Princeton University Press, Princeton, NJ, 1976.
4. J. Kohlas, and P.-A. Monney. *A Mathematical Theory of Hints: An Approach to the Dempster-Shafer Theory of Evidence*. Springer-Verlag, Berlin, Heidelberg, 1995.
5. R. R. Yager, and L. Liu, editors. *Classic Works of the Dempster-Shafer Theory of Belief Functions*. Springer-Verlag, Berlin, Heidelberg, 2008.
6. T. B. Sheridan. *Humans and Automation: System Design and Research Issues*. John Wiley & Sons, Hoboken, NJ, 2002.

7. B. Woolford, and F. Mount. *Handbook of Human Factors and Ergonomics*, 3rd edition, Chapter 34, pp. 929–944. John Wiley & Sons, Hoboken, NJ, 2006.
8. P. Eckart. *Spaceflight Life Support and Biospherics*, 2nd edition. Springer, New York, 2010.
9. S. Doll, and P. Eckart. *Human Spaceflight Mission Analysis and Design*, Chapter 17, pp. 539–573. McGraw-Hill, New York, 1999.
10. R. W. Humble. *Human Spaceflight Mission Analysis and Design*, Chapter 25, pp. 797–810. McGraw-Hill, New York, 1999.
11. C. M. McCleskey. Space shuttle operations and infrastructure: A systems analysis of design root causes and effects. Technical report, National Aeronautics and Space Administration, John F. Kennedy Space Center, FL, April 2005. NASA/TP-2005-211519.
12. J. R. Wilson. What's next for U.S. human spaceflight? *Aerospace America*, 50(1):24–31, 2012.
13. M. B. Abney, L. A. Miller, and T. Williams. Sabatier reactor system integration with microwave plasma methane pyrolysis postprocessor for closed-loop hydrogen recovery. In *Proceedings of the 40th International Conference on Environmental Systems*, Barcelona, Spain, July 11–15, 2010. AIAA.
14. National Aeronautics and Space Administration. *NASA Facts: International Space Station Environmental Control and Life Support System*. NASA, Huntsville, AL, May 2008.
15. S. Skogestad, and I. Postlethwaite. *Multivariable Feedback Control: Analysis and Design*, 2nd edition. Wiley-Interscience, Hoboken, NJ, 2005.
16. K. J. Astrom, and B. Wittenmark. *Adaptive Control*, 2nd edition. Dover Publications, Mineola, NY, 2008.
17. N. Hovakimyan, and C. Cao. *L1 Adaptive Control Theory: Guaranteed Robustness with Fast Adaptation*. Society for Industrial & Applied Mathematics, September 2010.
18. L. Vu, and D. Liberzon. Supervisory control of uncertain linear time-varying systems. *Automatic Control, IEEE Transactions on*, 56(1):27–42, 2011.
19. H. Lin, and P. J. Antsaklis. Stability and stabilizability of switched linear systems: A survey of recent results. *Automatic Control, IEEE Transactions on*, 54(2):308–322, 2009.
20. D. Liberzon. *Switching in Systems and Control*. Birkhäuser, Boston, MA, 2003.
21. J. P. Hespanha, D. Liberzon, and A. S. Morse. Overcoming the limitations of adaptive control by means of logic-based switching. *Systems & Control Letters*, 49(1):49–65, 2003.
22. M. Endsley, and D. G. Jones. *Designing for Situation Awareness: An Approach to User-Centered Design*, 2nd edition. Taylor & Francis Group, Boca Raton, FL, 2012.
23. G. Drayer, and M. Strefezza. A FAM-based agent for a ball and beam. In *IEEE-SOFA 2007. 2nd IEEE International Workshop on Soft Computing Applications*, pp. 89–94, Gyula, Hungary, and Oradea, Romania, August 21–23, 2007.
24. G. Drayer, and M. Strefezza. Integral control with error modulation in a FAM-based agent for a furuta inverted pendulum. In *2008 IEEE World Congress on Computational Intelligence*, pp. 1590–1597, Hong Kong, June 1–6, 2008.
25. G. Drayer, and A. Howard. A FAM-based switched control approach for the automation of bioregenerative life support systems. In *41st International Conference on Environmental Systems*, Portland, OR, July 2011. AIAA.
26. L. A. Zadeh. Toward a theory of fuzzy information granulation and its centrality in human reasoning and fuzzy logic. *Fuzzy Sets and Systems*, 90(2):111–127, 1997.
27. Y. Y. Yao. Granular computing: Basic issues and possible solutions. In *Proceedings of the 5th Joint Conference on Information Sciences*, vol. I, pp. 186–189, 2000.
28. Y. Y. Yao. The art of granular computing. In *Proceedings of the International Conference on Rough Sets and Emerging Intelligent Systems Paradigms*, pp. 101–112, 2007.
29. Y. Y. Yao. Perspectives of granular computing. In *Proceedings of 2005 IEEE International Conference on Granular Computing*, vol. 1, pp. 85–90, 2005.
30. G. Drayer, and A. Howard. Modeling and simulation of an aquatic habitat for bioregenerative life support research. *Acta Astronautica*, 93:138–147, 2014.
31. G. Drayer, and A. Howard. A granular approach to the automation of bioregenerative life support systems that enhances situation awareness. In *Proceedings of the 2nd IEEE Conference on Cognitive Methods in Situation Awareness and Decision Support*, March 2012.
32. L. A. Zadeh. Fuzzy sets. *Information and Control*, 8(3):338–353, 1965.
33. E. H. Ruspini. A new approach to clustering. *Information and Control*, 15(1):22–32, 1969.
34. M. Timmons, and J. Ebeling. *Recirculating Aquaculture*. North Eastern Regional Aquaculture Center, College Park, MD, 2007.

35. T. McCoy, S. Flint, J. Straub, D. Gazda, and J. Schultz. The story behind the numbers: Lessons learned from the integration of monitoring resources in addressing an ISS water quality anomaly. In *41st International Conference on Environmental Systems*, 2011.
36. R. C. Eberhart, J. Kennedy, and Y. Shi. *Swarm Intelligence*. Morgan Kaufmann, Burlington, MA, 2001.
37. A. Doan, R. Ramakrishnan, and A. Y. Halevy. Crowdsourcing systems on the world-wide web. *Communications of ACM*, 54:86–96, 2011.
38. N. R. Prestopnik, and K. Crowston. Gaming for (citizen) science: Exploring motivation and data quality in the context of crowdsourced science through the design and evaluation of a social-computational system. In *e-Science Workshops, IEEE International Conference on*, pp. 28–33, 2011.

# 26 Evaluating Image Fusion Performance

## *From Metrics to Cognitive Assessment*

### Zheng Liu and Erik Blasch

## CONTENTS

## 26.1 INTRODUCTION

Image fusion integrates information from multiple input images acquired by different imaging modalities or by the same modality operated in different conditions. Various applications on multisensor image fusion have been reported [1,2]. Pixel-level image fusion, also known as image-in image-out fusion, is to generate a composite image that should contain all the complementary information from the input images. Numerous pixel-level algorithms have been proposed for specific tasks and the fusion performance needs to be verified, assessed, validated, and compared. A comprehensive assessment will tell which fusion algorithm performs better for a specific set of

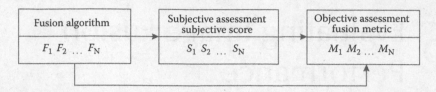

**FIGURE 26.1**    Fusion performance assessment and fusion metric validation. $F_i$, fusion algorithm; $S_i$, subjective score; and $M_i$, metric value.

sensors, targets, and environment data or applications [3]. Currently, the assessment is conducted with either subjective or objective methods. The subjective methods rely on human perception based measures whereas objective methods are based on computational models, which count the amount of image features, contents, or information transferred from inputs to the fused result [4]. These computational models are also known as fusion metric, which can reveal certain inherent properties of the fusion process or the fused image. However, multiple metrics sometimes may give contrary judgments and the same metric can even reach different conclusions for different data sets. Thus, a statistical analysis over all the data sets is necessary to determine the robust performance of a specific fusion algorithm [5].

Image fusion assessment includes the algorithm parameters, subjective scores, and objective metrics as illustrated in Figure 26.1. Both the objective and subjective assessment can be applied to assess the performance of a specific fusion algorithm. It is preferable that the scores and metrics correlate. On the one hand, extensive subjective assessment is not always feasible because of the high cost and the difficulties in the control of varied human factors (e.g., individual differences, biases, and sensing abilities). On the other hand, objective fusion metrics need to be validated with corresponding subjective scores over many operating conditions. Thus, statistical analysis can be performed to compare different algorithms over multiple data sets [6–10]. Specifically, different algorithms can be compared with a set of nonparametric statistical tests. With these tests, it is possible to tell which fusion algorithm performs better under certain conditions. When a new algorithm is proposed, it is good to have a baseline method from which to compare the current performance against other techniques in a similar scenario.

This chapter first describes the role and implementation of pixel-level fusion in Section 26.2.1. Section 26.2.2 presents the state-of-the-art of objective metrics and two trials on subjective assessment are described in Section 26.2.3. The role of natural imagery interpretability rating scales (NIIRSs) on fusion performance assessment is explored and discussed. The procedure to apply statistical approaches to analyze both subjective and objective fusion assessment data is illustrated in Section 26.2.5. Section 26.3 gives the experimental results on the assessment of fused night vision data sets. In Section 26.4, future research perspectives on cognitive assessment are highlighted and this chapter is concluded and summarized.

## 26.2    IMAGE FUSION PERFORMANCE ASSESSMENT

### 26.2.1    Image Fusion: Role and Implementation

Image fusion can be implemented at different levels, for example, pixel, feature, and decision levels as illustrated in Figure 26.2. The choice of fusion algorithm depends on the requirements of the specific application. Pixel-level fusion generates a composite from the input images, integrates the complementary information into the fused result, and reduces the uncertainty with the redundant information. Pixel-level fusion can be categorized into discriminative and combinative fusion. For discriminative pixel-level fusion, the fused image is a thematic map, which presents each pixel with predefined classes. Feature-level fusion combines the extracted features (e.g., lines, regions, shapes)

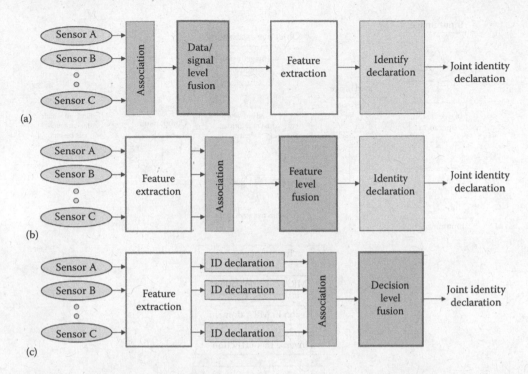

**FIGURE 26.2** Data fusion at three different levels. (a) Signal-level fusion, (b) feature-level fusion, and (c) decision-level fusion.

to determine the match of features in the fused images. An example is visual perception of a vehicle mapped to the location of the engine heat from the engine in an infrared image. Finally, decision-level fusion works with the outputs of the declarations of the image content [11,12].

In remote sensing, one application of image fusion is to place landscape in an image into categories, such as soil, forest, agriculture, water, and so forth, based on the spectral information or signature captured by different image modalities. Although the extent of the processing is at the decision or classification level for each pixel, the data resolution is still at the pixel level. The assessment can be done by comparing with the ground truth prepared manually by experts. With the established criteria in classification research, the fusion performance can be quantitatively evaluated.

The application of combinative fusion includes situation awareness and environmental perception, such as surveillance and driving assistant during the night. As illustrated in Figure 26.3, through the pixel-level fusion, both the foreground object and background scene are highlighted and enhanced in the fused image. Thus, abundant semantic context and information is available to interpret the scene presented in the fused image.

There are numerous algorithms proposed for pixel-level image fusion [1]. One implementation is in the multiresolution transform domain known as multiresolution analysis (MRA) based fusion as shown in Figure 26.3b. The MRA algorithms perform an image decomposition using a variety of structured image pyramids and wavelet transforms [1]. The key technique is the coefficient combination in the transform domain, which is a method of image fusion. Using these coefficients, the image has to be reconstructed using the inverse of the image pyramid or wavelet transform chosen for decomposition. Finally, a rendered fused image is available to the user and the optimized image is subject to user, application, and data availability needs. In this chapter, the fusion experiments on night vision images employ the MRA approaches. As shown in Figure 26.3a, the objective fusion metrics can be applied to assess the fusion performance while the subjective evaluation can be used to validate both the fusion algorithms and the fusion metrics.

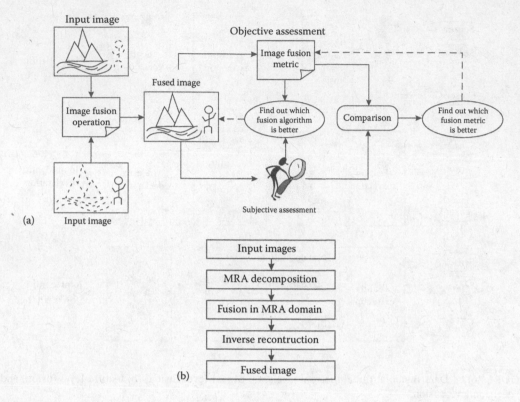

(a)

(b)

**FIGURE 26.3** Combinative image fusion. (a) Procedure of fusion and fusion performance assessment and (b) fusion with multiresolution analysis.

## 26.2.2 Objective Fusion Metrics

Objective assessment of image fusion builds a computational model that determines the fusion metric to evaluate the fusion performance. For most image fusion applications, there is no ground truth available for comparison. Thus, the objective assessment evaluates how much information is transferred from the input images to the fused image and the quantification of such transfer serves as the metric. Different approaches have been proposed for image fusion metrics [4]. These approaches can be classified into roughly four categories: information theory based metric, image feature based metric, image structural similarity based metric, and human perception inspired fusion metric. Details are described next for the four categories.

### 26.2.2.1 Information Theory Based Metrics

#### 26.2.2.1.1 Normalized Mutual Information ($Q_{MI}$)

Mutual information (MI) is a quantitative measure of the mutual dependence of two variables. The definition of mutual information for two discrete random variables $U$ and $V$ is

$$\text{MI}(U;V) = \sum_{v \in V} \sum_{u \in U} p(u,v) \log_2 \frac{p(u,v)}{p(u)p(v)} \tag{26.1}$$

where $p(u, v)$ is the joint probability distribution function of $U$ and $V$, and $p(u)$ and $p(v)$ are the marginal probability distribution functions of $U$ and $V$ respectively. In fact, MI quantifies the distance between the joint distribution of $U$ and $V$, that is, $p(u, v)$, and the joint distribution when $U$ and $V$ are independent, that is, $p(u)\, p(v)$. Mutual information can be equivalently expressed with joint entropy $\{H(U, V)\}$ and marginal entropy $\{H(U), H(V)\}$ of the two variables $U$ and $V$ as

$$\mathrm{MI}(U,V) = H(U) + H(V) - H(U,V) \tag{26.2}$$

where

$$H(U) = -\sum_u p(u)\log_2 p(u)$$

$$H(V) = -\sum_v p(v)\log_2 p(v)$$

$$H(U,V) = -\sum_{u,v} p(u,v)\log_2 p(u,v)$$

Qu et al. [13] used the summation of the MI between the fused image $F(i, j)$ and two input images, $A(i, j)$ and $B(i, j)$, to represent the difference in quality. The expression of MI-based fusion performance measure $M_F^{AB}$ is

$$M_F^{AB} = \mathrm{MI}(A,F) + \mathrm{MI}(B,F)$$

$$= \sum_{i,j}\left( h_{AF}(i,j)\log_2\frac{h_{AF}(i,j)}{h_A(i)h_F(j)} + h_{BF}(i,j)\log_2\frac{h_{BF}(i,j)}{h_B(i)h_F(j)} \right) \tag{26.3}$$

where $h_{AF}(i, j)$ indicates the normalized joint gray level histogram of images $A(i, j)$ and $F(i, j)$; $h_K(i, j)$ ($K = A$, $B$, and $F$) is the normalized marginal histogram of image $A$, $B$, or $F$, respectively.

One problem with Equation 26.3 is that it mixes two joint entropies measured at different scales. This may cause instability of the measure and bias the measure toward the source image with the highest entropy [14]. Thus, Hossny et al. [14] modified Equation 26.3 as

$$Q_{\mathrm{MI}} = 2\left[ \frac{\mathrm{MI}(A,F)}{H(A)+H(F)} + \frac{\mathrm{MI}(B,F)}{H(B)+H(F)} \right] \tag{26.4}$$

Hossny's definition is used in the experiments.

### 26.2.2.1.2  Fusion Metric Based on Tsallis Entropy ($Q_{TE}$)

Cvejic and Nava suggested using Tsallis entropy to define the fusion metric [15,16]. Tsallis entropy is a divergence measure of the degree of dependence between two discrete random variables. For the input image $A(i, j)$ and fused image $F(i, j)$, the Tsallis entropy is [15,16]

$$I^q(A,F) = \frac{1}{1-q}\left( 1 - \sum_{i,j}\frac{h_{AF}(i,j)^q}{h_F(j)h_A(i)^{q-1}} \right) \tag{26.5}$$

where $q$ is real value and $q \neq 1$. A quality metric of order $q$ can be defined as [15]

$$Q_{TE}^q = I^q(A,F) + I^q(B,F) \tag{26.6}$$

or as a normalized value [16]

$$Q_{TE} = \frac{I^q(A,F) + I^q(B,F)}{H^q(A) + H^q(B) - I^q(A,B)} \tag{26.7}$$

The use of Renyi entropy was also suggested by Zheng et al. [17]. However, the MI-based metric still needs a reference value to compare differences. It is not possible to tell in advance if a fused image with a given MI value is good or not, so a reference point is a must. Moreover, the MI-based approach is sensitive to impulsive noise and is subject to significant change in the presence of additive Gaussian noise.

### 26.2.2.1.3  Nonlinear Correlation Information Entropy ($Q_{NCIE}$)

For two discrete variables $U = \{u_i\}_{1 \leq i \leq N}$ and $V = \{v_i\}_{1 \leq i \leq N}$, the nonlinear correlation coefficient (NCC) is defined as [18]:

$$NCC(U,V) = H'(U) + H'(V) - H'(U,V) \tag{26.8}$$

which is similar to the definition of the mutual information between $U$ and $V$ in Equation 26.2. Considering NCC $(A,B)$ for image $A$ and $B$, the entropies are defined as

$$H'(A,B) = -\sum_{i=1}^{b}\sum_{j=1}^{b} h_{AB}(i,j)\log_b h_{AB}(i,j) \tag{26.9}$$

$$H'(A) = -\sum_{i=1}^{b} h_A(i)\log_b h_A(i) \tag{26.10}$$

$$H'(B) = -\sum_{i=1}^{b} h_B(i)\log_b h_B(i) \tag{26.11}$$

where $b$ is determined by the intensity level, that is, $b = 256$. A nonlinear correlation matrix of the input image $A(i,j)$, $B(i,j)$ and fused image $F(i,j)$ is defined as

$$R = \begin{pmatrix} NCC_{AA} & NCC_{AB} & NCC_{AF} \\ NCC_{BA} & NCC_{BB} & NCC_{BF} \\ NCC_{FA} & NCC_{FB} & NCC_{FF} \end{pmatrix} = \begin{pmatrix} 1 & NCC_{AB} & NCC_{AF} \\ NCC_{BA} & 1 & NCC_{BF} \\ NCC_{FA} & NCC_{FB} & 1 \end{pmatrix} \tag{26.12}$$

The eigenvalue of the nonlinear correlation matrix $R$ is $\lambda_i$ ($i = 1, 2, 3$). Therefore, the nonlinear correlation information entropy $Q_{\text{NCIE}}$ can be obtained:

$$Q_{\text{NCIE}} = 1 + \sum_{i=1}^{3} \frac{\lambda_i}{3} \log_b \frac{\lambda_i}{3} \tag{26.13}$$

### 26.2.2.2  Image Feature Based Metrics

Another type of assessment is implemented by measuring how the features are transferred from the input images to the fused one.

#### 26.2.2.2.1  Gradient Based Fusion Performance ($Q_G$)

Xydeas and Petrovic [19] proposed a metric to evaluate the amount of edge information, which is transferred from input images to the fused image. A Sobel edge operator is applied to get the edge strength of input image $A(i, j)$, $g_A(i, j)$ and orientation $\alpha_A(i, j)$.

$$g_A(i, j) = \sqrt{s_A^x(i, j)^2 + s_A^y(i, j)^2} \tag{26.14}$$

$$\alpha_A(i, j) = \tan^{-1}\left(\frac{s_A^x(i, j)}{s_A^y(i, j)}\right) \tag{26.15}$$

where $s_A^x(i, j)$ and $s_A^y(i, j)$ are the convolved results with the horizontal and vertical Sobel templates [19]. The relative strength ($G^{AF}$) and orientation values ($\Delta^{AF}$) between input image $A$ and fused image $F$ are

$$G^{AF}(i, j) = \begin{cases} \dfrac{g_F(i, j)}{g_A(i, j)}, & g_A(i, j) > g_F(i, j) \\[3mm] \dfrac{g_A(i, j)}{g_F(i, j)}, & \text{Otherwise} \end{cases} \tag{26.16}$$

$$\Delta^{AF}(i, j) = 1 - \frac{|\alpha_A(i, j) - \alpha_F(i, j)|}{\pi/2} \tag{26.17}$$

The edge strength and orientation preservation values can be derived:

$$Q_g^{AF}(i, j) = \frac{\Gamma_g}{1 + e^{\kappa_g\left(G^{AF}(i,j) - \sigma_g\right)}} \tag{26.18}$$

$$Q_\alpha^{AF}(i, j) = \frac{\Gamma_\alpha}{1 + e^{\kappa_\alpha\left(\Delta^{AF}(i,j) - \sigma_\alpha\right)}} \tag{26.19}$$

The constants $\Gamma_g$, $\kappa_g$, $\sigma_g$ and $\Gamma_\alpha$, $\kappa_\alpha$, $\sigma_\alpha$ determine the shape of the sigmoid functions used to form the edge strength and orientation preservation value [19]. Edge information preservation value is then defined as

$$Q^{AF}(i,j) = Q_g^{AF}(i,j)Q_\alpha^{AF}(i,j) \tag{26.20}$$

The final assessment is obtained from the weighted average of the edge information preservation values.

$$Q_G = \frac{\sum_{n=1}^{N}\sum_{m=1}^{M}\left[Q^{AF}(i,j)w^A(i,j) + Q^{BF}(i,j)w^B(i,j)\right]}{\sum_{n=1}^{N}\sum_{m=1}^{M}\left(w^A(i,j) + w^B(i,j)\right)} \tag{26.21}$$

where the weighting coefficients are defined as: $w^A(i,j) = [g_A(i,j)]^L$ and $w^B(i,j) = [g_B(i,j)]^L$, respectively. Here, $L$ is a constant value.

### 26.2.2.2.2  Image Fusion Metric Based on a Multiscale Scheme ($Q_M$)

Wang and Liu [20] proposed a metric, which is implemented with a two-level Haar wavelet as illustrated in Figure 26.4. The edge information is retrieved from the high- and band-pass

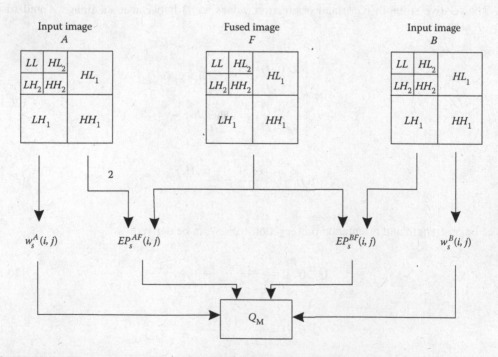

**FIGURE 26.4**  The implementation of a multiscale fusion metric.

components of the decomposition. At each level $s$, for input image $A(i, j)$ and fused image $F(i, j)$, there is

$$\psi_{H_s}^{AF}(m,n) = \exp\left(-\left|LH_s^A(m,n) - LH_s^F(m,n)\right|\right)$$

$$\psi_{V_s}^{AF}(m,n) = \exp\left(-\left|HL_s^A(m,n) - HL_s^F(m,n)\right|\right)$$

$$\psi_{D_s}^{AF}(m,n) = \exp\left(-\left|HH_s^A(m,n) - HH_s^F(m,n)\right|\right)$$

Then the global edge preservation value at scale $s$ can be derived as

$$EP_s^{AF} = \frac{\psi_{H_s}^{AF}(m,n) + \psi_{V_s}^{AF}(m,n) + \psi_{D_s}^{AF}(m,n)}{3} \tag{26.22}$$

A normalized performance metric weighted by $w_s^A(m,n)$ and $w_s^B(m,n)$ at scale $s$ is defined as

$$Q_s^{AB/F} = \frac{\displaystyle\sum_m \sum_n \left(EP_s^{AF}(m,n)w_s^A(m,n) + EP_s^{BF}(m,n)w_s^B(m,n)\right)}{\displaystyle\sum_m \sum_n \left(w_s^A(m,n) + w_s^B(m,n)\right)} \tag{26.23}$$

The high-frequency energy of the input images is used as a weight coefficient

$$w_s^A(m,n) = LH_{A_s}^2(m,n) + HL_{A_s}^2(m,n) + HH_{A_s}^2(m,n) \tag{26.24}$$

Similarly, $w_s^B(m,n)$ can be derived for image $B$. The overall metric is obtained by combining the measurement at different scales using [20]:

$$Q_M = \prod_{s=1}^{N}\left(Q_s^{AB/F}\right)^{\alpha_s} \tag{26.25}$$

where $\alpha_s$ is a constant to adjust the relative importance of different scales.

### 26.2.2.2.3 Image Fusion Metric Based on Spatial Frequency ($Q_{SF}$)

Zheng et al. [21] used "spatial frequency" to measure the activity level of an image $I(i, j)$ defined as follows:

$$SF = \sqrt{(RF)^2 + (CF)^2 + (MDF)^2 + (SDF)^2} \tag{26.26}$$

For an image $I(i, j)$, there are four gradients:

$$RF = \sqrt{\frac{1}{MN}\sum_{i=1}^{M}\sum_{j=2}^{N}\left[I(i,j) - I(i, j-1)\right]^2} \tag{26.27}$$

$$CF = \sqrt{\frac{1}{MN} \sum_{j=1}^{N} \sum_{i=2}^{M} \left[ I(i,j) - I(i-1,j) \right]^2} \qquad (26.28)$$

$$MDF = \sqrt{w_d \frac{1}{MN} \sum_{i=2}^{M} \sum_{j=2}^{N} \left[ I(i,j) - I(i-1,j-1) \right]^2} \qquad (26.29)$$

$$SDF = \sqrt{w_d \frac{1}{MN} \sum_{j=1}^{N-1} \sum_{i=2}^{M} \left[ I(i,j) - I(i-1,j+1) \right]^2} \qquad (26.30)$$

Here, RF, CF, MDF, and SDF are the four first-order gradients along four directions. Distance weight $w_d$ is $1/\sqrt{2}$. The four reference gradients are obtained by taking the maximum of absolute gradient values between input image $A$ and $B$ along four directions [21]:

$$Grad^D(I_R(i,j)) = \max\{abs[Grad^D(I_A(i,j))], abs[Grad^D(I_B(i,j))]\} \qquad (26.31)$$

where there is $D = \{H, V, MD, SD\}$, which represents horizontal, vertical, main diagonal, and secondary diagonal, respectively. With the reference gradients substituting the differences in Equations 26.27 to 26.30, the four directional references, $RF_R$, $CF_R$, $MDF_R$, and $SDF_R$, can be calculated. Thus, $SF_R$ can be derived from Equation 26.26. Finally, the ratio of SF error (metric $Q_{SF}$) is defined as

$$Q_{SF} = (SF_F - SF_R)/SF_R \qquad (26.32)$$

### 26.2.2.2.4  Image Fusion Metric Based on Phase Congruency ($Q_P$)

Zhao and Liu used the phase congruency, which provides an absolute measure of image feature, to define an evaluation metric [22,23]. In Ref. [22], the principal (maximum and minimum) moments of the image phase congruency were employed to define the metric, because the moments contain the information for corners and edges. The $Q_P$ metric is defined as a product of three correlation coefficients:

$$Q_P = (P_p)^\alpha (P_M)^\beta (P_m)^\gamma \qquad (26.33)$$

where $p$, $M$, $m$ refers to phase congruency ($p$) and maximum and minimum moments respectively and there are

$$P_p = \max\left( C_{AF}^p, C_{BF}^p, C_{SF}^p \right)$$

$$P_M = \max\left( C_{AF}^M, C_{BF}^M, C_{SF}^M \right)$$

$$P_m = \max\left( C_{AF}^m, C_{BF}^m, C_{SF}^m \right)$$

Herein, $C_{xy}^k$, $\{k|p, M, m\}$ stands for the correlation coefficients between two sets $x$ and $y$:

$$C_{xy}^k = \frac{\sigma_{xy}^k + C}{\sigma_x^k \sigma_y^k + C} \qquad (26.34)$$

$$\sigma_{xy} = \frac{1}{N-1} \sum_{i=1}^{N} (x_i - \bar{x})(y_i - \bar{y}) \qquad (26.35)$$

The suffixes $A$, $B$, $F$, and $S$ correspond to the two inputs, fused image, and maximum-select map. The exponential parameters $\alpha$, $\beta$, and $\gamma$ can be adjusted based on the importance of the three components [22].

### 26.2.2.3 Image Structural Similarity Based Metrics

The image similarity measurement is based on the evidence that the human visual system is highly adapted to structural information and a measurement of the loss of structural information can provide a good approximation of the perceived image distortion. Wang et al. [24] proposed a structural similarity index measure (SSIM) for image $A$ and $B$ defined as

$$
\begin{aligned}
\text{SSIM}(A,B) &= [l(A,B)]^\alpha [c(A,B)]^\beta [s(A,B)]^\gamma \\
&= \left( \frac{2\mu_A \mu_B + C_1}{\mu_A^2 + \mu_B^2 + C_1} \right)^\alpha \left( \frac{2\sigma_A \sigma_B + C_2}{\sigma_A^2 + \sigma_B^2 + C_2} \right)^\beta \left( \frac{\sigma_{AB} + C_3}{\sigma_A \sigma_B + C_3} \right)^\gamma
\end{aligned}
\qquad (26.36)
$$

where $\mu_A$ and $\mu_B$ are the average values of image $A(i,j)$ and $B(i,j)$, $\sigma_A$, $\sigma_B$, and $\sigma_{AB}$ are the variance and covariance respectively [24]. $l(A, B)$, $c(A, B)$, and $s(A, B)$ in Equation 26.36 are the luminance, contrast, and correlation components respectively. The parameters $\alpha$, $\beta$, and $\gamma$ are used to adjust the relative importance of the three components. The constant values $C_1$, $C_2$, and $C_3$ are defined to avoid the instability when the denominator are very close to zero. By setting $\alpha = \beta = \gamma = 1$ and $C_3 = C_2/2$, Equation 26.36 becomes

$$\text{SSIM}(A,B) = \frac{(2\mu_A \mu_B + C_1)(2\sigma_{AB} + C_2)}{\left(\mu_A^2 + \mu_B^2 + C_1\right)\left(\sigma_A^2 + \sigma_B^2 + C_2\right)} \qquad (26.37)$$

A previous version of this index is known as universal image quality index (UIQI) and is written as [25]:

$$
\begin{aligned}
Q(A,B) &= \frac{\sigma_{AB}}{\sigma_A \sigma_B} \frac{2\mu_A \mu_B}{\mu_A^2 + \mu_B^2} \frac{2\sigma_A \sigma_B}{\sigma_A^2 + \sigma_B^2} \\
&= \frac{4\sigma_{AB} \mu_A \mu_B}{\left(\sigma_A^2 + \sigma_B^2\right)\left(\mu_A^2 + \mu_B^2\right)}
\end{aligned}
\qquad (26.38)
$$

The following image structural similarity fusion metrics are based on these two definitions. The calculation of the quality indices is based on a sliding window approach, which moves from top left to bottom right. The SSIM and $Q$ value can be calculated locally, summed, and averaged to get the overall index.

### 26.2.2.3.1  Piella's Metric ($Q_S$)

Piella and Heijmans defined a three fusion quality index based on Wang's UIQI method [26]. Assume the local $Q(A, B|w)$ value is calculated in a sliding window $w$. There are calculation includes

$$Q_S = \frac{1}{|W|} \sum_{w \in W} \left[ \lambda(w) Q_0\left(A, F|w\right) + \left(1 - \lambda(w)\right) Q_0\left(B, F|w\right) \right] \tag{26.39}$$

$$Q_W = \sum_{w \in W} c(w) \left[ \lambda(w) Q_0\left(A, F|w\right) + (1 - \lambda(w)) Q_0\left(B, F|w\right) \right] \tag{26.40}$$

$$Q_E = Q_W(A, B, F)\, Q_W(A', B', F')^\alpha \tag{26.41}$$

where the weight $\lambda(w)$ is defined as

$$\lambda(w) = \frac{s(A|w)}{s(A|w) + s(A|w)} \tag{26.42}$$

Herein, $s(A|w)$ is a local measure of image salience. In Piella's implementation, $s(A|w)$ and $s(B|w)$ are the variance of image $A$ and $B$ within the window $w$ respectively. The coefficient $c(w)$ in Equation 26.40 is [26]:

$$c(w) = \frac{\max\left[ s(A|w), s(B|w) \right]}{\sum_{w' \in W} \left[ s(A|w'), s(B|w') \right]} \tag{26.43}$$

In Equation 26.41, $Q_W(A', B', F')$ is the $Q_w$ calculated with the edge images, that is, $A'$, $B'$, and $F'$ and $\alpha$ is a manually adjustable parameter to weight the edge-dependent information.

### 26.2.2.3.2  Cvejie's Metric $Q_C$

Cvejie et al. [27] defined a performance measure as:

$$Q_C = \sum_{w \in W} \mathrm{sim}\left(A, B, F|w\right) Q\left(A, F|w\right) + \left(1 - \mathrm{sim}\left(A, B, F|w\right)\right) Q\left(B, F|w\right) \tag{26.44}$$

where the function sim $(A, B, F|w)$ is [27]:

$$\mathrm{sim}\left(A, B, F|w\right) = \begin{cases} 0 & \text{if } \dfrac{\sigma_{AF}}{\sigma_{AF} + \sigma_{BF}} < 0 \\[2ex] \dfrac{\sigma_{AF}}{\sigma_{AF} + \sigma_{BF}} & \text{if } 0 \leq \dfrac{\sigma_{AF}}{\sigma_{AF} + \sigma_{BF}} \leq 1 \\[2ex] 1 & \text{if } \dfrac{\sigma_{AF}}{\sigma_{AF} + \sigma_{BF}} > 1 \end{cases} \tag{26.45}$$

The weighting factor depends on the similarity in spatial domain between the input images and the fused image. The higher the similarity between the input and fused image, the larger is the corresponding weighting factor.

### 26.2.2.3.3 Yang's Metric $Q_Y$

Yang et al. [28] proposed another way to use SSIM for fusion assessment:

$$Q_Y = \begin{cases} \lambda(w)\text{SSIM}(A,F|w)+(1-\lambda(w))\text{SSIM}(B,F|w), & \text{SSIM}(A,B|w) \geq 0.75 \\ \max\{\text{SSIM}(A,F|w),\text{SSIM}(B,F|w)\}, & \text{SSIM}(A,B|w) < 0.75 \end{cases} \tag{26.46}$$

The local weight $\lambda(w)$ is as defined in Equation 26.42.

### 26.2.2.4 Human Perception Inspired Fusion Metrics

#### 26.2.2.4.1 Chen–Varshney Metric $Q_{CV}$

The Chen–Varshney metric [29] consists of five steps as illustrated in Figure 26.5. Details are described below.

1. Extract edge information: The extraction of edge information is implemented by applying the Sobel edge detector to get the edge strength map $G_K(i,j)$.
2. Partition images into local regions: The images are divided into nonoverlapped local regions (windows).
3. Calculate local region saliency: The local region saliency is calculated as the summation of squares of edge intensities in the local region: $\lambda(A^W) = \sum_{w \in W} G_A(w)^\alpha$. Here, $\alpha$ is a constant.
4. Similarity measure in the local region: The measure is the mean squared value of the contrast sensitive function (CSF) filtered image $\hat{f}_A^W$.

$$D\left(I_A^W, I_F^W\right) = \frac{1}{|W|} \sum_{w \in W} \hat{f}_r^W(i,j)^2 \tag{26.47}$$

where $r$ refers to the input image A and B and $|W|$ is the number of pixels in local region $W$.

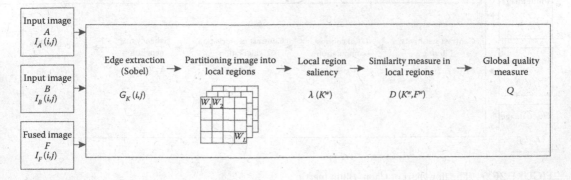

**FIGURE 26.5** Flowchart of Chen–Varshney metric.

5. Global quality measure: The global quality measure is the weighted summation over all of the nonoverlapping regions (windows).

$$Q_{CV} = \frac{\sum_{l=1}^{L} \left( \lambda \left( I_A^{W_l} \right) D \left( I_A^{W_l}, I_F^{W_l} \right) + \lambda \left( I_B^{W_l} \right) D \left( I_B^{W_l}, I_F^{W_l} \right) \right)}{\sum_{l=1}^{L} \left( \lambda \left( I_A^{W_l} \right) + \lambda \left( I_B^{W_l} \right) \right)} \tag{26.48}$$

#### 26.2.2.4.2  Chen–Blum Metric $Q_{CB}$

The flowchart of the Chen–Blum metric [30] is given in Figure 26.6. There are five steps involved in this algorithm.

1. Contrast sensitivity filtering: Filtering is implemented in the frequency domain. Image $I_A(i,j)$ is transformed into the frequency domain to get $I_A(m, n)$. The filtered image is obtained: $\tilde{I}_A(m,n) = I_A(m,n)S(r)$, where $S(r)$ is the CSF filter in polar form with $r = \sqrt{m^2 + n^2}$. In Ref. [30], there are three choices suggested for CSF, which include Mannos–Sakrison, Barton, and DoG filter.
2. Local contrast computation: Peli's contrast is defined as

$$C(i,j) = \frac{\phi_k(i,j) * I(i,j)}{\phi_{k+1}(i,j) * I(i,j)} - 1 \tag{26.49}$$

A common choice for $\phi_k$ would be:

$$G_k(x,y) = \frac{1}{\left( \sqrt{2\pi}\sigma_k \right)} e^{\frac{x^2 + y^2}{2\sigma_k^2}} \tag{26.50}$$

with a standard deviation $\sigma_k = 2$.

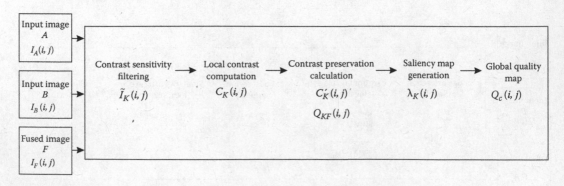

**FIGURE 26.6**  The flowchart of Chen–Blum metric.

3. Contrast preservation calculation: The masked contrast map for input image $I_A(i, j)$ is calculated as

$$C_A' = \frac{t(C_A)^p}{h(C_A)^q + Z} \tag{26.51}$$

Here, $t$, $h$, $p$, $q$, and $Z$ are real scalar parameters that determine the shape of the nonlinearity of the masking function [30].

4. Saliency map generation: The saliency map for $I_A(i, j)$ is defined as

$$\lambda_A(i, j) = \frac{C_A'^2(i, j)}{C_A'^2(i, j) + C_B'^2(i, j)} \tag{26.52}$$

The information preservation value is computed as:

$$Q_{AF}(i, j) = \begin{cases} \dfrac{C_A'(i, j)}{C_F'(i, j)} & \text{if } C_A'(i, j) < C_F'(i, j) \\ \dfrac{C_F'(i, j)}{C_A'(i, j)} & \text{otherwise} \end{cases} \tag{26.53}$$

Global quality map:

$$Q_{GQM}(i, j) = \lambda_A(i, j) Q_{AF}(i, j) + \lambda_B(i, j) Q_{BF}(i, j) \tag{26.54}$$

5. The metric value is obtained by average the global quality map, that is, $Q_{CB} = \overline{Q_{GQM}(i, j)}$.

## 26.2.3 Subjective Assessment

Although objective assessment is more attractive because of the low cost and less complexity in comparison with subjective assessment, it still needs to be verified and validated. Thus, subjective assessment still plays a key role in both the fusion performance assessment and fusion metric validation. Two trials on subjective fusion performance assessment were conducted [31,32]. However, the report on validation of objective fusion metrics is not available yet.

### 22.2.3.1 Case 1: Forced Choice Subjective Assessment

The first trial of subjective assessment is described in Ref. [31]. Passive, informal, preference tests were designed to compare the fusion algorithms with "visual inspection." In the tests, a wide variety of scenes and fusion scenarios were explored [31]. In total, 151 monochrome registered image pairs that were captured by a number of different sensors in real or realistic conditions were used in the tests. During the tests, human subjects were presented with a series of image sets consisting of two inputs and two fusion alternatives with these inputs. The subjects gave their individual preference to one or none of the fused images. For each image set $n$, the answers of subject $m$ for fusion schemes 1 and 2 are recorded as [31]:

$$P_{m,n} = \begin{bmatrix} P_{m,n}^0 \\ P_{m,n}^1 \\ P_{m,n}^2 \end{bmatrix} = \begin{cases} [1,0,0]^T, & \text{subject } m \text{ has no preference} \\ [0,1,0]^T, & \text{subject } m \text{ prefers fused image of scheme 1} \\ [0,0,1]^T, & \text{subject } m \text{ prefers fused image of scheme 2} \end{cases} \quad (26.55)$$

The recorded data are presented with $N \times M$ preference vectors, where $N$ and $M$ are the number of image sets and subjects respectively. These preference vectors are further used in three different ways. The first is simply to calculate the total preference score $T_i$ for schemes 1 and 2 with [31]:

$$T_i = \frac{1}{NM} \sum_{n=1}^{N} \sum_{m=1}^{M} P_{m,n}^i \quad (26.56)$$

The scheme with higher preference score has a better performance. The score $T_0$, if greatest, may indicate the equivalence of the two methods. Second, a scene preference score $S_i$ is derived by:

$$S_i = \frac{1}{N} \sum_{n=1}^{N} \begin{cases} 1, & \text{when } i = \underset{\forall j}{\arg\max} \left( \sum_{m=1}^{M} P_{m,n}^j \right) \\ 0, & \text{otherwise} \end{cases} \quad (26.57)$$

This score reflects the robustness of a fusion scheme to scene content. Third, individual preference score of fusion scheme is defined as:

$$I_i = \frac{1}{M} \sum_{n=1}^{M} \begin{cases} 1, & \text{when } i = \underset{\forall j}{\arg\max} \left( \sum_{n=1}^{N} P_{m,n}^j \right) \\ 0, & \text{otherwise} \end{cases} \quad (26.58)$$

This score provides a direct indication of the naturalness of fused images and may lead to an optimal fusion configuration in applications that involve lengthy display monitoring [31]. The three preference scores, that is, $T_i$, $S_i$, and $I_i$, perform relative fusion evaluation against another. It can be applied when only a few choices are considered. It is not practical when multiple fusion algorithms need to be compared and ranked over large data sets.

### 22.2.3.2   Case 2: Interpretability Subjective Assessment

Different from the first trial, which simply gives a preference to the "better" fused image, the assessment in the second trial was conducted by quantifying the interpretability of the fused image through human semantic segmentation [32]. Sixty-three human subjects were organized to perform a semantic segmentation on both the input and fused images. A gold reference is generated from the inputs and serves as the ground truth for the comparison with the fused image. The ground truth is compiled from users determining the pixels of interest, for example, where a target is, the image quality, and significant boundaries based on image contrast.

As illustrated in Figure 26.7, the subjects were asked to divide each image into pieces, which represents distinguishing something important in the image with manually drawn boundaries. The

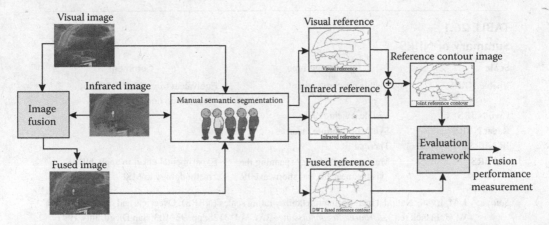

**FIGURE 26.7** Procedure to generate a "gold reference" from subjective assessment.

boundary map is then converted into a boundary mask image. The exact square two-dimensional Euclidean distance transform of the boundary map is first calculated using a square $3 \times 3$ structuring element. Then, the derived distance image is processed with a threshold operation to obtain the binary mask image. The mask images from all the subjects are summed and thresholded at a level corresponding to half the number of subjects contributing to the sum. Thus, the obtained result is called a *consensus binary mask image*. The logical union of the consensus binary mask images from visual and infrared inputs gives the reference mask image. A skeleton image containing boundaries of interest is derived with morphological operations and serves as the *reference contour image*. This procedure is applied to the input and fused images respectively. The contour from the fused image is then compared with the reference contour generated from input images. The comparison employs the precision-recall framework, where the precision ($P$) and recall ($R$) are defined as

$$P = \frac{\text{Number of correctly detected reference boundary pixels}}{\text{Total number of detected boundary pixels}} \tag{26.59}$$

$$R = \frac{\text{Number of correctly detected reference boundary pixels}}{\text{Total number of reference boundary pixels}} \tag{26.60}$$

The $F$-measure can be defined as

$$F = \frac{PR}{\alpha R + (1 - \alpha)P} \tag{26.61}$$

The relative cost $\alpha$ depends on the particular application. In the comparison, precision and recall are equally weighted. Thus, $\alpha$ is given a value of 0.5. Equation 26.61 changes to $F = 2PR/(R + P)$. A larger $F$-measure value between the fused result and "gold" reference indicates a better fusion performance. This approach can be applied to rank multiple fusion algorithms.

### 26.2.4 Nature Imagery Interpretability Rating Scales

The NIIRS was developed with a set of psychophysical scaling assessments to represent the interpretability of a given aerial image [33]. NIIRS quantifies the interpretability or usefulness of

**TABLE 26.1**

**Summary of NIIRS**

| Scale | Image Type | Comment |
|---|---|---|
| Visible NIIRS | Visible panchromatic | Equivalent to visible NIIRS, but focused on nonmilitary tasks |
| Civil NIIRS | Visible panchromatic | |
| Radar NIIRS | Synthetic aperture radar | |
| IR NIIRS | Thermal IR | |
| MS IIRS | Multispectral imagery spanning the visible, near IR, and shortwave IR | Experimental effort to apply NIIRS methodology to MSI |

*Source:*   J. M. Irvine. National imagery interpretability rating scales (NIIRS): Overview and methodology. In W. G. Fishell (ed.), *Airborne Reconnaissance XXI*, Vol. 3128, pp. 93–103, San Diego, July 1997.

imagery with 10 level scales (levels 0–9) [33–35]. Each level is defined by a set of information extraction tasks called criteria [33]. Table 26.1 lists the types of NIIRS [34]. The NIIRS definitions for visual, infrared (IR), multispectral (MS), and radar images are given in Table 26.2, where EO (electro-optic) represents visual.

In the application of surveillance and object recognition, the general image quality equation (GIQE) is employed to predict visible or IR NIIRS as a function of system design and operating parameters [33]. The NIIRS equations for the EO visual and IR sensors are given below [35]:

$$\text{NIIRS}_{EO} = 10.251 - a\log_{10}(\text{GSD}) + b\log_{10}(\text{RER}) - 0.656\text{H} - [0.344(\text{G/SNR})] \qquad (26.62)$$

$$\text{NIIRS}_{IR} = 10.751 - a\log_{10}(\text{GSD}) + b\log_{10}(\text{RER}) - 0.656\text{H} - [0.344(\text{G/SNR})] \qquad (26.63)$$

The parameters in the equation are

- GSD: geometric mean of the ground sample distance
- H: geometric mean height due to edge overshoot
- RER: geometric mean of the normalized relative edge response
- G: noise gain
- SNR: signal-to noise-ratio
  - 3.32 if RER = 0.9 and 3.16 if RER < 0.9
  - 1.559 if RER = 0.9 and 2.817 if RER < 0.9

NIIRS rating is a function of range or distance rather than target size, which is inherent in the definitions. When varied fusion algorithms are considered, this high-altitude imagery rating approach is not discerning enough to discriminate the differences between those fusion algorithms with only 10 discrete numbers. Usually, the comparison needs to be conducted between the fused and input images and images fused by different algorithms. In contrast, the probability of detection (POD) varies with a change in target size. The fused image can be assessed directly from its POD curve in comparison with the POD curves of the input images, when target detection is the major concern of the application.

In Refs. [33,36], the relationship between NIIRS and probability of detection, recognition, and identification was described and established. Given the number of cycles across the target's critical

## TABLE 26.2
## NIIRS Definitions for EO, IR, and Radar Images

| NIIRS Level | EO | IR | Radar | Resolution |
|---|---|---|---|---|
| 0 | Uninterpretable image | Uninterpretable image | – | – |
| 1 | Distinguish between taxiways and runways at large airfield | Detect large cleared areas | Detect lines of transportation | >9 m |
| 2 | Detect military training areas | Distinguish level of vegetation | Detect very large defensive berm | 4.5–9 m |
| 3 | Detect helipad based on configuration/markings | Detect driver training track | ID areas based on building pattern | 2.5–4.5 m |
| 4 | Recognize by general type (tracked/wheeled) | Detect individual thermally active vehicles | Detect a convoy | 1.2–2.5 m |
| 5 | ID by type large targets | Detect vehicles in revetment | Detect battery of towed artillery | 0.75–1.2 m |
| 6 | ID spare tire on medium sized truck | Distinguish between thermally active APC and tank | Distinguish between wheeled and tracked | 0.4–0.75 m |
| 7 | ID missile mount | ID missile transfer crane on transloader | Distinguish between medium tank and APC | 0.2–0.4 m |
| 8 | ID handheld SAM | Detect closed hatches | Distinguish between guns by overall configuration | 0.1–0.2 m |
| 9 | ID vehicle registration numbers | ID turret hatch hinges on armored vehicle | Detect gun tubes on SPAA gun | <0.1 m |

*Source:* B. Kahler and E. Blasch. Predicted radar/optical feature fusion gains for target identification. In *Proceedings of the IEEE 2010 National Aerospace and Electronics Conference (NAECON)*, pp. 405–412, July 2010.

*Note:* APC, armored personnel carrier; SAM, surface-to-air missile; SPAA, self-propelled anti-aircraft; SSM, surface-to-surface missile.

dimension $N$, the probability of success of the discrimination task is determined by the target transfer probability function (TTPF):

$$P(N) = \frac{(N/N_{50})^{27+07(N/N_{50})}}{1+(N/N_{50})^{27+07(N/N_{50})}} \tag{26.64}$$

$N_{50}$ is the 50% cycle criteria set for detection, recognition, or identification.

The POD curve of an IR sensor can be derived from the minimum resolvable temperature (MRT) models as illustrated in Figure 26.8. The basic procedure is as follows [33]:

1. Determine target parameters, for example, target height ($h$) and width ($w$).
2. Estimate the target-to-background temperature difference.
3. Find the spatial frequency on the MRT curve (Figure 26.8a), which matches the target apparent differential temperature on the target load line.

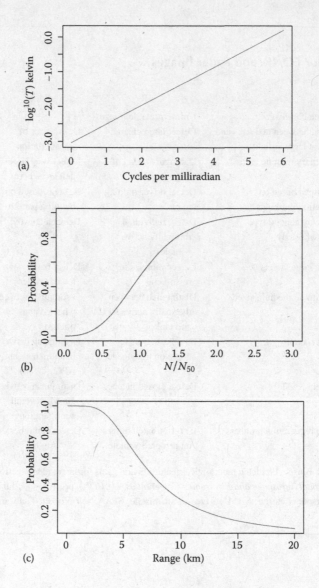

**FIGURE 26.8** Simulation of the process to derive the sensor POD curve. (a) Simulated minimum resolvable temperature curve, (b) simulated target transfer probability function, and (c) simulated probability of detection curve.

4. Resolve the number of cycle across the critical target dimension using

$$N = \rho \frac{d_c}{R} = \rho \frac{\sqrt{wh}}{R} \tag{26.65}$$

where

$\rho$ = maximum resolvable spatial frequency (cycles/milliradian)
$d_c$ = critical target dimension (meter)
$R$ = range from sensor to the target (kilometer)

5. Determine the POD using the TTPF function (Figure 26.8b) in Equation 26.64.

A typical POD curve is shown in Figure 26.8c.

The above process cannot be applied to image fusion directly. Given the input images and fused image, the POD curves for all the images can be derived. If the NIIRS of the input images is known, it is possible to get the NIIRS for the fused image. However, this still remains a topic for future research.

When the fusion does not discriminate different objects from input images, the POD curve may not be able to tell the differences between fusion algorithms. For example, the pixel-level fusion of visible and IR images highlights the object in the environment. Both the foreground and background need to be well presented in the fused image. In this case, the POD of object for the IR image and fused image could be approximately the same. Thus, POD will not be the best way to assess the performance of this type of image fusion, so we seek statistical methods for analysis. Statistical tests for subjective assessment can be used with NIIRS at each NIIRS level. If the subject is asked to assess the image fusion result at one NIIRS level, we can determine statistical score associated for that level.

## 26.2.5  STATISTICAL ANALYSIS OF FUSION PERFORMANCE ASSESSMENT DATA

Statistical tests can be employed to identify the performance differences between multiple fusion algorithms. The comparative tests provide statistical verification and validation of the fusion results. Usually, parametric tests assume the normality and equal variances of the data, which is not always satisfied in practice. Thus, nonparametric tests are needed to work with data that are not normal or of equal variance. As illustrated in Table 26.3, the Wilcoxon signed ranks test and the Friedman test are employed.

In fact, the comparison can be conducted in three different ways: (1) only two methods are compared; (2) one method is compared to all the others ($1 \times N$); and (3) all methods are compared to each other ($N \times N$) [6]. In the test, a null hypothesis ($H_0$) and an alternative hypothesis ($H_1$) are defined. The null hypothesis states that there is no effect or no difference between the algorithms, whereas the alternative hypothesis claims the presence of an effect or a difference between algorithms. As per the scientific method, one cannot prove $H_1$, but can disprove $H_0$, that is, there was no change in the image fusion results. A significance value $\alpha$ is used to determine at which level the hypothesis should be rejected [8]. The Wilcoxon signed ranks test and the Friedman test are used for a two-method comparison and a multiple comparison ($N \times N$), respectively.

### 26.2.5.1  Wilcoxon Signed Ranks Test

The Wilcoxon signed ranks test is a nonparametric alternative to the paired $t$-test and ranks the differences in performances of two algorithms for each data set [6,8,37]. It compares the ranks for the positive and negative differences.

Let $d_i$ be the difference between the $F$-measure values of two assessment approaches on $i$th out of $n$ problem or data set. The differences are ranked based on the absolute values. The use of average ranks is recommended to deal with ties. Let $R^+$ be the sum of ranks for the problems where the first assessment

**TABLE 26.3**

**Parametric and Nonparametric Statistical Tests**

| Statistical Tests | | Two Groups | $n$ Groups ($n > 2$) |
|---|---|---|---|
| Parametric test | Unpaired | Unpaired $t$-test | One-way ANOVA |
| (normality) | Paired | Paired $t$-test | Two-way ANOVA |
| Nonparametric test | Unpaired | Mann–Whitney $U$-test | Kruskal Wallis $t$-test |
| (no normality) | Paired | • Sign test | Friedman test |
| | | • Wilcoxon signed-ranks test | |

value is larger than the second and $R^-$ is the sum of ranks for the opposite. Ranks of $d_i = 0$ are split evenly among the sums; if there is an odd number of them, one is ignored [8]. The sum of ranks are defined as

$$R^+ = \sum_{d_i>0} \text{rank}(d_i) + \frac{1}{2} \sum_{d_i=0} \text{rank}(d_i) \tag{26.66}$$

$$R^- = \sum_{d_i<0} \text{rank}(d_i) + \frac{1}{2} \sum_{d_i=0} \text{rank}(d_i) \tag{26.67}$$

Let $T = \min(R^+, R^-)$ if $T$ is the smaller sum. When $T$ is less than or equal to the value of the distribution of Wilcoxon for $n$ degrees of freedom, the null hypothesis of equality of mean is rejected. For a larger number of data sets, the statistic

$$z = \frac{T - n(n+1)/4}{\sqrt{n(n+1)(2n+1)/24}} \tag{26.68}$$

is distributed approximately normally. With $\alpha = 0.05$, the null hypothesis will be rejected if $z$ is smaller than $-1.96$ [6]. In our case, the conclusion is that the two assessments are different.

### 26.2.5.2 Friedman Test with Post Hoc Analysis

#### 26.2.5.2.1 Friedman Test

The Friedman test is a nonparametric equivalent of the repeated-measures analysis of variance (ANOVA) that determines the difference between group means [6]. The Friedman test carries out a multiple comparison test to detect significant differences between two or more algorithms [8]. The algorithms are ranked for each data set. In the case of ties, average ranks are assigned. The Friedman test compares the average ranks of algorithms with the null hypothesis that all the algorithms are equivalent or behave similarly [6].

To calculate the test statistic, the original results are first converted to ranks. The detailed procedures are as follows [8]:

1. Collect results for each algorithm/problem pair.
2. For each algorithm/problem $i$, rank values from 1 (best result) to $k$ (worst result) as $r_i^j$ ($1 \le j \le k$).
3. Then obtain the final rank $R_j = \frac{1}{n} \sum_i r_i^j$ for each algorithm $j$.

The best algorithm should have the rank of 1. The Friedman statistic $\chi_F^2$ can be computed as

$$\chi_F^2 = \frac{12n}{k(k+1)} \left[ \sum_j R_j^2 - \frac{k(k+1)^2}{4} \right] \tag{26.69}$$

which is distributed according to $\chi_F^2$ with $k - 1$ degrees of freedom, when $n$ and $k$ are big enough. For a smaller number of algorithms and data sets, exact critical values have been computed [6,8]. Iman and Davenport proposed a better statistic, which avoids the undesirable conservative of $\chi_F^2$ [6,8,38]. The proposed statistic is [8]:

$$F_{\text{ID}} = \frac{(n-1)\chi_F^2}{n(k-1) - \chi_F^2} \tag{26.70}$$

which is distributed according to the $F$-distribution with $k - 1$ and $(k - 1)(n - 1)$ degrees of freedom. The table of critical values can be found in statistical books.

### 26.2.5.2.2 Post Hoc Analysis

The Friedman test detects the significant difference over the complete multiple method (or group) comparisons, but it does not tell which group. Thus, it is necessary to determine which pairs in the group have the significant differences. The Nemenyi test is used when all the algorithms are compared to each other. The performance of the two algorithms is significantly different if the corresponding average ranks differ by at least the critical difference (CD), which is given as

$$CD = q_\alpha \sqrt{\frac{k(k+1)}{6n}} \qquad (26.71)$$

where $q_{\alpha = 0.05, 0.10}$ is based on the studentized range statistic divided by $\sqrt{2}$. The critical value is adjusted for making $k(k - 1/2)$ comparisons. A table of critical values for the two-tailed Nemenyi test can be found in Ref. [6]. For a comparison of all algorithms, there will be $k(k - 1)/2$ hypotheses and the post hoc analysis can obtain a $p$-value that determines the degree of rejection of each hypothesis.

The $p$-value of every hypothesis can be obtained through the conversion of the rankings by using a normal approximation. The test statistic for comparing the $i$th and $j$th algorithm is [8]:

$$z = (R_i - R_j) / \sqrt{\frac{k(k+1)}{6n}} \qquad (26.72)$$

The $z$ value is used to find the corresponding probability from the normal distribution table, which is then compared with an appropriate $\alpha$ [6]. The value of $\alpha$ needs to be adjusted to compensate for the multiple comparisons. The Nemenyi test calculates the adjusted value of $\alpha$ in a single step by dividing it by the number of comparisons, that is, $k(k - 1)/2$ [39]. Therefore, the adjusted $p$-value for Nemenyi is $\min\{v; 1\}$, where $v = k(k - 1)p_i/2$.

The statistical tests in the following experiments were conducted with R [40] and the KEEL tools [41].

## 26.3  ASSESSMENT OF NIGHT VISION FUSION

### 26.3.1  EXPERIMENTS ON OBJECTIVE ASSESSMENT

The first part of the experiments is on the objective assessment. The night vision image pairs provided by Dr. Alex Toet were used [42]. The images are shown in Figure 26.9. Two fusion schemes were considered in the experiment, that is, direct fusion scheme and modified fusion scheme [43]. In the direct fusion, the fused image integrates the features from the IR and visible images, such as edges and boundaries. The fused image is not a natural image in either the visible spectrum or the infrared spectrum. In the modified fusion scheme, the original visible-spectrum image and the enhanced one, which is obtained from the IR image, are fused. Thus, the fused image is still within the spectrum of visible light. In other words, the direct fusion merges heterogeneous images while the modified one merges homogeneous images.

The following six representative multiresolution algorithms were employed to implement pixel-level image fusion: Laplacian pyramid (LAP) [44], gradient pyramid (GRAD) [45], ratio-of-lowpass pyramid (RoLP) [46], Daubechies wavelet four (DB4) [47,48], shift-invariant discrete wavelet (SIDW) [49], and steerable pyramid (STEER) [48]. The decomposition level is set to 4 and the fusion rule is the maximum selection for the high- and band-pass subimages (components) and average

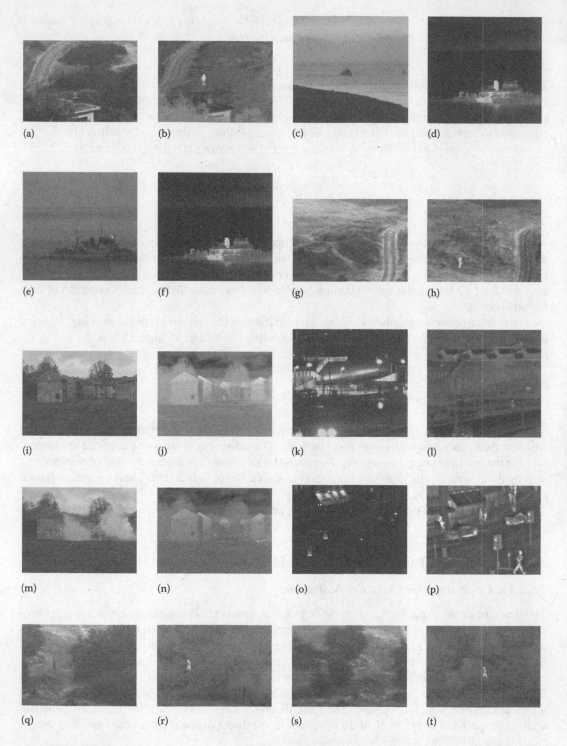

**FIGURE 26.9** Visible and infrared images used in this study. (a) UN camp (VI), (b) UN camp (IR), (c) b7118 (VI), (d) b7118 (IR), (e) b7436 (VI), (f) b7436 (IR), (g) Dune (VI), (h) Dune (IR), (i) octec02 (VI), (j) octec02 (IR), (k) e518a (VI), (l) e518a (IR), (m) octec21 (VI), (n) octec21 (IR), (o) quad (VI), (p) quad (IR), (q) trees 4906 (VI), (r) trees 4906 (IR), (s) trees 4917 (VI), and (t) trees 4917 (IR).

of the low-pass subimages (components). The Haar wavelet is used as the basis for SIDW and four orientational sub-bands along 0, 45, 90, and 135 degrees are implemented in the steerable pyramid decomposition.

Two fusion examples are given in Figures 26.10 and 26.11 respectively, where the original image pairs are from Figure 26.9a–d. Six multiresolution algorithms are implemented for the direct and

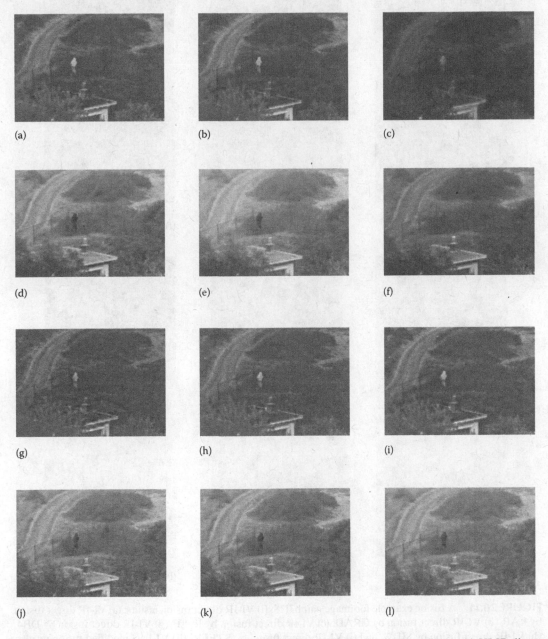

**FIGURE 26.10** A fusion example for image pair UN camp. (i) VI-IR direct fusion results: (a) VI-IR direct fusion by LAP, (b) VI-IR direct fusion by GRAD, (c) VI-IR direct fusion by RoLP, (g) VI-IR direct fusion by DB4, (h) VI-IR direct fusion by SIDW, and (i) VI-IR direction fusion by STEER. (ii) VI-EVI modified fusion results: (d) VI-EVI modified fusion by LAP, (e) VI-EVI modified fusion by GRAD, (f) VI-EVI modified fusion by RoLP, (j) VI-EVI modified fusion by DB4, (k) VI-EVI modified fusion by SIDW, and (l) VI-EVI modified fusion by STEER.

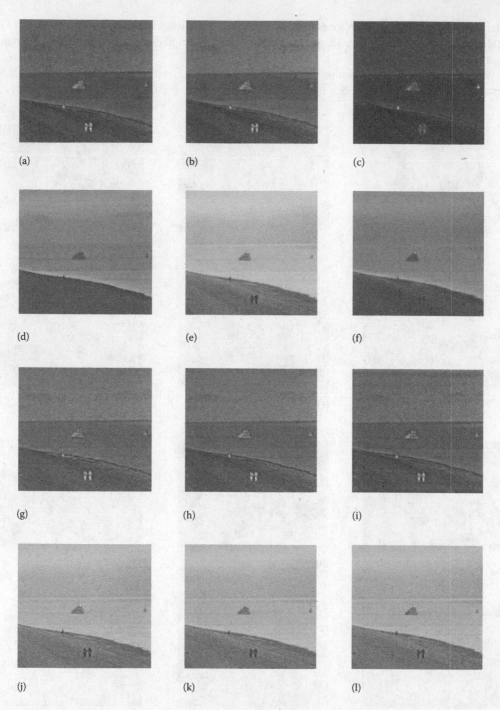

**FIGURE 26.11** A fusion example for image pair b7118. (i) VI-IR direct fusion results: (a) VI-IR direct fusion by LAP, (b) VI-IR direct fusion by GRAD, (c) VI-IR direct fusion by RoLP, (g) VI-IR direct fusion by DB4, (h) VI-IR direct fusion by SIDW, and (i) VI-IR direct fusion by STEER. (ii) VI-EVI modified fusion results: (d) VI-EVI modified fusion by LAP, (e) VI-EVI modified fusion by GRAD, (f) VI-EVI modified fusion by RoLP, (j) VI-EVI modified fusion by DB4, (k) VI-EVI modified fusion by SIDW, and (l) VI-EVI modified fusion by STEER.

modified image fusion schemes respectively. The differences between these two approaches and MRA fusion algorithms can be observed. For example, the person in the images has different contrasts with respect to the background.

Fusion metric values were computed with the approaches presented in Section 26.2.2. The correlation matrices of different fusion metrics for the two fusion schemes are plotted in Figures 26.12 and 26.13 respectively. The star on the top indicates the significance of the $p$-value for the correlation test. For the VI-IR direct fusion, the three largest values 0.789, 0.689, and 0.685 correspond to the correlation of $Q_{MI}$ and $Q_{NCIE}$, $Q_G$ and $Q_M$, $Q_S$, and $Q_C$ respectively. For the VI-EVI modified fusion, the largest values 0.854, 0.835, and 0.765 represent the correlation of $Q_{MI}$ and $Q_{NCIE}$, $Q_C$ and $Q_Y$, and $Q_G$ and $Q_C$. Obviously, $Q_{MI}$ and $Q_{NCIE}$ show a higher correlation in both cases as the two approaches are based on mutual information and are quite similar. The other metrics do not have the same correlation when applied to the results obtained by the two different fusion schemes. However, the fusion metrics with a higher correlation generally come from a same category. The correlation analysis reveals the similarities between the same types of fusion metrics.

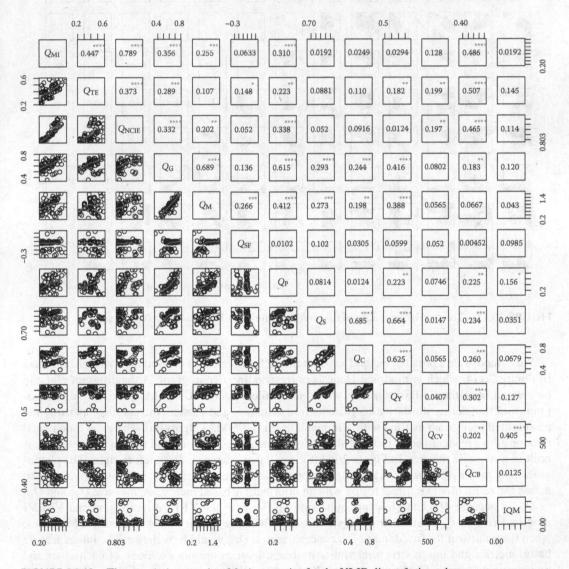

**FIGURE 26.12** The correlation matrix of fusion metrics for the VI-IR direct fusion scheme.

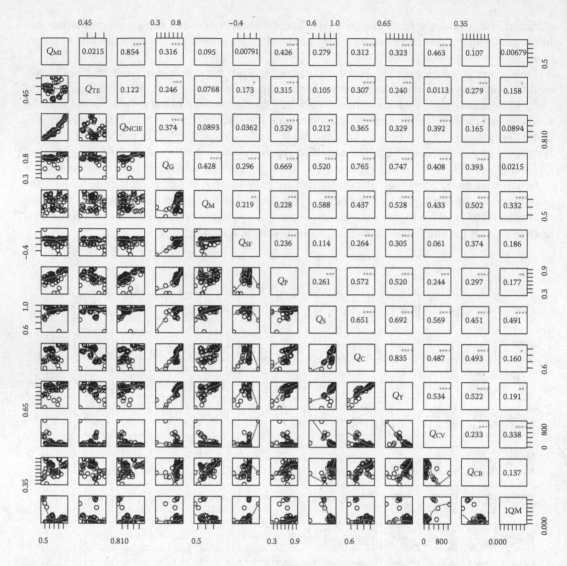

**FIGURE 26.13**   The correlation matrix of fusion metrics for the VI-EVI modified fusion scheme.

We demonstrate a possible method to select image fusion metrics derived from two phenograms in Figure 26.14, which is modified from its corresponding dendrogram. A dendrogram plot can be created from the similarity matrix with a tool called "DendroUPGMA" [50,51]. The dendrogram tool transforms similarity coefficients into distances and clusters the coefficients using the unweighted pair group method with arithmetic mean (UPGMA) algorithm. The local topological relationships are identified in order of similarity (Kendall correlation), and the phylogenetic tree is built in a stepwise manner. In Figure 26.14, the similarity increases from left to right. The phenogram may give a reference for choosing an appropriate fusion metric. For example, if we want to assess the result of VI-IR direct fusion with an information-based metric, we may use $Q_{TE}$ and $Q_{MI}$ or $Q_{NCIE}$, because $Q_{MI}$ and $Q_{NCIE}$ are quite similar. For both the VI-IR direct fusion and VI-EVI modified fusion, $Q_{CV}$ demonstrates a similarity with image quality measurement (IQM). Thus, when the quality of the fused image is a concern, metric $Q_{CV}$ would be a choice. The image feature based metrics and image structural similarity based metrics are not exclusive of each other and they group differently for the two fusion schemes. Nevertheless, image structural similarity based

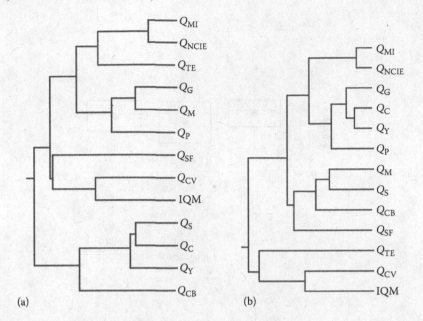

**FIGURE 26.14**  The phenogram for image fusion metrics. (a) VI-IR direct fusion and (b) VI-EVI modified fusion.

metrics $Q_C$ and $Q_Y$ are closely clustered. The reason why the two phenogram plots are differently structured is still not clear, but we speculate that it is partially due to the image modalities involved in the fusion process.

A left-to-right rule can be followed to choose a fusion metric. In Figure 26.14, a cutoff can be applied to a certain correlation value. For each cutoff, there are a varied number of clusters available. A fusion quality metric can be picked up from each cluster in favor of its characteristic, which may reflect the resemblance and similarity between the fused and the input images, or the clarity and perceptibility of the fused image.

## 26.3.2  EXPERIMENTS ON STATISTICAL ANALYSIS OF SUBJECTIVE AND OBJECTIVE ASSESSMENT RESULTS

The data sets used in these experiments contain image fusion assessment results by a group of subjects (user) and an expert (expert) [32]. The data included 10 sets of multisensor images were fused by three MRA algorithms: a pyramid-based (PYR) approach, a discrete wavelet transform (DWT), and a complex wavelet transform (CWT) [52]. For the objective assessment, we used the results in the previous experiments. Only 10 metrics, which give a value in the range [0, 1], were retained [4].

### 26.3.2.1  Test 1: User versus Expert Assessments

The first experiment investigates the difference of the assessments between the user and the expert in terms of the $F$-measure. Boxplots of the $F$-measure for different fusion algorithms by the user and expert are given in Figure 26.15a and b. The boxplot for the expert's and user's assessments is also given in Figure 26.15c for comparison. Both the median and mean values are highlighted. The data differences can be observed. The null hypothesis $H_0$ states that there is no difference between the assessments of the user and expert while the alternative hypothesis says that they are different. The Wilcoxon signed ranks test was performed. Table 26.4 gives the results at the significance level $\alpha = 0.05$. Herein, VS refers to the sum of ranks assigned to the $F$-measure differences between the user and expert with positive ($R^+$) or negative ($R^-$) sign. The $p$-value 0.241 is larger than 0.05, which

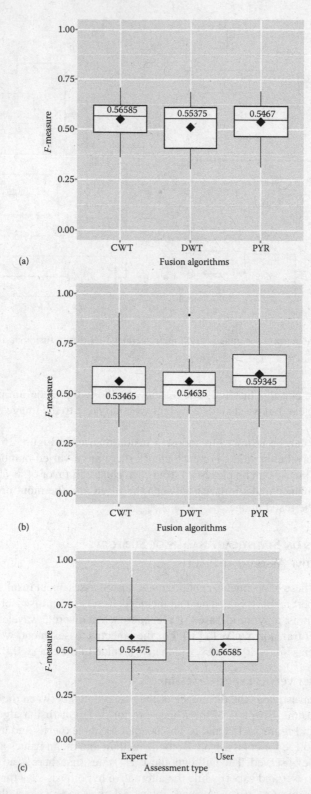

**FIGURE 26.15** The subjective assessments by (a) user and (b) expert. (c) Comparison between user's and expert's subjective assessments.

TABLE 26.4

**Results Obtained by the Wilcoxon Signed Ranks Test at a Level of Significance $\alpha = 0.05$**

| VS | $R^+$ | $R^-$ | $p$-Value |
|---|---|---|---|
| User versus expert | 290.0 | 175.0 | 0.2410 |

TABLE 26.5

**The Mean Value of the $F$-Measure**

| | PYR | DWT | CWT |
|---|---|---|---|
| User | 0.53718 | 0.51063 | 0.55032 |
| Expert | 0.60056 | 0.56420 | 0.56329 |

means the null hypothesis cannot be rejected, that is, there is no significant difference between the assessments from the user and expert.

The mean values of the $F$-measure are listed in Table 26.5. The maximum values are highlighted. The expert prefers the PYR while the user's favorite is CWT, when only the mean value is considered.

### 26.3.2.2 Test 2: PLIF Method Comparisons from User Assessments

In the second experiment, multiple comparisons were conducted for different fusion algorithms based on the users' assessments. The Friedman test tells whether the three approaches are similar or not. The test results and ranks by user's assessment are given in Table 26.6. However, this rank does not give much information about the performance of the algorithms. Unlike the classification application, the classifiers are evaluated for selection based on either the accuracy or the error rate. A higher rank refers to a higher accuracy or a lower error rate. Thus, it is an indication of the performance. For image fusion, the fusion performance value is evaluated. A higher rank can be reached by simply comparing the performance values. So, according to Table 26.6, a preference is given to PYR in terms of the average ranks from the Friedman test of users' assessments.

However, the $p$-value is larger than 0.05; the difference between these fusion algorithms is not significant in terms of users' assessments. As the difference is not statistically significant, there is no need to conduct a post hoc analysis, but the Nemenyi test results are listed in Table 26.7 for information. Nemenyi's procedure rejects those hypotheses that have an unadjusted $p$-value $\leq 0.016667$. Thus, all the hypotheses cannot be rejected according to the $p$-value in Table 26.7. The "major difference" is between the PYR and DWT by the users' assessments.

The test results on experts' assessments are given in Table 26.8. Again, no consequential difference observed at a significance level of 0.05. The difference between PYR and DWT is relatively more salient than between the other pairs, although all are trivial in terms of experts' assessments (see Table 26.9).

TABLE 26.6

**The Friedman Test of User's Assessments**

| Algorithm | User.PYR | User.DWT | User.CWT |
|---|---|---|---|
| Ranking | 1.7 | 2.4 | 1.9 |
| $\chi^2$ | 2.6 | Degree of freedom | 2 |
| $p$-Value | | 0.2725 | |

**TABLE 26.7**

**Adjusted *p*-Values for Subjective Comparison of Fusion Algorithms (by User)**

|   | Hypothesis | Unadjusted *p* | $p_{Neme}$ |
|---|---|---|---|
| 1 | User.PYR versus User.DWT | 0.117525 | 0.352575 |
| 2 | User.DWT versus User.CWT | 0.263552 | 0.790657 |
| 3 | User.PYR versus User.CWT | 0.654721 | 1.964163 |

**TABLE 26.8**

**The Friedman Test of Experts' Assessments**

| Algorithm | Expert.PYR | Expert.DWT | Expert.CWT |
|---|---|---|---|
| Ranking | 1.7 | 2.1 | 2.2 |
| $\chi^2$ | 1.4 | Degree of freedom | 2 |
| *p*-Value | | 0.4966 | |

**TABLE 26.9**

**Adjusted *p*-Values for Subjective Comparison of Fusion Algorithms (by Expert)**

|   | Hypothesis | Unadjusted *p* | $p_{Neme}$ |
|---|---|---|---|
| 1 | Expert.PYR versus Expert.DWT | 0.263552 | 0.790657 |
| 2 | Expert.DWT versus Expert.CWT | 0.371093 | 1.11328 |
| 3 | Expert.PYR versus Expert.CWT | 0.823063 | 2.46919 |

### 26.3.2.3 Test 3: Fusion Metric Comparisons

In the third part of the experiment, 10 objective fusion metrics are considered. Figure 26.16 shows the metric values for three algorithms. The Friedman test ($N \times N$) is performed and a *p*-value = $0.0247 < 0.05$ is obtained, which means there is a significant difference in the test. The CWT gains the highest rank 1.3 among the three. PYR and DWT are quite similar.

To understand where the differences happen, the Nemenyi test was conducted and results are given Table 26.10. Nemenyi's procedure rejects those hypotheses that have an unadjusted *p*-value = $0.013906 \leq 0.016667$ (see Table 26.11). Thus, the null hypothesis for DWT and CWT is rejected. In other words, the two algorithms are significantly different in the opinion of objective metrics while the differences among other pairs are trivial.

The results differ between the user and the fusion methods. The assessments from the users and the expert are quite the same, that is, no significant statistical difference is observed. This is understandable and reasonable, as many of the images presented to these groups had already been enhanced for targeting. However, there is no strong evidence to support the significant difference between the three fusion algorithms from the users' assessments. In other words, the three fusion algorithms perform equally well or the human subjects could not tell the differences through the designed evaluation process. There are several potential explanations for users not being able to distinguish between methods. First, in the subjective assessment, the input images, and fused result of the same set were assessed by different subjects. For example, the inputs of image set A are

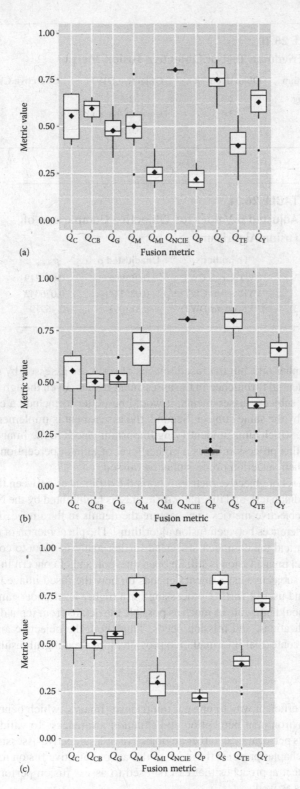

**FIGURE 26.16** The objective assessments with fusion metrics. (a) PYR, (b) DWT, and (c) CWT.

**TABLE 26.10**

**The Friedman Test of Objective Fusion Metrics**

| Algorithm | Objective.PYR | Objective.DWT | Objective.CWT |
|---|---|---|---|
| Ranking | 2.3 | 2.4 | 1.3 |
| $\chi^2$ | 7.4 | Degree of freedom | 2 |
| $p$-Value | | 0.0247 | |

**TABLE 26.11**

**Adjusted $p$-Values for Objective Comparison of Fusion Algorithms**

| | Hypothesis | Unadjusted $p$ | $p_{Neme}$ |
|---|---|---|---|
| 1 | DWT versus CWT | 0.013906 | 0.041719 |
| 2 | PYR versus CWT | 0.025347 | 0.076042 |
| 3 | PYR versus DWT | 0.823063 | 2.46919 |

assessed by subject number one, but the fused images of set A are assessed by subject number two. This may introduce potential turbulence in the assessment results. Even though the "gold reference" is a combination of all subjects' assessments, it would be better to include a complete assessment of a specific image set by the same subject. Second, the assessment is implemented through human semantic segmentation, which aims at identifying important pieces in the image and the results are user dependent. Thus, this process mimics a higher level of human perception on object detection while some details in their selection process may be missed.

On the other hand, a set of objective metrics reports the differences between the three fusion algorithms through the Friedman test, and the major difference is highlighted by the Nemenyi test's result. The reason is that the objective metrics may capture the details in the image. The weights on such details highlight the differences between fusion algorithms. The phenomenon of differences in image fusion between the user and the computer still need further analysis. How to conduct the subjective assessment on pixel-level image fusion is still an open question and of concern in Level 5 information fusion [53]. In fact, the subjective assessment depends on how the fused image is used, for example, a visualization to the end user or further processing for a higher level understanding. Generally, the image fusion method should integrate as much as possible the details from input data such as collected metadata, pixel normalizations, and mission needs. Thus, both the objective and subjective assessments can benefit from contextual information to provide more meaningful results.

## 26.4  SUMMARY

Image fusion offers an efficient way to present multisensor images, which benefits diverse applications such as robust environment perception and situation awareness. To validate the fusion algorithm and understand its performance, either an objective or a subjective assessment approach needs to be carried out. This chapter presents the state-of-the-art of objective fusion metrics and describes two subjective assessment approaches that can be used to assess fusion performance and validate objective fusion metrics as well.

The statistical tests make it possible to compare fusion algorithms over multiple image data sets in terms of individual or multiple metrics for a given NIIRS level. The statistical tests introduced here for image fusion can also be applied to validate the fusion metrics over multiple data sets and fusion algorithms. There are two scenarios where the statistical analysis can be applied. In the first case, when the subjective data are available, all the algorithms will be compared with subjective data. Either pairwise or multiple comparison ($1 \times N$ or $N \times N$) can be conducted with the objective data. When there is a new fusion algorithm coming in, the subjective assessment needs to be repeated, which is costly and not practical. The second scenario uses the objective fusion metric. In this case, individual or multiple fusion metrics can be used. The conclusion could be a new fusion algorithm that outperforms another algorithm in terms of a metric or a combination of multiple metrics for a given scenario. Statistical analysis can avoid the confusion caused by opposite results over varied data sets.

To validate a new fusion algorithm, a comparison with other fusion algorithms is performed using a series of fusion metrics to conduct verification. However, the validation issue exists in the objective fusion metrics also against the mission needs and robust performance requirements. A similar procedure with statistical analysis can be applied to compare the objective fusion metrics. When the subjective ground truth is available, we can learn which metric better matches the user's needs. When there is no such reference, we can only tell the differences between those metrics, but not which is better. However, the fusion algorithm will also affect the fusion metric. Therefore, the fusion metric study needs to be carefully and clearly defined as well against the operating conditions, mission needs, user objectives, and control of the image fusion parameters.

Future work includes three areas. The fused imagery [32] benchmark data set with human subjective assessment will be refined and available for further analysis. Also, we seek methods for user-defined objective scores and are developing an interpretability rating scale for image fusion results. Finally, we will induce quantifiable errors in the imagery, such as blurring, to determine the balance between the user experience and choice of image fusion methods to deal with challenging scenarios.

# REFERENCES

1. R. S. Blum, and Z. Liu, editors. *Multi-Sensor Image Fusion and Its Applications*. Signal Processing and Communications. Taylor and Francis, Boca Raton, FL, 2005.
2. Y. Zheng, W. Dong, and E. P. Blasch. Qualitative and quantitative comparisons of multispectral night vision colorization techniques. *Optical Engineering*, 51(8):087004, 2012.
3. B. Kahler, and E. Blasch. Sensor management fusion using operating conditions. In *Proceedings of National Aerospace and Electronics Conference*, Fairborn, OH, July 2008.
4. Z. Liu, E. Blasch, Z. Xue, J. Zhao, R. Laganière, and W. Wu. Objective assessment of multiresolution fusion algorithms for context enhancement in night vision: A comparative study. *IEEE Transactions on Pattern Analysis and Machine Intelligence*, 34(1):94–109, 2012.
5. Z. Liu, and E. Blasch. Statistical analysis of the performance assessment results for pixel-level image fusion. In *17th International Conference on Information Fusion*, pp. 1–8, Salamanca, Spain, July 2014.
6. J. Demšar. Statistical comparisons of classifiers over multiple data sets. *Journal of Machine Learning Research*, 7:1–30, 2006.
7. S. Garcia, A. Fernandez, J. Luengo, and F. Herrera. Advanced nonparametric tests for multiple comparisons in the design of experiments in computational intelligence and data mining: Experimental analysis of power. *Information Sciences*, 180(10):2044–2064, 2010.
8. J. Derrac, S. Garcia, D. Molina, and F. Herrera. A practical tutorial on the use of nonparametric statistical tests as a methodology for comparing evolutionary and swarm intelligence algorithms. *Swarm and Evolutionary Computation*, 1(1):3–18, 2011.
9. S. Garcia, A. Fernandez, J. Luengo, and F. Herrera. A study of statistical techniques and performance measures for genetics-based machine learning: Accuracy and interpretability. *Soft Computing*, 13(10):959–977, 2009.
10. J. Luengo, S. Garcia, and F. Herrera. A study on the use of statistical tests for experimentation with neural networks: Analysis of parametric test conditions and non-parametric tests. *Expert Systems with Applications*, 36(4):7798–7808, 2009.

11. E. Blasch, E. Bosse, and D. Lambert. *High-Level Information Fusion Management and Systems Design*. Artech House, Norwood, MA, 2012.

12. S. Gupta, K. P. Ramesh, and E. P. Blasch. Mutual information metric evaluation for PET/MRI image fusion. In *Aerospace and Electronics Conference, 2008. NAECON 2008. IEEE National*, pp. 305–311, July 2008.

13. G. Qu, D. Zhang, and P. Yan. Information measure for performance of image fusion. *Electronics Letters*, 38(7):313–315, 2002.

14. M. Hossny, S. Nahavandi, and D. Vreighton. Comments on "Information measure for performance of image fusion." *Electronics Letters*, 44(18):1066–1067, 2008.

15. N. Cvejic, C. N. Canagarajah, and D. R. Bull. Image fusion metric based on mutual information and Tsallis entropy. *Electronics Letters*, 42(11):626–627, 2006.

16. R. Nava, G. Cristobal, and B. Escalante-Ramirez. Mutual information improves image fusion quality assessments. SPIE News Room, September 2007.

17. Y. Zheng, Z. Qin, L. Shao, and X. Hou. A novel objective image quality metric for image fusion based on Renyi entropy. *Information Technology Journal*, 7(6):930–935, 2008.

18. Q. Wang, Y. Shen, and J. Jin. Performance evaluation of image fusion techniques, Chapter 19. In T. Stathaki, editor, *Image Fusion: Algorithms and Applications*, pp. 469–492. Elsevier, Amsterdam, 2008.

19. C. S. Xydeas, and V. Petrovic. Objective image fusion performance measure. *Electronics Letters*, 36(4):308–309, 2000.

20. P. Wang, and B. Liu. A novel image fusion metric based on multi-scale analysis. In *Proceedings of IEEE International Conference on Signal Processing*, pp. 965–968, 2008.

21. Y. Zheng, E. A. Essock, B. C. Hansen, and A. M. Haun. A new metric based on extended spatial frequency and its application to DWT-based fusion algorithms. *Information Fusion*, 8(2):177–192, 2007.

22. J. Zhao, R. Laganiere, and Z. Liu. Performance assessment of combinative pixel-level image fusion based on an absolute feature measurement. *International Journal of Innovative Computing, Information and Control*, 3(6A):1433–1447, 2007.

23. Z. Liu, D. S. Forsyth, and R. Laganiere. A feature-based metric for the quantitative evaluation of pixel-level image fusion. *Computer Vision and Image Understanding*, 109(1):56–68, 2008.

24. Z. Wang, A. C. Bovik, H. R. Sheikh, and E. P. Simoncelli. Image quality assessment: From error measurement to structural similarity. *IEEE Transactions on Image Processing*, 13(1):600–612, 2004.

25. Z. Wang, and A. C. Bovik. A universal image quality index. *IEEE Signal Processing Letters*, 9(8):81–84, 2002.

26. G. Piella, and H. Heijmans. A new quality metric for image fusion. In *Proceedings of International Conference on Image Processing*, Barcelona, 2003.

27. N. Cvejic, A. Loza, D. Bul, and N. Canagarajah. A similarity metric for assessment of image fusion algorithms. *International Journal of Signal Processing*, 2(3):178–182, 2005.

28. C. Yang, J. Zhang, X. Wang, and X. Liu. A novel similarity based quality metric for image fusion. *Information Fusion*, 9:156–160, 2008.

29. H. Chen, and P. K. Varshney. A human perception inspired quality metric for image fusion based on regional information. *Information Fusion*, 8:193–207, 2007.

30. Y. Chen, and R. S. Blum. A new automated quality assessment algorithm for image fusion. *Image and Vision Computing*, 27:1421–1432, 2009.

31. V. Petrovic. Subjective tests for image fusion evaluation and objective metric validation. *Information Fusion*, 8:208–216, 2007.

32. A. Toet, M. A. Hogervorst, S. G. Nikolov, J. J. Lewis, T. D. Dixon, D. R. Bull, and C. N. Canagarajah. Towards cognitive image fusion. *Information Fusion*, 11(2):95–113, 2010.

33. R. G. Driggers, J. A. Ratches, J. C. Leachtenauer, and R. W. Kistner. Synthetic aperture radar target acquisition model based on a national imagery interpretability rating scale to probability of discrimination conversion. *Optical Engineering*, 42(7):2104–2112, 2003.

34. J. M. Irvine. National imagery interpretability rating scales (NIIRS): Overview and methodology. In W. G. Fishell, editor, *Airborne Reconnaissance XXI*, vol. 3128, pp. 93–103. San Diego, CA, July 1997.

35. B. Kahler, and E. Blasch. Predicted radar/optical feature fusion gains for target identification. In *Aerospace and Electronics Conference (NAECON), Proceedings of the IEEE 2010 National*, pp. 405–412, July 2010.

36. R. G. Driggers, P. Cox, and M. Kelley. National imagery interpretation rating system and the probabilities of detection, recognition, and identification. *Optical Engineering*, 36(7):1952–1959, 1997.

37. P. Dalgaard. *Introductory Statistics with R. Statistics and Computing*, 2nd edition. Springer, New York, 2008.

38. R. L. Iman, and J. M. Davenport. Approximations of the critical region of the Friedman statistic. *Communications in Statistics—Theory and Methods*, 9(6):571–595, 1980.

39. B. Trawinski, M. Smetek, Z. Telec, and T. Lasota. Nonparametric statistical analysis for multiple comparison of machine learning regression algorithms. *International Journal of Applied Mathematics and Computer Science*, 22(4):867–881, 2012.

40. R Core Team. *R: A Language and Environment for Statistical Computing*. R Foundation for Statistical Computing, Vienna, Austria, 2013.

41. J. Alcalá-Fdez, L. Sánchez, S. García, M. J. del Jesus, S. Ventura, J. M. Garrell, J. Otero, C. Romero, J. Bacardit, V. M. Rivas, J. C. Fernández, and F. Herrera. KEEL: A software tool to assess evolutionary algorithms to data mining problems. *Soft Computing*, 13(3):307–318, 2009.

42. Image fusion website. Available at http://www.imagefusion.org, August 2009.

43. Z. Liu, and R. Laganiere. Context enhancement through infrared vision: A modified fusion scheme. *Signal, Image and Video Processing*, 1(4):293–301, 2007.

44. E. H. Adelson, C. H. Anderson, J. R. Bergen, P. J. Burt, and J. M. Ogden. Pyramid methods in image processing. *RCA Engineer*, 29(6):33–41, 1984.

45. P. J. Burt, and R. J. Kolczynski. Enhanced image capture through fusion. In *Proceedings of 4th International Conference on Image Processing*, pp. 248–251, 1993.

46. A. Teot. Image fusion by a ratio of low-pass pyramid. *Pattern Recognition Letters*, 9:245–253, 1989.

47. H. Li, B. S. Manjunath, and S. K. Mitra. Multisensor image fusion using the wavelet transform. *Graphical Models and Image Processing*, 57(3):235–245, 1995.

48. Z. Liu, K. Tsukada, K. Hanasaki, Y. K. Ho, and Y. P. Dai. Image fusion by using steerable pyramid. *Pattern Recognition Letters*, 22:929–939, 2001.

49. O. Rockinger, and T. Fechner. Pixel-level image fusion: The case of image sequences. *SPIE*, 3374:378–388, 1998.

50. S. Garcia-Vallve, J. Palau, and A. Romeu. Horizontal gene transfer in glycosyl hydrolases inferred from codon usage in *Escherichia coli* and *Bacillus subtilis*. *Molecular Biology and Evolution*, 16(9):1125–1134, 1999.

51. S. Garcia-Vallve, and P. Puigbo. Dendroupgma: A dendrogram construction utility, June 2010. Available at http://genomes.urv.cat/UPGMA/.

52. J. J. Lewis, R. J. O'Callaghan, S. G. Nikolov, D. R. Bull, and N. Canagarajah. Pixel- and region-based image fusion with complex wavelets. *Information Fusion*, 8(2):119–130, 2007.

53. E. Blasch. Introduction to Level 5 fusion: The role of the user. In *Handbook of Multisensor Data Fusion: Theory and Practice*, 2nd edition, pp. 503–535. CRC Press, Boca Raton, FL, 2008.

# 27 A Review of Feature and Data Fusion with Medical Images

*Alex Pappachen James and Belur V. Dasarathy*

## CONTENTS

## 27.1 INTRODUCTION

The fusion techniques that utilize multiple feature sets to form new features that are often more robust and contain useful information for future processing are referred to as feature fusion [1]. The term data fusion is applied to the class of techniques used for combining decisions obtained from multiple feature sets to form global decisions [2]. Feature and data fusion interchangeably represent two important classes of techniques that have proved to be of practical importance in a wide range of medical imaging problems.

There has been a significant growth in the amount of scientific literature on the fusion of medical images in general since the last decade [3,4]. This largely reflects the wider importance gained in the use of medical images and multiple imaging modalities in the clinical assessment of organ conditions. In addition, the noninvasive nature of medical imaging makes it an alternative to classical techniques of drug-induced patient assessment or invasive measurement techniques. Medical images of human organs and cells from different modalities indicate different types of features and details. The use of multiple images can reveal a wide range of useful information that is not otherwise visible from a single image modality. However, going through the details in an individual modality one at a time can lead to significant time lags and requires multiple levels of expertise, making this an expensive process for the patient and the health service provider. Multimodal and multisensory imaging systems can reduce the overhead to information processing through feature and data fusion techniques to improve the overall operational efficiency.

A large variety of imaging modalities are in use today, such as magnetic resonance imaging (MRI) [5–21], computerized tomography (CT) [8,13,15,18,20,22–31], positron emission tomography (PET) [17,32–44], single-photon emission computed tomography (SPECT) [7,8,10,26,28,30,32–34,45–62], and ultrasound (US) [22,41,63–76]. Among others, they largely find applications in the study of the brain [7,11,32–35,45,46,77–99], breast [28,62,73,100–111], prostate [25,41,52,54,56,63–68,70,72,112–127], and lungs [43,44,49,128–135].

The field of medical image fusion is faced with the problems of veracity, velocity, and volume of the data that require faster and efficient processing of information. This review chapter provides an overview of information fusion techniques making use of feature and data fusion principles that find application in medical image computing and analysis. The aim of this chapter is to provide a collective view of the applicability and progress of information fusion techniques in medical imaging useful for clinical studies [3,4,22,136–144].

## 27.2 FEATURE-LEVEL MEDICAL IMAGE FUSION METHODS

We organize the methodological developments in medical image fusion methods into those that rely on feature-level processing and those that work at decision-level fusion. Feature-level fusion often aids in improving image quality and extracts newer features that are otherwise difficult to find in the original set of features.

Feature-level fusion between images is challenged by the problem of interimage variability such as pixel mismatches (scale, rotations, shifts), missing pixels, image noise, resolution, and contrast. The inaccuracies in feature representations can lead to poor fusion performance and lesser robustness of the feature representation. In addition, this also means that wrong feature representations can lead to wrong conclusions (increased false positives and false negatives) that reduce the reliability of medical image analysis in clinical settings.

Figure 27.1 is a summary of the major medical image fusion methods that are used individually and in combination for solving clinically relevant medical imaging computing problems.

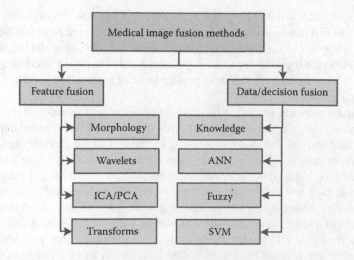

**FIGURE 27.1**  The classification tree for the major list of medical image fusion methods.

### 27.2.1 MORPHOLOGICAL OPERATORS AND FILTERS

Morphological operators make use of the connectedness between pixels either to improve the spatial arrange of the pixels or to distort them to extract useful features from the subset of spatially localized pixel features. The filters designed with morphological operators have been successfully applied in the problem of diagnosis of brain conditions to analyze and identify tumors [32,77,91]. The morphological operators are used for fusing the images from multiple modalities such as CT and MR [77,78], with a varied degree of success. The success of these operators depends on the size and design of the structuring operator that invariably controls the opening and closing operations in morphological filtering. Among many, the major operators used for fusion are averaging, morphology towers, K-L transforms, and morphology pyramids. The advantage of the morphological operators results from their simplicity and ability to parallelize for high-speed implementations, while the drawbacks largely result from the high dependence of pixel intensities.

### 27.2.2 WAVELET-BASED FEATURE FUSION

Wavelet transforms have the ability to compress the details of the images through their coefficients and to separate the fine and coarse details from one another. Because of the ability to represent the different properties of the image through coefficients, the impact of noise on the image would be reflected in one coefficient or another. This opens up the possibility to use wavelets to retain only those coefficients that are stable. Such coefficients from different features can be fused together to form more robust representations of the images [5,8,13,27,29,31,145–171]. In effect, the idea of the wavelet fusion is to inject good features from one image to another and in the process remove the problematic ones. Substitution, addition, aggregator functions, and data-driven models all form the methodological process of injection. Although the coefficients do show a compressed processing, the fused output image is optimized for maximum resolution and target quality. The high resolution of the input images can lead to increased computational complexity, whereas a combination of high- and low-resolution images for fusion can make the problem of feature level fusion challenging in terms of robustness. Examples of the application of wavelets include image pseudo coloring [85], improving the resolution of the images through super-resolution techniques [8], diagnosis with medical images [27,145,152,172], lifting schemes [173], image segmentation [146], planning for radiotherapy treatment using 3-D conformal mapping [154], and color visualization for labeling [167].

### 27.2.3 WAVELET-BASED HYBRID FEATURE FUSION

The features obtained from the wavelet feature fusion techniques have been used along with other feature extraction methods to improve the robustness of the wavelet-based fusion approaches. Neural networks, considered an excellent candidate for dimensionality reduction and feature extraction, have been employed along with fusion rules set by wavelet operators to implement medical image fusion [145,151,172]. Several combinations of the operators exist that have been combined along with wavelet operators to improve the robustness of the features. Some examples are combinations with support vector machines [150], the use of wavelet-texture measure [27], wavelet combined with magnetic resonance angiogram (MRA) [152,153], the use of wavelet-self adaptive operator [155], wavelet-resolution with entropy [156,158], nonlinear approach with properties of wavelet-shift invariant imaging [157], independent component analysis (ICA) combined with wavelet [174], wavelet and edge features [161], wavelet with a genetic approach [162], wavelet combined with contourlet transform [168], hybrid of neuron networks with fuzzy logic and wavelets [169], and wavelet entropy [171].

### 27.2.4 COMPONENT ANALYSIS TECHNIQUES

Several dimensionality reduction methods exist that can reduce the large feature set to a smaller subset of algebraically transformed features. The idea of extracting components from the images has been widely explored via ICA [97,174] and principal component analysis (PCA) [175–178]. Like wavelet coefficients, the derived feature coefficients from these techniques can be used to reconstruct the image with only a small number of feature coefficients. They find application in higher resolution and large volume imaging such as volumetric medical imagery [179]. A multimodal image fusion based on PCA using the intensity-hue-saturation (IHS) transform has been shown to preserve spatial features and required functional information without color distortion [178].

### 27.2.5 TRANSFORM-BASED APPROACHES

There are different mathematical transforms on features that can enhance the performance of the image fusion. For example, the combination of complex contourlet transform with wavelet has been shown to result in robust image fusion [176,177]. Transform-based methods are also applied for liver diagnosis [50], risk factor fusion [180], prediction of multifactorial diseases [180], parametric classification [180], local image analysis [181], and multimodality image fusion [168,176,177,182–184]. Possibilistic clustering methods show improvement over the fuzzy c-means clustering and have a wide range of application in registration stages of image fusion. Some of the applications of possibilistic clustering include tissue classification [87], diagnosis of brain conditions [34,185], and automatic segmentation [88].

## 27.3 DATA FUSION METHODS IN MEDICAL IMAGING

### 27.3.1 KNOWLEDGE IN DATA FUSION METHODS

Even more often, it becomes quite a difficult premise to replace the expertise of the medical practitioner in improving and validating the computer-aided analysis of the medical images in segmentation of the regions of interest, labeling and updating the points of interest, and re-registering the images. A high level of domain-specific knowledge is required to specify the type of image and region of interest, which leads to a range of practical applications in the image analysis concerned with region segmentation [79], microcalcification diagnosis [186], classification of tissues [89], diagnosis of brain-related condition [89], classifiers for fusion [107], breast cancer and tumor detection [107], and delineation and recognition of anatomical parts of the brain [79].

### 27.3.2 DATA FUSION WITH ARTIFICIAL NEURAL NETWORKS

Artificial neural networks (ANNs) represent a set of decision processing models inspired from the working of the human neural network. The neural networks consist of a weighted addition of inputs followed up with decisions at each of its nodes and further layers of neuron nodes acting as decision aggregates to global decisions. Because each node processes information from the group of input pixels, the network can learn and make decisions in modular levels. This makes it useful for a wide range of decision fusion applications that involve feature generation and classification [187], generic data fusion [145,186,187], various applications specific to image fusion [103,145,151,188–192], identification and diagnosis of microcalcification [186], breast cancer detection [103,109,193], data-driven medical diagnosis [145,172,191], cancer diagnosis [194], natural computing methods [195], and classifier fusion [193].

### 27.3.3 DATA FUSION WITH HYBRID ARTIFICIAL NEURAL NETWORKS

The combination of ANNs with other fusion techniques results in hybrid-ANN methods. They usually combine feature-level decisions and fusion-level decisions with the neural network training algorithms to improve image fusion performances. The major group of techniques includes wavelets combined with the neural network [145,151,172], neural networks combined with fuzzy logic [190,192], combinations of fuzzy logic with a genetic-neural network [195], and support vector machines (SVMs) combined with ANN and Gaussian mixture model (GMM) [193].

### 27.3.4 DATA FUSION WITH FUZZY LOGIC

The fuzzy approach to decision making allows for a greater level of flexibility in the grouping of features and decisions utilizing a wide set of fuzzy set operators and membership functions for image-based decision fusion algorithms [10,18,32,34,38,88,91,93,95,162,169,190,192,195–204]. They find applications in diagnosis of brain conditions [32,34,91,196], treatment of cancer [38], image integration and segmentation [38,88], maximization of mutual information [10], deep brain stimulation [93], segmentation of brain tumors [95], feature fusion and image retrieval [197,198], weighted entropy calculations in images [197], multimodal analysis and image fusion [162,190,199], ovarian cancer detection and diagnosis [200], sensor-oriented image fusion [201], natural computing methods [195], and gene expression [202,203].

### 27.3.5 DATA FUSION WITH HYBRID FUZZY LOGIC

The optimal selection of feature sets, membership functions, and fuzzy operators remains an open problem. Similar to other hybrid approaches, fuzzy decisions can be combined with other fusion approaches to obtain hybrid-fuzzy fusion algorithms. Common examples of hybrid-fuzzy fusion methods are fuzzy-neural network [190,192], fuzzy logic combined with genetic-neural network-rough set [195], fuzzy logic with statistical probability measures [202], and fuzzy logic combined with neural networks and wavelets [169].

### 27.3.6 SVM CLASSIFIER-BASED APPROACHES

Decision fusion is a straightforward operation when it comes to the majority of classifiers, as they inherently need to make local and global decisions to classify patterns. Most classifiers rely on thresholds to make a decision, whereas others go with statistical approaches; nonetheless ranking the scores and selecting the most likely one forms the core idea of asserting the presence or absence of a pattern. SVMs are a parameter-driven approach of detecting feature closeness and removing outliers for determining the class of the patterns. The ability to make decisions at local levels in the images is used in the process of decision fusion. Some of the applications of SVMs as a tool for image fusion include cancer diagnosis [194,205], classifier fusion [108,193,205], breast cancer diagnosis and treatment [108,193], image fusion [150,206], content-based image retrieval [207,208], tumor segmentation [206], gene classification [209], and feature fusion [208].

### 27.3.7 HYBRID SVM CLASSIFIER-BASED APPROACHES

The SVMs can be combined along with other fusion algorithms and techniques to improve processing speed and to work with better representations of low-dimensional feature vectors. These hybrid SVM methods include SVM combined with wavelets [150], SVM with adaptive similarity measures [207], SVM-data fusion [206], and SVM combined with ANN and GMM [193].

## 27.4 DISCUSSION AND CONCLUSIONS

Image fusion studies with medical images face several challenges having a significant impact in the field of medical diagnostics and monitoring. The wide-range use of information and communication technologies in the health sciences during the last decade has increased trust in technology for image analysis as an essential tool. However, there is hardly any imaging modality that can capture all the possible mechanisms required to reveal the conditions under study. This necessitates the use of multimodal imaging techniques; however, they are limited by the significant footprint it takes on computational and human resources to improve the efficiency of decision processing and clinical conclusions.

The technological challenges with image fusion are manifold, including sensor-level errors, imaging noise, interimage variabilities, motion artifacts, contrast variations, and interimage resolution mismatches. Many of these issues also make the automated co-registration and normalization process between the images a difficult problem to solve. They become even more serious issues in real-time imaging systems where high-speed sampling along with increased imaging accuracy is essential to ensure accuracy and reliability of image fusion methods.

Feature-level fusion methods are affected by the imaging quality and the natural variability of the modality. The importance of improving the image formation methods necessitates careful attention when designing feature fusion techniques, as they constitute the primary reason for the robustness of fusion techniques across a wide range of imaging conditions. Noise estimation is another important area that is growing in significance to improve the signal quality before fusion techniques can be applied. The processing speed of the large-volume feature fusion algorithms can be improved by the practical realization of algorithms in field-programmable gate array (FPGA) or graphical processing units. They could in the future find practical applications in real-time monitoring, telemedical diagnosis, and surgery.

The decision-level fusion methods require a good set of features at most times to ensure high reliability of fusion. Because they are highly dependent on the underlying data structure, they are generally referred to as a data fusion technique. The computational complexity of a majority of decision fusion techniques increases nonlinearly with any linear increase in feature size. In the future, this can be a serious challenge, as the convenience of data-driven processing requires a large volume of images for processing to ensure high accuracy and reliability. Although the data-driven techniques can lead to robust fusion rules, the trust of the users plays a major role in the adoption of data-driven techniques in mainstream health systems.

Overall, both feature and data fusion techniques have made promising progress in the practical domains of medical diagnosis and analysis. This is evident from the large number of algorithmic and medical studies that make use of automated medical image fusion techniques. Future progress could very well depend on developing techniques that are well tested across realistic case studies and scenarios across a large collection of data. This also requires a large-scale standardization of data sets to compare techniques that can be considered reliable to be used in clinical settings. The major methods that have been shown to be useful for feature and data fusion include wavelet transforms, neural networks, ICA/PCA, fuzzy logic, morphology methods, SVMs, and their combinations. Further major progress that is required is the miniaturization of medical devices with increased processing capability and reliability. The ability of these devices to make use of modern communication technologies also plays a major role in the sustainable use of the fusion algorithms.

## REFERENCES

1. M. Ulug, and C. L. McCullough, "Feature and data-level fusion of infrared and visual images," in *AeroSense'99*. International Society for Optics and Photonics, 1999, pp. 312–318.
2. T. Peli, M. Young, R. Knox, K. K. Ellis, and F. Bennett, "Feature-level sensor fusion," in *AeroSense'99*. International Society for Optics and Photonics, Orlando. FL, April 5, 1999, pp. 332–339.
3. A. P. James, and B. V. Dasarathy, "Medical image fusion: A survey of the state of the art," *Information Fusion*, vol. 19, pp. 4–19, 2014.
4. B. V. Dasarathy, "Editorial: Information fusion in the realm of medical applications—A bibliographic glimpse at its growing appeal," *Information Fusion*, vol. 13, no. 1, pp. 1–9, 2012.
5. Q. Guihong, Z. Dali, and Y. Pingfan, "Medical image fusion by wavelet transform modulus maxima," *Optics Express*, vol. 9, no. 4, pp. 184–190, 2001.

6. A. Taleb-Ahmed, and L. Gautier, "On information fusion to improve segmentation of MRI sequences," *Information Fusion*, vol. 3, no. 2, pp. 103–117, 2002.

7. M. Aguilar, and J. R. New, "Fusion of multi-modality volumetric medical imagery," in *Information Fusion, 2002. Proceedings of the Fifth International Conference on*, IEEE, Annapolis, MD, July 8–11, 2002, vol. 2, pp. 1206–1212.

8. R. Kapoor, A. Dutta, D. Bagai, and T. S. Kamal, "Fusion for registration of medical images—A study," in *Applied Imagery Pattern Recognition Workshop, 2003. Proceedings. 32nd*. IEEE, Washington, DC, October 15–17, 2003, pp. 180–185.

9. M. Vermandel, N. Betrouni, G. Palos, J.-Y. Gauvrit, C. Vasseur, and J. Rousseau, "Registration, matching, and data fusion in 2D/3D medical imaging: Application to DSA and MRA," in *Medical Image Computing and Computer-Assisted Intervention-MICCAI 2003*. Springer, New York, 2003, pp. 778–785.

10. C.-H. Huang, and J.-D. Lee, "Improving MMI with enhanced-FCM for the fusion of brain MR and SPECT images," in *Pattern Recognition, 2004. ICPR 2004. Proceedings of the 17th International Conference on*, IEEE, Cambridge, UK, August 23–26, 2004, vol. 3, pp. 562–565.

11. A.-S. Capelle, O. Colot, and C. Fernandez-Maloigne, "Evidential segmentation scheme of multi-echo MR images for the detection of brain tumors using neighborhood information," *Information Fusion*, vol. 5, no. 3, pp. 203–216, 2004.

12. Y.-M. Zhu, and S. M. Cochoff, "An object-oriented framework for medical image registration, fusion, and visualization," *Computer Methods and Programs in Biomedicine*, vol. 82, no. 3, pp. 258–267, 2006.

13. A. Wang, H. Sun, and Y. Guan, "The application of wavelet transform to multi-modality medical image fusion," in *Networking, Sensing and Control, 2006. ICNSC'06. Proceedings of the 2006 IEEE International Conference on*. IEEE, 2006, pp. 270–274.

14. O. Tanaka, S. Hayashi, M. Matsuo, M. Nakano, H. Uno, K. Ohtakara, S. Okada, H. Hoshi, and T. Deguchi, "Effect of edema on postimplant dosimetry in prostate brachytherapy using CT/MRI fusion," *European Journal of Cancer Supplements*, vol. 5, no. 4, p. 292, 2007.

15. S. F. Nemec, P. Peloschek, M. T. Schmook, C. R. Krestan, W. Hauff, C. Matula, and C. Czerny, "CT–MR image data fusion for computer-assisted navigated surgery of orbital tumors," *European Journal of Radiology*, vol. 73, no. 2, pp. 224–229, 2010.

16. S. Daneshvar, and H. Ghassemian, "MRI and PET image fusion by combining IHS and retina-inspired models," *Information Fusion*, vol. 11, no. 2, pp. 114–123, 2010.

17. H. Park, C. R. Meyer, D. Wood, A. Khan, R. Shah, H. Hussain, J. Siddiqui, J. Seo, T. Chenevert, and M. Piert, "Validation of automatic target volume definition as demonstrated for [11]C-choline PET/CT of human prostate cancer using multi-modality fusion techniques," *Academic Radiology*, vol. 17, no. 5, pp. 614–623, 2010.

18. J. Teng, S. Wang, J. Zhang, and X. Wang, "Fusion algorithm of medical images based on fuzzy logic," in *Fuzzy Systems and Knowledge Discovery(FSKD), 2010 Seventh International Conference on*, IEEE, Yantai, China, August 10–12, 2010, vol. 2, pp. 546–550.

19. E. Faliagka, G. Matsopoulos, A. Tsakalidis, J. Tsaknakis, and G. Tzimas, "Registration and fusion techniques for medical images: Demonstration and evaluation," in *Biomedical Engineering Systems and Technologies*. Springer, New York, 2011, pp. 15–28.

20. B. Hentschel, W. Oehler, D. Strauß, A. Ulrich, and A. Malich, "Definition of the CTV prostate in CT and MRI by using CT–MRI image fusion in IMRT planning for prostate cancer," *Strahlentherapie und Onkologie*, vol. 187, no. 3, pp. 183–190, 2011.

21. C. Tsien, W. Parker, D. Parmar, D. Hristov, L. Souhami, and C. Freeman, "The role of MRI fusion in radiotherapy planning of pediatric CNS tumors," *International Journal of Radiation Oncology Biology Physics*, vol. 45, no. 3, pp. 188–189, 1999.

22. J. Greensmith, U. Aickelin, and G. Tedesco, "Information fusion for anomaly detection with the dendritic cell algorithm," *Information Fusion*, vol. 11, no. 1, pp. 21–34, 2010.

23. M. Uematsu, A. Shioda, H. Taira, Y. Hama, A. Suda, J. Wong, and S. Kusano, "Interfractional movements of the prostate detected by daily computed tomography (CT)-guided precise positioning system with a fusion of CT and linear accelerator (focal) unit," *International Journal of Radiation Oncology Biology Physics*, vol. 54, no. 2, p. 13, 2002.

24. H. Fukunaga, M. Sekimoto, M. Ikeda, I. Higuchi, M. Yasui, I. Seshimo, O. Takayama, H. Yamamoto, M. Ohue, M. Tatsumi et al., "Fusion image of positron emission tomography and computed tomography for the diagnosis of local recurrence of rectal cancer," *Annals of Surgical Oncology*, vol. 12, no. 7, pp. 561–569, 2005.

25. R. Cambria, F. Cattani, M. Ciocca, C. Garibaldi, G. Tosi, and R. Orecchia, "CT image fusion as a tool for measuring in 3D the setup errors during conformal radiotherapy for prostate cancer," *Tumori*, vol. 92, no. 2, p. 118, 2006.

26. R. J. Ellis, H. Zhou, D. A. Kaminsky, P. Fu, E. Y. Kim, D. B. Sodee, V. Colussi, J. P. Spirnak, C. C. Whalen, and M. I. Resnick, "Rectal morbidity after permanent prostate brachytherapy with dose escalation to biologic target volumes identified by SPECT/CT fusion," *Brachytherapy*, vol. 6, no. 2, pp. 149–156, 2007.

27. K. Yuanyuan, L. Bin, T. Lianfang, and M. Zongyuan, "Multi-modal medical image fusion based on wavelet transform and texture measure," in *Control Conference, 2007. Chinese*. IEEE, 2007, pp. 697–700.

28. A. P. Pecking, W. Wartski, R. Cluzan, D. Bellet, and J. Albérini, "SPECT–CT fusion imaging radio-nuclide lymphoscintigraphy: Potential for limb lymphedema assessment and sentinel node detection in breast cancer," in *Cancer Metastasis and the Lymphovascular System: Basis for Rational Therapy*. Springer, 2007, New York, pp. 79–84.

29. C. Shangli, H. Junmin, and L. Zhongwei, "Medical image of PET/CT weighted fusion based on wavelet transform," in *Bioinformatics and Biomedical Engineering, 2008. ICBBE 2008. The 2nd International Conference on*. IEEE, 2008, pp. 2523–2525.

30. J. L. Alberini, M. Wartski, V. Edeline, O. Madar, S. Banayan, D. Bellet, and A. P. Pecking, "Molecular imaging of neuroendocrine cancer by fusion SPECT/CT," in *From Local Invasion to Metastatic Cancer*. Springer, New York, 2009, pp. 169–175.

31. Y. Liu, J. Yang, and J. Sun, "PET/CT medical image fusion algorithm based on multiwavelet transform," in *Advanced Computer Control (ICACC), 2010 2nd International Conference on*, IEEE, Shenyang, Liaoning, China, March 27–29, 2010, vol. 2, pp. 264–268.

32. C. Barillot, D. Lemoine, L. Le Briquer, F. Lachmann, and B. Gibaud, "Data fusion in medical imaging: Merging multimodal and multipatient images, identification of structures and 3D display aspects," *European Journal of Radiology*, vol. 17, no. 1, pp. 22–27, 1993.

33. J. Julow, T. Major, M. Emri, I. Valalik, S. Sagi, L. Mangel, G. Németh, L. Tron, G. Várallyay, D. Solymosi et al., "The application of image fusion in stereotactic brachytherapy of brain tumours," *Acta Neurochirurgica*, vol. 142, no. 11, pp. 1253–1258, 2000.

34. V. Barra, and J.-Y. Boire, "A general framework for the fusion of anatomical and functional medical images," *NeuroImage*, vol. 13, no. 3, pp. 410–424, 2001.

35. H. Lee, and H. Hong, "Hybrid surface-and voxel-based registration for MR-PET brain fusion," in *Image Analysis and Processing–ICIAP 2005*. Springer, New York, 2005, pp. 930–937.

36. J. A. Marquez, A. Gastellum, and M. A. Padilla, "Image-fusion operators for 3D anatomical and functional analysis of the brain," in *Engineering in Medicine and Biology Society (EMBS 2007). 29th Annual International Conference of the IEEE*. IEEE, 2007, pp. 833–835.

37. H. Lee, J. Lee, G. Kim, and Y. G. Shin, "Efficient hybrid registration method using a shell volume for PET and high resolution MR brain image fusion," in *World Congress on Medical Physics and Biomedical Engineering, September 7–12, 2009, Munich, Germany*. Springer, New York, 2010, pp. 2326–2329.

38. R. Wasserman, R. Acharya, C. Sibata, and K. Shin, "A data fusion approach to tumor delineation," in *Image Processing, 1995. Proceedings, International Conference on*, vol. 2. IEEE, 1995, pp. 476–479.

39. C. Anderson, M. Koshy, C. Staley, N. Esiashvili, S. Ghavidel, Z. Fowler, T. Fox, F. Esteves, J. Landry, and K. Godette, "PET-CT fusion in radiation management of patients with anorectal tumors," *International Journal of Radiation Oncology Biology Physics*, vol. 69, no. 1, pp. 155–162, 2007.

40. A. C. Riegel, A. M. Berson, S. Destian, T. Ng, L. B. Tena, R. J. Mitnick, and P. S. Wong, "Variability of gross tumor volume delineation in head-and-neck cancer using CT and PET/CT fusion," *International Journal of Radiation Oncology Biology Physics*, vol. 65, no. 3, pp. 726–732, 2006.

41. A. Venkatesan, J. Kruecker, S. Xu, J. Locklin, P. Pinto, A. Singh, N. Glossop, and B. Wood, "Abstract no. 155: Early clinical experience with real time ultrasound-MRI fusion-guided prostate biopsies," *Journal of Vascular and Interventional Radiology*, vol. 19, no. 2, pp. S59–S60, 2008.

42. Y. Nakamoto, K. Tamai, T. Saga, T. Higashi, T. Hara, T. Suga, T. Koyama, and K. Togashi, "Clinical value of image fusion from MR and PET in patients with head and neck cancer," *Molecular Imaging and Biology*, vol. 11, no. 1, pp. 46–53, 2009.

43. J. F. Vansteenkiste, S. G. Stroobants, P. J. Dupont, P. R. De Leyn, W. F. De Wever, E. K. Verbeken, J. L. Nuyts, F. P. Maes, and J. G. Bogaert, "FDG-PET scan in potentially operable non-small cell lung cancer: Do anatometabolic PET-CT fusion images improve the localisation of regional lymph node metastases?" *European Journal of Nuclear Medicine*, vol. 25, no. 11, pp. 1495–1501, 1998.

44. E. Deniaud-Alexandre, E. Touboul, D. Lerouge, D. Grahek, J.-N. Foulquier, Y. Petegnief, B. Grès, H. El Balaa, K. Keraudy, K. Kerrou et al., "Impact of computed tomography and [18]F-deoxyglucose coincidence detection emission tomography image fusion for optimization of conformal radiotherapy in non-small-cell lung cancer," *International Journal of Radiation Oncology Biology Physics*, vol. 63, no. 5, pp. 1432–1441, 2005.

45. J.-D. Lee, B.-R. Huang, and C.-H. Huang, "A surface-projection MNI for the fusion of brain MR and SPECT images," *Biomedical Engineering: Applications, Basis and Communications*, vol. 18, no. 4, pp. 202–206, 2006.

46. M. C. Dastjerdi, A. Karimian, H. Afarideh, and A. Mohammadzadeh, "FMDIB: A software tool for fusion of MRI and DHC-SPECT images of brain," in *World Congress on Medical Physics and Biomedical Engineering, September 7–12, 2009, Munich, Germany*. Springer, New York, 2009, pp. 741–744.

47. K.-P. Lin, and W.-J. Yao, "A SPECT-CT image fusion technique for diagnosis of head-neck cancer," in *Engineering in Medicine and Biology Society, 1995, IEEE 17th Annual Conference*, IEEE, Montreal, Quebec, Canada, September 20–24, 1995, vol. 1, pp. 377–378.

48. N. Tomura, O. Watanabe, K. Omachi, I. Sakuma, S. Takahashi, T. Otani, H. Kidani, and J. Watarai, "Image fusion of thallium-201 SPECT and MR imaging for the assessment of recurrent head and neck tumors following flap reconstructive surgery," *European Radiology*, vol. 14, no. 7, pp. 1249–1254, 2004.

49. S. Katyal, E. L. Kramer, M. E. Noz, D. McCauley, A. Chachoua, and A. Steinfeld, "Fusion of immunoscintigraphy single photon emission computed tomography (SPECT) with CT of the chest in patients with non-small cell lung cancer," *Cancer Research*, vol. 55, no. 23 Supplement, pp. 5759s–5763s, 1995.

50. T. Chung, Y. Liu, C. Chen, Y. Sun, N. Chiu, and J. Lee, "Intermodality registration and fusion of liver images for medical diagnosis," in *Intelligent Information Systems, 1997. IIS'97. Proceedings*, IEEE, Grand Bahama Island, Bahamas, December 8–10, 1997, pp. 42–46.

51. J. Li, and K. F. Koral, "An algorithm to adjust a rigid CT-SPECT fusion so as to maximize tumor counts from CT VOI in I-131 therapies," in *Nuclear Science Symposium Conference Record, 2001 IEEE*, IEEE, San Diego, CA, November 4–10, 2001, vol. 3, pp. 1432–1436.

52. D. B. Sodee, A. E. Sodee, and G. Bakale, "Synergistic value of single-photon emission computed tomography/computed tomography fusion to radioimmunoscintigraphic imaging of prostate cancer," in *Seminars in Nuclear Medicine*, vol. 37, no. 1, pp. 17–28, 2007.

53. M. Aguilar, J. R. New, and E. Hasanbelliu, "Advances in the use of neurophysiologically-based fusion for visualization and pattern recognition of medical imagery," *Information Fusion, 2003. Proceedings of the Sixth International Conference of*, IEEE, July 8–11, 2003, vol. 2, pp. 860–867.

54. B. Fei, Z. Lee, D. T. Boll, J. L. Duerk, J. S. Lewin, and D. L. Wilson, "Image registration and fusion for interventional MRI guided thermal ablation of the prostate cancer," in *Medical Image Computing and Computer-Assisted Intervention-MICCAI 2003*. Springer, New York, 2003, pp. 364–372.

55. B. Fei, Z. Lee, D. T. Boll, J. L. Duerk, D. B. Sodee, J. S. Lewin, and D. L. Wilson, "Registration and fusion of SPECT, high-resolution MRI, and interventional MRI for thermal ablation of prostate cancer," *Nuclear Science, IEEE Transactions on*, vol. 51, no. 1, pp. 177–183, 2004.

56. M. Krengli, M. Dominietto, S. Chiara, C. Barbara, R. Marco, I. Eugenio, K. Irvin, B. Frea et al., "Study of lymphatic drainage by SPECT-CT fusion images for pelvic irradiation of prostate cancer," *International Journal of Radiation Oncology Biology Physics*, vol. 63, p. S305, 2005.

57. M. Krengli, A. Ballarè, B. Cannillo, M. Rudoni, E. Kocjancic, G. Loi, M. Brambilla, E. Inglese, and B. Frea, "Potential advantage of studying the lymphatic drainage by sentinel node technique and SPECT-CT image fusion for pelvic irradiation of prostate cancer," *International Journal of Radiation Oncology Biology Physics*, vol. 66, no. 4, pp. 1100–1104, 2006.

58. C.-C. Tsai, C.-S. Tsai, K.-K. Ng, C.-H. Lai, S. Hsueh, P.-F. Kao, T.-C. Chang, J.-H. Hong, and T.-C. Yen, "The impact of image fusion in resolving discrepant findings between FDG-PET and MRI/CT in patients with gynaecological cancers," *European Journal of Nuclear Medicine and Molecular Imaging*, vol. 30, no. 12, pp. 1674–1683, 2003.

59. T. Denecke, B. Hildebrandt, L. Lehmkuhl, N. Peters, A. Nicolaou, M. Pech, H. Riess, J. Ricke, R. Felix, and H. Amthauer, "Fusion imaging using a hybrid SPECT-CT camera improves port perfusion scintigraphy for control of hepatic arterial infusion of chemotherapy in colorectal cancer patients," *European Journal of Nuclear Medicine and Molecular Imaging*, vol. 32, no. 9, pp. 1003–1010, 2005.

60. C. Beneder, F. Fuechsel, T. Krause, A. Kuhn, and M. Mueller, "The role of 3D fusion imaging in sentinel lymphadenectomy for vulvar cancer," *Gynecologic Oncology*, vol. 109, no. 1, pp. 76–80, 2008.

61. Z. Zhao, L. Li, F. Li, and L. Zhao, "Single photon emission computed tomography/spiral computed tomography fusion imaging for the diagnosis of bone metastasis in patients with known cancer," *Skeletal Radiology*, vol. 39, no. 2, pp. 147–153, 2010.

62. H. Iwase, Y. Yamamoto, T. Kawasoe, and M. Ibusuki, "Qs106. Sentinel lymph node biopsy using SPECT-CT fusion imaging in patients with breast cancer and its clinical usefulness," *Journal of Surgical Research*, vol. 151, no. 2, p. 287, 2009.

63. E. Holupka, I. Kaplan, E. Burdette, and G. Svensson, "Ultrasound image fusion for external beam radio-therapy for prostate cancer," *International Journal of Radiation Oncology Biology Physics*, vol. 35, no. 5, pp. 975–984, 1996.

64. I. Kaplan, E. Holupka, and M. Morrissey, "MRI-ultrasound image fusion for $^{125}$I prostate implant treatment planning," *International Journal of Radiation Oncology Biology Physics*, vol. 42, no. 1, p. 294, 1998.

65. L. Beaulieu, D. Tubic, J. Pouliot, E. Vigneault, and R. Taschereau, "Post-implant dosimetry using fusion of ultrasound images with 3D seed coordinates from fluoroscopic images in transperineal interstitial permanent prostate brachytherapy," *International Journal of Radiation Oncology Biology Physics*, vol. 48, no. 3, p. 360, 2000.

66. L. Taylor, J. Beaty, J. Enderle, and M. Escabi, "Design of a simple ultrasound/CT fusion image fusion solution for the evaluation of prostate seed brachytherapy," in *Bioengineering Conference, 2001. Proceedings of the IEEE 27th Annual Northeast*. IEEE, 2001, pp. 57–58.

67. B. C. Porter, L. Taylor, R. Baggs, A. di Sant'Agnese, G. Nadasdy, D. Pasternack, D. J. Rubens, and K. J. Parker, "Histology and ultrasound fusion of excised prostate tissue using surface registration," in *Ultrasonics Symposium, 2001 IEEE*, vol. 2. IEEE, 2001, pp. 1473–1476.

68. D. B. Fuller, H. Jin, J. A. Koziol, and A. C. Feng, "CT–ultrasound fusion prostate brachytherapy: A dynamic dosimetry feedback and improvement method. A report of 54 consecutive cases," *Brachytherapy*, vol. 4, no. 3, pp. 207–216, 2005.

69. N. Patanjali, M. Keyes, W. J. Morris, M. Liu, R. Harrison, I. Spadinger, and V. Moravan, "A comparison of post-implant US/CT image fusion and MRI/CT image fusion for $^{125}$I prostate brachytherapy post implant dosimetry," *Brachytherapy*, vol. 8, no. 2, p. 124, 2009.

70. F. Dube, A. Mahadevan, and T. Sheldon, "Fusion of CT and 3D ultrasound (3DUS) for prostate delineation of patients with metallic hip prostheses (MHP)," *International Journal of Radiation Oncology Biology Physics*, vol. 75, no. 3, pp. S327–S328, 2009.

71. A. Rastinehad, J. Kruecker, C. Benjamin, P. Chung, B. Turkbey, S. Xu, J. Locklin, S. Gates, C. Buckner, M. Linehan et al., "MRI/US fusion prostate biopsies: Cancer detection rates," *The Journal of Urology*, vol. 185, no. 4, p. e340, 2011.

72. N. Papanikolaou, D. Gearheart, T. Bolek, A. Meigooni, D. Meigooni, and M. Mohiuddin, "A volumetric and dosimetric study of LDR brachytherapy prostate implants based on image fusion of ultrasound and computed tomography," in *Engineering in Medicine and Biology Society, 2000. Proceedings of the 22nd Annual International Conference of the IEEE*, vol. 4. IEEE, 2000, pp. 2769–2770.

73. F. Arena, T. DiCicco, and A. Anand, "Multimodality data fusion aids early detection of breast cancer using conventional technology and advanced digital infrared imaging," in *Engineering in Medicine and Biology Society, 2004. IEMBS'04. 26th Annual International Conference of the IEEE*, IEEE, San Francisco, CA, September 1–5, 2004, vol. 1, pp. 1170–1173.

74. A. Hakime, F. Deschamps, E. G. M. De Carvalho, C. Teriitehau, A. Auperin, and T. De Baere, "Clinical evaluation of spatial accuracy of a fusion imaging technique combining previously acquired computed tomography and real-time ultrasound for imaging of liver metastases," *Cardiovascular and Interventional Radiology*, vol. 34, no. 2, pp. 338–344, 2011.

75. B. Hadaschik, T. Kuru, C. Tulea, D. Teber, J. Huber, V. Popeneciu, S. Pahernik, H.-P. Schlemmer, and M. Hohenfellner, "Stereotactic prostate biopsy with pre-interventional MRI and live US fusion," *The Journal of Urology*, vol. 185, no. 4, p. e924, 2011.

76. A. J. Walker, B. J. Spier, S. B. Perlman, J. R. Stangl, T. J. Frick, D. V. Gopal, M. J. Lindstrom, T. L. Weigel, and P. R. Pfau, "Integrated PET/CT fusion imaging and endoscopic ultrasound in the preoperative staging and evaluation of esophageal cancer," *Molecular Imaging and Biology*, vol. 13, no. 1, pp. 166–171, 2011.

77. S. Marshall, and G. Matsopoulos, "Morphological data fusion in medical imaging," in *Nonlinear Digital Signal Processing, 1993. IEEE Winter Workshop on*. IEEE, 1993, pp. 6.1_5.1–6.1_5.6.

78. G. Matsopoulos, S. Marshall, and J. Brunt, "Multiresolution morphological fusion of MR and CT images of the human brain," in *Vision, Image and Signal Processing, IEEE Proceedings*, vol. 141, no. 3. IET, 1994, pp. 137–142.

79. H. Li, R. Deklerck, B. De Cuyper, A. Hermanus, E. Nyssen, and J. Cornelis, "Object recognition in brain CT-scans: Knowledge-based fusion of data from multiple feature extractors," *Medical Imaging, IEEE Transactions on*, vol. 14, no. 2, pp. 212–229, 1995.

80. D. Dey, D. G. Gobbi, P. J. Slomka, K. J. Surry, and T. M. Peters, "Automatic fusion of freehand endoscopic brain images to three-dimensional surfaces: Creating stereoscopic panoramas," *Medical Imaging, IEEE Transactions on*, vol. 21, no. 1, pp. 23–30, 2002.

81. A. Viola, T. Major, and J. Julow, "The importance of postoperative CT image fusion verification of stereotactic interstitial irradiation for brain tumors," *International Journal of Radiation Oncology Biology Physics*, vol. 60, no. 1, pp. 322–328, 2004.

82. H. Eldredge, A. Doemer, D. Friedman, and M. Werner-Wasik, "Improvement in optic chiasm contouring for RT planning in patients with brain tumors using CT/MP-RAGE MRI fusion as compared to the routine T1-weighted MRI image," *International Journal of Radiation Oncology Biology Physics*, vol. 75, no. 3, p. S246, 2009.

83. S. A. Kuhn, B. Romeike, J. Walter, R. Kalff, and R. Reichart, "Multiplanar MRI–CT fusion neuronavigation-guided serial stereotactic biopsy of human brain tumors: Proof of a strong correlation between tumor imaging and histopathology by a new technical approach," *Journal of Cancer Research and Clinical Oncology*, vol. 135, no. 9, pp. 1293–1302, 2009.

84. X. Yong, S. Eberl, and D. Feng, "Dual-modality 3D brain PET-CT image segmentation based on probabilistic brain atlas and classification fusion," *Image Processing (ICIP), 2010 17th IEEE International Conference on*, IEEE, September 26–29, 2010, pp. 2557–2560.

85. C. Kok, Y. Hui, and T. Nguyen, "Medical image pseudo coloring by wavelet fusion," in *Engineering in Medicine and Biology Society, 1996. Bridging Disciplines for Biomedicine. Proceedings of the 18th Annual International Conference of the IEEE*, vol. 2. IEEE, 1996, pp. 648–649.

86. M. C. Erie, C. H. Chu, and R. D. Sidman, "Visualization of the cortical potential field by medical imaging data fusion," in *Visual Information and Information Systems*. Springer, New York, 1999, pp. 815–822.

87. V. Barra, and J.-Y. Boire, "Quantification of brain tissue volumes using MR/MR fusion," in *Engineering in Medicine and Biology Society, 2000. Proceedings of the 22nd Annual International Conference of the IEEE*, vol. 2. IEEE, 2000, pp. 1451–1454.

88. V. Barra, and J.-Y. Boire, "Automatic segmentation of subcortical brain structures in MR images using information fusion," *Medical Imaging, IEEE Transactions on*, vol. 20, no. 7, pp. 549–558, 2001.

89. W. Dou, S. Ruan, Q. Liao, D. Bloyet, and J. Constans, "Knowledge based fuzzy information fusion applied to classification of abnormal brain tissues from MRI," in *Signal Processing and Its Applications, 2003. Proceedings. Seventh International Symposium on*, vol. 1. IEEE, 2003, pp. 681–684.

90. R. Gorniak, E. Kramer, G. Q. Maguire Jr., M. E. Noz, C. Schettino, and M. P. Zeleznik, "Evaluation of a semiautomatic 3D fusion technique applied to molecular imaging and MRI brain/frame volume data sets," *Journal of Medical Systems*, vol. 27, no. 2, pp. 141–156, 2003.

91. I. Bloch, O. Colliot, O. Camara, and T. Géraud, "Fusion of spatial relationships for guiding recognition, example of brain structure recognition in 3D MRI," *Pattern Recognition Letters*, vol. 26, no. 4, pp. 449–457, 2005.

92. A. Villéger, L. Ouchchane, J.-J. Lemaire, and J.-Y. Boire, "Assistance to planning in deep brain stimulation: Data fusion method for locating anatomical targets in MRI," in *Engineering in Medicine and Biology Society, 2006. EMBS'06. 28th Annual International Conference of the IEEE*. IEEE, 2006, pp. 144–147.

93. A. Villéger, L. Ouchchane, J.-J. Lemaire, and J.-Y. Boire, "Data fusion and fuzzy spatial relationships for locating deep brain stimulation targets in magnetic resonance images," in *Advanced Concepts for Intelligent Vision Systems*. Springer, New York, 2006, pp. 909–919.

94. R. A. Heckemann, J. V. Hajnal, P. Aljabar, D. Rueckert, and A. Hammers, "Multiclassifier fusion in human brain MR segmentation: Modelling convergence," in *Medical Image Computing and Computer-Assisted Intervention–MICCAI 2006*. Springer, New York, 2006, pp. 815–822.

95. W. Dou, S. Ruan, Q. Liao, D. Bloyet, J.-M. Constans, and Y. Chen, "Fuzzy information fusion scheme used to segment brain tumor from MR images," in *Fuzzy Logic and Applications*. Springer, New York, 2006, pp. 208–215.

96. K. Yuan, W. Liu, S. Jia, and P. Xiao, "Fusion of MRI and DTI to assist the treatment solution of brain tumor," in *Innovative Computing, Information and Control, 2007. ICICIC'07. Second International Conference on*, IEEE, September 5–7, 2007, p. 620.

97. V. D. Calhoun, and T. Adali, "Feature-based fusion of medical imaging data," *Information Technology in Biomedicine, IEEE Transactions on*, vol. 13, no. 5, pp. 711–720, 2009.

98. F. Forbes, S. Doyle, D. Garcia-Lorenzo, C. Barillot, and M. Dojat, "Adaptive weighted fusion of multiple MR sequences for brain lesion segmentation," in *Biomedical Imaging: From Nano to Macro, 2010 IEEE International Symposium on*. IEEE, 2010, pp. 69–72.

99. W. Dou, A. Dong, P. Chi, S. Li, and J. Constans, "Brain tumor segmentation through data fusion of T2-weighted image and MR spectroscopy," in *Bioinformatics and Biomedical Engineering, (iCBBE) 2011 5th International Conference on*. IEEE, 2011, pp. 1–4.

100. C. P. Behrenbruch, K. Marias, P. A. Armitage, M. Yam, N. Moore, R. E. English, and J. M. Brady, "Mri–mammography 2d/3d data fusion for breast pathology assessment," in *Medical Image Computing and Computer-Assisted Intervention–MICCAI 2000.* Springer, New York, 2000, pp. 307–316.

101. K. G. Baum, K. Rafferty, M. Helguera, and E. Schmidt, "Investigation of PET/MRI image fusion schemes for enhanced breast cancer diagnosis," in *Nuclear Science Symposium Conference Record, 2007. NSS'07.* IEEE, vol. 5. IEEE, 2007, pp. 3774–3780.

102. G. M. Duarte, C. Cabello, R. Z. Torresan, M. Alvarenga, G. H. Telles, S. T. Bianchessi, N. Caserta, S. R. Segala, M. d. C. L. de Lima, E. C. S. de Camargo Etchebehere et al., "Fusion of magnetic resonance and scintimammography images for breast cancer evaluation: A pilot study," *Annals of Surgical Oncology,* vol. 14, no. 10, pp. 2903–2910, 2007.

103. H. Szu, I. Kopriva, P. Hoekstra, N. Diakides, M. Diakides, J. Buss, and J. Lupo, "Early tumor detection by multiple infrared unsupervised neural nets fusion," in *Engineering in Medicine and Biology Society, 2003. Proceedings of the 25th Annual International Conference of the IEEE,* vol. 2. IEEE, 2003, pp. 1133–1136.

104. Y. Chen, E. Gunawan, Y. Kim, K. Low, and C. Soh, "UWB microwave imaging for breast cancer detection: Tumor/clutter identification using a time of arrival data fusion method," in *Antennas and Propagation Society International Symposium 2006, IEEE.* IEEE, 2006, pp. 255–258.

105. Y. Chen, E. Gunawan, K. S. Low, S.-C. Wang, C. B. Soh, and L. L. Thi, "Time of arrival data fusion method for two-dimensional ultrawideband breast cancer detection," *Antennas and Propagation, IEEE Transactions on,* vol. 55, no. 10, pp. 2852–2865, 2007.

106. S. Ueda, H. Tsuda, H. Asakawa, J. Omata, K. Fukatsu, N. Kondo, T. Kondo, Y. Hama, K. Tamura, J. Ishida et al., "Utility of 18F-fluoro-deoxyglucose emission tomography/computed tomography fusion imaging (18F-FDG PET/CT) in combination with ultrasonography for axillary staging in primary breast cancer," *BMC Cancer,* vol. 8, no. 1, p. 165, 2008.

107. M. Raza, I. Gondal, D. Green, and R. L. Coppel, "Classifier fusion to predict breast cancer tumors based on microarray gene expression data," in *Knowledge-Based Intelligent Information and Engineering Systems.* Springer, New York, 2005, pp. 866–874.

108. M. Raza, I. Gondal, D. Green, and R. L. Coppel, "Classifier fusion using Dempster-Shafer theory of evidence to predict breast cancer tumors," in *TENCON 2006. 2006 IEEE Region 10 Conference.* IEEE, 2006, pp. 1–4.

109. Y. Wu, C. Wang, S. C. Ng, A. Madabhushi, and Y. Zhong, "Breast cancer diagnosis using neural-based linear fusion strategies," in *Neural Information Processing.* Springer, New York, 2006, pp. 165–175.

110. Y. M. Kirova, V. Servois, F. Reyal, D. Peurien, A. Fourquet, and N. Fournier-Bidoz, "Use of deformable image fusion to allow better definition of tumor bed boost volume after oncoplastic breast surgery," *Surgical Oncology,* vol. 20, no. 2, pp. e123–e125, 2011.

111. J. L. Jesneck, S. Mukherjee, L. W. Nolte, A. E. Lokshin, J. R. Marks, and J. Lo, "Decision fusion of circulating markers for breast cancer detection in premenopausal women," in *Bioinformatics and Bioengineering, 2007. BIBE 2007. Proceedings of the 7th IEEE International Conference on.* IEEE, 2007, pp. 1434–1438.

112. K. Kagawa, W. R. Lee, T. E. Schultheiss, M. A. Hunt, A. H. Shaer, and G. E. Hanks, "Initial clinical assessment of CT-MRI image fusion software in localization of the prostate for 3D conformal radiation therapy," *International Journal of Radiation Oncology Biology Physics,* vol. 38, no. 2, pp. 319–325, 1997.

113. R. J. Amdur, D. Gladstone, K. A. Leopold, and R. D. Harris, "Prostate seed implant quality assessment using MR and CT image fusion," *International Journal of Radiation Oncology Biology Physics,* vol. 43, no. 1, pp. 67–72, 1999.

114. L. Gong, P. S. Cho, B. H. Han, K. E. Wallner, S. G. Sutlief, S. D. Pathak, D. R. Haynor, and Y. Kim, "Ultrasonography and fluoroscopic fusion for prostate brachytherapy dosimetry," *International Journal of Radiation Oncology Biology Physics,* vol. 54, no. 5, pp. 1322–1330, 2002.

115. V. Servois, L. Chauveinc, C. El Khoury, A. Lantoine, L. Ollivier, T. Flam, J. Rosenwald, J. Cosset, and S. Neuenschwander, "Comparaison de deux méthodes de recalage d'images de scanographie et d'irm en curiethérapie prostatique. intérêt pour l'évaluation thérapeutique," *Cancer/Radiotherapie,* vol. 7, no. 1, pp. 9–16, 2003.

116. J. Crook, M. McLean, I. Yeung, T. Williams, and G. Lockwood, "MRI-CT fusion to assess postbrachytherapy prostate volume and the effects of prolonged edema on dosimetry following transperineal interstitial permanent prostate brachytherapy," *Brachytherapy,* vol. 3, no. 2, pp. 55–60, 2004.

117. T. Wurm, K. Eichhorn, S. Corvin, A. Anastasiadis, R. Bares, and A. Stenzl, "Anatomic-functional image fusion allows intraoperative sentinel node detection in prostate cancer patients," *European Urology Supplements*, vol. 3, no. 2, p. 140, 2004.

118. M. Moerland, I. Jurgenliemk-Schulz, and J. Battermann, "Fusion of pre-implant MRI and intra-operative US images for planning of permanent prostate implants," *Radiotherapy and Oncology*, vol. 75, p. S38, 2005.

119. D. Taussky, L. Austen, A. Toi, I. Yeung, T. Williams, S. Pearson, M. McLean, G. Pond, and J. Crook, "Sequential evaluation of prostate edema after permanent seed prostate brachytherapy using CT-MRI fusion," *International Journal of Radiation Oncology Biology Physics*, vol. 62, no. 4, pp. 974–980, 2005.

120. O. Tanaka, S. Hayashi, K. Sakurai, M. Matsuo, M. Nakano, S. Maeda, H. Hoshi, and T. Deguchi, "Importance of the CT/MRI fusion method as a learning tool for CT-based postimplant dosimetry in prostate brachytherapy," *Radiotherapy and Oncology*, vol. 81, no. 3, pp. 303–308, 2006.

121. S. Wachter, S. Tomek, A. Kurtaran, N. Wachter-Gerstner, B. Djavan, A. Becherer, M. Mitterhauser, G. Dobrozemsky, S. Li, R. Pötter et al., "11C-acetate positron emission tomography imaging and image fusion with computed tomography and magnetic resonance imaging in patients with recurrent prostate cancer," *Journal of Clinical Oncology*, vol. 24, no. 16, pp. 2513–2519, 2006.

122. R. J. Ellis, D. A. Kaminsky, H. Zhou, and M. I. Resnick, "Erectile dysfunction following permanent prostate brachytherapy with dose escalation to biological tumor volumes (BTVS) identified with SPECT/CT fusion," *Brachytherapy*, vol. 6, no. 2, p. 103, 2007.

123. J. Pouliot, A. J. Cunha, G. D. Reed, S. M. Noworolski, J. Kurhanewicz, I. Hsu, and J. Chow, "Multi-image fusions and their role in inverse planned high-dose-rate prostate brachytherapy for dose escalation of dominant intraprostatic lesions defined by combined MRI/MRSI," *Brachytherapy*, vol. 8, no. 2, pp. 113–114, 2009.

124. M. Aoki, A. Yorozu, and T. Dokiya, "Evaluation of interobserver differences in postimplant dosimetry following prostate brachytherapy and the efficacy of CT/MRI fusion imaging," *Japanese Journal of Radiology*, vol. 27, no. 9, pp. 342–347, 2009.

125. A. Mesa, L. Chittenden, J. Lizarde, J. Lee, M. Nelson, J. Lane, A. Spitz, and K. Tokita, "A gold fiducial based CT/MRI fusion method for prostate treatment planning," *International Journal of Radiation Oncology Biology Physics*, vol. 78, no. 3, p. S375, 2010.

126. S. Kadoury, P. Yan, S. Xu, N. Glossop, P. Choyke, B. Turkbey, P. Pinto, B. J. Wood, and J. Kruecker, "Realtime TRUS/MRI fusion targeted-biopsy for prostate cancer: A clinical demonstration of increased positive biopsy rates," in *Prostate Cancer Imaging. Computer-Aided Diagnosis, Prognosis, and Intervention*. Springer, New York, 2010, pp. 52–62.

127. O. Ukimura, M. Desai, M. Aron, A. Hung, A. Berger, S. Valencerina, S. Palmer, and I. Gill, "2131 Elastic registration of 3D prostate biopsy trajectory by real-time 3D TRUS with MR/TRUS fusion: Pilot phatom study," *The Journal of Urology*, vol. 185, no. 4, p. e853, 2011.

128. M. Uematsu, A. Shioda, A. Suda, K. Tahara, T. Kojima, Y. Hama, M. Kono, J. R. Wong, T. Fukui, and S. Kusano, "Intrafractional tumor position stability during computed tomography (CT)-guided frameless stereotactic radiation therapy for lung or liver cancers with a fusion of CT and linear accelerator (focal) unit," *International Journal of Radiation Oncology Biology Physics*, vol. 48, no. 2, pp. 443–448, 2000.

129. M. Schmuecking, K. Plichta, E. Lopatta, C. Przetak, J. Leonhardi, D. Gottschild, T. Wendt, and R. Baum, "Image fusion of F-18 FDG PET and CT-is there a role in 3D-radiation treatment planning of non-small cell lung cancer?" *International Journal of Radiation Oncology Biology Physics*, vol. 48, no. 3, p. 130, 2000.

130. P. Giraud, D. Grahek, F. Montravers, M.-F. Carette, E. Deniaud-Alexandre, F. Julia, J.-C. Rosenwald, J.-M. Cosset, J.-N. Talbot, M. Housset et al., "CT and 18F-deoxyglucose (FDG) image fusion for optimization of conformal radiotherapy of lung cancers," *International Journal of Radiation Oncology Biology Physics*, vol. 49, no. 5, pp. 1249–1257, 2001.

131. Y. Nakamoto, M. Senda, T. Okada, S. Sakamoto, T. Saga, T. Higashi, and K. Togashi, "Software-based fusion of PET and CT images for suspected recurrent lung cancer," *Molecular Imaging and Biology*, vol. 10, no. 3, pp. 147–153, 2008.

132. W. Ge, G. Yuan, C. Li, Y. Wu, Y. Zhang, and X. Xu, "CT image fusion in the evaluation of radiation treatment planning for non-small cell lung cancer," *The Chinese-German Journal of Clinical Oncology*, vol. 7, no. 6, pp. 315–318, 2008.

133. A. Kovacs, J. Hadjiev, F. Lakosi, G. Antal, G. Liposits, and P. Bogner, "Tumor movements detected by multi-slice CT-based image fusion in the radiotherapy of lung cancer," *Lung Cancer*, vol. 64, p. S50, 2009.

134. J. Bradley, K. Bae, N. Choi, K. Forster, B. Siegel, J. Brunetti, J. Purdy, S. Faria, T. Vu, and H. Choy, "A phase II comparative study of gross tumor volume definition with or without PET/CT fusion in dosimetric planning for non–small-cell lung cancer (NSCLC): Primary analysis of radiation therapy oncology group (RTOG) 0515," *International Journal of Radiation Oncology Biology Physics*, vol. 75, no. 3, p. S2, 2009.

135. X. Xu, J. Deng, H. Guo, M. Xiang, C. Li, L. Xu, W. Ge, G. Yuan, Q. Li, and S. Shan, "Ct image fusion in the optimization of replanning during the course of 3-dimensional conformal radiotherapy for non-small-cell lung cancer," in *Biomedical Engineering and Informatics (BMEI), 2010 3rd International Conference on*, vol. 3. IEEE, 2010, pp. 1336–1339.

136. J. Navarra, A. Alsius, S. Soto-Faraco, and C. Spence, "Assessing the role of attention in the audiovisual integration of speech," *Information Fusion*, vol. 11, no. 1, pp. 4–11, 2010.

137. C. E. Hugenschmidt, S. Hayasaka, A. M. Peiffer, and P. J. Laurienti, "Applying capacity analyses to psychophysical evaluation of multisensory interactions," *Information Fusion*, vol. 11, no. 1, pp. 12–20, 2010.

138. J. Twycross, and U. Aickelin, "Information fusion in the immune system," *Information Fusion*, vol. 11, no. 1, pp. 35–44, 2010.

139. S. Wuerger, G. Meyer, M. Hofbauer, C. Zetzsche, and K. Schill, "Motion extrapolation of auditory-visual targets," *Information Fusion*, vol. 11, no. 1, pp. 45–50, 2010.

140. T. D. Dixon, S. G. Nikolov, J. J. Lewis, J. Li, E. F. Canga, J. M. Noyes, T. Troscianko, D. R. Bull, and C. Nishan Canagarajah, "Task-based scanpath assessment of multi-sensor video fusion in complex scenarios," *Information Fusion*, vol. 11, no. 1, pp. 51–65, 2010.

141. J.-B. Lei, J.-B. Yin, and H.-B. Shen, "Feature fusion and selection for recognizing cancer-related mutations from common polymorphisms," in *Pattern Recognition (CCPR), 2010 Chinese Conference on*. IEEE, 2010, pp. 1–5.

142. S. Tsevas, and D. Iakovidis, "Dynamic time warping fusion for the retrieval of similar patient cases represented by multimodal time-series medical data," in *Information Technology and Applications in Biomedicine (ITAB), 2010 10th IEEE International Conference on*. IEEE, 2010, pp. 1–4.

143. H. Müller, and J. Kalpathy-Cramer, "The Image CLEF medical retrieval task at ICPR 2010—Information fusion to combine visual and textual information," in *Recognizing Patterns in Signals, Speech, Images and Videos*. Springer, New York, 2010, pp. 99–108.

144. Z. R. Mnatsakanyan, H. S. Burkom, M. R. Hashemian, and M. A. Coletta, "Distributed information fusion models for regional public health surveillance," *Information Fusion*, vol. 13, no. 2, pp. 129–136, 2012.

145. Q. Zhang, W. Tang, L. Lai, W. Sun, and K. Wong, "Medical diagnostic image data fusion based on wavelet transformation and self-organising features mapping neural networks," in *Machine Learning and Cybernetics, 2004. Proceedings of 2004 International Conference on*, vol. 5. IEEE, 2004, pp. 2708–2712.

146. S. Garg, K. U. Kiran, R. Mohan, and U. Tiwary, "Multilevel medical image fusion using segmented image by level set evolution with region competition," in *Engineering in Medicine and Biology Society, 2005. IEEE-EMBS 2005. 27th Annual International Conference of the*. IEEE, 2006, pp. 7680–7683.

147. L. X. M. L. J. Wang, and S. Hui, "New medical image fusion algorithm based on second generation wavelet transform," in *Computational Engineering in Systems Applications, IMACS Multiconference on*, 2006.

148. W. Li, X. Zhu, and S. Wu, "A novel approach to fast medical image fusion based on lifting wavelet transform," in *Intelligent Control and Automation, 2006. WCICA 2006. The Sixth World Congress on*, vol. 2. IEEE, 2006, pp. 9881–9884.

149. H. Zhang, L. Liu, and N. Lin, "A novel wavelet medical image fusion method," in *Multimedia and Ubiquitous Engineering, 2007. MUE'07. International Conference on*. IEEE, 2007, pp. 548–553.

150. W. Anna, W. Jie, L. Dan, and C. Yu, "Research on medical image fusion based on orthogonal wavelet packets transformation combined with 2V-SVM," in *Complex Medical Engineering, 2007. CME 2007. IEEE/ICME International Conference on*. IEEE, 2007, pp. 670–675.

151. L. Xiaoqi, Z. Baohua, and G. Yong, "Medical image fusion algorithm based on clustering neural network," in *Bioinformatics and Biomedical Engineering, 2007. ICBBE 2007. The 1st International Conference on*. IEEE, 2007, pp. 637–640.

152. B. Alfano, M. Ciampi, and G. De Pietro, "A wavelet-based algorithm for multimodal medical image fusion," in *Semantic Multimedia*. Springer, New York, 2007, pp. 117–120.

153. X. Li, X. Tian, Y. Sun, and Z. Tang, "Medical image fusion by multi-resolution analysis of wavelets transform," in *Wavelet Analysis and Applications*. Springer, New York, 2007, pp. 389–396.

154. L. Bin, T. Lianfang, K. Yuanyuan, and Y. Xia, "Parallel multimodal medical image fusion in 3d conformal radiotherapy treatment planning," in *Bioinformatics and Biomedical Engineering, 2008. ICBBE 2008. The 2nd International Conference on.* IEEE, 2008, pp. 2600–2604.

155. Y. Licai, L. Xin, and Y. Yucui, "Medical image fusion based on wavelet packet transform and self-adaptive operator," in *Bioinformatics and Biomedical Engineering, 2008. ICBBE 2008. The 2nd International Conference on.* IEEE, 2008, pp. 2647–2650.

156. Z. Wencang, and C. Lin, "Medical image fusion method based on wavelet multi-resolution and entropy," in *Automation and Logistics, 2008. ICAL 2008. IEEE International Conference on.* IEEE, 2008, pp. 2329–2333.

157. B. Yang, and Z. Jing, "Medical image fusion with a shift-invariant morphological wavelet," in *2008 IEEE Conference on Cybernetics and Intelligent Systems,* 2008, pp. 175–178.

158. R. Singh, M. Vatsa, and A. Noore, "Multimodal medical image fusion using redundant discrete wavelet transform," in *Advances in Pattern Recognition, 2009. ICAPR'09. Seventh International Conference on.* IEEE, 2009, pp. 232–235.

159. Z.-S. Xiao, and C.-X. Zheng, "Medical image fusion based on an improved wavelet coefficient contrast," in *Bioinformatics and Biomedical Engineering, 2009. ICBBE 2009. 3rd International Conference on.* IEEE, 2009, pp. 1–4.

160. L. Chiorean, and M.-F. Vaida, "Medical image fusion based on discrete wavelet transform using JAVA technology," in *Information Technology Interfaces, 2009. ITI'09. Proceedings of the ITI 2009 31st International Conference on.* IEEE, 2009, pp. 55–60.

161. X. Zhang, Y. Zheng, Y. Peng, W. Liu, and C. Yang, "Research on multi-mode medical image fusion algorithm based on wavelet transform and the edge characteristics of images," in *Image and Signal Processing, 2009. CISP'09. 2nd International Congress on.* IEEE, 2009, pp. 1–4.

162. A. Das, and M. Bhattacharya, "Evolutionary algorithm based automated medical image fusion technique: Comparative study with fuzzy fusion approach," in *Nature Biologically Inspired Computing, 2009. NaBIC 2009. World Congress on,* December 2009, pp. 269–274.

163. Y. Yang, "Multimodal medical image fusion through a new DWT based technique," in *Bioinformatics and Biomedical Engineering (iCBBE), 2010 4th International Conference on.* IEEE, 2010, pp. 1–4.

164. B. Li, L. Tian, and S. Ou, "Rapid multimodal medical image registration and fusion in 3D conformal radiotherapy treatment planning," in *Bioinformatics and Biomedical Engineering (iCBBE), 2010 4th International Conference on.* IEEE, 2010, pp. 1–5.

165. M. Agrawal, P. Tsakalides, and A. Achim, "Medical image fusion using the convolution of meridian distributions," in *Engineering in Medicine and Biology Society (EMBC), 2010 Annual International Conference of the IEEE.* IEEE, 2010, pp. 3727–3730.

166. W. Xue-Jun, and M. Ying, "A medical image fusion algorithm based on lifting wavelet transform," in *Artificial Intelligence and Computational Intelligence (AICI), 2010 International Conference on,* vol. 3. IEEE, 2010, pp. 474–476.

167. M. Ciampi, "Medical image fusion for color visualization via 3D RDWT," in *Information Technology and Applications in Biomedicine (ITAB), 2010 10th IEEE International Conference on.* IEEE, 2010, pp. 1–6.

168. S. Rajkumar, and P. S. Kavitha, "Redundancy discrete wavelet transform and contourlet transform for multimodality medical image fusion with quantitative analysis," in *Emerging Trends in Engineering and Technology (ICETET), 2010 3rd International Conference on.* IEEE, 2010, pp. 134–139.

169. C. Kavitha, and C. Chellamuthu, "Multimodal medical image fusion based on integer wavelet transform and neuro-fuzzy," in *Signal and Image Processing (ICSIP), 2010 International Conference on.* IEEE, 2010, pp. 296–300.

170. S. Vekkot, "Wavelet based medical image fusion using filter masks," in *Trends in Intelligent Robotics.* Springer, New York, 2010, pp. 298–305.

171. J. Teng, X. Wang, J. Zhang, S. Wang, and P. Huo, "A multimodality medical image fusion algorithm based on wavelet transform," in *Advances in Swarm Intelligence.* Springer, New York, 2010, pp. 627–633.

172. Q. Zhang, M. Liang, and W. Sun, "Medical diagnostic image fusion based on feature mapping wavelet neural networks," in *Multi-Agent Security and Survivability, 2004 IEEE First Symposium on.* IEEE, 2004, pp. 51–54.

173. S. Kor, and U. Tiwary, "Feature level fusion of multimodal medical images in lifting wavelet transform domain," in *Engineering in Medicine and Biology Society, 2004. IEMBS'04. 26th Annual International Conference of the IEEE,* vol. 1. IEEE, 2004, pp. 1479–1482.

174. Z. Cui, G. Zhang, and J. Wu, "Medical image fusion based on wavelet transform and independent component analysis," in *Artificial Intelligence, 2009. JCAI'09. International Joint Conference on.* IEEE, 2009, pp. 480–483.

175. W. Hao-Quan, and X. Hao, "Multi-mode medical image fusion algorithm based on principal component analysis," in *Computer Network and Multimedia Technology, 2009. CNMT 2009. International Symposium on.* IEEE, 2009, pp. 1–4.

176. N. Al-Azzawi, H. A. M. Sakim, A. Wan Abdullah, and H. Ibrahim, "Medical image fusion scheme using complex contourlet transform based on PCA," in *Engineering in Medicine and Biology Society, 2009. EMBC 2009. Annual International Conference of the IEEE.* IEEE, 2009, pp. 5813–5816.

177. N. A. Al-Azzawi, H. A. M. Sakim, and A. Wan Abdullah, "An efficient medical image fusion method using contourlet transform based on PCM," in *Industrial Electronics & Applications, 2009. ISIEA 2009. IEEE Symposium on*, vol. 1. IEEE, 2009, pp. 11–14.

178. C. He, Q. Liu, H. Li, and H. Wang, "Multimodal medical image fusion based on IHS and PCA," *Procedia Engineering*, vol. 7, pp. 280–285, 2010.

179. C. Wang, and Z. Ye, "First-order fusion of volumetric medical imagery," *IEEE Proceedings-Vision, Image and Signal Processing*, vol. 153, no. 2, pp. 191–198, 2006.

180. J. Phegley, K. Perkins, L. Gupta, and J. K. Dorsey, "Risk-factor fusion for predicting multifactorial diseases," *Biomedical Engineering, IEEE Transactions on*, vol. 49, no. 1, pp. 72–76, 2002.

181. B. Escalante-Ramírez, "The Hermite transform as an efficient model for local image analysis: An application to medical image fusion," *Computers & Electrical Engineering*, vol. 34, no. 2, pp. 99–110, 2008.

182. Z. Zhang, J. Yao, S. Bajwa, and T. Gudas, "'Automatic' multimodal medical image fusion," in *Soft Computing in Industrial Applications, 2003. SMCia/03. Proceedings of the 2003 IEEE International Workshop on.* IEEE, 2003, pp. 161–166.

183. L. Yang, B. Guo, and W. Ni, "Multimodality medical image fusion based on multiscale geometric analysis of contourlet transform," *Neurocomputing*, vol. 72, no. 1, pp. 203–211, 2008.

184. Y. Wei, Y. Zhu, F. Zhao, Y. Shi, T. Mo, X. Ding, and J. Zhong, "Implementing contourlet transform for medical image fusion on a heterogenous platform," in *Scalable Computing and Communications; Eighth International Conference on Embedded Computing, 2009. SCALCOM-EMBEDDEDCOM'09. International Conference on.* IEEE, 2009, pp. 115–120.

185. L. Gupta, B. Chung, M. D. Srinath, D. L. Molfese, and H. Kook, "Multichannel fusion models for the parametric classification of differential brain activity," *Biomedical Engineering, IEEE Transactions on*, vol. 52, no. 11, pp. 1869–1881, 2005.

186. G. L. Rogova, and P. C. Stomper, "Information fusion approach to microcalcification characterization," *Information Fusion*, vol. 3, no. 2, pp. 91–102, 2002.

187. S.-H. Lai, and M. Fang, "Adaptive medical image visualization based on hierarchical neural networks and intelligent decision fusion," in *Neural Networks for Signal Processing VIII, 1998. Proceedings of the 1998 IEEE Signal Processing Society Workshop.* IEEE, 1998, pp. 438–447.

188. S. Constantinos, M. S. Pattichis, and E. Micheli-Tzanakou, "Medical imaging fusion applications: An overview," in *Signals, Systems and Computers, 2001. Conference Record of the Thirty-Fifth Asilomar Conference on*, vol. 2. IEEE, 2001, pp. 1263–1267.

189. W. Li, and X.-F. Zhu, "A new algorithm of multi-modality medical image fusion based on pulse-coupled neural networks," in *Advances in Natural Computation.* Springer, New York, 2005, pp. 995–1001.

190. Y.-P. Wang, J.-W. Dang, Q. Li, and S. Li, "Multimodal medical image fusion using fuzzy radial basis function neural networks," in *Wavelet Analysis and Pattern Recognition, 2007. ICWAPR'07. International Conference on*, vol. 2. IEEE, 2007, pp. 778–782.

191. Z. Wang, and Y. Ma, "Medical image fusion using M-PCNN," *Information Fusion*, vol. 9, no. 2, pp. 176–185, 2008.

192. J. Teng, S. Wang, J. Zhang, and X. Wang, "Neuro-fuzzy logic based fusion algorithm of medical images," in *Image and Signal Processing (CISP), 2010 3rd International Congress on*, vol. 4. IEEE, 2010, pp. 1552–1556.

193. D. Lederman, B. Zheng, X. Wang, X. H. Wang, and D. Gur, "Improving breast cancer risk stratification using resonance-frequency electrical impedance spectroscopy through fusion of multiple classifiers," *Annals of Biomedical Engineering*, vol. 39, no. 3, pp. 931–945, 2011.

194. M. Sehgal, I. Gondal, and L. Dooley, "Support vector machine and generalized regression neural network based classification fusion models for cancer diagnosis," in *Hybrid Intelligent Systems, 2004. HIS '04. Fourth International Conference on*, December 2004, pp. 49–54.

195. F. Masulli, and S. Mitra, "Natural computing methods in bioinformatics: A survey," *Information Fusion*, vol. 10, no. 3, pp. 211–216, 2009.

196. W. Dou, S. Ruan, Y. Chen, D. Bloyet, and J.-M. Constans, "A framework of fuzzy information fusion for the segmentation of brain tumor tissues on MR images," *Image and Vision Computing*, vol. 25, no. 2, pp. 164–171, 2007.

197. X. Tai, and W. Song, "An improved approach based on FCM using feature fusion for medical image retrieval," in *Fuzzy Systems and Knowledge Discovery, 2007. FSKD 2007. Fourth International Conference on*, vol. 2. IEEE, 2007, pp. 336–342.

198. W. Song, and T. Hua, "Analytic implementation for medical image retrieval based on FCM using feature fusion with relevance feedback," in *Bioinformatics and Biomedical Engineering, 2008. ICBBE 2008. The 2nd International Conference on*. IEEE, 2008, pp. 2590–2595.

199. Y. Na, H. Lu, and Y. Zhang, "Content analysis based medical images fusion with fuzzy inference," in *Fuzzy Systems and Knowledge Discovery, 2008. FSKD'08. Fifth International Conference on*, vol. 3. IEEE, 2008, pp. 37–41.

200. A. Assareh, and L. G. Volkert, "Fuzzy rule base classifier fusion for protein mass spectra based ovarian cancer diagnosis," in *Computational Intelligence in Bioinformatics and Computational Biology, 2009. CIBCB'09. IEEE Symposium on*. IEEE, 2009, pp. 193–199.

201. J. K. Avor, and T. Sarkodie-Gyan, "An approach to sensor fusion in medical robots," in *Rehabilitation Robotics, 2009. ICORR 2009. IEEE International Conference on*. IEEE, 2009, pp. 818–822.

202. G. N. Brock, W. D. Beavis, and L. S. Kubatko, "Fuzzy logic and related methods as a screening tool for detecting gene regulatory networks," *Information Fusion*, vol. 10, no. 3, pp. 250–259, 2009.

203. R. K. De, and A. Ghosh, "Linguistic recognition system for identification of some possible genes mediating the development of lung adenocarcinoma," *Information Fusion*, vol. 10, no. 3, pp. 260–269, 2009.

204. M. Bhattacharya, and A. Das, "Multimodality medical image registration and fusion techniques using mutual information and genetic algorithm-based approaches," in *Software Tools and Algorithms for Biological Systems*. Springer, New York, 2011, pp. 441–449.

205. I. Dimou, G. Manikis, and M. Zervakis, "Classifier fusion approaches for diagnostic cancer models," in *Engineering in Medicine and Biology Society, 2006. EMBS'06. 28th Annual International Conference of the IEEE*. IEEE, 2006, pp. 5334–5337.

206. N. Zhang, Q. Liao, S. Ruan, S. Lebonvallet, and Y. Zhu, "Multi-kernel SVM based classification for tumor segmentation by fusion of MRI images," in *Imaging Systems and Techniques, 2009. IST'09. IEEE International Workshop on*. IEEE, 2009, pp. 71–75.

207. M. M. Rahman, B. C. Desai, and P. Bhattacharya, "Medical image retrieval with probabilistic multiclass support vector machine classifiers and adaptive similarity fusion," *Computerized Medical Imaging and Graphics*, vol. 32, no. 2, pp. 95–108, 2008.

208. Y. Huang, J. Zhang, Y. Zhao, and D. Ma, "Medical image retrieval with query-dependent feature fusion based on one-class SVM," in *Computational Science and Engineering (CSE), 2010 IEEE 13th International Conference on*. IEEE, 2010, pp. 176–183.

209. G. Pavesi, and G. Valentini, "Classification of co-expressed genes from DNA regulatory regions," *Information Fusion*, vol. 10, no. 3, pp. 233–241, 2009.

# 28 Multisensor Data Fusion
## *Architecture Design and Application in Physical Activity Assessment*

Shaopeng Liu and Robert X. Gao

## CONTENTS

## 28.1 INTRODUCTION

### 28.1.1 FUSION FOR PHYSICAL ACTIVITY ASSESSMENT

Physical activity (PA) is defined as body movement generated by skeletal muscle actions that increase energy expenditure above resting levels [1]. Scientific studies have shown that engaging in physical activities on a regular basis by walking, jogging, sports, or muscle-strengthening activities in general is effective for improving health, level of fitness, and weight control, and preventing or lowering the risk of cardiovascular disease, diabetes, obesity, early death, certain cancers, and stroke [2]. Ultimately, it helps improve the quality of life. Accurate monitoring and assessment of PA is of significant interest to the research community and society at large. Whereas in the past PA was assessed by different laboratory-based methods, including direct [3] or indirect [4] calorimeters to quantify the intensity of activities, wearable activity monitors in the form of wearable sensors have recently become the device of interest for PA assessment, owing to the low subject burden and noninvasive nature. Combined with advanced data fusion analytics, such a system has emerged as a

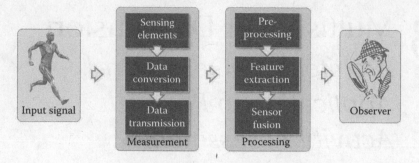

**FIGURE 28.1**    Architecture of multisensor data fusion for PA assessment.

method for accurate and reliable PA assessment under free-living environments [5–8]. This chapter highlights research on the multisensor data fusion technique applied to PA assessment.

### 28.1.2    ARCHITECTURE

PA assessment typically involves two main steps, as shown in Figure 28.1. Activities as "signals" are first captured by the **Measurement** system, which consists of *sensing elements*, *data conversion*, and *data transmission*. Sensing elements include one or more sensors, which usually are accelerometer(s) or physiological sensors, such as respiratory sensors. These sensing elements translate the physical activities into analog signals, which are in turn converted into digital data points in the data conversion step. The digitized data is then transmitted through either wired or wireless communication to computerized system for later **Processing**, which is the second main step. The digital sensor data will first be *preprocessed* (e.g., filtering, artifact removal, etc.) before *feature extraction*. Different features in the time, frequency, or other domains are extracted from the raw sensor data to characterize the original signals and reduce computational complexity and redundancy. These extracted features are then used as inputs to *data fusion* algorithms for PA assessment.

### 28.1.3    STRATEGIES

During the multisensor data fusion process, other than the fusion algorithms that will be described in detail in Section 28.3, a proper fusion strategy needs to be established first to ensure effective fusion performance. Typical fusion strategies include fusion across sensors, fusion across attributes, or fusion across domains [9].

#### 28.1.3.1    Fusion across Sensors

In this scenario, the multisensor data fusion is performed across a number of sensors that measure the same characteristics of the activity. For example, multiple accelerometers are placed at different locations of the human body to measure bodily motion.

#### 28.1.3.2    Fusion across Attributes

In this scenario, a number of sensors measure different characteristics (attributes) of the activity. For example, the fusion is performed across accelerometers that measure bodily motion and respiratory sensors that measure breathing rate and ventilation volume.

#### 28.1.3.3    Fusion across Domains (Time and Frequency)

In this scenario, the fusion is conducted across different feature domains (e.g., time and frequency) of one or more attributes. For example, the fusion is performed on the time-domain features of accelerometer and respiratory sensor data.

**TABLE 28.1**
**Sensors Used for PA Assessment**

| Types | Locations |
|---|---|
| Single accelerometer | Hip |
| Multiple accelerometers | Hip, chest, thigh (left and right), bottom of foot (left and right), wrist |
| Foot pressure sensor | Bottom of each foot (heel, metatarsal bones and big toe) |
| Heart rate (ECG) | Chest |
| Respiration | Rib cage, abdomen |
| Heat flux | Upper arm |
| Galvanic skin response | Upper arm |
| Skin temperature | Upper arm |

## 28.2 SENSOR DATA REPRESENTATION

### 28.2.1 SENSORS FOR PHYSICAL ACTIVITY MEASUREMENT

Various body-worn sensors, including accelerometers, foot pressure sensors, and physiological sensors such as respiration, heart rate, and skin temperature sensors, have been used for the PA assessment, as shown in Table 28.1. Among them, the accelerometer, uniaxial or triaxial, is one of the most commonly used wearable sensors for PA assessment, owing to its capability of detecting the quantity and intensity of human body movements at an affordable "expense"—relatively low sensor cost and low subject burden. Although simple in structure, the accelerometer can effectively detect both static and dynamic accelerations caused by changes in posture, body motion, or transition in motion patterns. Studies have been leveraging either single accelerometer for the PA assessment [10–14]. More details on using accelerometers for PA assessment can be found in previously published review articles [15–17].

For different types of activities that produce similar acceleration profiles but have different energy expenditure, accelerometer(s) alone cannot provide accurate PA assessment [18]. For example, walking at a certain speed may result in acceleration outputs similar to that of walking at the same speed while carrying a load or going uphill, although the energy expenditure is different. Often, data from the accelerometer will then be fused with data from additional accelerometers placed at various locations of the human body [18–20], or information from other types of sensors aiming to improve assessment performance. For example, foot pressure sensors have been fused with accelerometers to improve PA assessment performance. Sazonova et al. [21] developed a footwear-based device integrating one accelerometer and five force-sensitive resistors positioned at the heel, metatarsal bones, and the big toe. By using both the accelerometer and pressure sensor signals in the model, better physical activity energy expenditure (PAEE) prediction with a root mean squared error (RMSE) of 0.69 metabolic equivalents (METs) was achieved than that using only the accelerometer signal (RMSE of 0.77 METs). Sensors that measure human physiological responses, such as heart rate [22–24], respiration [18,25], or an armband [5,26,27] consisting of heat flux, galvanic skin response, and skin temperature sensors, have also been studied extensively.

### 28.2.2 FEATURE EXTRACTION

The raw signals from body-worn sensors during physical activities are often sampled over periods of time and contain a considerable number of data points. Although these raw data points have been directly used for PA assessment [28], signal features are usually extracted from the original sensor signals and used as inputs to sensor fusion algorithms. Time- and frequency-domain features

## TABLE 28.2
## Features Used in PA Assessment

**Features**

| | |
|---|---|
| Time domain | Data points of raw sensor signals |
| | Accelerometer counts (integral/sum of the signals over a period of time) |
| | Mean |
| | Standard deviation |
| | Coefficients of variation (CV) |
| | Peak-to-peak amplitude |
| | Percentiles (10th, 25th, 50th, 75th, 90th) |
| | Interquartile range |
| | Correlation between accelerometer axes |
| | Autocorrelation |
| | Skewness |
| | Kurtosis |
| | Signal power |
| | Log-energy |
| | Peak intensity (number of the signal peak appearances within a certain period of time) |
| | Zero crossings (number of times the signal crosses its median) |
| Frequency domain | Dominant frequency |
| | Amplitude of dominant frequency |
| | Signal entropy |

are commonly used for PA assessment, as shown in Table 28.2. Time-domain features are directly extracted from the time series of the original, raw sensor signals. A typical procedure for extracting time-domain features is to first divide a sensor signal into consecutive small windows. Then, the time-domain features are computed over each window. The size of the signal window may range from 0.25 to 60 seconds [29], but the determination of the window size usually depends on the specific application and the impact on the estimation accuracy. The window size can be fixed using a sliding window, or varied using an event- or activity-defined window. A more detailed review of the windowing techniques is discussed in Preece et al. [30].

Frequency-domain features are extracted from the coefficients obtained by performing spectral analysis, usually Fast Fourier Transform (FFT), on the original sensor signals. The values of the coefficients represent the amplitudes of the corresponding frequency components. In particular, the coefficient with the largest amplitude corresponds to the dominant frequency of the signal, which is an indicator of the activity intensity level. Both the dominant frequency [18,31] and its amplitude [31] have been commonly chosen as the frequency-domain features for PA assessment. Similar to time-domain features, measures of the probability distribution of the FFT coefficients have also been used. In addition, spectral entropy from the accelerometer signals has been studied for the estimation of activity intensity [31]. The spectral entropy measures the frequency concentration, and can be used as an indicator of the activity patterns [18]. For example, low spectral entropy corresponds to high-frequency concentration, which usually represents simple or uniform activity patterns such as treadmill locomotion. In contrast, high spectral entropy corresponds to low-frequency concentration, which indicates complex or nonuniform activity such as team sports, tennis, or self-paced walking.

In addition to the features computed directly from the sensor signals, subject demographic and anthropometric characteristics [19,21,29], such as age, height, weight, body mass index (BMI), body composition (e.g., fat mass, fat-free mass), gender, have also been considered for PA assessment.

### 28.2.3 FEATURE SELECTION

During a multisensor fusion study for PA assessment, it is often the case that quite a number of signal features have been included in the initial feature set. Therefore, it is important to remove irrelevant or redundant features and select "good" features to ensure an accurate multisensor fusion model that minimizes assessment errors and yields a general concept [32]. A two-step feature selection scheme is typically used for selecting optimal features for the PA assessment. An initial set of features is first selected through statistical analysis of the distribution of these features, and a more advanced feature selection scheme will then be applied to the initial feature set to determine the optimal features. In the authors' previous study [18], an initial set of 63 features (in both time and frequency domains) were selected via boxplots, which illustrate the distribution change among activities. A "good" feature is expected to have less distribution overlap between the activities, which is considered to be better for the differentiation of activities [33]. The 63 features were then evaluated by removing redundant features using the minimal-redundancy-maximal-relevance (mRMR) heuristic [34]. The mRMR method measures the relevance and redundancy of the feature candidates with the target class based on mutual information and selects a "promising" feature subset that has maximal relevance and minimal redundancy. A subset of 33 features with the best mRMR scores (difference between the relevance and redundancy) was selected for the PA assessment. More details are available in Ref. [18].

## 28.3 DATA FUSION ALGORITHMS

### 28.3.1 OVERVIEW OF CANDIDATES

The multisensor data fusion algorithms reported in past studies are of varying complexity, ranging from simple linear regressions to advanced machine learning techniques. These algorithms extract relationships and correlations, which are often hidden, between the sensor signals/features and physical activity type and/or the PAEE. Examples of such machine learning methods include hidden Markov models, quadratic discriminant analysis, artificial neural network, and support vector machines (SVMs). A list of recent multisensory data fusion methods for PA assessment and a detailed overview of these methods are provided in Ref. [35]. In the following sections, we will focus on one example of the methods: SVMs.

### 28.3.2 EXAMPLE: SVMS

SVMs transform the original multisensory signal features into a different and usually higher feature space/domain, where the relationship between the multisensor data and PA is modeled. To formulate a SVM multisensor data fusion model, assume a data set $\{x_i\}$ consisting of data measured by multiple sensors, where $x_i \in R^n$ ($n$ is the dimension of the input vectors) and $i = 1,\ldots, N$ ($N$ is the total number of data points). Within the duration of this data set, the subject is assumed to have engaged in two types of activity $\{y_i\}$, labeled as $-1$ and 1 ($y_i \in \{-1, 1\}$), respectively. Each data set $\{x_i\}$ can be associated with one of the two activity types $\{y_i\}$. To distinguish the activity type that the data set $\{x_i\}$ is associated with, a function $f(x)$ is assumed to exist that draws a separation decision boundary between the two activities. Data points above the boundary (when $f(x)$ yields a value $\geq 0$) belong to the activity labeled as 1, whereas data points below the boundary (when $f(x)$ produces a value $< 0$) are labeled as $-1$. Such a function $f(x)$ can be expressed as

$$\begin{cases} f(x_i) \geq 0 \Rightarrow y_i = 1 \\ f(x_i) < 0 \Rightarrow y_i = -1 \end{cases} \tag{28.1}$$

A drawback of such a separation boundary function built in the original feature space is that it is often a complex, nonlinear, and implicit function that is computationally demanding for determining activity types for each new data point added [36]. To overcome this limitation, the activities are assumed to be separable in an enlarged feature space with higher dimensionality than the original feature space [37] where a linear and explicit decision boundary—a hyperplane—can be formulated. The SVM algorithm first transforms the data $\{x_i\}$ from the original lower-dimensional space to a higher-dimensional space [37] via a transformation function $\phi$. A hyperplane $f'(x') = w^T x' + b = 0$, where $x' = \phi(x)$, is then built in the higher-dimensional space to separate the two activities. In this formulation, $w$ and $b$ are the weighing factors, and $x'$ is the transformed high-dimensional data. Similar to Equation 28.1, the hyperplane function for separating activities is expressed as

$$\begin{cases} f'(x_i') = w^T x_i' + b \ge 0 \Rightarrow y_i = 1 \\ f'(x_i') = w^T x_i' + b < 0 \Rightarrow y_i = -1 \end{cases} \tag{28.2}$$

The hyperplane is built such that it maximizes the distance (or margin) to the closest training data point of either activity. The process can be expressed as an optimization problem according to the following [37]:

$$\max_{w \in R^n, b \in R} D, \text{ subject to } y_i\left(w^T x_i' + b\right) \ge D, \forall i \tag{28.3}$$

where $D$ is the distance of the closest data point to the hyperplane and can be set as $1/\|w\|$ after normalization. For practical applications, when two activities may have overlapping (i.e., misclassified) data points in the feature space, slack variables $\xi = \{\xi_i\}$, where $i = 1,\ldots, N$, $\xi_i \ge 0$, are introduced, allowing certain points $x_i$ on the wrong side of the margin by an amount of $D\xi_i$. Points on the correct side of the margin are expressed with $\xi_i = 0$. Equation 28.3 can be rewritten as

$$\min_{w, \xi \in R^n, b \in R} \left\{ \frac{1}{2}\|w\|^2 + C \sum_{i=1}^{N} \xi_i \right\} \tag{28.4}$$

$$\text{subject to } \xi_i \ge 0, \, y_i\left(w^T \phi(x_i) + b\right) \ge 1 - \xi_i, \forall i$$

where $C$ is the cost parameter for those sample points misclassified by the optimal separating plane. The hyperplane decision function $f(x)$ is then determined as the following sign function ($\text{sgn}(t) = 1$ for $t \ge 0$, and $\text{sgn}(t) = -1$ for $t < 0$):

$$f(x) = \text{sgn}\left( \sum_{i=1}^{N} y_i \alpha_i \phi(x_i)^T \phi(x) + b \right) \tag{28.5}$$

From Equation 28.5, it is seen that the decision function can be determined without specifying the explicit form of the transformation $\phi$, but only the kernel function $K(x_i, x) = \phi(x_i)^T \phi(x)$ that computes inner products. Replacing the inner product in Equation 28.5, the decision function for distinguishing the two activities can be rewritten as

$$f(x) = \text{sgn}\left( \sum_{i=1}^{N} y_i \alpha_i K(x_i, x) + b \right) \tag{28.6}$$

When multiple activities need to be separated, a "one-against-one" approach can be taken for separating each set of two activities. For instance, to classify $d > 2$ activities, a SVM model can be built for each pair of activities from the training data to form a total of $d(d − 1)/2$ models. Using a new data point, each model will be tested, and a vote on which type of activity this data point should belong to will be cast. After all the models have been tested, the activity that has received the most votes will be identified as the activity that the new data point belongs to. Such a model built in a higher feature space is less computationally demanding for determining activity types for each new data point added than the model built using the original features, which is often a complex, non-linear, and implicit function. Various studies have successfully applied SVM [18,25,38] for estimating PAEE, and promising results have been reported.

### 28.3.2.1 PA Recognition—Classification

The recognition of activity types and validation of the developed multisensor data fusion model are generally performed via a two-step procedure. First, a training data set that consists of all the selected sensor signal features obtained from all subjects but one was constructed for building the multisensor fusion model, as well as selecting the best tuning parameters, which in the example of SVM are the cost parameter and kernel parameter. The tuning parameters are usually selected through cross validation (fivefold or more) to prevent overfitting. The parameters that yielded the highest recognition rate are then chosen during the process. On completion of training, the tuned model will be applied to the feature set of the subject that is left out of the training process to predict the activity type. Such a two-step procedure constitutes a "leave-one-subject-out" cross validation and is executed on each subject sensor data.

### 28.3.2.2 PAEE Estimation—Regression

The regression version of the SVM, support vector regression (SVR) [39], can be implemented to predict energy expenditure associated with each activity. Considering the broad range of intensity of the physical activities monitored, separate SVR fusion models are built by grouping activities of similar intensity level into four categories (e.g., sedentary, household, moderate locomotion, and rigorous activities) for improved the PAEE estimation accuracy. Data from any subject who has performed activities in a particular group are used in the training and testing process of the respective SVR model. Construction of the SVR models follows the "leave-one-subject-out" cross validation procedure. A two-tier estimation scheme is devised to estimate the PAEE. Specifically, the activity type of each 30-second data segment is first identified by the SVM model developed in the previous section and grouped into one of the four activity categories. Subsequently, the PAEE during the 30-second period is then estimated by the specific SVR model for that activity group.

## 28.3.3 COMPARISON OF FUSION STRATEGIES

To evaluate the developed multisensor data fusion method and compare different fusion strategies, experiments have been conducted by employing a total of 50 subjects who performed 13 types of activities of varying intensities. These 13 activities were separated into the following four categories, based on the intensity and similarity among the activities: (1) sedentary activity, (2) household and other activity, (3) moderate locomotion, and (4) vigorous activity. Detailed descriptions of the activities are provided in Ref. [18]. For the purpose of experimental organization, the activities were also separated into two routines, and selection of routine was balanced across subjects. Each subject completed one of the two routines while wearing a PA measurement device [40] that consists of two triaxial accelerometers placed at the hip and wrist, and one ventilation sensor secured to the abdomen (AB) at the level of umbilicus of the test subjects. The accelerometers measure the trunk and arm motions, while the ventilation sensor measures the expansion and contraction associated with breathing rate and volume representing the physiological response to bodily movement.

**TABLE 28.3**

**Fusion across Sensors for PA Assessment**

| Fusion across Sensors | Recognition Accuracy (%) | PAEE Estimation (RMSE) |
|---|---|---|
| Hip accelerometer | 75.8 ± 19.3 | 0.54 |
| Wrist accelerometer | 77.6 ± 14.7 | 0.51 |
| Ventilation sensor | 35.6 ± 14.5 | 0.72 |
| Hip + wrist accelerometers | 85.8 ± 14.1 | 0.54 |
| Hip accelerometer + ventilation | 78.3 ± 19.4 | 0.55 |
| Wrist accelerometer + ventilation | 80.1 ± 16.9 | 0.47 |
| Hip + wrist accelerometer + ventilation | 88.1 ± 10.1 | 0.42 |

**TABLE 28.4**

**Activity Type Recognition for Different Feature Sets**

| Feature Set | Recognition Accuracy (%) |
|---|---|
| Time-domain | 84.7 ± 16.7 |
| Frequency-domain | 66.5 ± 20.4 |

Table 28.3 compares PA assessment (both activity type recognition and PAEE estimation) by fusing different sensors. It can be seen that, for both activity type recognition and PAEE estimation, the performance is enhanced when more sensor data are fused in the models. For example, the average recognition accuracy has increased from 75.8%, 77.6%, and 35.6% for the single-sensor models respectively to 85.8%, 78.3%, and 80.1% for the dual-sensor models, and to 88.1% for the multiple-sensor model, respectively. The MET characterizes the energy expenditure associated with activities and was predicted in the study by using the SVR fusion model. The estimated MET values were then compared with the values measured by the respiratory gas exchange system along with the resting metabolic rate measurement. The estimation performance was assessed by the RMSE. The result also demonstrates high performance of PAEE estimation when fusing multiple sensors. For example, the PAEE was estimated with a RMSE of 0.42 when the model fused all three sensors.

Furthermore, the standard deviation of the recognition accuracies reveals the subject-to-subject variability. It is seen from Table 28.3 that the variability has decreased from 19.3% when using a single hip accelerometer to 14.1% and 16.9% when fusing two sensors (either the hip accelerometer with wrist accelerometer, or hip accelerometer with ventilation sensor), and to 10.1% when fusing all three sensors. It is concluded that when fusing more sensors, the fusion model becomes more generalizable than fusing data from fewer sensors, and thus the subject-to-subject variability can be effectively reduced.

Table 28.4 compares the fusion strategy among different feature domains (time and frequency) through the recognition of activity types. It is seen that time-domain features make a greater contribution than frequency-domain features to activity recognition (84.7% vs. 66.5%). Similar findings have also been reported in previous studies [41]. Noteworthy is that the fusion of both time- and frequency-domain features increases the recognition accuracy while reducing the subject-to-subject variability.

## 28.4  CONCLUSIONS

This chapter has presented a multisensor data fusion approach for assessing physical activities. In particular, an SVM-based multisensor fusion method for physical activity assessment has been introduced, which has demonstrated successful PA assessment through laboratory experiments. Two advantages of the algorithm have been verified: (1) improving the estimation accuracy of types and corresponding energy expenditure of physical activities and (2) reducing subject-to-subject variability in activity type recognition when multisensor signals were fused in the models. Research activities in new sensing technologies, feature selection, and advanced multisensor data fusion algorithms have the potential to further advance the state of science and engineering related to the assessment of energy expenditure and produce a valid, reliable, and accurate estimation of activity types and PAEE.

# REFERENCES

1. W.D. McArdle, F. Katch, and V. Katch, *Exercise Physiology: Nutrition, Energy, and Human Performance* (Philadelphia, PA: Lippincott Williams & Wilkins, 2001).
2. U.S. Department of Health and Human Services, *Physical Activity Guidelines Advisory Committee Report* (Washington, DC: U.S. Department of Health and Human Services, Office of Disease Prevention and Health Promotion, 2008), A2–4.
3. J. Webster, G. Welsh, P. Pacy, and J. Garrow, "Description of a human direct calorimeter, with a note on the energy cost of clerical work," *British Journal of Nutrition* 55 (1986): 1–6.
4. V. Diaz, P. Benito, A. Peinado, M. Alvarez, C. Martin, V. Salvo, F. Pigozzi, N. Maffulli, and F. Calderon, "Validation of a new portable metabolic system during an incremental running test," *Journal of Sports Science and Medicine* 7 (2008): 532–6.
5. M. Fruin, and J. Rankin, "Validity of a multi-sensor armband in estimating rest and exercise energy expenditure," *Medicine & Science in Sports & Exercise* 36 (2004): 1063–9.
6. D.L. Johannsen, M.A. Calabro, J. Stewart, W. Franke, J.C. Rood, and G.J. Welk, "Accuracy of armband monitors for measuring daily energy expenditure in healthy adults," *Medicine & Science in Sports & Exercise* 42 (2010): 2134–40.
7. D. John, and P. Freedson, "ActiGraph and Actical physical activity monitors: A peek under the hood," *Medicine & Science in Sports & Exercise* 44 (2012): S86–9.
8. K. Koehler, H. Braun, M. de Marées, G. Fusch, C. Fusch, and W. Schaenzer, "Assessing energy expenditure in male endurance athletes: Validity of the SenseWear armband," *Medicine & Science in Sports & Exercise* 43 (2011): 1328–33.
9. H. Mitchell, *Multi-Sensor Data Fusion: An Introduction* (New York: Springer, 2010).
10. D. Hendelman, K. Miller, C. Baggett, E. Debold, and P. Freedson, "Validity of accelerometry for the assessment of moderate intensity physical activity in the field," *Medicine & Science in Sports & Exercise* 32 (2000): 442–9.
11. C. Bouten, K. Koekkoek, M. Verduin, R. Kodde, and J. Janssen, "A triaxial accelerometer and portable data processing unit for the assessment of daily physical activity," *IEEE Transactions on Biomedical Engineering* 44 (1997): 136–47.
12. B. Najafi, K. Aminian, A. Paraschiv-Ionescu, F. Loew, C. Büla, and P. Robert, "Ambulatory system for human motion analysis using a kinematic sensor: Monitoring of daily physical activity in the elderly," *IEEE Transactions on Biomedical Engineering* 50 (2003): 711–23.
13. D.R. Bassett, A. Rowlands, and S.G. Trost, "Calibration and validation of wearable monitors," *Medicine & Science in Sports & Exercise* 44 (2012): S32–8.
14. A.G. Bonomi, G. Plasqui, A.H.C. Goris, and K.R. Westerterp, "Improving assessment of daily energy expenditure by identifying types of physical activity with a single accelerometer," *Journal of Applied Physiology* 107 (2009): 655–61.
15. K.Y. Chen, and D.R. Bassett, "The technology of accelerometry-based activity monitors: Current and future," *Medicine & Science in Sports & Exercise* 37 (2005): S490–500.
16. P. Freedson, D. Pober, and K.F. Janz, "Calibration of accelerometer output for children," *Medicine & Science in Sports & Exercise* 37 (2005): S523–30.
17. C.E. Matthews, "Calibration of accelerometer output for adults," *Medicine & Science in Sports & Exercise* 37 (2005): S512–22.
18. S. Liu, R.X. Gao, D. John, J. Staudenmayer, and P. Freedson, "Multi-sensor data fusion for physical activity assessment," *IEEE Transactions on Biomedical Engineering* 59 (2012): 687–96.
19. M.P. Rothney, M. Neumann, A. Béziat, and K.Y. Chen, "An artificial neural network model of energy expenditure using nonintegrated acceleration signals," *Journal of Applied Physiology* 103 (2007): 1419–27.

20. K. Zhang, P. Werner, M. Sun, F.X. Pi-Sunyer, and C.N. Boozer, "Measurement of human daily physical activity," *Obesity* 11 (2003): 33–40.
21. N. Sazonova, R.C. Browning, and E. Sazonov, "Accurate prediction of energy expenditure using a shoe-based activity monitor," *Medicine & Science in Sports & Exercise* 43 (2011): 1312–21.
22. S. Brage, N. Brage, P. Franks, U. Ekelund, and N. Wareham, "Reliability and validity of the combined heart rate and movement sensor Actiheart," *European Journal of Clinical Nutrition* 59 (2005): 561–70.
23. F. De Bock, J. Menze, S. Becker, D. Litaker, J. Fischer, and I. Seidel, "Combining accelerometry and HR for assessing preschoolers' physical activity," *Medicine & Science in Sports & Exercise* 42 (2010): 2237–43.
24. Z. Li, "Exercises intensity estimation based on the physical activities healthcare system," *Proceedings of the 2009 WRI International Conference on Communications and Mobile Computing* (January 6–8, 2009): 132–6, Yunnan, China.
25. D. John, S. Liu, J.E. Sasaki, C.A. Howe, J. Staudenmayer, R.X. Gao, and P.S. Freedson, "Calibrating a novel multi-sensor physical activity measurement system," *Physiological Measurement* 32 (2011): 1473–89.
26. D. Arvidsson, F. Slinde, S. Larsson, and L. Hulthén, "Energy cost of physical activities in children: Validation of SenseWear Armband," *Medicine & Science in Sports & Exercise* 39 (2007): 2076–84.
27. C. Bäcklund, G. Sundelin, and C. Larsson, "Validity of armband measuring energy expenditure in overweight and obese children," *Medicine & Science in Sports & Exercise* 42 (2010): 1154–61.
28. C.V.C. Bouten, K.T.M. Koekkoek, M. Verduin, R. Kodde, and J.D. Janssen, "A triaxial accelerometer and portable data processing unit for the assessment of daily physical activity," *IEEE Transactions on Biomedical Engineering* 44 (1997): 136–47.
29. I. Zakeri, A.L. Adolph, M.R. Puyau, F.A. Vohra, and N.F. Butte, "Application of cross-sectional time series modeling for the prediction of energy expenditure from heart rate and accelerometry," *Journal of Applied Physiology* 104 (2008): 1665–73.
30. S.J. Preece, J.Y. Goulermas, L.P.J. Kenney, D. Howard, K. Meijer, and R. Crompton, "Activity identification using body-mounted sensors—A review of classification techniques," *Physiological Measurement* 30 (2009): R1–33.
31. A.G. Bonomi, A.H.C. Goris, B. Yin, and K.R. Westerterp, "Detection of type, duration, and intensity of physical activity using an accelerometer," *Medicine & Science in Sports & Exercise* 41 (2009): 1770–7.
32. M. Dash, and H. Liu, "Feature selection for classification," *Intelligent Data Analysis* 1 (1997): 131–56.
33. J. Parkka, M. Ermes, P. Korpipaa, J. Mantyjarvi, J. Peltola, and I. Korhonen, "Activity classification using realistic data from wearable sensors," *IEEE Transactions on Information Technology in Biomedicine* 10 (2006): 119–28.
34. H. Peng, F. Long, and C. Ding, "Feature selection based on mutual information: Criteria of max-dependency, max-relevance, and min-redundancy," *IEEE Transactions on Pattern Analysis and Machine Intelligence* 27 (2005): 1226–38.
35. S. Liu, R. Gao, and P. Freedson, "Computational methods for estimating energy expenditure in human physical activities," *Medicine & Science in Sports & Exercise* 44 (2012): 2138–46.
36. V. Vapnik, *Statistical Learning Theory* (New York: John Wiley & Sons, 1998).
37. T. Hastie, R. Tibshirani, and J. Friedman, *The Elements of Statistical Learning* (New York: Springer, 2009).
38. M. Rumo, O. Amft, G. Tröster, and U. Mäder, "A stepwise validation of a wearable system for estimating energy expenditure in field-based research," *Physiological Measurement* 32 (2011): 1983–2001.
39. A. Smola, and B. Schölkopf, "A tutorial on support vector regression," *Statistics and Computing* 14 (2004): 199–222.
40. S. Liu, R. Gao, and P. Freedson, "Design of a wearable multi-sensor system for physical activity assessment," *Proceedings of the IEEE/ASME International Conference on Advanced Intelligent Mechatronics* (July 6–9, 2010): 254–9, Montreal, ON, Canada.
41. S. Preece, J. Goulermas, L. Kenney, and D. Howard, "A comparison of feature extraction methods for the classification of dynamic activities from accelerometer data," *IEEE Transactions on Biomedical Engineering* 56 (2009): 871–9.

# 29 Data Fusion for Attitude Estimation of a Projectile
## *From Theory to In-Flight Demonstration*

*Sébastien Changey and Emmanuel Pecheur*

## CONTENTS

## 29.1  NAVIGATION OF PROJECTILE AND SPECIFIC FEATURES

### 29.1.1  STATE OF THE ART FOR ATTITUDE ESTIMATION

Guided ammunitions are today a necessity in the military domain. The fundamental requirement of new kinds of weapons is the greatest lethality with minimal collateral damage. To meet this requirement, guidance, navigation, and embedded control systems are critical. To design guided ammunition, its embedded system needs to know in real time its angular position (attitude), velocity, and ideally its position. Several articles have dealt with attitude estimation of spacecraft (Shuster et al. 1982; Bar-Itzhack and Oshman 1985; Psiaki et al. 1990; Crassidis et al. 2007), but these methods cannot be easily applied to gun-launched projectiles because of their nonanalogous behavior. Some projectiles are stabilized by high-speed spinning movement, so we need to use specific ballistic models (Lieske and McCoy 1964; McCoy 2001) to design the navigation system. The main problem in using these models lies in the strong nonlinearities of the evolution equations. As a solution to this problem, recent work applies linear theory to projectile modeling (Hainz and Costello 2005) but only allows us to do an impact point prediction. Hence, it is very useful for control design but not for projectile localization. That is why we need to use nonlinear state estimation methods.

State estimation theory for nonlinear dynamical systems has been the subject of several research activities and remains an open research area (Ciccarella et al. 1995; Reif and Unbehauen 1999; Boutayeb 2004). The main motivation is that most physical processes are described by complex nonlinear differential equations. Therefore state estimators are central for control design, diagnosis, or supervision of this kind of system. Without being exhaustive, one of the standard methods for estimator synthesis consists of using a nonlinear change in coordinates to bring the original system into a linear one (or pseudo-linear one). Unfortunately, in practice and for several multi-input multi-output processes, this approach is rarely applied. When the feasibility conditions are not matched to establish canonical forms, an easy and workable technique consists of using the famous extended Kalman filter (EKF) in both deterministic and stochastic contexts. In spite of the extensive use of this method, there are two main drawbacks. The first one is high sensitivity to initializations or perturbations (in particular for hard nonlinearities) in the sense that performances may decrease significantly for a small variation of the initializations or in the presence of small perturbations. On the other hand, very few results have been established for stability analysis.

Indeed, if the convergence mechanism of the Kalman filter in the linear case is well understood by now, stability analysis is far from being solved when nonlinear models are considered. The main results in this field were developed earlier by Song and Grizzle (1995).

In this chapter, we provide a contribution to projectile attitude estimation using an EKF estimator. Even if the dynamical model has numerous nonlinearities, the results obtained on real applications are promising.

### 29.1.2  WHY CLASSICAL INERTIAL MEASUREMENT UNITS CANNOT BE USED

For technical and economical reasons, the only sensors that could be embedded on a projectile are accelerometers and magnetometers, so classical inertial measurement units (IMUs) based on rate gyroscope cannot be used. For projectile applications, the sensors must be gun hardened to high accelerations (in our case 15,000 $g$) and they must be robust to the projectile's speed (700 m/s). Second, the embedded algorithm must satisfy high performances in terms of computational requirements for real-time implementation according to very high dynamics and dynamical calibration of the sensors. Because of the high accelerations, the calibration of the sensors can change during the flight, so we need to do a real-time calibration during the flight.

## 29.2 DATA FUSION FOR ATTITUDE ESTIMATION

### 29.2.1 NOTATION

According to Euler's rotation theorem, any attitude may be described by three angles. With this global representation, the orientation of the velocity vector is not defined. Therefore another representation with five angles is necessary (Figures 29.1 and 29.2) (Fleck 1998).

### 29.2.2 EMBEDDED SENSORS

To compute in real time the attitude and position of the projectile, inertial sensors are embedded inside the projectile, as shown in Figure 29.3.

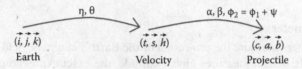

**FIGURE 29.1** Five angles are necessary.

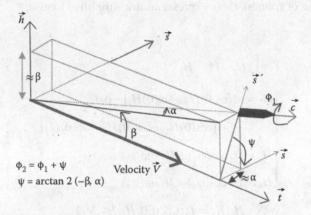

**FIGURE 29.2** Attitude definition.

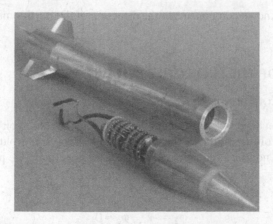

**FIGURE 29.3** Instrumented electronics embedded into the projectile; realized by ISL-STC department.

#### 29.2.2.1 Accelerometers

The three accelerometers measure the specific force along the three axes of the projectile. By using simplification (low amplitude of the attitude angle, simplification of second-order terms, etc.), the accelerometer measurements along radial axis $\vec{a}$ and $\vec{b}$ are expressed as

$$\begin{cases} A_a = A_{env}(\alpha\cos(\varphi_2) - \beta\sin(\varphi_2)) \\ A_b = -A_{env}(\alpha\sin(\varphi_2) - \beta\cos(\varphi_2)) \end{cases} \tag{29.1}$$

where $A_{env}$ is a function of velocity $V$, angular rotation $\omega_c$, and aerodynamic coefficients.

#### 29.2.2.2 Magnetometers

The three magnetometers measure the projection of the Earth's magnetic field along the three axes of the projectile. As the sensor measures global attitude, the exact expressions of the observations (written from the rotation matrix [Fleck 1998]) are composed by trigonometric expressions of the five considered angles. Because of the low amplitude of the attitude angles $(\alpha, \beta)$ and the low amplitude of $\eta$ (by definition of frames), these expressions are simplified to give

$$Y = \begin{bmatrix} H_c & H_a & H_b \end{bmatrix}^T$$

$$\begin{cases} H_c = \alpha[\cos(\theta)H_j - \sin(\theta)H_i] - \beta H_k \\ \qquad + [-\eta\cos(\theta)H_k + \cos(\theta)H_i + \sin(\theta)H_j] \\ H_a = H_1\cos(\varphi_2) + H_2\sin(\varphi_2) \\ H_b = H_2\cos(\varphi_2) - H_1\sin(\varphi_2) \end{cases} \tag{29.2}$$

$$\text{with} \quad H_1, H_2 = f(\alpha, \beta, \eta, \theta, H_i, H_j, H_k)$$

where $H_i, H_j, H_k$ are the known components of the direction of the Earth's magnetic field at the location of the experiment.

### 29.3 MODELING OF THE PROJECTILE BEHAVIOR

#### 29.3.1 THEORY AND EQUATION

All the attitude estimation algorithms presented in this chapter are based on EKF. These algorithms need an evolution model of the attitude of the projectile. The complete evolution model is described in Fleck (1998). The two nonlinear equations ususally used in ballistic are presented below.

$$\ddot{\varphi}_2 = -k\frac{D}{V}\dot{\varphi}_2 \approx 0 \tag{29.3}$$

$$\xi'' + (a_1 - i\,b_1)\xi' + (a_2 - i\,b_2)\xi = a_3 - i\,b_3 \quad \text{with } \xi' = \frac{D}{V}\dot{\xi} \tag{29.4}$$

where $D$ is the constant diameter of the projectile; $V$ is the velocity; $a_i$, $b_i$, and $k$ are functions of velocity $V$, altitude $y$, angular rotation $\omega_c$, and angles $\eta$ and $\theta$, depending on mechanical parameters; and $\xi$ is a complex variable, describing attitude, which can be approximated by $\alpha - i\beta$ because of the range (milliradians) of these angles.

In this way, real and imaginary parts of the mechanical equation of ballistics can be written to obtain state space evolution equation (state dimension, $n = 6$):

$$\dot{X} = F(X(t),t) = A(X(t),t)X(t) + B(X(t),t) \tag{29.5}$$

with

$$X = \begin{bmatrix} \alpha(t) & \beta(t) & \dot{\alpha}(t) & \dot{\beta}(t) & \varphi_2(t) & \dot{\varphi}_2(t) \end{bmatrix}^T$$

$$A = \begin{bmatrix} 0 & 0 & 1 & 0 & 0 & 0 \\ 0 & 0 & 0 & 1 & 0 & 0 \\ -\dfrac{V^2 a_2}{D^2} & \dfrac{V^2 b_2}{D^2} & -\dfrac{V a_1}{D} & \dfrac{V b_1}{D} & 0 & 0 \\ -\dfrac{V^2 b_2}{D^2} & -\dfrac{V^2 a_2}{D^2} & -\dfrac{V b_1}{D} & -\dfrac{V a_1}{D} & 0 & 0 \\ 0 & 0 & 0 & 0 & 0 & 1 \\ 0 & 0 & 0 & 0 & 0 & -k \end{bmatrix} \tag{29.6}$$

$$\text{and} \quad B = \begin{bmatrix} 0 & 0 & \dfrac{V^2 a_3}{D^2} & \dfrac{V^2 b_3}{D^2} & 0 & 0 \end{bmatrix}^T$$

### 29.3.2 Simulation of the Model Adapted to Real-Time Estimation

The theoretical model is generally enough to simulate a large-caliber projectile, such as a 155-mm projectile (spin stabilized) and a small-caliber projectile such as a 30-mm projectile (fine stabilized). These two kinds of projectile are presented in Figure 29.4. All the simulations and results presented in this chapter are for the small one.

(a)

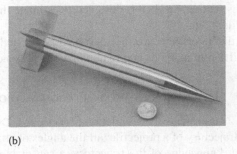

(b)

FIGURE 29.4 Two different kinds of projectiles: (a) 155-mm caliber projectile (40 kg) and (b) 30-mm caliber projectile (0.4 kg).

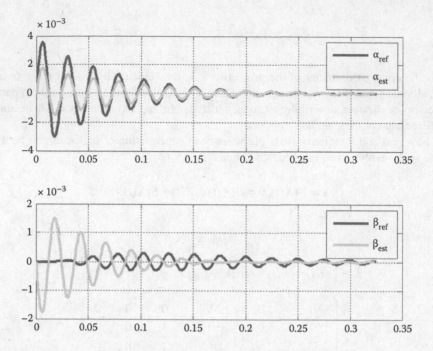

**FIGURE 29.5** Comparison between the complete model and the simplified model (to be used in the estimation algorithm).

By comparing the complete model with the simplified model adapted to the estimation, it is easy to see what the estimation algorithm needs to compensate to make the prediction converge to the real values. Figure 29.5 compares the simulation of the complete model at a sample time of 0.0001 second ($\beta_{ref}$) with the simulation of the simplified model at a sample time of 0.001 second ($\beta_{est}$).

The simplified model is precise enough to simulate the projectile behavior: phase delay and amplitude are due to false initial conditions. The goal of the filter is to use the simplified model and all the measurements to make the estimation fit the reality, without knowledge of the initial conditions.

## 29.4 ESTIMATION SIMULATION RESULTS

According to the sensors used onboard the projectile and the evolution model presented in the previous section, an EKF is designed to estimate the roll angle, the attitude, and the position of the projectile.

All the algorithms are first designed and tested in simulation; by this means, it is easy to test many different methods and to quantify the precision of the estimation.

### 29.4.1 DESCRIPTION OF THE SIMULATION METHOD

Figure 29.6 presents the simulation process. A 6 DOF (6 degrees of freedom) model simulates the whole trajectory of a projectile: all the angles of attitude and the 3-D position.

From knowledge of the trajectory, a sensor block generates the measurements of the sensors embedded onboard. This block simulates the nominal sensor model, the dynamics of the sensors, and the defaults identified during previous experiments. Then, different estimation algorithms are

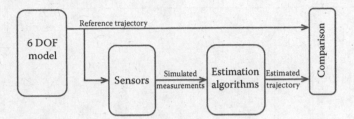

**FIGURE 29.6**  Simulation process.

tested. To quantify the results, the trajectory estimated is compared with the reference trajectory used to compute the measurements.

In the next section, we present some simulation results of the estimation algorithm based on EKF, developed by using the equation presented in Sections 29.2 and 29.3.

### 29.4.2  SIMULATION RESULTS

This section presents some simulation results on the navigation algorithms. In a first step, only the roll angle is estimated by the use of radial magnetometers. Then the attitude is estimated from radial accelerometers and the axial magnetometer. Eventually the position is computed by the use of all the sensors.

#### 29.4.2.1  Estimation of the Roll Angle from Magnetometers

The simulation results of the method depicted in Figure 29.7 are presented in the text that follows. The filter algorithm starts at time $t = 0.3$ s when the roll rate has reached its steady state. At the top of Figure 29.8, we compare the reference angle with the estimation angle: One revolution seems to be enough to make the estimation converge. The figure at the bottom focuses on the estimation error: after 0.1 s the estimation error stays underneath $1°$.

No direct measurement of the roll angle is available on the experiment system. Thus, to test and validate our method, we compare the magnetometer measurements and their estimates. The magnetometer estimations $H_{a\,est}$ and $H_{b\,est}$ are computed using the estimation state space vector. Figure 29.9 shows simulation results for the estimations of the scale factors ($Amp_A$ and $Amp_B$) and the offsets ($Offset_A$ and $Offset_B$) of the two embedded magnetometers. These four estimates meet the reference values. At the top of Figure 29.9, the estimation of the magnetometer measurements $H_{a\,est}$ and $H_{b\,est}$ converges to the simulated references signals $H_{a\,ref}$ and $H_{b\,ref}$; the simulation shows that only one revolution is enough to provide good accuracy.

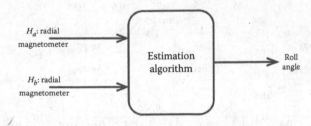

**FIGURE 29.7**  Estimation of the roll angle algorithm.

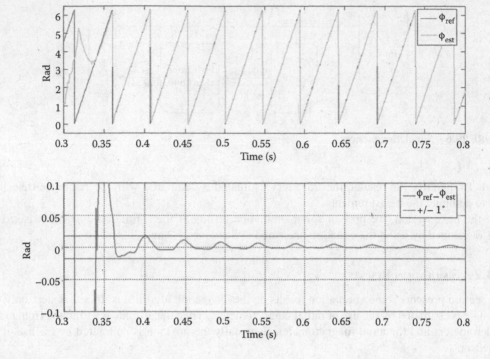

**FIGURE 29.8** Roll angle estimation.

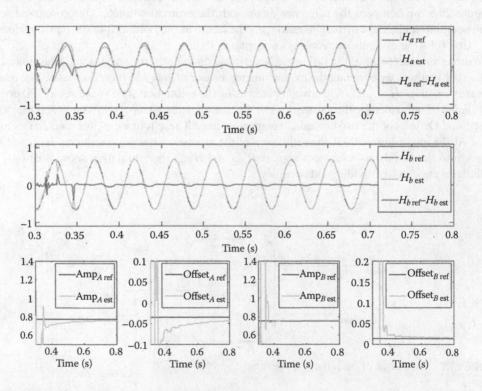

**FIGURE 29.9** Magnetometer measurement estimation.

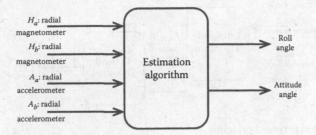

**FIGURE 29.10**  Estimation of the algorithm of the attitude angles.

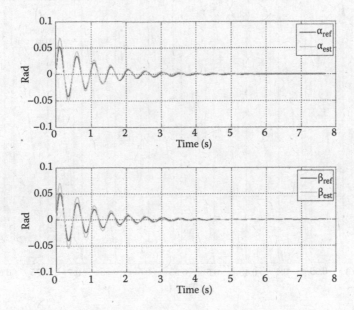

**FIGURE 29.11**  Estimation of attitude angles.

### 29.4.2.2  Estimation of the Attitude from Magnetometers and Accelerometers

The simulation results using the principle presented in Figure 29.10 are plotted in Figure 29.11.

Figure 29.11 shows that the estimation of very small attitude angles is possible. The estimated values are about 0.05 rad (3°).

## 29.5  IN-LAB EXPERIMENT

After the validation of the algorithms in simulation, some experiments have been developed in the laboratory. In this section, we present the lab demonstration of the real time estimation of the roll angle of a projectile.

### 29.5.1  Description of the Experiment

To validate the estimation of the roll angle in real time, the projectile is mounted on a test bench to generate the roll rate. A light-emitting diode (LED) is installed on the projectile. A camera is placed perpendicularly to the projectile's main axis. As the principle is described in the Figure 29.12, the algorithm computes in real time the roll angle of the projectile using the two radial magnetometer

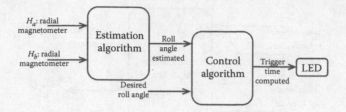

**FIGURE 29.12**   In-lab experiment principle.

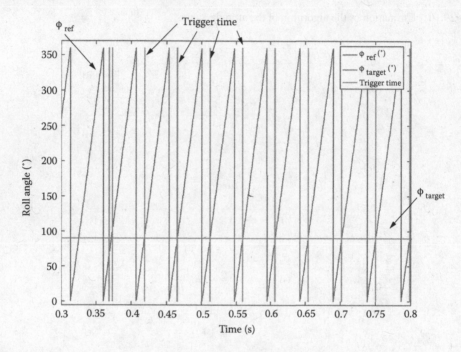

**FIGURE 29.13**   Trigger time computed for a target roll angle of 90°.

measurements. A second algorithm computes the trigger time of the LED to switch it on when the LED faces the camera.

The results of the computation of the trigger time are presented in Figure 29.13. As a new predicted trigger time is computed, the new value is sent to the projectile to update the previous value. Considering the 22 Hz roll rate and the 1 ms sample time, the trigger time can be updated up to 40 times per revolution.

The results presented in Figure 29.13 shows that only the first predicted trigger time is wrong, due to the delay imposed by the convergence of the algorithm.

## 29.5.2   IN-LAB RESULTS

To observe the LED, we used a high-speed camera with an acquisition speed up to 250,000 frames per second. The rotation frequency we reached was about 950 Hz. An algorithm giving the luminous intensity of each picture has been computed to determine pictures that the LED is switched on. This also leads to the corresponding trigger times of the LED.

Figure 29.14 represents the luminous intensity of the picture. Using this result the accuracy of the algorithm is estimated beneath 0.5° for a roll rate of 50 Hz, and beneath 5° for a roll rate of 1000 Hz (it is not technically possible to measure a better accuracy).

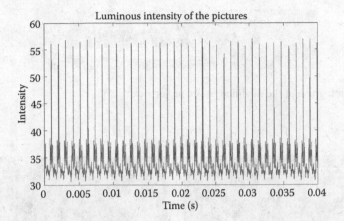

**FIGURE 29.14** Mean value of each picture.

## 29.6 IN-FLIGHT EXPERIMENT

This section discusses the validation of the algorithm under real conditions. The first subsection presents the real-time estimation of the roll angle of a projectile using embedded magnetometers and a digital signal processor (DSP). The second subsection will present the estimation of the attitude angle of a projectile based on data recorded during the flight of the projectile.

### 29.6.1 IN-FLIGHT REAL-TIME ESTIMATION OF THE ROLL ANGLE

To validate the roll angle estimation algorithm in-flight, an electronic module has been developed and gun hardened to be used into the projectile. The experiment and the in-flight results are presented in the next subsections.

#### 29.6.1.1 Description of the Experiment

The in-flight test was performed at the French-German Research Institute of Saint-Louis (ISL) open range test site over a flight distance of 450 m. Figure 29.15 shows a schematic of the experimental setup. The initial velocity was about 540 m/s. Videos of the flights were recorded by the "trajectory tracker" over a distance of about 90 m.

The projectile used for the spin rate estimation tests was a 90-mm smoke projectile modified for the purpose. The forebody of the projectile designed at ISL was made out of aluminum. The afterbody was taken from a 90-mm smoke projectile including six fins and screwed onto the forebody. To induce a roll rate, three of the fin tails were canted at an angle of 9°. The total mass of the projectile was 4.5 kg for a total length of 452 mm. A photograph of the model is shown in Figure 29.16.

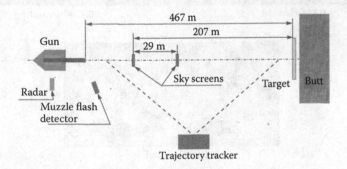

**FIGURE 29.15** Schematic of the free flight experimental setup.

**FIGURE 29.16** Photograph of the modified smoke projectile.

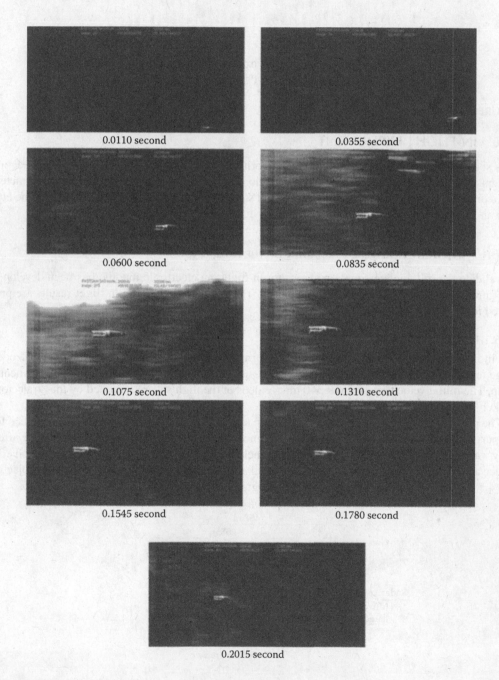

| 0.0110 second | 0.0355 second |
| 0.0600 second | 0.0835 second |
| 0.1075 second | 0.1310 second |
| 0.1545 second | 0.1780 second |
| 0.2015 second | |

**FIGURE 29.17** Video snapshots of the roll angle estimation.

### 29.6.1.2 In-Flight Results

The projectile was tracked by a high-speed camera. The target roll angle had been set so that the LED was switched on when it faced the camera. Figure 29.17 shows some snapshots extracted from the video. These represent the nine ignitions of the LED.

By analyzing the in-flight acquisition, it is very difficult to evaluate the accuracy of the estimation. However, and according to the laboratory experiments, we can assume that the accuracy of the algorithm is about 0.5°.

### 29.6.2 REAL-TIME ESTIMATION OF THE ATTITUDE'S ANGLE

In this section, the navigation algorithm is tested on real data, recorded during the flight. As there is no direct measurement of the attitude angles during the flight, the validation has to be done by using the redundancy of the different measurements.

### 29.6.2.1 Description of the Experiment

An electronic module was designed and gun-hardened to be embedded into the projectile. Figure 29.18 shows the electronic module composed of three magnetometers and three accelerometers. All the data are recorded during the flight using a telemetry system.

### 29.6.2.2 In-Flight Results

The method employed to validate the attitude is represented in Figure 29.19.

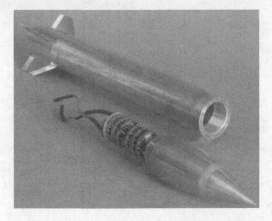

**FIGURE 29.18** Instrumented electronics embedded into the projectile: realized by ISL-STC department.

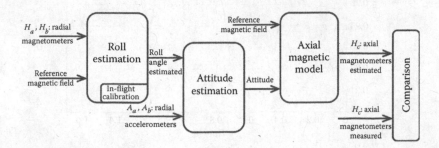

**FIGURE 29.19** Description of the validation algorithm.

First, the two radial magnetometers are used to estimate the roll angle in the "roll estimation" block. The "in-flight calibration" sub-block represents the estimation of the parameters of the two radial magnetometers in the estimation algorithm. Then, the attitude angles are estimated from the roll angle and from the two radial accelerometers ("attitude estimation" block). Furthermore, the axial magnetometer signal is reconstructed by using the three estimated angles. This *axial magnetometer reconstructed* signal can be compared to the *axial magnetometer measurement*.

The two blocks ("roll estimation" and "attitude estimation") are based on EKF. The details of these filters are given in (Changey et al. 2004, 2009, 2011).

Figures 29.20 and 29.21 present the estimation of the attitude angle of the projectile. These estimations cannot be compared with reference angles; this is due to the fact that there is no direct

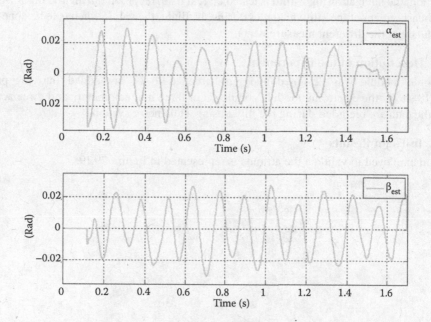

**FIGURE 29.20** Attitude estimation extracted from radial accelerometers.

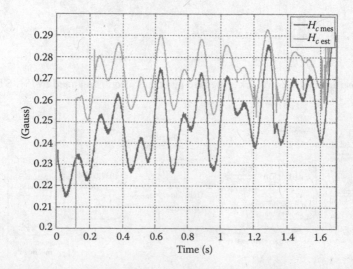

**FIGURE 29.21** Modeling of the axial magnetometer.

measurement of the attitude angle at the ISL proving ground. As is well described in Figure 29.19, the axial magnetometer measurement has not been used in this computation. Therefore, it will be used to validate the estimation.

The validation in Figure 29.21 consists of the comparison of the axial magnetometer measurement signal with its reconstruction based on the three estimated attitude angles. The differences are due to problems in the calibration of the sensors.

## 29.7 CONCLUSION

The simulations and the estimations are close enough to conclude that it is possible to compute a real-time estimation of the attitude of a projectile based on low-cost magnetometers and low-cost accelerometers measurements, using data fusion.

## REFERENCES

Bar-Itzhack, I., and Oshman, Y. (1985). Attitude determination from vector observations: Quaternion estimation. *IEEE Transactions on Aerospace and Electronic Systems*, 21(1):128–136.

Boutayeb, M. (2004). Synchronization and input recovery in digital non-linear systems. *IEEE Transactions on Circuits and Systems II: Express Brief*, 51(8): 393–399.

Changey, S., Fleck, V., and Beauvois, D. (2004). Non linear filtering for attitude estimation with magnetometer sensor. *ICNPAA*. Timisoara, Romania.

Changey, S., Pecheur, E., and Wey, P. (2009). Real time estimation of supersonic projectile roll angle using magnetometers: In-lab experimental validation. *2nd IFAC Workshop on Dependable Control of Discrete Systems*. Bari, Italy.

Changey, S., Pecheur, E., Wey, P., and Sommer, E. (2011). Real time estimation of projectile roll angle using magnetometers: In-lab experimental validation. *4th European Conference for Aerospace Sciences (EUCASS)*. St. Petersburg, Russia.

Ciccarella, G., Dalla Mora, M., and Germani, A. (1995). A robust observer for discrete time nonlinear. *System & Control Letters*, 24: 291–300.

Crassidis, L., Markley, F., and Cheng, Y. (2007). Survey of nonlinear attitude estimation methods. *Journal of Guidance Control and Dynamics*, 30(1): 12–28.

Fleck, V. (1998). *Introduction à la balistique exterieure*. Saint Louis: ISL.

Hainz, C., and Costello, M. (2005). Modified projectile linear theory for rapid trajectory prediction. *Journal of Guidance Control and Dynamics*, 28(5): 1006–1014.

Lieske, R., and McCoy, R. (1964). *Equations of Motion of a Rigid Projectile*. Aberdeen Proving Gound, MD: Army Ballistic Research Lab.

McCoy, L. (2001). *Modern Exterior Ballistics*. Atglen, PA: Schiffer Publishing.

Psiaki, M. L., Martel, F., and Parimal, K. P. (1990). Three axis attitude determination via Kalman filtering of magnetometers data. *Journal of Guidance*, 13(3): 506–514.

Reif, K., and Unbehauen, R. (1999). The extended Kalman filter as an exponential observer for nonlinear systems. *IEEE Transactions*, 47(8): 2324–2328.

Shuster, M., Lefferts, E., and Markley, F. (1982). Kalman filtering for spacecraft attitude estimation. *AIAA 20th Aerospace Sciences Meeting*, vol. 232. Orlando, FL.

Song, Y., and Grizzle, J. W. (1995). The extended Kalman filter as a local asymptotic observer for nonlinear discrete-time systems. *Journal of Mathematical Systems, Estimation and Control*, 5: 59–78.

# 30 Data Fusion for Telemonitoring
## Application to Health and Autonomy

*Céline Franco, Nicolas Vuillerme, Bruno Diot,*
*Jacques Demongeot, and Anthony Fleury*

## CONTENTS

## 30.1 INTRODUCTION

In an aging society facing the explosion of chronic (degenerative) and lifestyle-related diseases, it is generally accepted that e-health/telemedicine constitutes the best alternative to meet the increasing healthcare needs. On the one hand, healthy people would like to age-in-place well and maintain their healthy status and, on the other hand, chronic patients do not want their condition to override their daily living and yearn to preserve their quality of life. Moreover, the trend toward self-management, in addition to the therapeutic education of the patient contribute to the development of solutions dedicated to home-based or ambulatory monitoring. These solutions may allow physiological and/or behavioral monitoring from a network of wearable and/or environmental sensors.

This chapter is divided as follows. Section 30.2 is focused on an in-place network of sensors for monitoring activity at home. Data collected from both environmental (infrared sensors, contact switches, microphones) and embedded sensors (kinematic unit) were merged and used to detect activities of daily living (ADLs) and pathological situations observable in neurological degenerative diseases such as perseveration (i.e., repeating a task pathologically) and drift in nycthemeral rhythm. Section 30.3 is developed around the fusion of the built-in inertial sensors of a smartphone (3-D accelerometer, gyroscope and magnetometer). A smartphone-based biofeedback system was developed for self-measurement, training, and rehabilitation. It offers a range of derived applications, among them the application iBalance for postural capacities assessment and improvement.

## 30.2   HABITAT INTELLIGENT POUR LA SANTÉ (HIS): FROM SENSORS IN-HOME TO BEHAVIOR ANALYSIS

### 30.2.1   INTRODUCTION AND STATE-OF-THE-ART

Smart homes are a special field of pervasive computing. These environments are enriched by the presence of miniature sensors to capture information about the resident and his surroundings while being invisible. They were first developed to improve comfort, security, and remote control of home appliances. Their field of application has been extended to healthcare over the past 20 years [1–4]. Since then, projects of "health smart homes" have become numerous throughout the world (Table 30.1).

However, most of the smart home projects have stayed untapped and only a few of them were deployed in real environments. This may be accounted for by practical considerations—cost of the installation, difficulties to adapt it to an existing and nonstandardized environment—and technical ones—sensitivity of health data, lack of security against undesirable access or use.

To address practical limitations, off-the-shelf sensors such as inertial sensors (accelerometers, gyroscopes) or physiological sensors embedded into wearable devices, infrared sensors for presence detection, contact switches, microphones, or sensors installed directly onto objects [5] for the surroundings must be favored. Although the sensors of the last group give only access to low-level data, they may be sufficient to infer behavior. The choice of the sensors was discussed in different studies [6–8] depending on the interest to end-user (well-being, comfort, security, compensation of cognitive disabilities, therapeutic delivery, monitoring, etc.), but the most important application related to health is ADL recognition [9,10]. Indeed, an alteration into the proceedings of ADL—difficulties to perform, decrease in the level of activity, and disturbance of their circadian rhythm [11,12]—may indicate a likely degeneration and highlight signs of frailty and/or degenerative disease. In this chapter, we focus particularly on context-aware smart homes, that is, equipped with a network of sensors allowing detection of the resident location and activity.

### 30.2.2   CHALLENGES OF CONTEXT-AWARE SMART HOMES

To be reliable and accurate, ADL recognition necessitates a large network of various sensors capable of dealing with heterogeneous multimodal data. These data sometimes may be redundant, providing more consistency to the observation and sometimes missing or inconsistent, raising new questions. In addition, they are usually noisy and reflect phenomena occurring at different time scales. These features make their fusion challenging and many techniques of advanced data processing were proposed to deal with it [13,14].

Before fusing data, a preprocessing step is used most of the time. Indeed, even if raw data seem to be more informative, it turns out that reducing the dimension in a first stage provides more reliable results. The two main approaches consist of selecting the most relevant features (e.g., nearest neighbors) and extracting new feature variables with techniques such as principal component analysis, independent component analysis, or linear discriminant analysis. Moreover, there are many different ways to perform an ADL, yielding important intraindividual and interindividual variability.

To estimate new features, two approaches may be distinguished: knowledge-driven and data-driven techniques. In the first case, an expert defines logical rules between potential events from a priori knowledge on which the decision is based. This approach includes event calculus, temporal relation among events, fuzzy logic, and so forth. Although a logical approach enjoys the comfort and the rigor of a formal framework that is efficient on simulated data, it encounters difficulties while treating noisy and uncertain data such as real ones. Therefore the most well-spread techniques come from machine learning and rely on probabilities (Table 30.2).

**TABLE 30.1**

**Brief Review of Health Smart Homes with Activity Recognition Capacity**

| Reference | Project | Location | Use | Sensors | Fusion Algorithm |
|---|---|---|---|---|---|
| [15] | Aware Home | Georgia Tech, USA | Ubiquitous computing of ADLs | Ultrasonic sensors, RF technology, video, floor sensors | Hidden Markov model (HMM) |
| [16–18] | Mav Home | Texas, USA | Artificial intelligence Robotics Prediction and decision on activity | 60 X10 devices plugged into the home electric wiring system | Decision tree based on a finite-order Markov model |
| [19] | An intelligent room | Tokyo, Japan | Track any change of the system | Sensors attached to beds, floors, tables, and switches | Summarization algorithm |
| [11,20,21] | Habitat Intelligent pour la Santé (HIS) LILAS | Grenoble, France | Monitor activity and physiological parameters | Infrared sensor Contact switches Wearable devices Video Microphone | Statistical predictive algorithm Support vector machine (SVM) HMM |
| [13,22] | DOMUS Sweet Home project | Grenoble, France | Monitor activities | Infrared sensors, contact sensors, video camera (for annotation only), microphones | Markov logic network HMM Conditional random field (CRF) SVM Random forest |
| [23–25] | House of Matilda Gator Tech | Florida, USA | Location detection Optimize comfort and safety | Ultrasound, pressure sensors on floor, smart devices with sensors and actuators | HMM CRF Energy patterns |
| [26,27] | Housen project Placelab | Massachusetts, USA | Annotation of 84 ADLs Test and evaluation of ambient-assisted living tools | 214 sensors: temperature, light, humidity, water, gas flow, door switches, item use sensor, camera, microphone | Naïve Bayesian classifier (NBC) |
| [28] | Adaptation of a normal flat | Amsterdam, Netherlands | Activity recognition Annotations made by the resident via a speech recognition module | 14 state-change sensors: sensors placed onto microwave, fridge and toilet flush | Temporal probabilistic model HMM CRF |
| [9,29–31] | Casas | Washington, USA | Discover user behavior patterns | Motion, contacts doors, temperature, electricity, and water consumption | Frequent and periodic activity miner NBC HMM CRF |
| [32] | Core Lab | Taiwan | Location-aware activity recognition system | Power usage, contact, pressure, location, motion detection | NBC |
| [33] | GERHOME | Nice, France | Monitor activities | Water and electrical sensor, contactors pressure sensors and video | Event recognition |

## TABLE 30.2
### Main Approaches for Activity Recognition

| | Probabilistic Approach | Geometric Approach | Binary Classifier |
|---|---|---|---|
| Single-frame approach | *Naïve Bayesian* [34,35] <br> *Gaussian mixture model* [36] <br> Logistic classifier <br> Parzen classifier | Support vector machine <br> [13,20,35] <br> Nearest mean <br> Nearest neighbor [34,35] <br> Artificial neural network [37–39] | *Binary decision tree* <br> *(C4.5)* [34,35] <br> *Random forest* [13] |
| Sequential approach | Hidden Markov model [15,28,40] <br> Conditional random field [28] <br> Statistical relational learning [13] | | |

### 30.2.3 THE HIS PLATFORM

Since its development in 1994 in Grenoble, the multimodal platform, HIS, has been dedicated to the development and evaluation of simple and noninvasive devices for promoting aging-in-place and maintenance of autonomy. It takes the shape of a life-size flat located in the Faculty of Medicine of Grenoble and hosts different projects, among which is a unit dedicated to ADL recognition (Figure 30.1). It was deployed into the geriatric units of the hospital of Charles Foix (Ivry-sur-Seine) and la Grave (Toulouse) as well as into two independent living communities (Grenoble) and will be under the denomination Logement Intelligent pour la Longévité, l'Autonomie et la Santé (LILAS).

One field of investigation into the HIS is related to the recognition of the ADLs and monitoring of behavior patterns from localization data obtained by means of noninvasive sensors. Initially, the system was tested using simulated data of the space occupation [11]. This methodology was extended to the monitoring of various physiological parameters (weight, blood pressure, cardiac rhythm, temperature, glycemia, arterial oxygen saturation) [41]. Afterwards, a system of analysis

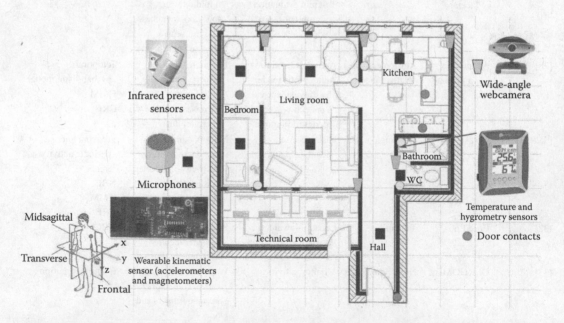

**FIGURE 30.1** Description of the HIS and distribution of the sensors.

adapted to the heterogeneous data provided by the HIS was developed to detect critical situations and included an unsupervised algorithm of temporal pattern recognition. A normal profile is designed from a learning corpus and important disparities trigger alarms [42,43]. Studies were then focused on the detection and identification of postural transitions and walking periods from accelerometric data [44] and refined by merging actimetric and localization data with a Bayesian model [45]. During the last decade, previous work was applied to recognition and classification of high-level activities (ADLs) in real conditions. For this purpose, a more complete kinematic unit was developed (3-D accelerometer and 3-D magnetometer) as well as associated algorithms for posture and level of activity detection. These actimetric data were used jointly with location data and sound data to reach the detection of seven different ADLs. This approach was validated on a corpus of real data gathered into two independent living communities during the AILISA project [46]. Recently, a monitoring system of the daily life habits of the resident was developed. It is based on a learning process and is able to classify the ADLs performed [20,21,47] and trigger alarms in case of deviations from the established profile [12].

For this project, the HIS was equipped with both surrounding and embedded sensors: infrared sensors (room occupation), contact switches on doors (use of the fridge, the cupboard and the dresser), temperature and hygrometry sensors into the bathroom, seven microphones (sound and speech), and a wearable kinematic sensor including a 3-D accelerometer and a 3-D magnetometer (postural and locomotor activities) [21] (Figure 30.2). Note that five wide-angle webcams were used

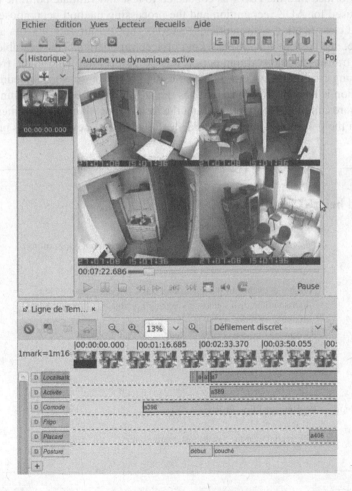

**FIGURE 30.2** Snapshot of the experimenter's interface (in French).

only for annotation purposes. Data collected were divided into 3-minute frames. A classification step based on a multiclass SVM was then used to match each frame to one of the following ADLs: (1) hygiene, (2) toilet use, (3) eating, (4) resting, (5) sleeping, (6) communication, and (7) (un)dressing. Fifteen healthy subjects (32 ± 9 years) proceeded to an experiment of daily living. They were asked to perform these seven activities, at least once in any order they chose, without time limitation. The duration of the experiment ranged from 00:23:11 hours to 01:35:44 hours.

Features were selected from the different sensors before classification using principal component analysis resulting in the features described in Table 30.3. A classification algorithm based on a multiclass SVM with a Gaussian kernel [20] was then applied on a 3-minute-long sliding window. To get better results, prior knowledge about the temporal or spatial context was introduced by associating logical rules between either the time of the day or the location with an ADL (e.g., eating at lunchtime, eating in the kitchen) [48]. All the data were used during the learning process to adjust the parameterization of the model. The method was then validated through the "leave-one-out" cross-validation technique.

It turned out that both temporal (90.1%) and hybrid (86.8%) approaches improved the level of ADL recognition of the generic approach (86.2%) while the spatial approach made it worse (78.9%). This last result may be accounted for by the poor sensibility of the infrared sensors and raises the question of the need to add more contactors placed on daily-life items to their setup placed on daily-life items (Table 30.4).

This study performed into the HIS was conducted to test and validate both the network of non-invasive sensors used for ADL recognition and the processing method decomposed into a feature selection step and classification with multiclass SVMs. The surrounding sensors used are off-the-shelf, noninvasive, and easy to adapt to any kind of accommodation. The wearable sensor used to monitor postural transitions and walking period was developed and hand-made from raw components. In daily living conditions, we may consider using kinematic sensors embedded into a smartphone. This solution is largely affordable and has the advantage of being communicative and able to get access to more complete information about the resident location and movements without possible confusion in the presence of multiple residents [49–51]. The next section is precisely focused

## TABLE 30.3
## Feature Selected before Classification

| Sensor | Features Selected | ADL |
|---|---|---|
| Kinematic unit | Percentage of time spent in various postures (standing, lying, sitting) and while walking | Sleeping<br>Resting<br>Eating |
| Microphones | Number of events for the nine sound classes: speech, door locking, door shutting, dish sounds, phone ringing, object falling, glass breaking, shouting | Communication<br>Cooking<br>Resting |
| Infrared sensors | Percentage of room occupation and number of detection | Sleeping<br>Cooking<br>Hygiene<br>Bowel movement<br>Resting |
| Contact switches on doors | Percentage of time open | Cooking<br>Dressing |
| Temperature and hygrometry | Differential measure for the last quarter hour | Hygiene |

**TABLE 30.4**
**Percentage of Good Classification of the Seven ADLs Considered**

| Activity | Generic | Spatial | Temporal | Hybrid |
|---|---|---|---|---|
| Sleeping | 98.0 | 69.4 | 100.0 | 98.0 |
| Resting | 78.1 | 78.1 | 83.6 | 78.1 |
| Dress/undress | 80.0 | 73.3 | 86.7 | 80.0 |
| Eating | 97.8 | 97.8 | 97.8 | 97.8 |
| Toilet use | 81.3 | 87.5 | 87.5 | 87.5 |
| Hygiene | 71.4 | 71.4 | 78.6 | 71.4 |
| Communication | 80.0 | 65.0 | 85.0 | 80.0 |
| Global | 86.2 | 78.9 | 90.1 | 86.8 |

*Note:* All the approaches except the generic one include *a priori* knowledge.

on a smartphone application for self-monitoring of postural stability, training, and rehabilitation. In this way, the smartphone is placed at the center of the sensor network of the home. It is used not only as a network of sensors but also as an active hub able to monitor, assess, and contribute to the preservation of the user's abilities; store information; and communicate with both the user and his medical care providers.

## 30.3 SMARTPHONE DATA ANALYSIS AND INTERPRETATION

### 30.3.1 INTRODUCTION AND STATE-OF-THE-ART

For decades, research on the use of microelectromechanical system (MEMS) sensors such as accelerometers, magnetometers, or gyroscopes has provided results on various domains such as fall [52,53], balance, and gait or movement analysis [54]. These sensors are proposed for general use into specifically designed materials such as Movea [55], XSens [56], ACTIM6D [57], and so forth. However, the cost of these last devices significantly exceeds the cost of their embedded sensors.

Today, accelerometers are integrated into most electronic devices such as cameras to automatically correct the orientation of the picture. Nintendo has introduced them into its Wii gamepad to change the way the player is living the experience. They made use of the player's movements instead of receiving controls from buttons. The same approach was adopted by developers of smartphones. Since the democratization of these devices that include MEMS sensors as standards, it has become possible to control the reaction of an application through the movements of the phone (and not only through its orientation).

The sensors embedded in these smartphones are similar to the those in on-the-shelf inertial measurement units (IMUs). The main difference in terms of movement recognition is the quality of the computation to obtain angles from the raw values. In addition, some IMU, such as XSens, may provide multiple points of measurement. Such a feature may be obtained with multiple smartphones but requires synchronizing gathered data in real time. Concerning the former point, the computational power available is largely sufficient to use standard algorithms such as Kalman filtering, with a large bandwidth (comparing to human movements) [58]. Therefore, the primary idea of the following work is to replace the different devices previously used by only one commonly available around the world: the smartphone. Interestingly, as a communication device, it can also send all its data through different networks. In addition, it can also be connected to supplementary devices such as vibrators or headsets (with commands and microphones) to expand its monitoring possibilities to active reeducation.

Possibilities are unlimited to obtain powerful and efficient devices that can be used by everybody at home for medical and nonmedical purposes (such as "quantified-self" applications that are now available and largely used for characteristics self-monitoring [59]).

### 30.3.2  SMARTPHONE SETUP AND ANGLE DATA COMPUTATION

Current smartphones generally contain a rapid ARM-based microprocessor, with sufficient frequency (under 700 MHz is rare) and enough RAM (1 GB can be found in the most popular phones). They also, as previously stated, generally include a complete IMU. The number of sensors depends on the quality of the smartphone (and its price range). All of the phones are at any rate at least equipped with an accelerometer, which is required by the three most common operating system to handle the orientation of the screen. Those containing a GPS (a large part of them) are also equipped with a magnetometer to determine the orientation on the map. Finally, the most equipped one (less and less rare as time goes on) will include a gyroscope for gameplay. The operating system of the smartphone includes variables to determine which kinds of sensor are available. Knowing this, it is possible to adapt the computation to the material.

The first task, to make possible any analysis of posture or activity, is to determine the movement of the phone brought by the user. For this, we have at our disposal three different kinds of sensors:

- An accelerometer, giving the acceleration of the phone on the three axes of the sensor. This acceleration is composed of both acceleration due to gravity and that due to the movement of the phone. Using the first one, we can induce the angle of the phone with this acceleration vector that is present all over the Earth with almost the same orientation and value.
- A magnetometer, giving the projection of the ambient magnetic field on the three axes of the sensor. This ambient field is composed of the Earth one, due to the center of Earth

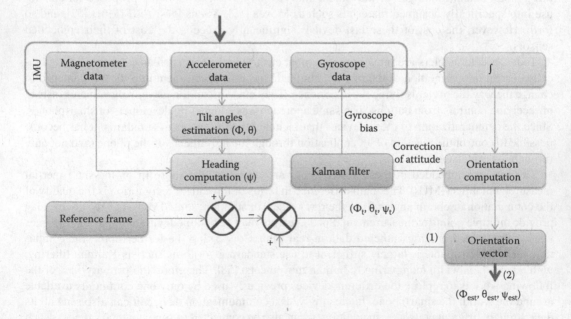

**FIGURE 30.3**  Definition of the axes on the phone and computation of the different angles on these axes.

activity, and of the local one, due to any magnetic materials around. Giving the orientation and value of the Earth one, we can determine the 3-D orientation of the phone. However, compared to other sources, this value is very weak and so this sensor can be noisy. The operating system of the phone uses it mostly for orientation, comparing to the north direction.

- The last one is a gyroscope that gives the angular velocity on each axis. This sensor is one of the most efficient to determine the movement of the phone considering a first basis orientation, but its major drawback is the need to integrate the signal, which leads to the appearance of a drift responsible for the loss of the user's real position.

Figure 30.3 shows the orthonormal basis created by the phone on the left and the computation of the angles on the right. From the three kinds of sensors, a Kalman-based fusion algorithm is used to evaluate angles and correct the drift and measurement errors of the different sensors, thereby considering the nine available dimensions. In this way, we have now, included in our smartphone, a complete IMU comparable with the ones from companies such as XSens or McRoberts.

In the next section, we detail a project we led using sensors embedded into the smartphone, from data fusion to application dedicated to postural monitoring and reeducation.

### 30.3.3 A Complete System for Assessing, Monitoring, and Training Balance Abilities of the Person

Auditory, tactile, or visual feedback has been proven to improve the posture of a person in daily living after a set of exercises over a long time. From the sensor's values fused and recovered on the smartphone as angles, we can infer the position of the person and then check or propose some corrections on his/her balance. A first application to postural monitoring and reeducation has been proposed in Ref. [60] and is depicted in Figure 30.4. In this work, we propose a complete system with the following capabilities:

- It can measure the angle of the user's trunk during a set of exercises proposed by a member of the medical staff.
- For instance, in some exercises, a headset, connected to the smartphone, provides the user with a warning feedback (an audio biofeedback [ABF]) in case of unstable position: If the person is not in a steady position and is on the left, a sound appears in the left ear until the person moves sufficiently to the right to correct his or her position, and respectively on the right.
- The application is standalone, as it will guide the person to perform the complete protocol using the text-to-speech ability of the phone to translate the medical staff proposition of exercises. It can communicate the results, presented to the user as a report on the execution of the protocol, to a medical staff linked to the patient.

The application has been designed to allow programming a sequence of exercises adapted to the person. It has been shown in Refs. [60] and [61] that the ABF is correctly used and has a good influence on the posture of the person, in respectively young adults and older persons. Figure 30.4b shows that the feedback is used to reduce the duration of instability in the more challenging position (tandem). Other figures show also that the movements and their frequency are largely reduced. As illustrated in the user's interface, it is possible to use other sensory modalities for the biofeedback depending on the user's preferences or abilities. For instance, a large set of experiments about the use of a feedback on the user's tongue for navigation or for postural reeducation has also been launched [62]. This application will feature in further work extended to gait monitoring and rehabilitation to complete the possibilities of the smart home.

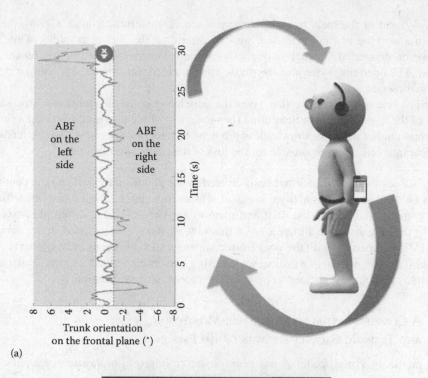

(a)

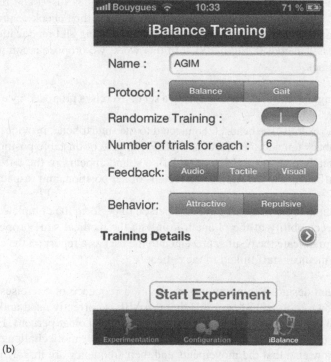

(b)

**FIGURE 30.4** Architecture of the application iBalance (a) and the graphical user interface (b).

## 30.4  CONCLUSION

In this chapter, we first presented a state-of-the-art of context-aware smart homes with activity recognition capacities and associated fusion algorithms. We detailed our progress over time on this field of research through the HIS platform and a network of noninvasive and on-the-shelf sensors. More than only sensing, our final goal is to interpret data to infer the resident's behavior in his or her home, and then to detect early signs related to the appearance of a disease.

In the second part, data from a smartphone were acquired and analyzed for rehabilitation purposes. In these two applications, balance and equilibrium abilities are assessed and corrected automatically by the use of biofeedback from the smartphone.

The smartphone could also be used, in addition to for exercises and rehabilitation at home, for a continuous analysis of the movements of the person. That would permit integrating it as a supplementary sensor in the smart home.

## ACKNOWLEDGMENTS

This work was supported in part by funding by IDS company, the French national program "programme d'Investissements d'Avenir IRT Nanoelec" ANR-10-AIRT-05, and Institut Universitaire de France.

## REFERENCES

1. M. Chan, E. Campo, D. Estève and J.-Y. Fourniols, "Smart homes—Current features and future perspectives," *Maturitas*, vol. 64, no. 2, pp. 90–97, 2009.
2. T. Gentry, "Smart homes for people with neurological disability: State of the art," *NeuroRehabilitation*, vol. 25, no. 3, pp. 209–217, 2009.
3. G. Demiris and B. K. Hensel, "Technologies for an aging sóciety: A systematic review of 'smart home' applications," *Yearbook of Medical Informatics*, pp. 33–40, 2008.
4. M. Alam, M. B. I. Reaz and M. A. M. Ali, "A review of smart homes 2014: Past, present, and future," *Systems, Man, and Cybernetics, Part C: Applications and Reviews, IEEE Transactions on*, vol. 42, no. 6, pp. 1190–1203, 2012.
5. P. Rashidi and A. Mihailidis, "A survey on ambient-assisted living tools for older adults," *IEEE Journal of Biomedical and Health Informatics*, vol. 17, no. 3, pp. 579–590, 2013.
6. D. J. Cook and L. B. Holder, "Sensor selection to support practical use of health-monitoring smart environments," *Data Mining and Knowledge Discovery*, vol. 1, no. 4, pp. 339–351, 2011.
7. D. Ding, R. A. Cooper, P. F. Pasquina and L. Fici-Pasquina, "Sensor technology for smart homes," *Maturitas*, vol. 69, no. 2, pp. 131–136, 2011.
8. T. Tamura, "Home geriatric physiological measurements," *Physiological Measurement*, vol. 33, no. 10, pp. R47–R65, 2012.
9. H. Fang, L. He, H. Si, P. Liu and X. Xie, "Human activity recognition based on feature selection in smart home using back-propagation algorithm," *ISA Transactions*, vol. 53, no. 5, pp. 1629–1638, 2014.
10. S. Chernbumroong, S. Cang and H. Yu, "A practical multi-sensor activity recognition system for home-based care," *Decision Support Systems*, vol. 66, pp. 61–70, 2014.
11. G. Virone, N. Noury and J. Demongeot, "A system for automatic measurement of circadian activity deviations in telemedicine," *Biomedical Engineering, IEEE Transactions on*, vol. 49, no. 12, pp. 1463–1469, 2002.
12. Y. Fouquet, C. Franco, B. Diot, J. Demongeot and N. Vuillerme, "Estimation of task persistence parameters from pervasive medical systems with censored data," *IEEE Transactions on Mobile Computing*, vol. 12, no. 4, pp. 633–646, 2013.
13. P. Chahuara, A. Fleury, F. Portet and M. Vacher, "On-line human activity recognition from audio and home automation sensors: Comparison of sequential and non-sequential models in realistic smart homes," *Journal of Ambient Intelligence and Smart Environments*, submitted for publication.
14. A. Mannini and A. M. Sabatini, "Machine learning methods for classifying human physical activity from on-body accelerometers," *Sensors*, vol. 10, no. 2, pp. 1154–1175, 2010.

15. C. D. Kidd, R. Orr, G. D. Abowd, C. G. Atkeson, I. A. Essa, B. MacIntyre, E. D. Mynatt, T. Starner and W. Newstetter, "The aware home: A living laboratory for ubiquitous computing research," in *Proceedings of the Second International Workshop on Cooperative Buildings, Integrating Information, Organization, and Architecture*, Pittsburgh, October 1 and 2, 1999.

16. S. Das, D. Cook, A. Battacharya, E. I. Heierman and T.-Y. Lin, "The role of prediction algorithms in the MavHome smart home architecture," *Wireless Communications, IEEE*, vol. 9, no. 6, pp. 77–84, 2002.

17. S. Das and D. Cook, "Designing smart environments: A paradigm based on learning and prediction," in *Pattern Recognition and Machine Intelligence*, vol. 3776, S. Pal, S. Bandyopadhyay and S. Biswas, Eds., Springer, Berlin and Heidelberg, 2005, pp. 80–90.

18. S. Das and N. Roy, "Learning, prediction and mediation of context uncertainty in smart pervasive environments," in *On the Move to Meaningful Internet Systems: {OTM} 2008 Workshops*, Monterrey, Mexico, November 9–14, 2008.

19. H. Noguchi, T. Mori and T. Sato, "Construction of network system and the first step of summarization for human daily action data in the sensing room," in *Knowledge Media Networking, 2002. Proceedings. IEEE Workshop on*, 2002.

20. A. Fleury, M. Vacher and N. Noury, "SVM-based multi-modal classification of activities of daily living in health smart homes: Sensors, algorithms and first experimental results," *IEEE Transactions on Information Technology in Biomedicine*, vol. 14, no. 2, pp. 274–283, 2010.

21. A. Fleury, M. Vacher, F. Portet, P. Chahuara and N. Noury, "A french corpus of audio and multi-modal interactions in a health smart home," *Journal on Multimodal User Interfaces*, vol. 7, nos. 1–2, pp. 93–109, 2013.

22. P. Chahuara, A. Fleury, F. Portet and M. Vacher, "Using Markov logic network for on-line activity recognition from non-visual home automation sensors," in *Ambient Intelligence*, vol. 7683, F. Paternò, B. de Ruyter, P. Markopoulos, C. Santoro, E. van Loenen and K. Luyten, Eds., Springer, Berlin and Heidelberg, 2012, pp. 177–192.

23. S. Helal, B. Winkler, C. Lee, Y. Kaddoura, L. Ran, C. Giraldo, S. Kuchibhotla and W. Mann, "Enabling location-aware pervasive computing applications for the elderly," in *Pervasive Computing and Communications, 2003. (PerCom 2003). Proceedings of the First IEEE International Conference on*, pp. 531–536, Dallas-Fort Worth, TX, March 23–26, 2003.

24. S. Helal, W. Mann, H. El-Zabadani, J. King, Y. Kaddoura and E. Jansen, "The Gator Tech Smart House: A programmable pervasive space," *Computer*, vol. 38, no. 3, pp. 50–60, 2005.

25. E. Kim, S. Helal and D. Cook, "Human activity recognition and pattern discovery," *Pervasive Computing, IEEE*, vol. 9, no. 1, pp. 48–53, 2010.

26. E. Tapia, S. Intille and K. Larson, "Activity recognition in the home using simple and ubiquitous sensors," in *Pervasive Computing*, vol. 3001, A. Ferscha and F. Mattern, Eds., Springer, Berlin and Heidelberg, 2004, pp. 158–175.

27. S. Intille, K. Larson, E. Tapia, J. Beaudin, P. Kaushik, J. Nawyn and R. Rockinson, "Using a live-in laboratory for ubiquitous computing research," in *Pervasive Computing*, vol. 3968, K. Fishkin, B. Schiele, P. Nixon and A. Quigley, Eds., Springer, Berlin and Heidelberg, 2006, pp. 349–365.

28. T. van Kasteren, A. Noulas, G. Englebienne and B. Kröse, "Accurate activity recognition in a home setting," in *Proceedings of the 10th International Conference on Ubiquitous Computing*, Seoul, South Korea, September 21–24, 2008.

29. D. J. Cook and M. Schmitter-Edgecombe, "Assessing the quality of activities in a smart environment," *Methods of Information in Medicine*, vol. 48, no. 5, pp. 480–485, 2009.

30. D. J. Cook, "Learning setting-generalized activity models for smart spaces," *IEEE Intelligent Systems Magazine*, vol. 2010, no. 99, p. 1, 2010.

31. D. J. Cook, A. S. Crandall, B. L. Thomas and N. C. Krishnan, "CASAS: A smart home in a box," *Computer (Long Beach, CA)*, vol. 46, no. 7, pp. 62–69, 2013.

32. C.-H. Lu and L.-C. Fu, "Robust location-aware activity recognition using wireless sensor network in an attentive home," *Automation Science and Engineering, IEEE Transactions on*, vol. 6, no. 4, pp. 598–609, 2009.

33. N. Zouba, F. Bremond and M. Thonnat, "Multisensor fusion for monitoring elderly activities at home," in *Advanced Video and Signal Based Surveillance, 2009. AVSS '09. Sixth IEEE International Conference on*, Genova, Italy, September 2–4, 2009.

34. L. Bao and S. S. Intille, "Activity recognition from user-annotated acceleration data," in *Pervasive Computing*, vol. 3001, A. Ferscha and F. Mattern eds., Springer, Berlin, 2004, pp. 1–17.

35. N. Ravi, N. Dandekar, P. Mysore and M. L. Littman, "Activity recognition from accelerometer data," in *Proceedings of the 17th Conference on Innovative Applications of Artificial Intelligence*, vol. 3, Pittsburgh, PA, July 9–13, 2005.

36. F. R. Allen, E. Ambikairajah, N. H. Lovell and B. G. Celler, "Classification of a known sequence of motions and postures from accelerometry data using adapted Gaussian mixture models," *Physiological Measurement*, vol. 27, no. 10, pp. 935–951, 2006.

37. A. Khan, Y.-K. Lee, S. Lee and T.-S. Kim, "A triaxial accelerometer-based physical-activity recognition via augmented-signal features and a hierarchical recognizer," *Information Technology in Biomedicine, IEEE Transactions on*, vol. 14, no. 5, pp. 1166–1172, 2010.

38. H. Zheng, H. Wang and N. Black, "Human activity detection in smart home environment with self-adaptive neural networks," in *Networking, Sensing and Control, 2008. ICNSC 2008. IEEE International Conference on*, Hainan, China, April 6–8, 2008.

39. M. Ermes, J. Parkka, J. Mantyjarvi and I. Korhonen, "Detection of daily activities and sports with wearable sensors in controlled and uncontrolled conditions," *Information Technology in Biomedicine, IEEE Transactions on*, vol. 12, no. 1, pp. 20–26, 2008.

40. A. Sundaresan, A. R. Chowdhury and R. Chellappa, "A hidden Markov model based framework for recognition of humans from gait sequences," in *Image Processing, 2003. ICIP 2003. Proceedings. 2003 International Conference on*, Barcelona, Catalonia, Spain, September 14–18, 2003.

41. G. Virone, B. Lefebvre, N. Noury and J. Demongeot, "Modeling and computer simulation of physiological rhythms and behaviors at home for data fusion programs in a telecare system," in *Enterprise Networking and Computing in Healthcare Industry, 2003. Healthcom 2003. Proceedings. 5th International Workshop on*, Santa Monica, CA, June 6–7, 2003.

42. F. Duchêne, C. Garbay and V. Rialle, "Learning recurrent behaviors from heterogeneous multivariate time-series," *Artificial Intelligence in Medicine*, vol. 39, no. 1, pp. 25–47, 2007.

43. G. Virone, "Assessing everyday life behavioral rhythms for the older generation," *Pervasive and Mobile Computing*, vol. 5, no. 5, pp. 606–622, 2009.

44. P. Barralon, N. Vuillerme and N. Noury, "Walk detection with a kinematic sensor: Frequency and wavelet comparison," in *28th Annual International Conference of the IEEE Engineering in Medicine and Biology Society*, New York, Aug. 30–Sept. 3, 2006.

45. P. Barralon, N. Noury and N. Vuillerme, "Classification of daily physical activities from a single kinematic sensor," in *Proceedings of the 27th Annual International Conference of the Engineering in Medicine and Biology Society IEEE-EMBS 2005*, New York, Aug. 30–Sept. 3, 2006.

46. A. Fleury, N. Noury, M. Vacher, H. Glasson and J.-F. Serignat, "Sound and speech detection and classification in a health smart home," in *Engineering in Medicine and Biology Society, EMBS 2008. 30th Annual International Conference of the IEEE*, Vancouver, Canada, August 21–24, 2008.

47. N. Noury, P. Barralon, N. Vuillerme and A. Fleury, "Fusion of multiple sensors sources in a smart home to detect scenarios of activities in ambient assisted living," *International Journal of E-Health and Medical Communications (IJEHMC)*, vol. 3, no. 3, pp. 29–44, 2012.

48. A. Fleury, N. Noury and M. Vacher, "Introducing knowledge in the process of supervised classification of activities of daily living in health smart homes," in *e-Health Networking Applications and Services (Healthcom), 2010 12th IEEE International Conference on*, Lyon, France, July 1–3, 2010.

49. Y.-S. Lee and S.-B. Cho, "Activity recognition using hierarchical hidden Markov models on a smartphone with 3D accelerometer," in *Hybrid Artificial Intelligent Systems*, E. Corchado, M. Kurzyński, and M. Woźniak Eds., Springer, Berlin and Heidelberg, 2011, pp. 460–467.

50. D. Anguita, A. Ghio, L. Oneto, X. Parra and J. Reyes-Ortiz, "Human activity recognition on smartphones using a multiclass hardware-friendly support vector machine," in *Ambient Assisted Living and Home Care*, vol. 7657, J. Bravo, R. Hervás and M. Rodríguez, Eds., Springer, Berlin and Heidelberg, 2012, pp. 216–223.

51. M. Shoaib, S. Bosch, O. D. Incel, H. Scholten and P. J. Havinga, "Fusion of smartphone motion sensors for physical activity recognition," *Sensors*, vol. 14, no. 6, pp. 10146–10176, 2014.

52. N. Noury, A. Fleury, P. Rumeau, A. K. Bourke, G. O. Laighin, V. Rialle and J. E. Lundy, "Fall detection—Principles and methods," *Conference Proceedings of IEEE Engineering in Medicine and Biology Society*, vol. 2007, pp. 1663–1666, Lyon, France, August 23–26, 2007.

53. M. Mubashir, L. Shao and L. Seed, "A survey on fall detection: Principles and approaches," *Neurocomputing*, vol. 100, pp. 144–152, 2013.

54. B. Najafi, K. Aminian, A. Paraschiv-Ionescu, F. Loew, C. J. Büla and P. Robert, "Ambulatory system for human motion analysis using a kinematic sensor: Monitoring of daily physical activity in the elderly," *IEEE Transactions on Biomedical Engineering*, vol. 50, no. 6, pp. 711–723, 2003.

55. P. Jallon, S. Bonnet, M. Antonakios and R. Guillemaud, "Detection system of motor epileptic seizures through motion analysis with 3D accelerometers," *Conference Proceedings of the IEEE Engineering in Medicine and Biology Society*, vol. 2009, pp. 2466–2469, Minneapolis, Minnesota, September 2–6, 2009.

56. D. Roetenberg, H. Luinge and P. Slycke, Xsens MVN: Full 6DOF human motion tracking using minia-ture inertial sensors, Xsens Technologies, Enschede, Netherlands, 2009.

57. A. Fleury, N. Noury, M. Vacher, "A wavelet-based pattern recognition algorithm to classify postural transitions in humans," in *EUSIPCO 2009, 17th European Signal Processing Conference*, Glasgow, Scotland, August 24–28, 2009.

58. S. Madgwick, An efficient orientation filter for inertial and inertial/magnetic sensor arrays, University of Bristol. Available from: http://www.x-io.co.uk/res/doc/an_efficient_orientation_filter_for_inertial _and_inertialmagnetic_sensor_arrays.pdf, 2010.

59. S. Mann, "Humanistic computing: 'WearComp' as a new framework and application for intelligent signal processing," *Proceedings of the IEEE*, vol. 86, no. 11, pp. 2123–2151, 1998.

60. C. Franco, A. Fleury, P. Y. Gumery, B. Diot, J. Demongeot and N. Vuillerme, "iBalance-ABF: A smart-phone-based audio-biofeedback balance system," *IEEE Transactions on Biomed Engineering*, vol. 60, no. 1, pp. 211–215, 2013.

61. A. Fleury, Q. Mourcou, C. Franco, B. Diot, J. Demongeot and N. Vuillerme, "Evaluation of a Smartphone-based audio-biofeedback system for improving balance in older adults—A pilot study," *Conference Proceedings of IEEE Engineering in Medicine and Biology Society*, vol. 2013, pp. 1198–1201, Osaka, Japan, July 3–7, 2013.

62. N. Vuillerme, N. Pinsault, O. Chenu, J. Demongeot, Y. Payan and Y. Danilov, "Sensory supplementation system based on electrotactile tongue biofeedback of head position for balance control," *Neuroscience Letters*, vol. 431, no. 3, pp. 206–210, 2008.

# 31 Sensor Data Fusion for Automotive Systems

*Max Mauro Dias Santos*

## CONTENTS

## 31.1 INTRODUCTION

A vehicular system is a component of a global transportation system with the capacity for connection and integration with the surrounding environment and contains entities such as pedestrian, infrastructure, and other components.

Today, automobiles are provided with a set of sensors and actuators in stand-alone configuration for the purpose of controlling subsystems, and in next-generation vehicles these are combined to support modular functions with a high level of integration among them. Thus, they provide the capacity of data fusion with control algorithms, bringing better performance and having an important role in the functioning of the whole system.

The use of new methodologies and paradigms for design of sensor fusion for large-scale utilization is needed so these will be systems with a high level of complexity to meet the requirements of new features. We can consider them as cyber-physical systems in which the use of these sensors is being applied exponentially in vehicles according to what is demanded by actors such as the users, environment, and government to improve performance, safety, comfort, connectivity, environment, and pedestrian aspects.

A transducer is a device that converts a signal in one form of energy to one in another form of energy. Energy types include electrical, mechanical, electromagnetic (including light), chemical, acoustic, and thermal. Although the term *transducer* commonly implies the use of a sensor/detector, any device that converts energy can be considered a transducer. A sensor is used to detect a parameter in one form and report it in another form of energy, often an electrical signal, and an actuator accepts energy and produces movement (action). The energy supplied to an actuator might be electrical or mechanical (pneumatic, hydraulic, etc.). An electric motor and a loudspeaker are both actuators, converting electrical energy into motion for different purposes.

Sensors applied to automobiles are defined as transductor devices that operate as components of a control system. Therefore, a wide range of available sensors for automotive applications are available that encompass a diversity of technologies that are able to measure physical aspects such as temperature, level, image, movement, and other relevant signals of the real world.

In the process of data acquisition by these sensors, consideration is given to providing data with information in different formats for a database. Any necessary pre-processing would occur at the application level where the intelligent and decision control strategies are embedded and have the capacity to generate outputs for the functions of advanced driver assistance.

The sensors can be connected at the electronic control unit (ECU) or at the electronic units named application specific integrated circuit (ASIC), which are distributed in an infrastructure of network communication transmitting frames with signals to be processed for upper application layers during sending and receiving.

Data collection by sensors will be performed at the exterior and interior of a vehicle, in the form of information and requirements, allowing their integration and processing a posteriori. However, the system designer should consider various aspects of sensors in the vehicle because the analytical model for the system architecture as a whole system is such as to ensure sensor data fusion with a view to making it possible to process data acquired at the application level that are in different formats and representations.

Communication among functions should be integrated and accomplished transparently at the application level, thus satisfying a modular system architecture based on the components with properly constructed interfaces.

Sensor data fusion plays an important role in current and future vehicular systems, as the sensors can remain connected with the environment in which they are inserted. The application of new sensors with advanced and innovative technologies alone is not sufficient without the utilization of enhanced techniques of signal processing such as data fusion methods and algorithms.

The operation of a stand-alone sensor cannot overcome certain physical limitations when new functions are to be supported. For instance, its sensor operation is realized with a limited range, the field of view at the environment, and the noise concomitant with the operation modes. Therefore when information comes from combining data with different formats that come from different sensors, this is characterized as a complex system that broadens the sensoring area around the vehicle that is covered by sensors and provides reliability for the whole system in case of a sensor failure.

It is very important to adopt a functional model for sensor data fusion for vehicles. Thus, according to this model the data processing would be divided according to the following levels of application: signal, object, situation, and application.

Hence, a storage and system manager is required for the levels of communication and exchange of data so that the functional model is able to support different architectures for the implementation of sensor fusion. These architectures could be categorized as centralized, distributed, and hybrid, which can provide both advantages and disadvantages depending on target system.

In the data fusion process the main focus is on object and situation refinement levels, which refer to the state estimation of objects and the relations among them. Discrimination among these levels is correspondingly also made by using the terms low- and high-level fusion instead of an object and situation's refinement.

There are several vehicular applications in which data fusions coming from many different sensors are necessary and these should be divided according to target application, such as for the applications on the intersection's safety and vehicle driveability in longitudinal and lateral orientations.

There are current trends to exploit also the wireless communications in vehicles, providing an interaction among vehicles and infrastructures, expanding the driver field view. Suppose, for instance, vehicles can talk to each other, forming ad hoc networks that may be useful in future applications to cover more safety cases than possible so far due to physical limitations of on-board sensors. In this way the horizon of electronics and software leads to awareness of the driver that can be extended even for some kilometers away without real vision. Several ongoing research projects are focusing on the design of efficient protocols and architectures for vehicular ad hoc networks in the standardized way for this kind of vehicular communication.

## 31.2 AUTOMOTIVE SENSORS

The automotive industry is facing new challenges in the sensors domain that should be considered as a set of several distributed transductors among a distributed computer architecture with the purpose of measuring the physical quantities in different data formats, leading to a complex system that in turn requires advanced and algorithm techniques for data fusion [1–3].

Extensive progress in the area of automotive sensors has been made in the last few years, bringing benefits for vehicles, as a simple switch with two states is able to recognize the driver/passenger's intention until a digital camera is able to record images or movement with algorithms that recognize objects, an environment, or a situation requiring a decision. These features thereby provide improvements at the level of performance, efficiency, safety, environment protection, comfort, driver assistance, connectivity, and other aspects related to transport.

For passive and active safety, we have passive systems such as airbag systems and active systems such as antilock brake systems (ABS) and electronic stability program (ESP), as well as other well-known examples, that encompass functions in the general context of improving driving stability and providing better performance, comfort, safety, and other regulated features.

For the automotive domain, sensors are essential and important components of innovations in automobiles that use electronic control systems. They are defined as "devices that transform (or transduce) physical quantities such as pressure or acceleration (called measurands) into output signals (usually electrical) that serve as inputs for control systems" [1].

It was not long ago that the first automotive sensors were composed only of discrete devices used to measure oil pressure, fuel level, coolant temperature, and so forth. For instance, in the late 1970s, microprocessor-based automotive engine control modules were phased in to satisfy federal emissions regulations, with significant improvements in vehicle perfomance.

In control applications at the engine and powertrain, the quantity and the diversity of sensors used increased exponentially since they were first introduced; their use can be seen today in modern vehicles with electrified propulsion (electrical and electrical hybrid vehicles).

Normally, the sensor for automotive applications must have combined/total error of less than 3% over its entire range of operating temperature and measurand change, including all measurement errors due to nonlinearity hysteresis, temperature sensitivity, and repeatability [4,5].

Moreover, even though hundreds of thousands of sensors may be manufactured sequentially on a large scale, calibrations of each sensor must be interchangeable within a 1% range of tolerance; the requirements of environmental operation are also very stringent, with temperature variation on the order of 40 to 125°C (engine compartment), vibration sweeps up to 10 $g$ for 30 hours, drops onto concrete floor (to simulate assembly mishaps), electromagnetic interference and compatibility, and so on. They must, therefore, satisfy a difficult balance among accuracy, robustness, manufacturability, interchangeability, and low cost.

For the further development of sensors in the automotive industry, there are three substantial factors. The challenge is that "what is technically achievable" changes to "which customer benefit is achievable with cost effective approaches." This has a significant influence on the structure of electronic systems in the vehicle where the links between the systems can enable an increased performance of the systems, which may lead to lower cost.

For instance, consider the connection of the ESP with the restraint system that leads to an increase of safety. If the ESP detects an unstable driving condition such as skidding, an accident may occur. Now the airbag system can be switched to an alert mode and the belt tensioners activated. Headrests and backrests may be adjusted to the optimal position. Figure 31.1 shows a combination of passive and active safety systems that can provide functionalities for driver assistance [6].

There are several vehicular systems that incorporate similar sensors, so by linking these systems via bus communication interfaces it becomes possible to reduce the total number of sensors from the combination of all necessary sensors into one cluster. This simplifies the connection

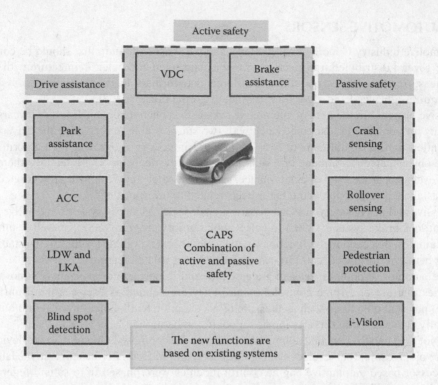

**FIGURE 31.1** The combination of active and passive safety systems to provide driver assistance.

of the systems and allows more functionality with lower costs for the automotive manufacturers: fewer connectors, less wiring, fewer assembly steps. Additional performance should be achieved by applying new sensors to these coupled systems. The sensors for the detection of the surrounding area of the car will identify critical situations that could lead to an accident and thus to proactively reduce the effects or even to avoid an accident by, for example, the execution of an automatic emergency braking.

The necessary sensors are based on video, infrared, or radar with the typical properties of real-time computing for their application. However, fusion of various sensor data provides a whole digital image of the vehicle environment, providing safety, comfort, and better performance.

A significant reduction in the volume and costs of sensors has been achieved with the increase of integration, thus becoming ubiquitous in various automotive applications. The original equipment manufacturer (OEM) automotive companies can integrate more intelligent vehicles, making them able to provide functionalities of autonomous cars. The smart car today can perform many driver-assistance tasks, such as providing guidance on how to avoid and prevent accidents and reduce the severity of accidents.

To perform these tasks, the vehicles have passive safety systems such as air bags and seat belts; active safety systems such as electronic stability control, adaptive suspension, and yaw and roll control; and driver assistance systems, including adaptive cruise control, blind spot detection, lane departure warning, drowsy-driver warning, and parking assistance. These systems require many of the same sensors as the autonomous car: ultrasonic sensors, radar, and image-vision cameras.

Thus, vehicles now use ultrasonic sensors to provide proximity detection for low-speed events, such as parallel parking and low-speed collision avoidance. For ultrasonic sensors, the detection works only at low speeds because it is able to sense acoustic waves; when the car is moving faster than a person can walk, the ultrasonic sensor is blind.

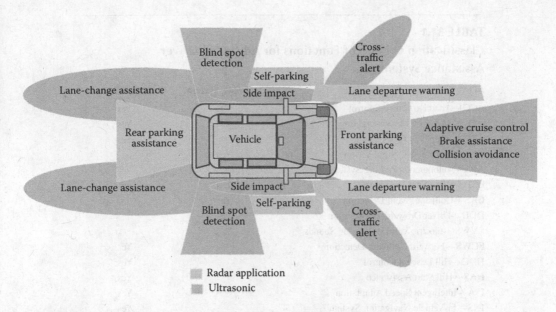

**FIGURE 31.2** Several ADAS systems using a diversity of sensors.

Although ultrasonic sensor technology is at a level of greater maturity and is less expensive than radar, car designers who care about the aesthetic appearance of a car are reluctant to have many openings with visible sensors on the vehicle's exterior. As a more powerful and more flexible technology, radar should begin to replace ultrasonic sensors in future designs, as shown in Figure 31.2.

Advanced driver assistance systems (ADAS) are becoming more popular and indeed necessary for current and future vehicular systems, with the ability to provide the driver, environment, and vehicle a better drivability and thereby improve the performance, safety, and comfort of the vehicle and environment involved.

ADAS are developed to automate, adapt, or enhance vehicle systems for safety and better driving. For instance, the safety features are designed to avoid collisions and accidents by offering technologies that alert the driver to potential problems, or by implementing safeguards and taking over control of the vehicle. Adaptive features may automate lighting, provide adaptive cruise control, automate braking, incorporate GPS or traffic warnings, connect to smartphones, alert the driver to other cars or dangers, keep the driver in the correct lane, or show what is in blind spots.

ADAS technology can be based on vision or camera systems, sensor technologies, car data networks, vehicle-to-vehicle (V2V), vehicle-to-infrastructure (V2I), and other related function in the system.

Most systems can be classified as either those that only inform the driver or those that provide actions that affect the vehicle itself. Therefore, they are classified as warning systems and reaction systems respectively. Table 31.1 presents a generic classification of the main function with significant applications for ADAS in vehicles.

## 31.3 SENSOR DATA FUSION

Sensor data fusion is the combination of sensory data acquisition from disparate sources with different formats such that the resulting information is in some sense better represented than would be possible when these sources were used individually in the stand-alone configuration, which can result in information that is more accurate, more complete, or more dependable, or referenced at the result of a broader view, such as vision systems. Examples of sensors are thermocouple, magnetic, accelerometers, photoelectrical, radar, sonar, TV cameras, GPS, and others [7,8].

**TABLE 31.1**

**Classification Criteria of Functions for Advanced Driver Assistance Systems**

| ADAS Function | Warning System | Reaction System |
|---|---|---|
| ACC—Adaptive Cruise Control | | Yes |
| AFLS—Advanced Front-Lighting System | | Yes |
| ALC—Adaptive Light Control | | Yes |
| ANV—Automotive Night Vision | Yes | |
| APS—Autonomous Parking System | | Yes |
| BSD—Blind Spot Detection | Yes | |
| CAS—Collision Avoid Detection | Yes | |
| DDD—Driver Drowsiness Detection | Yes | |
| EVWS—Electric Vehicle Warning Sounds | Yes | |
| FCWS—Forward Collision Detection | | Yes |
| HDC—Hill Descent Control | | Yes |
| HAS—Hill Start Assistance | | Yes |
| ISA—Intelligent Speed Adaptation | | Yes |
| INS—In-Vehicle Navigation System | | Yes |
| LDW—Lane Departure Warning | Yes | |
| LKA—Lane Keeping Assistance | | Yes |
| TSR—Traffic Sign Recognition | Yes | |
| VCS—Vehicular Communication System | Yes | |
| Customer Function by OEM | TBD | TBD |

There are several methods and algorithms for sensor data fusion that are able to cover a wide range of applications, such as industrial, aeronautic, automotive, and others. However, we can show some classical methods, for instance, central limit theorem; Kalman filter; Bayesian networks, Dempster–Shafer, and others. We now present some examples of sensor fusion calculations [9,10].

Let us consider an environment in which the temperature needs to be measured. Thus, there are two temperature sensors that measure the temperature at different points. The physical quantities of temperatures are defined by the variables $x_1$ and $x_2$, which denote two sensor measurements with noise variances $\sigma_1^2$ and $\sigma_2^2$, respectively. One way of obtaining the third variable based on the combined measurement and denoted $x_3$, is to apply the central limit theorem, which is also employed within the Fraser–Potter fixed-interval smoother. Equation 31.1 represents this approach analytically.

$$x_3 = \sigma_3^2\left(\sigma_1^{-2}x_1 + \sigma_2^{-2}x_2\right) \tag{31.1}$$

where $\sigma_3^2 = \left(\sigma_1^{-2}x_1 + \sigma_2^{-2}x_2\right)^{-1}$ is the variance of the combined estimate and it can be seen that the fused result is simply a linear combination of the two measurements weighted by their respective noise variances.

Another method to fuse two measurements of the sensor signal is to use the optimal Kalman filter in which the data are generated by a first-order system and $P_k$ denotes the solution of the filter's Riccati equation [11]. By applying Cramer's rule within the gain calculation, the filter gain is given by Equation 31.2.

$$L_k = \left[\frac{\sigma_2^2 P_k}{\sigma_2^2 P_k + \sigma_1^2 P_k + \sigma_1^2\sigma_2^2} \quad \frac{\sigma_1^2 P_k}{\sigma_2^2 P_k + \sigma_1^2 P_k + \sigma_1^2\sigma_2^2}\right] \tag{31.2}$$

By inspection, when the first measurement is noise free, the filter ignores the second measurement and vice versa. That is, the combined estimate is weighted by the quality of the measurements.

Sensor data fusion is defined as the capacity of processing information based on data acquisition from several different dispersed sensors, making it possible to compute an estimate of the states of the system and take decisions a posteriori with warning or direct actions over the vehicle. The resulting estimate is in some form better than it would be if the sensors were used as stand-alone and provide a better representation of the environment, with greater accuracy, reliability, accessibility, and safety integrity.

Furthermore, in some cases it may be possible to obtain the resulting estimate only by using data from different types of sensors; Figure 31.3 illustrates the basic concept of the sensor fusion framework. Initially, many systems were defined as stand-alone systems with one or several sensors transmitting information to only a single application and with no relationship between the data acquired. Using the sensor fusion approach, it is possible to remove one sensor from the system and still run the same tasks without function degradation, or then add new applications without the need to add new sensors.

The automotive system tends to use functions with a high level of integration at the level of sensors and actuators wherein the requirements of new features are requested and defined by users, the environment, and government to satisfy progress in performance, safety, comfort, connectivity, environment, and pedestrian aspects.

The settings of sensors in a distributed system can be divided into the following categories: complementary sensors, competitive sensors, and cooperative sensors. This division of settings is based on the role of each sensor in relation to others in the sensed environment, and for any type of configuration belonging to a category of networks. A fourth configuration called independent sensors was created that comprises networks of multiple sensors not belonging to the other three mentioned categories.

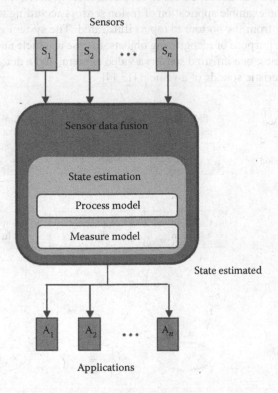

FIGURE 31.3  Main componentes of a sensor fusion framework.

Distributed systems for computer systems with sensors can be configured at different levels of complexity and architectures, making use of different modes of communication and requiring specific computational complexity yet merging data. For each network configuration it is still necessary to draw distinctions and choose specific methods for processing and combining data. The following arepossible configurations between the sensors for data fusion:

- *Complementary*: Sensors do not depend directly on one another but can be combined in a way that provides a more complete diagnosis of the phenomenon that is being observed.
- *Competitive*: Each sensor captures independent measurements of the same property and has two possible competitive settings: (1) measurement fusion from different sensors or (2) melting a single sensor measurement obtained at different instants of time.
- *Cooperative*: A cooperative sensor network uses information from many independent sensors. A stereoscopic vision is an example of a cooperative configuration of optical sensors by combination of two-dimensional images from two cameras positioned at slightly different points, which results in a three-dimensional image of the observed scene.
- *Independent*: An ad hoc sensor network is one that does not fit into one of three previous categories.

The aforementioned categories are not mutually exclusive, and many of the applications use more than one of these, thus setting a hybrid architecture. The use of video cameras for the monitoring of a specific area of interest is a good example of this. We can have areas that are covered by two or more cameras where information is competitive or cooperative, and there are significant areas covered by a single camera that constitute supplementary information. Figure 31.4 shows the different modes and configurations of sensor results achieved through the association of the aforementioned categories [12].

The use of sensor fusion can be carried out in several scenarios, and as an illustration, Figure 31.5 shows an adaptation of an example application of fusion sensors according to levels. The processing and control of data flow from the bottom to top as illustrated. The system is composed of five sensors that are used for the purpose of recognizing objects, in case a vehicle and two radars are operating at different frequencies; one infrared sensor; a video camera; and a detector of radio waves that can identify the characteristic sounds of a vehicle [13,14].

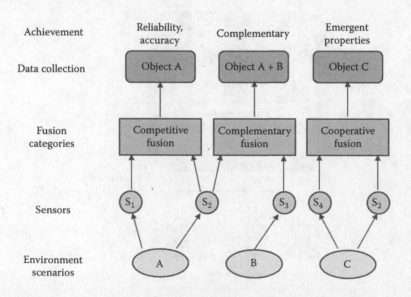

**FIGURE 31.4** Categories of sensor fusion: competitive, complementary, and cooperative.

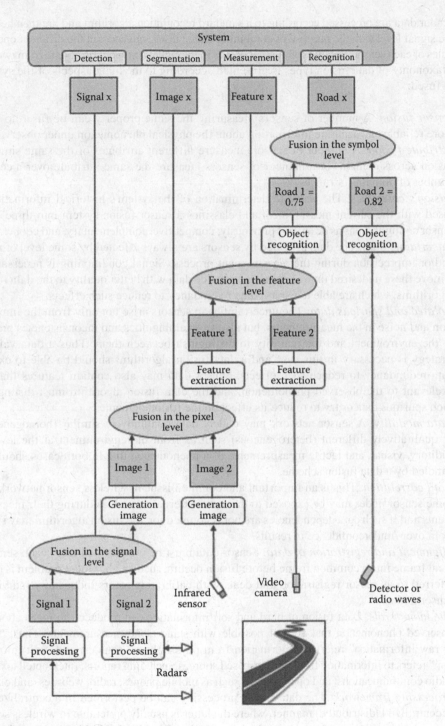

**FIGURE 31.5** Levels of fusion for object recognition.

The radar data are processed according to a standard recognition algorithm and are then held merging at the signal level with the purpose of combining data, taking into account the different operational frequencies of each sensor; finally a refinement occurs in the signal combination of data from two radars.

The taxonomy of data fusion types is presented according to the main aspects of the system for the data fused:

- *Sensor fusion*: A number of sensors measuring the same property can be fused to form more reliable and accurate information about the physical phenomenon under observation.
- *Attribute fusion*: A number of sensors measure different attributes of the same situation fusion across domains or a number of sensors measure the same attribute over a certain number of domains or ranges.
- *Fusion's categories*: The accurate determination of the system's historical information is fused with the current measurement and classifies a sensor fusion system into three basic sensor configurations as described previously: competitive, complementary, and cooperative.
- *Uncertainty data*: The data provided by sensors are always affected by some level of noise and/or imprecision during the measurement process. Signal conditioning is necessary to remove these undesired imprecisions and provide data with better quality to the data fusion algorithms, which are able to exploit data redundancy to reduce such effects.
- *Isolated and spurious data*: The uncertainties in sensors arise not only from the imprecision and noise in the measurements, but also from ambiguities and inconsistencies present in the environment and the inability to distinguish between them. Thus a data validity strategy is necessary in this case and a data fusion algorithms should be able to exploit data redundancy to reduce such effects. Sensor data may also contain features that are irrelevant to the observed phenomenon, and the data fusion algorithm must distinguish such spurious data or try to reduce its effect on the fusion outcome.
- *Data modality*: A sensor network may collect data qualitatively similar (homogeneous) or qualitatively different (heterogeneous) such as from the environment at the level of auditory, visual, and tactile measurements of a phenomenon that in both cases should be handled by a data fusion scheme.
- *Data correlation*: This is an important and common issue in wireless sensor networks, as some sensor nodes may be exposed to the same external noise bias during their measurements and if such data dependencies are not accounted for, the fusion algorithm may suffer from over-/underconfidence in results.
- *Alignment and registration of data*: Sensor data must be transformed from each sensor's local frame into a common frame before fusion occurs; such an alignment problem is often referred to as sensor registration and deals with calibration errors induced by individual sensor nodes.
- *The human role*: Data fusion of hard and soft information can produce estimates about an observed phenomenon that are not possible with stand-alone measurements. Thus, "hard or raw information" refers to information from physics-based sources, and "soft information" refers to information from human-based sources, including reports; intercepted text and audio communications; and open sources such as mobile phones, radio, websites, and others.
- *Processing framework*: The data fusion processing can be performed in a centralized or decentralized (distributed) manner, where the latter is usually preferable in wireless sensor networks as it allows each sensor node to process locally collected data. This is much more efficient compared to the communication burden required by the centralized approach when all measurements have to be sent to a central processing node for fusion.
- *Operational timing*: The area covered by sensors may span a vast environment composed of different and varying aspects, and in the case of homogeneous sensors, the operating frequency of the sensors may be different. A well-designed data fusion method should incorporate multiple time scales to deal with such timing variations in data.

- *Static versus dynamic environment*: The phenomenon under observation may be time invariant or varying with time. In the latter case, it may be necessary or useful for the data fusion algorithm to incorporate a recent history of measurements into the fusion process and the frequency of variations also must be considered in design or selection of the appropriate fusion approach.

In the real world, sensory data are generally imperfect, that is, uncertain, incomplete, imprecise, inconsistent, or ambiguous, or some combination of these. Initially, probability theory was used to deal with almost all kinds of imperfect information and afterward probabilistic techniques such as grid-based models, Kalman filtering, and sequential Monte Carlo simulation demonstrated better performance with the most common data fusion techniques. Later authors proposed alternative techniques such as fuzzy logic, interval calculus, and evidential reasoning.

Sensor data fusion systems can be found in a wide range of applications that have adopted the same functional model of data fusion developed in 1985 by the U.S. Joint Directors of Laboratories (JDL) Data Fusion Group [4,15,16]. The aim of this group was to develop a model that would help theorists, engineers, managers, and users of data fusion techniques to have a common understanding of the fusion process and its various levels. Since then, the model has been constantly revised and updated; that described in Figure 31.6 is from the 1998 revision.

In Figure 31.6, sensors are the primary step of the sensor data fusion and are at local, distributed, or external locations. The process of sensor data fusion is composed of five levels from treatment of the signals to refinement procedures and it is also considered a database system [17].

- *Level 0*: Measurement of the physical quantities by sensors that provide the raw data for the preprocessing (pixel/signal-level processing)
- *Level 1*: Estimation and prediction of entity states on the basis of inferences from observations
- *Level 2*: Estimation and prediction of entity states on the basis of inferred relations among entities
- *Level 3*: Estimation and prediction of effects on situations of planned or estimated/predicted actions by the participants
- *Level 4*: Acquisition of adaptive data acquisition and the processing related to resource management and process refinement (Table 31.2)

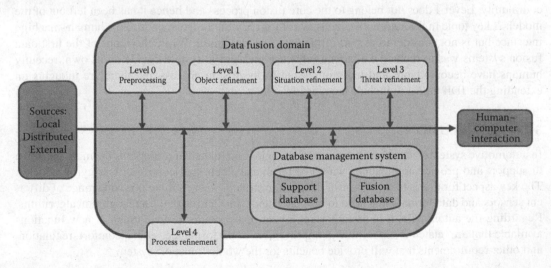

**FIGURE 31.6** Block diagram of the process model of data fusion.

**TABLE 31.2**

**Description of the Sensor Data Fusion Framework**

| Items | Description |
|---|---|
| Sources | These are sensors that provide signals containing information at a variety of levels in accordance with physical quantities measured. |
| Preprocessing (Level 0) | Preprocessing is an important step because it enables the data fusion process to concentrate on the data most pertinent to the current situation as well as reducing the data fusion processing load via data prescreening and allocating data to appropriate processes. |
| Object refinement (Level 1) | This level combines information such as locational, parametrization, and identity to achieve representation of individual objects. It is composed of four key functions: (1) transform data to a consistent reference frame and with units; (2) estimate or predict object position, kinematics, or related attributes; (3) assign data to the objects permitting statistical estimation; and (4) refine estimates of the objects identity or their classification. |
| Situation refinement (Level 2) | This level is responsible for attempts to develop a contextual description of the relationship between objects and observed events in order that this processing determines the meaning of a collection of entities and incorporates environmental information, a priori knowledge, and observations. |
| Threat refinement (Level 3) | Threat refinement projects from the current situation into the future, making inferences about possible threats, friendly and enemy vulnerabilities, and opportunities for equipment operations. Threat assessment is especially difficult because it deals not only with computing possible engagement outcomes, but also with assessing an enemy's intent based on knowledge about the pervasive environment in which the equipment is inserted. |
| Process refinement (Level 4) | This is a meta-process with three key aspects: (1) monitor the real-time and long-term data fusion performance; (2) identify information required to improve the multilevel data fusion product; and (3) allocate and direct sensor and sources to achieve mission goals. |
| Database management system | Database management is the most extensive auxiliary function and is required to support data fusion because of the variety and amount of managed data, as well as the need for data retrieval, storage, archiving, compression, relational queries, and data protection. |
| Human–computer interaction | In addition to providing a mechanism for human entry and communication of data fusion results to users, it provides the Man–Machine Interface (MMI). |

Figure 31.5 depicts how this model can be applied in multisensor automotive safety systems and corresponding revised JDL model, with a small adaptation. According to the automotive fusion community, Level 4 does not belong to the core fusion process and hence it has been left out of the model. A key topic in the automotive industry is Level 5, which corresponds to the human–machine interface but is not considered as part of the data fusion domain. While the scope of the first data fusion systems was to replace the human inference and let the system decide on its own, recently humans have become more and more important in the fusion process and there are thoughts on extending the JDL model to include humans in the loop.

## 31.4 CONCLUSION

In automotive systems, advances in sensor technologies and related areas have as the main objective to support and provide applications with a better management and control efficiency of vehicles. The key aspect for this is to adopt sensor data fusion strategies when there is a wide range of different sensors and data formats that need to be related and consider the pattern recognition algorithms. Regarding the automobile, it is an important aspect that provides a wide range of new functions available that are related to environment, improvement of performance, safety, comfort, regulation, and other requirements that will provide benefits for the whole transport system.

Data fusion for the different advanced vehicle sensors is at the center of this effort and tends to provide important aspects related to the perception of the environment. However, it should still be considered an ongoing investigation when there are many vehicles cooperating with each other. Thus, we can regard two types of communication between vehicles: communication of road-to-vehicle and vehicle-to-vehicle. These aspects always seek to improve and expand the properties of safety and comfort of vehicles.

We presented the state of the art in sensor data fusion for automotive applications, showing that this is a relatively new discipline in the automotive research area, compared to signal, image, or radar processing. The initial and the revised JDL functional fusion models, applicable to the automotive industry, have been highlighted. It can be stated that a central fusion architecture that uses more sensor data at the processing level is able to deliver higher performance.

## REFERENCES

1. R. C. Luo, and M. G. Kay. A tutorial on multisensor integration and fusion. In *16th Annual Conference of IEEE Industrial Electronics Society*, Vol. 1, pp. 707–720, November 27–30, 1990, Pacific Grove, CA.
2. D. L. Hall, and J. Llinas. An introduction to multisensor data fusion. *Proceeding of the IEEE* 85(1):6–23, 1997.
3. J. Llinas, and D. L. Hall. An introduction to multi-sensor data fusion. In *Proceedings of the 1998 IEEE International Symposium on Circuits and Systems*, Vol. 6, pp. 537–540, May–June 1998, Monterey, CA.
4. W. J. Fleming. Overview of automotive sensors. *IEEE Sensors Journal* 1(4):296–308, 2001.
5. O. Schatz. Recent trends in automotive sensors. In *Proceedings of IEEE Sensors*, pp. 236–239, October 24–27, 2004.
6. M. Darms, F. Foelster, J. Schmidt, D. Froehlich, and A. Eckert. Data fusion strategies in advanced driver assistance systems. *SAE International Journal Passenger Car—Electronic Electrical Systems* 3(2):176–182, 2010.
7. C. Stiller, F. P. León, and M. Kruse. Information fusion for automotive applications—An overview. *Special Issue on Information Fusion for Cognitive Automobiles* 12(4):244–252, 2011.
8. M. Kutila, P. Pyykonen, K. Kauvo, and P. Eloranta. In-vehicle sensor data fusion for road friction monitoring. In *IEEE International Conference on Intelligent Computer Communication and Processing (ICCP)*, pp. 349–352, August 25–26, 2011, Cluj-Napoca.
9. D. L. Hall, and S. A. H. McMullen. *Mathematical Techniques in Multisensor Data Fusion*, 2nd ed. Artech, Norwood, MA, 2004.
10. M. Stampfle, D. Holz, and J. C. Becker. Performance evaluation of automotive sensor data fusion. In *Proceedings of the 8th International IEEE Conference on Intelligent Transportation Systems*, pp. 590–595, September 13–16, 2005, Vienna, Austria.
11. L.-W. Fong. Multi-sensor data fusion via federated adaptive filter. In *2010 International Symposium on Computer Communication, Control and Automation*, pp. 209–212, May 5–7, 2010, Tainan, Taiwan.
12. X. Zhang, L. Xu, J. Li, and M. Ouyang. Real-time estimation of vehicle mass and road grade based on multi-sensor data fusion. In *IEEE Vehicle Power and Propulsion Conference (VPPC)*, pp. 1–7, October 15–18, 2013, Beijing.
13. F. Sandblom, and J. Sorstedt. Sensor data fusion for multiple configuration. In *IEEE Intelligent Vehicle Symposium (IV)*, pp. 1325–1301, June 8–11, 2014, Dearborn, MI.
14. T.-D. Vu, O. Aycard, and F. Tango. Object perception for intelligent vehicle application: A multi-sensor fusion approach. In *IEEE Intelligent Vehicle Symposium (IV)*, pp. 774–780, June 8–11, 2014, Dearborn, MI.
15. A. N. Steinberg, C. L. Bowman, and F. E. White. Revisions to the JDL DF model, sensor fusion: Architectures, algorithms, and applications. *Proceedings of SPIE* 3719:430–441, 1999.
16. J. Tian, J. Chen, L. Dou, and Y. Zhang. The reasearch of test and evaluation for multisensor data fusion systems. In *Proceedings of the 4th World Congresso n Intelligent Control and Automation*, pp. 2104–2108, June 10–14, 2002, Shangai, P. R. China.
17. N. A. Tmazirte, M. E. El Najjar, C. Smaili, and D. Pomorski. Dynamical reconfiguration strategy of a multi-sensor data fusion algorithm based on information theory. In *IEEE Intelligent Vehicle Symposium (IV)*, pp. 896–901, June 23–26, 2013, Gold Coast, Australia.

# 32 Data Fusion in Intelligent Traffic and Transportation Engineering

## Recent Advances and Challenges

*Nour-Eddin El Faouzi and Lawrence A. Klein*

## CONTENTS

## 32.1 INTRODUCTION

Transportation systems require reliable information for monitoring and managing operations to maximize the safety and efficiency of the highway system. As connected and autonomous vehicles proliferate, providing traffic flow, collision avoidance, and dangerous-condition warning information is becoming both a major challenge and a major opportunity for the public agencies and private companies that support Intelligent Transportation Systems (ITS). Simultaneously, the spread of Bluetooth® and Internet Protocol (IP)-based (cellular and Wi-Fi) communications technologies has increased travelers' proclivity for accurate road traffic information. Traffic sensors that monitor traffic flow at a given point are often ineffective in supplying the data required by modern transportation management systems. Other sources of data, such as surveillance cameras, global positioning system (GPS), cell phone tracking, probe vehicles, and license plate readers, are increasingly used to supplement the information provided by conventional measurement systems. In addition, traffic management agencies normally archive traffic flow data by time-of-day, day-of-week, month, season, and recurring special events. This offline information, together with sensor real-time data, is often found to be useful in predicting traffic trends.

Multisource data may be complementary in nature and, if this is the case, multisource data fusion can be applied to produce a better interpretation of the observed situation by decreasing the uncertainty present in individual source data. The fusion of multiple sources is correctly perceived as a well-adapted answer to the operational needs of traffic management centers and traffic information providers, allowing them to achieve their goals more effectively. The primary aim of this chapter is to acquaint readers with the most significant applications of data fusion (DF) to ITS and to indicate directions for future research in this area [1].

The chapter is organized into ten sections. Section 32.2 describes the development of sensor and data fusion as a technique for combining data and information from disparate sources and how the methods can be adapted to traffic management. The importance of the early incorporation of sensor and data fusion into the overall systems design of an ITS transportation network is introduced in Section 32.3. Section 32.4 explores sensor and data fusion architectures suitable for ITS applications while Section 32.5 examines alternative fusion models and architectures. Opportunities and challenges of ITS data fusion are discussed in Section 32.6, whereas Section 32.7 presents examples of data fusion applications to traffic management. Suggestions for how to select a data fusion algorithm for an ITS application are described in Section 32.8. Section 32.9 explores the ongoing need for data fusion research and Section 32.10 contains conclusions and suggestions for future research topics.

## 32.2  SENSOR AND DATA FUSION HISTORICAL DEVELOPMENT

Data fusion is concerned with

1. The representation of information within a computational database, particularly the information gained through data fusion
2. The presentation of this information in a way that supports the required decision processes when a human operator or decision maker is involved

Data fusion is not the goal or end. The goal is to provide a control system, that is, a machine or a human, the information necessary to support making automated or semi-automated decisions, such as in ITS applications where vehicle systems or operators may have to take corrective actions to ensure safety.

Several definitions of sensor and data fusion are found in the literature. The Joint Directors of Laboratories (JDL) model, perhaps the most widely cited, defines data fusion as "a multilevel, multifaceted process dealing with the automatic detection, association, correlation, estimation, and combination of data and information from single and multiple sources to achieve refined position and identity estimates, and complete and timely assessments of situations and threats and their significance" (Refs. [2,3], p. 1). The Institute of Electrical and Electronics Engineer (IEEE) Geoscience and Remote Sensing Society's definition is "the process of combining spatially and temporally-indexed data provided by different instruments and sources in order to improve the processing and interpretation of these data." The University of Skövde provides a definition in terms of information fusion as "the study of efficient methods for automatically or semi-automatically transforming information from different sensors and different points in time into a representation that provides effective support for human or automated decision making" (Ref. [4], p. 5). These definitions provide different insights into the role of sensor and data fusion. Their existence is a reflection of the diverse applications for sensor and data fusion.

The terms *data fusion* and *sensor fusion* are often used interchangeably. Strictly speaking, data fusion is defined as in the preceding text. Sensor fusion, then, describes the use of more than one sensor in a configuration that enables more accurate or additional data to be gathered about events or objects that occur in the observation space of the sensors. More than one sensor may be needed to completely and continually monitor the observation space for a number of reasons. For instance, some objects may be detected by one sensor but not another because of the manner in which signatures are generated, that is, each sensor may respond to a different signature-generation phenomenology. The signature of an object may be masked or otherwise hidden with respect to one sensor but not another; or one sensor may be blocked from viewing objects because of the geometric relation of the sensor to the objects in the observation space, but another sensor located elsewhere in space may have an unimpeded view of the object. In this case, the data or tracks from the sensor with the unimpeded view may be combined with past information (i.e., data or tracks) from the other sensor to update the state estimate of the object [5].

Data fusion is widely applied in diverse fields in civilian and military applications such as surveillance and reconnaissance, wildlife habitat monitoring, and detection of environment hazards [6–9]. Several methodologies have been proposed in the literature for the purpose of multisensor fusion and aggregation under heterogeneous data configurations. Owing to the different types of sensors that are used and the heterogeneous nature of information that needs to be combined, different data fusion techniques are being developed to suit the applications and data. These techniques are drawn from a wide range of areas including artificial intelligence, pattern recognition, statistical estimation, and others. The traffic engineering field has naturally benefited from this abundant literature. For instance, independent of specific application a variety of techniques are used ranging from a sample arithmetic mean to more complex DF approaches. More precisely, a three-way split could be suggested:

- Statistical approaches based on weighted combinations; multivariate statistical analysis; and its newest form, the data mining engine [10]. Among statistical techniques, the arithmetic mean is the simplest used for information combination. This approach is not suitable when the information at hand is not exchangeable or when estimators or classifiers have dissimilar performance [11–13].
- Probabilistic-based approaches such as the Bayesian approach with Bayesian network and state-space models [14]. Maximum likelihood methods and Kalman filter-based DF [15,16], possibility theory [17], evidential reasoning, and evidence theory [18–20] are widely used for multisensor DF. This latter technique can be viewed as a generalization of the Bayesian approach [19–21].
- Artificial intelligence encompassing neural networks and artificial cognition. In many applications, the neural network approach serves both as a tool to derive classifiers or estimators and as a fusion framework for classifiers and estimators [10,12].

Although the application of DF techniques to complex systems modeling is not new [3,22,23], there is a growing interest in their use in transportation systems. Road traffic management can be considered as a field where the benefits expected from the application of DF techniques are abundant. However, the benefits come with challenges in assessing feasibility, effectiveness, and usefulness of such approaches [24–26]. In the traffic engineering literature, the interest in DF is quite new and it coincides with the advent of ITS. The first paper that mentions DF for an ITS application was by Sumner in the early 1990s [27], where he acknowledges the importance of DF in enhancing the effectiveness of traveler information systems. Other literature exists regarding the application of DF in engineering and ITS [25,28–30].

Many of the data fusion models and processing techniques originally developed by the U.S. Department of Defense (DoD), namely the JDL model, to support the identification and tracking of military objects can be used today to aid traffic management on streets and highways [2,31–34]. The JDL data fusion model consists of a hierarchy of five processing levels with a potential sixth one. Level 0 deals with the preprocessing of data from the contributing source. It may normalize, format, order, batch, and compress input data [6,32]. It may even identify subobjects or features in the data that are used later in Level 1 processing. For traffic management, Level 1 processing concerns the combining or fusing of data from all appropriate sources, including real-time point and wide-area traffic flow sensors, transit system operators, toll data, cellular telephone calls, emergency call box reports, probe vehicle and roving tow truck messages, commercial vehicle transmissions, and roadway-based weather sensors [29,30]. It is here that the data fusion algorithms discussed in Section 32.7 are implemented. Level 2 processing identifies the probable situation causing the observed data and events by combining the results of the Level 1 processing with information from other sources and databases. These sources may include highway patrol reports and databases, roadway configuration drawings, local and national weather reports, anticipated traffic mix, time-of-day traffic patterns, construction schedules, and special event schedules. Level 3 processing assesses the traffic flow patterns and other data with respect to the likely occurrence of a traffic event (e.g., traffic congestion, incident, construction or other preplanned special event, fire, or police action) that impacts traffic flow. Level 4 processing seeks to improve the entire data fusion process by continuously refining predictions and assessments, and evaluating the need for additional sources of information. Sometimes a sixth level is added to address issues concerned with enabling a human to interpret and apply the results of the fusion process. The DF methods investigated in traffic literature involve basic functions such as temporal and spatial alignment of input data, data association, and data mining for knowledge extraction. The latter is also one of the potential objectives of multisource information fusion [35].

## 32.3 SENSOR AND DATA FUSION AS AN INTEGRAL PART OF SYSTEMS ENGINEERING IN ITS

Systems engineering is an interdisciplinary approach that should be utilized to develop complex systems including those that incorporate sensor and data fusion into ITS. Systems engineering focuses on defining customer needs and functionality early in the development cycle, documenting requirements and system design options, and then proceeding with the subsystem and component-level designs, and system validation. Systems engineering views the system from the points of view of all the stakeholders, that is, the owners, operators, managers, maintainers, and users of the system.

Risks are addressed as early as possible in the design process to keep their cost impacts as small as possible. Technology choices are made at the last possible moment so that they do not drive design decisions and to allow the possibility of incorporating the latest technological advances into the system build. Interfaces are focused on and specified as part of the system design to help ensure a smooth integration of individual subsystems, components, and processes. The systems engineering process is especially critical when developing systems to accommodate new concepts such as connected and autonomous vehicles that will operate together with older vehicles and legacy traffic management systems that do not incorporate the newer technologies and features.

The systems engineering process is illustrated in the Vee diagram of Figure 32.1. This particular version of the diagram is representative of one used for ITS projects that are part of a larger regional or even national architecture [36]. It demonstrates how early phases of a project directly affect end-of-project tasks. Accordingly, sensor and data fusion, if applicable, should be designed into the system at the earliest stages of the development process. Their inclusion and further development proceeds into all levels of the concept exploration and design phases, including benefits analysis, high-level system design, and lower-level subsystems and component designs. It is particularly important that the standards and data transfer protocols that specify the interfaces between subsystems and components be identified to ensure a smooth flow of data across these boundaries.

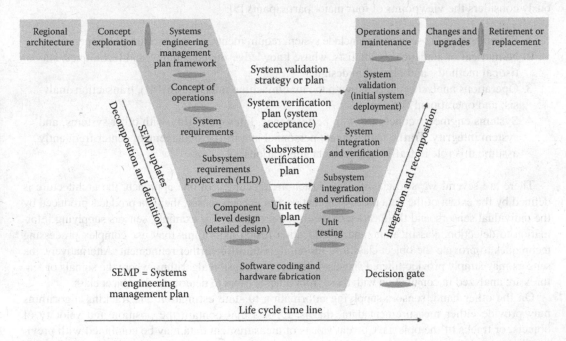

FIGURE 32.1 Vee model of systems engineering.

Some highlights of the Vee model are

- Emphasis on developing and using a systems engineering management plan (SEMP), stakeholder involvement, and validation of requirements and output products
- Depicting the relationship between the system requirements (left side) with the realization of the end product (right side)
- Importance of beginning verification planning when requirements are first defined at every level
- Mandating definition and control of the evolving baseline at each phase of the project through the use of a configuration management plan, risk management plan, and other requirements and validation test documents

## 32.4   SENSOR AND DATA FUSION ARCHITECTURES SUITABLE FOR ITS APPLICATIONS

An architecture is a structure of components, their relationships, and the principles and guidelines governing their design and evolution over time [37]. An architecture

- Identifies a focused purpose
- Facilitates user understanding and communication
- Permits comparison and integration
- Promotes expandability, modularity, and reusability
- Achieves most useful results with least development costs
- Applies to the required range of situations

The selection of a data fusion architecture requires an overall system perspective that simultaneously considers the viewpoints of four major participants [5]:

1. System users, whose concerns include system requirements, user constraints, and operations
2. Numerical or statistical specialists, whose knowledge includes numerical techniques, statistical methods, and algorithm design
3. Operations analysts concerned with the man/machine interface (MMI), transaction analysis, and operational concepts
4. Systems engineers concerned with performance, interoperability with other systems, and system integrity. Traffic management personnel at a traffic management center frequently assume this role in traffic management applications

There are several ways to classify data fusion architectures. In one approach, the architecture is defined by the extent of the data processing that occurs in each sensor, the data products produced by the individual sensors, and the location of the fusion processes. For example, sensors supplying information to detection, classification, and identification fusion algorithms may use complex processing techniques to provide the object class to a fusion algorithm for further refinement. Alternatively, the sensors may simply provide filtered signals or features to a fusion algorithm, where the signals or features are analyzed in conjunction with those from other sensors to determine the object class.

On the other hand, sensors supplying information to state estimation and tracking algorithms may provide either measurement data, that is, reports that contain the position and velocity of objects, or tracks of the objects. Current values of measurement data may be combined with previously obtained data to generate new tracks or the current data may be used to update preexisting tracks using Kalman filtering. These processes can occur in the individual sensors or at a central processing node, depending on the architecture. If the sensors supply tracks, the tracks can be associated with preexisting tracks residing in individual sensors or at a central processing node.

The tracking or state estimation associated processes of correlation and association are defined as follows. In correlation, tracks and measurement data from the same or different sensors are compared to determine candidates for association. Association combines the tracks and measurement data that are matched during correlation to enhance and update detection, classification, and tracking of the objects of interest. Cueing, another process shown in the architecture diagrams below, provides feedback of threshold, integration time, search area, and other changes in data processing and search parameters to the sensors based on the results of the fusion process.

One nomenclature used to describe data fusion architectures is sensor-level fusion (also referred to as autonomous fusion, distributed fusion, and post-individual sensor processing fusion), central-level fusion (also referred to as centralized fusion and pre-individual sensor processing fusion), and hybrid fusion, which uses combinations of the sensor-level and central-level approaches. They differ in the extent of the data processing that occurs in a sensor, sensor output types (e.g., features or object or event declarations), and the bandwidth needed to transmit the sensor outputs to a location where the fusion process occurs [5]. An alternative fusion architecture lexicon employs the terms pixel-level, feature-level, and decision-level fusion.

## 32.4.1 SENSOR-LEVEL FUSION

With sensor-level fusion, each sensor detects, classifies, identifies, and estimates the tracks of potential targets or objects of interest before data entry into the fusion processor. The target report contains at least three types of information: the classification or identity of the detected objects, a measure of how well the sensor believes it has classified or identified the objects, and the location of the objects in a reference coordinate system. The fusion processor combines the information from the sensors to improve the classification, identification, or state estimate of the objects.

Figure 32.2 depicts a sensor-level fusion architecture that is optimal for detecting and classifying objects if the sensors use independent signature-generation phenomena to develop information about the identity of objects in the field of regard, that is, they derive object signatures from different physical processes and generally do not cause a false alarm on the same artifacts [38]. The sensor footprints and data must also be spatially and temporally registered with respect to each other to ensure that the sensor signatures are characteristic of events or objects at the same spatial location and time instant [5]. Registration may be a simple task when the signatures arise from different information channels in the same sensor (e.g., reflectivity and range data from a laser radar or multispectral data from a multispectral or hyperspectral infrared or visible wavelength sensor). Registration is more difficult when information from spatially separated sensors is combined.

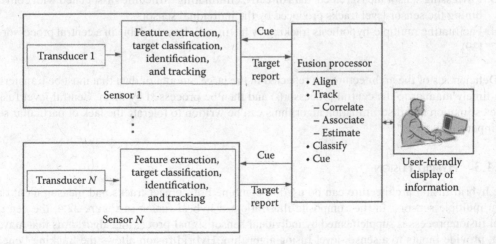

**FIGURE 32.2**  Sensor-level fusion architecture.

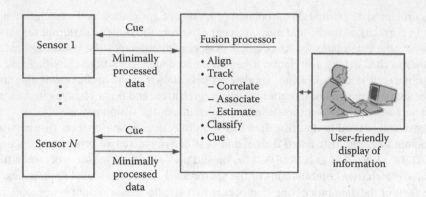

**FIGURE 32.3**   Central-level fusion architecture.

## 32.4.2 CENTRAL-LEVEL FUSION

Figure 32.3 illustrates a generalized central-level fusion architecture. Central-level fusion algorithms are generally more complex and must process data at higher rates than in sensor-level fusion, because the centralized architecture is designed to operate on the minimally analyzed data output by each sensor. The central-level fusion algorithm examines input data for object features or attributes that aid in tracking and discriminating among objects.

When the central-level fusion architecture is utilized for detection, classification, and identification of objects, each sensor may send minimally processed data to the fusion processor. Minimal processing includes operations such as filtering and baseline estimation. In state estimation and tracking fusion, the sensors typically supply measurement data, although sensor-generated tracks may also be sent to the fusion processor.

When used for state estimation, central-level fusion is optimal as it proves more effective than sensor-level fusion in estimating or predicting the future position of the object. Blackman observes that the increased tracking accuracy is due to a combination of effects:

1. Processing all the data in one place
2. Forming the initial tracks based on observations from more than one sensor, thus eliminating tracks established from partial data received by the individual sensors
3. Processing sensor measurement data directly, eliminating difficulties associated with combining the sensor-level tracks produced by the individual sensors
4. Facilitating multiple-hypothesis tracking by having all data available in a central processor [39]

Deficiencies of the architecture are reflected in the large amount of data that must be transferred in a timely manner to the central processor(s) and then be processed by them. Central-level fusion state estimation and discrimination algorithms can be written to tolerate the lack of particular sensor inputs.

## 32.4.3 HYBRID FUSION

The hybrid fusion architecture can be used to combine both object tracks and measurement data from multiple sensors. In the composite illustration of hybrid fusion in Figure 32.4, the central-level fusion process is supplemented by individual-sensor signal-processing algorithms that may, in turn, provide inputs to a sensor-level fusion algorithm. Hybrid fusion allows the tracking benefits of central-level fusion to be realized utilizing sensor measurement data and, in addition, allows

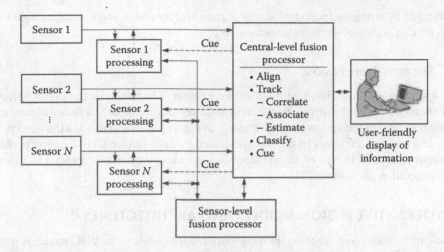

**FIGURE 32.4** Hybrid fusion architecture.

sensor-level fusion of object tracks computed by the individual sensors. Global track formation that combines the central- and sensor-level fusion tracks occurs in the central-level processor. Hybrid fusion can also be used to support object attribute classification when the signature data are not truly generated by independent phenomena. In this case, minimally processed data are sent to a central processor where they are combined using a fusion algorithm that detects and classifies objects in the field of view of the sensors. The limitations of hybrid fusion include the increased processing complexity and possibly increased data transmission rates.

Many hybrid architectures are application specific. Hybrid fusion can manifest itself in the form of hierarchical and distributed architectures. A hierarchical architecture contains fusion nodes arranged such that the lowest level nodes process sensor data and send the results to higher level nodes to be combined. Neyman–Pearson and Bayesian formulations of the distributed sensor detection problem for parallel, serial, and tree data fusion topologies are discussed by Viswanathan and Varshney [40]. Fixed superior-subordinate relationships do not exist in a fully distributed architecture. Each node can communicate with other nodes subject to connectivity constraints. The communication can be adaptive and dependent on the information content and requirements of the individual nodes. Significant savings in communication resources are achieved when the higher level nodes collect processing results periodically.

### 32.4.4 Pixel-Level Fusion

In pixel-level fusion, minimally processed data from different sensors, or sensor channels within a common sensor, are combined at the pixel or resolution-cell level of the sensors using a central-level fusion architecture. Little, if any, preprocessing of the data occurs.

An example of pixel-level fusion is found in LANDSAT imagery used to detect diseased crops or identify a particular crop. Identification is not made using the individual spectral bands of data, but rather the information from all bands is combined in a pixel-level fusion process before the scene is identified.

### 32.4.5 Feature-Level Fusion

Feature-level fusion is characteristic of either a central-level or sensor-level fusion architecture. Features are extracted from each sensor or sensor channel and combined into a composite feature, representative of the object in the field of view of the sensors. An example of a composite feature is one constructed by stringing individual sensor feature vectors end to end (concatenation) to form a longer vector that serves as the input to a classifier.

### 32.4.6 DECISION-LEVEL FUSION

Decision-level fusion is associated with sensor-level fusion. The results of the initial object detection, classification, and state estimate by the individual sensors are input to a fusion algorithm. Final classification occurs in the fusion processor using an algorithm that combines the detection, classification, and position attributes of the objects located by each sensor. Classification performance is suboptimal compared to that of feature-level fusion unless the sensors respond to independent signature-generation phenomena [38].

## 32.5 ALTERNATIVE FUSION MODELS AND ARCHITECTURES

In robotics applications and others employing similar functionality, the JDL model is not appropriate [41]. Therefore, other fusion frameworks have been developed for these applications. These architectures may be applicable to the autonomous driving scenarios being pursued by the United States in its Connected Vehicle Program and in other countries where vehicle-to-infrastructure, infrastructure-to-vehicle, and vehicle-to-vehicle communications will become more prevalent in the decades to come. The first of the robotics architectures is based on meta, algorithmic, conceptual, logical, and execution architectures. The second is structured on four independent design dimensions: centralized versus decentralized, local versus global interaction of components, modular versus monolithic, and heterarchical versus hierarchical.

### 32.5.1 META ARCHITECTURE

The meta architecture is a set of high-level considerations that strongly characterize the system structure. The selection and organization of the system elements may be guided by aesthetics, efficiency, or other design criteria and goals, for example, system and component comprehensibility, modularity, scalability, portability, interoperability, (de)centralization, robustness, and fault tolerance [41].

### 32.5.2 ALGORITHMIC ARCHITECTURE

The algorithmic architecture is a specific set of information fusion and decision-making methods. These methods address data heterogeneity, registration, calibration, consistency, information content, independence, time interval and scale, and relationships between models and uncertainty.

The algorithmic architecture contains three main fusion concepts: belief fusion, utility fusion, and policy selection. Belief fusion is achieved by communicating all beliefs to neighboring platforms. A belief is defined as a probability distribution of the world state space. Utility fusion occurs by separating the individual platform partial utility into the team utility of belief quality and local utilities of action and communication. The limitation is that the potential coupling between individual actions and messages is ignored because the utilities of action and communication remain local. Communication and action policies are chosen by maximizing expected values. The objective of the selected approach is to achieve point maximization for one particular state.

### 32.5.3 CONCEPTUAL ARCHITECTURE

The conceptual architecture contains the granularity and functional roles of components such as mappings from algorithmic elements to functional structures. The data types in this architecture include beliefs that are current world beliefs, plans that are future planned world beliefs, and actions that are future planned actions. The definition of component roles leads to a natural partition of the system.

Information fusion is achieved through the definition of four component roles for each data type, namely source, sink, fuser, and distributor. Connections between distributors form the backbone of a sensor network and the exchanged information is in the form of their local beliefs. Similar considerations determine component roles for the decision making and the system configuration tasks [41].

## 32.5.4 Logical Architecture

The logical architecture is designed from the conceptual architecture. It consists of detailed canonical component types (i.e., object-oriented specifications) and interfaces to formalize intercomponent services. Components may be ad hoc or regimented. Other concerns include granularity, modularity, reuse, verification, data structures, and semantics. Communication issues include hierarchical versus heterarchical organization, shared memory versus message passing, information-based characterizations of subcomponent interactions, pull–push mechanisms, subscribe–publish mechanisms, and so forth. Control involves both the control of actuation systems within the multisensor fusion system, as well as control of information requests and dissemination within the system, and any external control decisions and commands.

## 32.5.5 Execution Architecture

The execution architecture defines the mapping of logical components to execution (runtime) elements such as processes and shared libraries. Execution elements include internal or external methods of ensuring correctness of the code (i.e., that the environment and sensor models have been correctly transformed from mathematical or other formal descriptions into computer implementations), and also validation of the models (i.e., ensure that the formal descriptions match physical reality to the required extent).

## 32.5.6 Fusion in Terms of a Sensor-Computational Description

In another fusion architecture taxonomy found in robotics applications, the choice is among four independent design dimensions: centralized versus decentralized, local versus global interaction of components, modular versus monolithic, and heterarchical versus hierarchical [41]. The most common combinations are

- Centralized, local interaction, and hierarchical. An example of this combination is found in the subsumption multisensor architecture that defines behaviors as the basic components, and employs a layered set of behaviors to embody one monolithic program. Subsumption is defined in terms of two concepts, namely $C_1$ and $C_2$. We say that $C_1$ is subsumed by $C_2$ (i.e., $C_1 \subseteq C_2$) if $C_2$ is more general than or equivalent to $C_1$.
- Decentralized, global interaction, and heterarchical. An example is the blackboard system that uses logical sensors in the form of modular agents that post results to a blackboard. Other implementations of this combination of design dimensions include multisensor fusion systems that are capable of scene interpretation. They rely on the knowledge of sensor functionality, multiple viewing angles, and the recognition that some modeling approaches contain uncertainty and imprecision.
- Decentralized, local interaction, and hierarchical. An example of this architecture is the real-time control system (RCS), a cognitive architecture for intelligent control that uses multisensor fusion to achieve complex control [42]. RCS focuses on task decomposition as the fundamental organizing principle. It defines a set of nodes, each composed of a sensor processor, a world model, and a behavior generation component. Nodes communicate with other nodes, generally in a hierarchical manner, although across layer connections are allowed. The system supports a wide variety of algorithmic architectures.

- Decentralized, local interaction, and heterarchical. An example is an active sensor network (ASN) framework for distributed data fusion developed by Makarenko [43,44]. The distinguishing features of ASN are its commitment to decentralization, modularity, and strictly local interactions (this may be physical or by type). Decentralized implies that no component is central to the operation of the system, communication is peer to peer, and no central facilities or services are provided. These attributes produce a system that is scalable, fault tolerant, and reconfigurable. Local interactions mean that the number of communication links does not change with network size and that the number of messages should remain constant. These features make the system scalable and reconfigurable as well. Modularity supports interoperability as derived from interface protocols, reconfigurability, and fault tolerance as failure may be confined to individual modules.

## 32.6 OPPORTUNITIES AND CHALLENGES OF ITS DATA FUSION

Technological advances in road telematics (such as on-board electronic systems, vehicle localization mechanisms, telecommunications, and data processing) have expanded and improved the means of traffic data collection. Some of these developments are the availability of new sensor technologies and new architectures found in on-board vehicle equipment, enhanced roadside-mounted sensors that provide new data types or improved spatial resolution, and multiform data collection. Basic traffic flow data (volume, occupancy, and speed) needed by traffic operations personnel are typically obtained from sensors embedded in the pavement. The predominant sensor of this type is the inductive loop detector (ILD) that is able to measure temporal traffic flow characteristics at a given location. Other point sensors such as sonic and ultrasonic sensors, magnetometers, and Doppler microwave sensors can also be used to gather roadway network data. While these devices provide point data, they fail in measuring the spatial behavior of traffic flow. In addition, their deployment and maintenance costs become prohibitive when large-area coverage of a roadway network is required. Other fixed sensors with limited spatial capabilities have been developed and deployed to supplement loop detector data. These include visible and infrared spectrum video detection systems and surveillance cameras and multilane presence-detecting microwave radar sensors. The implementation of ITS applications and concurrent need for real-time and accurate data in support of various traffic management functions including incident detection, route guidance, and safety warnings found in connected vehicle applications, has shown the importance of having a complementary source of data for traffic flow parameter estimation.

One of these complementary data sources is probe vehicle data, also known as floating car data (FCD) and in its extended version as xFCD. With this technique, cars on the road shift from a passive attitude to an active one and act as moving sensors, continuously feeding information about traffic conditions to a traffic management center (TMC). More recently, cooperative systems research was performed where vehicles were connected via continuous wireless communication with the road infrastructure, exchanging data and information relevant to the specific road segment to increase overall road safety and enable cooperative traffic management [45]. Automatic vehicle identification (AVI) systems, based on different technologies, can be used as detection devices. These technologies include automatic vehicle tag identification, automatic license plate matching techniques, and GPS tracking and identification. With the advances in wireless communications and the spread of cellular phones, technical improvements in cellular positioning provide the opportunity to track cell phone equipped drivers as traffic probes. Many research studies have demonstrated the feasibility of using cell phones as traffic probes [46,47].

Therefore, a wide spectrum of data and heterogeneous sources of information are of potential use for a given traffic situation. As a result, many problems of traffic engineering become a typical data fusion problem. Indeed, some of the basic road traffic data gathered to meet operators needs have been made available through conventional fixed detectors. However, in the context of traffic operations where highly accurate information is needed, an ITS framework for instance, the information provided by fixed sensor data alone may not be sufficient except in some special situations such as for some network configurations or with closely spaced sensor coverage. A large-scale network is a situation where fixed sensor data alone cannot realistically provide the entire range of traffic information required by the TMC or traffic engineer. So, for traffic operations improvements, other sources of data are increasingly used to supplement the information provided by conventional measurement techniques. The purpose of DF is to produce an improved model or estimate of system parameters or events from a set of independent data sources. For traffic applications, the desired model is the state vector of the traffic phenomenon. These estimates may include statements about current or future vehicular speeds, mean speeds, travel time, vehicle classification, and similar topics of interest to travelers and traffic operators.

## 32.6.1 Data Fusion Algorithms for ITS Applications

Several data fusion algorithms are embedded in ITS applications. These include Bayesian inference, Dempster–Shafer evidential theory and some of its modifications, artificial neural networks, fuzzy logic, knowledge-based expert systems, and vehicle and pedestrian tracking based on the Kalman filter or extended Kalman filter (EKF), Monte Carlo techniques, and particle filters. The Bayesian and Dempster–Shafer approaches belong to the class of feature-based parametric algorithms. They directly map parametric data (e.g., features) into a declaration of identity. Physical models are not used. Artificial neural networks belong to the class of feature-based information theoretic techniques, which transform or map parametric data into an identity declaration. No attempt is made to directly model the stochastic aspects of the observables. Fuzzy logic and knowledge-based expert systems are examples of cognitive-based approaches that attempt to emulate and automate the decision-making processes used by human analysts. The Kalman filter and its nonlinear motion counterparts are examples of physical models since the kinematics of the objects being tracked are modeled. Physical models replicate object discriminators, in this case position, velocity, and sometimes acceleration that are easily observable or calculable [5].

### 32.6.1.1 Bayesian Inference

Bayesian inference is a probability-based reasoning discipline grounded in Bayes' rule. When used to support data fusion, Bayesian inference belongs to the class of data fusion algorithms that use a priori knowledge about events or objects in an observation space to make inferences about the identity of events or objects in that space. Bayesian inference provides a method for calculating the conditional a posteriori or posterior probability of a hypothesis being true given supporting evidence. Thus, Bayes' rule offers a technique for updating beliefs in response to information or evidence that would cause the belief to change [5].

Another interpretation of Bayesian inference is provided as follows. Suppose it is necessary to determine the various likelihoods of different values of an unknown state $H_i$. There may be prior beliefs about what $H_i$ might be expected. These are encoded in the form of relative likelihoods in the prior probability $P(H_i)$. An observation $E$ is made to obtain additional information about the particular $H_i$. These observations are modeled as a conditional probability $P(E|H_i)$ that describes, for each $H_i$, the probability of $E$ given that a particular $H_i$ is true. The new likelihoods associated with the $H_i$ are computed from the product of the original prior information and the information gained by observation (the evidence). This is encoded in the posterior probability $P(H_i|E)$ that describes the likelihoods associated with $H_i$ given the evidence $E$. In a fusion process, the denominator $P(E)$ is used to normalize the posterior and is not generally computed [41]. Limitations of Bayesian inference include

- Need to determine prior probabilities (if these are unknown, the principle of indifference can be invoked where the initial values of the priors are set equal to each other)
- Knowledge of likelihood function values corresponding to each observation or sensor and hypothesis or object of interest

- Ability to discard past evidence or data once the posterior probability corresponding to the new piece of data is calculated
- Complexities that arise when multiple potential hypotheses and multiple conditionally dependent events are evaluated
- Mutual exclusivity required of competing hypotheses
- Inability to account for general uncertainty, that is, Bayesian inference does not have a convenient representation for ignorance or uncertainty—equates concepts of ignorance and belief when direct knowledge is lacking

### 32.6.1.2  Dempster–Shafer Evidential Reasoning

Dempster–Shafer evidential theory generalizes Bayesian inference to allow for uncertainty by distributing support for a proposition (e.g., that an object is of a particular type) not only to the proposition itself, but also to the union of propositions (disjunctions) that include it and to the negation of a proposition. Any support that cannot be directly assigned to a proposition or its negation is assigned to the set of all propositions in the hypothesis space (i.e., uncertainty). Support provided by multiple sensors for a proposition is combined using Dempster's rule. Bayesian and Dempster–Shafer produce identical results when all singleton propositions are mutually exclusive and there is no support assigned to uncertainty. A requirement of the Dempster–Shafer method is the need to define processes in each sensor that assign the degree of support for a proposition. Limitations of the original method include the inability to make direct use of prior probabilities when they are known, requirement for independence of information from each sensor, and the counterintuitive output sometimes produced when support for conflicting propositions is large. Several methods have been proposed to modify Dempster's rule through the use of probability transformations that better accommodate conflicting beliefs and, in some cases, through the use of prior knowledge and spatial information [5].

### 32.6.2  Artificial Neural Networks

Artificial neural networks are hardware or software systems that are trained to map input data into selected output categories. The transformation of the input data into output classifications is performed by artificial neurons that attempt to emulate the complex, nonlinear, and massively parallel computing processes that occur in biological nervous systems. Limitations of this approach include the need to determine the type of neural network, number of hidden layers and weights, and the availability of adequate training set data for the application.

### 32.6.3  Fuzzy Logic

Fuzzy logic employs fuzzy set theory to open the world of imprecise knowledge or indistinct boundary definition to mathematical treatment. It facilitates the mapping of system state-variable data into control, classification, or other outputs. There are four elements to a fuzzy system, namely fuzzy sets, membership functions, production rules, and a defuzzification mechanism. Fuzzy sets are the state variables defined in imprecise terms. Membership functions are the graphical representation of the boundary between fuzzy sets. Production rules (also known as fuzzy associative memory) are the constructs that specify the membership value of a state variable in a given fuzzy set. Membership can range from 0 (definitely not a member) to 1 (definitely a member). The production rules, which govern the behavior of the system, are in the form of If–Then statements. An expert specifies the production rules and fuzzy sets that represent the characteristics of each input and output variable. Defuzzification is the process that converts the result of the application of the production rules into a crisp output value, which is used to control the system.

### 32.6.4 KNOWLEDGE-BASED EXPERT SYSTEMS

Knowledge-based systems incorporate rules and other knowledge from known experts to automate the object-identification process. They retain the expert knowledge for use at a time when the human inference source is no longer available. Computer-based expert systems frequently consist of four components: a knowledge base that contains facts, algorithms, and a representation of heuristic rules; a global database that contains dynamic input data or imagery; a control structure or inference engine; and a human–machine interface. The inference engine processes the data by searching the knowledge base and applying the facts, algorithms, and rules to the input data. The output of the process is a set of suggested actions that is presented to the end user [5].

### 32.6.5 KALMAN FILTERING AND NONLINEAR MOTION SIMULATION TECHNIQUES

The Kalman filter provides a general solution to the recursive, minimum mean-square estimation problem within the class of *linear* estimators. It minimizes the mean-squared error as long as the object's dynamics and measurement noise are accurately modeled. As applied to a tracking problem, the filter estimates an object's state at some time, for example, the predicted time of the next observation, and then updates that estimate using noisy measurements. It also provides an estimate of object-tracking error statistics through the state error-covariance matrix [48].

The extended Kalman filter is a form of the Kalman filter applied when the state model or the observation model corresponding to the object of interest are nonlinear. The EKF linearizes about the current mean and covariance of the state using first-order Taylor approximations to the time-varying transition and observation matrices. The EKF requires knowledge of the state transition and observation matrices and the noise statistics of the nonlinear dynamical system. In the EKF, the state distribution is approximated by a Gaussian random variable (GRV), which is then propagated analytically through the first-order linearization of the nonlinear system. This can introduce large errors in the true posterior mean and covariance of the transformed GRV, which may lead to suboptimal performance and sometimes divergence of the filter.

There are alternatives to the EKF when the system of interest is highly nonlinear. The first of these employs an unscented Kalman filter (UKF), which operates on the premise that it is easier to approximate a Gaussian distribution than it is to approximate an arbitrary nonlinear function. Instead of linearizing using Jacobian matrices as in the EKF, the UKF uses a deterministic sampling approach to capture estimates of the mean and covariance with a minimal set of carefully chosen sample points. The state distribution is again approximated by a GRV, but is now represented by the minimal set of sample points. These sample points completely capture the true mean and covariance of the GRV, and when propagated through the nonlinear system, represent the posterior mean and covariance accurately to the third order (Taylor series expansion) for any nonlinearity. The EKF, in contrast, only achieves first-order accuracy. Remarkably, the computational complexity of the UKF is the same order as that of the EKF [49,50].

Monte Carlo methods are appropriate for problems where state transition models and observation models are highly nonlinear. In particular, multimodal or multiple hypothesis density functions are well suited for Monte Carlo techniques. Monte Carlo filter methods describe probability distributions as a set of weighted samples of an underlying state space. Monte Carlo filtering uses these samples to simulate probabilistic inference usually through Bayes' rule. Many samples or simulations are performed and by analyzing the statics of the samples, a probabilistic picture of the process being simulated can be discerned. Sequential Monte Carlo filtering is a simulation of the recursive Bayes' update equations using sample support values and weights to describe the underlying probability distributions [41].

Particle filters are a recursive implementation of the sequential Monte Carlo algorithm. They provide an alternative to Kalman filtering when dealing with non-Gaussian noise and nonlinear systems. They utilize a weighted ensemble of randomly drawn samples (particles) as an approximation of the probability density of interest. When applied within a Bayesian framework, particle filters are used to approximate the posterior probability of a system state as a weighted sum of random samples [51].

## 32.7 EXAMPLES OF DATA FUSION APPLIED TO ITS STRATEGIES

El Faouzi and Lesort [11] and Sethi et al. [52] published the first papers in which a practical traffic and transportation problem was addressed as a data fusion problem. For the last 15 years, various authors have made significant contributions to the field of DF in transportation systems. ITS indisputably offers a most relevant framework for DF and also the most challenging [24–26]. Other conventional problems in transportation modeling have also been concerned with multisource processing, namely planning problems, demand estimation, and traffic estimation [53]. A variety of functions are assigned to intelligent transportation systems to address traffic mobility and safety issues. These functions appear in subsystems that include specific strategies to enhance the effectiveness of the transportation system in light of changing travel and commuter patterns or travel demand management. Typical subsystems found in ITS are advanced transportation management systems (ATMS), automatic incident detection (AID) that is normally a part of ATMS, advanced traveler information systems (ATIS), advanced driver assistance systems (ADAS), and commercial vehicle operations (CVO). Each of these subsystems can gather information from different data sources. DF techniques can then be used to combine the information to yield a better decision or understanding of the situation at hand.

### 32.7.1 ADVANCED TRANSPORTATION MANAGEMENT SYSTEMS

Advanced transportation management systems provide a systems approach for roadway management that incorporates ITS technology in the planning, programming, and evaluation of transportation facilities to better enable them to respond to recurring transportation demand and congestion. The facilities they manage include freeways (ramp metering, information dissemination, managed lanes, and active traffic management), arterials (traffic signals, transit and emergency vehicle signal priority, and parking guidance), integrated corridors, road-weather systems, and transportation management centers [54].

Taylor, Meldrum, and Jacobson [55] report on the design and performance of a fuzzy ramp metering algorithm that provides metering rate control based on local occupancy and speed data. The input variables to the ramp metering fuzzy control system are mainline and ramp occupancy data and mainline speed data computed from the previous 20-second period data, except for ramp occupancy that uses additional past data samples. The output variable is the metering rate. Seventeen production rules were heuristically developed based on experience in operating the ramp metering system. The fuzzy centroid is used to defuzzify the fuzzy values represented by the logical products and consequent membership functions. The fuzzy ramp metering algorithm was evaluated with FRESIM (FREway SIMulation) over a section of freeway that did not have many alternative diversion routes. The evaluation was against the other ramp meter controllers built into the FRESIM model, namely clock metering, demand/capacity metering, and speed metering. The performance criteria were to maintain a reasonable ramp queue, maximize distance traveled, minimize travel time (or equivalently maximize average system speed), and minimize the average delay/distance traveled per vehicle for all vehicles in the system. The more demand exceeded capacity, the greater was the ability of the fuzzy controller to balance the conflicting demands of mainline efficiency and minimum ramp queues as compared to the other controllers. The balance provided by the fuzzy controller was attributed to the ramp occupancy inputs available to this controller but not to the others. Generally, fuzzy control produced a modest improvement over the other metering controllers and when no metering was used at all.

Niittymaki and Kikuchi [56] designed a fuzzy system to control pedestrian crossing at a signalized intersection in much the same manner as an experienced crossing guard who regulates the timing of pedestrian crossings. The instrumented pedestrian crossing is a two-lane, two-way roadway facility. Two vehicle sensors, 60 m apart, are located on each approach and a pedestrian push button is available on each side of the crossing. The vehicle sensors provide the number of approaching vehicles. The objectives of the fuzzy system are to minimize pedestrian waiting time by accommodating the pedestrian as soon as possible, minimize the delay to vehicular movement by not stopping the vehicle flow for an unreasonably long period, and maximize vehicle and pedestrian safety by preserving and passing groups of approaching vehicles. The input variables are cumulative waiting time of one or more pedestrians from the last signal change; the larger of the approach volumes (i.e., number of vehicles) measured by the two sensors; and the smaller of the discharge gaps (i.e., time headway between two vehicles) measured by the two sensors in seconds. The output variables are the decision to extend the current signal phase or to terminate the current phase. A series of 18 production rules was used to implement the system. Defuzzification is performed by selecting the decision corresponding to the output of the rule that contains the maximum value of the logical product.

In addition to the AID applications discussed in Section 32.7.2, transportation management centers apply DF to traffic state prediction and active traffic management (e.g., dynamic speed limits, hard shoulder running, and queue-end automatic detection).

## 32.7.2 AUTOMATIC INCIDENT DETECTION

Incident detection methods for automatic recognition of incidents, accidents and other road events requiring emergency responses have existed for more than three decades. Most of the developed and implemented algorithms rely on loop detector data. However, these algorithms work with mixed success. Recently, there has been renewed interest in incident detection algorithms partly because of the availability of new sensors and data sources. One of these sources is probe vehicles. Hence, AID belongs to the class of problems that can be solved by DF techniques.

Several data fusion techniques that support incident detection and traffic management are reported in the literature. The data fusion algorithms utilized include Dempster–Shafer inference, Bayesian inference, and voting logic. Most of these applications combine probe vehicle data with conventional traffic data. As an example of such work, Koppelman et al. [57] and Ivan et al. [58,59] developed an AID system using surveillance data from two different sources: fixed detectors (e.g., inductive loop detectors) and probe vehicles specially equipped to report link travel time. The neural network approach was considered and two strategies were tested. The first one combined observed traffic directly to determine whether or not an incident is occurring. In the second, separate incident detection algorithms individually preprocessed data from each source, reporting scores that are combined by the neural network. Different neural network representations were studied in Ivan [59] where the results showed that probe and detector-based incident detection on arterial networks considerable offered considerable promise for improved performance and reliability. Dempster–Shafer inference or evidential reasoning was also used to implement an operational AID system [60].

Thomas [61] investigated this problem from a multiple attributes decision-making viewpoint with Bayesian scores. The author proposed an approach that utilizes combinations of probe travel times, number of probe reports, and ILD-reported occupancies and volumes as the inputs. The results show that models based solely on probe data lack in performance due to excessive overlaps in class distributions, while models based on detector occupancies and vehicle counts by lane perform outstandingly. It was also noted that the use of probe data enhanced the performance of detector data-based models.

More recently, Klein [62,63] studied the application of Dempster–Shafer inference to traffic management in support of incident detection and the identification of other events of concern to traffic managers. The application of the Dempster–Shafer inference algorithm to incident detection and verification is illustrated with an example consisting of three possible events, where data are supplied from three different types of sources. The available information is combined using Dempster's rule and the most probable event is identified.

Incident detection algorithm fusion is another area where classification accuracy needs improvement. This subject was investigated by Cohen [64] using three aggregation schemes: a logical aggregation, a neural network fusion, and a veto procedure. From the validation step, which was carried on real-world data, results demonstrated that both logical aggregation and veto procedure outperform the single best algorithm.

### 32.7.3 NETWORK CONTROL

Data fusion techniques were also applied to road network control. In Refs. [65,66] the problem of constructing an adaptive online traffic control method for urban and freeway road networks was investigated. Mueck's model [65] determines the queue length from vehicle counts produced by detectors located close to the stop line and from signal timing information. Wang and Papageorgiou [66] explored traffic state estimation on freeways using the extended Kalman filter. Similarly, Friedrich et al. [67] introduced a new approach based on queuing theory models for real-time queue length determination. In this latter method, Mueck's model serves as a quasi-measurement with the Kalman filtering technique.

### 32.7.4 ADVANCED TRAVELER INFORMATION SYSTEMS

Advanced Traveler Information Systems (ATIS) are one of several ITS subsystems that offer users integrated traveler information. In ATIS, a variety of automatic data collection techniques are employed to assist in understanding traffic conditions and derive relevant indicators to support traveler guidance [24]. One form in which traveler information is presented is travel time, and a number of systems are available for its dissemination. In this context, travel time is used as a measure of impedance (or cost) for route choice strategies. Travel time is also used by network operators as an indicator of quality of service (QoS). This raises the problem of estimating travel times with an acceptable degree of accuracy, which is a particularly difficult task in urban areas as a result of theoretical, technical, and methodological issues. Thus, to find the traffic conditions that prevail on an urban road, the traffic sensors that are normally used to measure traffic conditions are almost ineffective. New measurement device proliferation (surveillance cameras, GPS, cell phone tracking, license plate readers) implies that alternate sources of data are increasingly available to supplement the information provided by conventional measurement techniques and have the potential to improve the accuracy of travel-time estimates. As a result, travel-time estimation becomes a typical DF problem.

Many authors have discussed the requirements for data fusion in the ADVANCE program [68–70]. ADVANCE was an in-vehicle application of ATIS that provided route guidance in real time in the northwestern portion of Chicago and its northwest suburbs. It used probe vehicles to generate dynamic travel-time information on expressways, arterials, and local streets. A general framework combined data from loop detectors and travel-time reports from probe vehicles using inference rules. Evaluations of the proposed algorithms found that probe data greatly improve static (archival average) link travel-time estimates by time of day. Dailey et al. [71] report a detailed description of a current data amalgamation (fusion) within ITS projects and the presentation of a new quantitative data fusion algorithm to estimate speed from volume and occupancy measurements.

El Faouzi et al. [11,13] proposed an estimation framework for real-time traffic conditions characterization based on multisource data. As an illustrative example, multisource travel-time estimation was implemented using two data sources: data from conventional loop detectors that deliver Eulerian data and probe vehicles that collect Lagrangian data. Travel-time measurements collected by license plate matching were considered as a reference and were used for validation purposes only. The data from the loop detectors was of a statistical nature and can be viewed as a distributed estimation problem: each source derives an estimator of travel time and the individual estimates are then combined according to a weighted mean strategy. The weights were derived from variance–covariance estimation errors. Results display the propensity of the proposed schemes for estimation accuracy improvement. More recently, evidence theory was used to solve the same problem [21,26]. In these contributions, travel time was separated into classes to formulate the estimation problem as a classification one. Various strategies for classifier fusion were proposed and their evaluation displayed some improvement in classification rate.

Abe [72] reported work on travel-time forecasting where data from automatic vehicle identification devices were used for the correction. Dynamic route guidance systems (DRGS) are also an area where DF has potential. Kühne [73] proposed a framework for fusing information from various sources within DRGS. Once again, the objective is travel-time estimation and prediction. The data sources consist of loop detectors, probe vehicles, and a QoS indicator with some exogenous information: information on road works and incidents. The proposed solution is based on a weighted mean scheme. The weights are derived according to the source reliability. Choi [74] and Choi and Chung [75] addressed the problem of generating travel time from loop detectors, probe vehicles, and video-camera sources. They proposed a fuzzy logic based approach that was evaluated using a theoretical example.

### 32.7.5 ADVANCED DRIVER ASSISTANCE

Passenger safety is considered a primary function of ITS and driver assistance techniques are being developed from this point of view. Tremendous progress has been made with regards to vehicle safety and driver assistance. Early safety approaches emphasized precaution and focused on passive devices such as seat belts, air bags, and lighting. In spite of the crash-related injury severity rate reduction, drivers demand even greater improvements in vehicular transportation safety. Research trends show the increased use of active safety devices that complement the traditional passive ones. ADAS and collision avoidance systems (CAS) are an illustration of such trends. The main objective of these systems is to provide a more reliable description of the traffic scene surrounding the vehicle to vulnerable road users in pre-crash situations and to systems such as adaptive cruise control (ACC) and other CAS. Simultaneous localization and mapping is a technique used to obtain the static map of the environment and the vehicle position on the map [76]. Automated highways are another research topic where DF is important. In addition, autonomous vehicles are gaining importance due to their potential use in hazardous and unknown environments and their development for personnel highway driving. In any case, the vehicle needs to sense its environment with an array of sensors and the sensory information needs to be used effectively to provide decision support. Challenges involved are the heterogeneous nature of the data and extracting relevant features from the measurements. Typically, sensors of different capabilities are utilized to gain complimentary information.

Simultaneous localization and tracking (SLAM) has been an active research area in robotics for the last ten years. SLAM consists of multiple parts: landmark extraction, data association, state estimation, state update, and landmark update. Because the individual steps can be realized with a multitude of algorithms, there is no universally accepted algorithm for SLAM. Detection and tracking of moving objects comprise another set of techniques for obtaining information about the dynamic environment in which the vehicle is operating [77,78]. Many commercial products are available whose purpose is to alert drivers to inadvertent lane changes. In such systems, artificial intelligence techniques are used with image processing tools to extract information from 2-D and 3-D cameras. Initially, edge-based lane detection techniques were utilized. However, because there is good contrast between the road and lane markings, a combination of image thresholding followed by a perceptual grouping of the edge points to detect the lane markers of interest was found to be useful [79–81]. In ARCADE [82], which uses more advanced techniques than simple edge detection, one-dimensional edge detection is followed by a least median squares technique for determining the curvature and orientation of the road. Individual lane markers are then directly determined by a segmentation of the row-averaged image intensity values. Frequency domain techniques for lane extraction are detailed in Kreucher [83].

Operational DF systems utilize several sensor systems that are complementary and redundant and a DF process that provides a fused description of the traffic scene. This fusion incorporates the data of the available sensors into a single description. The problem to solve here is data association and data assimilation, a process where sensor data are matched with an environment description that requires synchronization of the sensor data and the associated object state. Whenever there are multiple sensors used to detect multiple objects, there is a need to associate the measurements with the individual objects [6]. Once the sensor measurements are associated with appropriate objects, the next step is to remove the sensor bias. This procedure is called sensor registration. Finally, the state (e.g., position and velocity) of the object is estimated using fused sensor measurements. The Kalman filter, its variants, and more recently particle filtering become an essential tool to perform this step [84]. For example, Murphy [85] discussed the role of sensor fusion in vehicle guidance and navigation, and proposed general methods for fusing data, and sensor-fusion activities within a robot architecture. In their work, Pei and Liou [86] proposed a three-dimensional vehicle motion estimation by fusion of multisource information. Image point and line features were considered for fusion. Langheim et al. [87] investigated DF systems for ACC with stop-and-go phenomena, and Stiller et al. [88] reported a DF framework for obstacle detection and tracking.

### 32.7.6 CRASH ANALYSIS AND PREVENTION

Although there has been a steady reduction in the number of accidents, accidents continue to contribute heavily to losses in both human and economic terms. The reduced number of accidents could be due to multiple efforts: road infrastructure improvements, regulations on alcohol and speed, and improvements in vehicle design that contribute to vehicle safety. One way to explain the circumstances and characteristics of traffic accidents is to utilize available retrospective accident data such as traffic accident records. Taking this approach, Sohn and Shin [89] employed both neural network and decision-tree algorithms to find a classification model for road traffic accident severity (bodily injury or property damage) as a function of potentially related categorical factors. They noted that the classification accuracy of the individual algorithm was relatively low. Furthermore, Sohn and Lee realized that the accuracy could be increased by applying data fusion and ensemble algorithms [90]. The data ensemble method combines various results obtained from a single classifier fitted repeatedly on several bootstrap samples [91–94]. Three approaches were explored: classifier fusion based on the Dempster–Shafer algorithm along with the Bayesian inference and logistic model; data ensemble fusion based on arcing and bagging; and clustering based on the $k$-means algorithm. Empirical results indicate that a clustering-based classification algorithm works best for road traffic accident classification.

### 32.7.7 TRAFFIC DEMAND ESTIMATION

One of the most important problems in the field of transportation planning and control is the problem of origin-destination (OD) estimation from link counts. To decrease the cost of passenger surveys, traffic counts are undertaken on specific links of the transportation network. An estimation of a most likely OD matrix is then derived from the counts. In the last two decades a large number of models have been developed for OD estimation from link counts.

Some of the proposed schemes derived the OD matrices by combining data from different sources. An illustration of this class of problems is the dynamic OD estimation initiated by Cremer and Keller [95,96]. Further development along the same lines was pursued later by others [14,97]. Kalman filtering is commonly applied in this class of problems. Ben Akiva and Morikawa [98] explored OD estimation methods that combine different data sources (stated preference data and traffic measurements) and more recently, Lundgring et al. [99] described a method for adjusting time-dependent travel demand information by incorporating link flow observations. They utilized the structure of the given OD matrix, which is compounded from different sources, for making simple overall adjustments.

## 32.7.8 TRAFFIC FORECASTING AND TRAFFIC MONITORING

Traffic flow forecasting is of increasing importance to traffic surveillance and management. Many traffic flow prediction schemes were based on classic autoregressive models, especially time series techniques. Some authors approached this problem in the context of a Bayesian framework [100]. Others used Kalman filtering techniques [101] or neural networks and system identification [102], and more recently, a nonparametric paradigm was adopted that incorporated kernel techniques [103]. None of these proposals allows one to achieve highly accurate predictions except in some special situations (for some network configurations and/or with high sensor coverage). This is caused to some extent by traffic dynamics that cannot be formalized by a single procedure. Therefore, in the context of traffic operations where highly accurate forecasts are needed, one can obtain different forecasts of the same quantity (the important assumption here is that different predictors are measures of the same quantity or various aspects of the same item) by two or more different methods. The set of available methods may consist of alternative models, different forecasters, or a mixture of models and forecasters.

Often, the approach used is to find the single "best" predictor in some sense (most accurate values, most appropriate models of the underlying process, most cost-effective, etc.) among the available forecasting methods. Another approach consists of combining these individual forecasts. The idea of combining estimators instead of selecting the single best model has a long history and has generated intensive theoretical work since the seminal article of Bates and Granger [104,105]. This work demonstrated that the linear combination of several predictors from a single data set can outperform the individual predictors. Others have extensively investigated methodological and practical issues related to combining forecasts produced by different methods in various contexts with notable successes.

El Faouzi [106] provided a methodological framework to combine various forecasts of the same quantity. He derived two predictors based on nonparametric traffic flow using a kernel estimator and predicting scheme founded on a propagation of a lagged upstream traffic flow. The proposed combination strategies exhibit encouraging results. Data integration and data fusion were applied in other traffic flow forecasting research. Cremer and Schrieber [107] studied the integration of in-vehicle information and loop detector data. The core of the integration step was extended Kalman filtering. More recently, Sau et al. [108] investigated traffic monitoring and prediction with multisource data. A particle filter was used as the estimation technique in this approach. Choi [109] examined the problem of missing data estimation and proposed a framework for missing data inference based on evidential reasoning.

## 32.7.9 ACCURATE POSITION ESTIMATION

In modern transportation systems, information about the position and the orientation of the vehicle should be accurate. Inertial navigation systems (INS) are one of the earliest forms of navigation techniques. INS, which rely on the principle of dead-reckoning, have a potential problem of integration drift caused by the accumulation of small errors in the measurement of acceleration and angular velocity into progressively larger errors in velocity. These errors are compounded into still greater errors in position. In the last few decades, GPS, initially developed as a military navigation aid, has gained a wide acceptance in civilian navigation systems. GPS functionality is based on three major components: satellites orbiting the Earth, control and monitoring stations on Earth, and GPS receivers owned by users. GPS satellites broadcast signals from space that are picked up and identified by GPS receivers. Each GPS receiver then provides three-dimensional location (latitude, longitude, and altitude) plus the time [110]. When the satellite signals are blocked by tall buildings or degraded by electromagnetic interference, GPS outage occurs. In such situations, due to the lack of reference signals, the estimation of position is impossible and the device ceases to work. DF can effectively be used to combat the drawbacks of both techniques. The benefits of using GPS with an INS are that the INS may be calibrated by the GPS signals and that the INS can provide position and angle updates at a quicker rate than GPS. For highly dynamic vehicles such as missiles and aircraft, INS fills in the gaps between GPS positions. In addition, GPS may lose its signal while the INS continues to compute the position and angle during the period of lost GPS signal. Thus, the two systems are complementary and are often employed together.

One of the earliest approaches in GPS–INS integration is to use Kalman filtering. For instance, in Ref. [111], a decentralized filtering strategy is developed for the GPS–INS integration. A Kalman smoother is used to integrate the differential range and phase measurements with the data from INS. Variants of the Kalman filter are often employed to improve integration. A constrained unscented Kalman filter algorithm was proposed in Ref. [112] to fuse differential GPS, INS (gyro and accelerometer), and digital maps to localize vehicles for ITS applications. For the Kalman filter to operate, accurate stochastic models of the sensors are required. Such requirements are often difficult to achieve and resulted in the application of artificial intelligence techniques for GPS–INS integration. Different types of artificial neural networks are used to combine the GPS and INS information. For example, multilayer perception and radial basis function neural networks are successfully used for GPS–INS integration [113,114]. GPS–INS integration is performed using adaptive neuro-fuzzy techniques in Ref. [115]. Such systems mimic vehicle dynamics by training the neural network modules during the availability of GPS signals.

Table 32.1 presents a summary of the data fusion algorithms and architectures that have been used to date to support ITS applications or strategies.

## TABLE 32.1
## Data Fusion Algorithms and Architectures Found in ITS

| Application | Data Fusion Algorithm | Architecture |
| --- | --- | --- |
| Ramp metering | Fuzzy logic | Sensor level |
| Pedestrian crossing time | Fuzzy logic | Central level |
| Automatic incident detection | Artificial neural network | Sensor level |
| Automatic incident detection | Bayesian inference | Sensor level |
| Automatic incident detection | Dempster–Shafer | Sensor level or decision level |
| Travel time | Inference rules | Sensor level |
| Travel time | Dempster–Shafer | Sensor level |
| Travel time | Weighted mean of several travel-time estimators. Weights are a function of the variance or covariance of the estimators | Sensor level |
| Travel time | Weighted mean where the weights are a function of the data source reliability | Sensor level |
| Travel time | Fuzzy logic | Sensor level |
| Vehicle and object tracking | Kalman filter | Central level |
| Lane departure warning | Image processing using edge detection and extraction of other features | Pixel level |
| Traffic state estimation | Extended Kalman filter | Central level |
| Crash analysis and prevention | $k$-means algorithm | Sensor level or decision level |
| Traffic forecasting and monitoring | Bayesian inference | Sensor level |
| Traffic forecasting and monitoring | Artificial neural network | Sensor level |
| Traffic forecasting and monitoring | Kalman filter | Central level |
| Traffic forecasting and monitoring | Extended Kalman filter | Central level |
| Traffic forecasting and monitoring | Kernel estimator | Central level |
| Traffic forecasting and monitoring | Particle filter | Central level |
| Vehicle position estimation | Unscented Kalman filter | Central level |
| Vehicle position estimation | Artificial neural network | Central level |

## 32.8 DATA FUSION ALGORITHM SELECTION

How does one know which data fusion algorithm or technique to use in a given application? A starting point is to evaluate the choice of algorithm and its performance based on the degree to which the technique makes correct inferences (a potential shortcoming of the original Dempster–Shafer theory) and the availability of required computer resources and algorithm input parameters. The selection process also seeks to identify algorithms that meet the following goals:

1. Maximum effectiveness: Algorithms are sought that make inferences with maximum specificity in the presence of uncertain or missing data. Required a priori data such as probability density distributions and probability masses are often unavailable for a particular scenario and must be estimated within time and budget constraints.
2. Operational constraints: The selection process should consider the constraints and perspectives of both automatic data processing and the analyst's desire for tools and useful products that are executable within the time constraints posed by the application. If the output products are to be examined by more than one decision maker, then multiple sets of user expectations must be addressed.
3. Resource efficiency: Algorithm operation should minimize the use of computer resources (when they are scarce or in demand by other processes), for example, CPU time and required input and output devices.
4. Operational flexibility: Evaluation of algorithms should include the potential for different operational needs or system applications, particularly for data driven algorithms versus alternative logic approaches. The ability to accommodate different sensors or sensor types may also be a requirement in some systems.
5. Functional growth: Data flow, interfaces, and algorithms must accommodate increased functionality as the system evolves.

### 32.8.1 PREREQUISITE INFORMATION FOR ALGORITHM SELECTION

Many of the Level 1 object refinement data fusion algorithms are mature in the context of mathematical development. They encompass a broad range from numerical techniques to heuristic approaches such as knowledge-based expert systems. Practical real-world implementations of specific procedures (e.g., Kalman filters and Bayesian inference) exist. Algorithm selection criteria and the requisite a priori data are still major challenges, however.

Applying classical inference, Bayesian inference, Dempster–Shafer evidential theory, artificial neural networks, fuzzy logic, and Kalman filtering data fusion algorithms to vehicle (and event) detection, classification, identification, and state estimation requires expert knowledge, probabilities, or other information from the analyst or data fusion specialist in the form of

- A priori probabilities and likelihood functions
- Probability mass
- Neural-network type, numbers of hidden layers and weights, and training data sets
- Membership functions, production rules, and defuzzification method
- Object kinematic and measurement models, process noise, and model transition probabilities (when multiple state models are utilized)

Table 32.2 shows examples of the prerequisite information typically required to utilize Bayesian inference, Dempster–Shafer evidential theory, artificial neural network, fuzzy logic, and Kalman filtering algorithms. Data fusion algorithm selection and implementation is thus dependent on the expertise and knowledge of the data analyst (e.g., to develop production rules or specify the artificial neural network type and parameters), analysis of the operational situation (e.g., to establish the

## TABLE 32.2

### Information Needed to Apply Bayesian Inference, Dempster–Shafer Evidential Theory, Artificial Neural Networks, Fuzzy Logic, and Kalman Filtering Data Fusion Algorithms to Vehicle Detection, Classification, Identification, and State Estimation

| Data Fusion Algorithm | Required Information | Example |
|---|---|---|
| Bayesian inference | A priori probabilities $P(H_i)$ that the hypotheses $H_i$ are true | Using archived sensor data or sensor data obtained from experiments designed to establish the a priori probabilities for the particular scenario of interest, compute the probability of detecting an object given that data are received by the sensor. The a priori probabilities are dependent on preidentified features and signal thresholds if feature-based signal processing is used or are dependent on the neural network type and training procedures if an artificial neural network is used. |
| | Likelihood probabilities $P(E|H_i)$ of observing evidence $E$ given that $H_i$ is true as computed from experimental data | Compare values of observables with predetermined or real-time calculated thresholds, number of desired object-like features matched, quality of feature match, etc., for each object in the operational scenario. Analysis of the data offline determines the value of the likelihood function that expresses the probability that the data represent an object of type $a_j$. |
| Dempster–Shafer evidential theory | Identification of events or objects $a_1, a_2, \ldots, a_n$ in the frame of discernment $\Theta$ | Identification of potential objects of interest, geological features, and other objects that can be detected by the sensors or information sources at hand. |
| | Probability masses $m$ reported by each sensor or information source (e.g., sensors and telecommunication devices) for individual events or objects, union of events, or negation of events | $m_{S1} = \begin{bmatrix} m_{S1}(a_1 \cup a_2) = 0.6 \\ m_{S1}(\Theta) = 0.4 \end{bmatrix}$ <br><br> $m_{S2} = \begin{bmatrix} m_{S2}(a_1) = 0.1 \\ m_{S2}(a_2) = 0.7 \\ m_{S2}(\Theta) = 0.2 \end{bmatrix}$ |
| Artificial neural networks | Artificial neural network type | Fully connected multilayer feedforward neural network to support object classification. |
| | Number of hidden layers and weights | Two hidden layers, with the number of weights optimized to achieve the desired statistical pattern capacity for the anticipated training set size, yet not unduly increase training time. |
| | Training data sets | Adequate to train the network to generalize responses to patterns not presented during training. |
| Fuzzy logic | Fuzzy sets | Object identification using fuzzy sets to specify the values for the input variables. For example, five fuzzy sets may be needed to describe a particular input variable, namely very small (VS), small (S), medium (M), big (B), and very big (VB). Input variables for which these fuzzy sets may be applicable include length, width, ratio of dimensions, speed, etc. |

*(Continued)*

**TABLE 32.2 (CONTINUED)**

**Information Needed to Apply Bayesian Inference, Dempster–Shafer Evidential Theory, Artificial Neural Networks, Fuzzy Logic, and Kalman Filtering Data Fusion Algorithms to Vehicle Detection, Classification, Identification, and State Estimation**

| Data Fusion Algorithm | Required Information | Example |
|---|---|---|
| | Membership functions | Triangular, trapezoidal, or other shapes. Lengths of bases are determined through offline experiments designed to replicate known outputs for specific values of the input variables. |
| | Production rules | If-Then statements that describe all operating contingencies. Heuristically developed by an expert based on experience in operating the object identification system or process. |
| | Defuzzification method | Fuzzy centroid computation using correlation-product inference. |
| Kalman filter | Kinematic and measurement models of the objects of interest | $x_{k+1} = F x_k + J u_k + w_k$, $z_{k+1} = H x_{k+1} + \varepsilon_{k+1}$, where $F$ is the known $N \times N$ state transition matrix, $J$ is the $N \times 1$ input matrix that relates the known input driving or control function $u_k$ at the previous time step to the state at the current time, $H$ is the $M \times N$ observation matrix that relates the state $x_k$ to the measurement $z_k$, and $w_k$ and $\varepsilon_k$ represent the process and measurement-noise random variables, respectively. |
| | Process noise covariance matrix | For a constant velocity object kinematic model, the covariance matrix is $$Q = q \begin{bmatrix} \frac{1}{3}(\Delta T)^3 & \frac{1}{2}(\Delta T)^2 \\ \frac{1}{2}(\Delta T)^2 & \Delta T \end{bmatrix}$$ where $q$ = variance of the process noise, and $\Delta T$ is the sampling interval. |
| Interacting multiple models | Object kinematic models, current probability of each model, and the model transition probabilities | Model transition probabilities given by $$\mu_k^j = \frac{\lambda_k^j \mu_{k=0}^j}{\sum_{l=1}^{r} \lambda_k^l \mu_{k=0}^l}$$ where $r$ is the number of models, $\lambda_k^j$ is the likelihood function of the measurements up to sample $k$ under the assumption that model $j$ is activated, and $M^j$ is the event that model $j$ is correct with prior probability $\mu_{k=0}^j$. |

*Source:* L. A. Klein. *Sensor and Data Fusion: A Tool for Information Assessment and Decision Making*, 2nd ed., Vol. PM 222. Bellingham, WA: SPIE, 2012.

number of kinematic models needed), applicable information stored in databases (e.g., to calculate the required prior probabilities or likelihood functions), types of information provided by the sensor data or readily calculated from them (e.g., sufficient information to compute probability masses), and the ability to adequately model the state transition, measurement, and noise models required by Kalman filter techniques.

## 32.9   ONGOING NEED FOR DATA FUSION RESEARCH

There is an ongoing need for data fusion research in many areas. Several of these are explored below.

### 32.9.1   RELIABILITY AND CREDIBILITY OF QUALITY OF INPUT DATA TO THE FUSION SYSTEM

Approaches are needed to assess the data fed into the fusion system and to calculate the degree of confidence in the data in terms of their reliability and credibility [116,117]. Khaleghi et al. [51] state that the most notable work in this area is the standardization agreements (STANAG) of the North Atlantic Treaty Organization (NATO) [118]. Here an alphanumeric rating system is used to combine a measurement of the reliability of the source of information with a measurement of its credibility, where both measures are evaluated using the existing knowledge. STANAG recommendations are expressed using natural language statements, which make them imprecise and ambiguous. The proposed formalism is based on the observation that three notions are required in an information evaluation system: The number of independent sources supporting the information, their reliability, and knowledge that the information may conflict with some available or prior information.

More recently, Cholvy has extended his work to incorporate the notion of degree of conflict, in contrast to merely conflicting or nonconflicting information [119]. Nonetheless, current formalism is still not complete as there are some foreseen notions of the STANAG recommendations, such as total ignorance about the reliability of the information source, that are not being considered. Another important aspect of input information quality is the rate at which information is provided to the fusion system. The information rate is a function of many factors, including the revisit rate of the sensors, the rate at which data sets are communicated, and the quality of the communication link [120]. The effect of the information rate is particularly important in decentralized fusion settings where imperfect communication is common.

### 32.9.2   ASSESSING PERFORMANCE OF THE FUSION SYSTEM USING MEASURES OF PERFORMANCE

Performance evaluation has different and possibly conflicting dimensions that may be difficult to capture in one comprehensive and unified measure. One can argue that performance assessment should be multifaceted accounting for not only the measurement of the extent of achieving the fusion goals, but also the amount of effort and resources spent to accomplish this task. Therefore, a comprehensive measurement of performance (MOP) might become too abstract and fail to properly reveal all dimensions of system performance. The alternative is to deploy a set of MOPs suited to the application [51].

#### 32.9.2.1   Adaptive Nature of MOPs

A fair indicator of fusion performance may require MOPs that are adapted over time or according to the given context and situation. Thus, the more difficult the evaluation scenario, the more challenging it becomes for the fusion system to maintain the desired performance level.

#### 32.9.2.2   Choice of MOPs Is Dependent on the Scenario in Which the Fusion System Operates

A multitarget system consisting of a finite set of vectors is fundamentally different from a single target system consisting of a single vector. This is due to the appearance and disappearance of targets (e.g., vehicles and pedestrians) over time, which causes the number of states to vary with time.

Likewise, multiple sensor systems have more parameters to evaluate than single sensor systems. One example is the assessment of the data and track association process that is integral to the multitarget problem.

Metrics can be computed for individual targets or over an ensemble of targets. Individual target metrics include track accuracy, track covariance consistency, track jitter, track estimate bias, and track continuity. Ensemble target metrics include average number of missed targets, average number of extra targets average track initiation time, completeness history, and cross-platform commonality history [51].

### 32.9.3 GROUND TRUTH

Ground truth is not usually known and yet many of the currently used MOPs require knowledge of the ground truth. This is the most common and serious issue that still needs to be addressed. A potential solution is to develop objective performance measures, that is, those that are independent of ground truth or human subjective evaluation.

### 32.9.4 COMMERCIAL OPERATING SYSTEM AND DATABASE MANAGEMENT SYSTEM SUITABILITY

Commercial operating systems (OS) and database management systems (DBMS) may be ill suited to ITS real-time requirements for sensor data processing and control of critical safety-related functions. A restriction of commercial database management systems is that they are designed for flexibility of application rather than real-time or fast-time processing [121]. Accordingly, database management for data fusion is still difficult to implement for the following reasons:

- Existence of large and varied databases with numerous records and record formats
- Support of rapid updates for incoming sensor data and fusion results
- Support of rapid retrievals for human analysts and automated fusion processes such as data association
- Need to provide flexible and user-friendly interfaces
- Requirement to maintain data integrity in real-time under rapid receipt of sensor data, intense human interactions, and asynchronous, out-of-sequence, and false sensor reports, and so forth
- Need to accept both fixed format and free-text message formats under multiple protocols

### 32.9.5 DESIGN FOR WORST-CASE SCENARIOS

ITS systems should be designed for worst-case scenarios such as occur in automated driving applications where delays at critical times are unacceptable. Rapid prototyping is the best solution for estimating data-processing requirements. Guidelines for rapid prototyping include using the target machine if possible, using prototype software in the required language, and driving the analysis with the worst-case load.

## 32.10 CONCLUSIONS AND RECOMMENDATIONS FOR FUTURE RESEARCH TOPICS

The application of data fusion to various transportation applications has been demonstrated for at least two decades and has given rise to an emergent field that is somewhat in its infancy. This chapter has described the state-of-the-art and practice of sensor and data fusion of traffic data from various sources. For the applications reported, DF techniques appear promising. However, these encouraging results should not conceal the challenges that still remain before any operational widespread deployment of DF in the transportation field occurs. These include obtaining data with the necessary accuracy to make the application effective, dynamic and real-time issues associated with data quality as traffic flow changes, the need to process the data in real time, and the development of methods to combine sensor or hard data with human-generated or soft data [1,51]. The benefits of DF will become more apparent as the number of successful and practical DF applications increases in the transportation field. It is certain, however, that there are real opportunities for additional DF applications in road transportation systems. Prospects include the increased collection of useable data from sources other than roadside sensors installed for traffic management and surveillance. Wireless technologies, which offer (1) the potential of easier reporting and access to customized information (e.g., cooperative systems with vehicle-to-vehicle, vehicle-to-infrastructure, and infra-structure-to-vehicle) and (2) the ability of tracking individual vehicles and obtaining information collected by FCD and xFCD, will enrich the available information that characterizes the traffic situation, and will certainly accelerate needs for operational DF systems.

# REFERENCES

1. N.-E. El Faouzi, H. Leung, A. Kurian, Data fusion in intelligent transportation systems: Progress and challenges—A survey. *Information Fusion*, 12:4–10, 2011.
2. O. Kessler, *Functional Description of the Data Fusion Process, Technology Report for the Office of Naval Technology Data Fusion Development Strategy*, Naval Air Development Center, Warminster, PA, November 1991.
3. E. Waltz, J. Llinas, *Multisensor Data Fusion*. Artech House, Norwood, MA, 1990.
4. H. Boström, S. Andler, M. Brohede, R. Johansson, A. Karlsson, J. van Laere, L. Niklasson, M. Nilsson, A. Persson, and T. Ziemke, *On the Definition of Information Fusion as a Field of Research*. University of Skövde Technical Report HS-IKI-TR-006, Skovde, Sweden, 2007.
5. L.A. Klein, *Sensor and Data Fusion: A Tool for Information Assessment and Decision Making*, 2nd ed., vol. PM 222. SPIE, Bellingham, WA, 2012.
6. D.L. Hall, *Mathematical Techniques in Multisensor Data Fusion*. Artech House, Norwood, MA, 1992.
7. J. Mnyika, H. Durrant-Whyte, *Data Fusion and Sensor Management: A Decentralized Information Theoretic Approach*. Ellis Horwood, London, 1994.
8. C. Harris, A. Bailey, T. Dodd, Multisensor data fusion in defense and aerospace. *Journal of Royal Aerospace Society*, 162(1015):229–244, 1998.
9. B.V. Dasarathy, Sensor fusion, potential exploitation: Innovative architectures and illustrative applications. *Proceedings of IEEE*, 85:24–38, 1997.
10. J. Han, J. Kember, *Data Mining Concepts and Techniques*. Morgan Kaufmann, Burlington, MA, 2000.
11. N.-E. El Faouzi, J.-B. Lesort, Travel time estimation on urban networks from traffic data and on-board trip characteristics. In *Proceedings of the Second World Congress on ITS*, Yokohama, Japan, 88–93, 1995.
12. S. Hashem, Optimal linear combinations of neural networks. *Neural Networks*, 10:599–614, 1997.
13. N.-E. El Faouzi, Heterogeneous data source fusion for impedance indicators. In *IFAC Symposium on Transportation Systems*, vol. 3, Chania, Greece, 1375–1380, 1997.
14. I. Okutani, The Kalman filtering approaches in some transportation and road traffic problems. In *Proceedings of the 10th ISTTT*, N. Gartner and N. Wilson (eds.), Elsevier, Amsterdam, 397–416, 1987.
15. D. Huang, H. Leung, EM-IMM based land-vehicle navigation with GPS/INS. In *IEEE International Conference on Intelligent Transportation Systems*, Washington, DC, October 2004.
16. D. Huang, H. Leung, An expectation maximization based interactive multiple model approach for collaborative driving. *IEEE Transactions on Intelligent Transportation Systems*, 6:206–228, 2005.
17. D. Dubois, H. Prade, *Possibility Theory*. Plenum Press, New York, 1988.
18. A.P. Dempster, Upper and lower probabilities induced by multivalued mapping. *Annals of Mathematical Statistics*, 38:325–339, 1967.
19. A.P. Dempster, A generalization of Bayesian inference. *Journal of the Royal Statistical Society, Series B*, 30:205–247, 1968.
20. G. Shafer, *Mathematical Theory of Evidence*. Princeton University Press, Princeton, NJ, 1976.
21. N.-E. El Faouzi, L.A. Klein, O. De Mouzon, Improving travel time estimates from inductive loop and toll collection data with Dempster–Shafer data fusion. In *Transportation Research Record*, vol. 2129. Transportation Research Board, National Research Council, Washington, DC, 73–80, 2009.

22. H. Durrant-Whyte, Sensor models and multisensor data fusion. *Journal of Robotics Research*, 6:97–113, 1988.
23. Y. Zhou, H. Leung, P.C. Yip, An exact maximum likelihood registration algorithm for data fusion. *IEEE Transactions on Signal Processing*, 45:1560–1572, 1997.
24. *Data Fusion for Delivering Advanced Traveler Information Services*. U.S. DoT and ITS Joint Program Office, Federal Highway Administration, Washington, DC, 247–254, 2003.
25. N.-E. El Faouzi, *Data Fusion: Concepts and Methods*, Monograph. INRETS (INRETS Edition), Lyon, France, 2000.
26. N.-E. El Faouzi, Travel time estimation via evidential data fusion. *Recherche Transports Sécurité*, 68:15–30, 2000.
27. R. Sumner, Data fusion in PathFinder and TravTek. In *Proceedings of the Vehicle Navigation and Information Systems Conference: VNIS'91*, 71–75, IEEE, 1991.
28. M. Abidi, R. Gonzalez (eds.), *Data Fusion in Robotics and Machine Intelligence*. Academic Press, San Diego, CA, 1992.
29. L.A. Klein, *Sensor Technologies and Data Requirements for ITS*. Artech House, Boston, 2001.
30. N.-E. El Faouzi, Multiform traffic data collection and data fusion in road traffic engineering. In *Proceedings of the Multisource Data Fusion in Traffic Engineering Workshop*, vol. 87, N.-E. El Faouzi (ed.). INRETS Publishing, Bron, France, 2003.
31. F.E. White Jr., Joint directors of laboratories data fusion subpanel report: SIGINT session. In *Technical Proceedings of the Joint Service Data Fusion Symposium I*, 469–484, 1990.
32. A.N. Steinberg, C.L. Bowman, F.E. White Jr., Revisions to the JDL data fusion model. *Proceedings of SPIE*, 3719:430–441, 1999.
33. H. Leung, J. Wu, Bayesian and Dempster–Shafer target identification for radar surveillance. *IEEE Transactions on Aerospace and Navigational Electronics*, 36:432–447, 2000.
34. E. Lefebvre, *Advances and Challenges in Multisensor Data and Information*. NATO Security through Science Series: Information and Communication Security, vol. 8, 2007.
35. B.V. Dasarathy, Information fusion, data mining, and knowledge discovery. *Information Fusion*, 1:1, 2003.
36. *Systems Engineering Guidebook for Intelligent Transportation Systems, Ver. 3.0*. U.S. Department of Transportation, Federal Highway Administration, California Division and California Department of Transportation, Sacramento, CA, November 2009. Available at www.fhwa.dot.gov/cadiv/segb (accessed April 13, 2015).
37. C4ISR ITF Integrated Architecture Panel, *C4ISR Architecture Framework, Version 1.0*. CISA-0000-104-96, June 7, 1996. Available at http://www.fas.org/irp/program/core/c4isr.htm.
38. G.S. Robinson, A.O. Aboutalib, Trade-off analysis of multisensor fusion levels. In *Proceedings of the 2nd National Symposium on Sensors and Sensor Fusion*, vol. II. GACIAC PR 89-01. IIT Research Institute, Chicago, 21–34, 1990.
39. S.S. Blackman, Multiple sensor tracking and data fusion, Chapter 7. In *Introduction to Sensor Systems*, S.A. Hovanessian (ed.). Artech House, Norwood, MA, 1988.
40. R. Viswanathan, P.K. Varshney, Distributed detection with multiple sensors: Part I—Fundamentals. *Proceedings IEEE*, 85(1):54–63, 1997.
41. H. Durrant-Whyte, T. Henderson, Multisensor data fusion, Chapter 25. In *Handbook of Robotics, Part C—Sensing and Perception*, B. Siciliano, O. Khatib (eds.). Springer, New York, 2008.
42. J. Albus, RCS: A cognitive architecture for intelligent multi-agent systems. In *Proceedings of the IFAC Symposium on Intelligent Autonomous Vehicles*, Lisbon, 2004.
43. A. Makarenko, *A Decentralized Architecture for Active Sensor Networks*. Ph.D. Thesis, University of Sydney, Sydney, 2004.
44. A. Makarenko, A. Brooks, S. Williams, H. Durrant-Whyte, B. Grocholsky, A decentralized architecture for active sensor networks. In *Proceedings of the IEEE International Conference on Robotics Automation*, New Orleans, LA, 1097–1102, 2004.
45. U.S. Department of Transportation Research and Innovative Technology Administration and ITS Joint Program Office. Available at http://www.its.dot.gov (accessed June 26, 2014).
46. J.L. Ygnace, *Travel Time/Speed Estimates on the French Rhone Corridor Network using Cellular Phones as Probes: STRIP Project*. Final Report, European Community Research Program SERTI, INRETS-LESCOT 0201, Lyon, December 2001.
47. Y. Youngbin, R. Cayford, *Investigation of Vehicles as Probes using Global Positioning System and Cellular Phone Tracking*. PATH Working Paper, Berkeley, CA, 2000.
48. Y. Bar-Shalom, T.E. Fortmann, *Tracking and Data Association*. Academic Press, Orlando, FL, 1988.

49. E.A. Wan, R. van der Merwe, The unscented Kalman filter for nonlinear estimation. In *Adaptive Systems for Signal Processing, Communications, and Control Symposium 2000*, AS-SPCC, 153–158, 2000.

50. J.J. LaViola Jr., A comparison of unscented and extended Kalman filtering for estimating quaternion motion. In *Proceedings of the 2003 American Control Conference*, vol. 3, 2435–2440, 2003.

51. B. Khaleghi, A. Khamis, F.O. Karray, S.N. Razavi, Multisensor data fusion: A review of the state-of-the-art. *Information Fusion*, 14:28–44, 2013.

52. V. Sethi, N. Bhandari, F. Koopelman, J. Schofer, Arterial incident detection using fixed detector and probe vehicle data. *Transportation Research*, 2:99–112, 1995.

53. J. Sussman, *Introduction to Transportation Problems*. Artech House, Norwood, MA, 2000.

54. ITS ePrimer. Available at http://www.pcb.its.dot.gov/ePrimer.aspx (accessed June 20, 2014).

55. C. Taylor, D. Meldrum, L. Jacobson, Fuzzy ramp metering: Design overview and simulation results. In *Transportation Research Record*, vol. 1634. Transportation Research Board, National Research Council, Washington, DC, 10–18, 1998.

56. J. Niittymaki, S. Kikuchi, Application of fuzzy logic to the control of a pedestrian crossing signal. In *Transportation Research Record*, vol. 1651. Transportation Research Board, National Research Council, Washington, DC, 30–38, 1998.

57. F. Koppelman, V. Sethi, J. Ivan, *Calibration of Data Fusion Algorithm Parameters with Simulated Data*. ADVANCE Project, Technical Report, TRF-ID-152, The Transportation Research Center, Northwestern University, Evanston, IL, March 1994.

58. J.N. Ivan, J.L. Schofer, F.S. Coppelman, L.L.E. Massone, Real-time data fusion for arterial street incident detection using neural networks. In *Transportation Research Record*, vol. 1497. Transportation Research Board, National Research Council, Washington, DC, 27–35, 1995.

59. J.N. Ivan, Neural network representations for arterial street incident detection data fusion. *Transportation Research C*, 5(3/4):245–254, 1997.

60. S.C. Byun, D.B. Choi, B.H. Ahn, H. Ko, Traffic incident detection using evidential reasoning-based data fusion. In *Proceedings of the 6th World Congress on ITS*, Toronto, Canada, November 1999.

61. N.E. Thomas, Multi-sensor multivariate and multiclass incident detection system for arterial streets. In *Proceedings of the 16th International Symposium on Transportation and Traffic Theory (ISTTT)*, J.-B. Lesort (ed.). Pergamon, Elsevier, Lyon, 315–340, 1996.

62. L. Klein, Dempster–Shafer data fusion at the traffic management center, Transportation Research Board. In *79th Annual Meeting*, Paper No. 00-1211, Transportation Research Board, Washington, DC, 2000.

63. L.A. Klein, P. Yi, H. Teng, Decision support system for advanced traffic management through data fusion. In *Transportation Research Record*, vol. 1804. Transportation Research Board, National Research Council, Washington, DC, 173–178, 2002.

64. S. Cohen, Fusion of incident detection algorithms. In *Proceedings of the Multisource Data Fusion in Traffic Engineering Workshop*, vol. 87, N.-E. El Faouzi (ed.), 109–117, 2003.

65. J. Mueck, Estimation methods for the state of traffic at traffic signals using detectors near the stop-line. *Traffic Engineering and Control*, 2002.

66. Y. Wang, M. Papageorgiou, Real-time freeway traffic state estimation based on extended Kalman filter: A general approach. *Transportation Research B*, 39:141–167, 2005.

67. B. Friedrich, R. Minciardi, Data fusion techniques for adaptive traffic signal control. In *IFAC Symposium on Transportation Systems*, Tokyo, Japan, 53–55, 2003.

68. A. Tarko, N. Rouphail, Travel time fusion in ADVANCE. In *Pacific Rim TransTech Conference: A Ride into the Future*, Seattle, WA, July 25–28, 1993.

69. P. Thakuriah, A. Sen, *Data Fusion for Travel Time Prediction: A Statement of Requirements*. Advance Working Paper Series, vol. 22, Task TRF-TT-07. Urban Transportation Center, University of Illinois, Chicago, 1993.

70. S. Berka, X. Tian, A. Tarko, *Data Fusion Algorithm for ADVANCE*. ADVANCE Working Paper 48, University of Illinois, 1995.

71. D.J. Dailey, P. Harn, P.-J. Linet, ITS data fusion, *Research Project T9903, Task 9: ATIS/ATMS Regional IVHS Demonstration*. Report WA-RD 410.1. Washington State Department of Transportation, 1996.

72. A. Abe, The correction of the forecasting travel time by using AVI data. In *Proceedings of the I.V.H.S.*, 1997.

73. R.D. Kühne, Data fusion for DRG systems. In *IFAC Symposium on Transportation Systems*, vol. 3, Chania, Greece, 1997.

74. K. Choi, Data fusion for generating the link travel time with insufficient data sources. In *Proceedings of the I.V.H.S.*, 1997.

75. K.Y. Choi, Y. Chung, Travel time estimation algorithm using GPS probe and loop detector data fusion. In *Proceedings of the 80th TRB Annual Meeting*, Paper No. 01-0374, Transportation Research Board, Washington, DC, 2001.

76. H. Durrant-Whyte, T. Bailey, Simultaneous localization and mapping (SLAM): Part I the essential algorithms. *Robotics and Automation Magazine*, 13:99–110, 2006.

77. Z. Hu, H. Leung, M. Blanchette, Evaluation of data association techniques in a real multitarget radar tracking environment. *Proceedings of SPIE*, 2561:509–518, 1995.

78. C. Wang, C. Thorpe, A. Suppe, Ladar-based detection and tracking of moving objects from a ground vehicle at high speeds. In *IEEE Intelligent Vehicles Symposium (IV2003)*, June 2003.

79. J. Tsao, R.W. Hall, S.E. Shladover, Design options for operating automated highway systems. In *Proceedings of the IEEE-IEE Vehicle Navigation Information Systems Conference*, 494–500, 1993.

80. S.K. Kenue, LANELOK: Detection of lane boundaries and vehicle tracking using image-processing techniques–Parts I and II. In *SPIE Mobile Robots IV*, 1989.

81. K.C. Kluge, *YARF: An Open-Ended Framework for Robot Road Following*. Ph.D. Dissertation, Carnegie Mellon University, Pittsburgh, PA, 1993.

82. K.C. Kluge, Extracting road curvature and orientation from image edge points without perceptual grouping into features. In *Proceedings Intelligent Vehicle Symposium*, 109–114, 1994.

83. C. Kreucher, S. Lakshmanan, LANA: A lane extraction algorithm that uses frequency domain features. *IEEE Transactions on Robotics and Automation*, 15:343–350, 1999.

84. Y. Bar-Shalom, X.R. Li, T. Kirubarajan, *Estimation with Applications to Tracking and Navigation: Theory, Algorithms, and Software*. Wiley, New York, 2001.

85. R.R. Murphy, Sensor and information fusion for improved vision-based vehicle guidance. *IEEE Intelligent Systems*, 13(6):49–56, 1998.

86. S.-C. Pei, G. Liou, Vehicle-type motion estimation by the fusion of image point and line features. *Pattern Recognition*, 31:333–344, 1998.

87. J. Langheim, A.J. Buchanan, CARSENSE—Sensor fusion for DAS. In *Proceedings ITS in Europe*, Lyon, 2002.

88. C. Stiller, J. Hipp, C. Rossig, A. Ewald, Multisensor obstacle detection and tracking. In *Proceedings 1998 IEEE International Conference on Intelligent Vehicles*, 451–456, 1998.

89. S.-Y. Sohn, H.W. Shin, Data mining for road traffic accident classification. *Ergonomics*, 44:107–117, 2001.

90. S.Y. Sohn, S.H. Lee, Data fusion, ensemble clustering to improve the classification accuracy for the severity of road traffic accident in Korea. *Safety Science*, 41:1–14, 2003.

91. L. Breiman, Bagging predictors. *Machine Learning*, 26:123–140, 1996.

92. L. Breiman, Arcing classifiers. *Annals of Statistics*, 26:801–849, 1998.

93. L. Breiman, Random forests. *Machine Learning*, 45:5–32, 2001.

94. D.H. Wolpert, Stacked generalization. *Neural Networks*, 5:241–259, 1992.

95. M. Cremer, H. Keller, Dynamic identification of OD flows from counts at complex intersections. In *Proceedings of the 8th International Symposium on Transportation and Traffic Theory (ISTTT)*, Toronto, Canada, University of Toronto Press, 1981.

96. M. Cremer, Determining time dependent trip distribution in a complex intersection for traffic responsive control. In *Proceedings on Control in Transportation Systems*, IFAC-IFIP-IFORS, Baden-Baden, Germany, 1983.

97. K. Ashok, M. Ben-Akiva, Dynamic O–D matrix estimation and prediction for real-time management systems. In *Proceedings of the 12th ISTTT*, 465–484, 1993.

98. M. Ben-Akiva, T. Morikawa, Methods to combine different data sources and estimates OD matrices. In *Proceedings of the 10th ISTTT*, Cambridge, MA, 459–481, 1987.

99. J.T. Lundgren, A. Peterson, S. Tengroth, Methods for pre-adjusting time dependent origin–destination matrices. In *Proceedings 10th World Congress and Exhibition on Intelligent Transportation Systems and Services*, Madrid, Spain, November 16–20, 2003.

100. J.F. Harrison, F. Stevens, A Bayesian approach to short-term forecasting. *Operational Research Quarterly*, 22:341–362, 1971.

101. I. Okutani, J. Stephanedes, Dynamic prediction of traffic volume through Kalman filtering theory. *Transportation Research B*, 18:1–11, 1984.

102. P.C. Vythoulkas, Alternative approaches to short term traffic forecasting for use in driver information systems. In *Proceedings of the 12th ISTTT*, Berkeley, CA, 485–506, 1993.

103. N.-E. El Faouzi, Nonparametric traffic flow prediction using kernel estimator. In *Proceedings of the 16th International Symposium on Transportation and Traffic Theory (ISTTT)*, J.-B. Lesort (ed.). Pergamon, Elsevier, Lyon, 41–54, 1996.

104. J.M. Bates, C.W.J. Granger, The combination of forecasts. *Operational Research Quarterly*, 20(4): 451–468, 1969.
105. W.J. Granger, Combining forecasts: Twenty years later. *Journal of Forecasting*, 8:167–173, 1989.
106. N.-E. El Faouzi, Combining predictive schemes in short-term traffic forecasting. In *Proceedings of the 17th International Symposium on Transportation and Traffic Theory (ISTTT)*, A. Ceder (ed.). Pergamon, Elsevier, Jerusalem, 471–487, July 26–28, 1999.
107. M. Cremer, S. Schrieber, Monitoring traffic load profiles with heterogeneous data sources configurations. In *Proceedings of the 16th International Symposium on Transportation and Traffic Theory (ISTTT)*, J.-B. Lesort (ed.). Pergamon, Elsevier, Lyon, 41–54, 1996.
108. J. Sau, N.-E. El Faouzi, A. Ben Aïssa, O. De Mouzon, Particle filter-based realtime estimation and prediction of traffic conditions. In *Proceedings of the ASMDA*, Chania, Crete, Greece, May 29–June 1, 2007.
109. D.B. Choi, On multisensor data fusion using attributes association for intelligent traffic congestion information inference. In *Proceedings of the I.V.H.S.*, 1989.
110. M.S. Grewal, L.R. Weill, A.P. Andrews, Global positioning systems. *Inertial Navigation, and Integration*. John Wiley, Hoboken, NJ, 2004.
111. M. Wei, K.P. Schwarz, Testing a decentralized filter for GPS/INS integration. In *Proceedings IEEE Position Location and Navigation Symposium*, Las Vegas, NV, 429–435, March 1990.
112. W. Li, H. Leung, Constrained unscented Kalman filter based fusion of GPS/INS/digital map for vehicle localization. In *IEEE 2003 International Conference on Intelligent Transportation Systems*, Shanghai, China, 1362–1367, October 2003.
113. N. El-Sheimy, K. Chiang, A. Noureldin, The utilization of artificial neural networks for multi-sensor system integration in navigation and positioning instruments. *IEEE Transactions on Instrumentation and Measurement*, 55:1606–1615, 2006.
114. R. Sharaf, A. Noureldin, A. Osman, N. El-Sheimy, Implementation of online INS/GPS integration utilizing radial basis function neural network. *IEEE Aerospace and Electronic Systems Magazine*, 20: 8–14, 2005.
115. A. Hiliuta, R. Landry, F. Gagnon, Fuzzy corrections in a GPS/INS hybrid navigation system. *IEEE Transactions on Aerospace and Electronic Systems*, 2:591–600, 2004.
116. L. Cholvy, Information evaluation in fusion: A case study. In: *Proceedings of the International Conference on Information Processing and Management of Uncertainty in Knowledge-Based Systems*, 993–1000, 2004.
117. V. Nimier, Information evaluation: A formalization of operational recommendations. In *Proceedings of the International Conference on Information Fusion*, 1166–1171, 2009.
118. *STANAG 2022: Intelligence Reports*. North Atlantic Treaty Organization (NATO), Brussels, December 1992.
119. L. Cholvy, Modelling information evaluation in fusion. In *Proceedings of the 10th International Conference on Information Fusion*, IEEE, Quebec, Quebec, 1–6, 2007.
120. A.E. Gelfand, C. Smith, M. Colony, C. Bowman, Performance evaluation of decentralized estimation systems with uncertain communications. In *Proceedings of the International Conference on Information Fusion*, 786–793, 2009.
121. D. McDaniel, G. Schaefer, Real-time DBMS for data fusion. In *Proceedings of the 6th International Conference on Information Fusion*, 2003.

# 33 Application of Multisensor Data Fusion for Traffic Congestion Analysis

Shrikant Fulari, Lelitha Vanajakshi,
Shankar C. Subramanian, and T. Ajitha

## CONTENTS

## 33.1 INTRODUCTION AND BACKGROUND

The exponential growth of personal vehicles, combined with an increase in trips and trip lengths, has resulted in acute road traffic congestion in most metropolitan cities around the world. A few reasons for congestion are higher demand rates, inadequate infrastructure, accidents, and construction. However, the most common reason is the traffic capacity supplied by traffic facilities being close to or less than the traffic demand. In recent years, the focus of congestion reduction has shifted from infrastructure and capital-intensive transportation strategies to more balanced and sustainable transportation solutions such as using Intelligent Transportation Systems (ITS).

In this regard, measuring congestion is a primary task and several congestion indicators are commonly used. These congestion indicators can be broadly categorized as speed-based indicators, temporal/delay based indicators, spatial indicators, service level/capacity indicators, reliability indicators, and economic cost/efficiency indicators [1]. Generally, these indicators are derived from three broad classes of measurements: location-based measurements (such as vehicle count and flow), temporal/speed indicators extrapolated or derived from the former (such as link travel time and delay), or spatial indicators (such as density and queue length). However, location-based measurements, being restricted to specific locations, cannot provide a complete picture of the condition/state of a road section to explain congestion. Hence, spatial variables such as density, travel time, and delays are more suitable to quantify congestion. Of these, traffic density is considered the primary measure to quantify congestion [2,3].

Traffic density is a macroscopic variable that indicates the number of vehicles inside a section of roadway. However, the direct measurement of traffic density involves aerial photography, which in most cases is impractical and expensive. In addition, most of the popular automated traffic sensors are location based in nature. Hence, density is usually estimated from location-based variables such as flow and speed. The most direct way to get density from location-based sensors is the input–output method [4]. This method uses the conservation of vehicles equation, adds the number of vehicles entering the section to the initial number of vehicles inside the section, and subtracts the number of vehicles exiting the section to provide the number of vehicles present in the section at the next instant/time interval. However, it has the limitation that the initial number of vehicles inside the section should be known, information that is typically difficult to obtain. Another issue is the sensitivity of this method to the accuracy of the count data collected [4–6]. Hence, researchers have explored other options to estimate density and a few of them are detailed in the text that follows.

Gazis and Knapp [7] proposed that traffic density can be estimated using time series analysis of flow and speed measurements at the entry and exit points of a road section. Nahi and Trivedi [8] derived a recursive estimator for traffic density in a section of roadway based on conservation of vehicles and using the Kalman filter. Gazis and Szeto [9] designed a density estimation scheme for multilane roadways using the Kalman filtering technique. The real-time estimation of traffic density was carried out by Sun et al. [10] using the cell transmission model. However, most of the aforementioned studies were reported from homogeneous and lane-based traffic conditions. There are only limited studies that explored density estimation under heterogeneous traffic conditions [11,12]. Overall, such an estimation of spatial variables from the location-based data has limitations, as the spatial variations are less effectively captured. The present study attempts to address this limitation by fusing location-based traffic data with limited spatial data available from Bluetooth® sensors.

Data fusion is a broad area of research in which data from several sensors are combined to provide comprehensive and accurate information [13]. Different terminologies are used related to data fusion such as information fusion, sensor fusion, multisensor data fusion, and multisensor data integration [14]. Information fusion encompasses theory, techniques, and tools employed for exploiting the synergy in the information acquired from multiple sources such that the resulting decision or action is in some sense better than what would be possible if any of these sources were used individually. Sensor fusion is a subset of information fusion that fuses data from sensors. It is inclusive in a broader terminology of data fusion depending on the data source. When different sensors are used to obtain certain data and when the data are fused, it can be termed multisensor data fusion.

There are different methods to fuse data. Van Lint and Hoogendoorn [15] have classified them into four main groups: estimation methods such as weighted averaging and Kalman filter technique, classification methods such as clustering algorithms, inference models such as Bayesian inference, and Dempster–Shafer evidential theory methods and artificial intelligence methods such as artificial neural networks (ANNs). In traffic engineering field, data fusion is used mainly to estimate parameters that cannot be estimated using a single sensor and to expand the spatial and temporal coverage while capturing the traffic characteristics; the important ones are discussed in the text that follows.

Kwon et al. [16] proposed a linear regression model for travel time prediction by combining both loop detector and probe vehicle data. Ivan [17] used the ANN technique to detect traffic incidents on signalized arterials using travel time data from loop detector and probe vehicle data. Choi and Chung [18] used fuzzy regression and the Bayesian pooling technique for estimating dynamic link travel time in congested urban road networks using global positioning system (GPS) and detector data. Chu et al. [19] used loop detector and probe vehicle data to estimate travel time using a model-based approach based on the Kalman filter technique. Anand et al. [20] used location-based sensor data and spatial data from probe vehicles fitted with GPS units for traffic density estimation. Anusha et al. [21] used location-based flow data and sparse travel time data obtained using GPS-equipped probe vehicles to estimate total travel time in an urban road stretch. In this study, a multisensor data fusion approach involving fusion of location-based and spatial variables using a model-based estimation method is attempted.

Traffic flow models are of two types depending on the level of detail in representing the state of traffic: microscopic and macroscopic. In microscopic models, each individual vehicle's motion is modeled, whereas in macroscopic models, the aggregate behavior of the traffic stream is considered. Microscopic models are comparatively difficult to calibrate owing to the effect of human behavior and hence are not typically preferred for real-time applications.

Macroscopic models can again be classified into continuum models and noncontinuum models. Continuum models are developed mainly using the analogy with the flow of liquids, where traffic flow is treated as analogous to flow of compressible liquids or gases. Many of these approaches need to make several assumptions about the traffic system to justify the analogy. There have been criticisms on the continuum models based on the fact that the number of vehicles in a typical section of roadway does not justify it being modeled as a continuum. Another issue was that continuum models allow the two-way propagation of disturbances, which is unrealistic in traffic [22–24]. In a noncontinuum approach to traffic flow modeling, only a finite number of vehicles is assumed to be present in a section of roadway during the analysis interval. This assumption overcomes several setbacks associated with the continuum approach and hence is adopted in the current study.

The use of such models for the fusion of location-based and spatial data for density estimation has not been reported under Indian conditions and is the main contribution of this study. In this regard, a mathematical model that describes the traffic flow process based on the conservation principle and steady-state traffic stream equation was developed. The flows entering and leaving the road section under study are the inputs to this model and the average section travel time is the output from this model. The density was estimated using data from two sources: a fixed location source (video) and a spatial source (Bluetooth), which should adequately capture its temporal and spatial characteristics. The traffic density and the space mean speed (SMS) were taken as the state variables. The state equations describing the temporal evolution of the state variables were developed from the principle of conservation of vehicles and a traffic stream model. The extended Kalman filter (EKF), which is a popular estimation scheme for nonlinear dynamic systems, was used along with the developed mathematical model to estimate the state variables. The estimated density values were then compared with the actual density obtained from the field to corroborate this model-based estimation scheme.

## 33.2 METHODOLOGY

As detailed earlier, a noncontinuum macroscopic dynamic traffic flow model based on the lumped parameter approach was used for density estimation using data fusion in this study. In a lumped parameter approach, the physical system under study is divided into lumped regions/domain, and within each region, the aggregate characteristics such as velocity, pressure, density, and so forth may vary with time, but are assumed to be uniform over the region. When this is applied to roadways, within a small section of roadway, the spatial variation of traffic variables (such as density, speed, etc.) is neglected and it is assumed that the variables depend only on time. A reasonable section length for this assumption to hold good may be around 1 to 1.5 kilometers, which is the usual spacing between automated data collection sensors. The section length ($L$) in this study was also selected in this range. To apply this procedure to longer roadways, the section can be divided into subsections of lengths in this range. The lumped parameter approach results in the governing equations of the model being ordinary differential equations (in the continuous time domain) and ordinary difference equations (in the discrete time domain), as detailed in Section 33.2.1.

### 33.2.1 MODEL FORMULATION

Consider a typical road segment as shown in Figure 33.1. The number of vehicles inside the section per unit length (density, $\rho$) and the SMS were considered as the macroscopic state variables to describe the state of traffic inside this section. The first governing equation, which is for density,

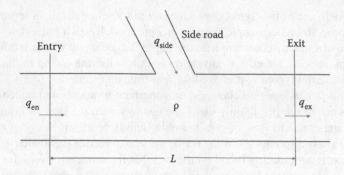

**FIGURE 33.1** Schematic diagram of a typical road section.

was formulated based on the conservation of vehicles inside the section (Figure 33.1). Let $N_{(k)}$ denote the number of vehicles inside the section at the $k$th instant of time. Then, the conservation of vehicles inside the section for a time step of $h$ can be represented as

$$N_{(k+1)} = N_{(k)} + h(q_{en(k)} - q_{ex(k)} + q_{side(k)}),$$ (33.1)

where $q_{en(k)}$ is the flow rate at which vehicles are entering into the section, $q_{ex(k)}$ is the flow rate at which vehicles are exiting from the section, and $q_{side(k)}$ is the net flow rate at which the vehicles are entering into the section from the side road in the time interval $(k, k + 1)$.

Dividing Equation 33.1 by the length of the section ($L$) results in

$$\rho_{(k+1)} = \rho_{(k)} + \frac{h}{L}(q_{en(k)} - q_{ex(k)} + q_{side(k)}),$$ (33.2)

where $\rho_{(k+1)}$ denotes the density inside the section at the $(k + 1)$th instant of time.

The second governing equation is a dynamic SMS equation formulated by incorporating the appropriate speed–density relationship developed for the specific traffic under study. The equation was derived as a dynamic equation based on the motive of minimizing the error ($e$) between the speed values estimated using the steady-state speed–density relation $v(\rho)$ developed and the observed values of $v$, that is $e = v(\rho) - v$. The time evolution of this error was hypothesized to be governed by the following equation, where the parameter $a$ was selected to be positive:

$$\frac{de}{dt} = -ae(t).$$ (33.3)

This equation is a linear homogeneous ordinary differential equation (ODE) and it is well known that its unique solution is $e(t) = e(0)\exp(-at)$ [25], where $e(0)$ is the initial error (can be either positive or negative), which will converge to zero with time. Although there may be other choices for describing the time evolution of the error function, an exponentially decaying error function (an exponential function is a very commonly used function in many phenomenological studies) has been chosen in this study as its performance will be comparable to that of any alternate choice. Discretizing this equation using a time step $h$ and Equation 33.2, the dynamic speed equation (i.e., the equation governing the evolution of speed) was obtained as

$$v_{(k+1)} = v_{(k)} + ah(v(\rho) - v_{(k)}) + \frac{h}{L}\frac{d(v(\rho))}{d\rho}(q_{en(k)} - q_{ex(k)} + q_{side(k)}).$$ (33.4)

Thus, the general formulation of the modified model is represented by Equations 33.2 and 33.4. Now, the appropriate steady-state speed–density relationship can be incorporated in Equation 33.4. The methodology adopted for the development of the stream model is discussed in Section 33.2.2.

## 33.2.2 Traffic Stream Model Development

Traffic stream models provide relationships among the three basic traffic variables—speed, flow and density—for steady-state conditions. Such speed–density relationships or equivalently, flow-density or speed-flow relationships, are essential components in most macroscopic models. These relationships are normally obtained through empirical observations and are location dependent. In general, stream models can be classified into two categories: single-regime models and multiregime models.

The single-regime models assume a unique relationship for representing the entire regime of traffic including the uncongested and congested regimes. Various single-regime models that explain speed–density relationship include Greenshield's linear model, Greenberg's logarithmic model, Underwood's exponential model, Drake's exponential model, and Pipes' generalized model [4]. However, in reality, a single relationship may not be enough, as the behavior of drivers is entirely different in uncongested and congested conditions. Hence, multiregime models have been used to improve the single-regime models by introducing models for representing different regimes of traffic. Popular models under this category are Edie's model, Drake's multi-regime models, and so forth [4]. In the present study, a multiregime model developed for the study site by Ajitha and Vanajakshi [26] was used. The best fitting model was reported to be a combination of Drake's exponential model, for the uncongested regime, and Pipes' model for the congested regime. A trust region based optimization algorithm [27] was used to estimate the parameters.

The error in fitting the aforementioned models to the data was quantified using the mean absolute percentage error (MAPE) given by

$$\text{MAPE} = \left[ \frac{1}{N} \sum_{i=1}^{N} \frac{|x_{\text{est}} - x_{\text{obs}}|}{x_{\text{obs}}} \right] * 100, \tag{33.5}$$

where $x_{\text{est}}$ and $x_{\text{obs}}$ are the estimated and observed values of the variable under consideration respectively, and $N$ is the total number of observations. Figure 33.2 shows the best fitting two-regime speed–density model illustrating the scatterplot of the set of observed values and the fitted curve.

Table 33.1 shows the corresponding model fitting results and related statistics. It can be seen that the MAPE value for speed was around 12%. Thus, this model, which uses Drake's exponential model in the uncongested regime and Pipes' model in the congested regime, can be considered as a good fit for the traffic under consideration. This model was also able to provide the complete set of key traffic stream parameters, which are given in Table 33.1. The obtained empirical speed–density relationships for uncongested and congested regimes respectively are shown in Equation 33.6 and 33.7.

$$v = v_f \exp\left[ -0.5\left( \frac{\rho}{\rho_{\text{cr}}} \right)^2 \right], \quad \text{for } 0 \le \rho \le \rho_{\text{cr}}, \tag{33.6}$$

$$v = \frac{v_{\text{cr}}}{\left( \frac{\rho_j}{\rho_{\text{cr}}} - 1 \right)} \left( \frac{\rho_j}{\rho} - 1 \right), \quad \text{for } \rho_{\text{cr}} \le \rho \le \rho_j. \tag{33.7}$$

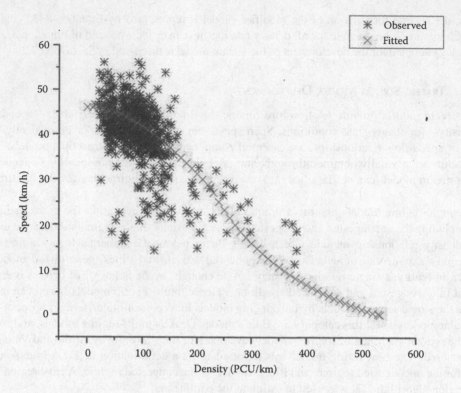

**FIGURE 33.2**   Best fitting two-regime speed density model.

**TABLE 33.1**

**Two-Regime Model Fitting Results and the Related Statistics**

| Model | MAPE (%) | Free Flow Speed ($V_f$) km/h | Critical Speed ($V_{cr}$) km/hr | Critical Density ($\rho_{cr}$) PCU/km | Jam Density ($\rho_j$) PCU/km |
|---|---|---|---|---|---|
| Drake's + Pipes' for Speed–Density | 11.78 | 46 | 34.2 | 227 | 633 |

Thus, the model equations can be summarized as

$$\rho_{(k+1)} = \rho_{(k)} + \frac{h}{L}(q_{en(k)} - q_{ex(k)} + q_{side(k)}),$$   (33.8)

$$v_{(k+1)} = v_{(k)} + ah\left\{v_f \exp\left[-0.5\left(\frac{\rho}{\rho_{cr}}\right)^2\right] - v_{(k)}\right\}$$

$$-\left(\frac{h}{L}(q_{en(k)} - q_{ex(k)} + q_{side(k)})\left[\frac{v_f \rho_{(k)}}{\rho_{cr}^2}\exp\left[-0.5\left(\frac{\rho_{(k)}}{\rho_{cr}}\right)^2\right]\right]\right),$$   (33.9)

for $0 \le \rho \le \rho_{cr}$,

$$v_{(k+1)} = v_{(k)} + ah\left[\left[\left(\frac{v_{cr}}{\left[\frac{\rho_j}{\rho_{cr}} - 1\right]}\left[\frac{\rho_j}{\rho_{(k)}} - 1\right]\right) - v_{(k)}\right] - \left(\frac{h}{L}(q_{en(k)} - \rho_{(k)}v_{ex(k)} + q_{side(k)})\left(\frac{v_{cr}}{\left[\frac{\rho_j}{\rho_{cr}} - 1\right]}\left[\frac{\rho_j}{\rho_{(k)}^2}\right]\right)\right)\right],$$

(33.10)

for $\rho_{cr} \le \rho \le \rho_j$.

Using this nonlinear model along with the extended Kalman filter (EKF), the model-based estimation scheme was designed as detailed in Section 33.3.

## 33.3 ESTIMATION USING THE KALMAN FILTERING TECHNIQUE

The Kalman filter [28] is a popular tool for recursive estimation of variables that characterize a system (these variables are usually referred to as "state variables"). The Kalman filter is a model-based estimation scheme that takes into account the stochastic properties of the process disturbance and the measurement noise. The process disturbance and the measurement noise are assumed to be independent of one another, white, and normally distributed with zero mean. The Kalman filter works like a predictor corrector algorithm, that is, it first predicts an a priori estimate of the state variables using the system model and the state estimate from the previous time interval, and then corrects the same using measurements to obtain an a posteriori state estimate. The Kalman filter has been widely used in many disciplines including the field of transportation [29–31]. The Kalman filter is used when the governing equations of the system are linear. When the system model is nonlinear, the extended Kalman filter (EKF) is typically used, and is discussed here.

Consider a nonlinear system whose model is given by

$$\mathbf{x}_{(k+1)} = \mathbf{f}(\mathbf{x}_{(k)}, \mathbf{u}_{(k)}, \mathbf{w}_{(k)}),$$

(33.11)

$$\mathbf{z}_{(k)} = \mathbf{g}(\mathbf{x}_{(k)}, \mathbf{v}_{(k)}),$$

(33.12)

where $\mathbf{f}$ represents the nonlinear function that relates the state at time step $k$ to the state at time step $k + 1$. Similarly, $\mathbf{g}$ is the nonlinear function that relates the state to the measurement. The preceding equations can be linearized using Taylor's series expansion to result in Equations 33.13 and 33.14

$$\mathbf{x}_{(k+1)} = \tilde{\mathbf{x}}_{(k+1)} + \mathbf{A}\left(\mathbf{x}_{(k)} - \hat{\mathbf{x}}_{(k)}^{+}\right) + \mathbf{W}\mathbf{w}_{(k)},$$

(33.13)

$$\mathbf{z}_{(k)} = \tilde{\mathbf{z}}_{(k)} + \mathbf{H}(\mathbf{x}_{(k)} - \tilde{\mathbf{x}}_{(k)}) + \mathbf{V}\mathbf{v}_{(k)},$$

(33.14)

where $\tilde{\mathbf{x}}$ and $\tilde{\mathbf{z}}$ are obtained using Equations 33.15 and 33.16.

$$\tilde{\mathbf{x}}_{(k+1)} = \mathbf{f}\left(\hat{\mathbf{x}}_{(k)}^{+}, \mathbf{u}_{(k)}, 0\right),$$

(33.15)

$$\tilde{\mathbf{z}}_{(k)} = \mathbf{g}(\tilde{\mathbf{x}}_{(k)}, 0).$$

(33.16)

Here, $\mathbf{A}$ is the matrix of the partial derivative of $\mathbf{f}$ with respect to $\mathbf{x}$, $\mathbf{W}$ is the matrix of the partial derivative of $\mathbf{f}$ with respect to $\mathbf{w}$, $\mathbf{H}$ is the matrix of the partial derivative of $\mathbf{g}$ with respect to $\mathbf{x}$, and $\mathbf{V}$ is the matrix of the partial derivative of $\mathbf{g}$ with respect to $\mathbf{v}$.

Now, the following recursive algorithm is used to obtain the estimate of the state variables:

1. The a priori estimate in the $(k + 1)$th interval of time is obtained through

$$\hat{\mathbf{x}}_{(k+1)}^{-} = \mathbf{f}\left(\hat{\mathbf{x}}_{(k)}^{+}, \mathbf{u}_{(k)}\right).$$

(33.17)

2. The a priori error covariance in the $(k + 1)$th interval of time is obtained through

$$\mathbf{P}_{(k+1)}^{-} = \left(\mathbf{A}\mathbf{P}_{(k)}^{+}\mathbf{A}^{T} + \mathbf{W}\mathbf{Q}\mathbf{W}^{T}\right).$$

(33.18)

3. The Kalman gain $\mathbf{K}_{k+1}$ is calculated through

$$\mathbf{K}_{(k+1)} = \mathbf{P}_{(k+1)}^{-}\mathbf{H}^{T}\left(\mathbf{H}\mathbf{P}_{(k+1)}^{-}\mathbf{H}^{T} + \mathbf{V}\mathbf{R}\mathbf{V}^{T}\right)^{-1}.$$

(33.19)

4. Then, the a posteriori state estimate is calculated through

$$\mathbf{x}_{(k+1)}^{+} = \mathbf{x}_{(k+1)}^{-} + \mathbf{K}_{(k+1)}\left(\mathbf{z}_{(k+1)} - \mathbf{g}\left(x_{(k+1)}^{-}\right)\right).$$

(33.20)

5. Finally, the a posteriori error covariance is obtained through

$$\mathbf{P}_{(k+1)}^{+} = \left((\mathbf{I} - \mathbf{K}_{(k+1)}\mathbf{H})\mathbf{P}_{(k+1)}^{-}\right).$$

(33.21)

These five steps (Equations 33.17 to 33.21) are repeated for each time interval to obtain the estimate of the state variables using EKF.

In this study, the density of vehicles ($\rho$) expressed in passenger car units/km (PCU/km) and the average SMS ($v$) in km/h were taken as the state variables of interest. The output variable was the average section travel time. The rates at which vehicles enter into the section from upstream and from the side road in PCU/h were provided as the inputs to the estimation scheme. Here, the inputs used, namely the flow and speed, were collected using location-based sensors, and the measurement, travel time, was obtained from Bluetooth sensors, making it a multisensor data fusion method.

Thus, the variables $\mathbf{x}$, $\mathbf{u}$, and $z$ are

$$\mathbf{x}_{(k)} = \begin{bmatrix} \rho_{(k)} \\ v_{(k)} \end{bmatrix}, \quad \mathbf{u}_{k} = \begin{bmatrix} q_{\text{en}(k)} \\ q_{\text{side}(k)} \end{bmatrix}, \quad z_{(k)} = tt_{(k)}.$$

The parameter $\mathbf{h}$ was obtained by taking the partial derivative of the measurement equation with respect to the state as

$$\mathbf{h} = \begin{pmatrix} 0 \\ \dfrac{-L}{v_{(k)}^{2}} \end{pmatrix}.$$

The parameters $\mathbf{W}$ and $V$ were assumed as

$$\mathbf{W} = \begin{bmatrix} h & 0 \\ 0 & h \end{bmatrix}, \quad V = 1.$$

For the present case, the function $\mathbf{f}$, relating the state $\mathbf{x}$ at $(k+1)$th instant to the $k$th instant in the free-flow regime, was obtained as

$$\mathbf{f} = \begin{bmatrix} \rho_{(k)} + \dfrac{h}{L}(q_{\mathrm{en}(k)} - q_{\mathrm{ex}(k)} + q_{\mathrm{side}(k)}) \\[2mm] v_{(k)} + ah \left\{ v_f \exp\left[ -0.5\left(\dfrac{\rho}{\rho_{\mathrm{cr}}}\right)^2 \right] - v_{(k)} \right\} \\[2mm] + \left( \dfrac{h}{L}(q_{\mathrm{en}(k)} - q_{\mathrm{ex}(k)} + q_{\mathrm{side}(k)}) \right) \left[ \dfrac{v_f \rho_{(k)}}{\rho_{\mathrm{cr}}^2} \exp\left[ -0.5\left(\dfrac{\rho_{(k)}}{\rho_{\mathrm{cr}}}\right)^2 \right] \right] \right) \end{bmatrix}. \tag{33.22}$$

Similarly, in the congested regime, $\mathbf{f}$ was obtained as

$$\mathbf{f} = \begin{bmatrix} \rho_{(k)} + \dfrac{h}{L}(q_{\mathrm{en}(k)} - q_{\mathrm{ex}(k)} + q_{\mathrm{side}(k)}) \\[2mm] v_{(k)} + ah \left[ \left( \dfrac{v_{\mathrm{cr}}}{\left[\dfrac{\rho_j}{\rho_{\mathrm{cr}}} - 1\right]} \left[ \dfrac{\rho_j}{\rho_{(k)}} - 1 \right] \right) - v_{(k)} \right] \\[2mm] - \left( \dfrac{h}{L}(q_{\mathrm{en}(k)} - q_{\mathrm{ex}(k)} + q_{\mathrm{side}(k)}) \right) \left( \dfrac{v_{\mathrm{cr}}}{\left[\dfrac{\rho_j}{\rho_{\mathrm{cr}}} - 1\right]} \left[ \dfrac{\rho_j}{\rho_{(k)}^2} \right] \right) \end{bmatrix}. \tag{33.23}$$

The data collected for this current study for implementation and evaluation of the above developed estimation scheme are discussed below.

## 33.4 STUDY STRETCH AND DATA COLLECTION

### 33.4.1 STUDY STRETCH

The study stretch identified for the present study was in the Rajiv Gandhi road in the city of Chennai, India. It is a six-lane roadway, with three lanes in each direction. For the present study, only one direction of traffic was considered. Figure 33.3 presents a location map showing the selected study stretch. Three pedestrian footbridges were available at the three locations labeled as A, B, and C along this study stretch, as shown in Figure 33.3. A side road for vehicles to enter into the section was present between B and C. The total length of the section AC is 1.738 km. The analysis was also carried out separately for the subsections AB and BC with section lengths of 1 kilometer and 0.738 kilometer respectively. This was due to the presence of a signalized intersection downstream of the point C that was causing the generation of long queues during red signals, especially during the peak hours. Hence, the analysis was carried out separately for these sections, to study the performance of the proposed scheme for sections of varying lengths as well as traffic conditions.

### 33.4.2 DATA COLLECTION

The present study required flow, speed, and travel time data. Of these, flow and speed data were collected using the videographic technique. However, extraction of travel time by using video cameras

**FIGURE 33.3** Location map showing selected study stretch. (From Google Maps.)

was very tedious and time consuming because it involves reidentification of vehicles in multiple videos. Hence, Bluetooth sensors were used to obtain the travel time data. The density data required to corroborate the results were obtained using the input–output method, which was explained in Section 33.1.

Three video cameras were installed at points A, B, and C as shown in Figure 33.3. Video data were collected on four different days that included both peak and off-peak periods, which were used for fitting the stream model. The data from the side roads were collected manually. To obtain the initial number of vehicles in the section for the input–output method, photographs of the study sections were taken at the start of video recordings. Bluetooth sensors were placed at the same locations where video cameras were placed and travel time data were collected on two out of the four days. The data from these two days were used to evaluate the performance of the scheme.

Data extraction from the videos was carried out manually. Five classes of vehicles—two wheelers (2Ws), three wheelers (auto rickshaw) (3Ws), light motor vehicles (including cars) (LMVs), and heavy motor vehicles (HMVs)—were considered to incorporate heterogeneity. To account for the lack of lane discipline, the data from all the three lanes of the roadway were aggregated and analyzed as a single lane. Data were extracted for every 1-minute interval. Data quality checks were carried out by visual observation and obvious outliers were removed. From these data, the traffic variables of interest namely flow, SMS and density were calculated as follows.

### 33.4.2.1 Flow

The traffic volumes in terms of vehicle counts in every 1-minute interval were converted into per hour flow values at the classified level for the five different categories of vehicles and then aggregated to a common unit using the standard passenger car equivalent unit (PCU) values recommended by the Indian Road Congress [32]. Table 33.2 illustrates the values of PCU used in this study. The average traffic composition of different vehicles present in the study stretch were observed to be around 48%, 6%, 35%, 4%, and 7% of TWs, 3Ws, LMVs, and HMVs, respectively.

**TABLE 33.2**
**Recommended PCU Values of Vehicles on Urban Roads by IRC**

| | Equivalent PCU Values | |
| --- | --- | --- |
| | Percentage Composition of Vehicle Type in Traffic Stream | |
| Vehicle Type | 5% | 10% and Above |
| Fast Vehicles | | |
| 1. Two-wheeler motorcycle or scooter | 0.5 | 0.75 |
| 2. Passenger car, pick-up van | 1.0 | 1.0 |
| 3. Auto-rickshaw | 1.2 | 2.0 |
| 4. Light commercial vehicle | 1.4 | 1.2 |
| 5. Truck or bus | 2.2 | 3.7 |
| 6. Agricultural tractor trailer | 4.0 | 5.0 |
| Slow Vehicles | | |
| 1. Cycle | 0.4 | 0.5 |
| 2. Cycle rickshaw | 1.5 | 2.0 |
| 3. Tonga (horse-drawn vehicle) | 1.5 | 2.0 |
| 4. Hand cart | 2.0 | 3.0 |

### 33.4.2.2 Space Mean Speed

The speed values extracted from video were spot speeds of individual vehicles and need to be converted to SMS. This was carried out by calculating the harmonic mean of spot speeds, which is a surrogate measure of SMS [33], as

$$v = \frac{1}{\frac{1}{N}\sum_{i=1}^{n}\frac{1}{v_{(i)}}},$$

where, $v$, $v(i)$, and $N$ represent SMS, spot speed of $i$th vehicle, and total number of vehicles respectively. The SMSs were calculated for the five categories of vehicles separately for every 1-minute interval. At the aggregate level, a weighted average value was computed considering the proportion of different categories of vehicles as the corresponding weight.

### 33.4.2.3 Density

The density values were determined using the input–output method [4]. The initial number of vehicles present inside the sections required for the input–output method was measured by taking aerial pictures of the sections at the start of the data collection. Then, the number inside the section at any interval is calculated by adding the number entering and subtracting the number exiting the section in that interval from the number inside the section in the previous interval.

## 33.5 IMPLEMENTATION AND EVALUATION OF ESTIMATION SCHEME

The implementation of the proposed scheme was done separately for sections AB, BC, and AC. The traffic variables required for the implementation of the proposed scheme included flow through the entry and exit points and side roads of the study stretches, spot speeds from which SMS was subsequently calculated, and travel time from Bluetooth. Using these data and assuming the initial values of state variables, the estimation scheme was implemented. The entire analysis was carried out at the aggregate level using the PCU converted data considering the five vehicle classes. The estimated density values were compared with the density values obtained using the input–output method. The mean absolute percentage error (defined in Equation 33.5) was used to quantify the errors. Sample results for sections AC and AB for day 1 are shown in Figures 33.4 and 33.5 respectively. The errors quantified for all the three sections are tabulated in Table 33.3.

From the results presented in Table 33.3, it can be observed that the MAPE values for density estimation for all the three sections for all traffic conditions were below 20%. According to Lewis' scale of interpretation of estimation accuracy [34], any forecast with a MAPE value of less than

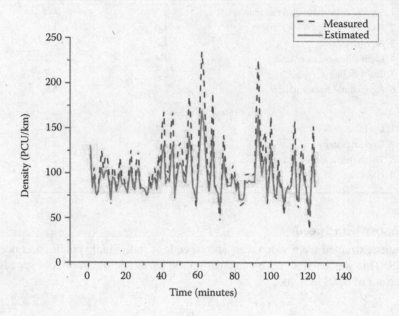

**FIGURE 33.4**   Comparison of measured and estimated density for section AC for day 1.

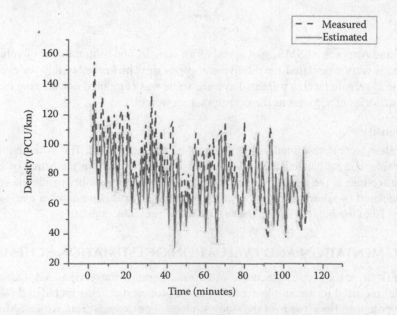

**FIGURE 33.5**   Comparison of measured and estimated density for section AB for day 1.

**TABLE 33.3**
**MAPE for Density Estimation Using the Model-Based Estimation Scheme**

| Serial No. | Section | Section Length | Day and Duration | Mean Absolute Percentage Error (%) |
|---|---|---|---|---|
| 1 | AC | 1.738 km | Day 1 (2 hours) | 11.9 |
| 2 | AB | 1 km | Day 1 (2 hours) | 10.7 |
| 3 | BC | 0.738 km | Day 1 (2 hours) | 15.7 |
| 4 | AC | 1.738 km | Day 2 (1 hour) | 15.3 |
| 5 | AB | 1 km | Day 2 (1 hour) | 13.6 |
| 6 | BC | 0.738 km | Day 2 (1 hour) | 13.4 |

10% can be considered highly accurate, 11% to 20% as good, 21% to 50% as reasonable and 51% or more as inaccurate. According to this, the results obtained can be classified as good. This shows that the model-based density estimation scheme with data fusion approach is a promising approach.

## 33.6 CONCLUDING REMARKS

Estimation of traffic density is one of the essential requirements for congestion analysis and its mitigation. The application of such methodologies for urban arterials may prove very useful for congestion management but limitation in data availability is one of the setbacks. The available data sources, both location-based and spatial, have limitations when it comes to the estimation of spatial variables such as density. Whereas the location-based sensors fail to capture the variations over distance, the spatial ones can collect information only from a sample of vehicles, which are equipped with identifiers. Hence, dependence on more than one source is preferred. The data fusion approach is one solution that is widely used in such cases and when implemented properly, proves to be a very cost-effective and viable choice.

In the present study, the data fusion approach was used for congestion analysis where traffic density was selected as the congestion indicator. A model-based approach was used and a dynamic macroscopic model was developed for density estimation. The data sources used for obtaining the required traffic variables were video cameras and Bluetooth sensors. Flow and speed were obtained from video and travel time data were obtained from Bluetooth sensors. The data were fused and used in the model-based estimation scheme to obtain density estimates. These were then compared with actual density values for evaluation and the results were found to be promising. These estimated/predicted density values can be used as a direct indicator of congestion and can be used to provide real-time congestion information to the road users so that they take well-informed travel decisions. Thus, this estimation scheme can ultimately be part of an overall congestion prediction and mitigation system.

## REFERENCES

1. C. Philippe, and W. John. *Managing Urban Traffic Congestion*. In European Conference of Ministers on Transport. Paris: OECD Publishing, 2007.
2. *Highway Capacity Manual* (HCM). Washington, DC: Transportation Research Board, National Research Council, 2000.
3. *Highway Capacity Manual* (HCM). Special Report 209. Washington, DC: Transportation Research Board of the National Academies, 2010.
4. A. D. May. *Traffic Flow Fundamentals*. Upper Saddle River, NJ: Prentice Hall, 1990.
5. L. Vanajakshi, and L. Rilett. Loop detector data diagnostics based on conservation-of-vehicles principle. *Transportation Research Record*, 1870(1):162–169, 2004.

6. L. Vanajakshi. *Estimation and Prediction of Travel Time from Loop Detector Data for Intelligent Transportation Systems Applications.* College Station, TX: Texas A&M University, 2004.

7. D. Gazis, and C. Knapp. On-line estimation of traffic densities from time-series of flow and speed data. *Transportation Science*, 5:283–301, 1971.

8. N. E. Nahi, and A. N. Trivedi. Recursive estimation of traffic variables: Section density and average speed. *Transportation Science*, 7:269–286, 1973.

9. D. Gazis, and M. Szeto. Design of density-measuring systems for roadways. *Transportation Research Record*, 495:44–52, 1974. Washington, DC: Transportation Research Board.

10. S. Sun, L. Mufioz, and R. Horowitz. Mixture Kalman filter based highway congestion mode and vehicle density estimator and its application. *Proceedings of the American Control Conference*, 3:2098–2103, 2004.

11. A. S. Padiath, L. Vanajakshi, and S. C. Subramanian. Estimation of traffic density under Indian traffic conditions. In *Transportation Research Board Annual Meeting CD-ROM*. Washington, DC: Transportation Research Board of the National Academics, 2010.

12. T. Ajitha. *A Model Based Approach for Real Time Estimation of Traffic Density under Indian Conditions.* Chennai, India: Indian Institute of Technology Madras, 2012.

13. D. Dailey, P. Harn, and P. Lin. *ITS Data Fusion*. ITS Research Program, Final Research Report. Washington State Transportation Commission, Seattle, WA, 1996.

14. L. Snidaro, I. Visentini, and G. Foresti. Data fusion in modern surveillance. In D. Ramagnino, P. Monokosso, and L. Jain (eds.), *Innovations in Defence Support Systems—3 Intelligent Paradigms in Security*, pp. 1–17. Berlin and Heidelberg: Springer Science+Business Media, 2011.

15. J. Van Lint, and S. Hoogendoorn. A robust and efficient method for fusing heterogeneous data from traffic sensors on freeways. *Computer Aided Civil and Infrastructure Engineering*, 24:1–17, 2009.

16. J. Kwon, B. Coifman, and P. Bickel. Day-to-day travel time trends and travel time prediction from loop detector data. *Transportation Research Record*, 1717:120–129, 2000.

17. J. Ivan. Neural network representations for arterial street incident detection data fusion. *Transportation Research Part C*, 5(3/4):245–254, 1997.

18. K. Choi, and Y. Chung. A data fusion algorithm for estimating link travel time. *Journal of Intelligent Transportation Systems: Technology, Planning, and Operations*, 7(3–4):235–260, 2002.

19. L. Chu, J. Oh, and W. Recker. Adaptive Kalman filter based freeway travel time estimation. In *Transportation Research Board 84th Annual Meeting*, Washington, DC, January 9–13, 2005.

20. A. Anand, L. Vanajakshi, and S. C. Subramanian. Traffic density estimation under heterogeneous traffic conditions using data fusion. In *IEEE Intelligent Vehicles Symposium (IV)*, Baden-Baden, Germany, June 5–9, 2011.

21. S. P. Anusha, A. Anand, and L. Vanajakshi. Data fusion based hybrid approach for the estimation of urban arterial travel time. *Journal of Applied Mathematics*, 2012:587913, 2012.

22. M. Papageorgiou. Some remarks on macroscopic traffic flow modeling. *Transportation Research Part A*, 32:323–329, 1998.

23. S. Darbha, K. R. Rajagopal, and V. Tyagi. A review of mathematical models for the flow of traffic and some recent results. *Nonlinear Analysis*, 69:950–970, 2008.

24. V. Tyagi, S. Darbha, and K. R. Rajagopal. A review of the mathematical models for traffic flow. *International Journal of Advances in Engineering Sciences and Applied Mathematics*, 1:53–68, 2009.

25. E. A. Coddington. *An Introduction to Ordinary Differential Equations.* Mineola, NY: Dover Publications, 1989.

26. T. Ajitha, and L. Vanajakshi. A traffic stream model for heterogeneous traffic conditions. *Proceedings of the ICE—Transport*, 167(4):240–247, 2012.

27. W. Forst, and D. Hoffmann. *Optimization–Theory and Practice.* New York: Springer Science+Business Media, 2010.

28. R. E. Kalman. A new approach to linear filtering and prediction problems. *Transactions of the ASME, Journal of Basic Engineering, Series D*, 82:35–45, 1960.

29. M. W. Szeto, and D. C. Gazis. Application of Kalman filtering to the surveillance and control of traffic systems. *Transportation Science*, 6:419–439, 1972.

30. Y. Wang, M. Papageorgiou, and A. Messmer. Real-time freeway traffic state estimation based on extended Kalman filter: A case study. *Transportation Science*, 41:167–181, 2007.

31. V. Tyagi, S. Darbha, and K. R. Rajagopal. A dynamical systems approach based on averaging to model the macroscopic flow of freeway traffic. *Nonlinear Analysis: Hybrid Systems*, 2:590–612, 2008.

32. Indian Road Congress (IRC). *Guidelines for Capacity of Urban Roads in Plain Areas.* New Delhi, India: IRC: 106, 1990.

33. F. L. Hall. Traffic stream characteristics. In N. H. Gartner, C. Messer, and A. K. Rathi (eds.), *Traffic Flow Theory: A State of the Art Report—Revised Monograph on Traffic Flow Theory*. Technical Report. Oak Ridge, TN: Oak Ridge National Laboratory, 1997.
34. D. L. Kenneth, and K. K. Ronald. *Advances in Business and Management Forecasting*. Bingley, UK: Emerald Books, 1982.

# 34 Consensus-Based Decentralized Extended Kalman Filter for State Estimation of Large-Scale Freeway Networks

*Liguo Zhang*

## CONTENTS

## 34.1 INTRODUCTION

Real-time knowledge about traffic conditions of urban road transportation systems is critical for traffic management and control. There are many well-established technologies for collecting vehicle speed and flux data, including loop detectors and automatic vehicle identification (AVI) sensors. However, equipped segments in the network are typically low and not representative of the urban network as a whole, which leaves the traffic conditions in most of the network unknown. Recently, novel ubiquitous sensing technologies have demonstrated the potential for unprecedented data collection at any spatial position of the large-scale road transportation networks; for example, dedicated probe vehicles, or simple portable radar speed detect guns embedded with wireless communication that constitute a wireless sensing network for vehicle mean speed measurement.

Road traffic estimation includes both traveling time estimate and traffic parameter estimate for any origin–destination pair of networks. The literature on travel time estimation using speed and position of probe vehicles has grown recently as the technology has become more available. Hunter et al. [1] present a probabilistic model of travel times in the arterial network, based on low-frequency taxi global positioning system (GPS) probes. Jenelius and Koutsopoulos [2] further consider the effect of explanatory variables on travel times using probe vehicle data.

Traffic parameter estimations mainly use the filtering approach with the prediction of the macroscopic traffic flow models. One of the most widely applied estimation methods is the discretization of the Lighthill–Whitham–Richards (LWR) model with an extended Kalman filter (EKF). Wang et al. [3] propose an extended Kalman filtering (EKF) to estimate the real-time traffic density and vehicle speed of a freeway link with a stochastic version of the METANET model. In Ref. [4], a particle filtering (PF) method is developed with a speed-extended stochastic cell transmission model (CTM). Sun et al. [5] present a solution to traffic density estimate with the sequential Monte Carlo algorithm that is a mixture Kalman filtering. Although previous work clearly demonstrates the possibility of extracting traffic parameters, these methods are all based on the general assumption that the boundary fluxes of the traffic networks are accessible or measurable.

A large disadvantage of the EKF is that it is too slow to perform in real time on the large-scale networks. For a larger scale freeway network that contains more than a few hundred measured cells, the complexity of EKF will make real-time calculations impossible on a normal computer, rendering the estimation method infeasible for large-scale online applications. To overcome this problem, a novel consensus-based decentralized EKF is proposed. The whole freeway is divided into several links under the structure of vehicle speed detecting networks. The logic of traffic flow dynamics (discretization LWR model) is used to correct only the state in the vicinity of the speed detectors. For the adjacent links, as the driving-out flux from the upstream link is equal to the driving-in flux of the downstream link, we consider a distributed fusion algorithm to connect and update the estimation of traffic parameters of the local links. Two key procedures are developed to perform the real-time distributed estimation of traffic state for large-scale freeway networks.

As a typical distributed parameter system, the traffic dynamics of freeway link are determined by both the current states and the boundary conditions. The boundary flux of the link is usually unknown or not available, as the freeway network is arbitrarily divided into subsystems with the real-time computation account. Therefore, we first extend the previous EKF on traffic parameter estimate by considering the jointed traffic density and boundary flux estimation of the freeway link. The proposed methodology is based on the simultaneous state and input estimate of Kalman filtering, which has been extensively applied in the recent literature [6,7]. From experimental test considerations, it would be a critical factor for the distributed estimate of the large-scale freeway networks. Second, we consider consensus average algorithm to fuse the boundary flux between the adjacent links, which are estimated with different measurement form upstream or downstream speed detectors, respectively. Stability and optimal estimation of consensus-based decentralized Kalmen filter were developed by Saber and Shamma [8] and Stankovic et al. [9]. A distinguishing advantage of the decentralized estimation is that real-time calculation is much faster than the classical EKF and it scales much better with the large-scale network size.

This chapter is organized as follows. The Godunov scheme for discretization of the LWR model is introduced in Section 34.2. Simultaneous state and input estimate of EKF filtering is developed in Section 34.3. Consensus-based decentralized EKF for the large-scale freeway networks is formulated in Section 34.4. In Section 34.5, an approximately five-mile freeway of Interstate 80 East (I-80E) in Alameda, Northern California is chosen to investigate the performance of the developed approach. Section 34.6 concludes the chapter.

## 34.2 GODUNOV SCHEME FOR LWR TRAFFIC FLOW MODEL

The first-order macroscopic (LWR) traffic flow model [10,11] formulates the relationship between the vehicle density $\rho(x, t)$ and the traffic flow $q(x, t)$, at the spatial position $x$ and time instant $t$ with the conservation law

$$\frac{\partial \rho}{\partial t} + \frac{\partial q}{\partial x} = 0. \tag{34.1}$$

Using the hydrodynamic of the flow-speed relation

$$v = \frac{q}{\rho}, \tag{34.2}$$

and further assuming that the speed measurement is available, then Equations 34.1 and 34.2 form a closed system to describe the traffic flow dynamics, as there are only two unknown variables $\rho$ and $q$ for two equations.

Based on the Smulders fundamental diagram [12], a static velocity–density relationship is given by the following piecewise function:

$$v = \begin{cases} \dfrac{v_c - v_f}{\rho_c}\rho + v_f, & \rho \le \rho_c \\[2ex] \dfrac{\rho_c v_c}{\rho_J - \rho_c}\left(\dfrac{\rho_J}{\rho} - 1\right), & \rho > \rho_c, \end{cases} \tag{34.3}$$

which contains four parameters that are specific to a link: free flow speed $v_f$, critical speed $v_c$, critical density $\rho_c$, and jam density $\rho_J$. These parameters also define the capacity of the link $q_{max} = v_c \rho_c$. Figure 34.1 shows the shape of this fundamental diagram.

Numerical solutions of Equations 34.1 and 34.2 could be found using the Godunov scheme, whereby each segment of the traffic link is discretized into cells with length $L_i$, $i = 1, 2,\ldots$, and time is discretized into interval with length $\Delta t$. According to the Courant–Friedrichs–Lewy condition, the numerical solution is stable as the length of each cell satisfies

$$L_i \le v_f \Delta t.$$

Given this space–time discretization, Equation 34.1 can be rewritten in the form

$$\rho_i(k+1) = \rho_i(k) + \frac{\Delta t}{L_i}\left(q_i^{in}(k) - q_i^{out}(k)\right) \tag{34.4}$$

where $\rho_i(k)$ is the vehicle density of cell $i$ at time $k$ and $q_i^{in}(k)$, $q_i^{out}(k)$ are vehicle flux entering and exiting cell $i$ during the time interval $[k\Delta t, (k + 1)\Delta t]$, respectively.

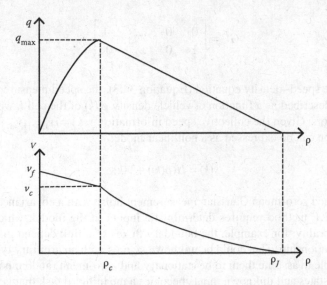

FIGURE 34.1  Traffic flow curves with Smulders fundamental diagram.

The traffic fluxes $q_i^{in}(k)$ between the cell borders are determined by comparing the available supply $S_i(k)$, the maximum flux that can enter the cell $i$, and the prevailing demand $D_{i-1}(k)$, the maximum flux that wants to exit the upstream cell $i - 1$, that is,

$$q_i^{in}(k) = \min\{D_{i-1}(k), S_i(k)\}. \tag{34.5}$$

These demand and supply functions are defined as

$$\begin{aligned} D_i(k) &= v(\rho_i)\rho_i, & S_i(k) &= q_{max}, & \rho_i(k) &\leq \rho_c \\ D_i(k) &= q_{max}, & S_i(k) &= v(\rho_i)\rho_i, & \rho_i(k) &> \rho_c. \end{aligned} \tag{34.6}$$

For the adjacent cells without the source terms, the driving-out flux of the upstream cell is equal to the driving-in flux of the downstream cell, that is,

$$q_i^{out}(k) = q_{i+1}^{in}(k), \quad i = 1, 2, \dots. $$

## 34.3   EKF FOR SIMULTANEOUS STATE AND INPUT ESTIMATION

In this section, a recursive EKF filter is introduced to estimate the state of the freeway link with unknown traffic flux at the upstream and the downstream boundaries.

The traffic state in the freeway link $j$ at time $k$ is uniquely described by the vector $\boldsymbol{\rho}(k) = [\boldsymbol{\rho}_1(k), \dots, \boldsymbol{\rho}_{n_j(k)}]^T$ of all densities $\rho_i(k)$ of cell $i$, $i = 1, \dots, n_j$. Based on the Godunov scheme for LWR model (Equation 34.4), the nonlinear state–space form of the traffic dynamics can be rewritten as

$$\boldsymbol{\rho}(k + 1) = F(\boldsymbol{\rho}(k)) + Gd(k) \tag{34.7}$$

where $d^T(k) = \left[ q_1^{in}(k), q_{n_j}^{out}(k) \right]$ is the unknown boundary driving-in and driving-out fluxes of the link, and

$$G^T = \begin{bmatrix} 0 & 0 & \dots & -1 \\ 1 & 0 & \dots & 0 \end{bmatrix}.$$

According to the speed–density equation (Equation 34.3), the speed measurement $v_l(k)$, $l = 1, \dots, m_j$, $m_j \leq n_j$, can be described as a function of vehicle density $\rho_l(k)$ of the cell $l$, where $m_j$ is the number of speed detectors. Given the collective speed information $y(k) = [v_1, \dots, v_{m_j}]^T$ for the link $j$, the observation equation can be expressed as a nonlinear model

$$y(k) = H(\boldsymbol{\rho}(k)) + v(k) \tag{34.8}$$

where $v(k)$ is assumed zero-mean Gaussian measurement noise with a covariance matrix $R(k)$.

The classical EKF method requires deterministic inputs in the model, which sometimes may not be the case in reality; for example, the boundary fluxes of a link depend mainly on the traffic supply–demand relationship. That could be unknown signals with an arbitrary type and magnitude, so it is not acceptable to assume them to be stationary and zero-mean random noise. In this regard, joint estimation of states and unknown inputs becomes a meaningful task that is often addressed as a constrained optimization problem.

We consider a recursive three-step filter of the form

$$\rho(k+1|k) = F(\rho(k|k)) \tag{34.9}$$

$$\hat{d}(k) = M(k)(y(k) - C(k)\rho(k+1|k)) \tag{34.10}$$

$$\bar{\rho}(k+1|k+1) = \rho(k+1|k) + G\hat{d}(k) \tag{34.11}$$

$$\rho(k+1|k+1) = \bar{\rho}(k+1|k+1) + K(k)(y(k) - C(k)\bar{\rho}(k+1|k+1)) \tag{34.12}$$

where

$$C(k) = \left.\frac{\partial H}{\partial \rho}\right|_{\rho(k+1|k)} \tag{34.13}$$

and $M(k)$, $K(k)$ can be determined from optimization.

The first step (Equation 34.9), which we call the time update, yields an estimate of the traffic density $\rho(k + 1|k)$, given measurements up to time $k$. The second step (Equation 34.10) yields the boundary flux estimate as the unknown input $\hat{d}(k)$. Finally, the third step (Equation 34.11), the so-called measurement update, yields an estimate $\rho(k + 1|k + 1)$ with the given measurement up to time $k + 1$ and input estimate $\hat{d}(k)$ in the second step.

The linearization matrix $A(k)$ of the traffic dynamic model (Equation 34.7) is used to correct the estimate,

$$A(k) = \left.\frac{\partial F}{\partial \rho}\right|_{\rho(k|k)}. \tag{34.14}$$

The estimate gain matrices $M(k)$, and $K(k)$ can be calculated using the error variance covariance equations as follows.

$$P(k+1|k) = A(k)P(k|k)A^T(k) + Q(k) \tag{34.15}$$

$$M(k) = \left(D^T(k)\tilde{R}(k)D(k)\right)^{-1} D^T(k)\tilde{R}^{-1}(k) \tag{34.16}$$

$$K(k) = P(k+1|k)C(k)\tilde{R}^{-1}(k) \tag{34.17}$$

$$\tilde{R}(k) = C(k)P(k+1|k)C^T(k) + R(k) \tag{34.18}$$

$$P(k+1|k+1) = (1 - M(k)C(k))P(k+1|k) \tag{34.19}$$

where $D(k) = C(k)G$, and $Q(k)$ is the process error covariance matrix in the linearization of the model (Equation 34.16).

A detailed description of the simultaneous state and input estimation with Kalman filter is provided in Refs. [6,7].

Compared with the classical EKF procedure, the simultaneous state and input estimation contains two more expensive operations: the inverse operation in Equations 34.16 and 34.17 that scales in the number of measurements and the matrix multiplications of Equation 34.10 that scales in the number of cells in the network. Once the EKF algorithm is applied to the entire freeway network, state vector $\rho(k)$ represents all cells in the entire network and $P(k|k)$ contains estimates of the covariance of the errors between all cells. Therefore, this procedure has one major concern: The calculation times can become very high.

## 34.4  CONSENSUS-BASED DECENTRALIZED EKF FILTERING

In this section, a consensus-based decentralized estimation method is proposed by using the logic of the network topology and fusing the estimation of the unknown input variables with local measurement. The new EKF implementation is much faster on larger scale online applications because it simplifies the matrix operations.

Stability and optimal estimate of the consensus-based decentralized Kalman filter have been extensively discussed in the recent literature. When a linearization of the nonlinear traffic flow model is made, we extend the aforementioned method to the EKF for simultaneous state and input estimate.

For adjacent two freeway links $j$ and $j + 1$ (see Figure 34.2), the traffic dynamics are formulated as

$$\rho_j(k+1) = F(\rho_j(k)) + G_1 q_{1,j}^{in}(k) - G_2 q_{n_j,j}^{out}(k), \tag{34.20}$$

and

$$\rho_{j+1}(k+1) = F(\rho_{j+1}(k)) + G_1 q_{1,j+1}^{in}(k) - G_2 q_{n_{j+1},j+1}^{out}(k), \tag{34.21}$$

where $G_1 = [1, 0,\ldots, 0]$ and $G_2 = [0, 0,\ldots, 1]$.

At the link boundary, the upstream driving-out flux of link $j$ is equal to the downstream driving-in flux of link $j + 1$, that is,

$$q_{n_j,j}^{out}(k) = q_{1,j+1}^{in}(k). \tag{34.22}$$

Therefore, we consider the consensus average algorithm to fuse the common flux, which is estimated using the local measurement observed from the upstream and the downstream links separately. To simplify the symbol, we denote $q_j^{out}(k)$, $q_{j+1}^{in}(k)$ as $q_{n_j,j}^{out}(k)$, $q_{1,j+1}^{in}(k)$ in our algorithm.

With the simultaneous state and input estimation, in the local EKF scheme, we have

$$\hat{q}_j^{out}(k) = M_j(k)(y_j(k) - C_j(k)\rho_j(k+1|k)), \tag{34.23}$$

$$\hat{q}_{j+1}^{in}(k) = M_{j+1}(k)(y_{j+1}(k) - C_{j+1}(k)\rho_{j+1}(k+1|k)). \tag{34.24}$$

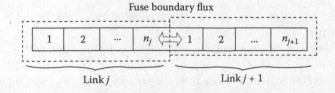

Fuse boundary flux

FIGURE 34.2  Fusing the boundary condition for consensus-based estimation.

where $y_j(k)$, $y_{j+1}(k)$ are local measurement of the upstream and the downstream links. Then input estimation is modified with the consensus average that

$$\hat{q}_j^{\text{out}}(k) = \hat{q}_{j+1}^{\text{in}}(k) = \theta_j \hat{q}_j^{\text{out}}(k) + \theta_{j+1} \hat{q}_{j+1}^{\text{in}}(k) \tag{34.25}$$

where the fusing weights satisfying $\theta_j + \theta_{j+1} = 1$. Finally, the new input estimation (Equation 34.25) is used by the local EKF scheme to update the state of the upstream and the downstream links $\rho_j(k + 1|k + 1)$ and $\rho_{j+1}(k + 1|k + 1)$, respectively. The preceding process is repeated for the next measurement $y_j(k + 1)$, and $y_{j+1}(k + 1)$.

The consensus-based decentralized EKF procedure has two major advantages to estimate the large-scale freeway networks: First, the matrix inverse and multiplication are performed on much smaller matrices so that the computation is thus very fast. Second, the freeway networks can be divided into any subsystems (local links) without considering the available measurement of the boundary flux. In fact, this characteristic is very suitable for large-scale and real-time applications.

## 34.5  EMPIRICAL STUDIES

In our experiment, the freeway network of interest is a section of Interstate 80 East, approximately five miles in length in Alameda, Northern California, as shown in Figure 34.3.

This section is instrumented with loop inductance detectors, which are embedded in the pavement along the mainline, high-occupancy vehicle lane, and off-ramps. The square points along the freeway link denote where loop detectors are installed. Each loop detector gives volume, speed, and occupancy measurements every 30 seconds. The utilized traffic data of 6 hours (7:00 am–1:00 pm) are collected from PeMS [13], which include the morning rush-hour congestion on October 20, 2013.

The freeway link is partitioned into six cells with lengths ranging from 0.65, 0.67, 0.82, 0.71, 0.73 to 1 mile, respectively. The fundamental diagrams are roughly calibrated using linear regression that uses one week of historical data from all detectors in this segment. The historical densities are computed for each cell using the occupancy divided by the $g$-factor, where the $g$-factor is the effective vehicle length for the detector. The four parameters of the Smulders fundamental diagram are validated as the free flow speed $v_f = 70$ mph, the critical density $\rho_c = 80$ veh/mile, the jam density $\rho_J = 320$ veh/mile, and the capacity volume $q_{\text{max}} = 4230$ veh/h, respectively, as shown in Figure 34.4.

As the speed measurements of each cell in this section are available, we can select some relevant measurement to build our speed detecting networks. The observation information acquired from portable speed detectors, such as using probe vehicles or radar speed guns, usually has larger measurement noise than loop inductance detectors. We assume the noise covariance of observation in the model in Equation 34.7 is $R(k) = 0.05$ (measurement noise is about 0.7 mph).

The purpose of our simulation is to estimate the traffic densities of the whole network using the speed measurement. Based on the consensus-based decentralized EKF method, this freeway

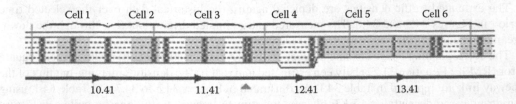

**FIGURE 34.3**  Freeway networks of I80-E, Alameda, CA with speed detectors (PeMS).

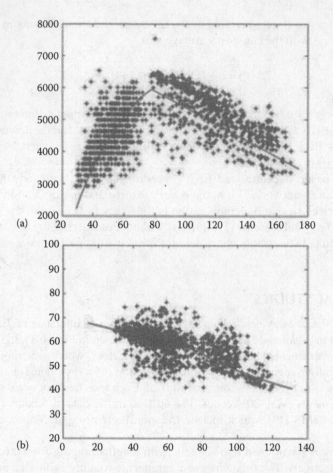

**FIGURE 34.4**  The fundamental diagrams of the freeway link calibrated from the traffic flow data collected on October 20, 2013. (a) Calibrated flow-density diagram. (b) Calibrated speed-density diagram.

network is divided into the upstream link 1, which includes cells 1, 2, and 3, and the downstream link 2, which includes cells 4, 5, and 6. Driving-in flux from the upstream boundary of cell 1 and driving-out flux of the downstream boundary of cell 6 are achieved from PeMS.

State vectors of two subsystems are $\rho_1 = [\rho_{1,1}, \rho_{2,1}, \rho_{3,1}]^T$ and $\rho_2 = [\rho_{1,2}, \rho_{2,2}, \rho_{3,2}]^T$, respectively. The synchronous mean speed measurement of cells 1 and 3 constitute the upstream observe vector $y_1(k) = [v_{1,1}, v_{3,1}]^T$, and speed measurement of cells 1 and 3 of cells 4 and 6 constitute the downstream observe vector $y_2(k) = [v_{1,2}, v_{3,2}]^T$. Boundary flux $q_3^{out}(k)$ (or $q_4^{in}(k)$) between cells 3 and cell 4 is considered as the unknown input of the two subsystems. The fusing weight $\theta_1$ and $\theta_2$ are selected as 0.5 in this simulation.

The estimated traffic densities are depicted against the historical data over the selected time period in Figures 34.5 and 34.6. The measured and estimated driving-out flux through the downstream boundary of cell 3 are depicted in Figure 34.7.

The corresponding mean absolute percent error (MAPE; Equation 34.26) and root mean square error (RMSE; Equation 34.27) between estimated and real traffic densities, driving-out flux of the freeway link are reported in Table 34.1. As illustrated in Figures 34.5 to 34.7 and Table 34.1, using consensus-based decentralized EKF filtering to estimate freeway traffic gives a quite satisfactory result.

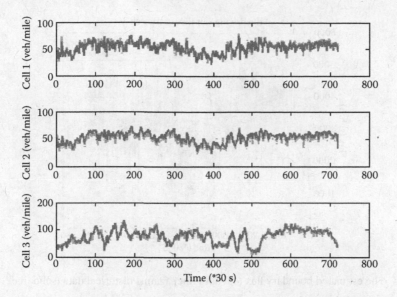

**FIGURE 34.5**    The estimated traffic densities (dot dash line) against historical data (solid line) of upstream link.

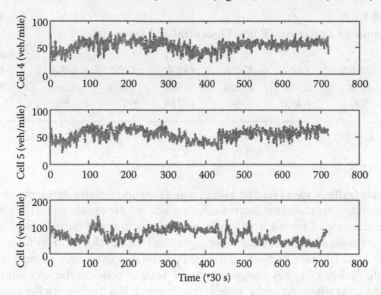

**FIGURE 34.6**    The estimated traffic densities (dot dash line) against historical data (solid line) of downstream link.

$$\text{MAPE} = \frac{1}{N}\sum_{k=1}^{N}\left|\frac{x(k|k)-x(k)}{x(k)}\right| \tag{34.26}$$

$$\text{RMSE} = \sqrt{\frac{1}{N-1}\sum_{k=1}^{N}(x(k|k)-x(k))^2} \tag{34.27}$$

where $x(k|k)$ represents estimated variable, $x(k)$ is the historical real data, and $N$ denotes total number of time sample steps.

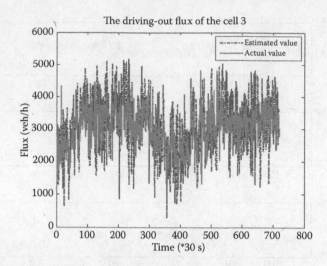

**FIGURE 34.7**  The estimated boundary flux (dot dash line) against historical data (solid line).

**TABLE 34.1**

**Performance of Consensus-Based Decentralized Estimation**

|        | Cell 1 | Cell 2 | Cell 3 | Cell 4 | Cell 5 | Cell 6 | Boundary Flux |
|--------|--------|--------|--------|--------|--------|--------|---------------|
| MAPE   | 0.0822 | 0.0437 | 0.2160 | 0.1938 | 0.0558 | 0.0988 | 0.1195        |
| RMSE   | 5.7141 | 5.4950 | 8.1905 | 6.6716 | 6.3787 | 5.9128 | 727.3017      |

## 34.6  CONCLUSION

Speed detection in traffic networks is easy to access and measure with the development of ubiquitous sensing technologies. This chapter presents a consensus-based decentralized EFK for real-time traffic parameter estimation of large-scale freeway networks with speed measurement. Simultaneous state and input estimate and consensus average algorithms are developed to perform the decentralized EKF procedure. The case studies highlight the potential of using only speed detecting data to monitor the state of urban freeway transport systems, without consider the accessibility of boundary flux when dividing whole networks. Hence, this approach has the potential of estimating traffic parameters of any segment of interest in large-scale transport networks.

## ACKNOWLEDGMENTS

This work was partially funded by NSFC (61374076) and Jinghua Scholar Project of Beijing University of Technology.

## REFERENCES

1. T. Hunter, T. Das, M. Zaharia et al. Large-scale estimation in cyberphysical systems using streaming data: A case study with arterial traffic estimation. *IEEE Transactions on Automation Science and Engineering*, 10(4):884–898, 2013.
2. E. Jenelius, and H. N. Koutsopoulos. Travel time estimation for urban road networks using low frequency probe vehicle data. *Transportation Research B: Methodological*, 53(4):64–81, 2013.

3.  Y. Wang, M. Papageorgiou, A. Messmer et al. An adaptive freeway traffic state estimator. *Automatica*, 45(1):10–24, 2009.
4.  L. Mihaylova, R. Boel, and A. Hegyi. Freeway traffic estimation within particle filtering framework. *Automatica*, 43(2):290–300, 2007.
5.  X. Sun, L. Munoz, and R. Horowitz. Mixture Kalman filter based highway congestion mode and vehicle density estimator and its application. In *Proceedings of the American Control Conference*, pp. 2098–2103, 2004.
6.  S. Gillijns, and B. De Moor. Unbiased minimum-variance input and state estimation for linear discrete-time systems. *Automatica*, 43(1):111–116, 2007.
7.  Y. Lu, L. Zhang, and X. Mao. Distributed information consensus filters for simultaneous input and state estimation. *Circuits, System and Signal Process*, 32(2):877–888, 2013.
8.  R. O. Saber, and J. Shamma. Consensus filters for sensor networks and distributed sensor fusion. In *Proceedings of the Decision and Control Conference*, pp. 6698–6703, 2005.
9.  S. S. Stankovic, M. S. Stankovic, and D. M. Stipanovic. Consensus based overlapping decentralized estimation with missing observations and communication faults. *Automatica*, 45:1397–1406, 2009.
10. M. J. Lighthill, and G. B. Whitham. On kinematic waves (II): A theory of traffic flow on long crowded roads. *Proceedings of the Royal Society of London A: Mathematical and Physical Science*, 229(1178):317–345, 1995.
11. P. I. Richards. Shockwaves on the highway. *Operations Research*, 4(1):42–51, 1956.
12. S. A. Smulders. Control of freeway traffic flow by variable speed signs. *Transportation Research B: Methodological*, 24(2):111–132, 1990.
13. PeMS Homepage. Available at http://pems.eecs.berkeley.edu/.

# Index

## A

ABF, *see* Audio biofeedback
ABS, *see* Antilock brake systems
ACC, *see* Adaptive cruise control
Accelerometers, 522, 527
Accumulated state densities (ASDs) and applications in
    object tracking, 295–329
  Bayesian tracking paradigm, 297–298
    filtering, 297
    fixed-interval retrodiction, 298
    likelihood function, 297
    prediction, 297
    retrodiction, 298
  data augmentation methods, 311–316
    ASDs and Q functions, 315–316
    expectation and maximization steps, 312–315
    noninformative measurement, 316
    probabilistic data association filter, 316
    probabilistic multiple hypothesis tracking, 316
  distributed ASD fusion, 316–323
    centralized Kalman filter, 322
    DASD filter implementation, 319–320
    exact DKF scheme, 317
    Gaussian posterior density, 317
    normalized root mean square error, 322
    numerical evaluation, 322–323
    relaxed evolution model, 318
    sensor likelihood function, 316
    sliding window mechanism, 320–321
  Gaussian ASDs (details of the proof), 325–328
  inverse of block matrices, 325
  notion of accumulated state densities, 298–305
    closed-form representation for ASDs, 299–302
    clutter, 303
    interacting multiple model filters, 302
    MHT/IMM filtering, 302–305
    mixture reduction techniques, 303
    multiple hypothesis tracking, 302
    Rauch–Tung–Striebel smoothing, 302
    well-separated objects, 303
  out-of-sequence measurements, 296, 305–311
    computational costs, 308–309
    continuous time retrodiction, 306
    information aging, 308
    modified Kalman update step, 307–308
    simulated example, 309–311
  probabilistic multiple hypothesis tracking, 296
  problem of tracking objects, 296–297
    distributed Kalman filter, 297
    expectation maximization methodology, 297
    track-to-track fusion, 297
  product formula for Gaussians in identical variables,
      324
  product formula for linearly conditional Gaussians, 324
  Schur complement, 325

Active sensor network (ASN), 574
Activities of daily living (ADLs), 535
Activity Recognizer Tool system, 11
Adaptive cruise control (ACC), 581
ADAS, *see* Advanced driver assistance systems
Additive uniform noise (AUN) error, 139
ADLs, *see* Activities of daily living
Advanced driver assistance systems (ADAS), 553, 581–582
Advanced transportation management systems (ATMS), 578
Advanced Traveler Information Systems (ATIS), 580–581
Agent interaction topology (AIT), 373
Agent opinions, 369–373
  agent interaction topology, 373
  agent state updates, 369–371
  belief revision, 371
  cautious CUE, 372
  characteristics of agent interactions, 372
  conditional update equation, 371
  convex sum formulation, 371
  discrete event-based time, 369
  modeling opinions and belief revision, 371–373
  probabilistic case, 372–373
  receptive CUE, 372
  spatial coupling, 373
  state models, 371
  temporal coupling, 373
AI, *see* Artificial intelligence
AID, *see* Automatic incident detection
AIT, *see* Agent interaction topology
Analysis of variance (ANOVA), 474
ANNs, *see* Artificial neural networks
ANOVA, *see* Analysis of variance
Antilock brake systems (ABS), 551
Application specific integrated circuit (ASIC), 550
Aquatic habitat model, 440, 449–450
Area sampling, 139
Artificial intelligence (AI), 64
Artificial neural networks (ANNs), 494, 576, 596
ASDs, *see* Accumulated state densities and applications in
    object tracking
ASIC, *see* Application specific integrated circuit
ASN, *see* Active sensor network
Asymmetric warfare-type problems, 8
Asynchronous iteration, 368
ATIS, *see* Advanced Traveler Information Systems
ATMS, *see* Advanced transportation management systems
Audio biofeedback (ABF), 543
AUN error, *see* Additive uniform noise error
Automatic incident detection (AID), 578
Automatic vehicle identification (AVI) sensors, 611
Automation, granular approach to, 438–444
  application to the model of the habitat, 440–442
  aquatic habitat, 440
  control laws and operating conditions, 441
  control signals, 440
  correspondence function, 440

623

Printed in the United States
by Baker & Taylor Publisher Services

Printed in the United States
by Baker & Taylor Publisher Services